ENERGY FROM BIOMASS

1st E.C. Conference

Proceedings of the International Conference on Biomass
held at Brighton, England, 4–7 November 1980.

ENERGY FROM BIOMASS

1st E.C. Conference

Edited by

W. PALZ

Commission of the European Communities, Brussels, Belgium

P. CHARTIER

Institut National de la Recherche Agronomique,
Département de Bioclimatologie, Versailles, France

and

D. O. HALL

University of London, King's College, London, UK

APPLIED SCIENCE PUBLISHERS LTD
LONDON

APPLIED SCIENCE PUBLISHERS LTD
RIPPLE ROAD, BARKING, ESSEX, ENGLAND

British Library Cataloguing in Publication Data

Energy from biomass.
  1. Biomass energy—Congresses
  I. Palz, W. II. Chartier, P.
  III. Hall, D. O. IV. Commission of the
  European Communities
  662'.6   TP360

ISBN 0-85334-970-3

WITH 198 TABLES AND 210 ILLUSTRATIONS

© ECSC, EEC, EAEC, Brussels and Luxembourg, 1981

Organization of the conference by: Commission of the European Communities, Directorate-General Research, Science and Education, Brussels in co-operation with the Department of Energy, London

Publication arrangements by: Commission of the European Communities, Directorate-General for the Information Market and Innovation, Luxembourg

EUR 7091

LEGAL NOTICE
Neither the Commission of the European Communities nor any person acting on behalf of the Commission is responsible for the use which might be made of the following information.

Printed in Great Britain by Galliard (Printers) Ltd, Great Yarmouth

F O R E W O R D

by the Executive Board

This international conference, the first held in Europe on biomass
technology, surely marks its coming of age among the renewable forms
of energy. For some European countries awareness of its promise has
been slow to develop; in these, biomass has been regarded as a marginal
source of energy that could at best make only a small contribution to
national energy supplies. The conference gave a clear response to this
view.

It made clear that biomass has much more than a marginal potential for
some European countries, e.g. Denmark or Eire. But, more importantly,
the world has to move away from undue dependence on individual sources
of energy towards a plurality of sources. The Commission's own programme
shows clearly that even biomass technologies whose individual contribution may be small in some countries can provide an aggregate impact on
European energy supplies that is strategically significant.

Even those countries that are dedicated to coal and nuclear strategies
as the main-stay of their future energy supplies may well come under a
variety of pressures to temper the scale on which these technologies
are used. Such pressures will stem not simply from environmental
issues, or from the strategic need to avoid undue dependence on imported energy; there is also a growing and deep rooted awareness within
Europe of the need to diversify into decentralised energy supplies.
Biomass provides a prime opportunity for so doing.

In spite of the promise of biomass it has to be recognised that its
successful development faces many problems and difficulties. Paradoxi-
cally these difficulties should be least in the short range future.
Biological waste products from industrial urban and rural areas are
already becoming economic as fuel savers. Furthermore, the problems and
rising costs of conventional waste diposal methods will steadily
emphasise the importance of alternative approaches. Nevertheless, these
short range developments will be limited in the total contribution

that biomass can make. Saturation of supplies will gradually set in.

When this saturation occurs, the collection and distribution infrastructure will be more developed than at present. This fact will make
it much more attractive in the medium term to grow crops to supplement
waste utilisation. These crops, whether trees or other plants, will
arise from the incentive to make economic use of waste ground, including the interplanting of secondary crops between main growing
seasons. One can expect increased yields of biomass crops which will
result in more biomass becoming available for energy conversion; this
will also reduce potential conflicts in land use. The best use of
biomass in each area will be determined according to its energy needs
and the alternative energy sources that are available. In some areas it
may be desirable to convert biomass into valuable liquid fuels, while
in other areas, traditionally simpler and more efficient uses may be
compatible with local needs.

The conference recognised that biomass holds a special importance for
the Third World countries. Indeed, they already make massive use of
it, though not necessarily in the optimal way. The exact way these
countries will develop the resource for optimum retrieval of energy
and soil nutrients, will depend upon the needs both of local society
and the ecosystem. There is a great diversity of possibilities, and
research will open up even more.

Generally, the diversity of biomass resources with numerous conversion
routes presents a confusing array of choices. It is just this array of
choices, however, that gives biomass a meaningfull role in satisfying
diverse energy requirements in different regions. In Europe, biomass
can make a small, but significant impact on our near term energy
requirements; with appropriate research and development, the energy
contribution from biomass will continue to grow in the future.

F.J.P. Clarke,   Harwell Laboratory, United Kingdom
A.S. Strub,   Commission of the European Communities, Brussels
T.K. Ghose,   Indian Institute of Technology, New Delhi
B. Berger,   Department of Energy, USA

Members of the Executive Board

CONFERENCE COMMITTEE

Executive Board

CLARKE, Dr F.J.P.
AERE Harwell, Oxon, UK

STRUB, Dr. A.S.
Commission of the European
Communities
Directorate General "Research,
Science and Education"
200, rue de la Loi
Brussels, Belgium

GHOSE, Prof. T.K.
Indian Institute of Technology,
Biochemical Engineering
Research Centre
Delhi, New Delhi, India

BERGER, Dr. B.
Department of Energy
600 East Street
NW, Washington DC, USA

Organising Committe

KING, Dr. G.H.
Energy Technology Support Unit
AERE Harwell, Oxon, UK

CHARTIER, Dr. P.
I.N.R.A.
Route de Saint Cyr
Versailles, France

HALL, Prof. D.O.
University of London
King's College
68, Half Moon Lane
London, UK

PALZ, Dr. W.
Commission of the European
Communities
Directorate General "Research,
Science and Education"
200, rue de la Loi
Brussels, Belgium

NICOLAY, Mr. D.
Commission of the European
Communities
Directorate General "Information
Market and Innovation" - JMO B4/072
P.O. Box 1907
Luxembourg, Grand Duchy

DUXBURY, Dr. J.K.
Department of Energy
Thames House South, Millbank,
London, UK

PREUVENEERS, Mr. C.J.A.
Education and Training Branch
AERE Harwell, Oxon, UK

BAYLISS, Mr. R.F.
Energy Technology Support Unit
AERE Harwell, Oxon, UK

Scientific Committee

BERESOVSKI, Mr. E.
UNESCO
Place de Fontenoy
Paris, France

CHENGUANG-QUIAN, Prof.
Bio Energy Laboratory
Chengdu Institute of Biology
Academia Sinica, Chengdu,
Sichuan Province
The People's Republic of China

FUKUI, Prof. S.
Department of Industrial Chemistry,
Kyoto University
Sakyo-ku, Yoshida,
Kyoto, Japan

HARRISON, Mr. J.D.L.
Energy Technology Support Unit
AERE Harwell, Oxon, UK

HANSEN, Dr. G.K.
Hydroteknisk Laboratorium
Den Kgl, Veterinaer - og
Landbohoejskole, Buelowsvej 13
Copenhague, Denmark

HUMMEL, Dr. F.
(Formerly CEC, Brussels)
8, The Ridgeway
Guildford, GU1 2DG, UK

HURAND, Mr. A.
Ministère de l'Agriculture
CNEEMA
Parce de Tourvoie
Antony, France

Scientific Committee (continued)

KINSELLA, Dr.E.
National Board for Science &
Technology
Shelbourne House, Shelbourne Road
Dublin, Ireland

LEQUEUX, Dr. P.
Commission of the European
Communities
Directorate General "Development"
200, rue de la Loi
Brussels, Belgium

LETTINGA, Dr. Ir. G.
Landbouw Hogeschool,
Wageningen Biotechnion,
De Dreijen, 12
Wageningen, The Netherlands

LICATA, Dr. R.
Via Ugo Ogetti 426
Rome, Italy

LIPINSKY, Dr. E.S.
Battelle Columbus Laboratories
505 King Avenue
Columbus, Ohio, USA

MARGARIS, Prof. N.S.
Laboratory of Ecology, Univer-
sity of Thessalonika
Faculty of Physics & Mathematics,
Thessalonika, Greece

MARTIN, Dr .W.
Commission of the European
Communities
Directorate General "Energy"
200, rue de la Loi
Brussels, Belgium

MAUREL, Mr. H.
Commission of the European
Communities
Directorate General "Agriculture"
200, rue de la Loi
Brussels, Belgium

McDIVITT, Dr. J.F.
UNESCO
Place de Fontenoy
Paris, France

MELLADO, Dr. L
INIA, Calle de Generale
San Jurjo 56
Madrid, Spain

MORANDINI, Prof. R.
Forestry Division, Istituto Speri-
mentale per la Selvicoltura
Viale Santa Margherita, 80
Arezzo, Italy

MUHS, Dr. H.J.
Institute of Forest Genetics &
Forest Tree Breeding
Sickerlandstr. 2
Grosshansdorf, F.R. Germany

NAVEAU, Prof. H.
Dept. de Chimie et de Physique
Appliquées
University of Louvain
Place Croix du Sud, 1
Louvain-la-Neuve, Belgium

QASEM, Prof. S.
Faculty of Agriculture,
University of Jordan
Amman, Jordan

SIREN, Prof. G.
Swedish University of Agricultural
Sciences
Uppsala, Sweden

TRINDADE, Dr. S.C.
Centro de Technologia Promon
Praie do Flamengo 154-120
Rio de Janeiro, Brazil

VAN UDEN, Prof. N.
Gulbenkian Institute of
Sciences
Oeirias, Portugal

WAGENER, Prof. K.
Technical University of Aachen
c/o KFA Jülich
P.O. Box 1913
Jülich, F.R. Germany

CHIEF CONFERENCE OFFICERS: DR. G.H.KING, DR. P. CHARTIER, PROF. D. HALL

DR. W. PALZ ADDRESSING THE PRESS

RECEPTION IN THE ROYAL PAVILION, HOSTED BY BRIGHTON BOROUGH COUNCIL

A LOOK INTO THE POSTER HALL

C O N T E N T S

Foreword v

Introduction 1

Summary report of the co-chairmen 2
P. CHARTIER, Centre National de la Recherche Agronomique,
Département de Bioclimatologie, Versailles, France and
D.O. HALL, University of London, King's College, London,
United Kingdom

OPENING SESSION

Introductory remarks 9
Dr. A. STRUB, Directorate General "Energy", Commission of the
European Communities, Brussels, Belgium

Opening speech 11
J. MOORE, Parliamentary Under Secretary of State for Energy,
London, United Kingdom

Opening speech 14
Dr. L. WILLIAMS, Director General for Energy, Commission of the
European Communities, Brussels, Belgium

Opening address 19
H. BONDI, Department of Energy, Natural Environment Research
Council, Swindon,Wilts, United Kingdom

Prospects for energy from biomass in the European Community 22
P. CHARTIER, Institut National de la Recherche Agronomique,
Versailles, France

Research and development on photochemistry and photobiology - 34
The European Commission's programme
D.O.HALL, University of London, King's College, London, United
Kingdom

SESSION I : AGRICULTURAL RESIDUES AND ENERGY CROPS

Summary of the discussions 47
N.S. MARGARIS, Laboratory of Ecology, University of Thessalonika,
Greece

Invited papers

Straw and animal residues available for energy   50
  Dr. F.P. REXEN, Bioteknisk Institut, Denmark

Energy crops - The case of Brasil   59
  S.C. TRINDADE, Centro de Tecnologia Promon-CTP, Rio de Janeiro,
  RJ-Brasil

Poster papers

Wetland energy crops: the productive potential of typha spp.,   75
phragmites communis and phalaris arundinacea in Minnesota
  N.J. ANDREWS and D.C. PRATT, Botany Department, University of
  Minnesota, USA

Natural vegetation as a renewable energy resource in Great   83
Britain
  G.J. LAWSON and T.V. CALLAGHAN, Institute of Terrestrial
  Ecology, Merlewood Research Station, Cumbria, United Kingdom

Selected natural and alien plant species as renewable sources of 90
energy in Great Britain - Experimental assessment and imple-
mentation
  T.V. CALLAGHAN, R. SCOTT, H.A. WHITTAKER and G.J. LAWSON,
  Institute of Terrestrial Ecology, Merlewood Research Station,
  Cumbria, United Kingdom

The productivity of catch crops grown for fuel   97
  S.P. CARRUTHERS, Department of Agriculture and Horticulture,
  University of Reading, United Kingdom

The allocation of land for energy crops in Britain   103
  R.G.H. BUNCE, Institute of Terrestrial Ecology, Merlewood
  Research Station, Cumbria, L.H. PEARCE, Forest Commission,
  Westonbirt Arboretum, United Kingdom and C.P. MITCHELL,
  Forestry Department, Aberdeen University, Scotland

The potential role of crop fractionation in the production of   110
energy from biomass
  L.G. PLASKETT, Biotechnical Processes Ltd, Tiverton, United
  Kingdom

The energetics of livestock manure management   118
  R.M. MORRIS, Open University, Milton Keynes, United Kingdom
  and S.B.C. LARKIN, N.C.A.E., Silsoe, United Kingdom

Production, distribution and energy content of agricultural   124
wastes and residues in the United Kingdom
  S.B.C. LARKIN, D.H. NOBLE and R.W. RADLEY, National College
  of Agricultural Engineering, Silsoe, Bedford, United Kingdom
  and R.M. MORRIS, The Open University, Milton Keynes, United
  Kingdom

Crop residue availability for fuel   131
  J.H. POSSELIUS and B.A. STOUT, Michigan State University,
  East Lansing, USA

Soil protection under maximum removal of organic matter   138
    CHASSIN, GUERIF, JUSTE, MONNIER, MULLER, REMY, STENGEL,
    TAUZIN, I.N.R.A., Agronomie et Science du Sol, France

Experimental study on continuous digestion of cellulose waste in a   144
combined reactor system
    F. ALFANI, Facoltà di Ingegneria, M. CANTARELLA and V. SCARDI,
    Facoltà di Scienze, Università di Napoli, Italy

The manure cellar as a source of heat   150
    O. TJERNSHAUGEN, Agricultural University of Norway

SESSION II: FORESTRY FOR ENERGY

Summary of the discussions   159
    G. SIREN, Swedish University of Agricultural Sciences, Uppsala,
    Sweden

Invited papers

Short-rotation forestry as a biomass source: an overview   163
    K. STEINBECK, School of Forest Resources, University of Georgia,
    Athens, USA

The outlook for energy forestry in France and in the European   172
Economic Community
    P. BOUVAREL, Department of Forest Research, I.N.R.A., France

Poster papers

Determination of yield of biomass from whole-tree harvesting of   181
early thinnings in Britain
    C.P. MITCHELL, J.D. MATTHEWS, C.G. MacBRAYNE and M.F. PROE,
    Department of Forestry, Aberdeen University, Scotland

Recovery of waste-products from wood-processing industry   187
    C. PARE and C. ROLIN, Institut National Polytechnique de
    Lorraine, Departement de Gestion des Entreprises,   Nancy,
    France

Biomass production with poplar   193
    H.A. VAN DER MEIDEN and H.W. KOLSTER, Institute for the
    Promotion of Industrial Wood Production, Wageningen, The
    Netherlands

Coppice willow for biomass in the United Kingdom   198
    K.G. STOTT, Long Ashton Research Station, University of Bristol,
    United Kingdom, G. McELROY and W. ABERNETHY, The Horticultural
    Centre, Loughgall, Co. Armagh, D.P. HAYES, Queen's University,
    Department of Agricultural Botany, Belfast, N. Ireland

Coppiced trees as energy crops   210
    M.L. PEARCE,Research and Development Division, Forestry
    Commission, Tetbury, Glos. United Kingdom

Method for the estimation of above-ground biomass and biomass   216
production in classical coppice and first results
  D. AUCLAIR and A. CABANETTES, I.N.R.A., Station de Recherches
sur la Forêt et l'Environnement, Centre de Recherches d'Orléans-
Ardon, Olivet, France

Coppice forests in Italy: their potential for energy   222
  G. SCARAMUZZI, A. ECCHER, S.A.F., Centro di Sperimentazione
Agricola e Forestale, Rome, O. CIANCIO, Istituto Sperimentale
per la Selvicoltura, Arezzo, Italy

Design of a harvester for 2-3 year old willow sticks   228
  J.W. DUFF and H.D. McLAIN, Loughry College of Agriculture and
Food Technology, Cookstown, Co Tyrone, N. Ireland

Short rotation forestry as a source of energy   232
  M. NEENAN and G. LYONS, An Foras Taluntais, Oak Park Research
Centre, Carlow, Ireland

Forest biomass as a source of energy in the UK - The potential   239
and the practice
  C.P. MITCHELL and J.D. MATTHEWS, Aberdeen University, Scotland

Short rotation biomass production of willows in New Zealand   244
  R.L. Hathaway and C.W.S. van Kraayenoord, National Plant
Materials Centre, Ministry of Works and Development, Palmerston
North, New Zealand

Forest energy crops from derelict and waste land   251
  V.N. DENNINGTON and M.J. CHADWICK, Department of Biology,
University of York, United Kingdom

Sélection d'eucalyptus adaptés aux conditions françaises en vue   257
de l'installation de taillis à courtes rotations
  J.N. MARIEN, Association Forêt-Cellulose - AFOCEL,
Montpellier, France

SESSION III: FERMENTATION TO ETHANOL AND BIOGAS

Summary of the discussions   261
  T.K. GHOSE, Indian Institute of Technology, New Delhi, India

Invited papers

Anaerobic digestion for energy saving and production   264
  G. LETTINGA, Department of Water Pollution Control, Agricultural
University, Wageningen, Holland

Ethanol - The process and the technology for production of   279
liquid transport fuel
  J. COOMBS, Tate & Lyle Ltd., Reading, United Kingdom

Poster papers

HF saccharification: the key to ethanol from wood?   292
D.T.A. LAMPORT, H. HARDT, G. SMITH, S. MOHRLOK, MSU-DOE Plant
Research Laboratory, M.C. HAWLEY, R. CHAPMAN and S. SELKE,
Department of Chemical Engineering, Michigan State University,
USA

Ethanol production by extractive fermentation   298
M. MINIER and G. GOMA, Institut National des Sciences Appliquées,
Toulouse, France

Agricultural waste treatment by means of ultrafiltration membrane   306
enzymatic reactors
L. GIANFREDA, Facoltà di Farmacia, and G. GRECO jr., Facoltà di
Ingegneria, Università di Napoli, Italy

l-thio-oligosaccharides as substrate analogs for investigations   312
in the field of polysaccharidases
M. BLANC-MUESSER, J. DEFAYE, H. DRIGUEZ and E. OHLEYER, Centre
de Recherches sur les Macromolécules Végétales, CNRS, Grenoble,
France

Degradation of cellulose with hydrogen fluoride   319
J. DEFAYE and A. GADELLE, Centre de Recherches sur les Macro-
molécules Végétales, CNRS, Grenoble, France and C. PEDERSEN,
The Technical University of Denmark, Lyngby, Denmark

Cellulase and β-glucosidase from clostridium thermocellum   324
excretion, localization and purification
N. CREUZET, N. AIT, P. FORGET and J. CATTANEO, Laboratoire de
Chimie Bactérienne, CNRS, Marseille, France

Continuous autohydrolysis, a key step in the economic conversion   330
of forest and crop residues into ethanol
J.D. TAYLOR, Stake Technology Ltd., Ottawa, Canada

Production of ethanol from wastes via enzymatic hydrolysis of   337
cellulose
P. ANDREONI, R. AVELLA, G. DI GIORGIO, M. LOPOPOLO, A. MANCINI,
Centro Studi Nucleari Casaccia, CNEN, Rome, Italy

Energy production from whey   344
G. MOULIN and P. GALZY, Chaire de Génétique et Microbiologie,
ENSAM-INRA, Montpellier, France

Separation of lignocelluloses into highly accessible fibre   348
materials and hemicellulose fraction by the steaming-extraction
process
J. PULS and H.H. DIETRICHS, Federal Research Center of Forestry
and Forest Products, Hamburg, Federal Republic of Germany

Ethyl alcohol production for fuel: energy balance   354
R.Y. OFOLI and B.A. STOUT, Agricultural Engineering Department,
Michigan State University, USA

A low energy system for the production of biomass ethanol   360
H.J. GROUT and M. ENGLISH, W.S. Atkins Group Consultants,
Epsom, Surrey, United Kingdom

The cellulolytic community of an anaerobic estuarine sediment   366
R.H. MADDEN, M.J. BRYDER and N.J. POOLE, Department of Microbiology, Aberdeen University, Scotland

The use of phase-plane analysis in the modelling of bio-metha-   372
nization processes in order to control their evolution
S.D. ANTUNES, Instituto Superior Tecnico, Lisboa, Portugal
and M. INSTALLE, Laboratoire d'Automatique et d'Analyse
des Systèmes and Laboratoire du Génie Biologique,
Université Catholique de Louvain, Louvain-la-Neuve,
Belgium

Application of microcalorimetry to the study of the fermentation   380
of lignocellulosic compounds
M.L. Fardeau, F. Plasse, J. Partos, J.P. Belaich, Laboratoire
de Chimie Bactérienne, CNRS, Marseille, France

Occurrence of cellulolysis and methanogenesis in various ecosystems   386
D. MARTY, J. GARCIN and A. BIANCHI, Ecologie et Biochimie
microbiennes du milieu marin, CNRS, ER 223, Université de
Provence, Marseille, France

Production of methane from freshwater macro-algae by an anaerobic   392
two step digestion system
C.M. ASINARI DI SAN MARZANO, H.P. NAVEAU and E.J. NYNS, Unit
of Bioengineering, University of Louvain, Louvain-la-Neuve,
Belgium

Operation of a laboratory-scale plug flow type of digester on pig   398
manure
F.R. HAWKES and J.R.S. FLOYD, Department of Science and D.L.
HAWKES, Department of Mechanical and Production Engineering, The
Polytechnic of Wales, Pontypridd,Mid-Glamorgan, United Kingdom

Fermentation to biogas using agricultural residues and energy   406
crops
D.A. STAFFORD and D.E. HUGHES, Department of Microbiology,
University College, Cardiff, United Kingdom

Horizontal-rotating-drum continuous fermentor   411
V.A. VASEEN, Avasco, Colorado, USA

Bioconversion of agricultural wastes into fuel gas and animal   416
feed
E. COLLERAN, A. BOOTH, M. BARRY, A. WILKIE, P.J. NEWELL and
L.K. DUNICAN, Department of Microbiology, University College,
Galway, Ireland

Proposals for the analytical modelling of a digester   423
E. LAVAGNO, P. RAVETTO and B. RUGGERI, Istituto di Fisica
Tecnica ed Impianti Nucleari, Politecnico di Torino, Italy

Computer aided design of anaerobic digesters for energy production   429
    D.L. HAWKES, Department of Mechanical and Production
    Engineering, The Polytechnic of Wales, Pontypridd, Mid
    Glamorgan, United Kingdom

A novel process for the anaerobic digestion of solid wastes   435
leading to biogas and a compostlike material
    B.A. RIJKENS, Institute for Storage and Processing of Agri-
    cultural Produce (IBVL), Wageningen, The Netherlands

Application of recent developments in anaerobic waste water   440
treatment to slaughterhouse wastes (Germany)
    G.D. LINDAUER and H. SIXT, Agrar- und Hydrotechnik GmbH,
    Essen, O. SCHULZE GmbH & Co KG, Gladbeck, Federal Republic of
    Germany

Findings of B.A.B.A.'s subcommittee to produce a code of   448
practice on safety in and around anaerobic digesters
    P.J. MEYNELL, British Anaerobic and Biomass Association Ltd.,
    Marlborough, Wilts., United Kingdom

An assessment of some commercial digesters in the United Kingdom   459
    P.J. MEYNELL, British Anaerobic and Biomass Association Ltd.,
    Marlborough, Wilts.,United Kingdom

Economic aspects of biogas-production from animal wastes   466
    W. KLEINHANSS, Institute of Farm Economics, Federal Research
    Center of Agriculture, Braunschweig, Federal Republic of
    Germany

Methane fermentation in the thermophilic range   472
    O. KANDLER, J. WINTER and U. TEMPER, Botanical Institute,
    University of Munich, Federal Republic of Germany

SESSION IV: THERMOCHEMICAL ROUTES TO GASEOUS AND LIQUID FUELS

Summary of the discussions   481
    R. OVEREND, National Research Council, Ottawa, Canada

Invited papers

Gasification - The process and the technology   485
    W.P.M. VAN SWAAIJ, Twente University of Technology,
    The Netherlands

The combustion, pyrolysis, gasification and liquefaction   496
of biomass
    T.B. REED, The Solar Energy Research Institute (SERI),
    Golden, Colorado

Poster papers

Energy from straw and woodwaste   509
    A. STREHLER, Bayerische Landesanstalt für Landtechnik,
    TU München, Freising, Federal Republic of Germany

Farm straw as fuel for power   516
  M.W. THRING and R.J. CROOKES, Department of Mechanical
  Engineering, Queen Mary College University of London,
  United Kingdom

Pyrolysis and combustion of wood in relation with its chemical   523
composition
  H. MELLOTTEE and J.R. RICHARD, CNRS, Centre de Recherches sur
  la Chimie de la Combustion et des Hautes Températures, Orleans,
  and B. MONTIES, INRA, Laboratoire de Chimie Biologique et
  de Photophysiologie, Grignon, France

On the influence of the different parts of wood on the production   529
of gaseous combustible products by pyrolysis
  J.R. RICHARD and M. CATHONNET, CNRS, Centre de Recherches sur
  la Chimie de la Combustion et des Hautes Températures, Orleans,
  France

The pyrolysis of tropical woods: the influence of their chemical   535
composition on the end products
  G. PETROFF and J. DOAT, Centre Technique Forestier Tropical,
  Nogent-sur-Marne, France

Gaseous fuel from biomass by flash pyrolysis   542
  S. CAUBET, P. CORTE, C. FAHIM, and J.P. TRAVERSE Laboratoire
  de Recherche sur l'Energie, Université Paul Sabatier, Toulouse,
  France

Fast pyrolysis/gasification of lignocellulosic materials at short   548
residence time
  X. DEGLISE, C. RICHARD, A. ROLIN and H. FRANCOIS, Laboratoire de
  Photochimie Appliquée, Université de Nancy I, Nancy, France

The conversion of biomass to fuel raw material by   554
hydrothermal treatment
  O. BOBLETER, H. BINDER, R. CONCIN, and E. BURTSCHER, Institute
  for Radiochemistry, University of Innsbruck, Austria

A process for thermo-chemical conversion of wet biomass into heat   563
energy
  H. HAVE, Jordbrugsteknisk Institut, The Royal Veterinary and
  Agricultural University, Taastrup, Denmark

Mass and energy balances for a two fluidised-bed pilot plant   569
which operates on wood fast pyrolysis
  X. DEGLISE, Université de Nancy I, Nancy, P. MORLIERE, TNEE
  (Tunzini-Nessi Entreprises d'Equipements), Argenteuil and Ph.
  SCHLICKLIN, Centre de Recherches de Pont-à-Mousson, France

Two new types of biomass gasifiers developed at C.N.E.E.M.A.   574
  J.F. MOLLE, Centre National d'Etudes et d'Expérimentation du
  Machinisme Agricole, France

Kinetic studies of pyrolysis and gasification of wood under   583
pressure with steam and oxygen
    A. DIVRY, P. DUBOIS and J.C. RENARD, Laboratoires de Marcoussis,
    Centre de Recherche de la Compagnie Générale d'Electricité,
    Marcoussis, France

Methanol catalytic synthesis from carbon monoxide and hydrogen   588
obtained from combustion of cellulose waste
    C. MASSON, A. BOURREAU, M. LALLEMAND, F. SOUIL and J.C. GOUDEAU,
    Université de Poitiers, St. Julien l'Ars, France

Process and equipment for the fluid bed oxygen gasification of   594
wood
    G. CHRYSOSTOME, Creusot-Loire, Le Creusot, France

Techno-economic evaluation of thermal routes for processing   598
biomass to methanol, methane and liquid hydrocarbons
    G. ADER, Ader Associates Ltd., Kent, A.V. BRIDGWATER and B.W.
    HATT, University of Aston in Birmingham, United Kingdom

Thermal processing of biomass to synthesis gas - a programme of   607
experimental investigations and design studies
    E.L. SMITH, B.W. HATT, G.A. IRLAM and A.V. BRIDGWATER,
    University of Aston in Birmingham, United Kingdom

Economics of combustion energy from crop residue   610
    F.J. HITZHUSEN, Department of Agricultural Economics, The Ohio
    State University, Columbus, Ohio, USA and M. ABDALLAH, Faculty
    of Administrative Sciences, Ryiadh University, Ryiadh, Saudi
    Arabia

Production of a low-Btu fuel gas by cocurrent gasification of   616
solid wastes
    J.J. HOS, A.A.C.M. BEENACKERS, F.G. VAN DEN AARSEN and W.P.M.
    VAN SWAAIJ, Twente University of Technology, Enschede, The
    Netherlands

SESSION V : NEW CONCEPTS IN FUELS BY BIOLOGICAL ROUTES

Summary of the discussions   625
    K. WAGENER, Technical University of Aachen, c/o KFA Jülich,
    Federal Republic of Germany

Invited papers

Prospects for abiological synthesis of biomass   627
    G. PORTER, FRS, The Royal Institution, London, United Kingdom

Algae as solar energy converters   633
    W.J. OSWALD, University of California, Berkeley, California,
    USA

Poster papers

Model system for the continuous photochemical production of
hydrogen peroxide mediated by flavins
  G. GOMEZ-MORENO, A.G. FONTES and F.F. DE LA ROSA
  Departamento de Bioquimica, Facultad de Biologia, Universidad
  de Sevilla and CSIC, Sevilla, Spain
    647

Renewable hydrocarbon production from the alga botryococcus
braunii
  C. LARGEAU, E. CASADEVALL and D. DIF, Laboratoire de Chimie
  Bioorganique et Organique Physique, ENSCP, Paris and
  C. BERKALOFF, Laboratoire de Botanique Cytophysiologie Végétale,
  ENS, Paris, France
    653

New concepts in solar biotechnology
  C. GUDIN, D. CHAUMONT and E. BERRA, Laboratory of Heliosynthesis
  BP, Lavera, France
    659

H_2 production by the photosynthetic bacteria, rhodopseudomonas
capsulata entrapped in alginate gels
  F. PAUL and P.M. VIGNAIS, DRF/Biochimie, Centre d'Etudes
  Nucléaires, Grenoble, France
    665

First experiments of production of macrophytes with waste water
and methanization of biomass
  M.-L. CHASSANY-DE CASABIANCA, Ecology and macrophytic production,
  CNRS, Montpellier and F. SAUZE, Station d'Oenologie, INRA,
  Narbonne, France
    672

Valorization of aquatic macrophytic biomass - Methane production -
Depollution and use of various by-products
  M.L. CHASSANY-DE CASABIANCA, CNRS, Montpellier, L. CODOMIER,
  Université de Perpignan, A. GELY, CNEEMA, Nîmes and F. SAUZE,
  INRA, Narbonne, France
    678

Growth of marine biomass on artificial structures as a renewable
source of energy
  J.G. MORLEY, Wolfson Institute of Interfacial Technology,
  University of Nottingham and J.M.JONES, University of Liverpool,
  Port Erin, Isle of Man, United Kingdom
    681

Mariculture on land - A system for biofuel farming in coastal
deserts
  K. WAGENER, Department of Biophysics, Technical University
  Aachen, Federal Republic of Germany
    685

Algal biomass from farm waste - a pilot plant study
  M.K. GARRETT and H.J. FALLOWFIELD, Department of Agricultural
  and Food Chemistry and Department of Agriculture for Northern
  Ireland, The Queen's University of Belfast, N. Ireland
    691

Photobiological production of fuels by microalgae
  S. LIEN, Solar Energy Research Institute (SERI) Golden,
  Colorado, USA
    697

Increasing biomass for fuel production by using waste luke-warm   703
water from industries
   C. PIRON-FRAIPONT, E. DUJARDIN and C. SIRONVAL, Photobiology
Laboratory, Liège University, Sart Tilman, Belgium

Algal fermentation, a promising step in biomass conversion   709
   K. KREUZBERG, Institute of Botany, University of Bonn, Federal
Republic of Germany

SESSION VI : IMPLEMENTATION - DEVELOPING COUNTRIES

Summary of the discussions   717
   E. DA SILVA, UNESCO, Paris, France

Invited papers

An Indian village agricultural ecosystem - Case study of Ungra   719
Village
- PART I - Main observations
   N.H. RAVINDRANATH, S.M. NAGARAJU, H.I.SOMASHEKAR,
   A. CHANNESWARAPPA, M. BALAKRISHNA, B.N. BALACHANDRAN and
   AMULYA KUMAR N. REDDY, with the assistance of P.N. SRINATH,
   C.S. PRAKASH, C. RAMAIAH and P. KOTHANDARAMAIAH, ASTRA, Indian
Institute of Science, Bangalore, India

- PART II - Discussion   727
   AMULYA KUMAR N. REDDY, ASTRA, Indian Institute of Science,
Bangalore, India

Fuelwood and charcoal in Africa   735
   E.M. MNZAVA, Ministry of Natural Resources and Tourism, Dar es
Salaam, Tanzania

Poster papers

A strategy for rural development in the Third World via biomass   752
resource utilisation
   M. SLESSER, C.W. LEWIS and I. HOUNAM, Energy Studies Unit,
University of Strathclyde, Glasgow, Scotland

The present use and potential for energy from biomass in Tanzania   758
   G. MAUER, University of Dar es Salaam, Faculty of Engineering,
Tanzania

Fractionation of biomass (sugar cane) for animal feed and fuel   763
   T.R. PRESTON, M. SANCHEZ and R. ELLIOTT, Facultad de Medicina
Veterinaria y Zootecnia, Universidad de Yucatan, Merida, Mexico

Biogas - Kenyan case   769
   P.N. KARIUKI DIC, Ministry of Energy, Nairobi, Kenya

Community biogas plants - Implementation in rural India   776
   R. ROY, Faculty of Technology, The Open University, Milton
Keynes, United Kingdom

A 4 to 6-MW wood gasification unit : scaling-up the gasification   782
technology into an unknown size category (case study Guyana)
   G.D. LINDAUER and T. KRISPIN, Agrar- und Hydrotechnik GmbH,
   Essen, Federal Republic of Germany

A design strategy for the implementation of stove programmes in   789
developing countries
   S. JOSEPH and Y.J. SHANAHAN, Intermediate Technology and
   Development Group Ltd., London, United Kingdom

The United Nations system and biogas activities   796
   E.J. DaSILVA, J.F. McDIVITT and V.A. KOUZMINOV, Unesco, Paris,
   France

SESSION VII : IMPLEMENTATION - DEVELOPED COUNTRIES

Summary of the discussions   813
   B. BERGER, Department of Energy, Washington

Invited papers

Alternative fuels from biomass and their use in transport   815
   W. BERNHARDT, Volkswagenwerk AG, Wolfsburg, Federal Republic of
   Germany

The French bioenergy programme   826
   H. DURAND, Commissariat à l'Energie Solaire, France

Planning for transport fuels from biomass - The New Zealand   838
experience
   G.S. HARRIS, New Zealand Energy Research and Development
   Committee, Auckland, New Zealand

Poster papers

Possibilities and limits of using biomass as substitutes for   854
exhaustible resources - A systems-analysis approach of producing
fuels from rape-seed in the Federal Republic of Germany
   T. BUEHNER and H. KOEGL, Federal Research Center of Agriculture,
   Braunschweig, Federal Republic of Germany

Biomass and hydrogen: an answer to the European liquids fuels   863
crisis in the 21st century
   M. MESSENGER, International Institute for Applied Systems
   Analysis, Laxenburg, Austria

Encouraging biomass use in California   870
   R.C. LANG, California Energy Commission, Sacramento, California

Optimization of an integrated renewable energy system in a dairy   876
farm
   L. BODRIA, G. CASTELLI, G. PELLIZZI and F. SANGIORGI, Istituto
   di Ingegneria Agraria dell'Università degli Studi di Milano,
   Italy

Heat supply system for the Community of Sent 883
  J. BUCHLI and J. STUDACH, IGEK Engineering Consultants,
  Switzerland

Realistic assessment of biomass energy contributions 890
  L.P. WHITE and L. PLASKETT, General Technology System Ltd. &
  Biotechnical Processes Ltd, Brentford, Middx., United Kingdom

Potential for energy cropping in Swedish agriculture 896
  K.G. BERGMAN, Swedish University of Agricultural Sciences,
  Uppsala, Sweden

Food and fibre or energy-implications of biomass energy for 903
Canada
  S. M. PNEUMATICOS, Conservation and Renewable Energy Branch,
  Energy, Mines and Resources, Ottawa, Ontario, Canada

The use of biogas for thermal, electrical & mechanical power 910
generation
  D.J. PICKEN and M.F. FOX, Leicester Polytechnic, United Kingdom

Generation of current from biogas 915
  E. DOHNE, Kuratorium für Technik und Bauwesen in der Landwirtschaft (KTBL), Darmstadt, Federal Republic of Germany

Studies of the potential for ethanol production from selected 922
biomass crops in temperate climates: an engineering approach
  C.E. DODSON, Helix Multi-Professional Services, S.R. MARTIN,
  Stone & Webster Engineering Ltd., London, United Kingdom

Use of 95%-ethanol in mixtures with gasoline 928
  A. SCHMIDT, Institute of Fuel Technology, Technical University
  Vienna, Austria

The utilization of sunflower seed oil as a renewable fuel for 934
Diesel engines
  J.J. BRUWER, B.v.D. BOSHOFF, F.J.C. HUGO, J. FULS, C. HAWKINS,
  A.N. v.d. WALT, A. ENGELBRECHT, Division of Agricultural
  Engineering, Department of Agriculture & Fisheries, Pretoria,
  and L.M. DU PLESSIS, Council for Scientific and Industrial
  Research, Pretoria, South Africa

The economics of improving octane values of gasoline with alcohol 941
additives
  P. JAWETZ, Independent Consultant on Energy Policy, New York,
  USA

Design and advance of the bioenergy in French sugar industry 947
  J.P. LESCURE, Institut de Recherches de l'Industrie Sucrière,
  Villeneuve d'Ascq, France

LIST OF PARTICIPANTS 954

INDEX OF AUTHORS 980

I N T R O D U C T I O N

In the last few years the interest and activity in methods for obtaining energy from biomass has expanded dramatically, in both developed countries and developing countries. Close examination is being given to the possibilities for using organic wastes, for growing specific plantations and crops, and to the technologies for converting the raw material to useful fuels.

As part of its expanding interest in fuels from biomass, the Commission of the European Communities in co-operation with the UK Department of Energy and assisted by an international Committee of experts, organised a Conference to discuss all aspects of the biomass technologies and their utilisation. The Conference brought together experts from many countries to present and discuss the most recent advances in research, development, demonstrations, design, manufacture, field testing and applications. It provided an international forum for the formal and informal exchange of new ideas and identification of problem areas.

The presentation of papers was considered in four main topics :
a) Resources
b) Conversion
c) New Concepts
d) Implemention

Keynote papers from invited specialists, who are leading experts in their fields, introduced each technical session. Contributed papers were presented in poster sessions where there was full scope for discussion. Rapporteurs finally presented an overview of specific topics discussed in the sessions.

The Conference was opened officially by Mr. John Moore, the Parliamentary Under-Secretary of State for Energy (U.K.). The participation by leading industrialists and administrators was particularly gratifying.

Dr. P. Chartier
Professor D.O. Hall
Conference Chairmen

SUMMARY REPORT OF THE CO-CHAIRMEN OF THE CONFERENCE

P. CHARTIER, France

D.O. HALL, U.K.

The co-chairmen used the Rapporteur reports and their own analysis of the highlights of the Conference to compile the following statements. They wish to emphasize the diversity of resources and conversion techniques available and the importance of implementing those systems which they think will be useful in developed and developing countries. Demonstrations are essential as well as appropriate Research and Development to optimize energy systems for both the short and the long term. Training and infrastructure adjustments are prime requirements for implementing biofuel schemes.

I - BIOMASS PRODUCTION FROM AGRICULTURE AND FORESTRY

1.1. Utilizing animal, crop and forest residues and introducing energy crops into conventional agricultural and forestry production needs careful system analysis including economics and ecological impacts.

1.2. Consideration of Fuel versus Food, Feed, Fiber and Chemicals open problems of competition for the residue feedstocks or for the land, on one hand, and to problems of regulation, on the other hand; regulation appears when local surpluses of food production exist (this is the case in the European Community for several products).

1,3. Ecologically acceptable solutions are needed to get an optimal sustainable yield ; the choice between furnaces or digestors for instance to produce heat from agricultural residues will be often influenced by the local soil and plant demands for organic matter and nutrients.

1.4. Continued Research and Development efforts on energy crops are required ; experimental trials must be carried out to solve problems of site evaluation, species selection, efficient fertilizer use and harvesting techniques ; the ecophysiological basis of biomass production must be simultaneously studied.

1.5. In Developing Countries, establishing energy crops for local use to reduce the pressure on natural vegetation due to collection of firewood is of prime importance to avoid desertification or flooding and other socioeconomic problems. It is essential that field training and trials at the local level is implemented.

1.6. Lignocellulosics are the most important feedstock for biofuel production ; the annual world production of lignocellulosics is more than ten times greater than those of sugars and starches ; the competition with food is thus limited to land use of more marginal lands and therefore of less serious consequence.

II - BIOMASS CONVERSION VIA BIOLOGICAL AND THERMOCHEMICAL ROUTES

2.1. Conversion techniques cannot be separated from local considerations about resources, infrastructure and end use applications.

2.2. Alternatives between biological and thermochemical routes for biomass conversion must be always considered bearing in mind the following guidelines which are only valid at present :

    - thermochemical routes are generally more efficient from an energetic point of view than the biological one's when the biomass is rather dry, which is often the case with lignocellulosic feedstocks.

    - the biological routes are generally more efficient in terms of nutrient and organic matter recycling and as far as anaerobic digestion is concerned offers the possibility of coupling depollution of liquid effluent with energy production.

2.3. As a liquid fuel, Methanol has at the present time better prospects than Ethanol in Developed and Developing Countries. The advantages of Methanol derive from the following considerations :

    - gasification of lignocellulosics is today a more efficient process than hydrolysis and fermentation,

    - substitution or mixture of Methanol ex-biomass with Methanol ex-fossil fuel such as coal and natural gas is possible ;

    - Cost analysis and Energy Balances are more favourable.

Methanol has nevertheless some drawbacks compared to Ethanol:

- Higher capital investment and more sophisticated technology are needed,

- Toxicity to human of Methanol could require its further conversion to Gasoline thus diminishing the overall performance of this route.

2.4. The fermentation route from sugars, starches and in the future lignocellulosics to higher alcohols such as Butanol is of prime importance to stabilize (with a few percentage) mixtures of Methanol or Ethanol and Gasoline.

2.5. Ethanol production from lignocellulosics must be kept as an option for fuel production ; moreover the hydrolysis R & D studies are very important for production of Feed and Chemicals (besides the fuel route).

2.6. The direct liquefaction of biomass by thermochemical means can also be kept as an option even though it is some time away from pilot plant demonstration.

2.7. Research and Development efforts must be increased in the fields of chemical engineering for gasifiers and biological engineering for anaerobic digestors. The present stages are :

(1) commercial viability exists for wood burning and Ethanol production from sugars and starches,

(2) more demonstrations are required for straw combustion, animal waste digestion and gasification of wood to produce power and electricity,

(3) pilot plants for Methanol production from lignocellulosics need to be constructed.

III - NEW CONCEPTS FOR BIOCONVERSION

3.1. The photosynthetic efficiency of plants for the conversion of solar energy seldom exceeds 1 % and for very good conditions with sugar cane and algae can attain 3 %. The practical maximum efficiency is probably about 6 % - on an annual basis for the total solar energy input.

Optimizing the overall photosynthetic efficiency of plants is thus of critical importance.

3.2. Photochemical and photobiological systems have advantages over plant systems due to the fact that they do not require arable land, water consumption, fertilizers and also their efficiencies of conversion could be higher. However as yet they are still at the early research stage but their potential is enormous.

3.3. The inherent physiological limitations in the efficiency of plants may be overcome by photobiologically mimicking the key reactions of plants e.g. water splitting for carbon fixation, production of ammonia and hydrogen to produce stored energy. An alternative approach is to design photochemical systems which split water with visible light to evolve H_2 (and O_2) and to fix carbon to organic compounds.

3.4. The use of algal ponds on non-agricultural lands has been demonstrated for example in California and Manila to be viable for disposal of urban, industrial and agricultural organic wastes. The yields of algae are about 60 tonnes dry weight per hectare and per year resulting in a 2 to 3 fold increase -via solar energy- of the original energy content of the organic wastes. The harvested algae can be fermented to biogas or also used as animal feed. Such algal ponds could be applicable in many of the sunny environments of the world wherever there are organic wastes, often polluting.

3.5. Recent research has shown that algae and photosynthetic bacteria can produce a range of valuable products such as food additives and pigments. In addition they have been shown to produce Hydrocarbons, Oil, Glycerol, Ethanol, Hydrogen gas, Ammonia, etc... under the appropriate conditions. The experience gained in the production of high value chemicals will be important in future systems where fuels may be the primary output. Increased efforts in this field are recommended because the benefits could be great.

IV - IMPLEMENTATION IN DEVELOPED AND DEVELOPING COUNTRIES

4.1. Regional and local analyses of the energy available from biomass resources to the end use of energy is of prime importance.

4.2. Overall energy balances of biofuel routes expressed as the quantity of displaced oil per unit area of land - in terms of energy or liquid fuel - is a very useful criteria to indicate the potential interest of the various routes. This analysis shows that the production of biomass per unit area of land (the bioproductivity) is often the more sensitive factor.

4.3. In Developed Countries, producing a liquid fuel such as Methanol and Butanol for the gasoline market and displacing oil for heating purposes in rural areas are the main targets.

4.4. In Developing Countries, biomass resource utilization including biofuels can be the basis of strategies for rural and even urban development as far as cooking, pumping and producing heat, power and electricity are concerned.

4.5. In Developing Countries, vegetable oils especially Palm oil are interesting candidates for direct use in diesel engines: using Ethanol from sugar or starch crops is only recommended when appropriate land is available (local decisions are imperative); Methanol could only penetrate the market by the year 1985-1990.

4.6. In Developing Countries, improvement of traditional techniques in the use of biofuels with an higher efficiency in order to reduce the pressure on the natural resources is a very important target. Gasification of lignocellulosics feedstocks to produce charcoal, electricity and heat is strongly recommended ; efficient stove and charcoal conversion programmes have to be reinforced ; biogas units are recommended especially when recycling of nutrients for agriculture is required ; the training of personnel and the development of appropriate infrastructures are crucial if biomass for energy schemes are to be successfully implemented.

4.7. More generally, once technical and economic bottlenecks have been identified and overcome, institutional adaptations, training and incentive procedures must be established.

OPENING SESSION

Chairman: Dr. A.S. STRUB, Commission of
the European Communities, Brussels

Introductory remarks
   Dr. A. STRUB, Directorate General "Energy", Commission of the
   European Communities, Brussels, Belgium

Opening speech
   J. MOORE, Parliamentary Under Secretary of State for Energy,
   London, United Kingdom

Opening speech
   Dr. L. WILLIAMS, Director General for Energy, Commission of the
   European Communities, Brussels, Belgium

Opening address
   H. BONDI, Department of Energy, Natural Environment Research
   Council, Swindon Wilts, United Kingdom

Prospects for energy from biomass in the European Community
   P. CHARTIER, Institut National de la Recherche Agronomique,
   Versailles, France

Research and development on photochemistry and photobiology -
The European Commission's programme
   D.O.HALL, University of London, King's College, London, United
   Kingdom

Podium at the Opening Session (from left to right): Dr. P. Chartier,
Sir Hermann Bondi, Mr. J. Moore, Dr. A. Strub, Dr. L. Williams,
Prof. D. Hall

INTRODUCTORY REMARKS

Dr. A. STRUB
Commission of the European Communities

Ladies and Gentlemen,

It is a great honour for me to chair the opening session of this
First European Conference on Energy from Biomass and to introduce
our eminent guest speakers who will welcome you here in Brighton.

This Conference is intended to bring together biomass experts from
many countries and we hope it will provide us all with a good picture
of the most recent advances in research, development and application
of biomass technologies for energy production.

It is a matter of fact that we in Europe - in common with many other
industrialized countries - have to make all possible efforts to reduce
our heavy dependence on oil as an energy source. This necessarily
implies that we have to call upon the so-called renewable energy
sources, even if sometimes their contribution is not as substantial
or as immediate as coal or nuclear energy.

The Commission of the European Communities is carrying out (and has
done since 1975), important R & D work in the field of biomass.
Some 36 research projects are under way within this programme, which is
managed by the Commission's Directorate XII, "Research, Science and
Education". You will hear more about this in Dr. Chartier's and
Prof. Hall's paper, to be delivered in the second half of this
morning session. I should also add here that since 1979 the Commission
has given financial support to larger scale demonstration projects,
including biomass. This demonstration scheme is managed by the
Commission's Directorate General XVII for Energy Matters.

Our energy R & D work normally involves close collaboration with
corresponding national programmes. When we at DG XII in Brussels

launched the idea of organizing this Conference, it seemed only natural and logical to call upon one of our powerful partners at the national level, to jointly organizing this important event here in Brighton. I would like to take this opportunity to thank our friends from the UK Department of Energy, in particular those from ETSU, for the marvellous job they have done.

But let me now invite Mr. John Moore to speak to us. Mr. Moore is, as you know, the Parliamentary Under Secretary of State for Energy of the United Kingdom. He has kindly agreed to open this Conference and I am sure he will present you with some very interesting views on the subject of biomass for energy production, – views by a politician who is responsible of a broad range of energy sources and energy matters in his country.

OPENING SPEECH

J. MOORE
Parliamentary Under Secretary of State for Energy

Ladies and Gentlemen,

my first duty is to bid you welcome

- firstly to this conference on Energy from Biomass;
- also to the UK and, in particular, to historic Brighton;
- and, to those of you from more distant parts of the world, welcome to
  the European Community.

It is gratifying to see the level of interest this conference has generated
attracting more than 500 delegates from all corners of the globe. I note
that some 30 countries are represented here. It is a truly international
conference and the level of participation is an indication of the current
interest in the technology and its application and in the wisdom of the
organisers, and in particular the Commission of the European Community,
in arranging such a Conference at this time.

This is the first major international scientific conference to be held on
biomass technology organised by the Commission of the European Community
and we are very pleased to be able to co-operate with the Commission in
its organisation. Within the Community biomass has come to the fore as a
possible major alternative energy source only in the last few years
although it should be remembered that it has been in use ever since man
felt the need for heat. Biomass is a dispersed resource and there are
institutional and practical difficulties in realising the contribution
it might make, and these are particularly apparent in the Community.

Using land to grow energy crops for biofuels is far from being the only use to which it can be put. The Community is a net importer of timber and in the UK we import a large part of our timber and more than half our food. Thus we have to take a critical look at the respective economics, which are constantly changing, before making any decision on land use. Even so we feel there is the potential for biomass to provide a few per cent of our total energy requirement. This, however, would need a commitment to large scale systems with considerable investments in research, development and demonstration in growing technologies and in conversion techniques. My Department is supporting a programme of work whose objective is to look more closely at these assessments and identify the technologies needed to realise the potential. Many of the projects in this programme are also supported by the European Commission and we look upon our participation in the Community programme, and the Community's participation in our programme, as an invaluable complementary exercise of benefit to both programmes.

During the course of this conference you will have the opportunity to hear something of the research which has been done since the Community's R & D programme started in 1975. This initially had a 4-year life. Dr. Strub, our Chairman from the European Commission for this session, established this programme so successfully that when the Commission suggested to the Council of Ministers in 1979 that there should be another 4-year programme building on progress from the first, the Council agreed. Closely following on this decision, but I am sure in no way linked, Dr. Strub and his Director General for Research, Science and Education, Dr. Günter Schuster, who unfortunately cannot be with us today, decided to ask us in the Department of Energy to join them in setting up the conference. We gladly agreed - especially when we discovered it would attract back for a while our ex-colleague from the Department of Energy, Mr. Leonard Williams, who is now the Director General for Energy in the Commission. We are doubly pleased to see him here as it emphasises the link between the objectives of energy policy and those of energy research, a link which we in the UK feel is very important.

However, this Conference is not merely looking at activities in the EEC. Just as sciene knows no boundaries, biomass knows no boundaries. Over half the people in the world currently use biomass for energy in the form

form of wood. The potential for biomass, particularly in developing countries, is very great, and would be particularly useful in those countries which have to import fossil fuels. It would ease chronic balance of payments problems and leave money available for other urgent needs. The availability of wood is of prime concern in promoting the use of biomass in developing countries.

The importance of biomass throughout the world is reflected in the amount of work being undertaken on it and some of this work you will be hearing about at this conference. This afternoon, for instance, we shall be hearing from Dr. Trindade from Sao Paulo, Brazil - one of the world's leading countries on this technology. It is not only experts in biomass who have heard of gasohol. The list of eminent scientists on the Scientific Committee for this conference, and where they come from, shows how truly international the topic is. In session VI the conference will be considering the implementation of biomass technology in the developing countries and this can be looked upon as a preview in this topic of the major Conference on New and Renewable Sources of Energy to be held under the United Nation's auspices in Nairobi nex August.

Ladies and Gentlemen, I hope you will find this conference a great opportunity for meeting fellow scientists working on the same topics as yourselves in different countries, for exchanging views and results of work and generally bringing yourself up to date with what is going on elsewhere. To communicate is an important part of a scientists work, not only with each other but to the outside world. The primary purpose of this conference is to provide a forum for an exchange between experts but I hope that it will also bring the subject to the attention of a wider audience. Sir Hermann Bondi, who was Chief Scientist at the Department of Energy until September and who is giving the keynote speech in a few minutes, is, I know, very conscious of the need for scientists to communicate. And, if you will forgive for a moment, I would like publically here to thank him for all he has done for us in the Department of Energy over the past 3 years during difficult times in the energy sector. It only remains for me to express the wish that you have a good and successfull conference. The organizers have provided the venue and the facilities, it is now up to you, the participants.

OPENING SPEECH

Dr. L. Williams
Director General for Energy
Commission of the European Communities

Introduction

I am pleased to have the opportunity of speaking to you today at this important and timely conference.

Ever since the first energy crisis of 1973, there has been an increasing interest in renewable energies in general and in recent years a growing awareness that biomass offers particularly promising prospects. So much so that over the next three years biomass is expected to double its share of the European Community's solar energy research and development programme.

I will not, however, intrude too much into the details of the Community's biomass programmes as these will be covered by Philippe Chartier later this morning. I should like rather to set these programmes into the broader context of Community energy policy and the international energy situation.

The Community's energy situation

Since 1973 the Community has been able significantly to reduce its dependence on imported oil thanks to energy saving, the North Sea, and increased use of gas, nuclear and coal. But our dependence is still high.

Last year oil accounted for nearly 55% of our total energy consumption and oil imports - 90% of them from OPEC - for 47%. The Community imported over 9 million barrels per day in 1979 - which is equivalent to almost 20% of the total free world oil supply - at a cost of over $ 70 bn (some 3.5% of our GDP). This year a smaller volume will cost us over $ 100 bn.

Community energy policy

There are four aspects to Community energy policy in response to this situation:

- reduced oil consumption;
- increased energy efficiency;
- diversification of our energy supply structure and development of indigenous resources; and
- encouragement of international co-operation on energy, with the other main consuming nations, the oil producers and the oil-importing developing countries.

By 1990 we want to see:

- the share of oil in total energy demand fall to around 40%;
- the ratio between the growth of energy demand and GDP fall to 0.7 or less;
- the contribution of coal and nuclear to electricity production rise to 70-75%;
- a greater role for renewable energy resources.

All this must take place alongside the implementation of sensible pricing policies for every fuel which reflect the real cost of suppplies and help to encourage both consumers and producers to take the right kind of investment decisions for the future.

Clearly, the need for action in each of these areas will vary from Member State to Member State. For example, the United Kingdom is already meeting the Community targets in two of the areas. Oil forms only 39% of the U.K.'s primary energy supply, while coal and nuclear fuel meet over 80% of the needs of electricity production. But these objectives provide a reference framework to guide national policies and an indication of the effort which the Community is making towards resolving the greater world energy problem. They give the Commission a basis for coordinating, stimulating and, if necessary, complementing national measures.

Energy R & D strategy

Energy research and development has a crucial role to play in all this:

- in energy conservation;
- in enhancing the scope for producing indigenous fossil fuels;
- in the promotion and application of nuclear fission energy;
- in the development of new energy sources (fusion, solar, etc.);
- in the development of substitutes for crude oil (for example, coal conversion);
- in the identification and development of substitutes for electricity (for example, hydrogen).

On the conservation side, we can already claim some success. In the 5 years from 1973 to 1978 the Community achieved more than 11% economic growth without increasing energy consumption. We believe that we are saving some 1.5 m b/d, the equivalent of $ 16 $\frac{1}{2}$ billion a year, mostly through national saving measures but also with the help of co-ordination at Community level. We are also meeting our commitments of the Strasbourg and Tokyo Summits to hold our oil imports at or below their 1978 level. But we are now at the stage where good house-keeping is not enough and the emphasis of effort must be on the development and introduction of new energy-efficient technologies.

On the supply side, our short term options are very limited indeed.

Community oil production might rise for a short period to $2-2\frac{1}{2}$ mb/d in the mid-1980s. Even then it would only represent 20% - 25% of our likely total of requirements. There is a need to ensure that the rate of exploration in the Community is maintained, indeed accelerated.

As for gas, Community production will peak before 1990. If gas is to go much above its present share of 17% in a larger market this can only be done by a three-fold increase in imports between now and the end of the decade.

This leaves us with <u>coal, nuclear and new energy sources</u>. Coal is the
Community's most abundant energy resource, yet it accounts for little
over 20% of total primary energy supply. Nuclear energy remains an area of
contention for energy forecasters. We believe that something over 12% of
our energy will come from nuclear sources by 1990 - the equivalent of 3 m
b/d of oil. We shall not be running nuclear-based economies, as some of
the opponents of nuclear power might have us believe. Yet, an increased
nuclear share will be very important. Without it, it is difficult to see
how we shall be able to avoid a serious risk of energy constraint on
reasonable levels of economic growth.

So what of the new sources, including biomass?

Given the length of time it takes for commercial development of any new
technology and given the scale of substitution which must take place
thereafter, we must accept that the contribution from new sources in the
next twenty years is bound to grow only at a modest pace. In the Commission
we doubt whether new sources of energy will be able to provide for the
Community more than the equivalent of 1-1.5 m b/d of oil (at the very
most 6-7% of our needs) by the year 2000. But that contribution will be a
vital one in bridging the gap between our overall demand for energy and
the available supply.

In the longer term, of course, there may be other possibilities. Energy
from fusion may possibly be available commercially by 2025 but that is,
unfortunately, quite some time in the future. The Community, the industrialised world and the Third World are going to have to pass through
some difficult moments before them.

Demonstration projects

I am optimistic that solutions to the energy problem will be forthcoming,
though whether they will come quickly enough is another matter. When I
look at the vast amount of money being spent on energy research within
the Community - currently well in excess of 2,500 MEUA (£1,500 m) - and
the diversity of effort, from photovoltaics to geothermal energy, from
nuclear safety to biomass, I am convinced that the important task of
laying the foundations for structural change is under way. But we have no

time to waste. We must pass from research and development of promising technologies to their commercial demonstration as quickly as possible.

This is the reason for the Community's programmes of support for demonstration projects in the energy field. We are now financing two separate programmes - one in energy saving and the other in the fields of solar energy, geothermal energy and coal gasification and liquefaction - each spread over a number of years.

60% of the finance allocated under our solar programme is devoted to biomass projects, including two major schemes - short rotation forestry in the Republic of Ireland and a French scheme for producing energy from sugar cane biomass on Reunion.

In the development of biomass we in the Community can learn a number of lessons from the developing countries. Those of you from the Third World will know of all the problems in the use of wood as a fuel. For many rural communities it remains the most important source of fuel for cooking and heating. One pressing problem is how to adapt technology to use it more efficiently and to avoid the risk of deforestation. In Western Europe wood was the major fuel in the middle of the last century but the eras of cheaper coal and oil reduced it to a marginal source. Today it supplies less than 1% of the Community's energy needs. But in our new energy circumstances we are rediscovering its importance in Europe. As we do so, we must draw on the experiences elsewhere.

Conclusion

Speeding up the introduction of appropriate biomass technology is a major technological, economic and social challenge to us all. I was struck by the conclusions of Sir George Porter at the conference on solar energy for development which the Commission held in Varese in 1979. Drawing on Linus Pauling he suggested that in the industrialised countries at least more technology is unlikely to increase happiness. Technology should therefore try more modestly to reduce unhappiness in the world. If here at this conference we can address our thoughts to answering how biomass can be developed to reduce the unhappiness of the world, the conference will be truly worthwhile.

OPENING ADDRESS

H. BONDI
Chief Scientist
Natural Environment Research Council

Energy should always be considered in the global context. Here the
striking fact is that the majority of mankind lives in developing
countries, and that in many of those the availability of food is by no
means continually assured. It must be the hope and indeed the expectation
of us all that there will be great advances in all those countries so
that the spectre of hunger and poverty will be banished. This will
require major developments in irrigation, in fertiliser supply, in
transport, in food storage, in food distribution. Note that every one of
these is an energy intensive business

Next, if we look at the leading developing countries then it is plausible
that in the next twenty or thirty years their advance will centre on
steel, on shipbuilding, on aluminium, perhaps on heavy chemicals (in just
the manner in which 30 or 50 years ago textiles was the way into industria-
lisation for countries). Again note that all those industries are energy
intensive.

By contrast, the growth industries in the advanced countries look like
being in computing, telecommunications, education, leisure industries,
entertainment, financial services, engineering consultancies. None of
these is energy intensive.

Thus we should not only expect but hope that global energy use will
increase rapidly, not because of growth in the industrialised countries,
but because all the developing countries from the poorest to the more
advanced will require greatly increased amounts of energy for progress.

Where is this energy to come from? The need is such that all potential
sources will have to be pressed into service, together with determined
measures to improve the efficiency of energy use. Therefore I detest

the term "alternative energy sources" which suggests that there is far more choice than we actually have. In particular it is always essential to keep the global picture in mind. When we in a nice country consider whether to sink a new coal mine or not, whether to build a nuclear power station or not, we must always remember that depriving ourselves of such sources of energy will inevitably increase our consumption of oil, reducing its availability and increasing its price for the poorest in the world, who so often need it most. I am not saying that this consideration always need to be decisive, but to make choices without keeping it in mind would certainly be very wrong. Naturally, the renewable (rather than "alternative") sources must be determinedly examined for their technical feasibility, safety, attractiveness and economics. Tidal energy has the rotation of the Earth as its primary source, geothermal energy the radioactivity of the Earth's interior, but all others derive from the Sun, a distant fusion reaction on whose steadiness we utterly depend. Even such small fluctuations as the geological record does not rule out could have grave implications for mankind.

The outstanding advantage of biomass among the solar sources of energy is its built-in storage. Whether in growing plants or in wastes, the choice of the time of conversion is largely up to us, and the resulting fuel itself can be stored, in sharp contrast to many other forms of utilising solar energy.

Its multivarious nature makes the examination of biomass for its technical feasibility, environmental impact and economic viability a substantial task. Among the economic constraints is the competition for land, for water, for nutrients, between food crops and energy crops, a constraint not applicable to the utilisation of wastes,where, however, the necessary gathering and transport can turn out to be an economic handicap.

One of the thrilling scientific questions is whether chlorophyll has been evolved under survival pressures to maximise efficiency, or whether other pressures have been dominant. Our chances of improving on the efficiency of chlorophyll depend on this.

There are many intricate questions on biomass. I am very pleased that my

organisation, the Natural Environment Research Council of the UK, is
energetically pursuing such researches under commissions from the EEC and
from the UK Department of Energy.

There are plenty of unresolved questions on biomass, and the organisers
of this conference are to be congratulated for having brought together,
at a timely moment, such a distinguished gathering from so many countries.
I am certain that the discussions, formal and informal, taking place here
will be of great help in advancing this important subject.

PROSPECTS FOR ENERGY FROM BIOMASS IN THE

EUROPEAN COMMUNITY

P. CHARTIER

Project Leader EC Programme on Energy from biomass (DG XII)

Institut National de la Recherche Agronomique,
Versailles, France

Summary

An Energy from Biomass Programme in the European Community has to be adapted to the various European regions which are generally characterized by a relatively low availability of land. The sources of biomass are first the crop residues, the animal wastes and the unused forest by-products. The introduction of energy crops at moderate level in conventional agriculture and forestry - perennials on available land and catch crops - could cover local energy needs by the year 2000 and provide in addition a small percentage of gasoline demand via methanol from wood or other ligno-cellulosic material. The potential contribution could be about 75 Mtoe by the year 2000 which represents 7,5% of the 1980 EC energy demand. The most promising routes have been identified. The utilisation of biomass to cover local energy demand for heat in rural areas in a more or less sophisticated way is a very strong competitor for the feedstocks. On the other hand, prospecting the feasibility of cultivating algae could give a long term response to an increase of the pressure over land use. As far as the liquid fuel demand is concerned, the methanol route appears more promising than the ethanol one (which could be devoted to chemical feedstocks). The cost of dry biomass varies from 35 US dollars for baled straw on farm to 90 US dollars per tonne for energy crop at factory. The various actions of the Commission in the biomass field are presented, the Research and Development Programme being more detailed.

1. INTRODUCTION

Energy from biomass in the European Community has to be adapted
to a situation characterized by a relatively low availability of land, a
high density of population (170 inhabitants per square kilometer) and a
high standard of energy consumption (4 tonnes of oil equivalent per capita). Nevertheless the still high and increasing level of crop yield
(4.7 tonnes per hectare of wheat grains for instance for the year 1978)
will offer some opportunity for energy plantation on land abandoned by
the conventional food farming, though new land requirements for wood production would have to be fulfilled. Presently the crop residues, the
animal wastes and the unused forestry by-products offer immediate opportunity to learn the technologies of using biomass and can fulfil by the
year 2000, 3 % of the 1980 EC energy demand (1200 Mtoe[1]).

So, the first step of biofuel development in European Community
(1975-1985) consists in using residues mainly to cover local energy needs
(low temperature heat especially) without any disturbance to agriculture
and forestry.

The second step (1980-1995) is the introduction of energy crops -
perennials on available lands and catchcrops - with low disturbance to
agriculture and forestry, to cover mainly local energy needs and a very
low percentage of gasoline demand.

The third step (1985-2000) corresponds to the development of energy plantations with some significant changes in agriculture and forestry,
as long as the community demands for food and wood can be satisfied and
the production permits some necessary international exchanges. The biofuels which correspond to the third step could satisfy a fraction of the
gasoline demand especially through methanol and a fraction of the solid
fuel market through dry wood chips and granulates.

The fourth step (beyond the year 2000) corresponds to the possible
emergence of new routes of bioconversion which could be a response to an
increasing pressure on agriculture and forestry lands. Algae production
is carefully studied for this reason in the frame-work of the EC R&D Energy
from Biomass programme.

(1) Mtoe = million tonnes of oil equivalent

2. BIOFUEL POTENTIAL IN THE EC 9 [1]

A first attempt of biofuel potential in the European Community has been made by WHITE and al (1980). The energy content of agriculture and forestry wastes is about 90 Mtoe. The energy content of biomass from energy crops and plantation schemes not involving major disturbance to forestry and agriculture has been estimated at 95 Mtoe. A lower envelope (table 1) has been defined afterwards (CHARTIER 1980) which accounts for various restraints to be applied to the potential (high dispersion, competitive utilization, soil humus requirement). The values quoted in table 1 must be considered as the upper limit of biomass feedstocks which could be used for energy by the year 2000 in each country. Nearly 40 Mtoe could be provided equally from crop residues, animal wastes and wood by-products. From 6 to 7 Mtoe are already produced via firewood and wood recycling for energy in industry. The energy crops or plantations could produce about 36 Mtoe : 28 Mtoe as dry lignocellulosic biomass such as short rotation forestry and 8 Mtoe as catch crops (wet biomass).

This contribution is strongly dependent upon land availability by the year 2000. It is envisaged (table 2) that the increased crop productivity and a better use of unproductive forest and marginal farming lands could offer some opportunities for increasing the area of productive forest (table 3) and for developing energy plantations and energy crops. The EC agriculture policy is characterized by a surplus of food which is funded at a high rate compared to the non member state one and a deficit of wood which is not supported. So that the future might be organized to fulfil the wood requirement and to regulate the surplus of food due to the predictable increase of crop and animal productivity by introducing energy plantations and energy crops in the agricultural and forestry systems. From 5 to 10 millions of hectares out of the 152 Mha could be used in this way by the year 2000.

3. BIOMASS FEEDSTOCKS AND CONVERSION ROUTES

The biomass feedstocks can be classified in :

- feedstocks available massively without any harvesting (animal manure, wood residues in industry, etc...) the cost of which being

(1) European Community, 9 member states

sometimes negative (polluting effect of sewage)

        - residues which have to be harvested before use such as
straw or unused coppice

        - energy crops which need land, husbandry and harvesting.

The cost of these feedstocks is higher for off-farm uses than for
on-farm ones (table 4) The energy content of one tonne of dry matter
being about equal to that of 0.4 tonne of oil equivalent it can be derived from table 4 that biomass is a raw material the cost of which is
often close to the price of coal.

The conversion routes from biomass to useful energy are numerous.
They can be classified according to the state of their technological
development and to their capacity to compete with fossil fuel especially
from the cost point of view (table 5). It appears that firewood and straw
bales are already cheaper than fuel for heating demand in rural area.
This conclusion has been drawn by a number of home users, as shown by
the increase in firewood consumption. The electricity generation using
air gasifiers is also competitive when the Diesel unit - in the range
50 KW to 1 MW - is used as reference. It is a conversion route especially
suitable for developing countries. In European conditions, the most competitive routes are mainly related to heat demand at relatively low temperature in rural areas. The cheapest fuel for mobile power route appears
to be methanol from wood ; compressed methane appears in second and
ethanol last. The comparison between methanol and ethanol as liquid fuel
or chemical has been made (table 6) ; the conclusion in either European
conditions is that methanol is more suitable for mobile power, and
ethanol for chemicals.

An increasing pressure on land by the first decades of the next
century could derive from an intensive utilization of biomass as fuel and
chemicals. This eventuality justifies a comprehensive set of R & D in the
algae field, though the present cost of this kind of biomass in an order
of magnitude higher than the terrestrial one (table 4). Little is known
about the practice of cropping algae, so that we have to wait for a lear-
ning curve that might reduce the future costs (table 4).

4. THE COMMISSION OF THE EUROPEAN COMMUNITY ACTIONS ON SYNFUELS FROM BIOMASS

The Research and Development programme of the Directorate General for Research Science and Education (DG XII) was the first and remains the most important action of the Commission about energy from biomass.

The EC fundings for demonstration as far as the feasibility studies of some particular bioconversion routes have been established previously by the R & D programme is under the responsibility of the Directorate General for Energy (DG XVII).

The impact on agriculture and forestry policy is carefully examined by the Directorate General for Agriculture (DG VI).

The utilization of the know-how for developing countries is funded by the Directorate General for Development (DG VIII).

The research and Development programme of the DG XII (10.7 millions of US dollars over 4 years [1]) is focussed on a set of actions which derived from the above considerations and on pilots for gaseification of wood to methanol. The various items are (table 7) :

    1) utilization of agricultural and forestry residues :
      11. assessing and harvesting the available ressources (59 F, 75 F, 78 I),
      12. anaerobic digestion to methane (10 UK, 15 EIR, 79 NL, 80 NL),
      13. thermal processing (2 UK, 4 DK, 6 DK, 71 F, 107 NL, 112 UK),
    2) production of biomass for energy :
      21. catch crops and energy crops (25 UK, 76 F, 96 UK),
      22. silviculture (13 UK, 67 UK, 69 IRL, 81 EIR, 116 F),
      23. algae (3 F, 7 F, 21 B, 34 D+I, 70 I, 105 UK),
    3) conversion of biomass to liquid fuel
      31. gaseification to synthetic gas (23 F, 24 D, 51 F, 91 F, 106 UK),
      32. catalytic hydrogenation of wood (19 B)
      33. cellulolysis by biological routes (60 I, 82 F, 87 NL).

[1] the projects are funded at 50% rate or lower, the rest being funded by the member states or the companies.

5. CONCLUSION

As long as the stress on energy resources is kept increasing, a growing part of the energy needs will have to be fulfilled by calling upon local resources everywhere in the world. Even in the European Community where the land availability is low compared to the energy demand such a programme can be significant. Moreover the production of synfuels from biomass can act as a regulator of the agriculture policy. Last but not least, the European know-how in the conversion technologies are available for developing countries which can find in synfuels from biomass a part of the response to the energy crisis.

REFERENCES

1. CHARTIER, P. 1979 - European Community Biomass Programme - UK ISES Meeting on Biomass for Energy. July 3rd. London.

2. CHARTIER, P. ; PALZ, W. 1980 - The Second European Community Programme on Energy from Biomass. Bio'Energy 80, Atlanta, April 1980 (under press).

3. CHARTIER, P. 1980 - Possibilités de Valorisation Energétique de la Biomasse dans la Communauté Européenne. Rapport préparé pour la Direction Générale de l'Energie de la Commission des Communautés Européennes. (unpublished).(The tables 1, 2, 3, 4 and 5 are taken from this work).

4. WHITE, L.P. ; PLASKETT, L.P. ; LOW, J.B. 1980 - Overview of opportunities for Energy from Biomass in the European Communities. W. PALZ and P. CHARTIER ed., Elsevier (under press).

5. X., 1978 to 1980 - Reports of the seminar of contractors of the EC Programme (DG XII). Versailles (F), Brussels (B), Taormina (I) and Amsterdam (NL). Unpublished papers.

TABLE 1 - Maximum Energy content of dry and wet crop residues, animal wastes (bedding included), forestry and wood wastes, dry and wet biomass from energy productions, which could be harvested as energy feedstocks by the year 2000 in the European Community (9 members states).

	Animal wastes(1) CH₄ Mtoe(3)	Wet crop residues(1) CH₄ Mtoe	Dry crop residues(2) Mtoe (3)	Forestry and wood wastes (2) Mtoe	Firewood (2) Mtoe	Total harvestable Mtoe	Energy production wet CH₄ Mtoe	Energy production dry Mtoe	Total Biofuels Mtoe
UK	2,0	0,1	1,1	0,3	0,2	3,7	1,0	7,2	11,9
EIR	0,6	ε	ε	ε	ε	0,6	0,1	2,4	3,1
DK	0,5	ε	1,1	0,1	ε	1,7	0,5	0,5	2,7
NL	0,6	0,1	ε	0,1	ε	0,8	0,1	ε	0,9
B&L	0,5	0,1	ε	0,3	0,3	1,2	0,2	ε	1,4
D	2,4	0,2	2,5	1,3	1,9	8,3	1,6	1,9	11,8
F	3,7	0,6	4,8	4,5	2,8	16,4	3,4	12,7	32,5
I	1,2	0,4	1,5	1,3	1,4	5,8	1,6	3,4	10,8
CE 9	11,5	1,5	11,0	7,9	6,6	38,5	8,5	28,1	75,1 (4)

(1) As net production of methane, 40 % of the production being used to heat the digestor

(2) Energy content of the solid fuel

(3) Straw used as bedding incorporated in animal wastes

(4) 6 % of the estimated consumption (EC 9) by the year 1985

TABLE 2 - Land use in the European Community (EC 9)

(millions of hectares)

	1980	Prediction 2000
Total Land Area	152	152
Utilisable Agricultural Land	87	80 (1)
Productive Forest Land (Timber)	19	26 (2)
Unproductive Forest Land & marginal farming land	23	14
Urban and Waste Land	23	25
Energy Plantation or Energy crop	0	7

(1) Hypothesis: - constant food consumption per capita
            - 260 millions of inhabitants by the year 1980,
              280 by the year 2000
            - self sufficiency for food without surplus
            - productivity improvment (R&D in agriculture): 1 % per year

(2) See table intitled "Wood Budget in the EC"

TABLE 3 - Wood budget of the European Community (EC 9)

(from the Forestry Policy in the European Community, 1978)

	1980	2000[1]	2030[1]
Consumption per capita (m^3)	0.85	0.85	0.85
Total consumption, firewood excepted, $(10^6 \ m^3)$	220	240	258
Forest extraction	72	91	191
Recycling	28	39	64
Imports	120	110	3
Productive forest land $(10^6 \ ha)$	19	26	26

(1) Hypothesis: - constant wood consumption per capita
            - 260 millions of inhabitants by the year 1980, 280 millions
              by the year 2000, 300 millions by the year 2030
            - self sufficiency for wood products by the year 2030
              except for tropical wood

TABLE 4 - Feedstock cost in the European Community

(per tonne of dry matter)

	FF 80	£ 80	US $ 80
- Firewood or baled straw on farm	140	14	35
- Cut straw on farm or baled straw off farm	190	19	48
- Cut straw at factory	240	24	60
- Wood chips from natural coppice at factory	250	25	63
- Dry energy crop, cut, on farm or about	270	27	68
- Wood chips from energy sylviculture, at factory	300	30	75
- Catch crops on farm or about	300	30	75
- Dry energy crop, cut, at factory	350	35	88
- Granulates, off factory	480	48	120
- Algae (1980)	2000/5000	200/500	500/1250
- Algae (2000)	450/650	45/65	110/160

TABLE 5 - Classification of the routes from Biomass to useful Energy

Cost of the fuel solution taken as 100 in 1980

(all taxes excluded)

Cost	Route	Energy
< 80	Firewood and baled straw on farm, heating and cooking	HEAT
80 to 120	Liquid manure to methane on farm, depollution cost deduced, heating and cooking	HEAT
120 to 150	Wood or straw, gasifier, electricity generation (reference Diesel, fed with fuel - less than 2 MW)	ELECTRICITY (1)
	Wood, boiler, electricity generation (reference boiler with fuel more than 5 MW)	ELECTRICITY (1)
	Straw for grain drying	HEAT
	Granulates, home heating	HEAT
	Wood chips, local heating unit	HEAT
	Wood chips to methanol	LIQUID FUEL
150 to 200	Solid manure to methane, on farm, heating and cooking	HEAT
	Solid manure to compressed methane (250 bars), local unit off farm	GASEOUS FUEL
	Liquid manure to methane on farm, depollution cost deduced, electricity	ELECTRICITY
	Sugar Beet (quota B) and Jerusalem Artickoke to Ethanol	LIQUID FUEL
200 to 300	Liquid manure to compressed methane on farm, depollution cost deduced	GASEOUS FUEL
	Sugar Beet (quota A) to Ethanol	LIQUID FUEL
	Catch crops to compressed methane, local unit off farm	GASEOUS FUEL
> 300	Solid manure to methane, on farm, electricity	ELECTRICITY

(1) Developing countries

TABLE 6 - Liquid fuel from biomass ethanol or methanol ?

- net productivity (toe ha^{-1}year^{-1})	M > E [1]
- economics	M > E
- potential of resources (biomass, coal, etc ...)	M > E
- pressure on agricultural land	M > E
- technical feasibility	M = E
- industrial development	E > M
- utilization as motor fuel	M = E

Conclusion : methanol for mobile power

[1] M > E means that methanol is more favourable than ethanol for the subject quoted in the first column.

TABLE 7 - Project E : List of contractors

Proposal n°	Organisations	Scientific Responsible
ESE/002/UK.	Queen Mary College University of London	Prof. M.W. THRING
ESE/003/F.	Ecole Nat. Sup. de Chimie Paris	Dr E. CASADEVALL
ESE/004/DK.	Royal Veterinary and Agricultural University	Prof. T. TOUGAARD PEDERSEN
ESE/006/DK.	Royal Veterinary and Agricultural University	Prof. T. TOUGAARD PEDERSEN
ESE/007/F.	Soc. Française des Pétroles BP	Dr C. GUDIN
ESE/010/UK.	University College, Cardiff	Dr D.A. STAFFORD
ESE/013/UK.	University of Aberdeen	Prof. J.D. MATTHEWS
ESE/015/EIR.	University College, Galway	Dr Emer COLLERAN
ESE/019/B.	Université Catholique de Louvain	Prof. B. DELMON
ESE/021/B.	Université Catholique de Louvain	Prof. E.J. NYNS Prof. H.P. NAVEAU
ESE/023/F.	Novelerg	Dr P. DUBOIS
ESE/024/D.	Imbert - Energietechnik GmbH und CO Kommanditgesellschaft	Dr. W.O. ZERBIN

TABLE 7 (continued)

Proposal n°	Organisations	Scientific Responsible
ESE/025/UK.	Natural Environment Research Council	Dr T.V. CALLAGHAN Dr S.W. GREEM
ESE/034/D.I.	Rheinisch-Westfälische Technische Hochschule Aachen/Instituto di microbiologia Firenze	Prof. Klaus WAGENER Prof. FLORENZANO
ESE/051/F.	Creusot-Loire	Dr G. CHRYSOSTOME
ESE/059/F.	C.N.E.E.M.A. - Antony	Dr. J. LUCAS
ESE/060/I.	Universita degli Studi di Napoli	Prof. F. ALFANI Prof. Liliana GIANFREDA
ESE/067/UK.	Forestry Commission	Dr D.R. JOHNSTON Dr M.L. PEARCE
ESE/069/EIR	Irish Peat Development Authority	M. J. HEALY
ESE/070/I.	C.S.A.R.E.	Prof. U. CROATTO
ESE/071/F.	C.N.E.E.M.A./ THIROUARD S.A.R.L. PROMILL / Alsthom-Atlantique	Dr J. LUCAS M. M. MORIN M. J. CARRASSE
ESE/075/F.	Inst. Nat. de la Recherche Agronomique	Dr P. CHASSIN Dr C. JUSTE Dr J.C. SOURIE
ESE/076/F.	INRA / C.N.E.E.M.A.	Dr M. ARNOUX Dr P. RAYNAUD
ESE/078/I.	Ente Nazionale Cellulosa e Carta	Prof. G. SCARAMUZZI
ESE/079/NL.	I.B.V.L. Wageningen	Dr B.A. RIJKENS
ESE/080/NL.	Agricultural University Wageningen	Dr G. LETTINGA
ESE/081/EIR.	An Foras Taluntais	Dr M. NEENAN
ESE/082/F.	National Institute of Applied Sciences of Toulouse	Prof. G. DURAND Dr. G. GOMA
ESE/087/NL.	Agricultural University Wageningen	Prof. W. PILNIK
ESE/091/F.	C.N.E.E.M.A. / Centre Techn. Forestier Tropical / Elf Aquitaine / Total Energie Développement / Creusot-Loire	Dr. J.F. MOLLE Dr. G. PETROFF M. R. BATTAIL M. J.P. MIVHZUC M. M. du CREST
ESE/096/UK.	University of Reading	Prof. C.R.W. SPEDDING
ESE/105/UK.	University of Liverpool University of Nottingham	Dr J.M. JONES Dr. D.M. DAVIS
ESE/106/UK.	Foster Wheeler Power Products Ltd	M. H.T. WILSON
ESE/107/NL.	Twente University of Technology	Prof. W.P.M. VAN SWAALJ
ESE/112/UK.	The University of Nottingham	B. WILTON
ESE/116/F.	I.N.R.A. Recherches Forestières	Dr J.F. LACAZE Dr E. TEISSIER du CROS

RESEARCH AND DEVELOPMENT ON PHOTOCHEMISTRY AND PHOTOBIOLOGY - THE
EUROPEAN COMMISSION'S PROGRAMME

D O HALL

Project Leader, EC Programme (DG XII, Solar Energy Project
Project D)
University of London King's College, London SE24 9JF

Summary

The production of hydrogen and other fuels by photochemical
photoelectrochemical or photobiological dissociation of water is an
important possibility for the utilization of solar energy. The general
objective is the study in natural and synthetic systems of oxidation-
reduction reactions and the photolysis of water to produce hydrogen or
other fuels. It includes the following areas: understanding of the
photoconversion mechanism; photochemical production of fuels and/or
electricity; improvement of hydrogen production via living cells,
construction of artificial systems based on photosynthesis models.

The following four areas were given preference in the selection of
25 contracts for the 2nd phase of the programme (1980-3):

1.   Conversion of light into electricity, hydrogen and/or fixed carbon
        by photochemistry, e.g. in semiconductor/electrolyte cells
        (Subject I - Photochemistry; Subject II - Photoelectrochemistry)

2.   Conversion of light into hydrogen or other fuels; study of photo-
        synthetic membranes; enzymes and their incorporation into matrices;
        (Subject IIIa - Photobiology mechanisms)

3.   Biological hydrogen production in algae and bacteria, including
        genetic adaptation and selection of new types of organisms and
        species
        (Subject IIIb - Photobiology cells and genetics)

4.   Construction of artificial systems by modelling of biological
        systems
        (Subject IV - Combined systems)

Cooperation between photochemists, photobiologists and physicists
both in academic and industrial laboratories is encouraged throughout the
programme.

A.   INTRODUCTION

The production of hydrogen or other fuels by the photochemical or
photobiological dissociation of water is one of the most attractive
possibilities for the utilization of solar energy. The substrate (water)
and the sunlight "interact" via a catalytic system to produce hydrogen (or
possibly fixed carbon or nitrogen instead, in newly studied systems). This
is a very new area of research having only been developed from discoveries
made in 1972 in the chemical and biological fields. The European
Commission was far-sighted in its formulation of a Solar Energy Programme
in 1974 to include this field of research. It is an area of basic directed
research which could have significant implications for solar energy conversion systems in the future. Already important discoveries and developments
are being made with increasing frequency and we in Europe are certainly in
the forefront of these developments.

Some references are given at the end of the article to give the
reader some assistance for further reading.

The compilation of this short report was with the help of Drs
A Mackor (Utrecht), D.von Wettstein (Copenhagen) and P Vignais (Grenoble).
I thank them very much.

B.   PHOTOCHEMISTRY AND PHOTOELECTROCHEMISTRY (Subjects I and II)

Following the rapid development of the dry photovoltaic cell for the
generation of electricity by sunlight, a new concept in recent years is the
wet photoelectrochemical (PEC) cell, consisting of a photosensitive semiconductor material in contact with liquid, normally aqueous, electrolyte
and a counter electrode. These cells can be used for the generation of
electricity by the addition of a suitable redox couple to the solution, or
they may produce hydrogen by photocatalytic water electrolysis. Current
materials for electricity production are GaAs or the so-called layered compounds like WSe_2; for water cleavage   the favourite materials so far
are the titanates (TiO_2/$SrTiO_3$). Carrying out this type of chemistry at
the interface of a solid semiconductor and a liquid is not an easy task
and a number of problems have been recognised. These problems exist in
three areas, viz. stability, spectral response and efficiency, and production and maintenance costs.

A second approach to this subject is the development of photochemi-

cal cells for hydrogen production by water cleavage, which differ from PEC
cells principally by the fact that in the former light absorption takes
place in solution, while in PEC cells a solid is the light-absorber.
Within Project D both directions are represented and the final solution
may well be a hybrid system of both principles.

In the first period (1975-79) of Project D we saw a rapid development of these areas all over the world. New options have been opened and
some of these have already disappeared. A large part of the work does not
belong to one discipline specifically, since most of the chemistry is taking place at interfaces and we find in many cases that existing theories
are not capable of predicting the new results. Therefore, basic inter-
disciplinary research and efforts aimed at application, go hand in hand.
Achievements in the past period include a liquid-junction solar cell for
the generation of electricity with 12% efficiency (based on the incoming
solar energy), a non-biased PEC cell for water-cleavage and an efficiency
for hydrogen production over 1% using a doped single-crystalline TiO_2-
electrode.

Very recently, a Swiss group has demonstrated that hydrogen and
oxygen can be produced in small quantities from a photochemical cell using
heterogeneous catalysts. In this system, light is absorbed by the transition-metal complex ruthenium tris (bipyridine), which then transfers an
electron to a relay compound, methylviologen (also called paraquat). In
the presence of a platinum catalyst, the latter compound produces hydrogen.
The oxidized ruthenium complex at the same time generates oxygen under the
influence of a ruthenium dioxide catalyst. In view of the limited stability
of transition-metal complex and catalysts and the rather low efficiency of
this system, the search for more suitable compounds and systems continues.

To stabilize and improve this effect, presently research in and
outside the EC is aimed at the use of micro-heterogeneous, colloidal
suspensions of the semiconductor materials such as TiO_2, as carriers for
these catalysts. French and other groups are working on an even simpler
system consisting of a metallized semiconductor in colloidal form. For
this purpose, strontium titanate is an attractive and stable material.
However, its visable absorbing properties have to be improved.

Another source of inspiration is the natural photosynthetic system
as one realizes that water splitting according to the following equation:-

$$2 \; H_2O \longrightarrow 2 \; H_2 + O_2$$

leads to both hydrogen and oxygen evolution; it may well be that the oxygen-producing part of the system, because of its corrosive oxidising properties, is the bottleneck. Therefore a study of model systems, having some resemblance to the oxygen-producing part of photosynthesis, will help us in selecting those photochemical systems in which an effective oxygen production takes place. In mimicking nature, one should also consider the photo(electro)chemical reduction of carbon dioxide to give methanol and other potential fuels. The feasibility of such an approach has recently been demonstrated. It widens the scope of the project by making use of the abundantly available CO_2 and N_2 for fixation to reduced compounds which are then valuable chemicals.

C.   PHOTOBIOLOGY-MECHANISMS (Subject IIIa)

All of our food, and much of our energy, comes ultimately from the utilisation of the sun's energy by green plants. Leaves and algae form sugars and other high energy substances by the process of photosynthesis. Inside the cells are the microscopic bodies called chloroplasts in which the conversion of solar energy into chemical energy takes place. The initial steps of this process involve the harvesting of the radiant energy by the green pigment chlorophyll, which is associated with proteins that are assembled together with lipids in a structure known as a thylakoid membrane.

The composition of this membrane is complex - the arrangement and size of many of the different components is being elucidated by a number of laboratories. In the part of the programme concerned with the mechanisms of photosynthesis they have studied the various components of this assembly, their chemical make-up, their functional properties, their ways of function-ing cooperatively and how they are constructed. For example, we now know that the first reduced (high energy) compound which can be identified in the membrane is an iron-sulphur protein which has much more energy in it than the hydrogen reduction.

The structure of the membrane can be examined for instance with the aid of the electron microscope, which reveals many small particles inside the membrane. These particles carry out the conversion of sunlight into chemical energy and are only 16 nm (1/60,000,000 cm) in diameter. Light is first collected by the light-harvesting chlorophyll-protein complex (LHC) which can now be isolated in a purified form. The initial stages of light capture and change separation occur in picoseconds (10^{-9} seconds) and result in a change separation across the membrane - this is the first stage of energy storage in the photosynthetic mechanism. This membrane is extremely efficient in functioning as a solar energy converter and works much better than man-made devices. If we can learn how the chloroplast membrane is able to convert sunlight into chemical energy, we should be able to design analogous artificial systems to do the job for us.

D.   PHOTOBIOLOGY - CELLS AND GENETICS (Subject IIIb)

One attractive method of solar energy conversion uses intact organisms as biological catalysts to produce hydrogen or ammonia at the expense of solar energy. Research with cyanobacteria (blue-green algae) and green algae has demonstrated that these organisms are capable of producing hydrogen and oxygen through the direct photodecomposition of water in a process that uses the system of photosynthesis already genetically engineered by nature. Photosynthetic bacteria, while unable to split water, have been shown to be capable of producing large quantities of hydrogen (without contaminating oxygen), or ammonia, with light and simple organic or inorganic substrates. Since suitable substrates are available as industrial wastes, solar energy conversion with photosynthetic bacteria might be used in conjunction with waste treatment.

Some of the ongoing research is directed to understanding, at the molecular level, the various processes of hydrogen production and the mechanisms involved in the water splitting reaction. Hydrogen production can be mediated by either hydrogenase or nitrogenase, and the properties of these enzymes from various sources is under investigation, including the regulation of enzyme synthesis and activity, and the relative oxygen stability. The production of reducing equivalents, and the flow of electrons to these enzymes, which may limit activity in some cases, is also an important area of research. These studies should lead to an understanding of the limits of these processes, and thus, to what can be done to optimize hydrogen production with the genetic types of organisms presently available. In addition, several of the projects are using genetic techniques to select mutants with an enhanced potential for hydrogen or

ammonia production. Examples of desirable genetically engineered characteristics include: hydrogenases with oxygen stability, increased cellular
hydrogenase content, and the ability to excrete newly fixed nitrogen as
ammonia.

Previous studies have partially established the physiological and
biochemical basis for H_2 and NH_3 production by photosynthetic organisms,
and have determined the necessary conditions for the maximum expression of
this activity. Future studies with whole organisms will maximise long
term H_2 and NH_3 production. In addition, the technical feasibility of
using photosynthetic organisms to produce H_2 (NH_3) at the expense of light
energy will be studied by building laboratory scale bioreactors with
immobilised cells.

E.   COMBINED SYSTEMS (Subject IV)

Hydrogen evolution from water   An exciting development in renewable
energy research was the discovery about seven years ago, that light could
split water into hydrogen and oxygen using membranes containing chlorophyll
with enzymes added to the water to act as catalysts. This was the first
step toward the development of a "biological" system that can turn solar
energy into a useful fuel, hydrogen. In those early days the process
continued for only minutes before the system "died" - a reflection of the
instability of biological components when removed from their natural
environment. Over the past few years a number of laboratories have prolonged the life of the system over 20-fold so that it carries on producing
hydrogen for 10 hours or more. At the same time the rate of hydrogen
production has gone up by an order of magnitude. Photochemists and photobiologists in industry and universities are interested in this problem,
because of the energy opportunities that solar hydrogen production from
water could open up and because photolysis has unique attributes that are
unmatched by any other known energy system: the substrate (water) is
abundant; the energy source (sunlight) is effectively unlimited; the
product (hydrogen) can be stored and is non-polluting; and the process is
completely renewable because when the hydrogen is "consumed" the substrate,
water, is regenerated. Another attraction of this system is that it
operates at normal ambient temperatures and does not involve toxic
intermediates.

The problems in constructing a stable biological system that will function for years rather than hours are enormous and may never be solved. However, if we could understand how the biological processes work and then possibly mimic them by constructing a completely synthetic system we might eventually be able to harness the solar energy that is available in temperate and hot climates. Such systems would use light at all intensities and temperatures; an attribute they would share with electricity-generating photovoltaic solar cells. Only the intensity of the light would determine the rate of hydrogen production. However, photobiological hydrogen production is still at the experimental stage and there has to be sustained and diverse research before it is ready for further development. Having said that, any "breakthrough", especially in the water-splitting reaction by photobiology or photochemistry could quickly change our idea of the future practicability of such a system.

With a mixture of membranes and enzymes under optimal conditions of pH (about 7) and temperature (about $25^{\circ}C$) hydrogen has been produced at a rate of about 50 micromoles per hour per milligram of chlorophyll. If scaled up this would correspond to a production rate of about one litre of hydrogen per hour per gram of chlorophyll. The approach is to replace the short-lived biological components with more stable synthetic chemicals that can do the same job.

Researchers have chemically synthesised a number of 4Fe-4S clusters which mimic the biological active centres of the enzymes and have shown that these iron-sulphur analogue compounds, some with small peptides attached, will replace ferredoxin (but unfortunately not hydrogenase) in the hydrogen producing system. We have only limited knowledge of how the hydrogenase works and how its action differs from that of the ferredoxins. However, we ultimately envisage replacing both these components in the chloroplast system with stable analogue compounds and possibly other compounds such as platinum.

Despite advances in the stabilisation of chloroplast membranes - for example, by immobilising the enzymes and encapsulating them in alginate gel films - chloroplasts aren't likely to be a part of any commercial solar energy conversion system. However, research into chloroplasts' composition and how they work may ultimately provide the basis of a synthetic process. What advances are likely in this area? The water-splitting reactions of the chloroplast membrane (Photosystem II) take place in a manganese-chlorophyll-protein complex; unfortunately we know little about the arrangement of the manganese ions and the mechanism of water splitting in living plants. Despite this lack of knowledge, two research groups seem to be close to splitting water with visible light.

From a practical point of view, it may be necessary to separate the light activated step of photosynthesis that evolves oxygen from the subsequent liberation of hydrogen in a dark reaction. A one stage system would generate a mixture of hydrogen and oxygen, and separation and collection over the whole area of any solar collector may be impractical. A two-stage system however, would consist of a first stage containing the photocatalytic system which generates a non-oxidisable carrier and evolves oxygen. Indeed, collection of the oxygen may prove economically important. The carrier is then circulated to the second stage, and this would consist of the hydrogen-evolving reaction, a dark reaction. After production of hydrogen from the carrier is recycled back to the first stage of the apparatus to be reduced in the light reaction, ready to take part in the cycle again.

So what are the short-term prospects for a photobiological/ photo-chemical system that produces hydrogen? That depends upon the progress made in understanding the science of the process. A crucial breakthrough in current research will come when we attain an efficient and stable system that mimics Photosystem II of chloroplast membranes; that is a process that splits water to evolve oxygen and provides protons and reducing equivalents. The second stage of such a system - the production of hydrogen gas - is well in hand as there are many alternative electron carriers and catalysts to produce hydrogen. However, the sensitivity of many of these components to oxygen may limit their practical use. An important requirement is a substance that can do the same task as hydrogenase, that is an oxygen stable redox and proton carrier.

Hydrogen evolution photoreactor   If we can do without the water splitting reaction and use reduced compounds such as ascorbic acid, EDTA or dithionite as the electron donor it is easy to obtain high rates of hydrogen production from a stable system. The photo-catalyst can be a pigment like a flavin or even the stable photosystem I of the chloroplast membrane. The electron and proton carriers used were dyes and hydrogenase. Stable enzymes and immobilized catalysts have also been used. Small photoreactors have been constructed which under the right conditions will evolve hydrogen at high rates, e.g. litres of H_2 per minute.

Artificial membranes   The construction of artificial membranes to replace the unstable biological membranes may not be easy but research is progressing. The purpose of one research project is to construct artificial inorganic membranes able to perform the photochemical decomposition of water. As membrane materials, various swelling layer-lattice, clay silicates (montmorillonites and hectorites) were used which are known to form stable intercalate compounds with a large variety of ionic complexes and polar molecules, such as water for instance. The membranes - or films - are obtained by sedimenting dispersed aqueous suspensions of silicate particles. Each particle is an arrangement of parallel clay sheets 9.6 $\overset{o}{A}$ thick, separated from one another by layers of the intercalated material. Typical membranes used in the study have a thickness of about 0.1 mm, a weight of about 3 mg/cm^2 and an apparent surface of more than 10 cm^2. Such a membrane consists of more than 10^4 superimposed sheets and has a true internal surface area available for chemical reaction of about 20 m^2. Due to the small disorder in the arrangement of the particles, it offers good porosity allowing gases to diffuse. Various electron and proton transfer complexes such as metal porphyrins, iron-sulphur compounds and inorganic metals such as Ru and Cr, complexed with dyes, have been incorporated into these thin, stable clay membranes. They seem very interesting membranes for future work.

Electric currents   The use of biological membrane components to generate electric potentials is an interesting aspect of solar energy conversion since this is an attempt to construct a solar cell. For example, the photosystem of red bacteria and the membrane protein of Halobacterium can be incorporated in to lipid vesicles or stacks of lipid membranes which can also include accessory dyes. These can be supported in millipore filters which separate two compartments - a potential difference is then generated upon illumination. Or, a net movement of salt ions such as Na can be induced, thus effecting desalination. Photovoltages have been demonstrated but the resistances so far have been too high to give a reasonable current.

A device has been constructed which uses the chloroplast membrane to generate a current by a photogalvanic cell. An open circuit potential of 220 mV was generated upon illumination of the combined half cells resulting in a current of 800 µA and a current density of 16 µA/cm^2 using platinum electrodes. One half-cell uses the water splitting reaction of the chloroplast membrane while the other half cell only uses the partial reaction (photosystem I) to generate the low potential. The half-cells are connected by an artificial membrane which is permeable to ions but not to the electron carrier (a dye). It may be possible to modify this very interesting galvanic cell to evolve H_2 and O_2 separately in each of the half cells, instead of generating a current.

F.   CONCLUSION

Thus we come back to the work reported in the beginning. Water can be used as a source of electrons and protons to produce hydrogen or can be used to generate a current. It is also possible to reduce CO_2 to organic compounds and nitrogen to ammonia - these are recent important developments Alternatively, reduced compounds can be upgraded in their energy to produce hydrogen upon illumination. Or, very simply, solar energy can be used to generate electricity using pigment complexes in solar cell type mechanism. The opportunities offered by chemical and biological components in a single or combined system are numerous and potentially important for solar energy conversion.

REFERENCES

1. J R Bolton and D O Hall (1979) Photochemical conversion and storage of solar energy. Ann.Rev.Energy 4, 353-401.

2. M Calvin (1979) Petroleum plantations and synthetic chloroplasts. Energy 4, 851-870.

3. J Connolly ed.(1980) Proc III Intl.Conf.Photochemical Conversion and Storage of Solar Energy, SERI, Golden, Col 80401 (in press).

4. D O Hall, M W W Adams, P Gisby and K K Rao (1980) Plant power fuels hydrogen production. New Scientist 86, 72-75.

5. K Kalyanasundaram, J Kiwi and M Gratzel (1978) Hydrogen evolution from water by visible light. Helvetica Chim Acta 61, 2720-2730.

6. M Kirch, J M Lehn and J P Sauvage (1979) Hydrogen evolution by visible light irradiation of aqueous solutions of metal complexes Helvetica Chim.Acta 62, 1345-1384

7. N N Lichtin (1980) Fixing sunshine abiotically. Chemtech (April) pp.252-260.

8. T Ohta ed(1979) Solar hydrogen energy systems. Pergamon Press Oxford, 264 pp

9. G Porter and M Archer (1976) In vitro photosynthesis. Interdisc. Sci Rev. 1, 119-143

10. P Weaver, S Lien and M Seibert (1980) Photobiological production of hydrogen. Solar Energy 24, 3-46.

SESSION I : AGRICULTURAL RESIDUES AND ENERGY CROPS

Chairman: Dr. A. WEISSMAN
                Bundesministerium für Ernährung,
                Federal Republic of Germany

Summary of the discussions

Invited papers

Straw and animal residues available for energy

Energy crops - The case of Brasil

Poster papers

Wetland energy crops: the productive potential of typha spp., phragmites communis and phalaris arundinacea in Minnesota

Natural vegetation as a renewable energy resource in Great Britain

Selected natural and alien plant species as renewable sources of energy in Great-Britain - Experimental assessment and implementation

The productivity of catch crops grown for fuel

The allocation of land for energy crops in Britain

The potential role of crop fractionation in the production of energy from biomass

The energetics of livestock manure management

Production, distribution and energy content of agricultural wastes and residues in the United Kingdom

Crop residue availability for fuel

Soil protection under maximum removal of organic matter

Experimental study on continous digestion of cellulose waste in a combined reactor system

The manure cellar as a source of heat

SUMMARY OF THE DISCUSSIONS

Session I   AGRICULTURAL RESIDUES AND ENERGY CROPS

Rapporteur :   N.S. MARGARIS, Greece

Speakers :   F. REXEN, Danemark
                  S. TRINDADE, Brazil

Chairman :   A. WEISSMAN, Germany

Poster Session : 11 papers presented

Three main subjects were presented :

Crop and Animal residues, Energy crops, Exploitation of natural (non-forest) systems.

1 CROP AND ANIMAL RESIDUES

Every attempt at energy exploitation from residues must be founded on data which describes production, distribution and seasonal availability related to economic, climatic and soil factors and energy balances. Papers dealing with the construction of regional models were presented here. Development of such models will undoubtedly help our understanding of the resources we have available.

The paper presented by Rexen on the availability of animal and crop residues in European Community showed not only the energy benefit we can derive, but also the problems arising from resource allocation. In particular the potential utilization of straw and wood for energy, as opposed to industrial use for the production of paper, particleboard etc, needs closer examination.

Data reported for livestock manure management provided very interesting aspects on nitrogen losses. According to Morris and Larkin these losses are equal to 10% of the current total usage of nitrogen fertilizer in the UK. Careful management will result not only in direct production of energy but also in considerable conservation since nitrogenous fertilizer production is an energy intensive process.

2 ENERGY CROPS

Important data was presented from Brazil concerning their programme of alcohol production. The experience gained will be very useful especially in other areas of the world with similar climatic and economic problems.

The land available for energy crops is a serious problem because of potential competition with agriculture. Therefore, the construction of land allocation models describing the existing situation in the U.K. (Bunce et al) was interesting as a first step.

Energy and food production can be integrated by the use of catch crops. Much new data, of course, is needed before energy production from catch crops can be a reality. Carruthers presented useful data on the productivity of some potential catch crops, dealing primarily with plants already used in agriculture. I think that the search for plants growing naturally in the agricultural lands between harvest and the sowing of the main crops will possibly provide more valuable results. For example during the summer in wheatfields of Greece and South Italy plants such as Inula grow very well during the dry summer months.

A serious problem with energy crops can be the removal of nutrients and organic matter from the system. The group from INRA, France, presented valuable data and I think we must have much more information in the near future concerning the flow of nutrients. Also, the regional availability of nutrient input through rain is important and should be studied.

3 NATURAL SYSTEMS

Many areas around the world are characterized as marginal. Grazing in such lands is the prevailing activity. In Europe, heathlands and mediterranean climate systems are two examples.

Establishment of energy crops creates not only economic but also ecological problems. High energy inputs, the use of fertilizers and pesticides as well as problems connected with monocultures (like sensitivity to insect attacks) are some of them. Natural systems, characterized with high species diversity, are much more stable and the understanding of their resilience to perturbations can guide us not only to likely economic benefits but also to ecologically acceptable solutions. Therefore, contributions like the one presented by Lawson and Callaghan on the utilization of natural systems must be taken seriously. Also, attempts like the one described by Andrews and Pratt concerning the use of wetlands for biomass production, instead of using peat surface mining, are more acceptable from the ecological point of view.

Personally I believe that the exploitation of natural systems for biomass production must be carried out with a balanced consideration of both ecological and energetic factors. For example, people are familiar with the experiments undertaken by Calvin in California in connection with Euphorbiaceae for hydrocarbon production. This however, may lead to problems already discussed concerning monocultures and land availability. In many parts of the mediterranean region we are fortunate to have such energy rich plants in ecologically balanced systems. I may point out here that the Euphorbiaceae species used in California were imported from the mediterranean region. It will be advantageous to work with these natural systems for biomass energy production.

STRAW AND ANIMAL RESIDUES AVAILABLE FOR ENERGY

F.P. Rexen

Summary

Energy from agricultural residues and wastes could make an important contribution to our present needs. In EEC the total amount of waste destroyed each year is more than 1.5 billion tons. One possibility is the manufacture of briquettes of waste materials as an efficient fuel. Straw is increasingly being utilized in farm straw furnaces and in industry. Animal manure may provide energy by anaerobic fermentation or by composting. Perhaps high quality residues - e.g. straw - is even better utilized as an alternative to petrochemical feedstock in a time with steeply rising raw material cost.

1. INTRODUCTION

Biomass is widely considered as an important alternative to the fossil energy sources, and by way of example experiments with growing of energy forests have been started in more countries.

However, an energy supply from energy plantations requires large, unused territories not being available in most EEC-countries. In return, there is a large production of various organic wastes in these countries. According to "Euroform no. 21/76" 1.5 billion tons of waste were destroyed in EEC in 1975. This amount is expected to increase 5 % per year.

A widespread application of organic by-products as an energy raw material presents technical as well as economical problems: The high moisture content in many of the products, the relatively low energy content of the biomass and problems as to harvest, storage, and transportation.

Nevertheless, the total, organic by-product quantity makes out a very large energy potential, which cannot be neglected in these energy-shortage times.

In Denmark, about 14 million tons of waste and by-products are produced every year but a considerable part is already used for various purposes. As an example, waste from the industry can only contribute moderately to an increase in the energy supply. On the other hand, a large part of the wastes from agriculture and households might be used as an energy raw material.

Table 1. Percentual distribution of waste on occupational group

	Part of the total waste lot, %
Households	6
Industry	6
Agriculture	82
Forestry	6

As appears from Table 1 the main part of the organic waste in Denmark results from the agriculture (about 82 %) but
also the households and forests deliver a large amount of waste.

Table 2. Yearly production of organic waste and by-products in
Denmark (1000 tons)

	Mainly cellulose-rich	Mainly starch-and sugar-rich	Mainly protein-rich	Mainly fat-rich	Mixed
Household	560	-	-	-	260
Industry	-	574	200	53	-
Agriculture	7340	290	-	-	4190
Forestry	800	-	-	-	-
Part of total waste lot	61 %	6 %	1,5 %	0.4 %	31.1 %

Table 2 states the waste according to composition. It appears that the main part is cellulose-rich, and this part
forms the most important part in the energy context.

2. WASTES FROM HOUSEHOLDS AND INDUSTRY

Presently, the wastes from industry is utilized most appropriately, which a.o. is due to the fact that this waste always is concentrated in rather large lots of a relatively uniform composition and at comparatively few places so that a
certain industrial processing is possible.

To this comes that the industrial waste normally is of a
more specific character than the wastes from households and
agriculture.

As mentioned, a considerable part of the 14 mill. tons of
waste is used for other purposes and it can be stated that the
industrial wastes only can contribute moderately to an increase of the energy supply. Then households and agriculture remain.

Most household waste is deposited at dumps but still more

district heating stations are built and heated by burning waste. The Danish district heating association states that the
yearly production of combustible household waste amounts to
about 2.3 mill. tons. From this, about 1.4 mill. tons are
burnt in about 30 refuse disposal plants. Every year, about
240 tera joule (10^{12}) are lead into the district heating network. This corresponds to 100,000 tons of oil equivalents per
year.

A few Danish firms have started some interesting experiments with production of fuel briquettes from household wastes
a.o. mixed with coal dust.

A briquette consisting of 60 % waste and 40 % coal dust
is supposed to contain about 4500 kcal per kg. Others plan to
produce fuel briquettes based on agricultural wastes, wood and
plastic wastes. Some experiments have already been carried out
with promising results.

3. WASTE FROM AGRICULTURE

The waste from agriculture is dominated by straw and animal residues.

Table 3. Approx. production of cereal straw in some European
countries. (EEC Yearbook of Agric. Statistics, 1976)

	Mill. tons/year	"surplus straw" mill. tons/year
Germany	21	?
France	28	5-6
Italy	10	?
Holland	1	?
Belgium	2	?
Great Britain	14	6
Ireland	1	?
Denmark	7	2
EEC total	85	?
Greece	3	?
Norway	1	?
Sweden	6	?

Table 3 indicates estimated figures for the straw production (rye, wheat, oats, and barley) in some European countries. It appears from the figures that the production is enormous in most of the mentioned countries, but it is an open question: How great a part of the straw lot is available for energy production?

In many countries the equipment for straw collection is not sufficient but this problem can be solved in time, although many farmers are of the opinion that the straw ought to remain in the field (ploughing under) for maintenance of the humus content. To this comes that the straw to a great extent is used for feed and bedding, just as a minor part is used in the industry.

It is estimated that France has a straw surplus of 5-6 mill. tons/year, whereas in England about 6 mill. tons are supposed to be applicable as energy raw materials. In Denmark, a potential of max. 2 mill. tons is expected. It has not been possible to state figures for the straw surplus in the other European countries.

The straw can be used directly as a solid fuel or be transformed into a liquid or gaseous fuel by pyrolysis processes or fermentation.

To-day, about 20,000 straw furnaces are installed in Denmark with an average capacity of 10-20 tons straw/year. This means that 200-400,000 tons of straw can be burnt in Danish straw furnaces presently. Such lots are hardly used, however, as many of the furnaces also are used for burning of wood and other kinds of organic wastes. In oil saving this means 50-100,000 tons/year corresponding to almost 1 % of our yearly fuel oil import.

In case all Danish farms invested in straw furnaces it might be possible to utilize about 3 mill. tons straw/year (150,000 farms) saving about 700,000 tons of oil (about 5 % of the fuel oil import). From this must be deducted the energy used by collection and burning the straw.

It is not realistic to imagine that the agriculture will

install straw furnaces for the simple reason that the straw surplus is very unevenly distributed - from district to district as well as from farm to farm.

At farms with large livestock production, the energy demand is very great but also the demand for straw feed and bedding, whereas a typical grain farm with a large straw surplus has a somewhat smaller energy demand. The trade with straw for burning purpose will hardly reach a greater extent for economic reasons.

4. ANIMAL RESIDUE

The animal residue makes out the largest organic by-product lot in the EEC.

Table 4. The yearly production of livestock manure in the EEC countries, mill. tons/year[x]

	Cattle	Swine	Total
West Germany	24.9	3.6	28.5
France	40.7	2.2	42.9
Italy	14.6	1.6	16.2
Holland	8.5	1.3	9.8
Belgium	5.2	0.8	6.0
Luxemburg	0.4	0.02	0.42
Great Britain	24.1	1.4	25.5
Ireland	10.3	0.2	10.5
Denmark	5.3	1.4	6.7
EEC	134.0	12.6	146.6

x) Estimated from the number of animals on the basis of the following figures (Vergara, 1978):
Cattle: 1723 kg manure (dry matter)/year
Swine: 182 - -   - -   -

As can be seen from Table 4, each year about 150 mill. tons of manure dry matter are produced in the 9 EEC countries.

Therefore, livestock manures could be a prime source for energy conversion. Once dried manure may be burnt, and when submitted to anaerobic digestion, methane gas is produced. But of course not all the manure produced is available for biomass conversion. Many livestocks are pastured, and when the livestock is dispersed on pasture land it might be somewhat unpractical to collect the manure. It has been estimated that if all the manure produced in Denmark is transformed into biogas, about 1 milliard m^3 gas can be produced, corresponding to about 600,000 tons of oil equivalents (about 4 % of the fuel oil consumption).

The Danish state has erected 3 biogas plants at different sites in the country. These plants have operated experimentally for 2 years, and the temporary conclusion is that the plants technically operate satisfactory, whereas the gas production is less than expected and the economy is not very good. Also private persons have established biogas plants, but the production is still very modest.

Just as by the straw it will not be possible to produce biogas from the complete manure lot, also because the plants must have a certain livestock basis to be profitable. The Danish consultant Carl Bro has stated that there ought to be at least 25 heads of cattle per biogas unit.

It also ought to be mentioned that the biogas plants still are at the experimental stage so that it presently is difficult to say whether the farmers ought to invest in such installations.

5. WOOD

Wood from forests, windbreaks, plantations, etc. makes out a considerable, potential energy source.

The Danish Forest Research Institute has calculated that if the marginal wood ressources are utilized, 1.6 mia. m^3 wood can be collected for burning purposes. From these 1.6 mill. m^3, 1.2 mill. m^3 come from the forests. The remaining originates from windbreaks and the wood industry.

The Institute has calculated that the energy content in the waste wood corresponds to 260-316,000 tons of heavy fuel oil. A complete utilization demands assistance from specially trained workers. It is estimated that 1000 extra forest labourers are needed.

The price setting at the wood market is presently very strange. It seems more profitable for the forest owners to sell the wood as fuel wood than as industry wood. These abnormal price relations are not durable and do probably express that the prices not yet have become steady. Another reason is, that there are government grants of up to 40 % to replace existing oil burning boilers. It is also these conditions that make straw burning so popular in Denmark.

One could imagine, however, that from a social point of view it might be more appropriate to use the wood and the straw for production of e.g. particle boards and cellulose. We could also ask the question whether it would be more appropriate to utilize the energy content as pyrolysis-gas or -oil.

The Danish Ministry of Energy has worked out a report on renewable energy in Denmark. In this report it is suggested that one third of Denmark, which will not be supplied with natural gas or district heating, might be heated by renewable energy sources. It has been calculated that 40 % of the heat demand and a part of the electricity demand can be covered in this area by installation of $\frac{1}{2}$ mill. different plants (corresponding to one plant per household).

The investments will amount to 28.5 billion Dkr. (2.2 billion £). This means that 6 % of the gross energy consumption in Denmark can be replaced by renweable energy.

Table 5 states the calculations from the Ministry of Energy.

Table 5. Investment and economy for plants for renewable energy

Type of plant	Number of plants, 1000	Investm. mill. Dkr.	Net saving mill.Dkr./ year	Pay-back time, years	Internal interest, %/year
Straw furnace	130	4550	163	28	-3
Wood furnace	50	1750	62	28	-3
Biogas	20	5500	157	35	-5
Wind mills	100	7500	374	20	0
Sun collecter	150	8250	86	96	-12
District heating plants	1	1000	56	18	1
Total	450	28550	898	31	-4

It is obvious that on a basis of common commercial thinking it will not for the time being be profitable to invest in renewable energy, but other conditions such as saving of currency and supply security also are important factors. Also a heavy increase of the oil price will alter the calculations positively.

6. CONCLUSION

Finally, I should like to mention that I find it encouraging that the state grants financial support for establishment of plants for renewable energy but in this connection we ought to consider that e.g. a strong subvention of energy plants for wood and straw will result in problems for basic industries, such as particle board industry and cellulose industry, which already now are concerned about the raw material supply.

ENERGY CROPS - THE CASE OF BRASIL

S.C. TRINDADE

Centro de Tecnologia Promon-CTP
Rio de Janeiro, RJ-Brasil

SUMMARY

The current world economic disarray has taken its heaviest toll on developing countries highly dependent on imported fossil fuels. High imbalance of payments, heavy foreign debt combine in Brasil with high inflation and fast economic growth. Consequently, the country has since 1975 resorted increasingly to expensive domestic sources of liquid fuels aiming at saving scarce foreign exchange. Brasil's energy vectors, composed of source-conversion-form-transmission-distribution-storage-end-use show the energy problem to be one of liquid fuels for transportation and industry. An agro-industrial system generating fuels could be generalized as a biomass energy factory producing solid, liquid and gaseous fuels. In this context, the net energy ratio - NER concept must be reassessed to emphasize the production of marketable high grade fuels from domestic lower grade energy inputs. Brasil's biomass energy-factory systems are in decreasing order of current importance: (i) ethanol from food crops (PROALCOHOL), (ii) methanol from wood (CESP program) and (iii) ethanol from wood (COALBRA program). In the future all biomass energy programs may merge into a broad based PROBIOMASS, including vegetable oils, to assure the long term success of biomass fuels based on diversity of feedstocks. PROALCOHOL is the only large scale biomass energy program in the world having an impact on a national scale. In 1980 virtually all gasoline consumed in Brasil contains 20% of ethanol by volume. In addition, by year end some 200 thousand new or converted cars will be consuming neat ethanol only (96% ethanol, 4% water by volume). Social, economic and environmental consequences of PROALCOHOL are also discussed.

1. INTRODUCTION

Developing countries highly dependent on imported fossil fuels are paying the heaviest tool in the midst of a prolonged world economic crisis. Brasil for instance relies on petroleum to supply 45% of her 118 million TOE annual primary energy requirements. Over 85% of crude is imported. Coupled with metallurgical coal imports, the country's dependence on foreign energy supplies reaches over 40% of her primary energy consumption (1).

Heavy reliance on imported fossil fuels at prices which have been increasing over and above world inflation has gradually led Brasil to a gloomy economic situation. Increasing cumulative imbalances of payments have contributed to a large foreign debt (over US$ 50 billion). On the other hand, domestic inflation that had been gradually decreasing through 1973 has come back to levels (over 100% per annum) that cannot be managed even with Brasil's indexation systems. In addition, the economy persists in growing at relatively high rates (over 5% per annum).

The oil import bill will almost certainly top US$ 11 billion in 1980 which coupled with the service of the foreign debt will virtually consume all export earnings. Vital imports such as raw materials, food, machinery, fertilizers, etc. rely on additional borrowing and foreign capital investment in the Brasilian economy.

Consequently, Brasil is increasingly resorting, since 1975, to expensive but domestic sources of liquid fuels which could contribute to saving scarce foreign exchange.

2. BRASIL'S ENERGY PICTURE

The complexity of energy issues requires a systemic analytical approach which recognizes the various energy vectors composed of source-conversion-form-transmission(transportation) -distribution-storage-end-use as exemplified in Figure 1 (2).

Ultimately, energy vector configurations aim at satisfying specific transportation, industrial, commercial and residential demands for energy. For example, the demand to consider in transportation is the ton-kilometers or passenger-kilometers services to be provided by appropriate energy vector configurations.

The Brasilian energy problem is mainly one of liquid fuels for transportation and industry. Consequently, the key energy vectors to consider are those which include liquid fuels as the required form of energy from a variety of sources (petroleum, biomass, etc.) to satisfy, as end-use, transportation and industrial requirements.

Major primary energy contributors in Brasil are indicated in Figure 2 (1,3) and are petroleum, hydroelectricity, biomass and coal.

As shown in Table I (4) the main derivatives of the crude oil barrel in Brasil are fuel oil (32% by volume), Diesel oil (28%) and gasoline (27%). Virtually all gasoline and most Diesel oil are consumed in transportation, whereas most fuel oil is used in industry. Electricity in Brasil is mainly hydro (over 90% of installed capacity) complemented by minor coal-based and oil-based generation. There is a nuclear power program underway based on a bilateral agreement with the Federal Republic of Germany. The first nuclear power plant, prior to the above mentioned agreement, is expected to start up in 1981.

TABLE I

BRASILIAN SECTORIAL CONSUMPTION OF OIL DERIVATIVES

- % of Crude Oil Barrel -

SECTOR	FUEL OIL	DIESEL	GASOLINE	TOTAL
TRANSPORTATION	1.0	21.0	25.0	47.0
INDUSTRY	27.0	3.5	1.0	31.5
ELECTRICITY	2.0	0.5	–	2.5
OTHER	2.0	3.0	1.0	6.0
TOTAL	32.0	28.0	27.0	87.0

Ref. (4)

SYSTEM'S VIEW OF ENERGY VECTORS

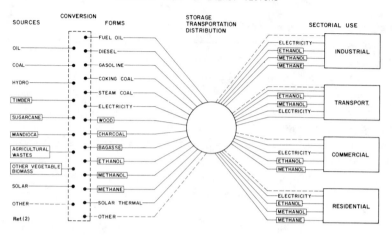

Fig. 1

BRASILIAN PRIMARY ENERGY PROFILE

1967 — 1978

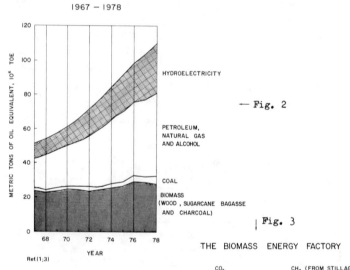

— Fig. 2

Fig. 3

THE BIOMASS ENERGY FACTORY

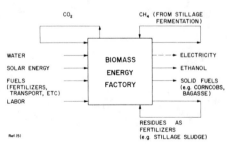

Biomass in Brasil is composed mainly of wood, charcoal and sugarcane bagasse. Ethanol is reported jointly with petroleum statistics. Coal is a relatively minor contributor to primary energy in Brasil.

3. BIOMASS ENERGY FACTORY AND THE NER MYTH

An agroindustrial system generating fuels could be generalized as a biomass energy factory producing solid, liquid and gaseous fuels, as depicted in Figure 3(5). In this context an agroindustrial system containing an ethanol distillery could be described as a biomass energy factory that

.consumes solar energy, water, carbon dioxide, labor, fuels (e.g. fertilizers);

.to produce solid fuels (i.e. sugarcane bagasse, corncobs, etc.), liquid fuels (i.e. ethanol), gaseous fuels (i.e. methane from stillage fermentation);

.and recycles carbon dioxide (from ethanol fermentation), methane (from stillage fermentation burned to raise steam under a boiler), residues (used as fertilizers, for instance).

NER- net energy ratio measures the ratio of the energy content of streams leaving (outputs) an energy conversion system with previously established boundaries, such as shown in Figure 3, and the energy content of streams entering the system (inputs).

$$NER = \frac{\text{Energy content of outputs}}{\text{Energy content of inputs}}$$

NER can be interpreted as a measure of the energy benefits-cost ratio of a conversion process.

Sugarcane and mandioca ethanol-producing agroindustrial systems are equivalent in terms of energy (NER=4.5). Exclusion of the system's wood lands from the calculation drastically reduces the NER for ethanol from mandioca (NER=1.0) (6).

NER analysis has been perhaps over-emphasized in the assessment of energy conversion systems. For instance, it

should not matter to a given country if ethanol is produced
with a low NER, provided that imported energy inputs are
minimized. The emphasis on NER analysis should be given to the
ratio of high grade fuels to lower grade domestic inputs. In
this context, to produce ethanol using wood or domestic coal
to raise steam is better than burning Diesel oil under a boiler
to distill ethanol. However, conventional NER does not
discriminate among wood, domestic coal and Diesel oil.

In conclusion, in spite of its imperfections, economic
analysis employing social prices and discount factors still is
the best tool for decision-making in the evaluation of energy
conversion projects (7).

4. BRASIL'S BIOMASS ENERGY PROGRAMS

There are three institutionalized biomass energy programs
currently in Brasil, namely:

- PROALCOHOL (ethanol from food crops such as sugarcane,
  mandioca, etc.);
- CESP's Methanol Program (based on wood gasification)
- COALBRA's Wood Ethanol Program

They all fall within the previously mentioned concept of
biomass energy factory systems.

PROALCOHOL, instituted in 1975, has continued a long
tradition in Brasil that started in the 1930's when ethanol was
used to hedge the important sugar industry from international
price fluctuations. Since 1975, however, Government has
provided, in addition to market guarantees, subsidized
financing for new or expanded ethanol distilleries.

Methanol from gasification of eucalyptus wood is a 5-year
US$ 80 million feasibility program being developed by CESP, the
energy utility of the powerful State of São Paulo (8). Ethanol
from acid hydrolysis of eucalyptus wood is an incipient
Brasilian federal program (Ministry of Agriculture's Forestation
Development Institute, IBDF), funded through a Government
corporation COALBRA, still in its infancy.

The use of agricultural feedstocks to make fuels brings

immediately to light the issue of competition of land and water for food, feed, fiber and fuel. This issue is of real concern even in countries such as Brasil and the U.S.A. which could grow crops at a surplus level (9).

In Brasil, PROALCOHOL's incentives provided to ethanol production at a time of low sugar prices (prior to 1980) have induced, to a minor degree, expansion of sugarcane agriculture at the expense of food cropland.

An analysis of the food versus fuel dilemma is illustrated in Figure 4. Selected countries are plotted according to their relative agriculture and energy self-sufficiency (10). It becomes clear that only those countries with agricultural surplus and energy deficit such as Brasil will tend to produce biomass fuels. The analysis could be extended to consider only the agricultural surplus or deficit of specific feedstocks for energy conversion. This would probably sharpen the identification of countries which are prone to embark on biomass energy producing systems on a large scale.

It is possible, in the case of Brasil, U.S.A. and other countries, that energy use of agricultural feedstocks may reinforce the agricultural base to supply the expanding demand for food, feed, fiber and fuel.

Consequently, to assure the long term implementation of biomass fuels based on diverse feedstocks, Brasil may expand PROALCOHOL into PROBIOMASS. Such an enlarged program would be broad based to include, in addition to current alcohol fuels program, vegetable oils, charcoal, direct use of wood, etc.

In particular, decoupling ethanol production from sugarcane only will be essential to irreversibly establish ethanol as a major liquid fuel in Brasil. The diversity of feedstocks, particularly in relation to food crop feedstocks, could in addition contribute to soften the food versus fuel issue. Finally, in the long run cellulosic materials will tend to predominate universally as biomass fuels feedstocks.

5. PROALCOHOL

5.1 Background

Brasil is undoubtedly the world leader in the production
and use of ethanol as a fuel. As such, PROALCOHOL is the only
large scale biomass energy program in the world with an impact
on a national scale.

In 1980 virtually all gasoline consumed in Brasil
contains 20% ethanol by volume. In addition, by the year end
some 200 thousand new or converted cars will be consuming neat
ethanol only (96% ethanol, 4% water by volume).

The 1973 sudden escalation of crude oil prices have led
to a change in Brasilian policy relative to ethanol production
and use as shown in Table II (5). Ethanol, which since the
1930's had been used as a hedge against sugar price variations,
became in 1975 an energy commodity on its own right.

TABLE II

ENERGY DECISION-MAKING AND PROALCOHOL

- POLITICAL PROCESS

- BIOMASS REQUIRES LAND, SUNSHINE AND WATER TO SUPPLY
  MARKETS WITH FOOD, FEED, FIBER AND FUEL

- FUEL ALCOHOL : BRASILIAN TRADITION SINCE 1930

- 1975 PROALCOHOL = LOW SUGAR PRICE + HIGH PETROLEUM PRICE

Ref.(5)

The Government supports PROALCOHOL in the following way:
market guarantees for all ethanol produced; administered price
of ethanol pegged to gasoline price at gas station; subsidized
loans for implementation of agricultural base (sugarcane,
mandioca, etc.) and alcohol distilleries; incentives for
hydrated ethanol users. Major accomplishments of PROALCOHOL
are listed in Table III (5).

The status of PROALCOHOL relative to ethanol production
as of June 1980 is shown in Table IV (11).

From the viewpoint of ethanol fuel utilization,Table V
illustrates the problems encountered in the Brasilian
experience with Otto engines. In addition, ethanol fuel
performance in Diesel engines and boilers is estimated (12).

TABLE III

BRASIL'S PROALCOHOL PROGRAM

LANDMARKS

1975   PROGRAM LAUNCHED
PRODUCTION (1974)   :   0.6   x $10^6 m^3$
ALCOHOL IN GASOHOL   : (0.2   x $10^6 m^3$)
GASOLINE DISPLACEMENT   :   1 %

1977   PRODUCTION   :   1.4   x $10^6 m^3$
ALCOHOL IN GASOHOL   : (0.6   x $10^6 m^3$)
GASOLINE DISPLACEMENT   :   5 %
SÃO PAULO STATE GASOHOL REACHES 20% ALCOHOL
TESTS WITH NEAT ALCOHOL POWERED CARS

1978   PRODUCTION   :   2.4   x $10^6 m^3$
ALCOHOL IN GASOHOL   : (1.4   x $10^6 m^3$)
GASOLINE DISPLACEMENT   :   11 %
50 GASOHOL BLENDING CENTERS BRASIL WIDE

1979   PRODUCTION   :   3.7   x $10^6 m^3$
GASOLINE DISPLACEMENT   :   14 %
4,000 CONVERTED FLEET VEHICLES ON NEAT ALCOHOL
NEAT ALCOHOL SOLD IN GAS STATIONS

1980   PRODUCTION FORECAST   :   4.0   x $10^6 m^3$
GASOLINE DISPLACEMENT   :LARGER THAN 20 %
200,000 NEW AND CONVERTED NEAT ALCOHOL CARS
AVAILABLE TO INDIVIDUAL CONSUMERS

1985   PRODUCTION FORECAST   : 10.7   x $10^6 m^3$

1987   PRODUCTION FORECAST   : 14.0   x $10^6 m^3$

Ref. (5)

TABLE IV

BRASIL'S PROALCOHOL PROGRAM

CURRENT STATUS   (JUNE,1980)

	NUMBER OF DISTILLERIES	PRODUCTION CAPACITY (10^6 m^3)
CROP YEAR 1979/80	198	3.80
UNDER IMPLEMENTATION	73	2.50
TOTAL	271	6.30

Ref. (11)

FOOD VS FUEL-ENERGY AND AGRICULTURAL
SELF-SUFFICIENCY RATIONS FOR SELECTED COUNTRIES
- 1976 -

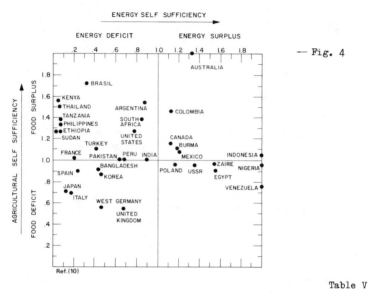

— Fig. 4

Table V
↓

ETHANOL   AS   A   FUEL

EQUIPMENT (CONVENTIONAL FUEL)	ETHANOL UTILIZATION	KEY POINTS	LOWER HEATING VALUE RATIO (PETROLEUM FUEL/ETHANOL)	VOLUMETRIC
OTTO ENGINE (GASOLINE)	• BLENDS (20% VOL. ETHANOL 80% VOL. GASOLINE)	- MATERIALS COMPATIBILITY - PHASE SEPARATION - DRIVEABILITY	1.5	1.0 - 1.1
	• NEAT ETHANOL (96°GL)	- ENGINE CONVERSION - COLD START - DISTRIBUTION NETWORK	1.6	1.2 - 1.3
DIESEL ENGINE (DIESEL OIL)	• BLENDS (MAX. 7% VOL. ETHANOL)	- MATERIALS COMPATIBILITY - PHASE SEPARATION		
	NEAT ETHANOL (PLUS ABOUT 10% CETANE IMPROVER ADDITIVE, INCREASE OF COMPRESSION RATIO, CONVERSION TO OTTO CYCLE OR INCLUSION OF "HOT POINT")	- MATERIALS COMPATIBILITY - COST AND AVAILABILITY OF ADDITIVES - DISTRIBUTION NETWORK	1.7	1.6 - 1.7
	• DUAL SYSTEM (DOUBLE INJECTION/CARBURETOR AND DIRECT INJECTION, EMULSIFIERS, ETC.)	- CONVERSION COST - DISTRIBUTION NETWORK - MULTI-FUEL		
BOILERS (FUEL OIL)	• BLENDS • NEAT ETHANOL • DUAL FUEL SYSTEM	- MATERIALS COMPATIBILITY - LOW COMPETITIVENESS	1.9	1.9 - 2.0

Ref.(12)

5.2 Environmental Impacts

A quick estimate shows that the 10.7 million m^3 of ethyl alcohol forecasted for production in 1985 will generate roughly 126 million m^3 of stillage. If not suitably recovered or treated, that volume could have a serious negative impact on the environment.

On the average, the pollution potential in terms of BOD in two liters of untreated stillage discharged into inland or coastal waters equals the sewage produced by one person/day. This rate provides an idea of the dimensions of the challenge inherent in the predicted volume of stillage, roughly equal to the sanitary waste (in BOD terms only since stillage contains no pathogenic micro-organisms) of 63 million inhabitants concentrated in Brasil's alcohol-producing regions.

It is pointed out elsewhere (13) that stillage may be considered as a raw material rather than as a pollutant. Its organic and mineral contents can feasibly provide useful products such as fertilizers, feed additives and methane fuel.

Air pollution effects of burning ethanol (as well as methanol) on a large scale in mobile (autos) and stationary sources require considerable attention, so far lacking in Brasil.

5.3 Social Impacts

In principle and in the long run PROALCOHOL should bring about improvements in rural employment, local and national income distribution and regional development, food, feed and fiber supply, etc.

However, in the short run the consequences of the fast implementation of PROALCOHOL have in many instances been quite the contrary of those mentioned above.

Since there was urgency in providing a partial substitute for gasoline, and given that the sugarcane cultivation is quite concentrated in the State of São Paulo, the incremental alcohol production has come primarily from that State.

The entry into the alcohol producing market of enterpreneurs from other industrial sectors has been quite slow. Therefore incremental production is concentrated in the few hands of the well established sugar mill and alcohol distillery owners.

PROALCOHOL's original subsidized financing resulted in minor displacement of food crop areas by sugarcane cultivation in both São Paulo as well as in Northeast Brasil.

Despite the shortcomings of the near term, it is reasonable to expect that as PROALCOHOL develops and new enterpreneurs get into the picture and particularly as independent distilleries become the suppliers of the majority of the alcohol produced in Brasil from a variety of agricultural feedstocks (sugarcane, mandioca, sweet sorghum, and cellulosic materials in the long run) the expected social benefits will materialize. That will however require political decision and Government monitoring.

The magnitude of PROALCOHOL will put a severe strain on the technical and managerial resources of Brasil. This applies to both the agricultural and industrial aspects of alcohol production. Consequently, urgent training programs are required to develop the skills necessary to carry the program to its fullest extent.

The positive impact that the utilization of ethanol as partial replacement for gasoline will have on Brasil's balance of payments could be considerable. However, it will not suffice as the main motive behind the PROALCOHOL program.

By way of illustration, if the scenario envisioned for 1985 is considered - wherein out of 10.7 million m^3 of alcohol 3 million m^3 of anhydrous alcohol replace an equal volume of gasoline and 6.1 million m^3 of 96°GL alcohol displace 4.7 million m^3 of gasoline (1.3 m^3 of hydrated alcohol equivalent to 1 m^3 of gasoline) - the 7.8 million m^3 of gasoline economized will amount to 28.8 million m^3 (181 million bbl) of crude oil. If the other petroleum fractions associated with gasoline production could also be replaced by domestic fuels - an extreme situation that would avoid importing those 28.8 million m^3 of crude oil - the corresponding foreign exchange savings calculated at the price of US$ 201.2/m^3 (US$ 32/bbl) would be US$ 5.8 billion in 1985. That total however is about 10% of Brasil's current foreign debt.

At US$ 32/barrel of crude it would only pay to make ethanol from sugarcane if sugar would sell below US$ 200/t in the international market. That was the price of sugar a couple of years ago. Current price (August,1980) is bordering US$800/t. Consequently, the need to disengage alcohol production from sugarcane only is confirmed, as well as the political decision in Brasil to make an expensive domestic fuel from biomass, notwithstanding the ups and downs in sugar prices.

In conclusion, it should be mentioned that currently (August,1980) regular gasoline sells for US$ 690/m^3(US$2.60/gal) at gas stations,where neat ethanol is sold for US$330/m^3 (US$1.25/gal). Gasoline is actually a blend with 20% anhydrous ethanol. The price of the latter at distillery gate is US$303/m^3 (US$1.15/gal). Consequently, quite a margin is realized when anhydrous ethanol is sold for the price of pure gasoline at gas stations. This, plus the heavy taxes on gasoline, help fund Brasil's energy mobilization fund which finances most alternative energy projects in the country.

6. <u>CONCLUSIONS</u>

PROALCOHOL's ambitious goals for the next ten-year period reveal a host of opportunities, including:

● The growing degree of energy independence to be achieved by replacing imported, non-renewable fuels.

● The growing effect brought on by the availability of alcohol, in terms of alcohol-derived chemicals to gradually supplant intermediate petrochemicals.

PROALCOHOL will encounter the following short and medium-term obstacles:

● The sizeable investments required, which will demand careful planning by the Government agencies administering the program and granting subsidized financing in competition with all other alternative energy sources.

• The implementation of independent distilleries at the level envisioned, and diversification of agricultural feedstocks in order to ensure PROALCOHOL autonomy by making alcohol supply less susceptible to international sugar price variations.

• The need for project planning and management commensurate with the scope of the new ventures, in order to ensure their viability. This is particularly true with respect to the development of agriculture to supply feedstocks to ethyl alcohol plants.

• The need for concurrent development of alternative fuels to replace Diesel and fuel oil and provide total petroleum displacement without generating gasoline surpluses.

• Adequate solutions at the national level to ensure air pollution control of ethanol derived exhaust gases and disposal and recovery/utilization of the stillage produced and thus prevent negative impacts on the environment.

Finally, since Brasil is the world leader in biomass energy utilization there is great interest worldwide in investigating the transferability of the Brasilian experience to other countries.

At first glance, the Brasilian ethanol fuel experience is non-transferable, given the very peculiar set of circumstances that led to PROALCOHOL. A deeper look into this issue however reveals that concepts, rather than specifics, could be transferred.

Analysis of alcohol fuels vectors is a first step in planning any national alcohol program. In essence, as far as alcohol fuels programs are concerned, other countries can learn from Brasil the following:

• Political decision is required to replace petroleum with more expensive but domestic alternatives such as Ethanol.

• Political consensus necessary for political decision can be rationally based on the following:

    . Current petroleum prices are political, not economical;

    . Petroleum prices are climbing faster than biomass derived ethanol prices;

. Petroleum substitutes such as ethanol are not aimed at replacing oil instantly but rather gradually (e.g. 20% ethanol in gasoline);

. Petroleum will eventually be replaced by a host of fuels so that methanol, ethanol, other biomass and coal, and shale derived liquid fuels will all have their share of the market;

. Domestically produced fuels can hedge a country against the political use of foreign oil supplies. They could also provide employment in the country, income redistribution and balance of payment savings.

BIBLIOGRAPHY

(1)   BRASIL. Ministério das Minas e Energia. Modelo Energético Brasileiro. (The Brasilian Energy Model, Ministry of Mines and Energy), Nov. 1979.

(2)   TRINDADE, S.C. Fuel Alcohol-Suitability of Sugarcane as Feedstock. In:WORLD SUGAR RESEARCH ORGANISATION CONF., S.Francisco, Feb.26-27,1980. Proceedings.

(3)   BRASIL. Ministério das Minas e Energia.Balanço Energético Brasileiro.(The Brasilian Energy Balance, Ministry of Mines and Energy), 1978.

(4)   CARMO,P.F. Fontes Alternativas para Geração de Energia (Alternative Sources for Energy Generation). In:SIMPÓSIO DE FONTES ALTERNATIVAS PARA GERAÇÃO DE ENERGIA ELÉTRICA NA AMAZONIA (Symposium on Alternative Sources for Electricity Generation in the Amazon Region), Manaus, Jul.25-27, 1979.

(5)   TRINDADE, S.C. The Brasilian Alcohol Fuels Programs. In: ASPEN INSTITUTE FOR HUMANISTIC STUDIES ANNUAL WORKSHOP ON ENERGY R&D PRIORITIES.3.Aspen, Jul.12-18,1980.

(6)   GOLDSTEIN JR.,L. et al. Comparação de Alternativas de Conversão de Energia (Comparison of Energy Conversion Alternatives). In: CONGRESSO BRASILEIRO DE ENERGIA (Brasilian Energy Congress).1.R.de Janeiro,Dec. 1978.

(7)   YOKELL, M. Physical Efficiency and Economic Efficiency
      as Criteria for Ranking Systems. In: INTERNATIONAL
      CONFERENCE ON ENERGY USE MANAGEMENT, Tucson, Oct.1977.

(8)   CTP-Centro de Tecnologia Promon. Proposição do Programa
      Integrado da CESP para o Metanol (Integrated Methanol
      Program for CESP). Study conducted by CTP to CESP- São
      Paulo State Energy Utility. 1980.

(9)   BROWN, L.R. Food or Fuel: New Competition for the World's
      Cropland. Washington, D.C., Worldwatch Institute,1980.
      (Worldwatch Paper 35).

(10) RASK,N. Using Agricultural Resources to Produce Food or
      Fuel: Policy Intervention or Choice ? In: INTERAMERICAN
      CONFERENCE ON RENEWABLE SOURCES OF ENERGY.1. N.Orleans,
      1979. Proceedings.

(11) GOLDSTEIN JR.,L. et al. Fermentation Ethanol as a
      Petroleum Substitute. In: INTERSOCIETY ENERGY CONVERSION
      ENGINEERING CONFERENCE.15. Seattle, Aug.18-22,1980.

(12) CTP-Centro de Tecnologia Promon. Diagnóstico do Programa
      Tecnológico Industrial de Alternativas Energéticas de
      Origem Vegetal (Assessment of the Biomass Energy
      Technology Development Program, of the Brasilian
      Industrial Technology Secretariat-STI). Study conducted
      by CTP to the Brasilian Industrial Technology Foundation
      -FTI, 1979.

(13) CTP-Centro de Tecnologia Promon. Stillage: Technical &
      Economic Evaluation of Processes for Stillage Recovery as
      Distillery By-Product. Multiclient Study, R.de Janeiro,
      1979.

WETLAND ENERGY CROPS: THE PRODUCTIVE POTENTIAL OF TYPHA SPP.,
PHRAGMITES COMMUNIS AND PHALARIS ARUNDINACEA IN MINNESOTA.

N.J. ANDREWS and D.C. PRATT

Botany Department, University of Minnesota

Summary

The use of wetlands to grow energy crops has been the
focus of several research programs at the University of
Minnesota. Typha spp. looks particularly promising as an
energy crop for wetland biomass production; it grows naturally in monoculture, is very productive, is easily propagated
and has few insect pests. The standing crop (above and below
ground) in natural Typha stands usually ranges from 20 to 40
tons/hectare. Productivities up to 30 tons/hectare/year
have been obtained with Typha grown under managed conditions
on peat. Other wetland species being considered as potential
energy crops are Phragmites communis, Phalaris arundinacea,
Carex spp. and Scirpus spp.

There are 5.7 million hectares of peatland in the northern lake states, Minnesota, Michigan and Wisconsin. The
feasibility of establishing large scale peat gasification
plants in Minnesota is currently being studied by the utilities companies. Wetland energy crops could provide an
alternative to large scale peat mining and gasification or
be useful in the reclamation of peatlands that have been
partially or totally mined. Work is currently under way to
study the feasibility of establishing extensive stands of
Typha on Minnesota peatlands. These studies are part of a
comprehensive program at the University of Minnesota which
includes work on fermentation technology, harvesting
methods, land use considerations and environmental constraints.

1. The Biomass Option for Minnesota.

Energy demand in Minnesota is expected to exceed the supply of traditional fuels within the next decade (1). Reducing demand, increasing supplies of traditional fuels and developing new sources of energy are the three major alternatives. The Minnesota Energy Agency (MEA) has estimated that 93% of Minnesota's 1978 primary energy demand (1250 trillion BTU/yr) could be met by special energy crops such as Typha spp. (cattails), Phragmites communis (reed), Phalaris arundinacea (reed canarygrass), and short rotation woody crops (Salix spp., Alnus spp., and Populus spp.). The land area potentially available for growing special energy crops in Minnesota is estimated to be 4.4 million hectares, of which 2.9 million hectares are peatlands (Figure 1). Before any of these biomass crops can be planted on a large scale the feasibility of establishing energy farms must be determined. Some of the important factors which must be considered include: (1) the environmental impact of large scale energy farms, (2) the social and public policy implications associated with such development, (3) the associated economic impacts, and (4) the feasibility of obtaining economically viable yields of these crops over extensive areas. One timely public policy question concerns the competing or conflicting uses of land suited for biomass production. Two utility companies are interested in harvesting peat to fuel gasification plants. Wetland energy crops can be viewed either as an alternative to peat mining or as a reclamation effort after peat has been excavated.

The Wetland Biomass Program at the University of Minnesota is currently funded by the Minnesota Energy Agency. Included in the program are studies of wetland plant productivity, biochemical conversion of wetland plant biomass to fuels, harvesting technology, land use and environmental constraints on biomass development and the economics, social and public policy implications associated with large scale biomass energy development.

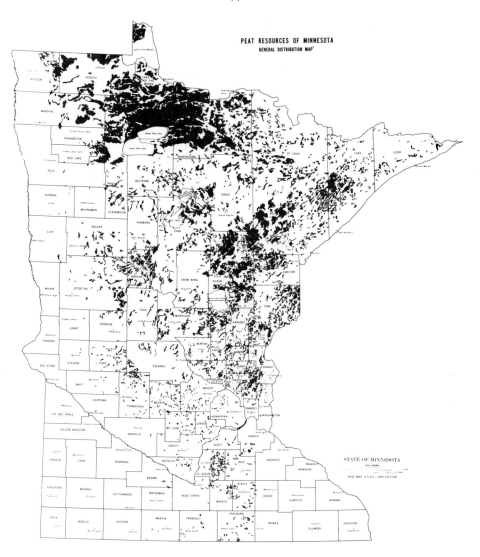

PEAT RESOURCES OF MINNESOTA
GENERAL DISTRIBUTION MAP

Fig. 1. * A preliminary generalized peat map compiled from
Minnesota soil atlas project information. 🝋 Peatland,
smallest delineation is approximately one square mile.

2. The Productive Potential of Typha spp.

Typha spp. is particularly promising as a wetland biomass crop; it grows naturally in monoculture, is very productive, is easily propagated and has few insect pests. Table
I presents standing crop of Typha spp. in natural stands in
Minnesota. Stands were sampled in late July and early August
when above ground standing crop is maximal (2). Above ground
yields range from 1.09-2.32 kg/m^2 (10.9-23.2 tons/ha). Below
ground biomass ranges from 2.32-4.72 kg/m^2. Estimating below
ground biomass is a particular problem; it is often difficult
to obtain samples from flooded areas and results are highly
variable. These yields are consistent with those reported in
the literature as summarized in Table II. Table III presents
a summary of yields obtained under managed conditions.
Annual productivity using either seeds or rhizomes ranges
from 1.11 to 3.08 kg/m^2. The annual productivity of Typha
using 1.5 m^2 paddies filled with peat plus fertilizer
(100:300:300 kg/ha, N:P:K) was approximately 28 t/ha. The
root to shoot ratio (R/S) increases from .2 to 1.0 after
early August as carbohydrates are translocated into the
overwintering rhizomes (19). Optimal harvest time will
depend on which component, the above or below ground, is of
most interest. The establishment of Typha stands on Minnesota peatlands is currently a major focus. In moving from
small scale paddy experiments to field trials optimal planting rates and techniques, fertilizer and lime application
rates and insect and weed control practices must be determined. Preliminary results from the first growing season
indicate that higher rates of fertilizer application and some
lime application will be necessary. Efforts to establish
stands using seed have been encouraging, and more extensive
field trials are planned for next season.

3. Other Potential Wetland Biomass Crops.

Phragmites communis, Carex spp., Phalaris arundinacea
and Scirpus spp. have been identified as potential wetland
biomass crops through a survey of the literature (Table III)
and a survey of the standing crop of these species in natural

Table I. Typha spp.: Standing Crop in Natural Stands in Minnesota (2, 3).

Species	Above Ground	Below Ground	Total	Location
Typha X glauca	2.32	2.40	4.72	Syre
	1.13	3.10	4.23	"
	1.33			
T. X glauca	1.30	1.02	2.32	Carlos Avery WMA*
	1.69			"
	2.05			"
T. X glauca	1.60			Sherburne NWR*
T. X glauca	1.52			Agassiz NWR
	1.63			"
	.67			"
T. X glauca	1.85			Lac Qui Parle WMA
T. X glauca	1.27			Roseau WMA
T. angustifolia	2.11			Eagle Lake WMA
	1.23			Walnut Lake WMA
T. latifolia	1.09			Moose Willow WMA

(Dry Weight, kg/m^2)

*WMA, Wildlife Management Area
NWR, National Wildlife Refuge

Table II. Summary of Wetland Plant Yields.

Species	Above Ground	Below Ground	Total	Location	Reference
Typha X glauca	1.44	2.65	4.09	Minn.	4
	1.68	2.96	4.64	Minn.	5
	1.36	--	--	N.Y.	6
	2.11	--	--	Iowa	7
T. latifolia	.43-2.25			S.E.(U.S.)	8
	1.40	.50	1.90	Wisc.	9
	.50-2.00	.20-1.40	.80-3.40	Czech.	10
Phragmites communis	1.11			Iowa	11
	.78			Iowa	12
	1.12			Czech.	10
Phalaris arundinacea	1.37			N.Y.	13
	.87			U.K.	14
	1.35			Wisc.	15
Carex atherodes	1.16			Minn.	16
	.94	.13	1.07	Wisc.	15
	.86	.16	1.02	N.Y.	17
	1.14			N.Y.	18
C. rostrata	.52			Minn.	16
Scirpus fluviatilis	.85	.43	1.28	Wisc.	15
	.45	1.38	1.83	Iowa	7

(Dry Weight, kg/m^2)

Table III. Typha spp.: Productivity under Managed Conditions (2, 3).

	Dry Weight, kg/m^2				
	Above Ground	Below Ground	Total	Net* Total	Date
Paddies, 1.5 m^2					
Rhizome	2.25	.46	2.71	2.48	Aug 4
	2.57	.63	3.20	2.92	Aug 18
	2.09	1.24	3.33	3.08	Sept22
	1.44	1.47	2.91	2.65	Oct 27
Seed					
1st season		.90	.40	1.30	1.30	Sept
2nd season	1.45	.82	2.27	1.98	Sept
Fertilizer Addition to					
Natural Stand					
fertilized	1.45	1.27	2.72		Sept
control	1.30	1.02	2.32		Sept
Establishment of Stands					
Minnesota Peatland					
planting density 9	.59	.55	1.14	1.11	Sept
rhizomes/m^2 25	.79	.82	1.61	1.53	Sept

*Total dry weight minus the dry weight of
 the planting material.

Table IV.   Potential Wetland Energy Crops: Standing Crop
             Estimates from Natural Stands in Minnesota (3).

Species	Shoot Dry Weight, kg/m^2	Density Shoots/m^2	Location
Monotypic Stands			
Carex atherodes	.79	209	Agassiz NWR
Phragmites communis	.82	63	Fort Snelling SP
	.75	41	Roseau WMA
	1.12	61	Agassiz NWR
Phalaris arundinacea	.76	388	Carlos Avery WMA
Scirpus validus	1.18		Agassiz NWR
S. acutus	1.39	460	Syre
Mixed Stands			
Carex atherodes			
C. stricta	.82	118	Fort Snelling SP
Spartina pectinata			
Carex trichocarpa	1.16	246	Fort Snelling SP

SP - State Park
WMA - Wildlife Management Area
NWR - National Wildlife Refuge

stands in Minnesota (Table IV). Several criteria are used in screening plants to identify potential wetland energy crops. Productivity must be 8-10 t/ha with possibilities for improvement, propagation methods using seeds or vegetative means should be known or easily developed, species must be adapted to the wetland habitat, and potential crop species must grow either in natural monoculture or in a mixed stand with species of similar harvesting requirements. Extensive work on developing Phragmites as a biomass crop is being carried out in Sweden (20, 21, 22). Phalaris is already a commercial crop; seed is readily available and management techniques have been investigated in some detail (23). Furthermore, its natural habitat is poorly drained areas, it appears to be quite resistant to insect attack, and mixed Phalaris-legume plantings to reduce fertilizer inputs look promising (23). Much less is known about Carex spp. and Scirpus spp., but the standing crop of several species fall within the 8-10 t/ha range. Yields can be increased with fertilizer additions (24) and the possibility of using a mixture of species with similar management and harvesting requirements presents an attractive alternative to the traditional monoculture approach.

Plans for the next growing season include establishing more extensive Typha stands (1-5 acres) using both rhizomes and seed, and small trial plots of Phragmites, Carex and Phalaris. Also, we will continue our studies of wetland plant productivity in natural stands.

References

1. Minnesota Energy Agency, 1980. Energy Policy and Conservation Biennial Report. MEA (Draft Version).
2. Pratt, D.C., 1978. Minnesota Energy Agency Report I: Cattails as an Energy Source.
3. Pratt, D.C., Bonnewell, V., Andrews, N.J. and Kim, J.H., 1980. Minnesota Energy Agency Report II: The Potential of Cattails as an Energy Source.
4. Bray, J.R., Lawrence, D.B. and Pearson, L.C., 1959. Primary Production in Some Minnesota Terrestrial Communities for 1957, Oikos 10, 38-49.
5. Bray, J.R., 1960. The Chlorophyll Content of Some Native and Managed Plant Communities in Central Minnesota, Can. J. Bot. 38, 313-333.

6. Bernard, J.M. and Fitz, M.L., 1978. Seasonal Changes in Above-ground Primary Production and Nutrient Contents in a Central New York Typha glauca Ecosystem, Bull. Torrey Bot. Club 106, 37-40.

7. van der Valk, A.G. and Davis, C.B., 1978. Primary Production of Prairie Glacial Marshes. In: Good, R.E. et al. (eds.). Freshwater Wetlands: Production Processes and Management Potential, Academic Press, N.Y.

8. Boyd, C.E. and Hess, L.W., 1970. Factors Influencing Shoot Production and Mineral Nutrient Levels in Typha latifolia, Ecology 51, 296-300.

9. Gustafson, T.D., 1976. Production, Photosynthesis and the Storage and Utilization of Reserves in a Natural Stand of Typha latifolia L., Ph.D. Thesis, Univ. Wisconsin, Madison.

10. Kvet, J. and Husak, S., 1978. Primary Data on Biomass Production Estimates in Typical Stands of Fishpond Littoral Plant Communities, In: Dykjova, D. and Kvet, J. (eds.) Pond Littoral Ecosystems, Springer-Verlag, N.Y.

11. Van Dyke, G.D., 1972. Aspects Relating to Emergent Vegetation Dynamics in a Deep Marsh, Northcentral Iowa, Ph.D. Thesis, Iowa State University, Ames.

12. van der Valk, A.G., 1976. Zonation, Competitive Displacement, and Standing Crop of Northern Iowa Fen Communities, Proc. Iowa Acad. Sci. 83, 50-53.

13. Wedin, W.F. and Helsel, Z., 1977. Plant Species for Biomass Production on Managed Sites, Proceedings of Biomass -- A Cash Crop for the Future, Kansas City, MO.

14. Pearsall, W.H. and Gorham, E., 1956. Production Ecology I. Standing Crops of Natural Vegetation, Oikos 7, 193-201.

15. Klopatek, J.M. and Stearns, F.W., 1978. Primary Productivity of Emergent Macrophytes in a Wisconsin Freshwater Marsh Ecosystem, American Midland Naturalist 100, 320-333.

16. Gorham, E. and Bernard, J.M., 1975. Midsummer Standing Crops of Wetland Sedge Meadows along a Transect from Forest to Prairie, J. Minn. Acad. Sci. 41, 16-17.

17. Bernard, J.M. and MacDonald, J.G., 1974. Primary Production and Life History of Carex lacustris, Can. J. Bot. 52, 117-123.

18. Bernard, J.M. and Solsky, B.A., 1970. Nutrient Cycling in a Carex lacustris Wetland, Can. J. Bot. 55, 630-638.

19. Linde, A.F., Janisch, T. and Smith, D., 1976. Cattail -- The Significance of its Growth, Phenology and Carbohydrate Storage to its Control and Management. Tech. Bulletin No. 94, Department of Natural Resources, Madison, Wisconsin.

20. Bjork, S. and Graneli, W., 1978. Energivass-Rapport Etapp I. Limnologiska Institutionen, Lund, Sweden.

21. Graneli, W., 1980. Energivass-Rapport Itapp II, Limnologiska Institutionen, Lund, Sweden.

22. Graneli, W., 1980. Energivass-Rapport Etapp III, Limnologiska Institutionen, Lund, Sweden.

23. Martin, G.C. and Heath, M.E., 1973. Reed Canarygrass, In: Heath, M.E. et al. (eds.) Forages, Iowa State University Press, Ames, Iowa.

24. Mason, J.L. and Mittimore, J.E., 1970. Yield Increases from Fertilizer on Reed Canarygrass and Sedge Meadows. Can. J. Plant Sci. 50, 257-260.

NATURAL VEGETATION AS A RENEWABLE ENERGY RESOURCE IN
GREAT BRITAIN

G.J. Lawson and T.V. Callaghan

Institute of Terrestrial Ecology

Merlewood Research Station

Grange-over-Sands

Cumbria LA11 6JU, UK

ABSTRACT

Current UK plant production could meet one third of the national
energy demand or 90% of the oil demand. Whilst it is not proposed, yet,
that large areas of agricultural land be diverted to energy production, it
is suggested that much of the 40% of rural Britain not used by intensive
agriculture or sylviculture could be devoted to energy crops. Natural
vegetation can, in some circumstances, provide higher yields than agriculture or forestry and photosynthetic efficiencies, comparable with those
of tropical species, have been observed.

Widespread species of natural vegetation could be harvested as an
opportunity energy crop without altering land use; other productive species
have a limited distribution and can be utilised only by establishing energy
plantations.

The energy content of terrestrial vegetation is reasonably constant,
and 18.3 KJ g^{-1} is an acceptable average.

Plant mineral content can be used to assess the total nutrient removal
in harvested vegetation, but it is difficult to predict the need for
nutrient replacement by fertiliser application without considering the
nutrient dynamics of each natural community.

The extent of several widespread species is assessed by a statistically
based sampling method.

The potential of a discrete area of land to sustain an energy conversion
unit is investigated using the county of Cumbria as an example.

1. INTRODUCTION

The possibility of energy farming is usually discussed in terms of agricultural or forestry crops, and is often related to countries with low population densities and ample supplies of underutilised land. Britain is overcrowded, sparsely forested and intensively farmed, so the potential for meeting a significant proportion of our primary energy demand (9576 PJ, 1980) from biofuels might too easily be discounted. Yet analysis of land use shows that 40% of rural land is not used for intensive agriculture or forestry, and surveys of the productivities of 'weed' species indicate that natural vegetation may often outyield the most productive agricultural or tree crops.

Two categories of natural vegetation energy crop can be considered: *OPPORTUNITY CROPS* (eg. *Pteridium aquilinum* - bracken and *Calluna vulgaris* - heather), which could be harvested from areas where they currently grow without altering traditional land use; and *DEDICATED CROPS*, which would involve establishing plantations of productive weed species (eg. *Reynoutria japonica* - Japanese knotweed) on fertile land. The first category, because it involves minimal competition with other land uses, should provide one of the first economically and socially acceptable forms of energy cropping.

Plant production in Great Britain is approximately 181×10^6 t yr^{-1} (1). This represents 3260×10^{15} J(PJ), which together with a possible 6-9 PJ from seaweed resources (2), is 34% of the 1980 UK primary energy demand or 90% of the oil demand. Thus the UK could not achieve energy self-sufficiency from biofuels without using the entire plant production from three times more rural land than actually exists. This is not a pessimistic conclusion however, because:

- each 1% of plant production converted into energy production would save some £160 million in oil imports;
- biofuels can, if required, be converted into high quality liquid fuels;
- biofuels potentially offer a major contribution towards energy self-sufficiency in rural and island communities (3).

2. PRODUCTIVITY OF NATURAL VEGETATION

Photosynthesis captures only 5.5% of incident solar radiation (4), which converts to a maximum possible yield of 134 t ha^{-1} yr^{-1} in southwest England and 89 t ha^{-1} yr^{-1} in northern Scotland. In practice however the product-

ivities of all types of vegetation are much lower than these maxima.
Agricultural yield does not exceed 25 t ha^{-1} yr^{-1} (5) (1.3% efficiency),
and, whilst some conifers may have total productivities as high as
36 t ha^{-1} yr^{-1} (1), the annual shedding of much of this production means
that harvestable yields of forests and coppice woodlands are unlikely to
exceed 20 t ha^{-1} yr^{-1} (1.0% efficiency) (6,7). It is interesting therefore
that some small stands of natural vegetation have yielded more than
30 t ha^{-1} yr^{-1} (maximum efficiency 2.2%) (8). These figures are higher than
could be expected for the same species growing in monocultures, but they
do indicate that British natural vegetation, in ideal circumstances, can
match the photosynthetic efficiencies of the most productive plants of the
tropics (9).

A survey of British scientific literature (10) and recent field
measurements (8) have indicated a variety of productive native and introduced species with potential as energy crops (Table I).

TABLE I

a) Selected Natural Vegetation Standing Crops Above 8 t ha^{-1}

SPECIES	STANDING CROP (t ha^{-1})	COUNTY
Grassland		
Dactylis glomerata (cock's foot)	9.1	Bucks
Deschampsia caespitosa (tufted hairgrass)	10.1	Highland
Calamagrostis canescens (bushgrass)	8.8	Cumbria
mixed chalk grassland	8.2	Bucks
Moorland		
Juncus effusus (soft rush)	8.0	Cumbria
Sphagnum sp.	9.6	Cumbria
Pteridium aquilinum (bracken)	14.1	Hants
" " "	10.0	Derbyshire
Marsh		
Typha latifolia (cat's tail)	11.2	Norfolk
" " "	10.7	Cumbria
Phragmites australis (reed)	39.7	Tayside
" " "	15.0	Norfolk
" " "	14.7	Tayside
" " "	13.5	Tayside
" " "	13.5	Devon
" " "	13.0	Cumbria
" " "	9.4	Norfolk
Phalaris arundinacea (reed-grass)	8.7	Cumbria
Saltmarsh		
Spartina anglica (cord-grass)	14.4	Merseyside

Waste ground

Reynoutria sachalinensis (giant knotweed)	37.5	Highland
Reynoutria japonica (Japanese knotweed)	25.3	Manchester
Epilobium hirsutum (great hairy willow herb)	15.3	Kent
Iris pseudacorus (yellow flag)	14.8	Kent
Chamaenerion angustifolium (rose-bay willow		
herb)	10.8	Derbyshire
" " "	8.0	Derbysire
Cirsium arvense (creeping thistle)	10.8	
Impatiens glandulifera (policeman's helmet)	10.5	Manchester
Urtica dioicia (stinging nettle)	9.7	Derbyshire
Petasites hybridus (butterbur)	8.2	Derbyshire

3. ENERGY CONTENT OF PLANT MATTER

An extensive literature review covering more than 200 species (1)
revealed very little variation in the energy content of terrestrial plants
(Table II). Considering the frequent errors involved in measuring plant
productivity, an energy content of 18.3 KJ g^{-1} dry weight appears an
acceptable standard figure.

TABLE   II   The Energy Contents of Plants

Groups of species	Mean energy content (KJ g^{-1} dry weight)	Standard error
Bryophytes	18.82	1.190
Pteridophytes	17.20	0.498
Evergreen trees	20.77	0.284
Deciduous trees	19.81	0.172
Coppiced trees	19.50	0.224
Deciduous shrubs	20.09	0.263
Herbaceous terrestrial species	18.75	0.260
Herbaceous aquatic species	12.92	0.820
Sedges	18.76	0.192
Grasses	17.77	0.112
Agricultural species	17.96	0.112
Horticultural species	18.85	0.429
Mean of all species	18.27	**0.112**

4. NUTRIENT CONTENT OF NATURAL VEGETATION

Two distinct questions relate to nutrients in natural vegetation:
- is it financially and energetically efficient to increase the yield of
dedicated energy plantations by applying fertiliser?;
- how long can we continue to harvest opportunity crops without replacing
nutrient losses by fertilising, and what replacement levels will be
necessary?
Neither question yet has a solution. Both are being investigated in small-
scale fertiliser trials (11), but partial answers to the latter have been

obtained by assessing the nutrients removed in annual harvesting (Table III) and by observing flows of nutrients to, from and within selected natural communities. This approach too is limited by lack of information, but initial attempts (10) indicate that:

- harvesting on moorlands will cause progressive nutrient impoverishment, although *Calluna* cutting could be carried out on a standard 15 year rotation with only the need for a small amount of P-fertiliser replacement;

- phosphorus is the most serious loss in hardwood coppice;

- many mires could be harvested continuously because drainage water replaces nutrient losses;

- saltmarshes have adequate supplies of most minerals but the prevention of autumn translocation of foliar nitrogen to below ground storage organs may diminish yields in subsequent yields.

TABLE (III) Nutrient Content of 24 Potential Energy Species Grouped into Broad Vegetation Categories

	N(%)	P(%)	K(%)	Ca(%)	Mg(%)
Woody shrubs	1.46	0.12	0.95	0.29	0.15
Herbs	3.30	0.35	2.29	2.66	0.42
Herbs of wet ground	1.74	0.13	2.20	1.10	0.24
Rushes	1.38	0.15	1.57	0.17	0.13
Grasses	1.92	0.21	2.08	0.45	0.21
Bracken	2.39	0.23	2.34	0.40	0.22

5. EXTENT AND DISTRIBUTION OF NATURAL VEGETATION

There are in the UK 67,680 km^2 of rough grazing, 14,260 km^2 of miscellaneous non-urban land and 3,670 km^2 of unproductive woodland (10). Together these categories constitute 38.5% of rural land. Much of this area is composed of only a few widespread species (Table IV) with *Calluna* (14,900 km^2) and *Pteridium* (3,200 km^2) being two obvious choices for energy crops.

TABLE (IV) The Percent of Rural Land in Britain Covered by the 10 most Abundant Species or Species Groups

SPECIES	%COVER	AREA(KM^2)
Calluna vulgaris (heather)	7.12	14,900
combined *Agrostis* (bent-grass)	5.14	10,800
Molinia caerulea (purple moor-grass)	3.33	7,000
Eriophorum/Trichophorum (cotton-grass/deer grass)	2.54	5,300
combined *Festuca*	1.97	4,100
combined *Poa* (meadow grass)	1.81	3,800
Dactylis glomerata (cock's foot)	1.56	3,300
Pteridium aquilinum (bracken)	1.54	3,200

Table (IV) continued.....

Nardus stricta (mat grass)	1.45	5,300
Vaccinium myrtillus (blaeberry)	0.65	1,400

Use could also be made of the productive species growing on Britain's 1,200 km² of derelict urban and industrial land, 1,800 km² of road verges, 3,755 km² of foreshore,1,096 km² of sports grounds and 900 km² of domestic lawns (10).

Spartina anglica, which does not appear to thrive in Scotland, is the only potential energy crop to have geographically limited distribution, although this may be largely due to its comparatively recent occurrence in this country.

6. ENERGY FROM NATURAL VEGETATION IN CUMBRIA

The ITE Land Classification of Cumbria (13) facilitates the mapping of vegetation within the country, and has been extended to map total plant energy production and used to site hypothetical catchments for different types of sizes of chemical conversion units (1). A similar approach is to examine the catchment size necessary to fuel a conversion unit if currently available crops of a single species were used as feedstock. With *Pteridium* a 500 m³ digestor could be supplied from a catchment of 8.7 km² (Figure 1). Naturally, for reasons of feedstock chemical quality and continuity of supply, such a digestor would not be supplied by a single species, but the approach illustrates a means of assessing the feasibility and planning of the siting of large digestors in rural areas. A thermal conversion unit, requiring the same annual dry matter intake (1,650 t) (14) would require a larger *Pteridium* catchment (14.5 km²), because the fronds would be harvested in autumn when they have dried out but also lost 40% of their peak weight. The same thermal unit fueled by *Calluna* would require 63.5 km².

7. CONCLUSION

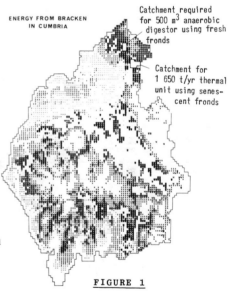

ENERGY FROM BRACKEN IN CUMBRIA

Catchment required for 500 m³ anaerobic digestor using fresh fronds

Catchment for 1 650 t/yr thermal unit using senescent fronds

FIGURE 1

Natural vegetation is shown to be both widespread, with many species ammenable to opportunity cropping, and productive, with several species being excellent candidates for dedicated energy crops. Many unknowns exist however, regarding the breeding, planting, management and harvesting of energy species and the availability of land for, and overall financial or energetic viability of, natural vegetation energy crops. The need for further research is overwhelming.

8. ACKNOWLEDGEMENT

This paper has been made possible by research grants from the Solar Biological Programme of the UK Department of Energy. The paper's contents represent solely the views of the authors, and are not necessarily shared by the Department of Energy.

REFERENCES

1. Callaghan, T.V., Millar,A., Powell,D. & Lawson,G.J. (1979). Carbon as a renewable energy resource in the UK. Report to UK Dept. of Energy from Institute of Terrestrial Ecology, Cambridge.
2. Newman,D.A. (1979). Aquatic plants as an energy crop in the UK. Report to the UK Dept. of Energy from University of Sheffield, Botany Dept.
3. Callaghan,T.V., Lawson,G.J., Scott,R. & Whittaker,H.A.(In press). Fuels from non-woody plants. In: Energy for Rural and Island Communities, edited by J.Twidell. Pergamon.
4. Hall,D.O. (1979). Solar energy use through biology - past, present and future. Solar Energy 22, 307-328
5. Loomis,R.S. & Gerakis,P.A. (1975). Productivity in agricultural eco-systems. In: Photosynthesis and Productivity in Different Environments, edited by J.P. Cooper, 145-172, Cambridge University Press.
6. Mitchell,C.P, Mathews,J.D., MacBrayne,C. & Proe,M. (1981).Determination of yield of biomass from whole-tree harvesting of early thinning in Britain. This volume.
7. Pearce, (1981). Coppiced trees as energy crops. This volume.
8. Callaghan,T.V., Scott,R. & Whittaker,H.A. (1980). The yield,development and chemical composition of some fast growing indigenous and naturalised British plant species in relation to management as energy crops. Report to UK Dept. of Energy from Institute of Terrestrial Ecology, Cambridge.
9. Loomis,R.E. & Williams,W.A. (1963). Maximum crop productivity - an estimate. Crop Sci., 3, 67-72.
10. Lawson,G.J., Callaghan,T.V. & Scott,R. (1980). Natural vegetation as a renewable resource in the UK. Report to UK Dept. of Energy from Institute of Terrestrial Ecology, Cambridge.
11. Callaghan,T.V., Lawson,G.J., Scott,R. & Whittaker,H.A. (1980). An exper-imental assessment of native and naturalised species of plants as re-newable sources of energy in Great Britain. In: EEC Energy Programme - Energy from Biomass. 4th Contractors Meeting,Amsterdam 18-19 Sept. 1980.
12. Bunce,R.G.H. (1980). The application of computer aided techniques for landscape description in Britain. In: Quantitative Landscape Mapping in Western Europe: a review of Experience in the United Kingdom,Switzerland West Germany,The Netherlands and Norway, edited by S.E. Bie.
13. Bunce,R.G.H. &, Smith,R.S. (1978). An ecological survey of Cumbria. Cumbria County Council and Lake District Special Planning Board. Kendal.
14. Wheatley,B.I. & A   G. (1979). Conversion of biomass to fuels by anaerobic digestion. Report to UK Dept. of Energy from Ader Associates, West Wickham, Kent.

SELECTED NATURAL AND ALIEN PLANT SPECIES AS RENEWABLE SOURCES
OF ENERGY IN GREAT BRITAIN - EXPERIMENTAL ASSESSMENT AND
IMPLEMENTATION

T. V. Callaghan   R. Scott   H. A. Whittaker and G. J. Lawson
Institute of Terrestrial Ecology (N.E.R.C.) Merlewood Research
Station, Grange-over-Sands, Cumbria, UK.

Summary

The yields of some naturally occurring plant species have been
assessed on a range of sites in North West Britain. Seasonal patterns
of production show differences between species and may allow phased
harvesting. Typical yields of 6 to 13 t ha^{-1} were found in continuous
stands though yields over 20 t ha^{-1} occurred in small patches of
Reynoutria species.

Choice of fuel conversion technology will be influenced by biomass
quality in relation to water content, organic fraction composition and
inorganic nutrient element concentration. Harvesting at peak biomass will
involve the nutrient depletion of the site whereas removal of senescent
material in Autumn should not significantly affect regrowth in subsequent
years.

Practical and economic limitations presently restrict the utilization
of potentially productive land over wide areas but harvesting machines
exist which are capable of dealing with dense vegetation on difficult
terrain and the survival potential of invasive weeds planted as dedicated
energy crops exceeds 98%. An estimate of £1.12 per barrel of oil equivalent has been derived for *Pteridium aquilinum* at the farm gate compared
with present expenditure of £120 ha^{-1} on its herbicidal control. Net
energy returns on opportunity harvesting of widespread species could be
high but the feasibility of managing the crops requires further detailed
investigation, especially with regard to maintenance of yield.

1. INTRODUCTION

Natural and semi-natural vegetation shows considerable potential as a renewable source of energy in the United Kingdom. It extends over 8.6×10^6 ha of land (40% of the rural area of the United Kingdom) and contains many productive species, some of which are more productive in their poor habitats than traditional crops under cultivation (1,2).

Extensive heather(*Calluna vulgaris*) moors or bracken (*Pteridium aquilinum*) fells could be harvested as "opportunity energy crops" without significantly disturbing traditional land use; alternatively species which are more productive than traditional crops could be planted as "dedicated energy crops" on areas of land devoted to energy fixation.

In order to manage an energy crop it is essential to understand the way in which the crop develops throughout a growing season and a knowledge of the seasonal changes in the chemical composition of biomass is important in determining the type of technology required to produce fuel. Also, it is necessary to know the composition of inorganic elements within harvested biomass so that replacement rates of nutrients can be estimated as it is essential that a perennial energy crop should give stable yields under a long-term harvesting regime. If, after these investigations, a species still shows potential it becomes important to investigate the technological and financial feasibility of managing the new energy crop.

2. SEASONAL DEVELOPMENT

A range of developmental pattern is shown by perennial plant species in the UK. *Petasites hybridus* (butterbur), a plant of damp woodland and wet meadows has herbaceous shoots which emerge in March and attain peak standing crop in mid-July (Fig. 1). Shoots of *Pteridium*, a widespread species of heath and woodland, do not emerge until mid-May but peak yield is obtained in the same month as *Petasites*. After peak standing crop senescent shoots of *Pteridium* may remain standing until the next year, whereas the shoots of *Petasites* collapse and quickly decompose on the ground surface during Autumn. *Spartina anglica* is a rhizomatous grass of low-level salt marshes. Spring growth is very slow but biomass continues to increase until mid-November. Slow senescence then follows (Fig. 1) and standing dead material persists well into the next year.

Flexible harvesting times during the senescent stages or a phased sequence of harvesting times staggered between different species exemplified by those in Fig. 1 would ensure a prolonged and fairly stable supply

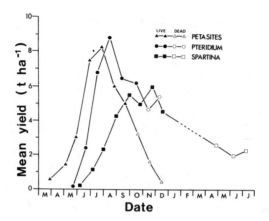

Fig. 1 Seasonal development of three potential energy crops in 1979.
of substrate to chemical conversion units, thereby alleviating physical
problems of storage and the biodeterioration of stored biomass.

3. CROP MANAGEMENT AND YIELD

Peak yields of the shoots of the species studied intensively varied
from 4.0 t ha^{-1} (*Filipendula ulmaria*) to 37.5 t ha^{-1} for *Reynoutria
sachalinensis*, an alien invasive weed of Great Britain (3). Most peak
yields were within the range of 6 to 13 t ha^{-1} and these yields were
achieved without cultivation or the addition of fertilizers.

Yields of *Pteridium* are particularly impressive as upland pastures
reclaimed from areas formerly under *Pteridium* yield *ca* 6.0 t ha^{-1} when
cultivated (4) and more recently studied areas of this weed species are
yielding up to 15.0 t ha^{-1}. As *Pteridium* occupies over 3,200 km^2 of Great
Britain, it is a potential opportunity crop of considerable significance.

Spartina showed peak yields of 6 t ha^{-1} (Figs. 1 & 2) at a poor site
but recent intensive studies at a more typical and extensive site show
yields of up to 16.8 t ha^{-1}. Although this species currently occupies
only 120 km^2 of British salt marshes, it is at an invasive stage and its
potential distribution on land currently of no economic use could be very
extensive.

Vegetation of waysides, amenity land and waste places often shows
high yields (eg. *Epilobium hirsutum* which yielded 12.9 t ha^{-1}) and should
be considered as a possible opportunity crop, especially as energy is at

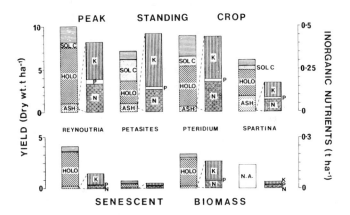

(Sol C: soluble carbohydrate, Holo = holocellulose, N.A. = not available)
Fig. 2 Yield and composition of potential energy crops at two stages.

present expended on control.

Impatiens glandulifera, an annual species, reaches a peak yield of
11.4 t ha^{-1} from seed within three months and is highly tolerant of
pollution. It could play an important role as a dedicated energy crop in
association with sewage treatment.

Reynoutria japonica (Japanese knotweed) R. sachalinensis and
Impatiens could be considered for planting as dedicated energy crops.
Both Reynoutria species showed yields over 20.0 t ha^{-1} in small stands,
though only 11.8 t ha^{-1} was achieved by R. japonica from an extensive
stand. Survival of plants from   rhizome fragments in our transplantation
trials exceeded 98%, so establishment of energy plantations could be easy.

A long term possibility is the replacement of some grasslands with
species like Reynoutria from which protein could be extracted (5) and the
residues used for fuel.

4. THE CONVERSION OF PLANTS TO FUEL

The quality of biomass will determine the type of conversion technology required to produce a usable fuel. Water contents, for example, are
important in determining the suitability of anaerobic digestion or thermal
methods of conversion whereas the contents of soluble carbohydrates and
protein, lignin and ash will influence the efficiency of anaerobic
digestion. If native species were harvested in summer at peak standing

crop, contents of energy-storing components (Fig. 2) and water would be
high (over 75% if fresh weight). Anaerobic digestion would probably be
the most suitable conversion method and species like *Petasites* and
Impatiens should digest very efficiently. Most herbaceous species should
digest efficiently, but *Spartina* possesses a high ash concentration (31%
of the dry weight). The actual efficiency with which these various
species digest remains to be assessed experimentally.

If harvesting were carried out during the senescent phase, water
contents and the concentrations of energy storing compounds would be low
(Fig. 2) and a thermal conversion process could most efficiently generate
fuel. *Petasites* and *Impatiens* could not be harvested under this regime
because of the autumnal collapse of their shoots, but the shoots of
Pteridium, Reynoutria, Spartina, Epilobium, Chamaenerion angustifolium and
Filipendula are readily harvestable during autumn and winter.

5. FERTILIZER REQUIREMENTS

Removing harvested biomass from site could affect the subsequent
regrowth of perennial species by removing inorganic and organic nutrients
which would normally be recycled by translocation from shoots to rhizomes,
leaching and decomposition. The continual removal of these elements from
the site will reduce subsequent yields at a rate dependent upon the size of
the nutrient pools held in the below ground biomass and soils, and the
inputs of nutrients through precipitation, rock-weathering and flooding
etc. As many native species grow on poor soils and use available nutrients
economically, significant response to artificial nutrient inputs may be
expected, especially under intensive cropping.

In general, plant matter shows peak concentrations of inorganic
nutrients (N,P and K) in spring and these decrease markedly over summer and
autumn as tissues age (2). The total amount of these elements in
harvested biomass is an interaction between their concentrations and the
dry weight of the biomass. This results in far lower contents of N, P and
K in senescent tissues than in healthy and younger tissues (Fig. 2). In
some species such as *Pteridium* and *Spartina*, nutrients appear to be translocated from shoots to rhizomes during senescence and contents of N, P and
K decrease faster than dry weight (6).

Biomass harvested at peak standing crop therefore removes far more
nutrients from site than if harvested when senescent. Dressings of N, P
and K required to replace nutrients removed in harvested biomass would be
greater than most typical lowland agricultural applications, if for

example, *Pteridium* were harvested at peak standing crop. However, if biomass were harvested when senescent, replacement levels of N, P and K would be very low. Annual nutrient inputs via flooding may be greater than nutrient removal (20 kg ha^{-1} of N, 2.5 kg ha^{-1} of P and 15 kg ha^{-1} of K) involved in cropping dead shoots of *Spartina anglica*. Thus, harvesting strategies aimed at removing senescent material would be preferable to those aimed at removing maximum yields in terms of nutrient replacement. Indeed, this approach has been used for many centuries in the harvesting of *Phragmites* (reed) for thatching.

6. ENERGETIC AND ECONOMIC FEASIBILITY OF HARVESTING

Areas of natural/semi-natural vegetation are currently unused for intensive agriculture because infertility and harsh climate preclude arable crops, mechanical harvesting is difficult even though the land is fertile (eg. salt marshes and freshwater marshes), or mechanical harvesting is uneconomic because the areas of land, often very fertile, occur in isolated small pockets (eg. road verges and stream sides). These areas will probably be the first to be available for energy cropping (7) and such cropping can be shown to be both technologically and economically feasible.

Maize harvestors could provide a precision chop of dedicated species (eg. *Reynoutria*) growing on flat stone-free land. Rotary slashers and flail mowers fitted with windrow of self-loading mechanisms can handle tougher vegetation (eg. *Pteridium* and *Calluna*) on stonier land. On steep slopes, mowers and slashers could be used with Swiss-designed low centre-of-gravity mini-tractors. Low ground pressure harvesters are already available for cropping marsh species (such as *Phragmites* (8)).

Whilst it is not yet possible to make accurate estimates of the energy input required to utilise opportunity crops of natural vegetation it can be predicted that this will be lower than most types of agriculture: cultivation and seeding are unnecessary, fertilizers will not be required initially and harvesting represents the only energy cost. The *Pteridium* cutting trials from which the financial costs below were updated (9) expended only 62.5 MJ ha^{-1} in fuel. Doubling this figure to allow for energy used in collection and transport suggests that the energy output/ input ratio of *Pteridium* harvesting may be around 1440.

Economic feasibility has been tentatively assessed for opportunity crops of *Pteridium*, *Calluna* and *Phragmites*. Harvesting costs are £1.12, £0.31 and £7.45 respectively per barrel of oil equivalent. None of these figures includes the cost of converting harvested plants into useful energy and the *Calluna* and *Pteridium* costs may be underestimates since they assume maximum yields and ideal ground conditions. Nevertheless, the comparative cheapness of energy from these species points to a real possibility of future exploitation, particularly when account is taken of the current cost of *Calluna* burning (£2.47 ha^{-1}) - reducing the cost of energy from *Calluna* to £0.26 per barrel oil equivalent - and chemical control of *Pteridium* (£120 ha^{-1}) - reducing the cost of energy from *Pteridium* to zero and saving the farmer and the state an additional £85 ha^{-1}.

7. ACKNOWLEDGEMENTS

This research was financed by the Solar Biological Programme of the UK Department of Energy and continuing research is also being financed by the Solar Energy Programme of the European Commission. The contents of this paper represent solely the views of the authors, which are not necessarily those shared by the Department of Energy.

8. REFERENCES

1. Callaghan, T.V., Millar, A., Powell, D. & Lawson, G.J. (1979). Carbon as a renewable energy resource in the UK. Report to UK Dept. of Energy from Institute of Terrestrial Ecology, Cambridge.
2. Lawson, G.J., Callaghan, T.V. & Scott, R. (1980). Natural vegetation as a renewable resource in the UK. Report to UK Dept. of Energy from Institute of Terrestrial Ecology, Cambridge.
3. Callaghan, T.V., Scott, R. & Whittaker, H.A. (1980). The yield, development and chemical composition of some fast growing indigenous and naturalised British plant species in relation to management as energy crops. Report to UK Dept. of Energy from Institute of Terrestrial Ecology, Cambridge.
4. Hill Farming Research Organisation (1979). Science and hill farming. Twenty-five years of work at the Hill Farming Research Organisation, 1954-1978. Midlothian, H.F.R.O., 184pp.
5. Plaskett, L.G. (1981). The potential role of crop fractionation in the production of energy from biomass. This volume.
6. Callaghan, T.V., Lawson, G.J., Scott, R. & Whittaker, H.A. (In press). Fuels from non-woody plants. In: Energy for Rural and Island Communities, edited by J. Twidell. Pergamon.
7. Lawson, G.J. & Callaghan, T.V. (1981). Natural vegetation as a renewable energy resource in Great Britain. This volume.
8. Björk, S. & Graneli, W. (1978). Energy needs and the environment. Ambio, 7, 150-156.
9. Scottish Machinery Testing Station (1952). Control of Bracken - Bowmont Water, Roxburghshire. S.M.T.S. Report on 1951 Operations (first year). Unpublished report, S.M.T.S. (Now S.I.A.E.) Bush Estate, Midlothian, Scotland. 8pp.

THE PRODUCTIVITY OF CATCH CROPS GROWN FOR FUEL

S.P. CARRUTHERS

Department of Agriculture and Horticulture,
University of Reading.

Summary

Agriculture, commonly seen as a consumer of energy, is here considered as a potential producer. There are a number of possible sources of energy within UK agriculture; the growing of a crop between the harvest and sowing of main crops (i.e. catch cropping) is one such source. Productivity, defined as net energy output as an available fuel, is determined by land-time availability, crop yield and efficiency of conversion to a fuel (i.e. anaerobic digestion to produce methane or fermentation and distillation to produce ethanol). A preliminary field experiment carried out in 1979 provided data on the yields and chemical composition of six species and drew attention to the practical problems associated with catch crop cultivation.

INTRODUCTION

Photosynthesis, originally responsible for the energy fixed in the most exploited fossil fuels, coal, oil and natural gas, is the process by which solar energy is captured by green plants and stored as carbon compounds; about ten times the world's annual energy consumption is captured each year by photosynthesis and stored in green plants. Shortages and rising prices of fossil fuels have provoked an interest in alternative, renewable energy sources; plant material represents one such·source.

AGRICULTURE AS A SOURCE OF PLANT MATERIAL FOR USE AS FUEL

Reasons for considering agriculture as a source of plant material for use as fuel include:
- considerable expertise in the capture of solar energy exists within agriculture;
- agriculture represents a significant component of U.K. land use (c.75%) and that of other EEC countries (48% - 81%);
- appropriate machinery and facilities for cultivating, harvesting and handling plant material are available on the farm;
- the farm may also represent an important outlet for the fuel.

Strategies: Assessment studies carried out at Reading and forming part of the Department of Energy's "Fuels from Biological Materials Programme" identified a number of strategies for obtaining plant material for fuels from within U.K. agriculture. These are summarized in Table I.

Degree of integration with food/feed production	Source	Time-scale	Real price of conventional fuels
	(1) Crop residues	Short-term	
	(2) Waste land on farms		
	(3) Catch fuel crops		
	(4) Dual purpose crops with food/feed and fuel component	Medium-term	
	(5) Mixed crops		
	(6) Dual purpose crops for fractionation		
	(7) Break fuel crops	Long-term	
	(8) Fuel crop plantations		

Table I: Possible sources of plant material for fuel within UK agriculture
After (1)

THE CATCH FUEL CROP SYSTEM

Background: Catch crops are grown in the time period available between the harvest and sowing of main crops. Traditionally catch crops
have been used to provide animal feed in the autumn - winter period;
using this opportunity to provide fuel is a strategy which should become
economically viable in response to moderate increases in the prices of
conventional fuels.

Advantages associated with growing catch crops for fuel include:
- relatively easy integration with current cropping regimes;
- minimal impact on traditional agricultural output;
- use of machinery already available on the farm;
- provision of fuel at a time when demand is high and material from other
  sources may not be available.

Desiderata: The catch fuel crop essentially seeks to use an available
resource i.e. land unoccupied for a particular time, to convert sunlight
energy to a usable fuel. It is important that the system:
- has a net output of fuel energy;
- is not wholly dependent on large inputs of non-renewable resources;
- has a minimum detrimental effect on other components of the rotation
  and, if possible, a substantial benefit.

The conversion process most suited to the wet, green feedstock produced by catch fuel cropping would be anaerobic digestion to produce methane. In the longer term, development of the technology associated with
the hydrolysis of cellulosic materials and the fermentation of the products
may result in the "ethanol route" becoming a more attractive option for
the conversion of catch crop material. There may be some synchrony between
the implementation of cath fuel crop systems and developments and improvements in efficiency of fermentation to ethanol.

DETERMINANTS OF PRODUCTIVITY

The productivity of the catch fuel crop system, defined as the net
energy output as an available fuel, is determined by:
- land-time availability;
- crop yield;
- efficiency of conversion to a fuel (i.e. methane or ethanol).

Land-time availability

Land-time availability, which is a function of area, length of time and time of year, can be assessed under current agricultural systems or considered in relation to a number of determining factors which may be subject to change and manipulation.

Assessments: The area of land and the date it is available, and hence the sowing date of the catch crop, is largely determined by the previous crop. The results of preliminary assessments are summarized in Table II.

Sowing date of catch crop	Previous crops	Area becoming available in the U.K.	
		Thousand hectares	%U.K. tillage
June	Early vegetables	66	1.4
End July- early September	Cereals Peas Winter rape Winter beans	4042	86
Mid-September onwards	Maize Potatoes Sugar beet Cereals	343	7.3

Table II: Land available for catch cropping
After (1)

Determining factors: Land-time availability is determined by the previous crop, the speed of harvest of the previous crop and by the promptness of establishment of the catch crop. There are a number of ways in which the effective amount of land-time can be increased either by releasing more land earlier in the year or by enabling a more rapid establishment of the catch crop. These include:

- alterations in main crop rotations;
- alterations in main crop species and varieties;
- whole crop harvesting of the main crop;
- undersowing and aerial sowing of the catch crop;
- direct drilling of the catch crop.

Crop yield

Crop yield is a function of the amount of solar radiation available and the efficiency of its fixation. The amount of solar radiation is indicated by land-time availability and will also be affected by location and weather. Other important environmental factors are temperature and soil moisture. Significant management factors include: crop species and variety, land preparation, fertilizing, seed rate and seed sowing method.

Species and variety: The type of crop that can be grown is largely determined by the sowing date as illustrated in Table III. Species can be selected from conventional agricultural species, non-U.K. crop species (e.g. Quinoa) or indigenous weeds (e.g. Sterile Brome).

Sowing date	Type of crop	Suitable species	Harvest date
June	Crops with large sinks (i.e. roots + stems) requiring longer growing period for full potential	Fodder beet Mangels Swedes Yellow turnips Marrow-stem kale Field bean	Oct-Nov
End July - Early September	Fast growing leafy or stemmy crop	Kale Rape Stubble turnip Fodder radish Mustard Forage pea Vetch	Oct-Nov
Mid-September	Crop able to overwinter and produce early spring growth	Barley Oats Rye Italian ryegrass	March

Table III: U.K. agricultural species suitable for use as catch crops

Conversion efficiency

The efficiency of conversion of plant material to a fuel is dependent on factors related to the conversion process and on the composition of the feedstock. There may be some correlation between ruminant digestibility (D-value) and the efficiency of conversion to methane by anaerobic digestion, but this has yet to be demonstrated experimentally.

1979 FIELD EXPERIMENT

A preliminary field experiment was carried out in 1979 to study the effect of species and growing period on catch crop yield and chemical composition and to draw attention to any practical problems associated with the growing of crops in the late summer - autumn period.

Results: Only the first sowing (7 August) produced significant yields; establishment of crops from later sowings (4 September, 3 October) was retarded or prevented by drought and low temperatures. The results are summarized in Table IV. Highest yields were achieved by fodder radish in 12 weeks; this species is suited to a short growing season, but senesces soon after flowering and is frost susceptible. Rape is more suited to a longer season and later in the year, showing some degree of frost tolerance.

Species	Mean max. D.M. yield of tops	D-value %
Fodder radish	5.11 t/ha in 12 weeks	69.48
Rape	4.66 t/ha in 16 weeks	75.33
Quinoa	4.37 t/ha in 16 weeks	59.66
Sterile brome	4.07 t/ha in 12 weeks	59.11
Kale	3.63 t/ha in 16 weeks	72.75
Forage pea	2.97 t/ha in 12 weeks	57.85

Table IV: Data from 1979 field experiment (crops sown 7 August)

Problems: The major problem experienced was late summer drought which delayed establishment of crops sown in September. Later-sown crops of some species also suffered frost damage.

CONCLUSIONS

Catch cropping has some potential as a source of fuel; this potential is likely to be higher in specific situations than in the U.K. as a whole and could be realized in response to moderate rises in fuel prices. Current research is aimed at gaining a clearer understanding of the determinants of crop yield and at investigating the methane production of various crops and, hence, at estimating the overall potential productivity of the system.

ACKNOWLEDGEMENTS

This work forms part of the Department of Energy's 'Fuels from Biological Materials Programe' which is co-ordinated by the Energy Technology Support Unit (ETSU). Thanks are due to the above for their financial support of the research. However, the views expressed in this paper are those of the author, and not necessarily those of the Department of Energy.

REFERENCE

(1) Spedding, C.R.W., D.M. Bather and L.A. Shiels (1979). An assessment of the Potential of U.K. Agriculture for Producing Plant Material for Use as an Energy Source. Unpublished report to the Department of Energy.

THE ALLOCATION OF LAND FOR ENERGY CROPS IN BRITAIN

R.G.H. Bunce[1], L.H. Pearce[2], and C.P. Mitchell[3]

1. Institute of Terrestrial Ecology, Merlewood Research Station

2. Forest Commission, Westonbirt Arboretum

3. Aberdeen University, Forestry Department

SUMMARY

The availability of land is a key factor in the future of energy crops in Britain. A stratification system is described that enables estimates from a small number of representative samples to be translated into national figures. The system uses environmental data to classify the land surface into strata, termed land classes, which may then be used as a basis for field sampling. The field data provide sufficient information on the sample squares to enable their potential for other uses, for example, single stem trees and coppice wood, to be estimated. These patterns can then be overlaid onto the existing land use in order to develop an optimal distribution of land for wood energy production, having account for specified constraints. The potential for wood energy crops can therefore be determined for Britain and possible areas of development identified. Linear programming is also applicable and could produce an integration of land use incorporating energy crops into the existing system.

1. INTRODUCTION

The availability of land is fundamental to the future of energy crops in Britain and the primary constraint is that of the current land use. All energy crops will have to compete with conventional, uses with the balance eventually being determined by relative profitability and a method is required for such comparisons. The present paper describes such an approach developed in relation to wood energy plantations, as compared with current land use and provides a link between actual areas of land and the strategic level. The method could also be applicable to other potential energy crops, such as natural vegetation,discussed elsewhere in this volume.

The method described below provides such a link by providing a stratification system to identify representative sample sites which can be studied in detail, and extrapolated to the whole of Britain. The present system has been developed from the application of computer methods to the classification of land through the analysis of environmental data from maps. The principle is that if stratification by environment alone shows sufficiently high correlations with biological parameters observed in the field, then the environment could be used for predictive purposes. The original study was carried out in a small area of Cumbria before being extended first to the Lake District National Park (Bunce, Morell and Stel (1975)) to Shetland, and then to the whole of Cumbria (Bunce and Smith (1978)). In the latter study high correlations were shown between the map stratification and the vegetation of the county. The method has now been extended to Britain and now includes soils and land use, as well as vegetation. The project has,(Bunce(1980) the following four phases:

Phase   I   The environmental analysis of the map data.

Phase   II   The use of the stratification to identify representative sample sites for field survey.

Phase III   The prediction of patterns for the whole of Britain from the sample field survey.

Phase   IV   The use of the framework to examine the potential of land for various uses.

The map data should contain sufficient information to enable an adequate summarisation of the environmental affinities of a given area of land. Accordingly a wide range of types of data are required and the following broad categories are used. (1) Topographic. (2) Human artefacts. (3) Geological. (4) Climatic. These data were recorded from a grid of 1228, 1 km squares.

The classification method used identifies indicator attributes at each stage of the analysis, and proceeds in an hierarchical fashion. The classes (termed land classes) produced are arbitrary divisions within the continuous variation of the British environment. Although there are no discontinuities between them, there is a greater variation between the classes than within them, and they thus provide a framework for the description of the land surface of Britain.

The second phase involved field visits to 8 squares drawn at random from each of the 32 land classes (ie. 256 squares) and records made of soils, vegetation and land use.

These data enable the land classes to be characterised in terms of biological factors observed in the field; 6,000 squares have been assigned to the 32 land classes and have been used to examine their distribution within Britain and to establish the proportion of the country covered by each class. National land use data are then derived as the product of the mean for sample squares in a particular land class and the total extent of that class. Similarly, estimates of potential can be converted to a British basis.

These results have been compared with the independent estimates of areas of land uses in Britain eg. those for crops by MAFF/DAFS. For the majority of crops eg. barley (2,117,000 ha land classification as opposed to 2,310,000 ha MAFF/DAFS). The estimates were within 5% but large discrepancies were present with minor crops eg rye 5,000 ha (land classification as opposed to 10,000 ha (MAFF/DAFS). Other comparisons have been made with factors such as urban land, and the estimates are also comparable. The studies suggest that the method can be used with some confidence to estimate parameters for which independent figures are not available and can also be used as a basis for monitoring changes in land use.

The field data provide a wide range of information on the ecology and environment of the sample squares and may be used as a basis for examining the potential of the particular area of land for uses other than the current pattern. The major advantage of the system is that it provides the framework for such a study to be carried out for the whole of Britain, and yet still is referable to particular areas of land.

2. METHODOLOGY

For the area of land in each of the 256 Land Classification sample squares the species best suited for wood production to the whole square (or part of it) was selected and its productivity estimated. For coppice, species selection was from member of the following genera; *Salix*, *Populus*, *Nothofagus*, *Alnus* and *Acer*. Choice of species for any piece of land in a square was made with reference to the following:
- a subjective interpretation of the geographic, topographic and climatic information incorporated in the classification system,
- the land form of the square under review (ie moisture receiving, sheltered; moisture receiving exposed, etc).

The productivity of the species suitable for a given site was estimated with reference to the optimum productivity of a species in the broad soil groups used in the survey. However, most sites will have sub-optimal conditions for growth and therefore maximum productivities were modified with reference to accumulated day degrees, summer soil moisture deficit, exposure, aspect, altitude and soil depth.

For single-stemmed trees the process of species selection and estimation of productivity was essentially the same as for coppice except that selection was from all tree species suitable for forestry in Britain. The area of land potentially available for single-stemmed trees is more extensive than for coppice species which are fairly site-demanding.

The distribution of the potential uses and the current land use within the squares can therefore be examined by progressively overlaying the maps. The areas of highest productivity can then be identified. Constraints according to existing land use can also be set.

Worked Example

In order to show the principles of the method an example of a single square is given overleaf.

The current land use has been simplified for the purposes of the example and has more detail in practice. In some cases the boundaries coincide because the units fit with current land use patterns. In other

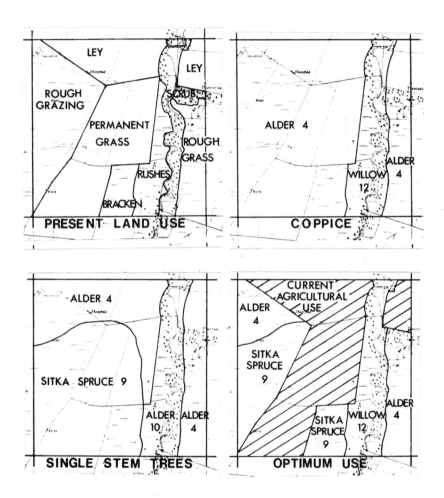

Figure 1. Comparison of a single 1 km square between current land use,
potential for coppice, potential for single stem trees and a
possible optimum use. All figures on the figure are t/ha/year.

cases, as with Sitka spruce in the single stem trees, a different environmental boundary was followed. For the optimum use it was assumed that the short term leys and permanent grass would remain in agricultural use. Elsewhere the highest yielding potential crops were assumed to be optimal.

The example simplifies the procedure which will be more complex when more land uses are involved and where the potential uses are more varied. Eight sample squares have been analysed from each land class and will be used as a basis for obtaining estimates of potential in Britain and its probable distribution pattern.

3. DISCUSSION

An important feature of the approach is that progressive refinements can be incorporated and new optimum used obtained. The refinements are of three main types:

1. Database: the base maps can be digitised and the overlaying of the potential uses carried out automatically Adjustments for factors such urban land and roads can then be made automatically. Other refinements could include access to peripheral land and ownership patterns.

2. Assumptions: the underlying assumptions between land uses and potential can be altered and their effect on the system examined. Gross margins in financial terms can also be used as a basis for comparison between potential energy crops and existing agricultural use, leading to the possibility of altering the margins of price between energy crops and conventional agriculture.

3. Optimisation: linear programming procedures could be applied to determine the optimal land use to achieve the highest yield. Preliminary work using a similar classification has been carried out in Cumbria (Bishop (1978)).

Although theoretically based, the approach therefore has the major advantage that it is linked to specific areas of land and enables estimates to be made for the whole of Britain from a small number of representative samples.

REFERENCES

Bishop,I. (1978). Land use in rural Cumbria: a linear programming model. Ph.D. Thesis University of Melbourne.

Bunce,R.G.H. (1980). The application of computer aided techniques for
landscape description in Britain In: Quantitative landscape mapping
in Britain. A review of experiences in the United Kingdom,
Switzerland, West Germany, The Netherlands and Norway. Edited by
Stein W. Bie. Norwegian Computing Center Publ. 658.

Bunce, R.G.H. & Smith,R.S. (1978). An ecological survey of Cumbria.
Kendal. Cumbria County Council and Lake District Special Planning
Board. 51p.

Bunce,R.G.H., Morell,S.K. and Stel,H.E. (1975). The application of
multivariate analysis to regional survey. J. Environ. Manage.
3, 151-165.

THE POTENTIAL ROLE OF CROP FRACTIONATION IN THE PRODUCTION OF ENERGY FROM BIOMASS

L. G. PLASKETT

Biotechnical Processes Ltd.,
Hillsborough House, Ashley, Tiverton, Devon EX16 5PA, U.K.

Summary

Whereas present projections indicate an extreme limit of availability of usable biomass fuels from agriculture of 45.8 M.t.o.e./yr for Europe (9), this figure can be increased severalfold by a scenario involving perennial non-woody energy plantation crops grown on land now used for supporting livestock. The principle would involve maximising overall biomass production, using the most vigourous plant genera, and then fractionating the crop to give high ME, high protein, low fibre fractions for use as feed and low ME, low protein, high fibre fractions for use as energy feedstock.

Theoretical analysis of the U.K. situation shows an extreme limit of usable fuel availability from grassland by this method of 19.8 M.t.o.e./yr, but a conceivable 47.5 M.t.o.e./yr from high-performing energy crop perennials replacing grassland, after taking animal feed requirements into account. Both figures are capable of extension by adding the area of rough grazing, and the limit of the scenario potential for the U.K., is estimated at 65 M.t.o.e. as usable fuel, or about 24% of projected 1985 energy requirements.

The required technology calls for mechanical separation of plant parts and tissues and juice expression. Reasons are given for considering the economic evaluation of such a venture to be completely different from that of fractionating grass or lucerne for wholly agricultural purposes.

1. INTRODUCTION

The majority of schemes proposed to date for producing usable fuels from biomass are restricted in scope and hence in their contribution to potential national or Community fuel supplies, because (i) they depend upon collecting residues from agriculture or forestry which are limited in supply or (ii) they depend upon utilizing agricultural land for the very limited times during which conventional agriculture leaves it unused, by planting catch crops or (iii) they depend upon establishing energy plantations on land unused by agriculture and, in densely populated countries, such land is either very unproductive or small in area. Total potential contributions from biomass fuels are discussed in detail elsewhere at this Conference (White & Plaskett, 1980), but an overall estimate of potential Community energy contribution from agricultural residues and catch crops would be 45.8 Mt oil equivalent/yr as converted fuel (Mt.o.e), though even this figure embodies certain assumptions that would demand major adjustments of agricultural practice in order to achieve it. The breakdown of the total in M.t.o.e./yr is; livestock wastes 8.2, crop wastes 16.3, catch crops 21.3.

These figures relegate biomass to a minor role in the supply of energy within the Community unless action can be taken to either intensify food production onto a much smaller land area, thereby freeing additional land for energy plantations, or, alternatively to maximise biomass production over a large proportion of the land presently used for agriculture alone by growing vigorous energy crop species and operating an integrated process for producing both feed and fuel from the increased biomass output. This paper concerns the second of these options for creating an increased potential for biomass fuels. The thesis offered is that biomass output from farmland could be far higher than at present and may be achievable by a radical alteration of crop species grown on land now devoted to the production of fodder and feeds, whether arable or not and whether it is grazed or cut. Crops used directly as human food do not lend themselves to substitution and are excluded from consideration. However, by processing selected portions of a vigorously growing energy crop it is reasonable to expect to generate products suitable as feed to either ruminant or non-ruminant animals, since the fibre content may be controlled by the processing operations. At the limit, therefore, all fodder and feeds, both Community-grown or imported, may be substitutable in this manner.

2. PRESENT GRASSLAND YIELDS AND FRACTIONATION

In the U.K. annual biomass yields per hectare from grassland used for supporting stock by grazing and/or conservation have been estimated to average 6.5t of dry matter (Cooper & Breese, 1971); typically about 60% of the gross energy of grass is metabolisable energy (ME), (MAFF, 1976). Therefore, of a probable gross energy yield of 120 GJ/ha/yr, approximately 72 GJ/ha/yr is ME. However, forage conservation and grazing are known to be quite inefficient, hence much less than this amount of ME is normally utilized. Forbes et al (1980) estimated the yield of utilized metabolisable energy (UME) from grass as 43.7 GJ/ha on dairy farms, 40.2 GJ/ha on non-suckler beef farms and 37.9 GJ/ha on suckler beef farms. The highest of these figures represents a 60% utilization of the ME in grassland yielding 6.5t/ha/yr dry matter, though it seems likely that on average dairy farms would yield more dry matter per hectare than the average grassland farm. A UME level of 43.7 GJ/ha/yr represents approximately 2.4t/ha/yr of metabolisable grass dry matter. From these considerations it appears that, if loss of grass dry matter could be prevented by avoiding the grazing damage and conservation losses now experienced and if the non-metabolisable element of the dry matter could be separated out by processing, then a 6.5t/ha/yr yield of grass dry matter would yield 2.4t/ha/yr of 100% utilizable livestock feed and 4.1t/ha/yr of energy feedstock. Even allowing for the inevitable losses in harvesting and processing, a major fuel component would clearly be available.

3. POTENTIAL GRASSLAND YIELDS AND FRACTIONATION

The above figures do not take into account the potential increase in grassland yield by increasing fertilizer input, which has been estimated for the U.K. by Wilkins et al (1981) to raise average yields to 10.9t/ha/yr with four-weekly cutting or to 15.7t/ha/yr with less frequent cutting (in each case N fertilizer is assumed applied to a level at which the incremental response to N is 10 kg dry matter/kg N). This would raise gross energy yield to 290 GJ/ha/yr, metabolisable energy yield to over 160 GJ/ha/yr at the higher of these production levels, figures that would lead to a gross energy production from the 1978 area of UK grassland (excluding rough grazings) of 46.5 Mt.o.e. of which, according to Green & Baker (1981) only some 6.8 Mt.o.e. would be required as metabolisable energy, completely utilized, to maintain the present ruminant animal stocks on their present rations of non-grass supplements. Since the total ME production would then be approximately about $3\frac{1}{2}$ times the feeding requirement for ruminants, there would obviously be scope for replacing imported feeds for ruminants and non-ruminants alike since, under a regime of exhaustive grass extraction, there need be no shortage of protein. If, for example, in practice, metabolisable energy from fractionated grass was 90% utilized and substitution of imported feeds demanded a 30% increase in the ME supplied from grass, then the ME demand would be 9.8 M.t.o.e. and the residual fuel component would amount to 36.7 M.t.o.e.; if this component also was collected with an efficiency factor of 0.9 and was converted to usable fuel with a 60% efficiency, the actual fuel yield would be 19.8 M.t.o.e. Obviously, major socio-economic, technical and financial problems would have to be overcome to make such a yield possible and the energy input for fertilizer must be taken into account; nonetheless, these calculations serve to illustrate the potential merit of integrating fuel production with farming, at least so far as the livestock sector is concerned. In this situation, grass becomes a perennial non-woody energy crop with a subsidiary use as animal feed.

4. ENERGY CROPS AND FRACTIONATION

It then becomes valid to enquire what plant genera, other than grasses, could be used in a similar way while giving a much higher yield of biomass per hectare. Callaghan et al (1978) and Lawson et al (1980) have listed genera such as Urtica, Pteridium, Polygonum, Impatiens, Gunnera, Epilobium etc., as being capable of annual dry matter yields from 10-40t/ha/yr (the highest value was 37.5t/ha/yr for Polygonum sachalinensis), while the genus Lavatera (L.arborea) is known to grow prolifically, from work done upon it as a potential textile fibre source in the late 1920's. It has not yet been shown that the highest yields are sustainable in successive years, nor have the establishment costs or fertilizer requirements been quantified. Nonetheless, the yields are impressive on account of their having been realized without genetic selection, without fertilizer or ground preparation and often on poor soils, such as may comprise rough grazing land. Their suitability for fractionation depends upon such factors as their macro-structure, their fibrousness and moisture content, protein content and relative freedom from toxic factors, at least in the ME-containing fractions.

Fractionation could comprise (i) mechanical separation of plant parts, such as leaves and shoots from course stems, or leaf veins from leaf par-

enchyma; the object here is to segregate high-fibre/low protein fractions
from low-fibre/high protein fractions that may be most suited to fuel use
or to ruminant or non-ruminant feed, and (ii) separation by dewatering
(which may be carried out either on whole crop or on fractions derived
from (i), which will produce a virtually fibre-free high protein juice and
a fibrous, low protein fraction. To date, studies on fractionation of
grass and lucerne have generally concentrated upon "creaming off" a modest
proportion of the juice and protein so as to leave the fibrous residue
with sufficient protein to act as ruminant feed. Hence, extraction has
been by no means exhaustive and the process equipment has had to be sized
to pass through a large volume of crop in relation to the protein extracted, a factor which has an adverse effect upon capital costs. With the introduction of the "fuel and feed" integrated concept, this constraint no
longer applies; crop may be expressed to the economic limit for recovery
of protein and ME, leaving a low protein, low moisture fibrous residue as
an energy feedstock. The juice solids can be expected to comprise 25-30%
of the crop solids and over 50% of the crop protein if the residue is
pressed to 50% moisture content; this residue could be used directly as
energy feedstock or subjected to water or alkali leaching for further recovery of ME and protein. The fibrous fraction may prove suitable for conversion to a solid fuel similar to refuse-derived fuels or could act as a
feedstock to thermal processes. Juices or juice fractions (e.g. deprot-
einised juice or juice expressed from moist stems, which are high in energy
and low in protein, are potential feedstocks to anaerobic digestion or
ethanol fermentation. Yeast and bacteria isolated from anaerobic digesters
and fermenters would be capable of re-cycle within the system as animal
feed protein. Hence, the overall production of true protein from the grass
or energy crop need not be restricted to that originally present in it as
harvested. A diagram illustrating these steps is given in Scheme 1.

One important criterion for a successful energy crop for fractionation would be the ease and effectiveness of separating ME from unmetabol-
isable energy by these measures. Some non-metabolisable energy will inevitably partition into the high ME/high protein fractions, while some protein
and ME will certainly be lost in the high fibre fuel fractions.

The separation of ME from non-metabolisable energy may well influence
the structure of animal populations, especially by lowering production
costs for pigmeat and making it no longer necessary to use ruminant animals
to obtain the benefit of the ME in foliage crops.

CONCEPTUAL SCHEME FOR COMBINED BIO-FUEL PRODUCTION AND LIVESTOCK FARMING
FROM GRASS OR OTHER PERENNIAL CROP, WHEN RUMINANT FEED IS NOT MANDATORY
(BASED ON JUICE EXPRESSION ALONE)

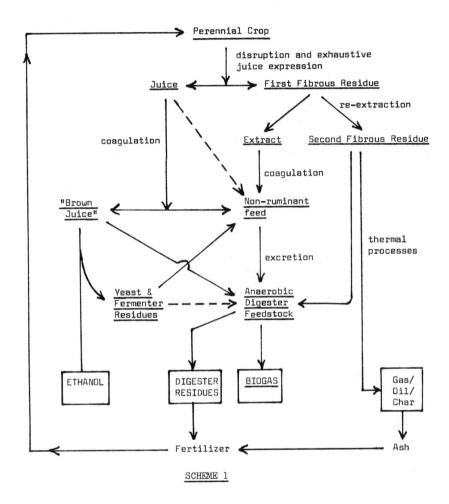

SCHEME 1

The extraction characteristics and the yields of the various fractions from potential non-woody energy crops are currently being investigated and projections are being made for processing costs and economics
for both grass and other genera in the integrated fuel/feed situation.

5. U.K. POTENTIAL FOR FRACTIONATED ENERGY CROPS ON PRESENT-DAY GRASSLAND

A 30t/ha/yr dry matter yield from energy crop perennials growing in place of present-day grassland would amount to 88.9 M.t.o.e. gross in the U.K., a probable 47.5 M.t.o.e. as converted fuel (60% efficiency) after allowing for animal feed demands. Such projections would clearly call for a major research and development effort to attain them, as well as a major programme of implementation, and, indeed, insuperable economic or technical barriers might be encountered. However, the projection serves to highlight perennial crops on present-day grassland and integrated fuel/feed schemes as possibly capable of transforming the biomass fuel concept from a minor contributor into a major one and, unless and until any obstacle to progress in this direction should appear, the subject appears to warrant a very important effort on all fronts. It is emphasised that potential biomass yields from crop wastes, catch crops, forestry and from plantation crops on non-agricultural land remain unaffected in this scenario, while livestock wastes would be diminished in total amount but increased in availability. Rough grazing land has not yet been taken into account. This is usually poorer land and it is not at present known what yield figure should be projected for energy crops grown upon it; if, for the sake of illustration, 15t/ha/yr were attained, the additional biomass yield would represent a projected additional 17.7 M.t.o.e./yr as converted fuel, giving a theoretical limit of about 65.2 M.t.o.e./yr for the U.K., equivalent to approximately 24% of projected U.K. 1985 energy demand (Palz & Chartier, 1981). A full comparison for the whole of Europe 9 would depend upon obtaining grassland statistics in a form other than that which is readily available; however, the limit of the Europe 9 potential usable fuel yield according to the energy crop/fractionation/feed and fuel scenario is expected to be in the region of 240 M.t.o.e/yr if grassland alone is considered, but considerably higher if the area now sown to cereals for stockfeed is also considered amenable to substitution. Under the latter conditions the extent to which the scenario increases the projected extreme limit of biomass fuel production from agriculture in Europe 9 beyond the 45.8 M.t.o.e. estimated as available from wastes and catch crops, approaches an order of magnitude.

The factors which can be expected to influence the financial appraisal favourably for this type of process, compared to the fractionation of lucerne for purely agricultural purposes (Wilkins et al, 1977) are (i) the

need to re-establish the crop only infrequently, (ii) much higher projected crop yields, (iii) higher degree of extraction of the crop combined with
retention of all or most of the juice solids (though Wilkins et al, 1979
did not consider extraction ratio a major factor), (iv) complete utilization of energy in the pressed crop fraction, (v) integration of crop processing and livestock feeding on the same site, instead of off-site sales
of feed.

REFERENCES

Callaghan, T.V., Millar, A., Powell, D & Lawson, G.J. (1979), "Carbon as a
            Renewable Energy Resource in the U.K. - a Conceptual Approach"
            Study undertaken for Energy Technology Support Unit, by
            Institute of Terrestrial Ecology

Cooper, J.P & Breese, E.L. (1971) "Plant Breeding : Forage Grasses and
            Legumes", in "Potential Crop Production", Ed. Wareing, P.F. &
            Cooper, J.P. Publ. Heinemann Educational Books.

Forbes, T.J., Dibb, C., Green, J.O., Hopkins,A & Peel, S (1980) "Factors
            Affecting the Productivity of Permanent Grassland; a National
            Farm Study", Grassland Research Institute and Agricultural
            Development Advisory Service

Green, J.O., & Baker, R.D. (1981) "Classification, Distribution and
            Productivity of U.K. Grasslands" in "Grassland in the British
            Economy", CAS Paper 10, Reading, Centre for Agricultural
            Strategy

Lawson, G.J., Callaghan, T.V., & Scott, R (1980) "Natural Vegetation as a
            Renewable Energy Resource in the U.K., " Study undertaken for
            Energy Technology Support Unit, by Institute of Terrestrial
            Ecology

MAFF (Ministry of Agriculture, Fisheries & Food), (1976), "Nutrient Allow-
            ances and Composition of Feedingstuffs for Ruminants", LGR 21

Palz, W & Chartier, P., (1981) "Energy from Biomass in Europe", Publ. Appl
            Sci. Publishers, in press

White, L.P & Plaskett, L.G. (1980) This Conference

Wilkins, R.J., Heath, S.B., Roberts, W.P., & Foxell, P.R. (1977) "A
            Theoretical Economic Analysis of Systems of Green Crop Fraction-
            ation" in "Green Crop Fractionation", British Grassland Society,
            Occasional Symposium No.9, Harrogate, Yorks, p 131

Wilkins, R.J., Morrison J & Chapman, P.F. (1981) "Potential Production
            from Grasses and Lagumes", in "Grassland in the British Economy"
            Ed. J.L.Jollans, CAS Paper 10. Reading : Centre for
            Agricultural Strategy.

THE ENERGETICS OF LIVESTOCK MANURE MANAGEMENT

R.M.Morris   S.B.C. Larkin
Open University   N.C.A.E.
Milton Keynes   Silsoe
U.K.   U.K.

Summary

Manure from housed livestock represents a major component of the biomass which is potentially available for energy production in the U.K. It also represents a source of plant nutrients which can be used by agricultural crops. Any losses of these nutrients from manures have to be made up using manufactured fertilisers, which have a relatively high energy and monetary cost. The losses of nutrients occurring as a result of different manure management practices, and the extent of these practices, were therefore examined.

Of the three major nutrients, nitrogen, phosphorus and potassium, nitrogen suffers the most severe losses, rising to well over half for the worst practices. Losses of nitrogen from manures in the U.K. are about one-third. This loss of nitrogen is equivalent to about 10% of the current usage of manufactured nitrogen fertiliser. The potential savings in nitrogen, and hence in support energy, through improved manure managements are examined. Anaerobic digestion appears to provide a substantial reduction in nitrogen losses as well as providing energy in the form of biogas. Some of the limitations to increased use of nutrients from livestock manures are discussed.

INTRODUCTION

Housed livestock in the U.K. void some 7m dry tonnes of faeces and urine per annum, (5), which represents a substantial proportion of the readily available biomass for energy production, with a gross energy content of approximately 100 PJ. At present 95% of this material is ultimately disposed by land spreading (11) after variable periods of storage. For intensive units with large numbers of animals on a small area, the problem of manure disposal without causing pollution of water courses or odour nuisance is paramount (14). For the less intensive mixed farm, residues may be regarded as a useful source of plant nutrients, or as a soil conditioner (3, 8). Nationally, it seems sensible wherever possible to try and recover the plant nutrients in livestock wastes, given the financial and energy costs of supplying these nutrients in manufactured fertiliser (6, 7). This paper investigates current livestock manure management practices and estimates the scope for improvement in nutrient recovery.

THE COMPOSITION OF LIVESTOCK WASTES

As voided, the faeces and urine from livestock contain varied amounts of the three major nutrients, nitrogen, phosphorus and potassium (N, P, K). There is a great deal of variation in published analyses for these nutrients, but Table (I) provides reasonably reliable averages for U.K. conditions derived from a range of sources, (e.g. 1, 9). Only the major housed species of livestock, cattle, pigs and poultry are considered.

TABLE (I)   The major plant nutrients in faeces and urine as voided by livestock (% dry weight).

Class of Stock	%N	%P	%K
Cows in milk	5.0	0.9	5.0
Dry cows	3.0	0.9	0.9
Growing beef	2.7	0.9	0.9
Pigs	6.2	2.4	3.9
Laying hens	4.5	2.5	1.9
Broilers	3.5	1.5	1.3

LOSSES IN STORAGE

After voiding by the animal, wastes may remain in situ or be moved to separate storage areas, as slurries or semi-solids, with or without additional bedding materials. Irrespective of the handling system, some losses of nutrients occur. Readily soluble compounds may be lost by leaching or run-off and nitrogen can also be lost by desorption of ammonia or by conversion of nitrogenous compounds into insoluble gaseous nitrogen by denitrifying bacteria. The exact sequence of breakdown depends on the conditions under which the waste is stored (1). Vanderholm, (15) Gracey, (4) Azevedo and Stout, (1) Summers and Bousefield, (13) and M.A.F.F. (9) provide estimates of the losses of nutrients under various handling and storage methods. Estimates derived from these sources are summarised in Table (II).

Table (II) Estimated percentage losses of plant nutrients from various storage and handling systems.

Handling System	N		P		K	
	range	mean	range	mean	range	mean
Slurry - unaerated	6-65	40	0-30	<5	n.i.	<5
aerated	10-90	66	n.i.	<5	n.i.	<5
digested	0-60	20	0	0	0	0
Semi solid, no litter	50-75	63	0	0	0	0
Semi solid ⎰outdoors	10-60	40	0-48	10	20-97	30
with litter⎱covered	10-40	20	0	0	0-5	<5

(n.i. - too little information available to provide an estimate)

Dairy cattle manures may be handled as liquid slurries, semi-solid slurries or as semi-solids with bedding. Current information (10) indicates that only about 2% is handled as a liquid, and 58% as a semi-solid with bedding. The slurried materials are usually stored in impervious lagoons or tanks, while semi-solids with bedding may be stored in the open, on more or less permeable surfaces. Conditions within stored slurries or semi-solids range from highly aerobic to almost totally anaerobic. Beef cattle manures are most likely to be stored in situ in covered or partially open yards, for up to six months, subject to limited drainage. The liquid may be absorbed by addition of up to 10 kg of straw or similar bedding material per beast per day (12). Conditions within the mass of manure are superficially aerobic with deeper layers almost completely anaerobic.

Pig manure is almost invariably handled as a slurry, with little bedding material. The slurry is stored in impervious tanks for periods up to a month, prior to disposal by landspreading. The store may be largely anaerobic, or it may be aerated to minimise odour and B.O.D. problems.

Poultry manure from caged layers is handled either as a semi-solid, as formed, or may be slurried with water for transport. It may be stored for up to a few months, usually under fairly aerobic conditions, prior to disposal by landspreading. The majority of broiler chickens are housed on litter (12), which is removed at the end of each production cycle. The resulting material is relatively dry and suffers almost no leaching loss prior to removal, but losses of nitrogen by denitrification may be quite high.

Table (III) summarises the handling systems used for manures from different types of livestock and estimates the nutrient losses entailed.
Table (III) Housing and manure handling systems for different livestock types and the estimated losses of nutrients entailed.

Livestock type	Handling System	% using Handling Systems	Nutrient losses %		
			N	P	K
Dairy Cattle	Solid + bedding	58	30	0	30
	semi-solid slurry	39	50	4	4
	liquid slurry	3	40	4	4
Beef	semi-solid + bedding	100	20	0	4
Pigs	scraped slurry	20	55	0	0
	liquid slurry	80	40	0	0
Layers	semi-solid	70	63	0	4
	liquid slurry	30	40	0	0
Broilers	semi-solid + litter	100	20	0	4

TOTAL NUTRIENTS PRESENT AND LOST

The total arisings of livestock wastes can be estimated for the various livestock classes (5). Combining these data with the information in Tables (I) - (III) enables an estimate to be made of the plant nutrients currently lost during storage, summarised in Table (IV).

For comparison the 1978 UK consumption of fertilisers was 997, 145 and 283 thousand tonnes of nitrogen, phosphorus and potassium respectively (2). In relation to these figures, the most serious loss is that of nitrogen, representing about 10% of current useage. Nitrogen is also the most energy consuming of the major nutrients used in the manufacture of fertilisers.

Table (IV) <u>Plant nutrients in livestock wastes (1) as voided by the</u>
<u>animals (2) lost during storage and handling</u> (Thousand tonnes)

Livestock type		N	P	K
Dairy cows and	(1)	99	18	91
heifers	(2)	37	6	17
Beef cows and	(1)	29	9	36
heifers	(2)	6	0	1
Beef feeders and	(1)	47	16	16
dairy replacements	(2)	9	0	1
Pigs	(1)	44	16	27
	(2)	19	0	0
Layers	(1)	41	15	12
	(2)	23	0	0
Broilers	(1)	22	10	8
	(2)	4	0	0
Total	(1)	282	84	190
	(2)	98	6	19

POSSIBLE IMPROVEMENTS IN MANURE HANDLING

For those manures handled as slurries, anaerobic digestion appears
to offer a dual benefit. The digestion releases some usable energy and,
under ideal conditions, results in no loss of nitrogen (13). Wider use
of this technique could save some 80 GJ per tonne of nitrogen saved (6),
although slurry spreading uses more energy than does the spreading of an
equivalent weight of nitrogen as manufactured fertiliser. (Lawson S.C.
personal communication). Use of digested residues does not in fact
increase the total quantity of manure to be spread, only its composition,
so this difference can be ignored. Another criticism of livestock
manures is that cropping and intensive livestock enterprises are geographically separated, so that transport costs are prohibitive. However,
a detailed analysis, reported in this volume (5) suggests that this is
not true except on a very broad scale.

The difficulty with organic manures is the uncertainty regarding
the availability of their nutrient content (3), which renders their
immediate value uncertain (9). Further work on this topic is obviously
desirable, as it is clear that readoption of techniques designed to

ensure retention of nitrogen in livestock wastes could make a worthwhile contribution to reducing the support energy needs of agriculture.

References

(1) Azevedo, J. & Stout P.R. 1974 Farm animal manures. Manual 44 California Agricultural Experiment Station Extension Service.

(2) Church, B.M. 1978. Use of fertilisers in England and Wales. Ann Rep. Rothamsted Exptl Sta. 1978 pt 2 131-6

(3) Cooke G.W. 1967. The control of Soil Fertility. Hafner, New York.

(4) Gracey, H.I. 1979 Nutrient content of cattle slurry and losses of nitrogen during storage. Exptl Husb. 35 47-51

(5) Larkin, S.B.C., Morris R.M., Noble D.H. and Radley R.W. 1980 Production distribution and energy content of agricultural wastes and residues in the United Kingdom. This volume.

(6) Leach G. 1976. Energy and food production. IIED, London.

(7) Lewis, D.A. & Tatchell, J.A. 1979. Energy in U.K. Agriculture J.Sci. Fd Agric. 30 449-57.

(8) Low A.J. 1973. Soil structure and crop yield. J. Soil Sci. 24 249-59.

(9) Ministry of Agriculture Fisheries and Food, 1979. Profitable utilisation of livestock manures. Booklet 2081, London.

(10) Ministry of Agriculture Fisheries and Food 1980. A.D.A.S. Statistical data: dairying systems in England and Wales. M.A.F.F. London.

(11) Pollock K.A. 1976. Using slurry as a fertiliser Soil and water, J. Soil & Water Man. Assoc. 4   3

(12) Staniforth A.R. 1979. Cereal Straw Clarendon Press.

(13) Summers R. & Bousfield S. 1980. A detailed study of piggery-waste anaerobic digestion. Agric. Wastes 2 61-78

(14) Royal Commission 1979. Agriculture and Pollution HMSO London.

(15) Vanderholm D.H. 1975. Nutrient losses from livestock waste during storage, treatment and handling. In: Managing livestock wastes A.S.A.E. Michigan.

PRODUCTION, DISTRIBUTION AND ENERGY CONTENT OF
AGRICULTURAL WASTES AND RESIDUES IN THE UNITED KINGDOM

S.B.C. LARKIN, R.M. MORRIS, D.H. NOBLE AND R.W. RADLEY

National College of Agricultural Engineering, Silsoe,
Bedford, MK45 4DT, England.
+The Open University, Milton Keynes, MK7 6AA, England.

Summary

Farm animal wastes and crop residues are a potentially important source of energy. To determine the extent of this potential and to aid the development of the resource, detailed information is required on the production, geographical distribution, seasonal variation, annual variability and likely future changes in production of wastes and residues. This information has been compiled for the United Kingdom and can be used to optimise the location and size of energy conversion plants.

Wastes from dairying, beef, pig, broiler and egg production and residues from cereals, sugar beet, potatoes, peas, beans, brassicas, carrots, hops and oil seed rape have been investigated. The mean level of production of each waste per unit area of crop or per unit of livestock for each age or weight category of each animal has been determined in terms of fresh weight, dry weight and energy content.

The numbers of animals and areas of crops have been taken from the June Agricultural Census for 1976. Results from the Census were reallocated to 5x5 km squares based on the Ordnance Survey National Grid system by computer. This made the data more easily manageable for further computation, analysis and interpretation. For each grid square the areas of crops and numbers of animals were combined with the figures obtained for waste production and time of year of production per unit areas of crop or per unit of livestock to give total waste production figures per month and per annum.

It was estimated that a total of around 20 million tonnes (dry and ashfree weight) of wastes were produced per year in the United Kingdom with an energy content of about 336 PJ.

From the results a clear picture of the production and distribution of all the major agricultural wastes and residues has been obtained. The suitability of different wastes for different conversion processes and the economics of the full operation from collection to energy production can now be considered.

1. INTRODUCTION

Farm animal wastes and crop residues are a potentially important source of energy in the United Kingdom. Other wastes that are potential energy sources such as industrial wastes and domestic refuse are concentrated in centres of population but agricultural wastes are distributed throughout the country. To determine the extent of the potential of agricultural wastes as an energy source and to aid the development of the resource,detailed information is required on the production, geographical distribution, seasonal variation and likely future changes in production of wastes and residues. This information has been compiled for the United Kingdom and can be used to optimise the location and size of energy conversion plants, minimising energy and economic costs of transporting the wastes.

2. WASTE AND RESIDUE ASSESSMENT

Wastes from dairying, beef, pig, broiler, turkey and egg production and residues from cereals, sugar beet, potatoes, peas, beans, brassicas, oil seed rape and hops have been investigated. In some vegetable crops significant quantities of the edible portion of the crop are wasted. Losses are caused by trimming to improve marketability, ploughing in of crops that cannot be sold economically, damage in harvesting, rotting in storage and frost damage. These wastes have also been investigated. The level of production of each waste or residue per unit area of crop or per animal has been determined in terms of fresh weight, dry weight and energy content (1).

Production of crop residues such as straw, potato haulms and Brussels sprout stems will vary annually and geographically. Factors such as the weather, soil type, fertiliser use, crop variety and variations in harvesting will all affect residue production. However, there is likely to be some correlation between the yield of the residue and the yield of the crop. For most crops the ratio of residue yield to crop yield was estimated from figures available in the literature or obtained from researchers specialising in the particular crop. Each of these ratios had been derived either from experiments on crop growth or surveys of yield in the field. The ratios were applied to crop yield data which is available on a county basis (2,3,4).

Yields of animal wastes were determined from a survey of the

literature which produced a reasonable number of direct measurements or
data from which waste production could be calculated. Daily waste production was estimated for animals of different age groups or weight classes
to match the information available on animal numbers. As wastes can
only be collected from housed animals and most cattle are only housed
during the winter months the average date of cattle housing in the Autumn
and turnout in the Spring for each county in the country was assessed from
information in the literature or obtained from the Agricultural Development
and Advisory Service (ADAS) or other contacts. The number of days of housing each month was then combined with the daily waste production figures to
produce monthly totals.

3. WASTE AND RESIDUE MAPPING

The numbers of animals and areas of crops were taken from the June
Agricultural Census. For all holdings above a minimum size and standard
man day requirement a census form has to be completed in June every year
covering about 140 items. The census forms are collated to produce the
total numbers of animals of each age or weight class and the total area
of each crop for each parish in the country. Parish results would be
difficult to use for planning of waste utilization so they were converted
to data for 5x5 km grid squares using the computer program CAMAP (5), in
which each parish is represented as a group of 1 km grid squares throughout which agriculture is assumed to be evenly distributed. The data for
blocks of 1 km grid squares was summed to give figures for all the 5x5 km
grid squares in the country. The grid square data on crop areas and animal
numbers was combined with the figures obtained on waste and residue yield
per animal or per hectare of crop, appropriate to the county covering the
square or the largest portion of it, to produce waste and residue production figures for every grid square in the country.

4. RESULTS

For each residue or waste the quantity produced per year and per
month, for months in which any of the waste is produced, has been calculated for each grid square in the country. Maps showing the density of
waste production have been produced which can be used to indicate areas
likely to be worth closer examination for plant location studies or other
purposes. Two examples are shown below. Each of these is for an indivi-

dual ADAS region. For a plant location study the actual data for individual grid squares would be used. If a study covers more than one region the data files can be combined.

4.1 Dairy Waste in the West Midlands Region

The West Midlands region, comprising the counties of Cheshire, Shropshire, Staffordshire, Hereford & Worcester, West Midlands and Warwickshire, has the highest densities of wastes arising from dairy cows, as shown in Fig. 1.

Fig. 1. Density of total annual production of dairy wastes in the West Midlands Region

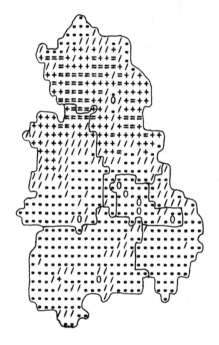

Key:
Tonnes dry matter
per 5x5 km grid
square.

0		0
.	>	0
/	>	500
+	>	1000
=	>	1500

The greatest density of dairy cattle waste production found in the region was 2270 tonnes dry matter per annum in one 5x5 km square near Crewe in Cheshire. Total waste production in the region was estimated at 797,000 tonnes per annum.

4.2 Cereal straw residues in the Eastern Region

Fig. 2 shows a similar map of total cereal straw dry matter production per annum in the Eastern Region which comprises the counties of

Norfolk, Suffolk, Essex, Cambridgeshire, Bedfordshire, Hertfordshire and part of Greater London. The highest total for one square was 7971 tonnes dry matter in the area surrounding Burnham-on-Crouch in Essex. Total annual straw production in the Eastern Region was estimated at 2,518,000 tonnes dry matter from wheat, barley, oats and mixed corn.

Fig. 2   Density of total annual straw production in the Eastern Region.

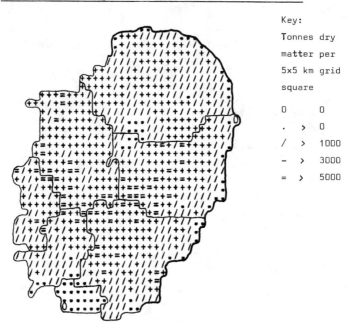

Key:
Tonnes dry
matter per
5x5 km grid
square

0	0
. >	0
/ >	1000
− >	3000
= >	5000

4.3 Total Waste and Residue Production

Total agricultural waste and residue production in Great Britain was estimated to be around 20 million tonnes dry weight with an energy content of around 336 PJ as shown in Table I. This compares with a total U.K. primary fuel consumption of 9241 PJ in 1979 (6).

Table I   Total Agricultural Waste Production and Energy Content

| | kT dry matter | | TJ | |
	England and Wales	Scotland	England and Wales	Scotland
Crop residues and wastes	11,843	1,466	207,594	26,190
Cattle wastes	3,701	836	49,531	11,287
Pig wastes	887	78	16,866	1,493
Poultry	1,354	167	20,238	2,472
Total	17,785	2,547	294,229	41,442

5. CONCLUSION

Agricultural wastes and residues can make a relatively small, but significant contribution to the energy requirements of the United Kingdom. Some of this energy is likely to be produced by conversion processes carried out on the farms on which the wastes are produced, possibly with some inputs from other farms in the area.   Other waste conversion processes are likely to be carried out on an industrial scale. The data described in this paper should be used, in combination with the results of studies on conversion processes and the suitability of different wastes for different conversion processes, to maximise the efficiency of the use of agricultural wastes as an energy source.

ACKNOWLEDGEMENTS

This work is funded by the Department of Energy and forms part of the Energy Technology Support Unit's programme on biological solar energy. We would like to thank Dr. G.H. King, Dr.K. Langley and Dr.O. Brandon of ETSU for their support.   The views expressed are those of the authors and not necessarily of the Department of Energy.

REFERENCES

1. Larkin, S.B.C., Morris, R.M., Noble, D.H., and Radley, R.W. (1981) Production and distribution of agricultural wastes in the United Kingdom and their potential for use as an energy source.   In Vogt, F. (ed) Energy Conservation and the use of Solar and other Renewable Energies in Agriculture, Horticulture and Fishculture. Pergamon, Oxford.(In press).

2. MAFF(1977) Agricultural Statistics, England and Wales, 1975.
    H.M.S.O., London.

3. MAFF (1979) Horticultural Crop Intelligence Reports on the Crops
    and Supplies of Fruit and Vegetables in England and Wales.
    Reports Nos. 589-609. MAFF, London.

4. MAFF (1980) Estimated yields and Production of Crops : England
    and Wales, Regions and Counties : 1979 harvest. MAFF Agricultural
    Censuses and Surveys Branch, Guildford.

5. Hotson, J. McG. (1978) CAMAP 6. Department of Geography,
    University of Edinburgh, Edinburgh.

6. Central Statistical Office (1980) Monthly Digest of Statistics
    April, 1980.

CROP RESIDUE AVAILABILITY FOR FUEL

J. H. Posselius and B. A. Stout

Department of Agricultural Engineering
Michigan State University
East Lansing, Michigan 48824 U.S.A.

Summary

A computer program has been developed for analyzing crop residue
removal restrictions as a function of soil type, field topography, climate
conditions, agricultural systems and energy balance. The paper reviews soil
organic matter requirements for continued crop production, describes the
computer program, and presents output of six sample runs of the program.
The scenarios analyze a 60 ha corn field with grain yields of 6000 kg/ha
field slopes where limited to a two percent grade. All input data were held
constant for the sample runs except the soil type. The clay loam soil was
the most tolerant to residue removal while a clay soil was the least
tolerant.

After numerous trials it has been concluded that no broad generaliza-
tion can be made regarding crop residue removal. Guidelines must be established for each individual field where crop residue removal is proposed.

1. INTRODUCTION

Crop residues are receiving much attention as a potential energy resource. Few scientists advocate total removal of crop residue from the soil
for it is recognized that crop residue is essential for soil erosion control and maintenance of productive capacity. The essential questions are--
to what extent can crop residue be removed without soil loss and reduced
productivity? And, can crop residue be grown, harvested, collected, compacted, transported, converted to more useful energy forms and utilized for
fuel while maintaining a positive energy balance?

A computer program[1] has been developed that takes into account all
relevant factors and calculates the amount of biomass available from each
individual field. It should be noted that the program is applicable only to
areas where climate, soil types, and agricultural practices have been cata-
gorized by the U.S. Soil Conservation Service, or similar organization for
use in the universal soil loss equation.

2. SOIL REQUIREMENTS FOR CONTINUED CROP PRODUCTION

The soil is probably man's greatest natural resource. It has been said
that human vanity can best be served by a reminder that whatever his accomplishments, his sophistication, his artistic pretensions, mankind owes his
very existence to a six-inch layer of topsoil. Nature builds soil slowly,
but it may be destroyed very rapidly. Deterioration in soil productivity is
usually associated with mismanagement; therefore, it is imperative that
crop residues be used for fuel only after basic soil maintenance requirements are met.

Before determining the portion of crop residues that may be taken from
the soil consider some of the functions of crop residues:

-provide surface protection.

-act as a storehouse of nutrients--nitrogen, phosphorus, etc.

-stabilize structure and improve tilth (wet aggregate stability).

-reduce bulk density.

-enhance infiltration and moisture retention.

-provide energy for microorganism activity.

-increase cation exchange capacity.

-release carbon dioxide.

---

[1]This computer program has been developed to work on a Texas TI-59
programmable calculator equipped with a printer.

With commercial fertilizers available at reasonable prices the need for organic matter has decreased from a nutritive standpoint (1). According to Tisdale and Nelson (2) "With the increasing use of commercial nitrogen it is not necessary nor wise to rely on the soil organic matter for high yields of a crop such as corn." Soil organic matter remains important, but mainly to prevent erosion and promote water infiltration and efficient water usage. It also plays a significant role in soil aggregation and tilth maintenance. It is these roles that the computer program will address.

Erosion Control

Although practices have been developed to reduce soil erosion (3), erosion occurs on much of the farmland. Some loss from erosion is unavoidable but there is a point where soil loss is sufficiently small that crop production will be maintained or perhaps increased through the years. "The 'soil loss tolerance' denotes the maximum level of soil erosion that will permit a high level of crop productivity to be sustained economically and indefinitely" (4).

Soil loss tolerances for the U.S. range from 4.5 metric tons/ha/yr to 11 t/ha/yr. Factors in determining these limits include soil depth, physical properties and other characteristics affecting root development, gully prevention, on-field sediment problems, seeding losses, soil organic matter deduction and plant nutrient losses (4).

Erosion is the effect of two forces--water and wind. Equations have been developed to predict the losses for each of these. The universal soil loss equation (USLE) is designed to predict long-time soil loss from particular field areas in specified cropping and management systems for water erosion. Widespread field use has substantiated its usefullness and validity for this purpose. The USLE equation is:

$$A = RKLSCP$$

Where;

A is the computed soil loss per unit area per year.

R is the rainfall and runoff factor. Components of R are based on rainfall intensity and duration.

K is the soil erodibility factor. Characteristics of K include soil texture, organic matter content, size and shape of aggregates, and the permeability of the least permeable soil layer.

L & S are the slope length factors. Both L and S are based on increased velocity as slope gradient and length increase.

C is the cover and management factor. The components of C include amount of vegetative cover, time of the year the field is plowed and planted, management of residues, amount and type of tillage, expected yields and crop rotation.

P is the support practice factor. The P value depends on soil conservation practices such as contouring, strip cropping or terracing.

The wind erosion equation, derived from the basic relationship between annual soil loss by wind erosion from a given field and the factors influencing it, is expressed as:

$$E = f(I', K', C', L', V').$$

Where;

E' is the computed soil loss per unit area per year.

I' is the soil erodibility index indicated by soil aggregate size and percentage of slope.

K' is the soil surface roughness factor based on the size of stable clods and surface ridges.

C' is the climate factor indicated by wind velocity and soil surface moisture.

L' is the length of unsheltered field width measured parallel to the prevailing wind.

V' is the vegetative cover factor, including the height of residue and the total amount.

Nutrient Maintenance

Commercial fertilizers can replace nutrients lost by crop residue removal. To estimate total nitrogen, phosphorus and potassium content of the crop residues, the amount of residue is multiplied by the average composition values. Larson, et al. (6) conclude that the nutritive value of the residues represents an appreciable portion of the total commercial fertilizers applied. Potassium and nitrogen levels in crop residues are relatively high. Phosphorus concentrations in residues are low.

It may be more economical to provide necessary nutrients via commercial fertilizers than from residues. Normally, if a leguminous crop is turned under, about 45 kg/ha of nitrogen is made available to the succeeding crop. This seldom provides the total nitrogen requirement. Moreover, it is usually cheaper to buy commercial fertilizers than to grow legumes for this purpose. When straw, corn stover or other crop residues low in nitrogen are incorporated into the soil, microorganism activity uses up most of the available soil nitrogen. If the roots constitute the only new residue source for humus maintenance few problems exist. But where large amounts of both tops and roots are present a sufficiently wide carbon-nitrogen ratio may cause nitrogen deficiencies during rapid decay in spring and early summer (1).

It may be possible to remove all above-ground organic matter and yet maintain or increase soil fertility by increasing fertilizer rates (1,2,6).

Soil Physical Properties

Crop residue function in soil maintenance is more than just erosion control and nutrient supplement. Residues reduce the bulk density of the soil, enhancing infiltration, moisture retention and respiration. Cation exchange capacity, aggregation and tilth are also increased. Unlike residues used for soil protection or nutrient maintenance, no equation or multiplier factor exists for determining the exact requirements needed to maintain ideal physical soil properties.

Figures relating to the soils physical characteristics vary significantly for each soil type, location, and various management practices. According to Allison (1) root residues represent almost the sole source of organic matter available for humus maintenance for a large portion of America's farming areas. The roots alone are usually inadequate to maintain humus content at high levels but will maintain the level commonly reached after 50 or more years of continuous farming, i.e., humus level 30 to 50% below virgin levels. The increased plant growth due to fertilization increases the amount of root residues which, in turn, keeps humus in many soils at an acceptable level.

3. A TOOL FOR ESTIMATING CROP RESIDUE AVAILABILITY FOR FUEL

Soil maintenance must be a prime consideration when crop residue removal is proposed. If soil needs can be met with partial removal of crop residues (along with adequate fertilization and other feasible chemical practices), there should be no objection to their removal. No one system has been available to the farmer giving the guidelines for residue removal. However, a computer program has been developed to determine the removable residues from each field and the expected net energy gain after inputs from crop component production (field tillage, planting, chemicals and their application), harvesting, and post-harvest processing, handling and transportation are considered. This program gives farmers and energy planners a tool for analyzing crop residue availability for fuel, feed or any purpose.

Computer Program Sections

This tool for analyzing residue management systems has six basic
sections:

1. A predicts the total above ground biomass yields and energy potential in
   the field.
2. B and B' cover residue needs and practices for water erosion control.
3. A' covers residue needs and practices for wind erosion control.
4. C, C' and D' predict available residues.
5. D addresses the energy balance.
6. E examines the transportation and handling system in regards to hauling
   capacities and maximum distance.

A step by step procedure, input data needed, and copies of the computer
program for computing crop residue availability are available from the
authors.

Program Validation

The program has been run for different scenarios on hypothetical and
actual fields in midwest U.S., with the results as diverse as the fields
and agricultural practices analyzed (Table 1). Basically it has proven to
be a bit more conservative than traditional methods of soil preservation.

4. CONCLUSIONS

Every field where crop residue removal is proposed should be analyzed.
If appropriate management practices are followed, partial residue removal
from some fields is feasible without undue risk of soil loss or other adverse effects. The computer program gives specifics for each individual
field (Table 1).

The energy available from crop residues will usually be far greater
than that required for production harvesting and handling; however, the
economics will probably limit the use of residues before the energy
considerations.

5. REFERENCES

1. Allison, F. E. (1973). Soil Organic Matter and Its Role in Crop Production.
   Elsevier Scientific Publishing Co. Amsterdam, London, New York.
2. Tisdale, S. L. and W. L. Nelson (1975). Soil Fertility and Fertilizers.
   MacMillan Publishing Co., New York.
3. Beasley, R. P. (1972). Erosion and Sediment Control. The Iowa State
   Univ. Press, Ames, IA 50010.

4. Wischmeier, W. H. and Smith, D. D. (1978). Predictions of rainfall erosion losses--a guide to conservation planning. U.S. Dept. of Agric., Agric. Handbook No. 537.

5. Woodruff, N. P. and F. H. Siddoway (1965). Soil Science Society Am. Proceedings Vol. 29 (5): 602-608.

6. Larson, W. E. (1976). Residues for soil conservation. Paper 9818. Sci. Journal Series.

Table 1. Sample Runs[1] for 60 ha Corn Field with Grain Yields of 6000 kg/ha.

Run	1	2	3	4	5	6
		Clay		Sandy	Loamy	
Soil Type	Clay	Loam	Loam	Loam	Sand	Sand
Above Ground Biomass Grain and Residue Dry kg	610,000	610,000	610,000	610,000	610,000	610,000
Residue Needed for Water Erosion Protection kg	245,000	168,000	134,000	118,000	67,000	67,000
Residue Needed for Wind Erosion Protection kg	150,000	163,000	180,000	190,000	180,000	200,000
Total Residue[2] Available kg	75,000	143,000	136,000	122,000	136,000	109,000
Net Energy[3] Gain kg	8.32×10^8	1.58×10^9	1.51×10^9	1.36×10^9	1.51×10^9	1.21×10^9

[1]Assumes all climatic, geographic, agricultural systems inputs the same for all six scenarios, only soil type different.

[2]Total residue available after soil requirements have been met, does not include grain.

[3]Net energy gain after production harvesting, and post harvest energy inputs accounted for, including chemical inputs and handling losses of 15%. Does not include off-farm transport.

SOIL PROTECTION UNDER MAXIMUM REMOVAL
OF ORGANIC MATTER

Institut National de la Recherche Agronomique
149, Rue de GRENELLE
75341 - PARIS CEDEX 07
FRANCE

CHASSIN, GUERIF, JUSTE, MONNIER, MULLER, REMY, STENGEL, TAUZIN.
I. N. R. A. - Agronomie et Science du Sol
Avignon, Bordeaux, Chalons-sur-Marne, Laon, Versailles - FRANCE

SUMMARY

In this paper, the author explains why several research
works on the subject "Energy Production using straw and soil protection"
have been conducted by nearly all the French I.N.R.A. (Institut National
de la Recherche Agronomique) teams whose major activity has been focused
on the soil organic matter (Avignon, Bordeaux, Chalons-sur-Marne, Dijon,
Laon, Versailles - Grignon).

After describing the purpose of these works and the methodology used, several results are presented concerning :

- the field experiments which have carried out for many
  years to study the effects of straw removal on soil properties.

- theoretical and experimental studies on the relation between the soil organic matter and the soil properties.

- the technical solutions designed to make up for the negative effects of massive straw removal on soil properties.

I - WHY RESEARCH ON "SOIL PROTECTION UNDER MAXIMUM REMOVAL OF ORGANIC
MATTER" HAS BEEN FINANCED BY THE C.E.E. AND I.N.R.A. (contacts n° 05376
ESF and 326781 ESF) WITHIN THE "SOLAR ENERGY - BIOMASS" PROGRAMME.

In the very large field of alternative energy research, many
people considered using crop residues for energy rather than for agronomy purposes (soil organic matter control). However, this project may be
realistic only if the energy source is renewable, i.e. if the removal of
crop residues does not produce a strong decrease in soil fertility, which
has not been proved as of today.

It is now well established that several soil properties are
altered in the presence of organic matter, for instance :

- dynamics of fertilizing substances and, particularly,
  of nitrogen.
- mechanical and physical properties (structure, porosity,
  structural stability...).
- physico-chemical properties (adsorption, water retention, cation exchange capacity...).

ALLISON's work (1973) provides more information on these
topics.

Besides, it is known that the soil organic matter content
necessarily drops if crop residues are not returned to the soil. Consequently, we may raise the following question : which will be the short
or long terme effects of such a technique on soil properties, thus on
soil fertility ?

The answer to this question may be provided in the "Organic
matter-fertility" field in which many scientists of all countries have
been working for a long time. It is therefore obvious that studies
conducted by I.N.R.A. may lead to results only in the mean term, but
with a most accurate timing. We will consider these aspects in the
second part of this paper.

II - <u>PURPOSE OF WORK CONDUCTED AT THE I.N.R.A.-PROCEDURE</u>.

        The studies done within the "soil protection under maximum removal of organic matter" programme are aimed at answering the following two questions :

    1°) - Can the fertility of soils bearing a succession of essentially cereal crops be maintained if nearly all their aerial parts are systematically removed ?

    2°) - If not, what are the remedial techniques or the minimum amounts of residues which should not be removed so that the negative effects of such a technique be compensated ?

        Unfortunately, such research is made difficult for two reasons :

+ complexity of systems under study : in spite of the large number of publications on soil organic matter, few theories have been widely accepted. Therefore, it is difficult to determine the functional relations between soil properties and organic matter properties.

+ kinetics of organic matter decay : it is impossible to obtain significant results in this field without undertaking long range tests. As the project of crop residue removal bears, a priori, no immediate consequences, it could be thought that sufficient time is available to rectify a possible decrease in fertility. It would be wrong to go ahead with such an analysis because it takes a long time to rectify defects due to not returning crop residues to soils, thus significantly decreasing the agricultural income.

        Due to such considerable difficulties, the I.N.R.A. decided to set up their research programme as follows :

+ Results of middle and long range tests conducted in France by agronomy scientists who, for various reasons (never based so far on using straw for energy), dealt with the long term consequences on soil fertility which might result from straw removal.

+ Using laboratory and field data to elaborate a model simulating the long term evolution of different fractions of organic matter which exert a significant effect on some soil properties.

+ Studying modes of action, so far rarely studied, of organic matter on soil mechanical properties and dynamics of trace-elements especially under a maize culture removing large amounts of zinc.

+ Determination of a technique largely compensating the non-return of fertilizing substances to soils by using ashes from straw combustion.

III - RESULTS.

All the results mentioned in this note are presented in detail in two reports written by C. JUSTE (INRA - 33140 Pont-de-la-Maye - FRANCE) for Commission of the European Communities, Directorate General Scientific and Technical Information and Information Management (LUXEMBURG).

31 - Results of long range experiments.

Results from experiments conducted in France suggest that straw incorporation in soil leads to an easier use of soil and a smaller risk of accidents rather than a significant increase in the optimum production capacity of soils (See report n° 3 "Soil protection under maximum removal", contract 05376 ESF).

32 - Estimated decay of organic matter contents.

Traditional concepts are contradicted by the following evidence : existence of a marked biodegradability of maïze stalks (REMY, MULLER) on the one hand, and unlikelihood of a "non decomposable" fraction of bound organic matter being proportional to the clay content (MONNIER and STENGEL) on the other hand.

The considerable effect of temperature on the in-vitro straw decaying rate (29 % at 10°C against 70 % at 25°C in a year) is likely to account for the relative lack of interest for high straw exportation shown by agronomers in northern countries. It is also to be noted that nitrogen supply decreases the carbon mineralization rate (REMY - MULLER).

33 - <u>Laboratory studies</u>.

MONNIER and GUERIF have shown that properties such as elasticity or sensitivity to soil compaction are strongly dependent on the presence of straw fragments. Straw incorporation in soil thus improves the bearing capacity and trafficability of soils.

Besides, JUSTE and TAUZIN have shown that straw decay in soils which are poor in one trace-elements (zinc) due to agricultural techniques, tends to compensate for this defect.

34 - <u>Study of the fertilizing value of ashes</u>.

The above two examples clearly show the necessity of subordinating the study of remedial techniques to the indeniable existence of straw action on soil properties. Bearing this in mind, JUSTE and TAUZIN yielded evidence that the supply of fertilizing elements by straw may be compensated for by restituting ashes produced during straw combustion. They also show in an undisputable way that the potassic fertilizing value of ashes is slightly or much higher than that of $K_2 SO_4$.

IV - <u>CONCLUSION</u>.

It can be noticed that the research programme includes three aspects :

- estimation of quantitative, qualitative and spatial evolution of organic matter as a function of the crop residue control.

- further studies concerning the action of organic matter on soil properties and behaviour.

- testing of remedial techniques.

Thus, we may think that this comprehensive knowledge will allow us to determine a certain number of rules to comply with so that the use of crop residues for energy purposes does not cause a strong decrease in the soil agricultural production capacity.

LITERATURE

ALLISON F.E. (1973) - Soil organic matter and its role in crop production. Elsevier Scientific Publishing Company - London 637 pages.

JUSTE C. (1977) - Protection des sols en régime d'exportation maximale. (Etude n° 3, contrat 05376 ESF) Commission des Communautés Européennes 48 pages.

JUSTE C. (1980) - Protection des sols en régime d'exportation maximale de matière organique. (Etude n° 1, contrat 326 78 1 ESF) Commission des Communautés Européennes 53 pages.

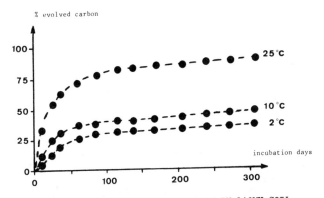

KINETIC OF CORN-STALK MINERALIZATION IN LAOMY SOIL

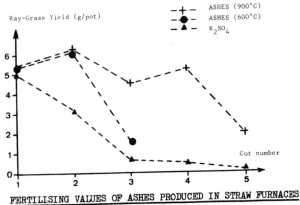

FERTILISING VALUES OF ASHES PRODUCED IN STRAW FURNACES

EXPERIMENTAL STUDY ON CONTINUOUS DIGESTION OF CELLULOSE WASTE IN A COMBINED REACTOR SYSTEM

F. ALFANI° and M. CANTARELLA[+], V. SCARDI[+]

° Istituto di Principi di Ingegneria Chimica, Facoltà di Ingegneria, Università di Napoli, P.le Tecchio, 80125 Napoli, Italy.

+ Cattedra di Chimica delle Fermentazioni e Batteriologia Industriale, Facoltà di Scienze, Università di Napoli, via Mezzocannone 8, 80135 Napoli Italy.

Summary

Cellulose as a novel energy source is attracting widespread attention also in connection with the problem of solid waste disposal. Besides the direct utilization as a fuel, cellulose waste can be converted into other useful products;e.g. it can be hydrolysed to glucose, which in turn can be fermented to ethyl alcohol. Cellulose digestion is nowadays performed almost generally by enzymic hydrolysis thanks to the availability of stable and very active enzyme preparations (cellulases) of microbial origin, such as the well-known cellulase from the fungus Trichoderma Viride. The first two components of cellulase complex are strongly adsorbed by cellulose substrate under the conditions which are optimum for the enzymic action. Furthermore, the end products of the enzymic digestion, namely cellobiose and glucose, are strong inhibitors. All these facts led us to consider the possibility of performing the cellulose digestion in a combined reactor system which permits continous production of glucose and removal of cellobiose by converting it into glucose.
In this communication preliminary experiments are discussed which indicate the role of the pretreatment on the enzymatic hydrolysis of commercial cellulose (Avicel) and olive wastes.

The utilization of cellulosic wastes as an alternative natural energy resource is matter of a large number of papers pubblished in the last years. In some circumstances it is suggested to reserve a certain amount of territory to produce crops which can be used as raw materials for energy production. This use of land is alternative to its traditional one that, of course, consists in producing foods for man and animals. In Brasil and U.S.A. it has been programmed that the 20% of gasoline consumption and the 10% of total oil consumption was supplied by alcohol obtained by sugar fermentation and oil extracted from cactus Euforbia, respectively.(1) These projects can be pursued in Countries where a relative large extension of rural area is at hand as compared to the population.

In table 1 (2) are summarized data for U.S.A., Brasil and Countries of the European Community. Brasil and U.S.A. have large extension of rural area, much larger than the ones strictly needed to satisfy the nutriment of population as evidenciated by the values reported in columns 2 and 3. The data in column 2 have been evaluated on the assumption of 0.3 tillable hectare per habitant that represents an average estimated worth for well industrialized agricultures. On the other hand, in all the European Countries, except France, the tillable territories are just enough to solve the problem of nourishment without considering that part of territory has to be reserved to produce fodder for cattle. Hence in Europe, in our opinion, only the sugar extraction from agricultural wastes and its fermentation to alcohol can contribute to reduce the energy deficit. A parallel source is represented by wastes of wood working. In table 1 are also reported the hectares planted with woods. In addition a possible source of cellulosic materials is represented by the excess of several crops which are each jear destroyed, for economic considerations.

Since one of the major cost in the processing of cellulosic wastes is represented by the collecting and transport of the materials from the fields to the plants, it seems particularly attractive the use of wastes by agricultural products, like grapes and olives, which have been preliminary processed to produce wineand olive oil, and are there fore dispo-

T A B L E   1

States	Territory ha x 10³	Territory needed for nutriment ha x 10³	Rural Area ha x 10³	Territory planted with wood and water surfaces ha x 10³
UNITED STATES	936.312	61.869	429.830	313.623
BRASIL	851.197	28.680	202.630	513.546
DENMARK	4.307	1.490	2.941	569
IRELAND	7.028	891	5.732	444
GREAT BRITAIN	24.482	16.671	18.568	2.348
HOLLAND	3.695	3.958	2.073	623
BELGIUM - LUXEMBURG	3.310	3.020	1.685	730
GERMANY	24.858	18.385	13.269	7.621
FRANCE	54.703	15.378	32.067	14.687
ITALY	30.126	17.009	17.524	7.034

sable in relative large amount near social cellars and oil-presses.

In table 2 (2) is illustrated the situation in Italy averaged on the last three years. The arboreous culture covers in Italy the 9.84 % of the country territory, grapes and olive cultures represent the 44.34 % and 35.57 % of this. In table 2 the amount of grapes reported in column 1 is only that relative to wine production, whilst in column 2 are indicated the quintals of grapes processed by social cellars and wine industries.

T A B L E   2

	Harvested q x 10⁶	Harvested for processing q x 10⁶	Residual generated q x 10⁶	Cellulose content q x 10⁶
Olives	34.59	33.79	13.28	5.31
Grapes	91.28	34.08	3.75	1.50

The quite large amounts of cellulose which are present in the grape and olive wastes, as reported in Table 2, clearly indicate the interest to investigate the utilization of such cellulose materials as energy resource and to perform economic evaluations on the process.

Our research, which is part of the project of the European Community Solar Energy Programme- Energy from Biomass, is mainly focused to obtain quantitative data on the enzymatic hydrolysis of cellulose, present in the wastes previously mentioned, to glucose.

As reported in the literature the cellulose, inside natural wastes, is surrounded by a lignin seal which precludes the enzymatic attack and is present in the cristallyne form which protects the internal bonds from hydrolysis (3,4). These obstacles to the enzymic attack can be removed, by making a pretreatment of the cellulosic substrates. Different techniques, which consist in a mechanical, thermal or chemical manipulation of natural wastes, are suggested (5,6). Once the cellulose is ava ble in the amorphous form, a subsequent enzymatic hydrolysis is easily performed. However the complete conversion of cellulose to glucose is precluded by the inhibition of glucose and cellobiose towards β-glucosidase (EC 3.2.1.21) and 1,4 β-glucancellobioidrolase (EC 3.2.1.91) respectively (7,8).

In order to overcome these obstacles, a membrane reactor which allows to remove continously the reaction products from the reacting mixture has been adopted. Moreover this reactor design allows a better utilization of enzyme since it can be recovered on the membrane (9,10) . The first step of our experiments consists in a study of acid and non toxic solvent attack of either insoluble commercial cellulose (Avicel and filter paper) and cellulosic residues in order to set the best operating conditions for this necessary pretreatment.

The effect of acid pretreatment is clearly shown by the results of early experiments reported in Fig 1 and 2. In figure 1 glucose and reducing groups concentrations as function of process time for two runs performed with 50 mg of Avicel powder and the same amount of Avicel kept for 3 hr in 1% by volume H_2SO_4 at 80°C, are plotted. The Avicel was placed

in a Amicon stirred membrane cell, 70 ml by volume, filled with 50 mM Na-
Acetate, pH 4.8, and the cell was continously fed with the same buffer
solution.In the case of the pretreated Avicel, the same procedure was
adopted, excepting for a neutralization step and washing in buffer solution prior to kinetic test. The reaction temperature is 45°C, and cellulase
(Trichoderma Viride) concentration is 33.2 mg (0.02 EU/mg) in the 70 ml
cell volume. The cellulase was preliminarly diafiltered in order to remove
the reducing groups which are present in the enzymic mixture. On the
other hand, in Fig. 2. the results of a Kinetic test performed in a batch
reactor using wastes of olive press- processing are reported. The concentrations of glucose and reducing groups are reported as function of time.
The experimental conditions are the following: 200 mg of olive wastes free
of oil were placed in 200 ml of Na-Acetate buffer 50 mM, pH 4.8, and kept
at 42°C in the presence of 33.2 mg of cellulase (Trichoderma Viride). The
reactor vessel is stirred, and at different time, samples of solution are
collected, filtered and tested following the Nelson and GOD methods.

A sample of olive wastes has been pretreated for 17 hr at 56°C in 1%
by volume H_2SO_4 solution in a stirred vessel. A Nelson analysis of the
final solution indicates the presence of 1.54×10^{-4} mol of reducing groups.
The GOD test shows that the glucose is not present in the solution, hence
the reducing groups are pentose and or pentose polymers produced by hemicellulose acid hydrolysis. As for the Avicel, the solution of olive wastes,
prior to the kinetic test, was neutralized and the filtrate washed with
distilled water until the reducing groups are no more detectable.
Comparison of glucose and reducing groups concentration curves for the two
cases of acid treated and untreated substrates indicates that the acid
pretreatment enhances the enzymatic attack.

ACKNOWLEDGEMENTS

Thanks are due to Dr. Albanesi D. and Dr. Vetronile A. for their help
in the experimental work.
This programm has received grant from CEE (Contrct Number ESE/060/I).

REFERENCES

1. G. Grassi Conferenza sull'uso dell'energia alternativa in agricoltura, Regione Lazio (Roma novembre 1979)

2. "Annuario di Statistica Agraria" Istituto Centrale di Statistica, 25, Roma 1979

3. M. Chang, J. Polymer Sci. Part C, 36, 343 (1971)

4. T.A. Hsu, M.R. Ladisch and G.T. Tsao, Chemtech p. 315 May 1980

5. R.G. Kelsey and F. Shafizadeh, Biotechnol. Bioeng. 22, 1025-1036 (1980)

6. G. Jayme and F. Lang, Methods Carbohydr. Chem. 3, 75 (1963)

7. C.S. Gong, M.R. Ladish, T. Tsao, Biotechnol. Bioeng. 19, 959 (1977)

8. L.E.R. Berghem, L.G. Pettersson, U.B. Asiö-Fredriksson, Eur. J.Biochem. 53,55 (1975)

9. M. Cantarella et al., Biochemical J. 179, 15 (1979)

10. F. Alfani et al., Chem. Eng. Sci. 34, 1213 (1979)

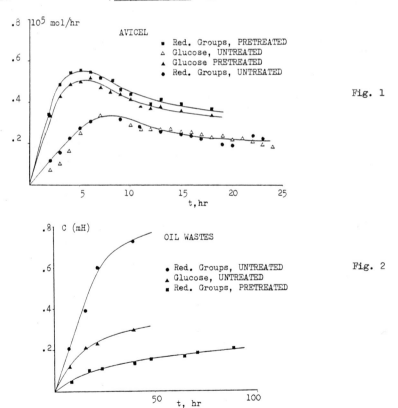

Fig. 1

Fig. 2

THE MANURE CELLAR AS A SOURCE OF HEAT

O. Tjernshaugen

Department of Farm Buildings, Agricultural University of Norway

Summary

Pilot- and full-scale experiments have been conducted in order to study the environmental impact of liquid composting in relation to the problem of energy recovery. Various types of aerators have been tested. The machine that combines the functions of aeration, stirring and loading has proved most suitable.

The composting process produces a manure temperature of 40-50°C. This heat is transferred via a heat exchanger made of 1" PEL plastic. The heat is easily extracted by using an air heater. The heat may be used for warming up dwellings as well as farm houses.

So far the experiments have indicated an effect factor of 4 - 4.5. Calculations show that 24 kWh per cow per day of energy may theoretically be recovered. In practice, however, 15 - 20 kWh per cow per day must be regarded a satisfactory result. An investment of ca 170 US dollars per cow is sufficient to obtain this result.

1. INTRODUCTION

Traditionally, livestock manure is a valuable source of plant nutrition. Ca 40% of nitrogen, phosphorous and potassium consumtion in Norway
is supplied by livestock manure.

In addition, livestock manure constitutes a considerable energy
source. Based on calculations of fuel value of dry manure, Norwegian livestock farms yearly produce manure equivalent to 6 billion kWh.

On the negative side is the fact that manure is easily transformed
from being an energy source to a pollution source.

The Department of Farm Buildings of the Agricultural University of
Norway therefore intends to develop a manure treatment procedure that is
favourable from an environmental point of view at the same time as the
energy potential of manure is taken into use. The effort is supported
financially by the Royal Ministry of Petroleum and Energy.

This paper will deal with some of the results that so far have been
achieved in recovering compost heat from slurry.

2. PRINCIPLES AND TECHNICAL PROCEDURES

Composting of slurry is accomplished by mechanical aeration. The
aerator ensures an optimal air distribution in the manure. The air distribution is more vital than air quantity and air temperature. Good air dispersion leads to aerobic bacterial growth. Decomposition process is roughly as follows:

Organic material + bacteria + $O_2 \rightarrow H_2O + CO_2$ + sulfate, nitrate,
etc. + new bacteria + energy.

This means that the aerobic bacteria breaks down the organic material
in the manure to relatively harmless end products under emission of heat.
This is the heat that we want to put into use. Several types of aerators
have been used in the experiments at the Department of Farm Buildings.
Figure 1 shows one of the most suitable aerators. This is a submerged
centrifugal pump with extra aerating equipment installed. This pump combines the functions of stirring, loading and aerating. The pump is placed
on a stand on the floor of the manure storage.

Figure 1: Submerged centrifugal pump with aerating equipment.
Heat exchanger in manure cellar.

3. WHAT IS ACHIEVED BY COMPOSTING

Composting of slurry gives:

    a)   strong reduction in odor and gas hazard

    b)   manure that is easy to handle

    c)   hygienization

Liquid composting releases certain amount of ammonia. Four years of experiments at the Soil Science Department have proved that this does not lead to crop reduction.

Liquid composting has previously been regarded as being energy con-sumptive. The results presented here prove that the composting process on the contrary gives a fair energy surplus.

4. PRACTICAL EXAMPLES OF ENERGY RECOVERY

Figure 2 shows a plant with an outdoor manure storage. The manure floats or is being scraped towards the mixing pit A. Frome here the manure is every day pumped to the treatment pit B. This is a concrete pit that holds at least one week's production of manure. The dimensions should be kept within maximum 8 meters in diameter and 6 meters in height. If daily manure amounts require larger dimensions, several treatment pits should be built. The walls and the ceiling of the pit must be insulated, k-values for wall 0.2 W/m^2.$^\circ$C and for ceiling 0.3 W/m^2.$^\circ$C.

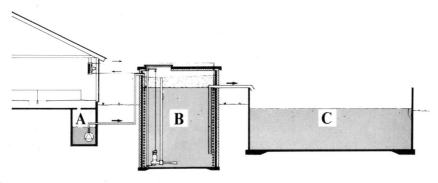

Figure 2:   Compost heat recovery in an outdoor plant.
A = mixing pit, B = treatment tank, C = manure storage.

As heat exchanger is installed a 1" PEL pipe of ca 8 cm length per m^3 storing volume. The heat exchanger is mounted on bars along the inside walls of the pit.

The aerator used in the example in Figure 2 is a submerged centrifugal pump, operated by a timer. Normal operation time is 10 - 15 minutes per hour. The energy demand of the largest aerator that was tested was 12 kW, of the smallest one, 1.5 kW.

Figure 3 shows a plant where the manure is stored in a cellar under the cattle barn. An aerator similar to the one shown in Figure 2 was used. The heat exchanger, however, is formed differently.

For both of the examples that are presented, the hot water is led to an air heater (radiator with fan) in the dwelling or in the farm building. This is a simple and efficient method of heat extraction. The aerobic process is conducted so that water temperature remains at 40 - 45°C

5. PRELIMINARY RESULTS

Preliminary results from 2 experimental plants are presented in Table 1. The results show that a certain treatment volume is required before the size of the surplus becomes interesting. Once a plant functions properly it should give 15 - 20 kWh per cattle x day.

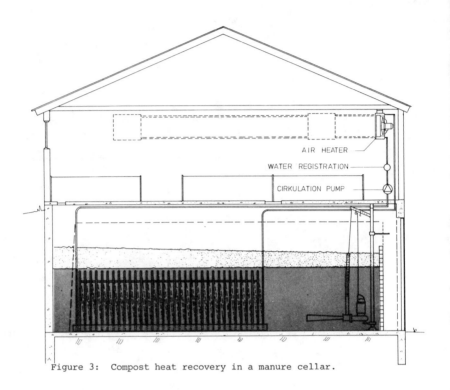

Figure 3: Compost heat recovery in a manure cellar.

Table 1. Experiments on recovery of compost heat from manure storage.
A: silo, B: cellar under the barn

	Herd A 24 cows	Herd B 30 sows
	Manure cellar 25 m^3	Cellar 250 m^3
Average temperature of the manure	44°C	41°C
Average temp. of circulating water from heat exchanger	43°C	40°C
Amount of water circulating	9.5 m^3/day	25.4 m^3/day
Energy budget		
Energy recovered	75 kWh/day	334 kWh/day
Energy consumed	17.5 -"-	79 -"-
Surplus	57.5 -"-	255 -"-
Effect factor	4.3	4.2

The equipment required for such plants costs approximately 3 - 4000 US dollars. Costs for actual manure storage room and ordinary pumping equipment not included. Such kind of equipment should in any case be available on any livestock farm.

6. FURTHER RESEARCH PLANS

The objective for further experiments is to increase the credit side and reduce the debet side of the budget.

Furthermore, a fresh air intake will be mounted in a research plant according to the so-called recirculating principle. This will be combined with extraction of compost heat (see Figure 4). The idea is that pre-heating of the fresh air will further increase the climatic advantages that the recirculating principle already implies.

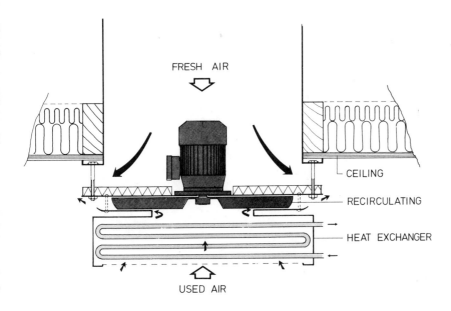

Figure 4: Combining compost heat recovery with fresh air intake.

SESSION II: FORESTRY FOR ENERGY

Chairman: Prof. R. MORANDINI
         Istituto Sperimentale per la
         Selvicoltura, Italy

Summary of the discussions

Invited papers

Short-rotation forestry as a biomass source: an overview

The outlook for energy forestry in France and in the European
Economic Community

Poster papers

Determination of yield of biomass from whole-tree harvesting of
early thinnings in Britain

Recovery of waste-products from wood-processing industry

Biomass production with poplar

Coppice willow for biomass in the United Kingdom

Coppiced trees as energy crops

Method for the estimation of above-ground biomass and biomass
production in classical coppice and first results

Coppice forests in Italy: their potential for energy

Design of a harvester for 2-3 year old willow sticks

Short rotation forestry as a source of energy

Forest biomass as a source of energy in the UK - The potential
and the practice

Short rotation biomass production of willows in New Zealand

Forest energy crops from derelict and waste land

Sélection d'eucalyptus adaptés aux conditions françaises en vue
de l'installation de taillis à courtes rotations

SUMMARY OF THE DISCUSSIONS

Session II FORESTRY FOR ENERGY

Rapporteur :	G. SIREN, Sweden
Speakers :	K. STEINBECK, USA
	P. BOUVAREL, France
Chairman :	R. MORANDINI, Italy
Poster Session :	11 papers presented

Daniel AUCLAIR on behalf of Pierre BOUVAREL described the general premises of energy production mainly in a forestry context within the EC and France. For economic reasons conventional forestry was the main interest. In the analysis of present land use attention was paid to the large area of underproducing coppice forests covering some 6.2 million/ha in France alone. In the forest improvement program to be implemented high priority will be given to conversion of these coppice forests into either forest for high-value timber production or high-yielding short rotation forestry plantations for various purposes including biomass production for energy. In the first instance however, proper utilization of existing resources would include better use of logging wastes, thinning removals and easily available industrial residues for energy production. The need to develop mechanized harvesting systems for mountainous terrain as well as promotion of cooperation of the 1.5M owners of the 6.8M ha small forest estates was stressed. Intensified forestry would raise the present forestry share of energy production from 3.3M toe (tonnes oil equivalent) to 7.5M toe.

Future use of the 3.0M ha of agricultural land that has been or should be taken out of the subsidized and heavily over-producing agriculture seems to be still under discussion. Here large scale introduction of intensive short-rotation plantations would be a practical solution to two problems :

a) the economic burden of agricultural over-production could be considerably reduced without demographically negative consequences

b) 5M ha energy plantations would contribute at least some 20 M toe per year to the energy resources of France. Compared with coal, the environmental consequences of wood based energy production would cause considerably less problem.

The paper of Dr Steinbeck concentred entirely on short rotation forestry for biomass as it is carried out in the USA. After defining the concept of this new form of intensified forestry and mentioning the eco-physiological basic premises he gave a condensed description of the agricultural measures necessary for achieving the relevant production goals.

Short-rotation plantations of broad-leaved species repeatedly harvested on cycles of less than 10 years regenerate the stand by sprouting from root-stocks. Because of carbohydrate reserves in the root-stocks,the high N-availability in the undisturbed rhisosphere, the sprouts (especially in the case of pioneer species) grow rapidly. The number of shoots per area unit is important for achieving optimum production. Fertilization and weed control are needed. Selection of the best species and clones for the prevailing ecological conditions requires elaborate screening programs. In the establishment of stands both biology and agricultural-forestry techniques have to consider the requirements of the harvesting system. In the USA the spacing is $\geqslant$ 1m2, rotation length $\geqslant$ 3 years. Prior-to-harvest energy analysis indicates an energy out-in ratio averaging 15 : 1. The input is mainly caused by heavy clearance operations of forest land and some nitrogen application. If abandoned farm land and sufficient biological nitrogen fixation were available the outcome would be different; ratios of 25 : 1 and higher appear feasible. According to Steinbeck, much remains to be learned in site evaluation, species selection and efficient fertilizer use. Genetic improvement and attainment of sustained high yields with minimum inputs are major biological research priorities.

The poster session dealt mainly with utilization of existing resources, short rotation plantations and mechanization of harvesting. Many of the contributions originated from research and pilot projects financed by EC. A sample of abbreviated titles gives a good picture of the dominant subjects : whole tree harvesting and early thinnings, biomass produc-

tion with poplars, potential and practice of forest biomass, short rotation forestry, coppice forestry has a source of energy, recovery of
waste-products, a harvester for 2-3 year old willow sticks.

Of all the information available only a few examples can be given here.
Total output of utilizable resources in the U.K including single stemmed energy plantations could contribute with some 5.4 M toe/yr by year
2000. In the Netherlands 10 tons   DM/ha/yr has been achieved in poplar
experiments whereas in the U.K a poplar clone in a nursery environment
has reached 20 tons DM/ha/yr over a 10-20 year period with an annual
harvesting cycle. Extrapolation to full scale is assumed to reduce the
yield to some 14 tons DM/ha/yr. Derelict land has produced more than
4 toe/ha/yr in U.K.

Two of the poster papers drew a special mention. The report on coppice
willow for biomass in the U.K (by Stott, McElroy et al) reveals some
exciting yield figures from experiments carried out on disadvantaged
soils in Northern Ireland. A three year old stand of S. aquatica
gigantea has produced more than 17.5 t DM/ha/yr. In Long Ashton the
Mullatin-variety has in the best case produced 47 t FW/ha/yr, which may
correspond closely to 20 tons D W. The motive for mentioning the
highest yield figures only is rather simple :

a) there is no reason to assume that the clones so far selected do
represent the best choice
b) ecophysical growth optimization measures based upon careful site-
analysis have been very moderate if at all carried out.

The other poster paper deserving attention came from New Zealand.
Biomass production of 50-60 tons FW/ha/yr has been achieved with a number of clones ; a new hybrid (S.matsudana/x/alba) being the best producing - average about 25 tons DM/ha/yr in a 2 year experimental plot.
The size of the plot excludes any major edge effect, in case this detail
should have been neglected. In a 6 year old experiment the average yield
of the seven best clones varied between 15-26 FW/ha/yr. Regarding fertilization no information was given ; annual rainfall is 1000 mm ; information on optimum spacing is not available.

The comments from the floor underlined the importance of continued research efforts all over the world in order to increase and improve the knowledge in growth related biosciences - especially for the benefit of developing countries. The necessity of excluding the edge effects from the average yield figures was mentioned. A proposal to express all yields in terms of oven dry weight per hectare and year (ODW $ha^{-1}yr^{-1}$) was put forward.

Concluding Remarks

It appears that in the present conditions wood utilisation should be planned and developed on an integrated basis : in the developed countries wood from the existing forests and from new fast growing plantations will be mainly used for industrial purposes. Residues from the utilisation of existing forests and industries will be available for energy purposes. This might give new economically interesting products to the market for small dimension wood. Moreover attention should be paid to research on unused wood in coppices, which are important in some countries.

Taking into account the knowledge already gained on the subject, short rotation plantations should be developed on a larger scale especially in many developing countries, where fuels for domestic uses are often scarce and firewood collection is destroying large areas of trees. Intensified vocational training at the growers level needs to be strongly emphasised. Strengthening applied research in the fields of genetics, tree physiology, site improvements and utilization seem essential to support the work presently being developed in the field of short rotation forestry.

SHORT-ROTATION FORESTRY AS A BIOMASS SOURCE: AN OVERVIEW

Klaus Steinbeck

School of Forest Resources, University of Georgia,
Athens, GA   30602   U.S.A.

Summary

Short-rotation forests are plantations of closely spaced, broadleaved trees which can be harvested repeatedly on cycles of less than ten years. After each harvest of the above-ground portions of the trees, the root-stocks remain intact and resprout, thus regenerating the stand. Because these rootstocks already exploit the rhizosphere fully and also contain carbohydrate reserves, the new sprouts grow rapidly. During plantation establishment, fertilization and weed control are usually needed and additional fertilization is necessary after the harvests.

Screening programs to evaluate the biomass production of various forest tree species are under way in Canada, France, Ireland, Sweden and the U.S. Plantation establishment techniques for cuttings, bare-rooted seedlings and container stock are being refined. Researchers in the U.S. tend to favor spacings which allot a growing area 1 m^2 or more to each rootstock and three year or longer harvesting cycles. Initial energy balances indicate that for the biomass production phase from stand establishment to just prior to harvest an output of 12 to 25 energy units for every invested unit can be expected.

1. TREES AS A BIOMASS SOURCE

Forest trees produce a long-neglected variety of raw materials in addition to the well known, traditional lumber and fiber products. These underutilized materials include lignin, proteins, sugars, vitamins, oils, gums and various other extractives. Cellulose can substitute for many petroleum derived feedstocks in the chemical industry, particularly in the manufacture of plastics. In addition, the actual as well as the potential use of trees as an energy source is currently of great interest in many parts of the world.

The increased utilization of the forest resource for this variety of purposes obviously has advantages and disadvantages. Among the most important positive, long-term advantages are that trees are renewable indefinitely, cause no net change in the atmospheric carbon dioxide levels when they are utilized, require relatively low energy inputs in manufacture and contain few pollutants. Disadvantages of increased utilization of trees include the accelerated removal of nutrients and organic matter from forest soils as well as high harvest and transport costs because the resource generally is scattered, bulky, and high in moisture.

At this point a comparison of trees with agricultural crops or other herbaceous sources of terrestrial plant biomass is perhaps in order. The perennial growth habit of trees offers flexibility in harvest timing within a year and also the option to accumulate large quantities of biomass in the field over several years. Forest trees make low nutritional demands on soils and protect them from erosion, therefore they can be grown on sites not suitable for agriculture. Most broadleaved and some coniferous species do not need to be replanted after each harvest because regeneration is assured by stump- or root-sprouts. Also, woody materials are less subject to deterioration in post-harvest storage than the more succulent, herbaceous plant tissues.

2. SHORT-ROTATION FORESTS DEFINED

Short-rotation forests are plantations of closely spaced, broadleaved trees which are harvested repeatedly on cycles of less than 10 years. The key to high growth rates lies in the rootstocks. Once planted and established, they remain in the ground after each harvest of the above-ground portions and then resprout. The emerging shoots (coppice) grow rapidly because the roots already have established access to soil water and nutrients and also contain stored carbohydrates which help sustain rapid regrowth rates.

Short-rotation forestry offers a relatively fast return on investment to the landowner. Therefore, it often will be an attractive alternate crop on fallow agricultural land. Compound interest charges and sometimes taxes also are decreased. The system lends itself to mechanization because the crop is uniform and relatively small. On the other hand, it can also be labor intensive and therefore be attractive for countries where labor is abundant and jobs scarce. In subtropical and tropical regions, short-rotation forests are compatible with agro-forestry practices. The raw material produced is suitable for conventional, reconstituted wood products like paper and particle board and holds promise as a source of protein and cellulose. It is a versatile energy source because it can be converted into a variety of gaseous, liquid, and solid fuels. Its most important asset in view of the shrinking forest landbase in the world, however, may well turn out to be that it permits increased production per unit of land area. Biomass yields from short-rotation forests depend on many factors, among them choice of tree species, site quality, length of harvesting cycle (rotation), and cultural practices. The objective of the remainder of this overview will be an assessment of the importance of some of the environmental and biological factors which affect yields. I make no claim to a thorough literature review and will base much of this assessment on personal experience with short-rotation test plantations in the state of Georgia in the Southeastern United States.

3. SPECIES SELECTION

The proper matching of the tree species to be grown on a particular site for specific end uses is the most important decision to be made prior to plantation establishment. Much remains to be learned in this area and species screening studies are under way in the peat fields of Ireland, wetlands of Sweden, and the better forest sites of Canada, France and the United States. In general, broadleaved tree species with rapid juvenile growth rates which do well on a variety of sites and sprout satisfactorily from the stumps are used. Other selection criteria include pest resistance, high genetic variability, ease of vegetative propagation and the ability to tolerate competition in plantations.

End use considerations may determine the species to be planted because it may possess certain desirable fiber, extractive, cellulosic, or energy content characteristics. Physical wood properties such as hardness, color, and machinability tend to be of little importance in species selection because the uses of short-rotation grown raw material almost always involve chipping or other fractionation, chemical digestion or outright conversion into liquids, gases or amorphous materials. In most cases the ability of a species to produce maximum tonnages on a particular site will be its most important criterion for selection. Paper (2, 7, 21) and particle board (17) have been successfully produced from coppice of various species. Livestock feed and vitamin supplements can be derived from finely ground leaves and small branches. Such fodder derived from Populus leaves compared favorably with that produced by alfalfa (6). The cellulosic composition of different species may be important when the coppice is to be used as a chemical feedstock or to be fermented to alcohol. Few opportunities, however, seem to exist to select hardwood species with high caloric contents per unit of mass (16), although more species need to be screened.

4. PLANTATION ESTABLISHMENT

Site preparation so thorough as to approach agricultural conditions is generally necessary for successful short-rotation plantation establishment. Intensive preparation helps control competing vegetation, results in higher water and nutrient availability, and facilitates equipment passage in planting and cultural operations.

Bare-rooted (23, 28) or containerized seedlings or cuttings (8, 12, 14, 15) can be planted with somewhat modified, off-the-shelf equipment or by hand during the dormant season. Fertilization is necessary on all but the best sites (26, 27) to aid the trees in capturing the site. Minerals can be broadcast, banded, or dropped directly into the slit opened by the planting machine. The two former methods result in increased weed competition but have the advantage of timing flexibility. Often surface application is delayed until the seedling roots have overcome the transplant shock either during the first summer or the following spring. Planting slit fertilization has not been thoroughly tested but holds promise. Apparently sufficient soil to prevent root burn falls into the slit to cover the fertilizer before the seedling is machine planted. Significant savings in fertilizer inputs could be achieved with slit application if further testing shows favorable results.

Several new herbicides are currently being tested in hardwood plantations, and no definitive recommendations are possible. In the Southeastern USA, a pre-emergence herbicide treatment just before budbreak is recommended (10), but often should be followed by harrowing between the rows in mid-growing season.

In summary, during the initial year in the field on a well prepared site, short-rotation forests minimally require the following operations: Planting, herbicide spray, fertilizer application and two harrowings between the rows. Additional harrowing may be needed in the second growing season if the tree crowns have not closed ranks.

5. SPACING, ROTATION AGE AND YIELDS

The spacing of the rootstocks, also called stools, should result in early and complete site occupancy above and below ground. Spacing should be determined by the sprout growth pattern of the species to be planted, the rotation length needed to produce a specified size or condition of the raw material, and the dimensions of the equipment available for the cultivation and harvesting operations.

The initial harvest should occur after the crowns have captured the site and shaded out much of the competing vegetation. Yields from this initial rotation will generally be lower than those expected from subsequent coppice rotations (13, 19, 30). Likewise, sprout growth in the first year after the initial harvest is also disappointingly low (22, 31) and it is not yet clear whether this is attributable to an immature rootsystem or to internal, physiological factors.

Very close spacings tend to produce higher coppice yields when combined with two and three year rotations (1, 8, 12, 13). It is difficult to define 'very close' because so many spacings have been tried. At this point I think of any arrangement that provides less than one square meter of growing space to each stool in temperate climates as very close. Annual harvests produce low coppice yields which decline with repetition (1, 22). The mass of annually coppiced American sycamore rootstocks was significantly lower than that of others harvested on two- and seven-year cycles (25) and this probably in part explains the low yields. Annual harvests also result in high nutrient removals, therefore this very short rotation option seems unrealistic.

The trend in American short-rotation forestry research is towards longer rotations and wider spacings. Mean annual increments are higher and yields are relatively independent of spacing because early full site occupancy is not as critical and the number of sprouts maintained on each rootstock is higher at wider spacings (8, 12, 13, 22). Spacings which provide from one to about 3.5 m^2 land area for each rootstock and rotations from three to eight years appear most promising at this point.

6. NUTRIENT REMOVALS

Short harvesting intervals and the complete removal of the above-ground tree portions will accelerate the nutrient drains from the site. The amount removed will vary with the tree species, site, and rotation age. The amount of nutrients removed by harvesting 3-year-old sycamore planta-tions grown on two contrasting sites in Kentucky were affected by site and fertilizer treatment (29). When sycamore was coppiced on 2-year-rotations in Georgia (20) and Mississippi (3), nitrogen removals were reported at 46.2 and 58.6, phosphorus at 5.8 and 14.4, potassium at 21.8 and 36.2, calcium at 35.1 and 40.7, and magnesium at 5.1 and 8.9 kg/ha/2 years, respectively, when harvested without leaves during the dormant season. In Mississippi these nutrient losses exceeded the gains estimated from weathering, precipitation, and other sources. Significant sycamore clone x nutrient treatment interactions have been reported (18), indicating that genetic selection for efficient nutrient utilization on various sites is possible.

7. ENERGY BALANCES

The potential of short-rotation forests for energy production has engendered considerable interest (5, 9) and initial attempts to evaluate biomass yields in terms of energy inputs needed to achieve various outputs have been made (4, 24, 30). In non-irrigated plantations, the highest inputs are needed for fuel to power bulldozers and tractors and for the manufacture of fertilizers. In one case (24), heavy site preparation con-sisting of piling of the vegetation and stumps left after a commercial clearcut and two passes with a wildland harrow consumed more than 90% of the fuel inputs. Much of the remaining 10% was accounted for by machine planting. Cultural operations such as disking, fertilizer spreading and herbicide application required little energy. Among the fertilizer inputs, nitrogen alone can account for more than 90% of the energy needs. These might be reduced by the use of nitrogen fixing tree species either in pure or mixed stands or by the use of leguminous, herbaceous cover crops. The latter would serve the dual purpose of augmenting the nitrogen supply and also controlling the weeds. For the biomass production of Populus 'Tristis #1' beginning with energy inputs needed for machinery and materials production all the way to energy outputs of coppice on the stump just prior to harvest, an energy out:in ratio of 16:1 has been reported (30). When American sycamore was grown on various rotations, energy inputs beginning with heavy site preparation to energy output also just prior to harvest ranged from 12 to 17 and averaged 15.3 energy out:in (24). Even though these two estimates are very close, ratios of 25:1 and higher appear feasible (11).

8. OUTLOOK

In the United States several large forest industries are interested in the short-rotation forestry system and initial operational plantings are being established. Much remains to be learned in site evaluation, species selection, and efficient fertilizer use, but hopefully the successes will outnumber the inevitable failures. Genetic improvement of the promising species and the achievement of sustained high yields with a minimum of inputs are biological research priorities. The development of harvesting machinery and conversion technology for liquid and gaseous fuels seem to represent the most pressing technological needs which confront short-rotation forestry.

References

1. Anderson, H. W. 1979. Time-related variation in the performance of a hybrid cottonwood minirotation as influenced by spacing and rotation length. Proc. N. Am. Poplar Council, Crystal Mtn., MI.: 35.
2. Barker, R. G. 1974. Papermaking properties of young hardwoods. Tappi 57(8): 107.
3. Blackmon, B. G. 1979. Estimates of nutrient drain by dormant-season harvests of coppice American sycamore. USDA For. Serv. Res. Note SO-245, Sou. For. Expt. Sta., New Orleans, LA 5 pp.

4. Blankenhorn, P. R., T. W. Bowersox, and W. K. Murphey. 1978. Recoverable energy from the forests. Tappi 61(4): 57.

5. Brown, C. L. 1976. Forests as energy sources in the year 2000: What man can imagine, man can do. J. For. 74: 7.

6. Dickson, R. E. and P. R. Larson. 1977. Muka from _Populus_ leaves: A high-energy feed supplement for livestock. In: Tappi For. Biol. Wood Chem. Proc., Tappi, Atlanta, GA.

7. Einspahr, D. W., J. R. Peckham and M. K. Benson. 1970. Fiber and pulp properties of triploid and triploid hybrid aspen. Tappi 53(10): 1853.

8. Ek, A. R. and D. H. Dawson. 1976. Actual and projected growth and yields of _Populus_ "Tristis #1" under intensive culture. Can. J. For. Res. 6(2): 132.

9. Fege, A. S., R. E. Inman, and D. J. Salo. 1979. Energy farms for the future. J. For. 77: 358.

10. Fitzgerald, C. H., R. F. Richards, C. W. Selden, and J. T. May. 1975. Three year effects of herbaceous weed control in a sycamore plantation. Weed Sci. 23:32.

11. Geyer, W. A. 1980. Personal Communication.

12. Heilman, P. and D. V. Peabody. 1980. Effect of close spacing and short harvest cycles on productivity of black cottonwood in intensive culture. Can. J. For. Res. In press.

13. Kennedy, H. E. 1975. Influence of cutting cycle and spacing on coppice sycamore yield. Research Note SO-193, South. For. Expt. Sta., For. Serv., New Orleans, LA. 3 pp.

14. Kennedy, H. E. 1977. Planting depth and source affect survival of planted green ash cuttings. Res. Note SO-224, South. For. Expt. Sta., New Orleans, LA. 3 pp.

15. McAlpine, R. G., D. Hook, and P. P. Kormanik. 1972. Horizontal planting of sycamore cuttings. Tree Plant. Notes 23(2): 5.

16. Neenan, M. and K. Steinbeck. 1979. Caloric values for young sprouts of nine hardwood species. For. Sci. 25(3): 455.

17. Rice, J. T. 1973. Particleboard from "Silage" sycamore - laboratory production and testing. For. Prod. J. 23(2): 28.

18. Steinbeck, K. 1971. Growth responses of clonal lines of American sycamore grown under different intensities of nutrition. Can. J. Bot. 49: 353.

19. Steinbeck, K., R. G. McAlpine, and J. T. May. 1972. Short rotation culture of sycamore: A status report. J. For. 70(4): 210.

20. Steinbeck, K., R. G. Miller, and J. C. Fortson. 1974. Nutrient levels in American sycamore coppice during the dormant season. Ga. For. Res. Paper 79. Ga. For. Res. Council, Macon, GA. 5 pp.

21. Steinbeck, K. and E. N. Gleaton. 1974. Young sycamore cuttings give promise in fine paper tests. Pulp and Paper 48(13): 96.

22. Steinbeck, K. and C. L. Brown. 1976. Yield and utilization of hardwood fiber grown on short rotations. Appl. Polymer Symp. 28: 393.

23. Steinbeck, K. and J. T. May. 1977. Artificial hardwood regeneration in the Southeast - Some aspects of planting stock quality. Proc. 4th N. Am. For. Biol. Workshop. State University of New York, Syracuse, N.Y. 218 pp.

24. Steinbeck, K. 1980. Short-rotation forests for energy production. Am. Chem. Soc. Proc., San Francisco, CA. In press.

25. Steinbeck, K. and L. C. Nwoboshi. 1980. Rootstock mass of coppiced Platanus occidentalis as affected by spacing and rotation length. For. Sci.: In press.

26. USDA Forest Service. 1976. Intensive plantation culture: Five years research. Gen. Tech. Rep. NC-21, N.C. For. Expt. Sta., St. Paul, Minn. 117 pp.

27. Wittwer, R. F., R. H. King, J. M. Clayton, and O. W. Hinton. 1978. Biomass yield of short-rotation American sycamore as influenced by site, fertilizers, spacing, and rotation age. S. Jour. Appl. For. 1: 15.

28. Wood, B. W., S. B. Carpenter and R. F. Wittwer. 1976. Intensive culture of American sycamore in the Ohio River valley. For. Sci. 22(3): 338.

29. Wood, B. W., R. F. Wittwer and S. B. Carpenter. 1977. Nutrient element accumulation and distribution in an intensively cultured American sycamore plantation. Plant and Soil 48: 417.

30. Zavitkovski, J. 1979. Energy production in irrigated, intensively cultured plantations of Populus "Tristis #1" and jack pine. For. Sci. 25(3): 383.

31. Zavitkovski, J. 1980. Personal Communication.

THE OUTLOOK FOR ENERGY FORESTRY IN FRANCE
AND IN THE EUROPEAN ECONOMIC COMMUNITY

P. BOUVAREL

Department of forest research, I.N.R.A., France

Summary

The nine countries of the European Community produce together 80 M m3 (million cubic meters) of wood, and import wood and wood products equivalent to 120 M m3. The demand will increase 2 % per year. So the possibilities for producing wood for energy will be limited by the competition with other uses. This competition exists also for sawmill wastes (25 M m3, of which two thirds are burned). The fuelwood included in the recensed production amounts 8 M m3, the non commercialized fuelwood is probably more than 20 M m3. The possibilities of increasing wood for energy relies to use of bark (8 M m3), collection of wastes in forest (depends on proper collection equipment and economic and energetic rentability) and energy plantations on bare lands (around 5 M ha available for forestry).

France and Italy have together 7 M ha of coppices ; the coppices over 30 years represents a reserve of biomass which could be rapidly used in case of emergency.

The french forests (14 million ha) produce 30 M m3 : 4 are already converted into energy. The estimations of the energetic potential available are (per year) :

- coppices over 30 years : 1,5 to 2 M t.o.e. (for 10 years)

- wood from early thinnings : 0,4 to 0,8 t.o.e.

- forest wastes : 0,3 to 0,7 M t.o.e.

- industrial wastes still available : 0,8 to 1 M t.o.e.

In total, 3 to 4,5 M t.o.e., plus 3 M t.o.e. already coming from fuelwood or burned industrial wastes. These figures are contingent to probable competition with other uses, already severe between needs for pulpwood and increasing demand for domestic fuelwood.

The program for short term energy plantations is still at the research level. The trend is : rotations over 8 years, use of broadleaved trees (poplars, alders, eucalypts) but also fast growing conifers in view to establish multiple use stands. On large areas, the average production will rarely exceed 9 t of dry matter per ha and per year.

The following is an expose dealing with the production of
wood utilizable for energy purposes with the exception of processing techniques and marketing channels. The situation described with respect to
France is based chiefly on the studies and evaluation made by the Forestry
Biomass Group of the Solar Energy Commission. The data relevant to the whole E.E.C. are taken, for the most part, from the report entitled "Political forestry within the Community" presented in December 1978. No reference
will be made to the particular policies of the member governments which
are dependent upon the specific situation of each country with respect to
national fossil fuel resources.

o o
o

In the nine countries of the E.E.C., forests cover 31 million
hectares. In five countries (Belgium, Germany, France, Italy and Luxemburg)
they cover more than 20 % of the territory. Average production amounts to
only 2,5 m3 per hectare/year and 3,5 m3 if we take into account only the
productive timber forests. It is lower than potential production which has
been estimated of the order of 5 m3 if still unproductive young forests
and too numerous old ones are taken into consideration, as well as mediocre coppices which represents approximately 7 million hectares in France
and in Italy. It should also be noted that the total commercial production
of about 30 million cubic meters does not take into account non-commercia-
lized fuel wood the volume of which is very difficult to estimate but may
exceed 20 million cubic meters.

Distribution of forests within the E.E.C.

	Total (1000 ha)	Forest (% of the territory)	Hectare per inhabitant
Belgium	615	20	0,06
Denmark	470	11	0,09
Germany	7 200	29	0,12
France	13 950	25	0,28
Ireland	330	4	0,09
Italy	6 300	21	0,12
Luxemburg	85	32	0,24
Netherlands	310	8	0,02
United Kingdom	2 020	8	0,04
	31 280	21	0,12

Sawmill refuse of indigenous and imported logs amount to 2,5 million cubic meters. About half is reused in paper manufacture with this increasing tendency. The remainder is usually burned to furnish power for the plant itself. The greater part of the 8 to 10 million cubic meters of bark remains unused even though attempts have been made to put them to use as horticultural compost of in making particle boards with non debarked timber. There are no overall statistics as yet on <u>forest refuse</u>. Their quantity and especially their economic availability vary greatly according to the type of forest and mode of exploitation.

Lastly, the black liquors from paper manufacture which represents expressed in units of energy, approximately 20 % of the biomass harvested when recovered and burned, furnishes more than half the energy for pulp plants.

<u>Wasteland has been estimated at 5 million hectares</u>. They could be planted with forest trees, especially to provide energy, without at the moment running any competition with agriculture.

This data would seem to indicate that there are very real possibilities for increasing the biomass production, whether short term, by adjusting felling to actual increase of growth and collecting residues, or longer term transformation of poor forests and replanting bare earth with fast-growing trees. An increase from 80 to 100 million cubic meters between nom and the year 2000 for commercial use seems a reasonable goal.

But the use for energy purposes of what currently appears to be available, and future increase of the same, must be prudently estimated. The E.E.C. produces 80 million cubic meters of wood , and <u>its net imports</u> in terms of equivalent roundwood amounts to 120 million cubic meters as compared to 40 million cubic meters in 1950. The value of the commercial deficit is second only to that of oil products. Increase in the demand for wood and derivative products is estimated at 2 % per year, higher than the anticipated increase in production. Under these conditions, it is more than probable that <u>competition between energy production and other uses will be extremely severe</u>.There is considerable concern already in certain countries whose industries use small wood in the paper and panel industries, in the face of the rising cost of fuel wood. These countries may have ti help support these industries. Furthermore, the major objective of the E.E.C. countries is the production of timber and a great deal of effort is being expended to use wood of small diameter (finger jointing, glued-laminated techniques among others) which will reduce the available quantity. This policy designed to economize on currency, is uncontestable. All technological improvement including basic and applied research with a view to developing uses of wood especially in construction, result in savings in energy, for wood is the material whose use and transformation requires the least energy, compared for example, to cement, aluminium and plastics.

Another fact complicates the situation. This is the great parcelling out of private forests which represent 60 % of the wooded area in the E.E.C. Only 50 000 out 3 000 000 private owners have forest of more than 50 hectares. This is likely to curb possible government intervention in mobilizing wood for energy purposes. Of course, the little forests already furnish and can furnish more fuel wood for demestic use but only in cases where the forests are combined with agricultural exploitation, representing, as the case may be according to country, one third or one half these forests, the remainder belonging to non-resident owners.

Taking the above limitations into consideration, the generally reserved attitude of the forestry authorities of E.E.C. countries, is understandable, and the contribution of existing forests to the production of energy should concern firstly waste wood currently left in the forest as well as industrial waste not yet put to use, including bark. But the latter presents processing problems. The use of such waste wood presupposes developing machinery to gather the wood and transform it into chips. Several such are already being tested, some have already been put into operation. Studies as to its profitability, in terms of economy and energy, have not yet been completed. The use of wood of small diamter from the first thinnings and of as yet unused sawdust depends on too many factors (energy cost, worldwide cost of wood for pulp and government policies toward the paper industry) for a prognostic to be possible. The situation is, however, quite different for countries such as France and Italy with large coppiced areas as we shall see when studying the situation in France.

In any case, and no matter what system of transforming wood into energy is used, the cost of preparing and transporting it will be determinant and will be doubt lead to the use of the forestry biomass in a short circuit within regions close to large forests.

Remain the short-term rotation plantations specifically for use in producing energy. In all countries with the exception of Ireland, everything is still in the research stage, at best in that of pre-development. Research is being done as to the various types of wood, the density of plantations, duration of turnover, the cost of silviculture, fertilization, harvesting machinery, etc. Results obtained from the first trials gave very high figures, sometimes more than 15 tons of dry matter per hectare/year for turnovers of 5 to 8 years. But these figures were obtained from trials on generally rich soil and small aeras. More recent trials on varying types of soil resulted in less optimistic evaluation. Considering the inevitable reduction involved when large aeras of hundreds, even thousand of hectares are under consideration instead of small experimental areas, it is reasonable to suppose that only rarely will a regular average production of 9 tons per hectare/year exceeded. Genetic selection of highly favorable new varieties could bring about a revision, within the next 10 years, of this figure to 12 tons approximately.

The availability of land in such highly populated countries as those of the E.E.C., will be an important factor in limiting plantations for energy purposes. Of course, if only the food requirements of these countries are taken into consideration, the current areas under cultivation may be adequate or, with probable increased productivity, actually produce a surplus. However, it is not unlikely that the E.E.C. will increase its agricultural exportation in which case there will be aggravated pressure on the lands, including those which are at present actually vacant, so that there will be growing competition between energy plantation and classic reforestation.

o°o

The situation in France is roughly that described for the whole E.E.C. except for coppice resources. Forests which cover 14 million hectares represent almost half the forest surface of the E.E.C. 30 million m3 of wood are harvested and commercialized every year or, in terms of the surface of exploited forest (9 million hectares) 3,3 m3 per hectare/year. The total increase is approximately 50 million m3 or the equivalent of 12 million t.o.e. (x) representing 7 % of the country's total energy consumption.

(x) t.o.e. : ton of oil equivalent

The surface covered by <u>coppices and coppices-with-standards-</u> -a traditional type of management favoring fire wood- amounts to about 5 million hectares. Production varies from 0,5 ton of dry matter/hectare/ year in the Mediterranean region to 5 tons in good chestnut coppices. Many of these coppices have not been exploited since the last war and <u>represent a considerable reserve of standing biomass</u>. A prudent estimate would be 1 million hectares of old coppice immediately exploitable, representing at least 50 million tons of standing dry matter. Exploitation over a period of 10 years would give an additional annual yield of 1,5 to 2 million t.o.e. The most important aspect of this resource is that it represents mobilizable "<u>first aid</u>" in case of a crisis arising, and would allow, for instance, putting into effect a program of exploitable energy-producing plantation with optima rotation which is rarely the shortest. This prudent estimate is well under the figure suggested by GUILLARD (6 million t.o.e. over 15 yea - rs) but it takes into consideration mobilization for standing wood-produ- cing energy of all existing coppices (225 million tons). Coppices on very sloping areas and those in the process of being converted into timber fo- rest cannot be exploited mechanically. Moreover, there is considerable competition for this type of wood for paper and boards.

In any case, coppices are France's principal forestry biomass resource. What was once considered a handicap is proving a lucky thing for the country as well as for those owners who, until recently, for lack of outlets, left their coppices unexploited. It is, however, ardently to be hoped that the revenue thus obtained be used to transform mediocre coppi- ces into productive forest land for example through plantation of conifers.

Mobilization of coppices has already begun without exercice of any particular pressure beyond the rapid increase in firewood prices due, perhaps, to precautionary buying of wood which will require at least two years to dry. This phenomenon is particularly prevalent in rural areas where in the last two years purchases of wood-burning domestic boilers ha- ve increased sevenfold.

The newfound importance of coppices as a source of energy jus- tifies research on the precise estimate of the biomass (based on data of the National Forestry Inventory, expressed in volume of wood of more than 7 cm diameter), and on techniques for improving productivity (choice of stems, rate of cuttings, results of mechanization and fertilization).

Asied from coppices, potential resources in forestry biomass are as follows :

- <u>Wood from thinning</u> : mostly the first thinning, which is rarely profitable, of young resinous stands (2 million hectares planted over the last 30 years). A total volume of from 15 to 35 million m3, over ten years would amount to 0,3 to 0,7 million t.o.e. per year. This is, however, a resource widely dispered and in small parcels. A small additional resource amounting to 0,1 million t.o.e. can be counted upon from winrows left in place after clearing to make way for the planting of rapidly growing species of trees.

- <u>Residue from exploitation left in the forest</u> : the total volume is estimated at 6-7 million m3, but a large part is already used for domestic heating. The remainder may amount 0,3 to 0,7 million t.o.e. per year but here again, we are talking about a highly dispersed resource, the harvesting cost of which would be determinant.

<u>Industrial waste woods</u> have the advantage of being concentrated in sawmills and factories. The overall estimate of 17 million m3 includes sawdust and bark. Slabs and edging are practically all used in paper manufacture. Industries which use their waste wood actually save 0,8 to 1 million t.o.e. ; this figure could be doubled by additional mobilization of waste wood currently neglected.

The annual production of 30 million m3 stipulated for France includes approximately 1,5 million m3 of firewood but this does not include autoconsumed firewood, the amount of which is difficult ot determine (of the order of 8 million m3).

<u>To summarize, the annual resources in available biomass in France</u> may be evaluated as follows :

- old coppices, exploited over 10 years : 1,5 to 2 M t.o.e.
- early thinning and winrows   : 0,4 to 0,8
- residues from forest exploitations   : 0,3 to 0,7
- industrial residues   : 0,8 to 1

<u>or a total of 3 to 4,5 million t.o.e.per year,</u> to be added to the <u>3 million</u> already used (fire for domestic heating, industrial residues used as fuel) and to the <u>0,3 million t.o.e. from the buring of blakc liquors of the paper plants.</u>

These estimates are prudently presented and take into account serious reservations about competition with other uses for wood valid in France as in the E.E.C. as a whole. We nite that GUILLARD advances figures of 7,5 to 8,5 million t.o.e. which certainly represents a maximum.

Experiments on short - term rotation energy plantation : the oldest of these are twelve years old. There is to date no really widescale plantation. The varieties used up until now are for the most part poplars, for we have large collections of clones in the process of selection. Further, plane-trees, eucalypts (in southwestern France but the risks of cold make very careful selction necessary). Research is just beginning on alders and coniferous trees likely sprout and to be used for coppices (*Sequoia sempervirens*).

Tests of poplars after 5 years have given, with 5000 feet to the hectare, 12 tons of dry matter per hectare/year on good soil. It would be imprudent to anticipate on an average, over large areas, more than 9 tons per hectare/year which might be increased to 12 tons with genetically improved material (within 5 to 10 years). The concept of the energy plantation is rapidly evolving and the following tendencies have been observed :

- prolonging the term of rotations : the biological and economic optimum lies between 8-10 years rather than at 5 years
- improving existing coppices to give comparable yields at lower prices
- softwood in dense plantations, cut at 15-20 years, could relieve short turnover coppices
- generally speaking, systems with flexible objectives are being sought. These would produce according to the conjunction of circumstances and needs, wood for energy as well as for paper and boards. Mixed plantations could involve a definitive lumber species and some intermediate lines of varieties exploitable at short term for energy.

Assuming the size of agricultural lands in France it is probable that the majority of energy plantations will be situated on terrain today considered as marginal, that is, of, at most, average fertility. Therefore, despite possible progress through research (selected species, mycorhization) our production estimates remain modest.

BIBLIOGRAPHY

- COMMISSARIAT A L'ENERGIE SOLAIRE (COMMISSION ON SOLAR ENERGY)

    Etude et recommandations pour l'exploitation de l'énergie
    verte - 1980

    (Study and recommendations for the exploitation of green
    energy - 1980)

- COMMISSION DES COMMUNAUTES EUROPEENNES (COMMISSION OF EUROPEAN COMMUNI-
    TIES)

    Politique forestière dans le Communauté - Bulletin des
    Communautés Européennes - 1979

    (Forestry policy within the Community - Bulletin of
    European Communities - 1979)

- GUILLARD, J.

    Wood for fuel in France - In "Wood for fuel" - Stockholm -
    1980

- HUMMEL, F.

    The likely impact of using wood for fuel on forest and the
    forest industry in the E.E.C. - In "Wood for fuel" -
    Stockholm - 1980.

° °
°   °

DETERMINATION OF YIELD OF BIOMASS FROM
WHOLE-TREE HARVESTING OF EARLY THINNINGS IN BRITAIN

C.P. MITCHELL, J.D. MATTHEWS, C.G. MacBRAYNE and M.F. PROE

Department of Forestry, Aberdeen University, St Machar Drive,
Aberdeen AB9 2UU, UK

Summary

At present,markets in Britain for small roundwood from early thinnings are depressed.   The small roundwood could be used for fuel and estimates suggest that 600 Kdt/yr, or in energy terms 12 PJ/yr, could be obtained by the year 2000.   The work reported was undertaken to confirm these estimates and to provide information about management of forests for energy.

Growth and biomass production of unthinned (approximately 10 to 20 year old) stands of ten species, which show potential as producers of forest biomass for fuel, have been studied on a range of sites of varying quality.

Preliminary results and an example of a table compiled to aid managers of forests for energy are given and implications for marketing examined. These results suggest that previous estimates of yield of forest biomass from whole-tree harvesting of early thinnings were conservative.   The use of small roundwood could make a small but valuable contribution to Britain's energy needs.

1. INTRODUCTION

At present, markets in Britain for small roundwood from early thinnings are depressed. However, whole trees from early thinnings could be used as a source of fuel. Estimates suggest that up to 600 Kdt/year, or in energy terms 12 PJ/year, could be obtained by the year 2000 from whole-tree harvesting of trees in early thinnings (1). The work reported here was undertaken to confirm these estimates and provide information for forest managers.

2. METHOD

Ten species were selected for their potential as producers of biomass and these are being examined across a range of ages up to approximately 20 years and growth rates for total above-ground biomass and its partition (Table I). The effect of geographical variation in site quality and on biomass production was countered using stratification by Yield Class (which is an indicator of site production potential). This study concerns yields from early thinnings so crops which have not reached thinning stage

Table I

Stratification of Sampling

Species	Age 10-14	Class 18-22
alder	LH*	LH
birch	LH	LH
Nothofagus	LH	LH
sycamore	LH	LH
Corsican pine	H	LH
Douglas fir	LH	LH
hybrid larch	LH	LH
Scots pine	H	LH
Sitka spruce	LH	LH
western hemlock	H	LH

*Yield Class: L-low; H-high

in the young age class were not included.

Tree crops on each site were characterized using standard mensurational parameters including basal area, stocking, top height. Selected trees were then sampled destructively to determine dry matter partitioning between stem, branches and foliage. The method was:-

- to define a mathematical relation between an easily measured tree parameter and dry weight per tree e.g. dry weight/tree = b(g) + a where g = basal area in m^2;

- to synthesize dry weight per hectare by summing the dry weight of the trees in the stand;

- to convert conventional yield tables in terms of volume to units of dry matter in order that the dry matter production of a crop can be given on an area basis by reference to either the mensurational tables or from easily measured forest parameters.

3. RESULTS AND DISCUSSION

Yields of biomass per hectare and per year are given in Table II.

Table II   Yields of Biomass for Ten Species

Species	Age (years)	Yield Class (m^3/ha/yr)	Biomass Yield (dry tonnes/ha)	Annual Biomass Yield (dry tonnes/ha/yr
Alder	12	6	29.6	2.5
	18	10	87.2	4.8
	22	14	211.4	9.6
Birch	25	6	100.0	4.0
	20	9	100.0	5.0
Nothofagus	24	11	145.0	6.0
Sycamore	28	4	93.0	3.3
	22	10	114.2	5.2
Corsican pine	19	12	96.2	5.1
	20	16	139.7	7.0
Douglas fir	14	20	78.8	5.6
	18	12	71.3	4.0
Hybrid larch	13	10	36.6	2.8
	20	10	112.4	5.6
	20	15	154.9	7.7
Scots pine	14	14	52.2	3.7
	18	8	62.7	3.5
	19	14	108.6	5.7
Sitka spruce	14	26	110.5	7.8
	22	14	108.7	4.8
	20	26	211.7	10.6
Western hemlock	15	28	174.9	11.7
	18	20	107.7	6.0

(Based on data collected up to September 1980)

Table III   Management Table: Sitka spruce; Yield Class 14, aged 22 years

MAIN CROP BEFORE THINNING

Number of Trees	Mean Tree Diameter (cms)	Mean Tree Volume to 7cm t.d. (m³)	Basal Area (m²/ha)	Stem Volume to 7cm t.d. (m³/ha)	Total Above-Ground Biomass (dry tonnes/ha)	Biomass of Tree Components (dry tonnes/hectare)						
						Stem			Branches			Foliage
						Total	Wood	Bark	Total	Dead	Live	
2806	11	0.0523	31.79	146.75	108.73	53.80	47.73	6.07	38.82	11.13	27.69	16.11

YIELD FROM 1ST THINNING (⅓ of main crop)

| 935 | 11 | 0.0388 | 10.60 | 48.92 | 36.24 | 17.93 | 15.91 | 2.02 | 12.94 | 3.71 | 9.23 | 5.37 |

Adopting a 5cm Top Diameter
Utilisation Standard: -

	MAIN CROP		THINNING YIELD	
Mean Tree Volume to 5cm t.d. (m³)	Stem Volume to 5 cm t.d. (m³/ha)	Total Stem Biomass (Dry tonnes/ha)	Stem Volume to 5 cm t.d. (m³/ha)	Total Stem Biomass (Dry tonnes/ha)
0.0589	165.27	58.92	55.09	19.64

Table IV   Marketing Options for First Thinning
                     (Sitka spruce; Yield Class 14, aged 22 years)

Option	Revenue (£/ha)
1. Whole crop for energy: 36.24 dt/ha @ £18.50/dt (see note a)	£670.44
2. Stems (to 7cm) for pulp: 48.92 m³/ha @ £11.75/m³ (see note b)	574.81
3. Stems (to 5 cm) for pulp: 55.09 m³/ha @ £11.75/m³	647.31
4. Stems (to 7cm) for pulp; remainder for energy 48.92 m³/ha + 18.31 dt/ha	913.55
5. Stems (to 5cm) for pulp; remainder for energy 55.09 m³/ha + 16.60 dt/ha	954.41

Notes: a) Value of 1 dry tonne of wood for energy.

b) Price (1978) of 1 m³ ob fresh conifer roundwood at mill.

c) These calculations assume that the consumer values quoted reflect comparative standing values i.e. costs of extraction, handling and delivery charges for energy production and mill roundwood are similar.

The data indicate that previous estimates of yield of forest biomass from whole-tree harvesting of early thinnings were conservative. We conclude that the use of small roundwood could make a valuable contribution to Britain's energy needs.

The results for one of the crops sampled (Sitka spruce; Yield Class 14, aged 22 years) have been presented in a table to aid management for energy (Table III). This table has been prepared in two stages; firstly, following present forestry practice in Britain in which stems to 7cm top diameter are harvested and secondly adopting a 5cm top diameter utilization standard. The yield from first thinning has been calculated assuming that one third of the crop is harvested non-selectively.

The implications of these results for this crop of Sitka spruce can be seen in the marketing options for first thinnings given in Table IV. The value of 1 dry tonne of wood for fuel has been derived from the energy equivalent price of coal at 1978 prices; i.e. 1 tonne of coal (27 GJ) costs £25; 1 tonne of wood (20 GJ) costs 25 x $\frac{20}{27}$ = £18.50.

With these assumptions and from a financial viewpoint an option in which stems are harvested to a 5cm top diameter for pulp and the remainder utilized for energy is the best. This takes no account of increased harvesting, marketing and utilization costs with this practice but does indicate that a potentially valuable additional market for small round-wood might be developed.

REFERENCES

1. MITCHELL, C.P. & MATTHEWS, J.D. (1979) The Potential of Forest Biomass as a Source of Energy with Special Reference to Trees with Normal Single-Stemmed Habit of Growth. Aberdeen University, Report to ETSU, Harwell.

RECOVERY OF WASTE-PRODUCTS FROM WOOD-PROCESSING INDUSTRY

C. PARE and C. ROLIN

Département de Gestion des Entreprises
de l'Institut National Polytechnique de Lorraine
1, rue Grandville - 54042 NANCY CEDEX - FRANCE

M. CASTAGNE (Director of Research)

Summary

METHOD OF APPROACH

1) The quantities of wastes have been assessed at every stage of timber transformation through a range of 150 interviews covering the whole wood industry and giving indications about the average output and the existing uses of wood residues.

2) Inquiries about existing and feasible processes in the following areas : energy producing, compound materials, compost, animal feeding, wood chemistry.

SIGNIFICANTS RESULTS

a) Existing processes : Except for barks and hard wood sawdust, most of other wastes are usefully converted : coniferous sawdust is exported to german manufactures of compound panels ; dry wastes are used for combustion ; long wastes from sawmill cuts are used for producing paper paste. Ground to the high prices of wastes and their incompressible transport costs, the applications on a short scale and involving a high added value, are only profitable.

b) Feasible processes : For dry wastes : development of compatible boilers with application on site for heating and drying ; For barks : isolation, antipollution treatments, soil improvements ; For saw dust : compressed blocks for combustion, wood and resin mouldings, compost, animal feeding.

c) Future development : The acute researches in wood chemistry are attending to issue on new by-products of barks.

1. INTRODUCTION

Le Département de Gestion des Entreprises de l'Institut National
Polytechnique de Lorraine a entrepris, pour le compte de la D.G.R.S.T.
et en collaboration avec le Groupement Interprofessionnel pour la Promotion et l'Economie du Bois en Lorraine, le Laboratoire d'Economie Fores-
tière de l'ENGREF et le Laboratoire de Photochimie de l'Université de
NANCY I, une étude dont l'objectif était l'évaluation des quantités de
déchets disponibles dans l'industrie du bois en Lorraine-Alsace, ainsi que
l'étude des différentes voies de valorisation existantes ou possibles, en
vue d'évaluer la faisabilité d'implantation des différentes filières
retenues. Nous présentons dans ces quelques pages les principaux résultats
de cette étude.

2. EVALUATION DES QUANTITES DE DECHETS DISPONIBLES PROVENANT DES INDUS-
TRIES DU BOIS

La méthode retenue a été l'interview directe qui, pratiquée auprès
de 150 industriels de la filière, a permis dans un premier temps de
déterminer les rendements matières, et donc les quantités de déchets
produites par unité (m3 grumes) de bois traité. Nous avons ainsi étudié
les scieries, en distinguant feuillus, de résineux, et leur production de
déchets, ceux-ci étant de trois types : dosses et délignures, sciures et
écorces.

- Les industries de seconde transformation (meubles, menuiseries,
charpente, emballage, divers) et les industries du panneaux, les déchets
produits pouvant être de deux types : déchets massifs et de faible
granulométrie.

- L'industrie de la pâte à papier ayant pour principal déchet, des
écorces.

Cette série d'interviews nous a, de plus, permis de faire le point des
utilisations actuelles de ces déchets ainsi que de leur prix sur le marché.

La dernière étape de cette première partie de l'étude a consisté à
étendre les résultats obtenus sur l'échantillon à l'ensemble des entreprises de la région étudiée. Nous avons pour cela utilisé les informations
concernant les scieries et leur production, que le Service Régional
d'Aménagement Forestier de Lorraine et d'Alsace nous ont communiqué et

DECHETS DE :	TOTAL	PANNEAUX	PATE A PAPIER	ENERGIE	EXPORT	DISPONIBLE
SCIERIE	tonnes à 70 % d'humidité environ par an					
Ecorces Résineux	66 280			9 %		91 %
Sciures Résineux	93 920	48 %		10 %	36 %	6 %
Dosses et délignures Résineux	280 630	15 %	73 %		12 %	0 %
Ecorces Feuillus	18 780			17 %		83 %
Sciures Feuillus	65 770	5 %		32 %	35 %	28 %
Dosses et délignures Feuillus	167 770	19 %	40 %	30 %		11 %
SECONDE TRANSFORMATION	m^3 à 20 % d'humidité par an					
Gros morceaux	68 000	← 20 % →		79 %		1 %
Faible granulométrie	55 000 m^3 non foisonnés	← 71 % →		21 %		8 %
PATE A PAPIER	tonnes humides par an					
Ecorces	6 000 + stock 300 à 500 000 m^3	/			30 %	70 %

les informations concernant le nombre de salariés des entreprises de seconde transformation et du panneau fournies par l'I.N.S.E.E.

Les résultats de cette étude, dont la fiabilité, mesurée par recoupements auprès des principaux acheteurs, et de l'ordre de 85 % en moyenne, sont présentés de façon synthétique dans le tableau de la page suivante. Il apparaît clairement sur ce tableau, que la majorité des déchets sont déjà exploités : dosses et délignures par les industries du panneau et de la pâte à papier ; sciures par les industries du panneau et en vue de produire de l'énergie ; déchets secs issus de la seconde transformation, pour l'autoproduction d'énergie.

Les prix de vente moyens relevés début 1980 étaient : 110 F/t de dosses et délignures résineux, 70 F/t de dosses et délignures feuillus, 30 F/t de sciures. Les coûts de transport étaient : 28 F/t à 57 F/t de dosses et délignures, 22 F/t à 49 F/t de sciures et écorces (rayons 60 et 200 km).

En conclusion, seules les écorces, et dans une moindre mesure, les sciures de feuillus, sont disponibles en quantité relativement importante et à des prix assez bas, au moment de notre étude.

3. RECENSEMENT DES FILIERES POSSIBLES DE VALORISATION DES DECHETS DE BOIS

L'objectif de cette deuxième partie de l'étude a été d'établir un inventaire des technologies de valorisation non traditionnelles, assorti d'un descriptif technique et économique si possible très précis de chaque procédé indiquant en particulier leur stade de développement : laboratoire, pilote, industriel. Ainsi, l'étude bibliographique réalisée auprès du CTB, du CSTB, de l'ENGREF et d'autres sources diverses, ainsi que l'examen des brevets internationaux et le contact direct avec des industriels français et étrangers, nous ont permis de distinguer deux voies principales de valorisation : la voie des matériaux à base de bois et celle de la chimie.

- "La voie de la production de matériaux" se divise en quatre sousfilières :

. Dans la catégorie des matériaux à base de bois et de ciment, nous avons pu distinguer les panneaux légers ciment-bois, caractérisés par de très bonnes qualités d'isolation phonique et thermique et une bonne résistance au feu ; les panneaux mi-lourds et les blocs ainsi que les panneaux lourds, intéressants pour leurs qualités mécaniques et de résistance au feu.

.On trouve ensuite les matériaux à base de bois et de résine avec un grand nombre de procédés utilisant des bois d'essence variable et des résines thermodurcissables ou thermoplastiques en quantité variable selon les résultats attendus aux applications nombreuses : automobile, meuble, ...

.Les matériaux à base de bois seulement concernent des procédés de compression sous très fortes pression et température dont la mise en oeuvre est délicate et coûteuse en énergie.

.Enfin, les particules de bois peuvent être traitées chimiquement ou thermo-mécaniquement en vue d'être associées à des liants ou en vrac, utilisables principalement comme isolants.

Pour ce qui concerne la "voie chimique", nous avons retenu les 8 sous-filières suivantes :

.La combustion, procédé connu, permettant l'obtention de vapeur ou d'eau chaude, en vue de la production de chauffage, de séchage ou de la production d'électricité.

.La pyrolyse permet l'obtention de charbon de bois et de gaz pauvre.

.La gazéification, donnant un gaz au pouvoir calorifique de 1 à 2 th/m3, difficilement transportable et au stockage onéreux, utilisable de préférence directement pour le chauffage ou le séchage des bois.

.La liquéfaction est un procédé encore au stade du laboratoire dont l'objet est la production d'huile lourde.

.La distillation qui permet d'obtenir des suppléments vitaminés pour le fourrage du bétail ou des huiles essentielles destinées à la pharmacie et à l'industrie des cosmétiques.

.L'hydrolyse et les voies de valorisation des produits d'hydrolyse : ration énergétique en aliments pour le bétail, protéines obtenues par fermentation, éthanol obtenu par distillation, furfural. Une expérience pilote se monte actuellement à l'INPL de NANCY en vue de développer ce procédé à l'échelon industriel.

.La fermentation aérobie/anaérobie consiste en une fermentation des matières organiques en vue d'obtenir du gaz carbonique, de la chaleur et de l'humus utilisable comme compost.

.Enfin, la densification permet la production de briquettes ou granulés destinés à la combustion.

4. ETUDE DE FAISABILITE

Les procédés présentés dans la deuxième partie étant tous "nouveaux", l'étude de leur implantation en France passe par l'examen de trois critères

principaux : la disponibilité de la matière première bois, en quantité et
en qualité ; l'ouverture du marché ; la rentabilité du procédé.

.Nous avons confronté ces quantités de bois, disponibles ou susceptibles
de le devenir, avec la taille probable des unités de traitement afin
d'avoir une idée des rayons d'approvisionnement moyens. Suite à cette analyse, seront retenus comme valables les procédés valorisant les sciures et
les écorces, de taille réduite (50 à 100 t bois brut/jour.maximum).

.D'autre part, nous avons décrit pour chaque sous-filière les atouts et les
freins à l'introduction des différents produits sur le marché.

.Enfin, pour chacune des sous-filières présentées en deuxième partie, nous
avons retenu un procédé susceptible d'être développé en France à court ou
moyen terme. Selon un schéma commun de calcul et sous certaines hypothèses,
nous avons déterminé quelques valeurs techniques et financières qui permettent d'éliminer les procédés non rentables et de classer les autres.

  Les principaux résultats sont les suivants :

- Techniques concernant les matériaux de construction : rentables dans
l'ensemble, mais problèmes d'approvisionnement en gros déchets et problème
de marché.

- Production de compost à base d'écorces : problème d'ouverture du marché.

- Utilisation des déchets secs sur place pour le chauffage et le séchage :
à favoriser.

- Production d'électricité faible puissance: intéressant pour l'industriel
producteur de déchets et consommateur à la fois d'électricité et de fluide
thermique. Concept de l'énergie totale.

- Valorisation des produits de l'hydrolyse : rentable car la mise en oeuvre
passe par une production multiproduits, bouclée.

- Grosse unité de production de méthanol, ou centrale électrique : non
rentable étant donné le coût d'achat, et de ramassage des déchets.

5. CONCLUSION

  Fortement concurrencée par d'autres applications (pâte à papier,
panneau), la valorisation des déchets des industries du bois est pénalisée
par une insuffisance quantitative des déchets et par un retard dans le développement des "technologies douces". Il serait important de poursuivre
ce travail par un recensement des quantités de petits bois, par l'étude
d'engins de ramassage de ces déchets en forêt, par le développement de pilotes industriels débouchant sur la fabrication en série de petites unités
de valorisation des déchets et par l'étude de la chimie fine du bois susceptible de déboucher sur des produits à plus forte valeur ajoutée.

BIOMASS PRODUCTION WITH POPLAR

H.A. VAN DER MEIDEN and H.W. KOLSTER

Stichting Industrie-Hout
(Institute for the Promotion of Industrial Wood Production)
Wageningen, The Netherlands

1. INTRODUCTION

The availability of wood in the main producing areas in the world has its limits; in some regions like Northern Europe these limits have been reached. Developing countries will need a rapidly increasing share of the world wood production. Consequently, importing regions like the E.C. must do their utmost to increase their own wood production on the shortest term possible. This need is underlined by the increasing interest of wood as a source of energy.

In this view the importance of fast growing species and very short rotations is evident. Such rotations offer another advantage; the shorter the period between planting and harvesting, in combination with attractive financial results, the more private land owners, farmers included, will be interested to contribute to the extension of the wood producing area.

The idea of mini-rotations is not a recent one. Especially in the Near East the exploitation of densely spaced poplar plantations at very short rotations is practised since long.

The realization of these plantations, however, evokes many questions with regard to site, clones, planting material, spacing, soil preparation, rotation, harvesting, regeneration, wood quality, possible uses of the product, etc. It is obvious, therefore, that a research program concerning mini-rotation plantations is wide-ranging, many-sided and therefore very costly. This is why this research ought to be co-ordinated internationally.

2. THE EXPERIMENTAL FIELDS

The Stichting Industrie-Hout started its research on mini-
rotation with poplars in the beginning of 1974. Up to now a number of
trial fields has been established, the oldest three being 6 to 7 years.
They were planted on different kinds of soil. The grassland soil was
ploughed and afterwards subsoiling in stripes was applied in order to
facilitate the planting of the one year old unrooted sets. These were
topped after planting; the length of this removed top was 40 to 60 cm. All
lateral branches were removed. The sets were soaked before planting; this
considerably reduces losses as appeared from earlier research. The choice
of the different spacings (2 to 3 m) was based on the aim to restrict the
costs of this sort of plantation and on the conviction that a reasonable
diameter of the trees has to be achieved in connection with harvesting
costs, wood quality (not too much bark) and wood price.
The rotations envisaged are 5 to 7 years.
The creation of a second generation after felling demands yet much research.

3. RESULTS

The growth in the three experimental fields concerned is mentioned in
figures 1 and 2. After 6 years the mean annual increment was still lower
than the current increment. This implicates that the optimal rotation from
the point of view of mass production has not yet been reached.
In experimental field Hummelo the influence of felling date on resprouting
of stumps was investigated between the middle of November 1979 and April
1980; no differences were found. In February 1980 some wood properties and
the weight of the trees were determined with some sample trees; the results
are mentioned in table I.
Part of this experimental field was felled in March 1980. Also because of
the small quantity harvested (25 tons) the real harvesting costs were high,
namely £ 37 per ton and the revenues of the chips produced low, namely £ 14
per ton fresh at the roadside. Based on a quantity of at least 500 tons the
chip price could be $ 20 to 25 while the total costs per ton fresh will
then be $ 20 to 26, specified as follows: Felling $ 7,5 to 10, skidding $ 5
to $ 5,5, chipping at roadside $ 7,5 to 10.
In 1980 sprouting of the stumps developed favourably.

4. CONCLUSIONS

1. It is technically possible to produce large quantities of poplar wood per hectare in relatively dense plantations (1600 - 2500 trees per ha) and at short rotations. Mean annual increment of 20 to 30 m3/ha can be realized with, at an age of 5 to 6 years, current increments of 35 to 50 m^3. This is an indication that the optimal rotation in the experiments is longer than 5 to 6 years.

2. In the type of plantations concerned the volume production is particularly interesting for judgement of growth development and optimal rotation and for making comparisons with other methods of wood growing. For the financial results, however, the weight of the wood produced is more important because the potential consumers of the wood produced (pulp, board, energy) are interested in dry matter and not in volume. Dry weight is not only determined by the stem volume but also by the percentage of branches and by the density of the wood. These factors vary considerably between clones, reasons why breeding research ought to give them much attention.

3. The relation between harvesting costs and revenues is not favourable. The financial result is further worsened by the costs of establishment and maintenance, about $ 18 per ton fresh, and by soil and other fixed costs, $ 20 per ton, interest included. The price of the green chips should be at least $ 60 per ton at the roadside in order to have costs and revenues equalized. The second generation if realized by sprouts could be cheaper. However, other exploitation methods with the same plantation type must be investigated, the production of roundwood instead of chips included.

4. Short rotation poplar growing according to the method described could offer good possibilities where soil and labour are cheap and wood is very scarce. Particularly when other energy sources are lacking or very expensive, the method offers already now interesting perspectives.

TABLE I   Weight and wood properties in experimental field Hummelo,
          spacing 2,5 x 2,0 m. Trees 6 years old,
          felled February 19th 1980.

A) Data per tree

	cv. Rap	cv. Dorskamp
Number of trees investigated	5	3
Diameter b.h. (cm)	13,0	13,8
Height (m)	15,9	13,6
Volume to the top (dm^3)	98,7	91,5
Thickness bark at 0,1 m height(mm)	4,3	5,7
Fresh weight (kg)		
stem	73,6	73,1
branches	5,3	21,7
total	78,9	94,8
% branches	6,7	22,0
Moisture content (% of dry)		
stem	144	144
branches	132	135
Nominal specific gravity		
stem	298	320
branches	347	345

B) Data per hectare

	cv. Rap	cv. Dorskamp
Fresh weight (tons)		
stem	147,2	146,2
branches	10,6	43,4
total	157,8	189,6
Dry weight (tons)		
stem	60,4	60,0
branches	4,6	18,4
total	65,0	78,4
Harvest 20/3/80(stem and branches)		
bone dry weight chips (tons)	59,2	76,1

figure 1

figure 2

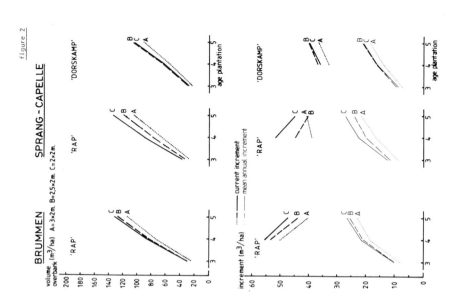

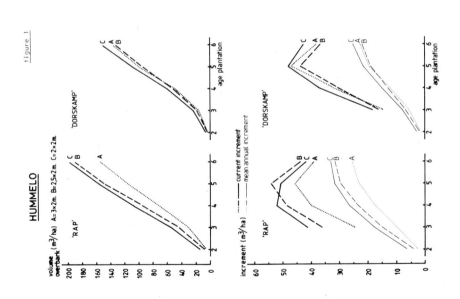

COPPICE WILLOW FOR BIOMASS IN THE U.K.

K.G. STOTT

Willows Officer, Long Ashton Research Station, University of Bristol,

Long Ashton, Bristol

G. McELROY and W. ABERNETHY

The Horticultural Centre, Loughgall, Co. Armagh, N.Ireland

D.P. HAYES

Department of Agricultural Botany, Queen's University, Belfast

Summary

The main objective was to determine maximum crop yields in the
shortest time on 'disadvantaged soils' in Northern Ireland.   We have
evaluated species, plant spacing, pre and post planting ground management,
cutting cycle, and the area of potential locations in the U.K.

S x 'Aquatica Gigantea' at densities of 0.5 - 1.0^{2}m, maintained weed-
free on a 3-year cutting cycle is currently preferred.   Its best
replicated yield on the poor surface water gley soils of Fermanagh identified as the most likely location in U.K. was 92.1 tonnes fresh weight of
3 year old sticks/ha - or at 57.5% dry matter, 17.65 tonnes D.M./ha/yr.
S x dasyclados and clones of S.viminalis are also promising.

Northern Ireland has the greatest potential in the U.K.   Willow
biomass at 25 tonnes fr.wt./ha/yr at a pulpwood value of £10/tonne rideside
(1977 prices) would be sufficiently more profitable than poor currently
under utilized stock grazing to release at least 20,000 ha. of disadvantaged land in Fermanagh and West Tyrone.   In England, 4,860 ha in
Somerset and 9,620 ha of Devon culm clays have the greatest potential.

Costings emphasised that mechanical harvesting will be necessary.
Currently, converting willow biomass into cattle feed is a more promising
market than pulp and cattle acceptance trials are in progress in Northern
Ireland.

1.   INTRODUCTION

Projected world shortages of wood pulp from conventional forestry led

in 1973 to the initiation of studies in Northern Ireland to evaluate novel

plant sources for the production of cellulosic pulps on 'disadvantaged

land' in the Province.   Work centred on the Horticultural Centre, Loughgall,

Department of Agriculture, N.Ireland and the Department of Agricultural

Botany, Queen's University.   At the Horticultural Centre the first object-

ive was to evaluate species.   Coppice basket willows - serviced since

1922 by a Willows Officer at Long Ashton Research Station were an immediate choice - leading to the collaborative studies here described. The main objective was to determine maximum crop yields in the shortest time, on disadvantaged soil types.

Initial trials at Loughgall showed that selections of willows from Long Ashton were preferable to New Zealand flax (Phormium tenax) and two species of bamboo. Hence trials concentrating on willow were established at Long Ashton and in Somerset but mostly in N.Ireland which we identified as possessing the greatest area of potential land in U.K. for willow biomass (1).

Since 1973 concern for supplies of energy from fossil resources has focussed world wide attention on alternative energy sources giving our trials a wider significance.

2. METHODS AND RESULTS

2.1 Selection

Of some 300 willows listed in the National Willow Collection at Long Ashton in 1971 (2), species of merit for the production of short rotation coppice biomass are described in Table I with notes on pertinent characteristics and the fresh weight yield of annual rods for the four years 1974-1977. Data for the most productive clones (12 stools only; 1 x 1 m spacing) is given together with country of origin. Because the reputed good pulping attributes of S x 'Aquatica Gigantea' were upheld (3, 4, 5),it predominates in our trials. In general this choice has been upheld by the trials of candidate willows detailed in Table II. These trials (100 plants/clone), are replicated only at Long Ashton but show that the close basket willow spacings of 0.7 x 0.35 m give enhanced early yields (Somerset), that the drier site at Long Ashton is less productive than the alluvial silt bottom land at Loughgall, and that S x dasyclados has performed well at all sites and S x 'Aquatica Gigantea' better than S. viminalis at Loughgall. Like S x dasyclados,the Swedish willow Q666 is probably a hybrid of S.viminalis x caprea x cinerea.

2.2 Factors affecting yield

Spacing. More precise data on yields result from the large replicated trial planted in February 1975 at Loughgall which compares S x 'Aquatica Gigantea' with a local Northern Ireland clone of S.viminalis, with and without nitrogen at 8 spacings using sizeable unit plots of 6 x 3 m.

Table III which averages the results for 4 annual harvests confirms previous results that close spacings enhance early yields. For the same density (e.g. 0.75^{2}m), asymmetric spacings (e.g. 3.0 x 0.25 m) are initially less productive than more regular ones (1.5 x 0.5 m). Given time, yield/ ha of annual coppice stabilizes irrespective of spacing (Table IV). Spacing is essentially a compromise between the need for early returns, the extra cost of high density plantings, the extent to which machinery needed in management dictates row width and the need to produce a certain sized product, (itself constrained by harvesting method) dictates cutting cycle.

In 1976 replicated trials using three preferred densities:- 0.25, 0.5 and 1.0^{2}m/plant were established jointly at Long Ashton and Loughgall. This range was thought likely to prove most appropriate for growing 3 yr coppice of S x 'Aquatica Gigantea' large enough and with an adequate fibre length (5) for pulp, but small enough to be mechanically harvested by the light machinery required on the land likely to become available. It was thought that a 3 year cutting cycle might give optimum annual increment (Table V) (6).

Nutrition – crop protection. Table V shows that on Loughgall soils 48 kg/ha of nitrogen improves yields. At this site, S x 'Aquatica Gigantea' is significantly more productive (mean 26.6 tonnes fresh weight/ha/yr) than S.viminalis Northern Ireland (15.1).

Whereas annual production of 'Aquatica Gigantea' (averaged over all spacings) is still increasing, that of 'N.Ireland' is declining, due at least in part to severe defoliation caused by rust, Melampsora larici- epitea (7). Larches (Larix spp.) – the alternate host, flank the site – S x 'Aquatica Gigantea' is not susceptible – nor as yet notably to other fungi and the many insects that attack willows (8).

Establishment. Two extension trials planted in February 1976 in County Fermanagh, N.Ireland test willows on the type of site most likely to be available and compare the effect of four ground management treat- ments on establishment and the effect of harvesting cycle on yield (7). The best yields have been obtained consistently where the ground was ploughed before planting. Killing the vegetation with paraquat prior to ploughing was slightly better at one site. In general unrooted willow cuttings are extremely susceptible to weed competition and establish best where weeds are completely eliminated (9).

Cutting. Economics dictate that biomass willows will have to be
harvested mechanically and it will not be possible to regulate so accur-
ately the height above the stool at which the rods are cut. Within limits
(here 0 - 5 cm) this does not appear critical:- Table IV shows that height
of cut did not affect yields of S.triandra once the stools were mature
(6 years) with a well developed stool and rooting system.

The trials in Fermanagh on typical surface water gleys show (Table
VI), that significantly greater mean annual yields (fresh and dry weight)
were obtained from a 3 year harvesting cycle (when first harvested 3
years from planting) than from a biennial harvest or from the mean yield
of the three annual harvests. (The highest yield was 92.1 tonnes/ha of
fresh material/3 year period, for a mean of 4 replicates, equivalent to
17.65 tonnes D.M./ha/yr.). Additional cutting cycle trials will mature in
1981 season at Long Ashton and Loughgall to check the above result that
the maximum mean annual increment requires at least a 3 year cycle and it
now appears prudent to attempt to manipulate one of these trials to obtain
data for cutting cycles up to 6 years.

2.3 Dry matter production - comparison with grass

Table VII quotes % dry matter of one(47) and 3 year old rods (57.5) and
shows that S x 'Aquatica Gigantea' at Castle Archdale gave a mean dry
matter yield of 15.8 tonnes/ha/yr which compares very favourably with the
dry matter yield from adjacent well managed grassland on the same Department of Agriculture Experimental Farm.

2.4 Derivation of location and area of potential land for willow
     Biomass in the U.K.

Initial willow yield data was promising enough to encourage the Paper
Industries Research Association in 1975 to commission a Coppice Willow
Pulpwood Feasibility Study (1), of which pages 66-116 detail the methods
used to estimate potential pulp willow land in U.K. and give the results
obtained. The methods are indicated semi-schematically in Table VIII with
results for England and Wales, and Table IX for Ireland. Areas were
located and quantified by referring first to the Soil Survey's Soil Water
Regime map for England and Wales by Thomasson (10). The wet-soil groups
2b, 3b and 3a were studied in more detail on the National Soil Map of the
Soil Survey (11). From this map it is possible to locate within the water

regime groups, soil groups suited to willows. Soil groups symbols 4, 6 –
Alluvial gleys – some 275,400 ha. were thought "most likely" out of a
"possible" of 855,600 ha. – about 8% of the land surface.

Additionally, to test a sample region, in the South West Region land
utilization maps were scrutinized for poor grade 4 and 5 land (12) lying
within the "Stagnogley"symbol 56. An estimated additional 175,200 ha.
were thought sufficiently poor to have potential for willows. This method
could be used for the large tracts of 56 present elsewhere in England and
Wales.

Within the South West, the most likely area (symbols 4, 6), the
Somerset Levels and Moors, were studied in detail by reference to the
Soil Classification Maps and appended description of their characteristics,
influence on agriculture and land use, (13, 14) to discover for one sample
location what proportion of the area was likely to be competitively suited
to willow.   Table VIII shows that of a total of 54,300 ha. – the most
likely soil series (Midelney, Fladbury, Allerton and Max) amounted to
17,800 ha. Discussions and visits with A.D.A.S. personnel compared existing land uses (stock raising) with projected pulp-willow profitability.
This was based on a realistic return of £250/ha/yr derived from 25 tonnes
fresh weight/ha/yr and pulpwood valued at £10/tonne rideside – (1977 prices).
On this basis 4,860 ha. were thought better suited to willows. This
approach should be followed for all locations but time only permitted a
questionnaire to appropriate A.D.A.S. Drainage Officers. From their
comments a reasoned assessment of 25,520 ha. of real potential in England
and Wales was derived.

Broadly similar methods were followed for Northern Ireland. The
National Soil Map (15), particularly the separate sheet for Northern
Ireland by McAllister and McConaghy (16) detailing the influence of soil
association on agriculture; a similar report for Eire (17), and a
detailed account of land use in Northern Ireland (18) were used.

The surface water gleys (Table IX) have the greatest potential for
willow for biomass and occupy 277,000 ha. in N. Ireland (plus 384,000 south
of the border in Eire).   Excluding land too poor for willows or higher
than 200 m, or in forest, 196,300 ha. – about one sixth of the Province
was identified.

Estimates of the proportion that might be released for willows were
obtained by extensive discussion with Agronomic, Soil and Drainage advisers

of the Department of Agriculture, Northern Ireland (19). Estimates ranged from 20,300 ha.(20% of disadvantaged land) to 45,000 ha. based on soils and land use, and 44,700 ha. based on comparative agricultural profitability (Table IX).

Advisers are confident that if willows proved reasonably profitable, 20% of the subsidized disadvantaged land in Fermanagh and West Tyrone would be immediately available with a further 20,000 ha. in the event of of success - all in a compact area around Enniskillen. In addition many small local areas - drumlin hollows, or wet pockets of grade C_2D_2 (18) particularly in the Lough Neagh - River Bann trough would be brought into production.

In England, whilst some 4,860 ha. in Somerset have real potential the only large block is 9,620 ha. of Culm Clays, poor wet soils with impeded drainage, in north Devon, ("Stagnogleys" symbol 51, 68). The problems of conventional agriculture in this area have been well documented by the Culm Clay Study Group (20).

For the whole of U.K, Fermanagh and West Tyrone is the most worthwhile area to pursue for willows for pulp, and probably therefore willows for biomass. Indeed, though the project was initiated for pulp, we are now more interested in the production of cattle feed pellets by a related multi-raw material process developed by Natural Fibres U.K. Ltd and B.P. Normally this process converts straw into cattle feed pellets but with modifications it can produce pellets suitable for the paper industry. Seven tonnes of 3 year-old sticks of S. x 'Aquatica Gigantea' were converted by this process by B.P. Nutrition Ltd, Grantham in April 1980 of which 2.5 tonnes of cattle feed pellets were returned for cattle acceptance trials at Greenmount Agricultural College, Northern Ireland.

Acknowledgements

We thank the Paper Industries Research Association for funding the Coppice Willow Pulpwood Feasibility Study and Mr.Trevor Dean, Natural Fibres U.K. Ltd for continuing advice and encouragement.

REFERENCES

1.   STOTT, K.G. (1977). Coppice Willow Pulpwood Feasibility Study.
          Report to Paper (Board, Printing, Packaging) Research Association
          (P.I.R.A.), Randalls Road, Leatherhead, Surrey. pp 122.

2.   STOTT, K.G. (1972). Check list of the Long Ashton collection of
          willows with notes on their suitability for various purposes.
          Rep. Long Ashton Res.Stn for 1971, 243-249.

3. McCOLLUM, I.N. (1975). Novel sources of cellulose. Ann.Rep.Hort. Centre, Loughgall. Dept.Agr.N.Ireland for 1974, 49-50.

4. McADAM,J.H. (1975). The production of cellulose for paper pulp in Northern Ireland with special reference to the use of willow (Salix species). M.Ag.Thesis, Dept.Agricultural Botany, The Queen's University, Belfast.

5. HAMER, R.J. (1975). Brief evaluation of the pulping potential of first year growth willow. Report to Queen's University, Belfast. PIRA, 9 pp. (Confidential report by Paper Industries Research Association to Dept.Agric. and Bot. Queen's).

6. STOTT, K.G. (1978 and 1979). Willows for paper pulp. Ann.Rep. Long Ashton Res.Stn for 1977, 37-38; for 1978, 50.

7. McELROY, G.H. (1979). Novel sources of cellulose. Ann.Rep. Hort. Centre, Loughgall. Dept.Agr. N.Ireland for 1978, 55-59.

8. STOTT, K.G. (1956). Cultivation and uses of basket willows. Q. J Forestry 50 (2), 103-112.

9. STOTT, K.G. (1980). Control of weeds in short rotation coppice willow. Proc.Conf.Weed Control in Forestry, 1980. Univ. Nottingham, Assoc. Applied Biologists, England, 33-44.

10. THOMASSON, A.J. (1975). Soils and field drainage. Tech. Monograph No.7. Soil Survey, Rothamsted Exp.Stn. Harpenden, Herts. 80 pp.

11. AVERY, B.W., FINDLAY, D.C. and MACKNEY, D. (1975). Soil map of England and Wales, 1: 1,000,000. Ordnance Survey, Southampton.

12. ANON. (1970). Land classification of England and Wales. MAFF Publication Branch, Government Buildings, Tolcarne Drive, Pinner, Middx. HA5 2DT.

13. AVERY, B.W. (1955). The soils of the Glastonbury district of Somerset (sheet 296). Mem. Soil Surv. Gt.Britain, Min.Ag.Fish and Fd, HMSO, 129 pp.

14. FINDLAY, D.C. (1965). The soils of the Mendip district of Somerset (sheets 279, 280). Mem. Soil Surv. Gt.Britain, Eng. and Wales, Ag.Res.Council, Rothamsted Exp.Stn, Harpenden. pp.204.

15. ANON. (1974). The soil map of Ireland, National Soil Survey, An Foras Taluntais, Ordnance Survey Office, Phoenix Park, Dublin.

16. McALLISTER, J.S.V. and McCONAGHY, S. (1968). Soils of Northern Ireland and their influence upon agriculture. Record of Agricultural Research 17, pt 1, Min.Agric.N.Ireland, 101-108.

17. GARDINER, M.J. and RYAN, P. (1969). A new generalised soil map of Ireland and its land-use interpretation. Ir.J.agric. Res.8.95-109.

18. SYMONS, L. (Ed.). (1963). Land use in Northern Ireland. Univ. of London Press Ltd, Warwick Square, London, EC4. 288 pp.

19. FURNESS, C.W. (1977). Farm management standards. Agriculture in Northern Ireland 51, No.9.

20. GREEN, C. (1976). The culm measures in Devon and Cornwall. Internal Report, MAFF, SW Region. 15 pp.

SELECTION   TABLE I

MEAN TONNES FRESH WT/HA AT LONG ASHTON 1974-1977, (NO. STOOLS SAMPLED)

1-YEAR-OLD RODS		MOST PRODUCTIVE VARIETY	ORIGIN

SALIX DAPHNOIDES — LARS 8/1

Tolerates dry sands, poor soils, acidic to pH 4.5. Few pests/diseases. Hardy. Produces few but long (2.5 - 3.0m) annual rods/coppice stool. Coppice vigour maintained beyond 4th year. Tree to 10m.

14.1 (60)   18.5   GB

S.TRIANDRA — BLACKMAUL X WHISSENDER

Rich moist lowland soils, not peats or more acidic than pH 5.5. Many pests/ diseases. Prefers mild climate. Moderate number 2m rods/coppice stool. Coppice vigour not maintained beyond 4th year. Tree to 7m.

16.8 (36)   21.0   GB

S.PURPUREA — JAGIELLONKA

Tolerates sands - peats - clays. pH to 5.0. Few pests/diseases. Hardy - frost resistant. Many small 1m rods/coppice stool. Coppice vigour not maintained beyond 3rd year. Tree to 6m.

17.6 (72)   19.3   P

S.ALBA, HYBRIDS WITH S.FRAGILIS — COERULEA WANTAGE HALL

Moist, fertile soils, often near running water. pH to 5.0. Few pests, some diseases. Fragilis and some hybrids notably hardy, frost and wind resistant. Moderate number 2m rods/coppice stool. Coppice vigour maintained beyond 6th year. Tree to 18-22m.

17.7 (130)   23.1   GB

HIGHEST COPPICE YIELDS

S.VIMINALIS and HYBRIDS

Moist heavy soils. Hybrids with cinerea, caprea group, tolerate peats and acidity to pH 4.5. Hardy, frost and wind resistant. Some pests, few diseases. Moderate number 3m rods/coppice stool. Coppice vigour maintained beyond 4th year. Tree to 8m.

22.8 (108)

Variety	Yield	Origin
VIMINALIS	31.8	GB
BOWLES HYBRID	27.0	GB
MULLATIN	47.3	DDR
DASYCLADOS	27.2	GB
S.X AQUATICA GIGANTEA	29.3	DK

SELECTION   TABLE II

	LONG ASHTON PLANTED 1976 0.7 x 0.7m 4th ANNUAL YIELD		SOMERSET PLANTED 1977 0.7 x 0.35m 3rd ANNUAL YIELD			LOUGHGALL PLANTED 1977 1.0 x 0.5m 3rd ANNUAL YIELD
	TONNES FRESH WT./ HA	NO. RODS/ STOOL	TONNES FRESH WT./ HA	NO. RODS/ STOOL	ROD(m) MEAN LENGTH	TONNES FRESH WT./ HA
S.X AQUATICA GIGANTEA 'GERMANY'	-	-	9	4.5	2.2	-
S.X AQUATICA GIGANTEA 'KORSO'	19	18.4	19	7.3	2.6	20.3
S.X REIFENWEIDE (viminalis x cinerea)	13	21.7	23	9.1	2.2	-
S.DASYCLADOS (viminalis x cinerea) x caprea	23	19.8	29	3.9	2.1	24.7
S.VIMINALIS 'NORTHERN IRELAND'	18	30.0	30	12.0	2.3	16.8
S.VIMINALIS LARS 4/16	18	15.3	31	7.6	2.5	17.9
S.VIMINALIS 'GIGANTEA' 'ENGLISH ROD'	- 12	- 25.0	32 -	7.5 -	2.6 -	- -
S.VIMINALIS 'BOWLES HYBRID'	18	15.4	35	6.5	2.7	15.0
S.VIMINALIS 'MULLATIN'	20	27.7	37	8.4	2.5	15.8

SPACING   TABLE III

S xAQUATICA GIGANTEA

PLANTED Feb.1975

LOUGHGALL HORTICULTURAL CENTRE

YIELD OF 1-YEAR-OLD RODS
FOR THE 4 YEARS,
1976-1979

AREA (^{2}m)	SPACING (m)	YIELD (Tonnes fr.wt/ha)
0.25	1.0 x 0.25	37.2
0.375	1.5 x 0.25	30.1
0.5	2.0 x 0.25	25.6
0.5	1.0 x 0.5	30.1
0.75	3.0 x 0.25	20.1
0.75	1.5 x 0.5	27.9
1.0	2.0 x 0.5	22.7
1.5	3.0 x 0.5	19.3
S.E.D.		0.86

SPACING IS A COMPROMISE BETWEEN THE NEED FOR EARLY MAXIMUM YIELD, PLANTING COSTS AND HARVESTING REQUIREMENTS.

207

CROP YIELD (TONNES FRESH WT/HA) WITH (+) AND WITHOUT (-) NITROGEN (45kg/ha) MEAN ALL SPACINGS

TABLE V — FERTILIZER

+	**
28.3	25.2̄

S.E.D. 0.43

YIELDS (TONNES/HA) IN FERMANAGH

TABLE VI

	CASTLE ARCHDALE	NEWTOWN BUTLER
ANNUAL HARVEST 1 YR-OLD-RODS Mean ann. yield for 1977-1979 Fr.wt.	21.9	17.0
Dry wt.	10.3	8.0
BIENNIAL HARVEST 2-YR-OLD RODS Mean ann. yield for 1977-1978 Fr.wt.	22.1	18.3
Dry wt.	11.1	9.2
TRIENNIAL HARVEST 3-YR-OLD RODS Mean ann. yield for 1977-1979 Fr.wt.	23.4	27.4
Dry wt.	13.5	15.8
S.E.D. Fr.wt.	0.52	0.56

CUTTING HEIGHT — TABLE IV

S. TRIANDRA PLANTED 1960 AT LONG ASHTON RESEARCH STATION: 5 SPACINGS:

0.6 x 0.18m
0.7 x 0.18m   0.7 x 0.36
0.76 x 0.18m   0.76 x 0.36
i.e.92,000 - 41,000 STOOLS/HA

YIELD IN TONNES FRESH WT/HA

	STOOLS PLANTED AND CUT AT GROUND LEVEL	STOOLS PLANTED AND CUT 2-5 CM ABOVE GROUND LEVEL
4th SEASON CROP,1963	16 (**)	20
6th SEASON CROP,1965	16 (n.s.)	17

CUTTING HEIGHT INITIALLY AFFECTS PRODUCTIVITY. FROM THE 3rd SEASON S.TRIANDRA YIELD AT THESE SPACINGS STABILIZED AT 15.8 TONNES FRESH WT/HA.

TABLE VII **ANNUAL DRY MATTER PRODUCTION
IN TONNES DM/HA/YR**

WILLOW v GRASS ON SURFACE WATER
GLEY SOILS. CASTLE ARCHDALE
EXPERIMENTAL STATION, FERMANAGH

	1978-79	1979-80	**MEAN**
PASTURE			
PERENNIAL RYEGRASS	9.2	9.5	**9.4**
IMPROVED PERMANENT	6.6	4.8	**5.7**
WILLOW			
S.x AQUATICA GIGANTEA			
1 YR-OLD RODS 47% DM FRESH WT.	10.4	11.0	**10.7**
3 YR-OLD RODS 57.5% DM FRESH WT.	-	15.8	**15.8**

TABLE VIII **POTENTIAL WILLOW BIOMASS AREA**

ENGLAND - WALES

HECTARES

Soil Water Regime	National Soil Maps: Soil Group No.	Most likely Somerset Soil Series	cf.profitability willows for pulp and existing Land use: ADAS Assessment	
2b	Alluvial Gley 4,6, 275,400	Somerset 54,300 ➔ 17,800 ➔	**4,860**	
		Rest 221,100 ➔	**8,060**	
		Assessment via local ADAS drainage Officers		
3b	Stagnogley 68,51, 120,000 ➔		**12,600**	
	69, 460,200 — — — — — — — ?			
Total	855,600 (8% land surface)		**25,520**	
3a	Stagnogley 56, 175,200	Speculative e.g.by reference to SW Region Land utilization Maps Grades 4 and 5		

TABLE IX   **IRELAND**

National Soil Map		Area land class B4 or below. Discount forest, Land > 200m	ASSESSMENT OF LIKELY WILLOW AREA (HA) FERMANAGH-TYRONE based on:			
			% area for willows – based on proportion of poor soils + Land use (McAllister)		Agricultural profitability Pulp willow v. Stock raising @ 1.2ha/beast	Subsidized disadvantaged land
(symbol)	(ha)		(%)	(ha)		
12	23,500	23,500	15	3500		
13	19,000	19,000	25	4900	of	of
15	115,500	34,800	20	7000	134,000 ha	147,000 ha
					Suppose 33% to willows	Suppose 20% to willows
17	119,000	119,000	25	29,900		
Total	277,000	196,300		**45,300**	**44,700**	**20,300**
Nearby Eire	386,000		25	96,500		

COPPICED TREES AS ENERGY CROPS

M. L. PEARCE

Forestry Commission - Research and Development Division

(Contractors to the U.K. Department of Energy and the
Commission of the European Communities)

Summary

A desk study (1978/79) by the author, indicated that dry matter production of above ground biomass from trees could attain levels of up to
20 tonnes ha^{-1} yr^{-1}. Using some broadleaved tree species managed on a
coppice system, it was predicted that these production levels could be
achieved over ultra short rotations. The author examines corroborative
evidence from some small scale inter-related silvicultural research, and
compares the target production figures with generally accepted photosynthetic efficiencies. A programme of research is described, wherein
production models will be constructed for a range of species, crop
spacings and rotation ages. Seven experiments will be established on a
range of potentially suitable site types, selected to reflect the categories of U.K. land which could become available for this kind of crop.
An evaluation of production costs will be attempted and the sites will be
offered to test harvesting machinery. The resulting biomass from each harvest cycle will be analysed for feedstock characteristics.

1. INTRODUCTION

Literature searches during 1978/79 by the author (1) revealed that many researchers throughout the northern temperate world were establishing production models for broadleaved coppice growth over short rotation periods. The predominant tree species used was Populus (or a Populus cultivar), with some examples from Salix and Acer. Accepting the diffi-culties of environmental variation, and extrapolation from experimental to field scale, the evidence suggested that production levels of up to 20 tonnes (dry) ha^{-1} yr^{-1} could be expected from the best sites in Britain. Production would include all above ground growth with the exception of leaves. Short term experimental work by Cannell (2) substantiated the theory that the biomass production from fully-stocked young hardwood stands is independent of planting density, and that the maximum production possible was 14 tonnes (dry) ha^{-1} yr^{-1}. However, all plants in these ex-periments were pruned down to a single stem, and the results may not re-flect the production curves expected from a true coppice (multi-stem) system.

Photosynthetic efficiencies - the measure of an organism's ability to utilise and store solar radiation - have been postulated, with world values of a maximum of 10% for the tropics, and without a controlled environment, as low as 2% in the north temperate zone (3). A production level of 20 tonnes (dry) ha^{-1} yr^{-1} equates with a photosynthetic efficiency of approx $1\frac{1}{2}$%. (Ref Figure 1).

Our experience of broadleaved coppice systems in Britain is limited to species which have been grown for specific market requirements (other than energy feedstock), and production levels have been below 10 tonnes (dry) ha^{-1} yr^{-1}. However, these levels have been sustained over long periods from the original stools (root systems) with rotation lengths (harvesting intervals) of from 7-16 years. Production levels from clonal Poplars in a nursery environment have reached 20 tonnes (dry) ha^{-1} yr^{-1} and been sustained on the same stools for 10-20 years with an annual harvesting cycle.

2. PROGNOSIS

The coppice system of tree growth for biomass production consists of the establishment of a tree crop by planting normal nursery stock or un-rooted cuttings in the case of Poplar or Willow. The growth after 2 or 3 years is removed and the stool (rootstock) produces multi-stems which can be harvested on a rotation cycle determined by the time at which the maximum mean annual increment (MMAI) and the current annual increment (CAI) coincide. Broadleaved tree species are expected to have an optimal rotation length of approx 3-5 years on this system, and can be harvested during the winter period after leaf fall. As the leaves contain a low proportion of the total biomass, but a high proportion of the absorbed nutrients, this harvesting method will help to maintain soil fertility. With appropriate harvesting techniques, suited to stem removal with minimal stool destruction, a stool life of several rotations of biomass harvesting can be expected. Optimum stool spacing - relative to species, must be researched, but rapid site colonisation by the coppice stems should eliminate competition from other vegetation, and keep weeding costs low. The candidate species for such a system, that meet the criteria of rapid growth from coppice, are clonal Poplar and Willow, Nothofagus (Southern Beech), Alder (for its ability to fix atmospheric nitrogen), and where the environment permits, Eucalyptus. On the basis of this prognosis, production curves were predicted (4) and a comparison with the more normal curves for timber production can be seen in Figure 2.

3.   RESEARCH PROPOSALS

Under a joint contract with the United Kingdom Department of Energy and the Commission of the European Communities, a series of 7 experiments will be established in Britain during the next 6 months. The objective will be to compile production models (Biomass tonnes (dry) ha^{-1}) for a range of species, a range of tree densities, and a range of rotation ages. The experiments will be established on sites which reflect the potential availability - namely (i) derelict (under utilised) woodland and (ii) marginal (under utilised) agricultural land, and will cover broad environmental zones. The period of the contract is 4 years, but it is hoped that substantive data will continue to be collected after that period. The potential damage which may be inflicted on the coppice stools by mechanised harvesting techniques, and the ultra-short rotation lengths, make it imperative that the health and longevity of the coppice stools be monitored over a long period.

The experimental design is of replicated plots of species and crop spacing, which will be split for testing rotation age. Sufficient plants will form a buffer to eliminate the effect of adjacent plots, whilst the remaining central plot stocking will be adequate to provide a statistically viable result. Plot replication will be adequate to test any site variation which may exist. A Nelder design plot will be established on each site to provide a relatively quick indication of the effect of stool density on production - with each of the circular plots being split for 4 species. The candidate species have been chosen to meet the criteria of maximum production on a range of edaphic site conditions. They are listed earlier in this paper, and experience suggests that they offer the greatest potential for coppice growth. The genus Eucalyptus has been little researched in Britain and its tropical and semi-tropical origins limit many of its species for use in the temperate world.

However, its potential growth rates are so high that it is felt desirable to include some species on sites which may be environmentally tolerable. Whilst it is generally acknowledged that the production from fully stocked plantations is independent of crop spacing, there is no hard evidence that it remains true for coppice systems over very short rotations. The growth profile of vigorous multi-stemmed and well rooted coppice stools may be dependent upon stool stocking when harvesting rotations of 5 years or less are undertaken. It is known that a wide variation of stem diameter arises from a coppice stool and that spatial colonisation by the coppice shoots is both vertical and horizontal. This rapid and vigorous growth from a well established stool suggests the necessity to test the effect on production of stool spacing. Evidence from existing and long lived coppice systems shows no fall off in production after many rotations of 7-16 year intervals. These experiments are designed to measure the effect on production of much shorter rotation lengths, when every rotation gained without loss of production, results in a reduction of production costs. Although it is not the subject of these contracts, the experimental plots will provide a "test bed" for harvesting machinery, and the resulting feedstock can be subjected to analysis by those involved with the conversion systems for final production.

4. PERSPECTIVE

There is little variation in the calorific value of oven-dry wood either between species, or the component parts of the tree and for the purposes of this paper a value of 20 GJ/tonne (dry) of woody material is assumed. It is acknowledged that moisture content and wood density vary with species and component tree parts, but production data has been reduced to a dry weight basis. Thus the target production level used in the prognosis will result in an energy production before conversion to fuel, of 400 GJ ha^{-1} yr^{-1}. This is double the current annual _per capita_ consumption of energy in the United Kingdom.

There are several alternative pathways for the conversion of woody biomass to fuel, and their associated energy input/output ratios must be considered. Using a guide figure of only a 50% conversion efficiency it can be seen that 1 hectare of land could theoretically produce the present _per capita_ energy consumption. It is of course unlikely that the form of energy in which this current consumption is demanded could be met entirely by woody biomass.

However, it is not difficult to envisage a scenario whereby much of the localised demand for low grade (non-transportable) energy could be met by the localised planting of coppice. It is with this scenario in mind that the sites selected for this R&D programme reflect the land which could become available without serious disturbance to current practice or environmental factors. Much of the current under-utilised and semi-derelict broadleaved woodland could be up-graded to coppice production - the sites are usually fertile and the interest ill defined. On many farms in lowland Britain there exists sufficient unused land of high fertility to provide the local low grade energy by coppice production.

As present energy sources either become too price sensitive or even non-existent, the need for a renewable source becomes imperative, and the current prognosis for coppice production is cost favourable. Using present day costs against the target production level (and accepting the conjectural nature of the harvesting costs) it is possible to project a production cost of £27 tce at the forest or farm gate - a sufficient incentive at least for the R&D programme.

REFERENCES

1. PEARCE, M. L. 1980 Coppiced Trees as Energy Crops. Report to the
   U.K. Dept. Energy.
2. CANNELL, M. G. R. 1980 Productivity of Closely-spaced Young Poplar on
   Agricultural Soils in Britain. Forestry. Vol. 53 No 1.
3. HALL, D. O. 1977 Solar Energy conversion through Biology. Fuel 1978
   Vol. 57 June.
4. EDWARDS, P. N. 1977 Short Rotation Mensuration. Forestry Commission.
   Internal Paper.

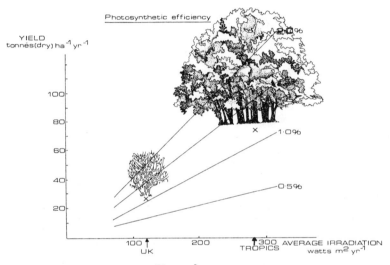

Figure 1

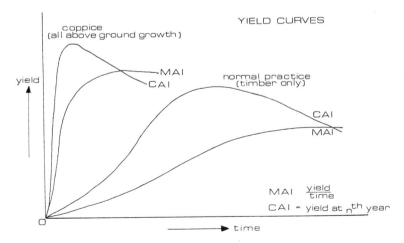

Figure 2

METHOD FOR THE ESTIMATION OF ABOVE-GROUND BIOMASS

AND BIOMASS PRODUCTION IN CLASSICAL COPPICE AND FIRST RESULTS

D. AUCLAIR and A. CABANETTES

I. N. R. A. Station de Recherches sur la Forêt et l'Environnement

Centre de Recherches d'Orléans - Ardon 45160 Olivet - FRANCE

SUMMARY

A study was undertaken to obtain simple methods of evaluation of coppice biomass. Dimentional analysis lead to the equation : $Y = a + b\ C^2 + \varepsilon$, where Y is the dry weight and C the girth at breast height. Neither did the height of the shoot nor the number of shoots per stool bring much additional precision. The sampling can be done shoot by shoot without considering the stools, about fifteen trees per plot are sufficient. In three pure stands and one mixed stand the parameters of the equation were different. Three methods for estimating productivity were compared, showing that the current annual increment is still higher than the mean annual increment in spite of the age of the stands (30 to 40 years). The results obtained in poor stands representative of the Centre of France were 67 to 109 metric tons of above-ground biomass per hectare, of which only 75 percent is available at present. The current annual increments are 2.7 to 5.6 t per hectare per year and the mean annual increments 1.9 to 4.1 t/ha/yr. A simple evaluation is given for the french forests using these results.

1 - INTRODUCTION

At present there exist in France over five million hectares of coppice, either as simple coppice or coppice with standards, which have been left over for the past thirty years or more. These may provide an important re-source of fuelwood for the future, but there is for the time being very little knowledge concerning the amount of available resource. It is also important to determine the possible production of such forests if they are to be used for biomass production.

In recent years the number of studies on forest biomass has been increasing rapidly (1). Many methods of biomass estimation have been used, and descri-bed by several authors (2, 3). But the coppice system presents particulari-ties which have to be taken into account in the sampling. The stools have several stems which may compete with one another. Each stool has had several rotations and the age of the stumps is often unknown. Few studies have been done on the particular coppice system (4, 5), and we have tried here to ob-tain simple methods of evaluation of above-ground coppice biomass. These have been described in more detail elsewhere (6).

2 - METHODS

Four different stands of approximately one-tenth of a hectare were sampled near Orleans : one 35 year old *Carpinus betulus* stand with a mean height of 11.8 m, one 25 year old *Betula pubescens* stand with a mean height of 11.4 m, one 40 year old *Quercus robur* stand with a mean height of 8.9 m, and one 30 to 35 year old mixed stand of *Castanea sativa, Quercus robur*, and *Betula verrucosa* of mean height 10.0 to 11.0 m.

Dimensional analysis was used to estimate the biomass of various above-ground compartments : total tree, cross-sections of 7 cm, 4 cm, 2.5 cm (approx 1"), "*stacked*" wood.

A total of 208 trees were sampled in the following way :
- stumps on which all shoots were sampled, with a varying number of shoots per stump,
- shoots of one given diameter class, issued from stumps with a va-rying number of shoots,
- shoots of various classes of diameter,
- shoots harvested and stacked in the "*usual*" way (2 m long and larger than 4 cm).

Each compartment of each sampled tree was weighed fresh and a subsample taken to the laboratory to be oven-dried at 105°c. A disk was sampled at breast height for age and production estimations.

The year's leaves were gathered on the ground at the end of autumn on one-square-meter plots.

The following independent variables for the regressions were measured : girth at breast height, length and height of the tree, the social status of the tree, the height of the first branch and the girth of the stem under the first branch.

On one subsample the branches smaller than 0.5 cm were weighed, as well as the bark on all compartments.

The biomass production was estimated by three methods :
- the addition of the mean productions of the different compartments (15 cm, 7 cm, 4 cm, 2.5 cm), a method described by KESTEMONT (7),
- the use of the derived equations relating biomass to girth as breast height and the measure of the last five annual growth rings,
- the total biomass divided by the age (= mean production).

3 - RESULTS

A stepwise multiple linear regression lead us to select the following equation :

(1) $Y = a + b\ C^2 + \varepsilon$

Y is the biomass or volume of a given compartment

C is the girth at breast height

The next independent variables selected were $C^2 \times L$ and $C^2 \times 1/N$, where L is the tree length and N the number of shoots per stump. They brought little additional precision compared to the amount of work they entail.

To evaluate the biomass of stools it was found that a regression on the stools was poorer than a regression based on the individual shoots. This is due to the small number of stools sampled, but to sample more stools would mean harvesting a much larger number of individual shoots.

The biomass of each shoot was not shown to be dependent on the stool (and the number of shoots on the stool), therefore the sampling can be very much simplified by being done shoot by shoot without considering the stools.

The optimum number of trees to be sampled was studied by comparing regressions based on 31, 15 and 7 observations. With fifteen trees sampled the confidence interval of the stand is lower than 3 per cent. This meets the conclusions of RIBE (8). This number amounts approximately to one stere, which is the apparent volume of one cubic meter (about $0.66 m^3$ real volume and 1/3 of a cord).

The variance of the residues of equation (1) is of the form var $\varepsilon = k C^4$. We therefore introduced a weight-factor $1/C^2$ into the equation.

The equations were compared for each site and each species. They were significantly different, showing an influence both of the site and the species.

The water content varies along the shoot as shown in figure 1. However in the stands studied here the conversion from fresh weight to dry weight can be done using only one sample per shoot at breast height.

<u>Figure 1</u>

Water content of the various compartments. The dotted line corresponds to $\emptyset < 0.5$, it is included in the last compartment.

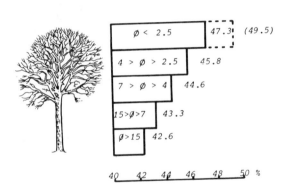

The evaluation of the productivity was dependent on the method used. The results were 40 to 60 per cent higher with the first method than with the second. The latter is more classical and has a more rigorous theoretical basis. The current annual increments thus estimated were 2.7 to 5.6 t/ha/yr. The last method gives an evaluation of the mean annual increment : 1.9 to 4.1 t/ha/yr.

The distribution of total above-ground biomass is shown in figure 2. The standing biomass was estimated as follows :

   . 109 t/ha in the Carpinus Stand

   . 80 t/ha in the Betula stand

· *77 t/ha in the mixed stand*

· *67 t/ha in the Quercus stand*

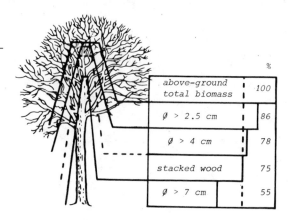

	%
above-ground total biomass	100
∅ > 2.5 cm	86
∅ > 4 cm	78
stacked wood	75
∅ > 7 cm	55

These results have been used by BOUCHON (9) to estimate the standing bio-
mass in the coppice and coppice-with-standards stands for the whole of
France. He used the data of the French national forest inventory, which
gives the volume of "larger timber" (larger than a diameter of 7 cm). The
use of a mean conversion factor of 1.82 to convert large wood data to to-
tal above-ground wood, and of a conversion factor of 0.55 to convert volu-
me to dry weight leads to a total result for french forests :

total area : 6 222 000 ha

total above-ground biomass : 225 000 000 metric tons

mean biomass per hectare : 36.2 t/ha

mean biomass increment : 1.6 t/ha/yr

LITERATURE CITED

(1) PARDE, J. *(1980). Forest biomass. Forestry Abstracts,* 41,343-362

(2) YOUNG, H. E. *(1976). Oslo biomass studies. College of life sciences and agriculture, University of Maine at Orono, USA, 302 p.*

(3) KEAYS, J. L. *(1971). Complete tree utilization. An analysis of the literature. Information report, Western forest products laboratory, Canada n° VP-X-69, VP-X-70, VP-X-71, VP-X-77, VP-X-79.*

(4) FORD, E. D. and NEWBOULD, P. J. *(1970). Stand structure and dry weight production through the sweet chestnut coppice cycle. J. Ecol. 58, 275-296.*

(5) DUVIGNEAUD, P. and KESTEMONT, P. *(1977). Productivité biologique en Belgique. Duculot, Paris-Gembloux, 617 p.*

(6) AUCLAIR, D. and METAYER, S. *(1980). Méthodologie de l'évaluation de la biomasse aérienne sur pied et de la production en biomasse des taillis. Oecol. applic. 1, 357-377.*

(7) KESTEMONT, P. *(1975). Biomasse, nécromasse et productivité aériennes ligneuses de quelques peuplements forestiers en Belgique. Thesis, Faculty of Sciences Free University of Brussels, 334 p.*

(8) RIBE, J. H. *(1979). A study of multi-stage sampling and dimensional analysis of "puckerbrush" stands. Bull. n° 1, The Complete tree Institute, University of Maine at Orono, USA, 108 p.*

(9) BOUCHON, J., DIVOUX, A., PARDE, J., TISSERAND, A. *(1980). L'estimation en bois total des taillis, et des houppiers des arbres "Chêne" et "Hêtre" à partir des données de l'Inventaire Forestier National. EEC preliminary report, project F. 203, contract 470.78.7. ESF.*

COPPICE FORESTS IN ITALY: THEIR POTENTIAL FOR ENERGY

G. SCARAMUZZI*, O. CIANCIO** and A. ECCHER*

* S.A.F.-Centro di Sperimentazione Agricola e Forestale, Roma (ENCC Group)
**Istituto Sperimentale per la Selvicoltura, Arezzo

Summary

Due to the area they cover (over 3.6 million hectares, equalling some 57% of the national forest surface) and the many complex problems they entail, coppice forests represent the major theme of Italian forestry.

The drastic fall in wood removals from Italian coppices occurred over the past three decades is discussed, and an estimate of the expectable coppice area economically harvestable and of the wood yield obtainable is given.

A research project for an enhanced utilization of Italian coppices promoted by the National Agency for Cellulose and Paper, including an assessment of their potential for industry and energy supply, is outlined.

1. INTRODUCTION

Due to the area they cover (over 3.6 million hectares, equalling some 57% of the national forest surface) and the many complex problems they entail, coppice forests represent the major theme of Italian forestry.

As a result of the crisis of vegetal fuels and the contemporary rural exodus, yearly forest removals from Italian coppice forests, that in 1947-48 exceeded 10 million m^3, have gradually decreased until reducing over the last period (1973-77) to less than 3 million m^3. Only during the last 3 years, a slight reversal of tendency has been recorded.

The need for an enhanced utilization of coppices has been repeatedly stressed, particularly over the last years because of the dramatic level attained by the national timber deficit and the supervened energy crisis. However, the high harvesting costs and the quality of the obtainable stock, until recently scarcely accepted or even refused by wood industries, have opposed the realization of concrete initiatives.

Today, the existing availability of an adequate harvesting machinery and the changed industry attitude make it possible to expect an economic utilization of many coppices left to age over rotation because of their negative stumpage value.

2. DISTRIBUTION AND COMPOSITION OF ITALIAN COPPICES

The regional distribution of the various coppice types in Italy is given in Table 1. The following regions show the largest coppice surface: Tuscany (18.4% of the national coppice area), Piedmont (10%), Emilia-Romagna (8.2%).

When referring the coppice area to the region's territory, Liguria proves the highest value (35%), followed by Tuscany (29%) and Umbria (28%). When regarding, on the other hand, the coppice surface as a percentage on the forest area, Umbria ranks first (92%), followed by Marches (83%), Emilia-Romagna (79%) and Tuscany (77%).

As concerns coppice composition, mixed stands prevail (53%). Among pure stands, a majority of oak coppices occurs, followed by beech and

Table 1 – Regional distribution of various coppice types in Italy (ISTAT, 1978).

region	simple ha	composite — hardwoods only ha	composite — hardw.and conifers ha	composite — sub-total ha	total ha	total %
Piedmont	283,683	60,314	18,938	79,252	362,935	10.0
Valle d'Aosta	4,229	975	2,957	3,932	8,161	0.2
Lombardy	173,144	71,151	33,161	104,312	277,456	7.7
Trentino–Alto Adige	87,523	1,278	4,678	5,956	93,479	2.6
Friuli–Venezia Giulia	35,487	28,113	2,241	30,354	65,841	1.8
Veneto	96,315	14,038	11,346	25,384	121,699	3.4
Liguria	148,070	22,374	21,919	64,293	192,363	5.3
Emilia–Romagna	278,009	16,354	1,059	17,413	259,422	8.2
Tuscany	432,954	204,316	28,334	232,650	665,604	18.4
Umbria	194,747	44,307	1,721	46,028	240,775	6.7
Marches	117,826	7,991	3,696	11,687	129,513	3.6
Latium	239,236	33,042	114	33,156	272,392	7.5
Abruzzi	73,673	47,676	315	47,991	121,664	3.3
Molise	24,721	25,193	67	25,260	49,981	1.4
Campania	185,666	7,173	––	7,173	192,839	5.3
Basilicata	72,580	2,701	25	2,726	75,306	2.1
Apulia	47,468	3,857	94	3,951	51,419	1.4
Calabria	114,865	34,111	3,496	37,607	152,472	4.2
Sicily	65,588	13,575	1,615	15,190	80,778	2.2
Sardinia	166,928	1,726	2	1,728	168,656	4.7
Italy	2,842,712	640,262	135,778	776,043	3,618,775	100
% on total	78.6	17.7	3.7	21.4	100	

chestnut stands. Mixed coppices largely prevail in Tuscany, Lombardy, Umbria and Marches, pure stands being more widespread in Emilia-Romagna, Piedmont and Latium. Beech is the most common species in Abruzzi, Trentino-Alto Adige, Veneto and Friuli; chestnut prevails in Piedmont, Liguria and Campania, whereas in all the remaining regions oak coppices prevail among pure ones.

Simple-mixed coppices (1,400,000 ha) also include those known under the name of 'Mediterranean maquis', amounting to some 900,000 hectares. These are mostly made up of degraded associations, showing a predominance of Ericaceae, that supply material of very poor technological value.

3. SILVICULTURAL AND GENERAL REMARKS

The drastic fall in wood removals from coppices has brought about as a consequence the over-aging of many stands as also the accumulation of considerable timber volumes.

At first sight, this would appear positive from the silvicultural standpoint, in view of a widespread conversion of coppices into high forest. On the contrary, different factors of disequilibrium occur, with impending danger for the forest perpetuity.

In this connection, it should be stressed that under Mediterranean environmental conditions the high functionality level of coppices lies at the basis of their spreading and multisecular tradition, as also of their intrinsic resistance to natural and anthropic adversities (fire, grazing, deforestation, etc.).

The new social tensions entailed by the increasing needs and the worsening of the energy crisis are presently stressing again the validity of coppice management, seen as the result of the man-environment inter action and integration in mountainous and hilly areas. Coppices provide, in fact, the highest rates of volume and value increment, as well as the highest labour employment per unit area.

Coppice forests utilization should, therefore, be conveniently enhanced. The possibility of an economic utilization is, however, affected by the degree and times of realization of a number of conditions, such as an adequate mechanization, that should also allow the complete harvesting of the entire above-ground biomass; a unitary management of forest compart ments of economic size; the creation of the required infrastructures, main ly a suitable network of forest roads; the planning of fellings, to balance timber offer and demand, with a view to reaching an adequate remuneration of wood production.

Mention should also be made that a part of the over-aged stands are susceptible of conversion into high forest and that a part of the degraded stands can be improved and transformed by thickening and conifers' addi- tion for being later converted in pure or mixed high forests. The above- quoted stands can be estimated will not exceed 20,000 hectares per year, on the whole. Thus, in the coming twenty years the coppice area might be reduced by 400,000 hectares at the most, decreasing from 3.6 to 3.2 million hectares.

Taking a mean rotation of 20 years, for the next two decades a yearly harvestable area of 160,000 hectares (3,200,000/20) can be estimated, which is absolutely unattainable under present conditions. In fact, about half this area consists of poor and hardly accessible stands, that despite me chanization advances still remain economically unutilizable.

From the remaining 80,000 ha, taking a prudential mean annual incre- ment of 3 m^3/ha, the total timber volume obtainable by traditional methods would amount at least to 4.8 million m^3/year, with a 60% increase as against present removals.

The supposed complete utilization of the above-ground wood biomass would allow the yearly production obtainable from coppices to increase to 7.2 million m^3.

4. OUTLINE OF E.N.C.C.'s EXPERIMENTAL PROJECT

In view of the ever more increasing difficulties met by the paper in dustry for raw material supply and in the belief of contributing to the solution of one of the most prominent problems of Italian forestry, the

National Agency for Cellulose and Paper (E.N.C.C.) has promoted a large re
search project for an enhanced utilization of coppices in Italy, to be car
ried out by its scientific and technical branches in cooperation with other
research organizations interested in the problem. The project has recently
been granted .a financial support by the Commission of the European Communities.

The research object is to find out technically and economically suitable harvesting methods in agreement with the safeguard of forest productiv
ity, and to assess the possible suitable uses of the obtainable biomass.
The trials are expected to cover different areas representative of the main
Italian coppice types, and will deal with silvicultural, harvesting and
technological aspects.

The silvicultural aspects are concerned with site characterization,
from the pedologic, climatic and vegetational point of view; stand charac
terization; forest production assessment; standards selection in relation
to the harvesting system used; machinery effects upon stumps and standards
vegetative conditions, on natural regeneration, on stability and evolution
of soil and vegetation.

The harvesting aspects include felling, processing, skidding and chip
ping trials of the obtained stock. The choice of the mechanical equipment
and of the harvesting system is made in relation to the stand type, geo-
morphology and machinery working capacities.

The technological trials aim at evaluating the suitability of the obtainable wood biomass for various end uses.

The project will include an evaluation of the expectable size of cop
pice biomass use for industry and energy, and of the conditions for its
technical and economic convenience.

DESIGN OF A HARVESTER FOR 2-3 YEAR OLD WILLOW STICKS

J W DUFF and H D McLAIN

Loughry College of Agriculture and Food Technology
Cookstown, Co Tyrone, N Ireland

Summary

A trailed harvester to cut and tie willow sticks into bundles weighing
approximately 30-40 kg was designed built and tested between 1977 and
1979. This machine demonstrated that it was possible to harvest 2-3
year old willow sticks into bundles but a number of problems were found
with the design. The conveying system was unreliable, the bundle
separation was inadequate, the bundle density was too low and the machine
was too heavy for operation on soft ground.

Based on the experience gained with the first machine a new harvester
has been designed and built. This machine is mounted on an agricultural
tractor to improve manoeuverability and make it suitable for use on
marginal land. A spring tine conveying system has been introduced and
the two stage packing system is designed to achieve the required bundle
separation. A density control unit has been incorporated into the tying
chamber. The whole design has been kept as simple, compact and robust
as possible. Field testing will be carried out during the winter of
1980/81.

Purpose

The Department of Agriculture for Northern Ireland has been engaged,
since 1974, in assessing the potential for growing short rotation willow
crops on marginal land. These experiments have been carried out by the
Department's Horticultural Centre at Loughgall. In conjunction with
this they required a mechanised harvesting system to be developed so
that if and when commercial willow production was economically viable
a means of harvesting would be available. In the Northern Ireland
situation it is considered that willow production would be as a
"cash crop", grown by farmers on marginal land. Harvesting would be
carried out by a contractor or by the farmer himself.

Aim

To design and build a mechanised harvesting system to cut willow sticks
up to 3 years old and to process the material into a form suitable for
transport from the field. Both harvester and transport system were to
be suitable for use on marginal land.

Design and Build-Phase 1

After consideration of the possible alternatives it was decided to design
a machine to tie bundles of sticks approximately 30-40 kg in weight. A
bundle gives a versatile package suitable for further processing in a
number of ways. It also permits natural drying of the crop. The
bundle weight chosen was considered to be the maximum which would be
easily handled. The bundles were to be bound by twines approximately
600 mm and 1700 mm from the base. It was decided to leave the design of
a transport system to a later date as bundles could be handled by existing
agricultural machinery, perhaps with some modification.

Machine design commenced during the Autumn of 1977 and after initial
trials with a reciprocating cutting mechanism a cutting system based on
twin circular saws was adopted and these performed satisfactorily.

To this cutting mechanism the remainder of the components of the machine
were added in stages. The final machine design could be divided into
four distinct operational units:-

Cutting Unit	- 2 circular saws
Conveying Unit	- 2 pairs of narrow conveyor belts
Packing Unit	- single and later twin rotating tines
Tying Unit	- 2 needles and knotters

As so little willow material of the correct age and row spacing (700 mm) was available for test purposes a technique for simulating willow of the required age was developed. Trials, which were completed by mid 1979, showed that it was possible to successfully cut and bundle willow sticks but a number of problems were found with the machine:-

1. The hydraulically driven circular saws operated satisfactorily though more power was consumed than expected.
2. The conveying system worked but was unreliable due to the narrow belts running off their pulleys.
3. The packing system worked but separation of the bundles was inadequate.
4. The tying units performed well but the bundles were of low density.
5. The machine was not sufficiently robust to deal with 3 year old material.
6. The trailed design would have been unsatisfactory in soft conditions and on hills.

The problems listed could not be easily overcome in the existing machine design. A new machine had to be built so the opportunity was taken for to make a complete reassessment of the design taking into account the valuable experience gained on the first machine.

Design and Build - Phase 2

The design philosophy adopted was to keep the machine as simple and compact as possible using existing proven technology where practical. This latter point was particularly important as design commenced in May 1980 and the machine had to be ready to cut a 3 year old crop during the winter of 1980/81.

Much time was spent at the design phase and each idea was tested in "mock-up" form to test its validity. The new harvester has the following features:-

1. Mounted on an agricultural tractor which is operated in reverse using modified controls. This enables the machine to operate on soft ground and greatly improves manoeuverability. The height of cut can also be easily adjusted.

2. Spring tined conveying units are used to feed the willow sticks to the circular saw. The speed of these do not have to be exactly matched to the forward speed of the harvester as is the case with belts.

3. A single circular saw of diameter 760 mm is used to cut the crop. This is easily driven off the tractor PTO and it also helps to convey the butts of the sticks.

4. A two-stage packing system is used to convey the material from the saw to the tying chamber. The first stage consists of rotating tine units approximately 500 mm and 1700 mm from the base of the sticks. Each unit consists of a double throw crankshaft with a tine on each pin. This system quickly clears the cut sticks from the saw blade and places them upright in an off-blade area ready for the second stage to take them through checks into the tying chamber. The second stage consists of a single rotating tine which is designed to provide the separation required to prevent the bundles being tied together.

5. The tying units of the first machine have been retained as they performed well.

6. A density control unit has been incorporated into the exit from the tying chamber. The bundle size is also variable.

7. The design of the machine has been kept as simple and compact as possible, with the distance between cutting and bundling being approximately 1200 mm. The frame has been designed to have sufficient strength to handle a 3 year old crop.

Conclusion

The first machine demonstrated that it was possible to develop a harvester to cut 2-3 year old willow sticks and tie them into bundles weighing approximately 40 kg. The second machine has been designed to eliminate the shortcomings of the first. The building of this machine is well advanced and it is planned to have it operating during the winter of 1980/81 and to complete the development by 1982.

SHORT ROTATION FORESTRY AS A SOURCE OF ENERGY

M. NEENAN and G. LYONS

An Foras Taluntais, Oak Park Research Centre,
Carlow, Ireland

Summary

   The term short rotation forestry applies to forest species which give rapid juvenile growth and are capable of re-growth from harvested stumps, through successive cutting cycles. As a biomass energy crop, it has many advantages, including: it's wide climatic suitability, adaptability to low grade habitats, ease of propagation and high yield potential.

   This paper discusses the main feasibility criteria for short rotation forestry (S.R.F.) energy production and evaluates it's potential impact in the Irish energy context. Results of recent E.E.C. funded research work are presented to establish current fuel yield levels, land availability and economic returns. Finally, the paper examines the application of S.R.F. biomass technology in the electricity, industrial and domestic energy utilisation sectors.

1. INTRODUCTION

At higher latitudes where shorter day-lengths co-incide with colder
weather, the use of some solar energy technologies is limited. In such
circumstances, photosynthetic production of energy is one of the more
promising possibilities. Since the efficiency of photosynthesis is about
the same in all C3 species, the basic problem in selecting an effective
biomass crop is that of trapping all of the useful solar radiation.

In the production of terrestrail biomass, there are three main
biological constraints:-

(i)   availability of a species which is adapted to soil and climatic con-
       ditions
(ii)   presence of a photosynthetic apparatus over most of the year
(iii) occurrence of a full leaf canopy throughout the growing season

There are, in addition, economic constraints to biomass energy production.
These include: cost competitiveness of the biomass fuel, biomass energy
output/input ratio, the possibility of alternative markets and the fact
that the price of some commodities is determined by political decisions
rather than market forces. This last factor would eliminate many candidate
biomass products, the price of which is determined under Common Agricultural Policy (C.A.P.) regulations.

In view of these limitations, the Irish Agricultural Research Institute (An Foras Taluntais) selected short rotation coppice forestry
(S.R.F.) as the most suitable biomass species for energy production (1).
Working under contract to D.G. Xll of the E.E.C., the Institute is investigating the possibility of establishing S.R.F. energy plantations for
electricity generation and also for domestic and industrial heating applications.

2. S.R.F. - ENERGY FEASIBILITY

The feasibility of producing energy from short rotation forestry
(S.R.F.) on a large scale is determined by four main criteria:

                    1. Fuel yield potential
                    2. Land availability
                    3. Economics
                    4. Market availability

2.1 Fuel yield potential

At present, 25 S.R.F. silvicultural trials are being conducted in Ireland, investigating soil type - species interactions, plant populations, cutting cycles and cultivation treatments. Results from these experimental plots indicate that a yield of 12.0 t/ha per year of dry matter is obtainable when the correct combination of these factors is employed. This is equivalent to a primary fuel yield (in gross calorific value terms) of 5.4 tonnes of oil equivalent (TOE) per hectare per year. At this yield level, 2.4 per cent of Ireland's land area could provide 10 per cent of it's current primary energy demand (8.6 MTOE in 1979).

It is worthwhile to quantify the useful energy potential of these biomass yields. In converting S.R.F. wood chips to steam and electricity, material of up to 30 per cent moisture content can be burned effectively in conventional boilers (2). Assuming realistic boiler and turbine efficiencies (for 30% m.c. fuel) (3), the useful energy output per planted hectare may be summarised as follows:

Heat energy output : 162 GJ/ha/annum

Electrical energy output : 19.35 MWh/ha/annum

Possible installed electrical capacity (@ 55% load factor) : 4.0 kw/ha

2.2 Land availability

The national soil survey has identified 31 individual soil types in Ireland. Three main groups of these are potentially suitable for S.R.F. energy plantations (4). These are presented in Table I, together with their estimated total energy potential.

TABLE I: Land availability for S.R.F. energy plantations

Soil type/ Topography	Area (M. hectares)	Potential Fuel Yield (MTOE/annum)
1. Mountain and hill land	1.45	7.83
2. Wet drumlin land	0.22	1.19
3. Organic soils	0.50	2.70
Total :	2.17 (31% of tot. land area)	11.72

Current land uses for these soil categories range from rough grazing of cattle and sheep to conventional silviculture. However, much of these areas have been or will be abandoned from agriculture due to low economic returns. Clearly, the terrain and isolation of many mountainous regions would render the land unsuitable for energy production on a large scale.

While the proportion of this land which could be made available for energy plantations depends on sociological and political factors, the above upper-bound estimate indicates that land availability is not a major constraint. Already, 80,000 hectares of peatland (organic soil category) are under State ownership for peat fuel production. On depletion of the peat resource, this area alone could provide 0.4 MTOE per year, or 4.7 per cent of current primary energy demands, under S.R.F. plantations.

2.3 Economics

The economics of S.R.F. energy production are sensitive to fuel yield, land price and availability, and to a range of plantation management parameters including : cutting cycle, plant population, cultivation practices and harvesting techniques. In addition, revenue from fuel sales will be largely dependent on the biomass energy value compared with other indigenous or imported primary fuels.

Recent work has shown (5) S.R.F. fuel production costs of the order IR £15 to IR £20 per tonne green weight delivered of wood chips. This is equivalent to a primary energy cost of IR £66 - IR £88 per TOE. Under an energy pricing policy based on fuel oil, these S.R.F. production costs would provide up to a 15 per cent internal rate of return for the energy plantation enterprise.

Figure 1 shows the variation of S.R.F. primary fuel production costs with biomass yield potential and cutting cycle. These costs are compared with 1979 primary fuel prices for imported oil and indigenous (machine) peat used in electricity generation. It is evident that, at current yield levels, S.R.F. is an economic alternative to oil, although marginally more expensive than native peat. However, with the projected decline in both oil and peat production over the coming decades, the economics of S.R.F. energy plantations should prove even more favourable.

2.4 Market availability

At a time of rapidly increasing global and national energy consumption, an assessment of market availability for renewable energy resources may seem unnecessary. However, the long lead-time required for implementation of energy supply alternatives is most often associated with deficiencies of the consumer (or market), rather than the production technologies. In this context, the well established technology of solid fuel combustion should enable an immediate penetration of S.R.F. wood fuel in the energy supply market.

In Ireland, the market potential for direct combustion of S.R.F. biomass occurs in three main utilisation sectors:

1. Electricity generation
2. Industrial process heat/steam
3. Space heating

Currently, peat burning electric power plant accounts for 15.5 per cent of the country's total installed electrical capacity (2890 MW). Under S.R.F. plantations, an area of 112,000 hectares could maintain this electrical capacity. Lalor has suggested (6) that the State owned peatlands (and associated network of power stations) should be developed for biomass energy production, as their peat reserves become exhausted over the next thirty years. These cut-over peatlands would provide over 70 per cent of the land required to maintain the above (15.5 per cent) electrical capacity.

While the infrastructure for electricity generation from biomass (peat) is already developed in Ireland, the use of S.R.F. wood fuel for process and space heating applications is a more energy efficient alternative. These two utilisation sectors alone account for 62 per cent of final energy deliveries (7). In 1977 terms, this was equivalent to 47 per cent of primary energy demand, or 3.5 MTOE (8), of which an estimated 1.5 MTOE was consumed for domestic heating. In these sectors, considerable potential exists for small-scale development of S.R.F. energy production, as a farm-based or co-operative enterprise. However, such developments will come, only as economic pressures are applied with the escalation in costs of the more convenient energy forms, gas, oil and electricity.

3. CONCLUSIONS

At present yield levels, S.R.F. biomass can satisfy the limiting land availability, economic and market criteria to provide a significant proportion of Ireland's energy requirements. However, further research is necessary to establish this technology, by increasing fuel yield potential and reducing production costs for both large and small scale applications. While the currently established modes of utilisation are electricity generation, domestic and industrial heating, the conversion of S.R.F. biomass to transportation fuels may become important during the coming decades.

In conclusion, S.R.F. biomass can make an impact on future national energy supplies only if an early commitment is given to greatly increase the rate of afforestation and to encourage the development of low grade land areas as energy plantations.

REFERENCES

(1) Neenan M., and Lyons G.,
Energy from Biomass : The Production of energy by photobiological methods. An Foras Taluntais report on contract Nos. 055-76-ESEIR December 1977.

(2) O'Connell L., and Gallagher P.,
'Combustion of green wood chips on a chain grate stoker'. Ceneca Agriculture Energy Conference, Paris, February 1980.

(3) Lunny F.,
Personal communication. Energy Division, National Board for Science and Technology, Shelbourne House, Shelbourne Rd., Dublin. October 1980

(4) Neenan M., and Gardner M.,
'Potential land availability for biomass' in Production of Energy from Short Rotation Forestry, M. Neenan and G. Lyons (eds.), An Foras Taluntais. October 1980. In press.

(5) Lyons G., and Vasievich M.,
'Economics of short - rotation forests for energy in Ireland'.

October 1980.   In preparation.

(6) Lalor E.,

Solar Energy for Ireland.   National Science Council report.
Stationary Office, Dublin, pp. 40 - 41, 1975.

(7) Henry E. W., and Scott S.,

A national model for fuel allocation - a prototype.   The Economic
and Social Research Institute, Paper No. 90, Dublin.   September
1977.

(8) Department of Industry, Commerce and Energy.

Energy in Ireland.   Dublin, 1978.

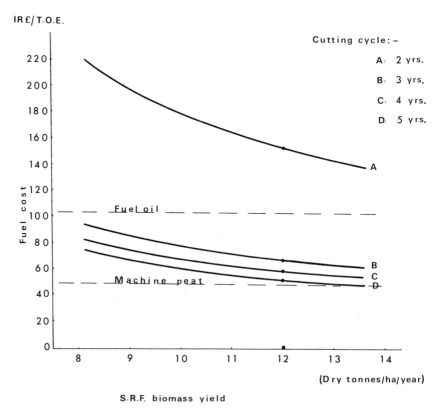

Fig:1   Comparison of primary fuel costs for oil, peat and S.R.F wood chips.

FOREST BIOMASS AS A SOURCE OF ENERGY IN THE UK;
THE POTENTIAL AND THE PRACTICE

C P MITCHELL and J D MATTHEWS

Department of Forestry, Aberdeen University,
St Machar Drive, Aberdeen, UK.

Summary

The purpose of the study was 1) to assess the feasibility of using forest biomass from single-stemmed trees as a source of energy in the United Kingdom and 2) to identify the silvicultural, managerial and economic problems of growing trees on short rotation for energy.

In UK forestry two existing sources of wood could be used directly as solid fuel or be converted into secondary gaseous or liquid fuels. These sources are forest residues and the produce from early thinnings. In future, supplies of forest biomass for fuel might be obtained from a third source, namely energy plantations.

Recoverable forest residues are estimated to be 1.5 Mdt in 2000 or, in energy terms, 0.7 MTOE. Whole-tree harvesting of early thinnings could yield 600 Kdt/yr by 2000 or, in energy terms, 0.3 MTOE/yr. Energy plantations using single-stemmed trees are expected to yield 8-12 dt/ha/yr on rotations of 12 to 20 years. The area of land suitable for energy plantations is estimated at 1.5 Mha which could realize 10 Mdt/yr or, in energy terms, 4.4 MTOE. Thus forest biomass from single-stemmed trees could produce 5.4 MTOE/yr by 2000.

In order to identify the problems of growing trees for energy forest energy plantations using ten species with good potential for biomass production in Britain will replace three types of land use:- under-productive agricultural land, old woodland and young forestry plantations. These will be cleared and the ground prepared and planted at 1 m spacing between plants. These plantations will be established in four geographic regions of Britain. Methods of harvesting single-stem and coppiced forest energy plantations as well as old woodland will be investigated.

The results of this assessment indicate that the use of forest biomass for energy in the UK is potentially viable. The results of the trials will lead to a series of recommendations for implementing short rotation forest energy plantations.

1. INTRODUCTION

In Britain very little natural forest cover remains. The majority of the 1.7 Mha of existing forest consists of man-made conifer plantations. The forest area of Britain per head of population is one of the lowest in Europe while consumption of wood is among the highest. Ninety two per cent of current annual consumption of wood and wood products is imported.

In British forestry there are two existing sources of wood which could be used directly as solid fuel or could be converted into secondary gaseous or liquid fuels (Table I).

Table I Sources of Forest Biomass for Energy

Source	Forestry System
1. Forest Residues	Existing plantings/ Existing management
2. Early Thinnings	Existing plantings/ New management/ Whole-tree harvesting
3. Energy Plantations	New plantings/ New management

Forest residues arise in the forest during harvesting. The present markets for early thinnings are the pulp and board industries but these markets are weak. Supplies of forest biomass for fuel might be obtained from energy plantations. This paper examines the potential of each of these sources and describes work proposed on the production of woody biomass from single-stemmed trees grown on short rotation.

2. FOREST RESIDUES

In the current harvesting of trees in Britain only the merchantable stem to 7 cm top diameter, comprising 55 per cent of the total biomass, is extracted from the forest. The remaining 45 per cent, consists of tops, branches and foliage (20 per cent), stumps and roots (20 per cent) and sawdust (5 per cent), normally is left in the forest. Not all residues generated by normal harvesting practice are recoverable. Twenty per cent of roots and stumps are potentially harvestable but this proportion is not expected to increase because of the steep terrain and inaccessibility of sites to suitable harvesting equipment. Thirty per cent of tops and branches could be harvested now and this could rise to 75 per cent by 2000 as harvesting machines and systems are improved.

Wood production forecasts for Britain are made by the Forestry Commission. When used in conjunction with the proportions given above these forecasts can be used to estimate the quantities of forest residues arising (Table II).

Table II   Forest Residues in United Kingdom

Forest Residues	Arising		Recoverable	
	1977 Estimate	2000 Projection	1977 Estimate	2000 Projection
	(Kt dry/ann)		(Kt dry/ann)	
Roots + Stumps	700	1600	140	320
Tops + Branches	700	1600	231	1200
	1400	3200	371	1520

In 1977 about 1.4 Mdt of forest residues were generated although only 370 Kdt could have been harvested. Due to the age structure of British forests and the intensive planting programme since 1950 the quantity of forest residues arising will more than double to 3.2 Mdt/yr by 2000. Of this 1.5 Mdt or, in energy terms, 0.7 MTOE are considered to be recoverable.

3. EARLY THINNINGS

Thinning is an essential part of the management of British plantations because the prime object of silviculture is to produce sawlogs. At present the market for early thinnings is weak and there is a tendency not to thin. An additional outlet for the produce of early thinnings could be fuel. Whole-tree harvesting of early thinnings could yield 300 Kdt/yr at present, rising to 600 Kdt/yr or, in energy terms, 0.3 MTOE, by the year 2000. As such material can command a similar price for energy as for existing markets then, if a strong fuel market were developed some of the 0.7 Mdt/ yr of short roundwood predicted for the year 2000 could find its way into the energy market rather than the pulp and board industries.

In order to provide more precise information to managers on yields from whole-tree harvesting of early thinnings biomass yield tables are being produced for ten species for energy (1).

4. ENERGY PLANTATIONS

Forest energy plantations are designed to produce biomass for energy
in the shortest possible time.   Most interest has centred around very short
rotations of less than 5 years using hardwood coppice.   Coppice production
in Britain tends to be limited to fertile lowland and sheltered sites.
The availability of this type of land is limited and the concept of short
rotation energy plantations has been extended to cover rotations of up to
20 years using single-stemmed trees which can be established on less fertile
sites in the lowlands and uplands of Britain.   Yields from complete-tree
harvesting of trees grown on rotation of up to 20 years range from 8 to
12 dt/ha/yr (Table III) - length of rotation depends on species and site.
A factor limiting

Table III   Estimated Yields from Selected Species on Short Rotation (dt/ha/yr)

Species	Location	
	Lowland sites	Upland Sites
Nothofagus spp	14	6
Douglas fir	11	6
Corsican pine	9	5
Sitka spruce	9	5
Western hemlock	9	5
ash	8	-
birch	9	7
Japanese/hybrid larch	8	7

development of energy plantations in Britain is availability of land.
Mitchell & Matthews (2) have estimated that 1.5 Mha of land is potentially
available of which 0.5 Mha comprises small areas of unused or under-used
land such as field and stream margins, rough grazing, unmanaged scrub
woodland in the lowlands, and 1 Mha of unused or under-productive marginal
agricultural land in the uplands.   This land if used for single-stem
forest energy plantations could produce 10 Mdt/yr or, in energy terms,
4.4 MTOE/yr.

In order to identify the problems of growing single-stem trees for
energy on such land forest energy plantations using ten species with
good potential for biomass production in Britain (birch spp, sycamore,
alder spp, hybrid larch, Nothofagus spp, Douglas fir, Corsican pine,
Scots pine, Western hemlock and Sitka spruce) are being established in
replicated trials in an experimental programme conducted by Aberdeen

University funded jointly by CEC and UK D. Energy. These plantations will replace three types of present land-use:- under-productive agricultural land, scrub woodland and young forestry plantations (the inclusion of the last category will permit the problems anticipated with second rotation energy plantations to be studied). These sites will be cleared prior to establishment, the ground prepared and planted at 1 m spacing between plants. The plantations will be established in four geographic regions of Britain:- NE Scotland, the border between England and Scotland, the border between England and Wales, and the SW peninsular of England. Methods of harvesting single-stem and coppiced forest energy plantations as well as old woodland will be investigated during the experimental programme.

5. CONCLUSIONS

The results of the assessment indicate that the use of forest biomass for energy in Britain is potentially viable for the following reasons;

- residues and early thinnings are two sources of biomass for fuel which exist now
- recoverable forest residues could yield 1.5 Mdt/yr (0.7 MTOE) by 2000
- biomass from early thinnings could yield 0.6 Mdt/yr (0.3 MTOE) by 2000
- forest energy plantations using single-stem trees could yield 10 Mdt/yr (4.4 MTOE)
- costs of all three sources are estimated to be about £1/GJ which makes them competitive with fossil fuels.

The results obtained from study of the trial forest energy plantation will lead to a series of recommendations for the implementation of short rotation forest energy plantation.

REFERENCES

1. MITCHELL, C P; MATTHEWS, J D; MacBRAYNE, C G & PROE, M F (1980). Determination of Yield of Biomass from Whole-tree Harvesting of Early Thinnings in Britain. These proceedings.

2. MITCHELL, C P & MATTHEWS, J D (1979). The Potential of Forest Biomass as a Source of Energy with special reference to trees with normal single-stemmed habit of growth. Aberdeen University, Report to ETSU, Harwell.

SHORT ROTATION BIOMASS PRODUCTION OF WILLOWS IN NEW ZEALAND

R.L.Hathaway and C.W.S. van Kraayenoord

National Plant Materials Centre,
Ministry of Works and Development,
Palmerston North, New Zealand

Summary

New Zealand needs feedstock for conversion to gaseous and liquid fuels and is ideally suited to the production of biomass - especially mini rotation coppicing systems of willows and poplars. Biomass production of 50-60 t/fr.wt./ha/yr have been achieved with a number of clones of species of S.purpurea, S.viminalis and the new S.matsudana x alba hybrids, of which NZ 1002 produced best as 2 year old rods- equivalent to 62.4 tonnes fr.wt/ha/yr. Dry weight was about 45% for these hybrids and 50-54% for the other species.

The use of biomass as a feedstock for conversion to gaseous and liquid fuels is of considerable interest to New Zealand, where the continued supply of imported fuel is uncertain, and costs of energy are escalating rapidly.

New Zealand is ideally suited to the production of biomass for this purpose, as plant growth rates are relatively rapid, compared to many other energy deficient countries, and sufficient land area is available to allow such an industry.

Although it is likely that any energy conversion industry utilising woody material would initially be based on Pinus radiata, there are many areas in New Zealand ideally suited to the production of deciduous hardwoods such as willows and poplars, which can produce a greater level of biomass per unit area (under suitable conditions) than softwoods. They have the added advantage that they can be grown in a short - or mini-rotation coppicing system, without the need to replant after each harvest.

In order to evaluate the potential of such a system, in particular the levels of biomass production attainable per hectare, measurements have been made of the annual and biennial production from established willow beds, and a trial established to determine the effect of planting density and harvesting cycle (one or two years) on production.

Methods and Materials

Clones included in the trials, and planting densities are given in the following sections. All harvests were made in June or July (southern winter) after leaf fall was completed, and before flushing commenced in the spring.

Green weights of stem and branch material were determined immediately after harvesting, and a sample taken for the determination of percentage dry weight and total dry matter production.

Prior to the 1980 harvest of the spacing trial, measurements were also made of the number of live shoots, height of tallest shoot, and diameter of largest shoot for each stool sampled.

Results

A. Production from established beds

(I) The first measurements were made in 1969, on a willow bed which had been cut annually for two years. Each of the clones for which yield was recorded occupied only a single row in the bed, with one half of each row being planted at 0.3 m spacing, and the remainder of the row at 0.6 m. Distance between rows was 1.8 m. The soil type was a recent alluvial sandy loam, and mean annual rainfall is 1,000 mm.

The yield of 50 stools of each clone at each spacing was recorded. Results are given in Table I.

Table 1: Production of one year coppice growth on three year-old roots at Massey Plant Materials Centre Nursery (1969). (tonnes/ha/yr).

Clone	Spacing 0.3 m x 1.8 m (18518 per ha.)		Spacing 0.6 m x 1.8 m (9259 per ha.)	
	Green wt	Dry wt*	Green wt	Dry wt*
S.caprea 'N'	27.3	15.2	16.6	9.2
S.daphnoides 'G'	35.5	19.7	34.3	19.1
S.discolor 'B'	45.5	25.3	30.2	16.8
S.purpurea 'Booth"	33.2	18.5	28.1	15.6
S.purpurea 'PMC"	36.4	20.2	27.8	15.5
S.viminalis 'Gigantea"	51.4	28.6	53.8	29.9

Further measurements of a wider range of clones were made in 1977 of one and two year coppice growth.

Coppice growth on six year old roots at the National Plant Materials Centre Nursery. Total fresh weight production of 10 stools in a 50 stool row of each clone was recorded. Spacing was 0.45 m within rows, and 1.50 m between rows.

Table II. Production of one and two year coppice growth on six-year old roots (1977).

	Rotation length (yr)	Green weight per stool (kg)	Green weight per ha. per yr (t)	% dry weight	Dry weight per ha. per yr. (t)
Acutifolia	1	1.80	26.40	56.7	15.0
Daphnoides 'G'	1	2.30	33.73	56.1	18.9
Discolor 'B'	1	2.55	37.40	48.7	18.2
Incana	1	1.30	19.07	56.7	10.8
Incana x Acutifolia	1	1.92	28.16	56.5	15.9
Purpurea 'Booth'	1	2.35	34.47	53.1	18.3
Purpurea 'Goldstones'	1	1.30	19.06	58.4	11.1
Purpurea 'Holland'	1	1.60	23.47	55.3	13.0
Purpurea 'Holland'	2	3.00	22.00	55.3	12.2
Purpurea 'Irette'	1	3.45	50.60	51.4	26.0
Purpurea 'PMC'	1	1.95	28.60	58.6	16.8
Purpurea 'Shultz'	1	1.00	14.67	63.1	9.3
Purpurea 'Shultz'	2	2.35	17.24	55.1	9.5
Purpurea 'Denmark'	2	4.90	35.93	56.5	20.3

B.   Trial to determine effect of planting density and harvesting cycle on biomass production of six salix clones.

The high biomass production recorded in the established beds prompted the establishment in 1977 of a replicated trial to compare the production of 6 clones at 5 spacings and 2 harvesting cycles. As well as the higher producing osiers and sallows for which yield had been measured in the established beds, two salix matsudana x alba hybrids produced at the National Plant Materials Centre which showed very rapid early growth rates were also included. The trial was planted using 25 cm cuttings in July 1977 and all shoot growth cut back to 5 cm above ground level in July 1978. One half of the trial (2 replicates) was harvested in 1979 after one years growth (Table III), and the whole trial harvested in 1980, giving further data on one year growth, and the first data on 2 year yields (Table IV).

Table III. Results of first year harvest from spacing trial.

	Spacing (m)			Tonnes/ha Green Wt)	
	0.3x0.3	0.3x0.6	0.3x1.2	0.6x1.2	1.2x1.2
S.Daphnoides 'G'	28.1	24.8	18.7	15.3	7.9
S.Purpurea 'Booth'	35.6	41.9	31.5	24.8	18.6
S.purpurea 'Irette'	43.9	41.6	47.1	37.2	26.4
S.viminalis 'Gigantea'	35.9	43.4	46.0	43.0	37.5
S.matsudana x Alba 'NZ1002'	37.8	37.5	46.7	46.1	41.0
S.matsudana x Alba 'NZ1046'	38.6	42.0	54.9	47.6	42.2

one year old rods

Table IV. Results of second year harvest from spacing trial.

Species/ Clone	Spacing (m)	Fresh wt[1] per ha (tonnes)	%age Dry wt	1 year Rotation Number of live shoots per stool	Height tallest shoot per stool(m)	Diam. of largest shoot per stool (cm)
Salix daphnoides 'G'	0.3x0.3	35.1	52.2	3.96	2.93	1.15
	0.3x0.6	29.4	53.9	5.09	3.31	1.48
	0.3x1.2	30.1	50.3	7.25	3.73	1.46
	0.6x1.2	17.8	49.9	9.85	3.83	1.48
	1.2x1.2	17.7	55.2	12.90	3.82	1.56
Salix purpurea 'Booth'	0.3x0.3	48.7	48.5	5.68	2.95	1.17
	0.3x0.6	51.2	48.6	12.00	3.10	1.34
	0.3x1.2	38.6	49.2	15.44	3.26	1.46
	0.6x1.2	29.7	47.4	25.25	3.28	1.53
	1.2x1.2	31.6	50.1	33.70	3.21	1.66
Salix purpurea 'Irette'	0.3x0.3	45.1	51.4	5.64	2.75	1.00
	0.3x0.6	49.1	52.2	9.05	3.07	1.20
	0.3x1.2	58.7	52.5	18.25	3.21	1.39
	0.6x1.2	44.7	53.9	31.95	3.32	1.53
	1.2x1.2	44.5	51.6	41.25	3.54	1.62
Salix viminalis 'Gigantea'	0.3x0.3	38.5	49.0	2.72	3.16	1.36
	0.3x0.6	39.2	48.9	4.74	3.53	1.78
	0.3x1.2	51.3	47.5	9.72	3.94	2.86
	0.6x1.2	42.0	46.5	17.55	3.95	2.91
	1.2x1.2	47.5	47.9	24.45	4.01	2.92

	Spacing (m)	Fresh wt[1] per ha (tonnes)	%age Dry wt.	Number of live shoots per stool	Height tallest shoot per stool(m)	Diam. of largest shoot per stool (cm)
Salix matsudana x Alba 'NZ1002'	0.3x0.3	47.4	41.4	2.78	3.00	1.34
	0.3x0.6	53.9	43.0	4.32	3.04	1.82
	0.3x1.2	56.4	43.8	6.75	4.13	2.44
	0.6x1.2	41.9	39.8	9.90	4.06	2.42
	1.2x1.2	45.1	43.8	16.85	4.27	2.81
Salix matsudana x Alba 'NZ1046'	0.3x0.3	52.0	45.7	2.57	3.00	1.20
	0.3x0.6	45.4	50.0	3.87	3.40	1.68
	0.3x1.2	45.2	47.8	7.38	3.46	1.86
	0.6x1.2	40.8	48.3	14.80	3.79	2.13
	1.2x1.2	49.6	46.1	25.69	3.93	2.25

[1] Stem and branch material only.

		two year old rods		2 year Rotation		
Salix daphnoides 'G'	0.3x0.3	54.3	49.1	3.41	3.18	1.44
	0.3x0.6	44.0	49.4	4.42	3.53	1.55
	0.3x1.2	43.8	50.6	4.8ε	4.21	2.00
	0.6x1.2	26.9	50.7	6.05	4.73	2.45
	1.2x1.2	24.5	52.1	5.30	4.51	2.4l
Salix purpurea 'Booth'	0.3x0.3	47.9	52.0	4.31	3.18	1.58
	0.3x0.6	47.8	48.6	3.34	3.83	2.02
	0.3x1.2	40.2	50.8	4.28	4.22	2.44
	0.6x1.2	26.2	52.4	5.40	4.73	3.19
	1.2x1.2	33.9	46.0	11.15	5.05	4.05
Salix purpurea 'Irette'	0.3x0.3	70.9	51.6	3.38	3.58	1.39
	0.3x0.6	73.4	54.3	5.40	3.77	1.70
	0.3x1.2	77.9	54.0	9.97	4.10	1.11
	0.6x1.2	67.8	55.6	13.13	4.54	2.41
	1.2x1.2	75.3	53.9	18.31	4.39	2.57
Salix viminalis 'Gigantea'	0.3x0.3	87.8	50.3	2.88	4.24	1.73
	0.3x0.6	77.2	50.7	3.75	4.60	1.92
	0.3x1.2	53.7	50.7	6.22	4.97	2.28
	0.6x1.2	79.7	51.3	10.75	5.46	2.53
	1.2x1.2	73.0	52.7	13.80	5.03	2.59

	Spacing (m)	Fresh wt[1] per ha (tonnes)	%age dry wt.	Number of live shoots per stool	Height tallest shoot per stool(m)	Diam. of largest shoot per stool (cm)
Salix matsudana x Alba 'NZ1006'	0.3x0.3	93.6	44.7	1.76	3.85	1.78
	0.3x0.6	124.8	45.6	2.02	4.69	2.72
	0.3x1.2	108.4	45.3	3.78	5.14	3.44
	0.6x1.2	112.0	45.6	6.10	5.87	4.58
	1.2x1.2	112.0	45.2	7.85	6.22	5.82
Salix matsudana x Alba 'NZ 1046'	0.3x0.3	96.3	47.5	1.41	3.69	1.86
	0.3x0.6	87.7	47.1	1.96	4.41	2.42
	0.3x1.2	82.2	47.0	3.16	4.58	2.91
	0.6x1.2	74.9	47.2	3.85	5.50	3.99
	1.2x1.2	94.9	49.6	5.95	5.97	5.29

[1] Stem and branch material only

Discussion and Conslusions

From these results it appears that biomass production levels of 50-60 t/ha/yr (green weight) can be readily achieved with a number of willow clones grown in a short rotation coppicing system on good soil types. This is considerably higher than the production of other tree species grown in a conventional sawlog rotation, although it is quite possible Eucalyptus species could produce similar levels of biomass when grown in very short rotations.

Spacing and harvesting cycle obviously have a large effect on production, and although not determined quantitatively, a clone x spacing x harvesting cycle interaction is evident.

When harvested annually, the higest producing clone is S.purpurea 'Irette' (58.7 t/ha/yr at 28,000 plants/ha in the spacing trial, Table 4). When harvested biennially however, maximum production for this clone was 39.5 t/ha/yr at the same spacing, only 67% of the annual harvest production.

The clone S.matsudana x Alba 'NZ 1002' produced best in the 2 year rotation, (62.4 t/ha/yr) although this was only slightly greater than the 56.4 t/ha/yr produced when harvested annually.

Planting density had a large effect on production when the crop was harvested annually, but much less effect when harvested biennially. This is important in terms of planting and harvesting costs, as less plants are required to obtain similar production levels. Rotations longer than two years may be even more economic, although the need for heavier harvesting machinery would have to be taken into consideration.

It should be pointed out that these trials were conducted on first class soils under near optimum conditions, and the biomass production obtained would be near the upper limit for the species. In New Zealand, sites suited to large scale production of willows for conversion to energy would probably be restricted to less fertile, high water table sites, the use of better soil types would depend on the economics of the system, and the return available to growers. These trials have shown the potential of willows for biomass production in New Zealand, but extensive further work will be required before any definite conclusions can be drawn as to the viability of growing willows for energy production.

FOREST ENERGY CROPS FROM DERELICT AND WASTE LAND

V.N. DENNINGTON and M.J. CHADWICK

Derelict Land Reclamation Research Unit, Department of Biology
University of York, York, United Kingdom

Summary

Land use competition in the United Kingdom means that it is
unlikely that large areas will be devoted to forest energy crops.
However, derelict land amounts to at least 71,000 ha to which may be
added 260,000 ha of associated waste and degraded land.   Much of this
has potential for forest energy crop plantations although some sites
comprise difficult substrates.   This study investigated tree growth
and standing crop on a range of sites as well as determining the weight
proportions between trunk, branches and leaves.   The chemical composition of leaves and wood was ascertained.   Average annual production
varied between 0.52 (Betula pendula on railway land) and 11.22 t/ha/yr
(Alnus glutinosa on pulverized fuel ash).   Harvesting a total crop of
4.0 t/ha could remove over 36 kg/ha of nitrogen as well as 1.5 and
over 12 kg/ha of phosphorus and potassium from the site.   If the crop
were taken after leaf fall these amounts could be significantly reduced.
Even so nutrient reserves could be rapidly depleted.   Coppicing combined with regular fertilizer applications and the use of nitrogen
fixing species offer a possible method of exploiting these sites for
energy crops.

1. INTRODUCTION

If forest crops are to be grown for energy in Britain this will require a considerable area of land. There is increasing competition already for land for agriculture, forestry, housing, services, industry and amenity uses. However, there exists in Britain a substantial area of derelict and waste land which is at present virtually unused and which may have a potential for the growth of energy crops.

2. DERELICT AND WASTE LAND IN BRITAIN

Government surveys identified 71 155 ha of land which fell within the official definition of dereliction (Table I). This land arises mainly from the mineral working and processing industries, rail and military closures and industrial decline. A number of local authorities have, however, carried out independent surveys covering a broader spectrum of derelict and waste land and have identified a larger area (Table II). This is due mainly to the inclusion of degraded sites such as neglected woodland, farmland and public open space, disused allotments, vacant sites, overgrown tips and spoil heaps and general waste land. This land is often associated with derelict land in urban and industrial areas. The total area of derelict and waste land in Britain may therefore be estimated at about 331 000 ha (Table III). 250 000 ha may have a potential for growing trees for energy purposes.

3. ASSESSING TREE GROWTH ON DERELICT AND WASTE LAND

During the last 20 years in Britain considerable expertise has been built up on the establishment and maintenance of trees on restored sites. Some of these sites were used in the present study to assess the yield of trees on derelict land.

Sites covering a range of substrates, and a wide climatological and geographical range from the North East to the South West of the country, were considered. The sites included two naturally overgrown areas of disused railway land and others on restored colliery spoil, pulverized fuel ash, a sand and gravel working, land contaminated by heavy metals and land degraded by atmospheric pollution. Trees included in the study cover a range of hardwood and softwood species.

Table I. The total area of derelict land in Britain

Country	Area of derelict land (ha)
England	43 273
Wales	14 478
Scotland	13 404
Total	71 155

Table II. Areas of derelict land from Government surveys and local authority estimates of derelict and waste land

Category	Location	Area ha	
		Derelict and waste land Local Authority data	Derelict land Government 1974 surveys
Industrial	Merseyside	3 517	529
	South York-shire	8 730	1 565
	Gwent	1 765	2 762
Urban	Southwark	341	27
	Tower Hamlets	193	0
Industrial/ Rural	Northumberland	2 413	2 411
	Derbyshire	2 419	1 797
Rural	Dorset	> 510	510

Table III. Estimated total area of derelict and waste land in Britain

Country	Area of derelict and waste land (ha)
England	200 000
Wales	59 000
Scotland	72 000
Total	331 000

The tree standing crop at each site was determined by a combination of destructive and non-destructive measurements. Tree numbers and size

distribution (diameter at breast height) were determined in sample
plots.   About 10 trees covering the size range were felled (at ground
level) and the fresh weight of trunk and branches + leaves determined
in the field.   The age of each tree was determined from annual rings.
Leaves were removed from sample branches and the fresh weights of
leaves and wood determined separately.   A ratio was then derived
relating the fresh weight of leaves to that of the total branch sample.
Sub-samples of trunk, branch and leaf components were returned to the
laboratory for determination of moisture content.

A second degree polynomial of the form $y = a + bx + cx^2$, where
y = fresh weight, x = diameter at breast height and a, b and c are
constants, was found to describe the relationship between total fresh
weight and diameter and trunk fresh weight and diameter for almost all
sites.   The curves differ between sites, emphasising the importance of
site characteristics in determining yield.   These curves, together with
the size distribution data, were used to determine the total fresh
weight of the standing crop and the relative contributions of trunk,
leaves and branches to the total (Table IV).   The data in Table IV show
considerable differences in the productivity of various substrates.
This is partly due to variations in tree density, age of standing crop
and climatic factors but substrate factors such as toxicity, acidity and
nutrient status also play an important part.

4. THE MANAGEMENT OF WASTE LAND SITES FOR ENERGY CROPPING

Tree crops for energy on derelict land could be conveniently
managed by coppicing.   A series of short rotation crops could be
harvested from a single planting.   This not only has economic
advantages but also minimises the disturbance by machinery on sites
where surfaces may be unstable or liable to compaction.   In a study of
12 species coppiced on a colliery site in West Yorkshire almost 100%
regeneration was obtained from the stumps of trees harvested 30 cm
above ground.   Of the hardwoods Alnus glutinosa, Salix caprea and
Populus alba produced the most shoots per stump.

Analyses of the heavy metal content of tree tissue in this study
suggest that concentrations are unlikely to present a significant
problem even in trees from contaminated sites such as the Lower Swansea

Table IV. The standing crop and average annual production on derelict and waste land sites

Species	Site	Substrate	Dry weight of standing crop t/ha				Average annual production t/ha/yr
			Trunk	Branches	Leaves	Total	
Alnus glutinosa	Mitchell	colliery spoil	19.31	3.40	2.41	25.12	1.67
	Choppington	colliery spoil	2.45	2.35	1.93	6.73	0.39
	Tir John	PFA	87.69	15.44	31.46	134.59	11.22
A. glutinosa + Salix caprea	Roundwood	railway land	13.60	7.58	4.03	25.39	2.30
Betula pendula	Rowsley	railway land	2.04	2.12	1.02	5.18	0.52
	Halewood	railway land	21.08	11.57	5.48	38.13	3.81
	Drakelow	PFA	27.07	11.50	6.72	45.29	4.12
Populus robusta	Drakelow	PFA	32.33	11.00	5.40	48.73	4.43
	Tir John	PFA	63.59	7.59	13.83	85.01	7.08
Larix leptolepis	Bedwas	colliery spoil	31.46	11.40	3.45	46.31	2.21
Pinus sylvestris	Kilvey Hill	degraded land	2.85	3.87	4.04	10.76	1.08
	Trentham	sand & gravel working	13.83	14.38	10.52	38.73	2.58
Pinus contorta	Lower Swansea Valley	contaminated land	37.00	21.55	13.11	71.66	5.12

Valley.   Many derelict sites have undergone severe disturbance and
have a poor nutrient status.   Repeated harvesting may significantly
deplete the soil nutrient pool and this study has examined the nutrient
export associated with harvesting coppiced crops (Table V).   A

Table V.   The export of major nutrients associated with harvesting a
crop of 4 dry tonnes/hectare at the end of the growing season

Species	Site	Substrate	Nutrient	Nutrient content kg/ha		
				Leaves	Wood	Total
Alnus glutinosa	Mitchell	colliery spoil	N	9.49	10.07	19.56
			P	0.62	0.17	0.79
			K	3.26	5.83	9.09
	Choppington	colliery spoil	N	26.38	9.55	35.93
			P	1.84	0.23	2.07
			K	8.95	7.29	16.24
	Tir John	PFA	N	28.52	7.67	36.19
			P	1.40	0.11	1.51
			K	6.92	5.45	12.37
	Roundwood	railway land	N	17.96	9.55	27.51
			P	1.26	0.48	1.74
			K	5.90	6.93	12.83
Betula pendula	Rowsley	railway land	N	8.43	9.77	18.20
			P	1.65	0.17	1.82
			K	5.28	5.93	11.21
	Halewood	railway land	N	7.70	9.85	17.55
			P	1.61	trace	1.59
			K	5.58	5.27	10.85
	Drakelow	PFA	N	5.82	7.87	13.69
			P	1.37	0.10	1.47
			K	4.51	5.05	9.56

considerable conservation of nutrients may be achieved by harvesting
after leaf fall since about half the N and K and 90% of the P of the
total crop occurs in the leaves.   Since the leaves represent on average
less than 20% of the total dry weight of the crop this conservation can
be achieved by a relatively small reduction in the total dry weight
harvested.   From limited evidence for colliery spoil and PFA sites it is
clear that repeated harvesting could only be sustained if soil nutrient
status is maintained by regular fertilizer additions and the use of
nitrogen fixing species.

SELECTION D'EUCALYPTUS ADAPTES AUX CONDITIONS
FRANCAISES EN VUE DE L'INSTALLATION DE
TAILLIS A COURTES ROTATIONS

(Breeding eucalyptus for short rotation coppice in France)

J.N. MARIEN
Association Forêt-Cellulose - AFOCEL
5 rue des Palombes
34000 Montpellier - FRANCE

Le but est d'obtenir des clones d'Eucalyptus à la fois résistants au froid (jusqu'à - 15° C) et à croissance rapide, pour les utiliser sous forme de taillis à courtes rotations. La production attendue de telles plantations est de 20 m^3/ha/an, soit 12 à 15 tonnes de matière sèche.

Pour cela, l'utilisation du phénomène d'hétérosis provoqué par l'hybridation interspécifique entre E. gunnii (résistant au froid) et E. dalrympleana (à croissance rapide) constitue la base du programme. Les graines obtenues sont semées en pépinière et triées après 2 hivers sur leur résistance au froid et leur vigueur juvénile. Les clones retenus sont alors rajeunis et propagés végétativement, soit par bouturage herbacé classique, soit in vitro. La dernière étape est constituée par la mise en place de tests clonaux.

Après la première campagne complète, on a obtenu les résultats suivants :

- nombre de plants avant gelées...................... 2 673   100 %

- nombre de plants résistants au gel.................   376   14 %

- nombre de plant après sélection sur la vigueur......   13   0,5 %

- gain sur la hauteur des 13 clones sélectionnés
  par rapport à la moyenne des plants résistants...... $\dfrac{218}{106}$ 206 %

SUMMARY

To get frost resistant Eucalypts for planting in France, we use interspecific hybridations. The seeds are sown and screened for juvenile growth and frost resistance in a high altitude nursery. The selected clones are propagated by vegetative propagation and compared in clonal tests.

SESSION III: FERMENTATION TO ETHANOL AND BIOGAS

Chairman: Prof. H. NAVEAU
           Biotechnology Unit, Catholic University
           of Louvain, Belgium

Summary of the discussions

Invited papers

Anaerobic digestion for energy saving and production

Ethanol - The process and the technology for production of
liquid transport fuel

Poster papers

HF saccharification: the key to ethanol from wood?

Ethanol production by extractive fermentation

Agricultural waste treatment by means of ultrafiltration membrane
enzymatic reactors

1-thio-oligosaccharides as substrate analogs for investigations
in the field of polysaccharidases

Degradation of cellulose with hydrogen fluoride

Cellulase and β -glucosidase from clostridium thermocellum
excretion, localization and purification

Continuous autohydrolysis, a key step in the economic conversion
of forest and crop residues into ethanol

Production of ethanol from wastes via enzymatic hydrolysis of
cellulose

Energy production from whey

Separation of lignocelluloses into highly accessible fibre
materials and hemicellulose fraction by the steaming-extraction
process

Ethyl alcohol production for fuel: energy balance

A low energy system for the production of biomass ethanol

The cellulolytic community of an anaerobic estuarine sediment

The use of phase-plane analysis in the modelling ob bio-metha-
nization processes in order to control their evolution

Application of microcalorimetry to the study of the fermentation
of lignocellulosic compounds

Occurrence of cellulolysis and methanogenesis in various ecosystems

Production of methane from freshwater macro-algae by an anaerobic
two step digestion system

Operation of a laboratory-scale plug flow type of digester on pig
manure

Fermentation to biogas using agricultural residues and energy
crops

Horizontal-rotating-drum continuous fermentor

Bioconversion of agricultural wastes into fuel gas and animal
feed

Proposals for the analytical modelling of a digester

Computer aided design of anaerobic digesters for energy production

A novel process for the anaerobic digestion of solid wastes
leading to biogas and a compostlike material

Application of recent developments in anaerobic waste water
treatment to slaughterhouse wastes (Germany)

Findings of B.A.B.A.'s subcommittee to produce a code of
practice on safety in and around anaerobic digesters

An assessment of some commercial digesters in the United Kingdom

Economic aspects of biogas-production from animal wastes

Methane fermentation in the thermophilic range

261

SUMMARY OF THE DISCUSSIONS

Session III   FERMENTATION TO ETHANOL AND BIOGAS

Rapporteur :   T.K. GHOSE, India

Speakers :   G. LETTINGA, Netherlands
                  J. COOMBS, U.K.

Chairman :   H. NAVEAU, Belgium

Poster Session : 32 papers presented

The presentation of LETTINGA reviewed the various possibilities of biomethanation of biomass via thermophilic, mesophilic and cryophilic bacteria. He dealt with the inhibition compounds present in the substrates
or produced during the process of biomethanation. He described the two
phase process involving liquefaction of complex substrates into volatile
fatty acids and bio-conversion of the acids into methane and carbon dioxide. Amongst the substrates referred to, stillage from ethanol fermentation was mentioned as the most important waste generated by the ethanol
process for use as an energy source. A number of digestor systems were
discussed, the drawbacks identified and the possible improvements indicated. Various experiences in this area were indicated. The important
aspects which came out of the paper included information on low digesta-
bility of substrates, low tolerance of volatile fatty acids by the microflora, slow digestion rates inherent in the process, low methane content
of the gas produced, and the disposal of the digested solids. During the
discussion, different views were expressed in favour of two-phase systems
(liquefaction and biomethanation) and single-phase systems in which both
these processes are carried out together. The speaker was in favour of
a single system while a number of participants felt the necessity, under
site specific circumstances, of having the process segregated into two
phases. This view was also supported by an argument in favour of

possibilities which exist for constructing digestors with cheap materials, thereby reducing the installation cost and running costs. Amongst the liquefaction methods mentioned by the speaker, were included enzyme liquefaction, chemical and thermochemical conversion, heat treatment and the gamma radiation with aeration, mechanical treatment like ball milling, etc. It was mentioned that thermophilic biomethanation of biomass residues carried out between 55-60°C provides a gas productivity up to 3.5 m^3 of gas per digestor volume per day as against 2.5 m^3 per day by mesophilic digestion at 30-35°C.

A comment was made that the expression of the amounts of volatile matter converted into methane needs to be expressed as the fraction which undergoes digestion, rather than the amount of volatile matter added to the system. Another important point raised by one participant was on the theoretical limit of a microbial cell in the digestion system to convert biomass into methane. It was revealed during the discussion that these aspects need to be studied along with other aspects of improvement in the reactor design and performance, the use of mixed substrates which are compatible with each other, and segregation of the system into acid forming and methane generating phases. There is also considerable scope for improving such systems in which microflora are fixed on inert solid supports which presently in most cases of residue conversion are provided by the biomass itself.

In COOMB'S presentation on ethanol production he expressed the opinion that methanol rather than ethanol would be the main approach for conversion of biomass into liquid fuel in Europe. During the course of the discussion several matters came up with respect to the availability of site specific raw materials for the conversion into ethanol ; some participants expressed the view that production of ethanol from starchy materials, cane molasses or lignocellulosics should be kept as options vis à vis methanol production from lignocellulosic materials because circumstances in various countries are so different that any one option may not be the preferred answer to the question of liquid fuel supply from biomass. It was also felt that ethanol production based on established knowledge can be easily implemented while methanol production systems presently require higher technology and capital inputs. The general opinion however was that a mix of both the processes should be considered. It was also felt that much more practical information and research was required on the lignocellulosic bioconversion into ethyl alcohol.

The session was supported by posters dealing with biomethanation of agriculture residues, wood and forest residues, saccharification of lignocellulosic substances, and ethanol production based on sugars, starches and lignocellulosics. Some of the poster papers like extraction of ethyl alcohol by non-toxic higher alcohols were discussed and hydroflouric acid treatment of lignocellulosics for the conversion into sugars. It was felt that these are important developmental inputs which require support if all the options are to be kept open for intermediate or long term development of technologies of biomass conversion into ethanol. A note of caution was expressed on the need of adequate understanding and application of the net energy ratio concept for conversion of biomass into fuels like methanol and ethanol. One important feature of biomass energy conversion processes was considered to be an integrated system which incorporates the uses of stillage from ethanol plants via production of algae and other aquatic plants which are then fermented to methane. An estimate was given that one billion cubic metres of biogas can be produced from the stillage of an ethanol plant producing one million cubic metres of ethanol per year.

The session concluded with a note of caution concerning the site specific nature of biomass substrate resources, the technology to be selected for the conversion of these resources into ethanol or methanol, and the existence of substancial scope in R&D&D on both chemical and biotechnology to determine appropriate selection of these conversion processes in the near future.

ANAEROBIC DIGESTION FOR ENERGY SAVING AND PRODUCTION

G. Lettinga

Department of Water Pollution Control
Agricultural University
De Dreijen 12
Wageningen

SUMMARY

A survey is presented of the present knowledge about the effect of important environmental factors on anaerobic digestion and of the state of art concerning the practical application of the process for both energy production and energy saving. Although additional research is still required in certain aspects, it may be foreseen that a significantly increased use will be made of the process in the near future both for wastewater treatment (energy saving + energy production) as well as for the mere production of energy.

1. INTRODUCTION

Although most of the available biomass has a fairly high energy content on a mass basis, the energy content on a volume basis frequently is very low, due to its high moisture content. This obviously is the case for domestic wastewater and many industrial effluents.

In order to be able to valorize wet biomass for energy generation, the conversion method to be choosen should have a low energy requirement relative to the gross energy produced, whereas the method should be inexpensive and simple, and applicable at both small as well as large scale. Considering these requirements anaerobic digestion at present looks the most appropiate conversion method, because the fuel produced (CH_4) escapes spontaneously from the system and the method is relatively cheap and simple.

Anaerobic digestion is certainly not a new development. As a matter of fact the method is already extensively applied since many years in municipal treatment plants for sludge stabilization. However, contrary to countries like China, the process thus far has not found extensive application for biogas production in the Western part of the world. It is only quite recently that in Western countries it has been recognized that anaerobic digestion also here has attractive potentials, both for energy saving as well as energy production.

2. THE ANAEROBIC DIGESTION PROCESS

Basically anaerobic digestion of complex organic matter consists of three successive steps (1, 10, 11), viz.:

a. hydrolysis, where organic polymers are hydrolized to their individual monomers as a result of enzymatic attack.

b. acid formation. The hydrolized compounds are converted intracellularly by a group of 'acid forming' bacteria to simple compounds, such as volatile fatty acids (VFA), CO_2, NH_3 and H_2.

c. methane fermentation. The simple compounds from the preceeding step are converted to methane and CO_2 by a group of stricktly anaerobic bacteria. As a matter of fact only a very limited number of substrates can be used directly by methane bacteria, viz. $CO_2 + H_2$, $CH_3 COOH$ and CH_3OH. The fermentation of the higher VFA and alcohols can only be accomplished by the aid of a group acetogenic bacteria (2).

In a well balanced digestion process these three separate steps proceed simultaneously.

Anaerobic digestion combines a high degree of waste stabilization with a very low production of bacterial matter. This holds especially for the growth of the methanogenic and acetogenic bacteria, consequently in the digestion of simple substrates like VFA and alcohols. A somewhat higher biomass yield is obtained in the digestion of more complex dissolved wastes, such as carbohydrates, due to the growth of the acidogenic bacteria in addition to the methanogenic and acetogenic bacteria.

A significantly larger quantity of solids remains in the digestion of complex solid organic matter, such as refuge, manure, sewage sludge and various types of plant material. This can be mainly attributed to the relatively high content of poorly and non-biodegradable (i.e. lignin-cellulose) matter in these materials. In conventional digestion processes the enzymatic attack is insufficient to hydrolyze these compounds within a reasonable period of time.

Important environmental factors

Temperature

Three temperature ranges can be distinguished i.e. thermophilic $(50-65^{\circ}C)$, the mesophilic $(20-45 ^{\circ}C)$ and the psychrophilic $(< 20 ^{\circ}C)$ range. Most experience exists for the mesophilic range.

Thermophilic digestion proceeds somewhat faster and is more complete than mesophilic digestion, but the process seems to be more sensitive under thermophilic conditions, e.g. for ammonia (3).

Conform thermophilic digestion relatively little is known about digestion under psychrophilic conditions. Results obtained with simple (VFA) and more complex substrates (e.g. dry solids of potato, domestic sewage) indicate that methanogenesis still proceeds at a reasonable rate under psychrophilic conditions, but that especially the hydrolysis step is affected by the temperature below 20 $^{\circ}C$.

Anaerobic digestion can resist sharp temperature fluctuations in the range 10-45 $^{\circ}C$, provided these fluctuations do not initiate other adverse environmental conditions. Temperatures in the range of 45-55 $^{\circ}C$ are quite detrimental for mesophilic cultures. This seems especially true for cultures which pass into the endogenic phase after a period of a high substrate level. No relevant information is available in the literature about the resistance of thermophilic cultures to temperature fluctuations.

pH and bicarbonate alkalinity are closely related factors. As acidic pH-conditions should be prevented to any price in cultures fermenting VFA, an adequate quantity of bicarbonate alkalinity should always be available. For cultures using merely methanol as substrate low pH-values (e.g. as low as pH 4.0) can be tolerated, provided methanol is fermented directly and not via the intermediate formation of VFA. Whether or not methanol is fermented directly seems to be related to the presence and availability of one or more trace elements (8).

Nutrients

An optimal performance of anaerobic digestion requires that all essential ingredients for bacterial growth are available (N, P, S and trace elements). The minimum requirements for N and P can be estimated from the growth yield and the bacterial composition. For cultures fermenting a mixture of VFA (sludge yield ~0.05) the minimum COD:N:P ratio is 1000:5:1, while for more complex substrates like carbohydrates (sludge yield ~ 0.15) it is 350:5:1. For most types of biomass this condition is met. In addition to NH_4^+-N and $PO_4^{'''}$-P some S should be present as SO_4^{2-}.

Toxic compounds

Anaerobic digestion is a rather sensitive process. Many compounds (e.g. $CHCl_3$, CH_2Cl_2, CCl_4, CN^- and free heavy metal ions) already are highly toxic at concentrations of approximately 1 ppm, while for others such as formaldehyde and H_2S this is the case at moderate concentrations (50-400 ppm).

Fortunately most of these compounds do not occur at toxic levels in most natural organic materials, and if so in many cases they may be eliminated as a result of precipitation reactions (i.e. heavy metals as sulphide precipitate) or stripped from the system by the biogas produced. Moreover, anaerobic cultures have the ability to acclimatize to a significant degree to most of these compounds.

As anaerobic digestion frequently is applied to concentrated feed stocks it should be recognized that some specific ingredients of the waste or metabolic intermediates or end products (e.g. NH_4^+-N) may reach critical levels. Although it has been postilated (10) that NH_4^+-N becomes toxic at concentrations beyond 3 g/l especially at pH-values above 7, a stable mesophilic digestion process can still be maintained at NH_4^+-N-concentrations as high as 5 g/l, provided the system has been allowed to acclimatize (e.g. 9).

3. ENERGY GENERATION and SAVING by ANAEROBIC DIGESTION

In applying A.D. for energy (CH_4) generation or energy saving it is of eminent importance to be able to conduct the process in simple, inexpensive and compact installations and in such a way that the smallest possible fraction of the gross energy produced is required for the operation of the process.

In order to arrive at a compact installation, the admissible loading rate of the digester should be as high as possible. Obviously in practice this can only be accomplished by maintaining a high concentration of viable bacterial mass in the digester. The conversion of the feedstock organics should therefore be as complete as possible and a high retention of viable biomass under conditions of high loading rates shoud be assured.

Unfortunately in various types of available complex feedstocks (i.e. refuge, sewage sludge, energy crops) non – or slowly biodegradable organics constitute up to 40-50 % of the total solids, and therefore these ingredients ultimately may contribute up to 90 % of the solids in the digested residues. On the other hand in the digestion of dissolved – more or less completely biodegradable feedstocks – a highly active sludge will be produced and the viable bacterial matter may constitute here up to 90 % of the digested solids. The mere practical difficulty in the latter case is to accomplish a high sludge retention, whereas in the digestion of complex solid feedstocks the main task is to improve the biodegradability of the organic matter.

In considering the digestion of complex feedstocks the application of a two-step process (i.e. with a separate liquefaction and methanogenesis step) might become beneficial, provided it can be shown that

    a. the overall investment and operation costs are comparable with that of the one-step process;

    b. the conversion of substrate organics to methane can compete with that of the more conventional approach.

It is obvious that little if anything will be gained when a high capacity methanogenesis step inherently is connected with the use of a relatively large reactor for liquefaction. As in the digestion of many complex feed-

stocks hydrolysis is the rate limiting step, optimalization of the
digestion process step desires an enhancement of the liquefaction step
in both approaches, the one and the two-step digestion process.

Thus far no evidence has been obtained that liquefaction is more
complete nor faster in using a separate liquefier. Within the framework
of the EEC-solar energy R&D programme attention will be paid to this
matter, especially to research concerning the enzymatic attack of more
complex feedstocks. Evidence of the potentials of enzymatic enhancement
of the liquefaction of cellulose residues has recently been reported (12).
For the digestion of 'solid feedstocks' such as straw and refuge an
attractive approach may be a 'solid bed liquefier', through which an
aqueous solution is percolated (i.e. preferentially the effluent of the
methanogenic step). Eventually this aqueous solution can be enriched with
appropiate enzymes.

Except by enzymatic and microbiological pretreatment a number of other
methods or combinations of methods are available to improve the biodegradability of biomass, e.g.:
Chemical pretreatment, e.g. with mineral acids or alkali, particularly
at elevated temperatures (thermochemical pretreatment) (13, 14).
Although these methods may lead to a substantial increase in the biodegradability, it is rather doubtful if they can be made economically
feasible, particularly for on farm application of digestion for energy
production purposes. The same applies for physical methods, such as heat
treatment and a combined process of aeration + β (or γ) irradiation (15).
Application of these methods seems to be limited to large scale digesters,
particularly if there is a need for a substantial reduction in the volume
of the solids (i.e. as in the case of sewage sludges), because in addition
to an improved biodegradability (up to 50 %) β-γ irradiation + aeration
results in an improved biodegradability of the sludge. In order to increase
the accessibility of the ligno-cellulosic complex, application of
mechanical methods such as ball milling may become attractive, because
they increase the rate of the digestion process.

3.1 Application of anaerobic digestion to low and medium strength
dissolved wastes

As mentioned before, in treating dissolved wastes the main task is
to achieve a high retention of viable biomass under conditions of high

organic and hydraulic loading rates. Up to recently such processes were
not available at full scale level, and consequently the use of anaerobic
digestion for treating low and medium strength wastes was rather hypothetical. Fortunately, the situation has changed in this respect considerably in the last years as a result of the development of new simple and
cheap processes such as the Anaerobic Filter (17) and the Upflow Anaerobic
Sludge Blanket (UASB-) process (4, 16). The UASB-process is applied at full
scale level since 1977. These methods may lead to a considerable save in
energy in waste treatment, i.e. up to 25 kWh (= 90 MJ).$PE^{-1}.year^{-1}$*
compared to the use of conventional aerobic activated sludge processes.
Moreover, anaerobic wastewater treatment also implies the production of
a substantial quantity of useful biogas, viz. up to 16 Nm^3 CH_4 (= 600 MJ).
$PE^{-1}.year^{-1}$ at a biodegradability of the organics of 70 %.

Applying anaerobic digestion as a waste treatment method, in various
cases fuel can be produced, whereas at the same time an effective and
cheap abatement of pollution is achieved without the use of high grade
energy. Therefore in this way waste may represent a source for fuel. As
an example the application of the process to the treatment of stillage
or bottomslops may be mentioned. The use of CH_4 produced from it may lead
to an energy saving of 30-40 % in the production of ethyl alcohol.
Considering that 10 m^3 CH_4 can be produced per m^3 stillage, the potential
methane production capacity from stillage in a country like Brasil with
an projected fuel alcohol production of approximately $4.5.10^6$ m^3/Y in
1986 amounts up to $0.6.10^9$ m^3/Y in 1986!

Of the processes developed thus far the UASB-process looks the most
elegant approach, because apart of its potentials the design and
construction of the process is simple (figure 1.) and the investment and
operation costs are relatively low.

Wastewater is flown upwards through a sludge blanket in the UASB-
concept, and gas and suspended solids are separated from the treated
liquid by means of a gas-solids separator device which is situated in the
upper part of the reactor. Contrary to ideas underlying convential
anaerobic wastewater treatment processes (viz. the Anaerobic Contact
process, figure 1.) any form of forced mixing can be omitted or kept at
a minimum (i.e. some mechanical mixing is required in the case of very
dilute wastes, at low loading conditions and if the waste contains a
high fraction of dispersed solids).

*PE = population equivalent (0.18 kg COD/day).

Full scale UASB-plants treating sugar beet waste (19) and potato processing waste can handle loading rates up to 15-20 kg $COD.m^{-3}.day^{-1}$ at treatment efficiencies as high as 95 %. Presumably even significantly higher loading rates can be applied, because in 6 m^3 pilot plant experiments loads up to 50 kg $COD.m^{-3}.day^{-1}$ at detention times as low as 4 hrs could be satisfactorily accomodated (18).

Recent experiments indicate that the UASB-process can also be applied for domestic wastes. In using a sugar beet sludge adapted more or less granular-sludge, up to 85 % COD-reduction has been achieved at 12 hrs detention time even under very low temperature conditions (5-8 ^{o}C during wintertime) (4).

Similar results have been reported for pilot plant experiments with the Anaerobic Filter (AF) (20). In the AF-process (see figure 1.) wastewater flows upwards through a reaction vessel which has been provided with some inert static support (e.g. gravel, coke, plastic media, etc.). The AF-concept originates from the early sixties, but thus far no reports appeared in the literature about full scale plants.

Although the potentials of the UASB-process look better and this process presumably is cheaper than the AF-process, its merits can usefully be applied in the UASB-process, i.e. probably an improved solids separation can be achieved in the UASB-concept by providing the settler compartment in the UASB-reactor with a static support (see figure 1.). Accordingly the AF-process can be improved by equiping it with a gas-solids separator above the filter.

A more recent approach, although still in the laboratory phase of research, is the 'Anaerobic Attached Film Expanded Bed' (AAFEB-) process (21). Here an upflow column filled with small (< 1 mm) pvc particles provide a large surface area for microbial attachment. In order to achieve good treatment efficiencies the process should be operated at a relatively high recycle ratio (superficial velocity: 3 m/hr).

In considering the results with the UASB-process some form of forced (intermitted) mixing also might suffice for this purpose. On the other hand 'Fluidized Anaerobic Sludge Bed' (FASB) systems (figure 1.) may become attractive for treating very dilute wastes (or medium strength wastes at low temperatures), once granular sludge such as has been cultivated in UASB-plants treating sugar beet and potato processing wastes is available (and can be maintained!) in sufficient quantities.

3.2 Anaerobic digestion of highly concentrated complex organic feedstocks

For an economic production of a high quantity of net energy from concentrated complex wastes, tentatively one-step digestion processes look more appropiate than processes with phase separation. However, this may change in the near future if it can be shown that phase separation leads to a cheaper and more effective process.

As mentioned before the best way to increase the loading potentials of a digestion process is to improve the biodegradability of the feed- stock components, because this will result in a sludge having a higher fractional content of viable biomass and consequently (at a certain sludge retention of the system) also to the pursued higher retention of viable biomass. Obviously another way to increase the quantity of viable biomass in the digester may be found in recycling a part of the digested solids after the solids concentration has been increased by subjecting the effluent stream to either thickening or centrifugation. The limits of sludge recirculation lay in technological difficulties and the relatively high costs of concentration processes. Obviously the same objections apply for concentrating the feedstock, which represents another way to increase the loading potentials of a digestion system. Moreover, above a certain DS-content - depending on the type and origine of the feedstock - there may also exist physiological restrictions to a further increase of the DS-content of the feedstock, viz. due to the fact that inhibitory intermediates or end products may also occur at increased concentrations. In this way more activity may be lost than is gained by increasing the content of the viable biomass.

A rather effective and more attractive method to increase the DS- content of a digester may be found in a minimization of mechanical mixing of the digester, the more because there is also evidence that the metabolic activity of bacteria improves if they can operate in a more or less quiescent environment. Therefore a minimization of mechanical mixing of the digester should be pursued. In practice this may be accomplished in digesters of a more or less plug flow type (22, 35).

Obviously the loading rate of a digester can also be increased by decreasing the digestion time. Dependent on the type of feedstock (biodegradability, DS-content, etc.) and the digestion temperature applied, minimum values for the digestion time are in the range of 5-10 days, below which the methane production will collaps (23, 36, a.o.). With respect to

the breakdown of malodorous compounds the process may already detoriate
at detention times in the range of 10-20 days (e.g. 23).

Research in the field of energy production from various types of
biomass (manure, refuge, agricultural residues and energy crops like
algae) has been significantly encouraged in recent years. Part of this
work is granted in the framework of the EEC solar energy programme, but
considerable grants are also provided now by various national governments.
The highest methane production rates reported for one-step pilot plant
digesters under mesophilic conditions are up to 2.5 $m^3.m_r^{-3}.day^{-1}$, while
for thermophilic conditions values up to 3.5 $m^3.m_r^{-3}.day^{-1}$ (24) have been
found. Whether or not such production rates can be achieved in full scale
on farm plants has still to be shown. Otherwise, it is not the gross gas
production rate that determines the feasibility of the process, but the
net methane yield that will be achieved. The net energy production of
a digester depends on a number of factors, viz.:

1. type of the feedstock, its VS-content but particularly its ligno-
   cellulose content (see table 1. and 2.).

2. feedstock temperature, because it is one of the main factors
   determining the heat requirements of the process. In moderate
   climate regions the VS-content of the feedstock should be at least
   4 à 5 % (25) will there be talk of a net methane production under
   severe winter conditions. This restriction on the process is
   considerably less, if the biogas is applied for electricity
   generation, because in that case sufficient waste heat is available.

3. type, size and operation of the digester.
   The net energy production will increase at increasing size and
   improved insulation   of the digester, especially at cold weather
   conditions. Moreover, a minimization of mechanical mixing (i.e. use
   of a plug flow digester!) may also result in an increased net
   energy production.
   Recent economic analyses indicate that for a maximum net energy
   production, mesophilic digestion processes should be operated
   5-10 ^{o}C below the optimum temperature of 35 ^{o}C (25).

In addition to these factors an economic generation of energy by
means of anaerobic digestion obviously also depends on:

1. The costs of the feedstock, which may vary from a negative price
   in the case of wastes (viz. if the use of anaerobic digestion

eliminates costly and environmentally sensitive disposal problems)
to positive values in the case of energy crops. At present in most
countries energy farming only looks attractive in the case of an
integrated harvesting of crops for both food and fuel, and where
the fuel component comes from the by-product of the crops. However,
there seems to exist already situations where the cultivation of
energy crops is economically feasible, e.g. in the South Island of
New Zealand (26).

2. Size of the digester.

At digester volumes above about 200 m^3, the capital costs of a
digester system will make up a smaller proportion of the costs
of the biogass (26).

3. Digester construction.

Digester constructions of the type of earthern reactor basins
provided with an underdrain system and plastic liner only cost
about one third of rigid wall tanks which at a 50-100 cow scale
may cost up to $ 100/m^3. Evidently an important factor in the
construction costs of digesters is whether or not there exists a
market for manufacturing prefabriquated fully assembled units (22).

4. Local retail prices for fuel and electricity, and the price at
which excess biogas, or electricity generated from it, can be sold.

4. CONCLUSIONS

In conclusion it can be stated that anaerobic digestion (AD) represents
a very attractive method both for the conversion of wet biomass in methane as
well as to combat environmental pollution. Application of the process to
the treatment of (mainly) liquid wastes will lead in the near future to
a considerable save in the use of high grade energy in wastewater treatment, and at the same time - and quite attractive - to the production
of energy from the pollutants.

Simple, cheap and effective methods are available now in the form of
the Anaerobic Filter (AF) and the Upflow Anaerobic Sludge Blanket (UASB)
process. These processes presumably will make anaerobic treatment even
feasible for domestic wastes at temperatures as low as 5-10 $^\circ$C. The AF
and UASB process, or combination of these processes, can also be applied
as second step in the digestion of complex solid feedstocks, after they have

been subjected to liquefaction in a 'liquefier'.

Concerning the application of A.D. to complex feedstocks for mere energy production, tentatively one-step digestion processes (especially of the plug flow type) look most appropiate. However, this may change in the future once it can be shown that processes with phase separation are more effective and/or cheaper. Both for one-step as well as for two-step digestion processes there is a condiderable need for cheap and effective liquefaction methods for poorly biodegradable compounds of the lignin-cellulose type and others.

5. LITERATURE

1. P.N. Hobson, et al, CRC Critical Reviews, June, 1974, 131.

2. M.J. Mc.Inerney, et al, Arch. Microbiol. 122, 129-135, 1979.

3. A.F.M. v. Velsen, et al, 1st Int. Symp. An. Digestion, Cardiff, 17-21 September, 1979.

4. G. Lettinga, et al, 35th Ind. Waste Conference, Purdue University, 13-15 May, 1980.

5. G.J. Stander, et al, 22nd Ind. Waste Conf., Purdue University, 1967.

6. L. v.d. Berg, et al, Can.J. Microbiol. 22, 1312, 1976.

7. A.F.M. v. Velsen, et al, Neth. J. Agric. Sci. 27, 255-267, 1979.

8. G. Lettinga, et al, Water Research 13, 725, 1979.

9. A.F.M. v. Velsen, Water Research 13, 995-999, 1979.

10. P.L. Mc.Carty, Public Works 95, 91-94, 1964.

11. E.J. Kirsch, R.M. Sykes, Progress in Ind. Microbiology, 9, 155-237, 1971.

12. L.D. Bullock, et al, I.G.T. Symp. 'Energy from Biomass and Wastes IV' 21-25 January, 1980, Florida USA, Lake Buena.

13. J.T. Pfeffer, et al, Bioeng. & Biotechnol. 16, 771, 1974; 18, 1179, 1976.

14. D.C. Stuckey, et al, Biotechnol Bioeng. Symp. 8, 219, 1978; J. Water Poll. Control Fed. 50, 73, 1978.

15. W.C. Gottschall, et al, I.G.T. Symp. 'Energy form Biomass and Wastes' Washington D.C., August 1978.

16. G. Lettinga, et al, Proc. 4th European Sewage and Refuge Symp., EAS Munich, 5-9 June, 1978, 226; 1st Int. Symp. An. Digestion, Cardiff, 17-21 September, 1979.

17. P.L. Mc.Carty, et al, J. Water Poll. Control Fed., 41, R160, 1969.

18. G. Lettinga, et al, Biotechn. & Bioeng. 22, 699-734, 1980.

19. K.C. Pette, et al, 35th Ind. Waste Conf., Purdue University, 13-15 May, 1980.

20. R.K, Genung, et al, Conference 'Biotechnology in Energy Production and Conservation, Gatlinburg, 10-12 August, 1978.

21. W.J. Jewell, et al, 52 Annual Water Poll. Control Fed. Conf., 11 October, 1979, Houston, Texas.

22. W.J. Jewell, et al, papers presented at 1st Int. Symp. An. Digestion Cardiff, 17-21 September, 1979.

23. A.F.M. v. Velsen, et al, Neth. J. Agric. Sci., 25, 151-169, 1977.

24. G. Sheleff, et al, 1st Int. Symp. An. Digestion, Cardiff, 17-21 September, 1979.

25. D.L. Hawkes, 1st Int. Symp. An. Digestion, Cardiff, 17-21 September, 1979.

26. D.L. Stewart, 1st Int. Symp. An. Digestion, Cardiff, 17-21 September, 1979.

27. P. Chynoweth, et al, I.G.T. Symp. papers 'Energy from Biomass and Wastes', Washington D.C., August, 1978.

28. K.K. Chin, et al, I.G.T. Symp. papers 'Energy from Biomass and Wastes', Washington D.C., August, 1978.

29. C.G. Golueke, et al, Applied Microbiol. 7, 219.

30. D.M. Eisenberg, et al, 1st Int. Symp. An. Digestion, Cardiff, 17-21 September, 1979.

31. M.P. Bryant, et al, 'Microbial Energy Conversion', Göttingen, 1976, 347, 1977.

32. Y.C. Converse, et al, Trans. Am. Soc. Agric., 20, (2), 336, 1977.

33. G.M. Patelunas, R.W. Regan, J. Environm. Eng. Div., 851-861, October, 1977.

34. A.G. Haskimoto, et al, 1st Int. Symp. An. Digestion, Cardiff, 17-21 September, 1979.

35. T.D. Hayes, et al, 1st Int. Symp. An. Digestion, Cardiff, 17-21 September, 1979.

36. E. Hindin, et al, J. Water Poll. Control Fed., 35, 501, 1963.

37. J. Maly, et al, J. Water Poll. Control Fed., 43, 641, 1971.

TABLE I – Yields of biogas obtainable from various crops and waste materials after 10 and 25 days of digestion at 37°C, using feedstocks containing 5 % TS (26)

Organic matter	Liters biogas.* per kg TS$_{added}$	
	HRT = 10 days	HRT = 25 days
oats	420	490
ground hay	390	480
maiz	290	470
chopped hay	330	460
poultry manure	290	430
wheat straw	265	400
lake weed	290	380
sugar beet tops	290	340
lucerne hay	280	330
grass	235	305
news paper	190	230
cattle manure	170	220
sheep manure	105	210

* methane content of the biogas: 60 – 70 %

TABLE II – Performance data for the digestion of various types of biomass

type of biomass	methane yield $m^3.kg\ VS^{-1}_{added}$	VS-destruction %	Conditions temp. °C	Conditions loading rate kg VS.m^{-3}.day^{-1}	Conditions detention time (days)	Ref.
raw kelp	0.31	48	35	1.6	18	27
water hyacinth	0.20	56	37	1.81	10	28
	0.25	66	37	0.91	20	
algae	0.25		35	1.44	30	29
	0.32		50	1.44	11-30	
algae	0.27	53	35	1.0	20	30
	0.32	56	35	0.67	28-30	
cattle waste	0.255	47	60	8.6	9	31
	0.171	35.8	60	26	3	
dairy manure	0.208	43.1	35	4.32	15	32
dairy manure	0.145-0.193	28-39	37.5	8.96-2.2	7.5-30	33
livestock wastes	0.22	44.2	55	3.4	20	34
	0.31	52.8	55	5.2	12	
	0.23	46.1	55	11.4	6	
	0.21	39.8	55	15	4	
cow manure*	a 0.145 / b 0.115	a 30.5 / b 24.5	35	10.6	10	35
	0.19 / 0.15	34.1 / 29	35	7.5	15	
	0.21 / 0.195	40.6 / 32	35	3.5	30	
piggery wastes	0.22	44.1	32	1.4	40	23
	0.24	48.9	32	2.7	20	
	0.22	39.2	32	3.6	15	
	0.20	38.3	32	4.5	12	
sewage sludge	0.40	77	37	0.80	33	36
	0.36	77	37	1.20	33	
	0.31	76	37	1.60	33	
	0.26	70	37	1.76	33	
	0.20	60	37	2.16	33	
	0.40	46	20	batch experiments		37
	0.40	49	30			
	0.41	50	50			

* a) plug flow digester; b) conventional digester

278

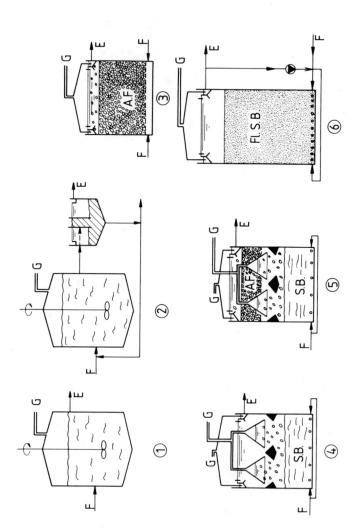

Figure 1. Anaerobic treatment processes: (1) conventional flow through digester, (2) anaerobic contact process, (3) anaerobic filter (AF), (4) upflow anaerobic sludge blanket (UAFB), (5) combined AF + UASB, (6) fluidized sludge bed. (G=gas, E= effluent, F= feed.

ETHANOL - THE PROCESS AND THE TECHNOLOGY FOR PRODUCTION

OF LIQUID TRANSPORT FUEL

J.COOMBS

Group Research & Development

Tate & Lyle, Limited

Summary

Ethanol produced from carbohydrate feedstocks (sugar crops, starch crops or agricultural and food process wastes) using yeast is of value as a liquid transport fuel in its own right or blended with petroleum. The process may be divided into the stages of substrate preparation, fermentation, distillation and, depending on use, dehydration, denaturation and/or blending. In particular, the technology of substrate preparation will vary with the nature of the feedstock, e.g. juice extraction for sugar crops, milling and hydrolysis for starch crops, milling separation and hydrolysis for cellulose derived from wood or crop residues, clarification and dilution for molasses, concentration for whey, etc. The resultant solution of fermentable carbohydrates (generally sucrose, glucose or fructose) may be fermented in a batch system or with recycling of the yeast (generally Saccharomyces cerevisae). Distillation to give 96% alcohol/water mix may be followed by dehydration using benzene or other organic solvent as entrainer. A particular problem may arise from the large volume of effluent produced. A power alcohol programme may be implemented for reasons of self-sufficiency, as a means of saving of foreign exchange, to use agricultural surpluses, as a means of regaining energy lost as crop wastes or as a means of converting a low grade fuel (e.g. poor quality coal) to one of higher value. Examples of existing programmes are cited and discussed briefly in terms of costs and energy balance. The justifications for and potential of power alcohol in Europe, and the opportunities for export of technology to other parts of the world, are briefly reviewed.

1. INTRODUCTION

Organic matter produced by current photosynthesis each year has an energy content in excess of that contained in the known reserves of oil(1). The problem with use of this organic material as a fuel is that it is often of low energy and physical density, is widely distributed, diverse in form, may contain a large percentage weight of water and is usually seasonal in availability. In addition, its use is not compatible with the use of technology which has been developed for oil based fuels. This is particularly true in the case of transport and agricultural machinery where the locomotive power is based on the use of the internal combustion engine.

The conversion of simple sugars to ethanol by yeast fermentation is, at present, the only proven technology by which plant material (biomass) can be converted to a more easily handled and stored fuel of higher energy content compatible with the internal combustion engine. In this paper the emphasis is put on the use of ethanol as a transport fuel since the needs for energy for heating, electricity generation or stationary power may be satisfied by direct combustion of biomass, or after thermal conversion to charcoal or gas(2) with high thermal efficiencies. However, the requirements for transport are more demanding since they are, in general, associated with the use of the internal combustion engine. Aspects of the use of alcohols as a motor fuel are discussed in more detail elsewhere(3), but in general terms can be summarised as follows.

Blends of petroleum plus 10 to 15% alcohol by volume (e.g. U.S.gasohol) can be used with little, if any, modification in a standard spark (Otto cycle) engine. In fact, addition of alcohol to unleaded petroleum will increase the research octane number by about 4 points. At higher levels (20-40%) modifications are necessary to the engines. These entail a modified carburretor, due to the lower stoichiometric air: fuel ratio, an alcohol resistant tank and fuel delivery system, a heated inlet manifold and a modified high compression cylinder head and a modified ignition system. The main disadvantage of alcohols as a fuel is their lower calorific value, resulting in a lower mileage per gallon. However, this is offset in part by the fact that alcohols will produce more power, have a higher thermal efficiency and will produce more power. In order to use alcohol alone as a fuel, special engines have been developed. Unfortunately, what is good for the Otto cycle engine is unsuitable for the compression ignition (diesel) type, in which the fuel requirements are dictated largely by fuel ignition quality. However, alcohols can be used as a supplementary fuel being aspirated into the diesel engine with normal diesel fuel at levels of 40 to 50% substitution.

At present, most alcohol being used as a fuel for motor transport is being used as a blend with petroleum. In order to blend an anhydrous alcohol is required, resulting in a higher cost of production.

The interest in the use of alcohol as a motor fuel has been rekindled (ethanol was used as a fuel in the development of the internal combustion engine and many countries used alcohol petroleum blends prior to the Second World War) by the increase in oil prices, and fear of the potential use of oil as a political weapon, since 1973. Since alcohol fuels may be manufactured from renewable indigenous feedstocks, they represent a source of fuel of particular interest to those countries which have to import oil and are thus being faced with problems of both shortages and increased prices resulting in severe balance of payment problems.

However, when looked at in detail, power alcohol does not always provide the instantaneous answer to such problems as is sometimes suggested. To be viable a power alcohol programme must show advantages in terms of net energy gain (or be fueled by a low cost indigenous energy source, such as wood or cheap coal), and be supported by some specific economic measures. This last point is of particular importance since, at present, using conventional accounting procedures, the costs of producing alcohol may exceed the costs of importing gasoline. On the other hand, the cost, in terms of hard currency of supporting loans for the capital installations for power alcohol production will, in general, be less than the hard currency cost of oil imports. Thus a power alcohol programme should result in a net saving of foreign exchange.

In general, the price of power alcohol produced by current technology will be high since the only proven process is the production of ethanol by yeast fermentation of solutions of simple sugars derived from starch crops (maize or cassava) or sugar cane. Until satisfactory methods of hydrolysing cellulose-containing material, such as wood and farm wastes, are developed, this will remain the case. As a result fuel alcohol costs are strongly related to world sugar and starch prices, hence fuel production may compete for land and other resources with food production. Thus the expected costs and benefits of a power alcohol programme will vary from country to country due to the complex interaction between such factors as alter- native uses and comparative values of feedstocks, by-products, land costs, labour costs, processing costs, effluent treatment, etc.

In general, the amount of energy expended in growing a crop, transporting it to a factory, processing to produce the fermentation substrate, fermentation, distillation, dehydration and effluent treatment, will be close to or greater than the amount of energy contained in the final alcohol produced. Hence, a net gain of solar energy trapped in the end product will only be achieved if part of the crop can be used to fuel the factory process. At present, the sugar cane-based power alcohol system is the only one which shows a significant net energy gain. This is because the fibrous parts of the cane (bagasse) may be used as a fuel. In addition, the cane-based system has the advantages that in many places the infra- structure and know-how already exists as a result of cane sugar production. In those countries that already produce significant quantities of sugar, imbalances may permit rapid initiation of power alcohol programmes at lower cost, due to local availability of excess molasses, excess cane or excess mill capacity. However, many countries, including most European ones, cannot grow cane and must look to alternative systems. The viability of these alternatives will depend on the provision of adequate feedstocks, the availability of technology for conversion to ethanol, and the availability of a suitable low cost fuel to power the process. These points are now considered in more detail.

2. BIOMASS FEEDSTOCKS FOR ALCOHOL PRODUCTION

The major feedstocks which could be used for alcohol production are shown in Table I. This table also indicates the current production and consumption of oil products. In addition to these crops major by-products are also available in the form of molasses (about 30×10^6t produced per annum; the EEC imported 3.3×10^6t in 1979) and milk solids (18×10^6t on a global basis with about 5.5×10^6t produced in the EEC). The figures for sugar and starch crops indicate harvested material, grains may contain over 80% starch, however, the content of fermentable material in

Table 1. (a) Total production (10^6 t) and yields (t.ha^{-1}) of starch and sugar crops
(F.A.O. statistics Ref.4); (b) Forest area (10^6 ha) and growing stock
(M^3 ha^{-1}) (Ref.5); (c) Petroleum production and consumption (Ref.6).

		World	Europe
(a)	Cereals total	1,580 (2.0)	167 (3.5)
	Wheat	441 (1.9)	94 (3.6)
	Rice	376 (2.6)	2 (4.5)
	Maize	362 (3.0)	71 (3.5)
	Barley	196 (2.1)	48 (4.2)
	Roots/tubers total	522 (11)	120 (21)
	Potatoes	272 (15)	119 (21)
	Cassava	119 (9)	–
	Sweet potatoes	100 (8)	0.1 (11)
	Sugar cane	781 (56)	0.4 (81)
	Sugar beet	289 (32)	141 (37)
	Sucrose	91	19.7
(b)	Forests	4,077 (120)	154 (103)
(c)	Petroleum production	3,074	88
	Oil consumption	3,076	714

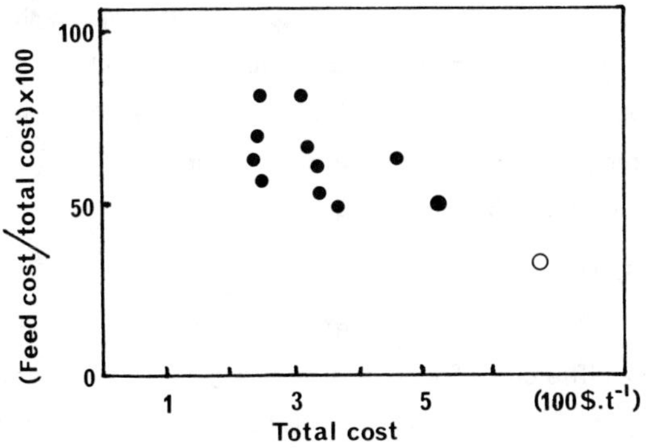

Figure 1. Scatter diagram showing % cost of feedstock as a function
of final selling price for ethanol from starch and sugar
crops (●) or wood (O) based on data in Refs. 18 and 22.

roots and tubers (20 to 30%) and sugar crops (16%) is, of course, much lower. Much of this material is required for food and animal fodder and is thus not available for use for alcohol production. The only feedstock available in a quantity comparable to that of oil is wood and other ligno-cellulosic residues.

Although in theory feedstocks for alcohol production may consist of any material which can be hydrolysed to produce fermentable sugars, at present there is no proven economical route for preparation of glucose syrups from ligno-cellulose. Hence, at present, power alcohol programmes are based on three main crops, sugar cane, maize and cassava. Cane-based systems offer major advantages over other systems, since where alcohol is the only product they may be completely autonomous in terms of fuel. Alternatively, distilleries based on sugar cane molasses may be fuelled using excess bagasse from sugar processing. In contrast, starch-based distilleries need an external source of fuel. Production of power alcohol from maize has a net energy ratio close to one. In the case of cassava, the energy value of the crop residues is insufficiently understood, and optimum cultivation, harvesting and feedstock preparation techniques have yet to be developed. In general, systems based on temperate grain crops, roots and tubers will also show a net energy ratio close to one.

Agricultural costs constitute the greatest proportion of the overall cost of any biomass fuel derived from purpose grown crops (Figure 1). It is also highly desirable to be able to supply feedstocks on a regular basis for a sustained period over the year to maximise use of capital equipment. This requires either that the feedstock crop be harvested for processing throughout the greater part of the year, or that it can be converted to a form suitable for storage.

3. FEEDSTOCK PREPARATION

Sugars may be extracted by mechanical expression of the juice in roller mills for sugar cane(7) or by diffusion as for beet(8), resulting in a feed containing 13 to 16% of fermentable sugars. On the smaller scale (9) starch crops may be dry milled to release the starch grains and produce a slurry which is then heated with a thermophilic bacterial amylase at $90^\circ C$ for thinning followed by saccharification (hydrolysis of the dextrins to glucose) by a glucoamylase at 50 to $60^\circ C$. Fibre may be removed either before or after saccharification. On a larger scale conventional wet

milling (10) may be used to produce a high purity starch which is then hydro-
lised either by enzyme/enzyme techniques or by acid/enzyme processes. In most
cases fermentations of maize will produce various by-products - animal feed, gluten
or oil, the value of which can be used to offset the cost of alcohol production.

4. FERMENTATION

Ethanol is produced by yeasts under anaerobic conditions according to the
equation:- $C_6H_{12}O_6 \rightarrow 2C_2H_5OH + 2CO_2$

On a glucose conversion basis the alcohol weight yield is around 50%, but
since a proportion of the sugar supplied is used for cell growth, maintenance energy
or converted to other metabolites, the actual yield is usually about 87 to 90% of
the theoretical giving a weight yield over around 45% on feed. However, the proportion of energy recovered in ethanol is over 90%. The regulation of ethanol
production is complex, the concentrations of substrate, oxygen and product(ethanol)
all affect yeast metabolism, and cell viability(11). Selection of suitable strains of
yeast with a higher tolerance for both substrate and alcohol concentrations has been
of particular importance in increasing yield. Where practicable, the level of fermentable solids generally used is in the region of 16 to 25% giving (in a batch system)
final alcohol concentrations of 7 to 12%.

Essentially, there are three methods of fermenting sugar feedstocks, namely
batch, batch recycle and continuous. In batch fermentation the substrate is fermented
out by a yeast inoculum which has been freshly cultured. The yeast passes out at
the end of the fermentation and consequently a new culture must be grown. This
continual propagation is costly in terms of substrate which could otherwise be converted into ethanol. Such processes also have long fermentation times. Both these
disadvantages can be circumvented by separating the yeast from the fermented mash
and returning to to the next fermentation. Hence, the effect of such batch recycle
methods is to decrease fermentation time and avoid yeast propagation except at start
up. Continuous fermentation is the subject of considerable research effort. However,
problems may be encountered in reaching a favourable balance between productivity
and substrate utilisation.

For efficient alcohol production these factors, productivity and sugar consumption
are of paramount importance. In many cases the two parameters oppose each other.

In batch fermentations complete sugar consumption means longer fermentation times and hence lower volumetric ethanol productivities. In continuous systems lowering dilution rate to increase substrate utilisation will again reduce productivity. Specific ethanol productivities for yeast range from less than one to about 1.8g per h ethanol per gram of cells(12). Higher productivities (2.5 to 3.8) have been recorded for the bacteria <u>Zymomonas</u> <u>mobilis</u>(13).

5. DISTILLATION

Distillation remains the only proven method of separating ethanol from the fermentation brew. An efficient still will produce a 96% ethanol/water mixture. Production of anhydrous alcohol requires a further distillation following addition of a suitable entrainer. As high an alcohol content as possible in the fermentation brew is desirable since steam consumption for distillation increases rapidly as the alcohol concentration is reduced. For example, a 10% alcohol content beer requires 2.25Kg steam/litre of 96% alcohol produced, whereas 5% beer requires in excess of 4Kg steam(7).

For double effect operation in a classical still design, the rectifying column works in double effect with the entrainer recovery column and the boiling column works with the dehydration column. The use of petroleum as an entrainer is of particular interest as it need not be removed from the alcohol, and the entrainer recovery column is eliminated. Power alcohol containing 10 to 15% of gasoline would also be effectively denatured.

6. EFFLUENT TREATMENT

The treatment of distillery slops probably represents the most serious problem to be faced in large scale alcohol production. Characteristically, stillage has a BOD of between 20,000 and 40,000 and a pH of about 4.5, but the exact composition depends on the substrate being used for fermentation. Obviously, the more pure the initial feedstock, the less residual material there is to appear in the stillage. Moreover, with yeast recycling techniques the majority of the yeast solids are removed and will not appear in the stillage effluent. There is no single solution to stillage treatment - the options which are currently available are as follows: (a) land disposal; (b)irrigation;(c) evaporation for animal feed;(d) evaporation for incineration; (e) activated sludge treatment;(f) lagonning; (g)anaerobic digestion.

All these methods, with the possible exception of direct disposal on land, are

expensive and the potash and phosphate content of the stillage is small. In
addition, there is a limit to the amount of slops which can be added to farm land
due to the changes in soil ion balance, water logging and effects of organic
materials on activities of soil microorganisms. Evaporation of slops from low purity
feedstocks such as molasses may offer useful and valuable by-products in terms of
an animal feed component or as a fuel for steam generation. Although, entailing
high capital cost, this is the most environmentally and socially desirable means of
slop disposal. Conventional activated sludge treatment of distillery wastes requires
that they are first diluted to reduce the BOD, hence anaerobic digestion appears
more promising. The use of effluents to dilute further quantities of feedstocks, such
as molasses, concentrated sugar syrups or raw sugar (or as imbibition water in the
case of cane juice fermentation) is of interest since it has the effect of reducing
the volume of effluent leaving the distillery. However, it also increases the concentration of non-fermentable solids in the fermenter which may effect the yeast
metabolism.

7. LIGNO-CELLULOSE AS FERMENTATION FEEDSTOCK

Because of the potential of large amounts of low cost feedstock offered by
wood and agricultural residues, considerable research is being carried out on the
production of ethanol from such materials. The problem with cellulose is that in
its natural state, in the plant cell wall, it exists as an insoluble complex with
lignin with the individual cellulose molecules orientated in microfibrils which
show varying degrees of crystallinity preventing attack during hydrolysis. The
plant cell wall also contains xylans and hemicelluloses which on hydrolysis yield
sugar which cannot be used by yeast. Possibilities exist for the fermentation of
these sugars by bacteria such as Klebsiella or Aeromonas to produce butandiol(14).
Alternatively, ligno-cellulose may be used as fermentation substrate for strains of
the anaerobic, thermophilic bacteria Clostridium thermocellum or C. thermosacchary-
olyticum. However, such systems are still under development. For conventional
yeast fermentation the wood or waste must be mechanically and chemically processed prior to cellulose hydrolysis. Milling is the most effective physical pre-treatment, but is expensive in terms of cost and energy. Delignification can be achieved using reagents such as sodium hydroxide, peracetic acid, sodium hypochlorite,
etc. Alkali treatment will swell and separate the fibres. Hydrolysis may be

accomplished using either mineral acids or biological methods using, for instance, enzymes from the fungi Trichoderma vivide. However, in spite of considerable research in the U.S., in particular,(14,15) no proven commercial process is current-ly available.

8. POWER ALCOHOL PROGRAMMES

The major power alcohol programme, is that in Brazil(16) based on cane grown on newly planted land with a view to off-setting large balance of payment deficiencies. In 1978/79 Brazil produced 3.4 billion l. of alcohol of which 2.7 were anhydrous. By 1985 they expect to produce between 8 and 12 billion l. Agreements have been signed to produce 250,000 ethanol fueled cars this year and a further 650,000 over the next two years. In addition, a further 270,000 cars will be converted to run on alcohol.

In the United State gasohol(10% ethanol blend) is produced from a crop surplus of maize, in part to offset balance of payments and, in part, to increase self-sufficiency in fuel production. Current production is around 120 million U.S. gallons p.a., and is expected to rise to over 500 million gallons in 1981, and 900 million gallons in 1982. This would replace 1.8% of the 50 billion gallons of unleaded petroleum.

Many other plants which have recently come on stream, or are planned, are based on molasses. Recent reports(17) include details of the factory in Zambia which will produce 1.5×10^6 l. p.a., the plant in Zimbabwe which has already produced 3×10^6 l. since it was opened in April, a 30,000 l. per day plant in Bacolad City in the Philippines, a 150,000 l. per day plant in Nicaragua, a 3,000 l. per day plant in Paraguay and a 20×10^6 l. p.a. plant at Kisumu in Kenya. In addition, the Philippines are expected to spend $\$2.8 \times 10^9$ to produce an annual output of 5.8×10^6 barrels of ethanol by 1989 by planting 270,000 ha of cane. This should replace over 30% of petrol requirements and will save $US477 million. Many other countries have initiated studies based on molasses, sugar cane, cassava, maize. In particular, New Zealand is looking at the potential of sugar beet.

9. ENERGY RATIOS AND COSTS

There are two factors which must be considered to determine whether it is worthwhile embarking on a specific route of fuel production from biomass. The first concerns the net energy ratio (NER) or the final yield of energy in useful products divided by total energy inputs, the second consideration is of the straight economics. Within the literature wide variations occur in values quoted for both NER and costs of production for ethanol. As far as NER is concerned, this is due in part to different approaches taken. Three different types of ratio can be recognised. In the first, all inputs are given an energy equivalent value and all outputs are taken as having an energy value. The second approach is to take the same total energy inputs (sometimes excluding energy costs of making machinery and factory if the distillery is based on an existing plantation and factory), but restrict the output to the energy content of the useable ethanol alone. A third approach is to apportion the energy inputs to the various excess end products, i.e. a certain input will be set against the production of ethanol and another against, for instance, the concentration of stillage for animal feed.

In spite of these **differences** it is clear that a net energy gain will occur in cases where fermentation and distillation is powered by the burning of crop wastes as in the case of sugar cane, or by the burning of wood obtained from close by as for a eucalyptus/cassava factory. The NER reported vary from about 2.4 to over 7. For most starch crops and sugar beet the value of NER is close to or below one, i.e. more energy is used than is produced. However, this may be worthwhile if the fuel source is, for instance, cheap coal of poor quality which is, in effect, converted to a high quality liquid fuel without the environmental problems associated with production of synthetic gasolines from coal. A very low NER of 0.1 to 0.2 is obtained for fermentation of cellulose using present techniques of hydrolysis.

Costs are equally variable. For instance, Sheehan(18) lists 39 estimates based on various starch crops, sugar crops, molasses and wood. These range from less than 10¢ per litre to over 60. However, a particular price is not related to a particular crop, for instance, costs quoted for sugar cane derived ethanol range from 14 to 60¢ per litre. Even these costs are affected by rapid rates of inflation in most countries; furthermore, many are based on paper studies. However, a 1979 costing of ethanol derived from sugar cane in Brazil(19) gave 30.5 US¢ for cane alcohol and 31.7 US¢ per litre for cassava derived alcohol. At the present price of corn in the U.S. ethanol can be produced at about $1.25 per US gallon, resulting in gasohol at a cost of 98¢ per gallon as compared with gasoline at 95¢. However, ethanol production is profitable because of the tax structure. The Federal Government has passed an exemption of gasoline excise tax on gasohol made with ethanol from farm products, for standard gasohol this is equivalent to $0.40 per gallon of ethanol. At present, the political economic justification is based on the fact that in order to maintain corn prices the government subsidises each bushel of corn not produced with one dollar. A bushel can be converted to $2\frac{1}{2}$ gallons of ethanol to be used in 25 gallons of gasohol. In addition, many States have passed special additional tax exemptions for gasohol varying from $0.04 to $0.07 a gallon.

10. POWER ALCOHOL FOR EUROPE

At present the cane-based system is the only one which does not have a need for an external source of fuel. Hence, in Europe an external source of fuel is needed (e.g. gas, oil, coal or wood). In general the net energy gain using beet or wheat will be small or negative. Such processes may be justified on the basis of agricultural support in times of surpluses or in terms of converting a poor grade solid fuel to a more useful product. A major problem with European crops is continuity of supply, large quantities of feed materials would have to be processed and stored to enable year round working, with a resultant increase in energy requirements. Imported molasses could be used, however, supplies may decrease (and prices rise) as more can producing countries initiate power alcohol programmes. In the same way agricultural surpluses will fluctuate, and hence costs of alcohol production may be linked to food prices - or worse, food prices may rise as fuel costs rise if food crops are used for production. These problems could, in part, be overcome if lignocellulose could be efficiently converted to ethanol. However, it is probable that the technology for production of methanol from wood, gas or coal will develop faster than that for ethanol in Europe at least. Processes have been described for converting methanol to petroleum (20) and ethanol (21). The cost of such products may be less than that of ethanol produced by fermentation of sugar or starch crops. Hence, the most realistic assumption is that production of fermentation ethanol for fuel use will not be as important in Europe as in many other regions of the world. In this case power alcohol programmes should be seen, in part, as an opportunity for the export of European technology and, in part, as a future source of imported fuel. For instance, substitution of power alcohol quotas for cane sugar quotas would reduce problems such as the sugar surplus in Europe and establish new markets for the less developed countries.

REFERENCES

1. Hall, D.O. and Coombs, J. (1979) Report No.5. p.2. The Watt Committee on Energy Ltd., London.

2. Van Swaaij, W.P.M. (1980). This volume.

3. Bernhardt, W. (1980). This volume.

4. Anon (1979) F.A.O. Production Yearbook. Statistical Series 22. Rome.

5. King, K.F.S. (1978) Resources of Organic Matter for the Future. St.-Pierre, L.E. (Ed.), p.35. Multiscience, Canada.

6. Moore, R.(Ed).(1979) Oil and Gas International Year Book, Financial Times.

7. Paturau, J.M.(1969) By-products of the Cane Sugar Industry, Elsevier,N.Y.

8. McGinnis, R.A. (1971) Beet Sugar Technology, 2nd Edition. Beet Sugar Dev., Foundation.Colo.

9. Anon (1980) Fuel from Farms. Report SERI/SP-451-519. U.S. Government Printing Office, Washington D.C.

10. Whistler, R.L. and Paschall, E.F. (1967) Starch: Chemistry and Technology, Vol.II. Industrial Aspects, Academic Press, N.Y.

11. Nord, F.F. and Weiss,S.(1958) In: The Chemistry and Biology of Yeasts, Cook, A.H.(Ed.) p.323. Academic Press. N.Y.

12. Bazua, C.D. and Wilke, C.R. (1977) Biotech.Bioeng. Symp.7. 105.

13. Rogers, P.L., Lee, K.J. and Tribe, D.E. (1980) Process. Biochem., 15,7.

14. Ladisch, M.R., Flickinger, M.C. and Tsao, G.T. (1979) Energy, 4, 263.

15. Cooney, C.L., Wang, D.I.C., Wang, S.D., Gordon, J. and Jiminez, M. (1979) Biotech. Bioeng., Symp.8, 103.

16. Anon (1980) F.O.Licht's International Molasses Report. 17, 83.

17. Anon (1980) F.O.Licht's International Molasses Report. 17, various.

18. Sheehan, G.J., Greenfield, P.F. and Nicklin, D.J. (1978) in Alcohol Fuels, pp.6-11. Inst.Chem.Eng., N.S.W. Branch, Australia.

19. Anon (1979) The Brazilian Alcohol Programme. Int.Mol.Rep.Special Edition. F.O. Licht, Germany.

20. Penick, J.E., Meisel, S.L., Lee, W. and Silvestri, A.J. (1978) in Alcohol Fuels, p.5. Inst.Chem.Eng., N.S.W. Branch, Australia.

21. Anon (1980) Chem. and Eng. News, 58,37.

22. Stewart, G.A., Gartside, G., Gifford, R.M., Nix, H.A., Rawlins,W.H.M. and Siemon, J.R. (1979) The potential for Liquid Fuels from Agriculture and Forestry in Australia. C.S.I.R.O., Australia.

HF SACCHARIFICATION: THE KEY TO ETHANOL FROM WOOD?

Derek T. A. Lamport, Haim Hardt, Gary Smith and Sharon Mohrlok

MSU-DOE Plant Research Laboratory,

and

Martin C. Hawley, Richard Chapman and Susan Selke

Department of Chemical Engineering,

Michigan State University, East Lansing, Michigan 48824 USA

Summary

Hydrogen fluoride attacks, via solvolysis, polysaccharides including cellulose rapidly at circumambient temperatures to yield glycosyl fluorides which are hydrolysed to the free sugars when the HF contains traces of water. We have reinvestigated this reaction with a mind to its use as a source of sugars as chemical feedstock for fermentation or synthesis. Using standardised wood chips of Big tooth Aspen (Populus grandidentata) exposed to HF under various conditions we determined: fluoride retention in the soluble and insoluble products, sugar recoveries, degree of sugar reversion, optimal post hydrolysis conditions, and fermentability of the products. Most of the fluoride retained after evacuation for 2 hrs. at 100°C, was in the water soluble fraction (ca. 4 mg F'/gram initial wood) with little in the insoluble lignin fraction (0.6 mg/gram lignin or 0.12 mg/gram initial wood). The wood ash content (ca. 0.7%) may account for some or all of the fluoride retained. Sugar recoveries were 90-95% of theoretical with only 2-3% remaining in the lignin fraction. Removal of HF led to extensive sugar repolymerisation. The reversion products were mainly water soluble oligosaccharides which gave >95% monomers after post hydrolysis for 1 hr. at 140°C in 50 mM sulphuric acid. After neutralisation of the post hydrolysis products with calcium carbonate, and filtration, yeast fermented the wood sugars, with ethanol as the major distillation product.

INTRODUCTION

Large amounts of waste cellulosic materials are available yet remain unused because of the difficulties involved in "cracking" the cellulose macromolecule. For example according to the recent Office of Technology Assessment report (1) acid hydrolytic processes for the production of glucose from cellulose are currently uneconomic, while microbiological methods present no clear cut advantages.

We have for sometime been interested in the structure of the growing (primary) cell wall, especially its hydroxyproline-rich glycoprotein component extensin (2). This led us to consider various methods of deglycosylation. When we considered a glycoprotein from a peptide chemist's point of view, the carbohydrate substituents could be regarded as protecting groups...removable by a deprotecting reagent such as anhydrous hydrogen fluoride (3) which leaves peptide bonds intact even at room temperature. HF however, cleaves glycosidic linkages with impressive speed and efficacy. We have therefore spent some time reinvestigating and extending much earlier "forgotten" reports (4) of wood saccharification via anhydrous HF, which apparently reached the pilot plant stage (5). Here we described results of our experiments using liquid HF and wood chips in the presence of known small amounts of water. We have determined fluoride retention, sugar recoveries, degree of reversion, post hydrolysis conditions, and fermentability of the products.

METHODS AND RESULTS

1. HF Solvolysis and Fluoride Retention.

We treated small wood chips (ca. 1x1x.2 cm) with liquid hydrogen fluoride in a Kel-F vacuum distillation apparatus manufactured by the Toho Kasei Company of Japan and marketed in the United States by Peninsular Labs. Inc., San Carlos, California.

In typical experiments we treated 1 gram samples of wood with HF (containing 0 to 20% water) for 1 hr at 0°C, and then removed the HF by evacuation into a calcium oxide trap. Maximal HF removal necessitated heating the reaction products (Fig. 1). Addition of water to these products then yielded a water soluble sugar fraction and an insoluble lignin fraction. The lowest fluoride content achieved for these two fractions, as determined with an Orion combination fluoride electrode, was ca. 4 mg/g and 0.1 mg/g wood for sugar and lignin fractions respectively under the following conditions: HF solvolysis 1 hr at 0 degrees, then HF evacuation at room temperature followed by evacuation for 2 hr. at 100°C.

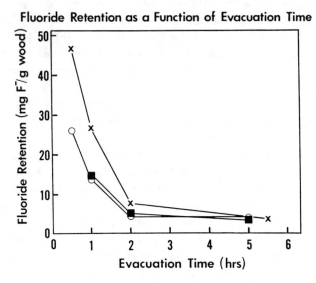

Figure 1. Evacuation was at 100°C.   x——x anhydrous HF; o——o HF
containing 5% H$_2$O; ■——■ containing 10% H$_2$O

We were able to detect sugar fluorides only when HF solvolysis ocurred
in the complete absence of water. Thus glucosyl fluoride was an HF solvolysis product in model experiments with pure cellulose (filter paper) treated
for 1 hr at 0°C with anhydrous HF followed by evaporative HF removal with
no heat applied, and then addition of water with a little calcium carbonate. After chromatography of the water soluble products on Whatman #1
(using the epiphase from a butanol/ethanol/water 4/1/5 mixture) we eluted
the appropriate area (Rf = 0.38 visualised with alkaline silver nitrate)
and after acid hydrolysis determined the glucose/fluoride ratio (found:
1.2). The Rf 0.38 spot gave only one major peak after trimethylsilyla-
tion and gas chromatography on a 12' x 0.2 mm 3% SP2100 column (running
conditions: 120 - 225°C at 4°/min, initial delay 4 min. Carrier He 40
mL/min). TMS-glycosyl fluoride had a relative retention time of 0.86 with
TMS-mannitol as the internal standard.

2. Sugar Recoveries
    HF solvolysis renders 70-80% of the wood soluble in water. We have

attempted to determine precise sugar recoveries by analysing both the water soluble and insoluble lignin residue for sugar content after further hydrolysis (dilute sulphuric or trifluoroacetic acid) or methanolysis with methanolic HCl. Analysis of the free sugars or methyl glycosides as their alditol acetates (6) or trimethylsilyl derivatives (7) allowed us to account for around 95% of the wood sugars, mainly as glucose and xylose, the bulk (ca. 95%) being in the water soluble fraction, although a small amount (ca. 2-3%) was trapped in the lignin fraction.

3. Degree of Sugar Reversion

Concentrated sugar solutions tend to polymerise when heated especially under acidic conditions, and 1-6 linkages predominate. Our experiments with HF treated wood chips show this clearly. HF solvolysis rendered most (>95%) of the wood sugars soluble, which appeared mainly as the monomers, when HF was removed by rapid evaporation and neutralisation. However for maximal HF removal we heated the reaction mixture after an initial flash evaporation. The sugar solvolysis products remained water soluble but judging from ethanol precipitability and gel filtration on Biogel P2 columns (Fig. 2) extensive reversion had occurred, leaving only 10-20% of the sugars as monomers.

4. Post Hydrolysis of Reversion Products

To ensure fermentability of the wood sugars we optimised post hydrolysis conditions which, for dilute sugar solutions (0.5% w/v) were 1 hr at 140°C with 50 mM sulphuric acid. These conditions converted >95% of the water soluble oligosaccharides to the monomers, as judged from (a) gel filtration on Biogel P2 columns before and after complete acid hydrolysis (b) direct gas chromatographic determination of monomers as their trimethylsilyl derivatives compared with total soluble saccharides determined after methanolysis as their trimethylsilyl-O-methyl glycosides. Methods (a) and (b) agreed well.

5. Fermentability of the HF Solvolysis/Post Hydrolysis Products

As a "benchmark" experiment we determined whether or not yeast (Saccharomyces cereviseae) could ferment the wood sugars obtained as described above. After the sulphuric acid post hydrolysis we neutralised with calcium carbonate which also ensured a very low fluoride level, then added a simple salt mixture, sterilised and inoculated with yeast. After

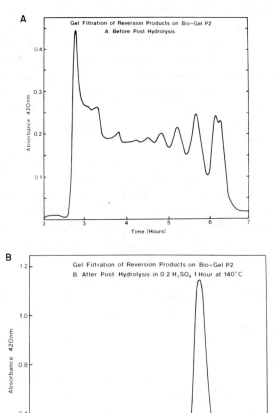

Figure 2. After HF solvolysis we loaded 0.5-2mg of the water soluble
wood sugars onto a 180 x 1.3 cm P2 (-400 mesh) column and
monitored the eluate with an automated system using an
orcinol/sulphuric acid reagent. A. Before post hydrolysis
B. After post hydrolysis in 0.2M H_2SO_4

appropriate incubation we identified ethyl alcohol (via gas chromatography
on Supelco Carbopack B/5% carbowax 20M isothermally at 85°C) as the major
volatile product.

DISCUSSION

These experiments demonstrate the biochemical feasibility of the wood to ethanol route via HF solvolysis. The economic feasibility depends on factors such as HF recovery and recycling, fluoride retention in both soluble and insoluble fractions, ability to deal with reversion products, and quality of the lignin produced. Precise determination of HF recovery is in the planning stage. Fluoride retention is known quantitatively but not qualitatively...the ash content (ca. 0.7%) of our wood may account for most of the fluoride retained although we cannot yet exclude the possibility of some fluorination of the lignin residues. Overall fluoride contamination does not seem to be a severe problem because of the ease with which fluoride can be removed, and because yeast can tolerate surprisingly high fluoride levels. Oligosaccharide reversion products arise especially during the HF removal heating stage and are probably inescapable, but can be dealt with effectively by a post hydrolysis yielding >95% sugar recovery At the moment we have no data on the quality of the residual lignin, for example whether it has undergone condensation reactions, or whether it still remains reactive and could serve as a suitable feedstock, e.g. for the Noguchi hydrogenation process. This could have an important bearing on the overall economic feasibility of HF saccharification which at the moment is very attractive particularly in view of the excellent sugar yields.

ACKNOWLEDGEMENTS:

We thank Dr. Andrew Mort for helpful discussions and Ms. Joann Lamport for help at all stages. This work is supported by US DOE Contract DE-AC02-76ER01338.

REFERENCES:

1. Congress of the United States Office of Technology Assessment (1980) "Energy from Biological Processes," 1-195.
2. Lamport, D.T.A. & Northcote, D.H. (1960) Nature 188, 665-666.
3. Mort, A.J. & Lamport, D.T.A. (1977) Anal. Biochem. 82, 289-309.
4. Fredenhagen, K. & Cadenbach, G. (1933) Angewandte Chemie 46, 113-117.
5. Luers, H. (1938) Holz Roh. und Werkstoff 1, 342-344.
6. Albersheim, P., Nevins, D.J., English, P.D. & Karr, A. (1967) Carbohyd. Res. 5, 340-345.
7. Bhatti, T., Chambers, R.E. & Clamp, J.R. (1970) Biochim. Biophys. Acta 222, 339-347.

ETHANOL PRODUCTION BY EXTRACTIVE FERMENTATION

M. MINIER and G. GOMA

Département de Génie Biochimique et Alimentaire, ERA-CNRS n° 879,
Institut National des Sciences Appliquées, Avenue de Rangueil,
F 31077 Toulouse CEDEX

Summary

Ethanol production is characterised by the production of an inhibitory
metabolite which depress the reaction rate and stops the ethanol production
when the concentration reachs the range of 10-15 % V/V. The ideal technology of ethanol production must use plug flow reactor to delete the ethanol
effect, and, remove the inhibitory product during the fermentation: extractive fermentation.
     The proposed technology uses both plug flow reactors and liquid-liquid
extraction to achieve continuously the extractive fermentation of ethanol.
     The solvent used for liquid-liquid extraction is dodecanol. The study
of the inhibitory effect of aliphatic alcohol of different carbon number is
achieved. The inhibitory effect increases with carbon number; no growth is
observed in the presence of alcohols which have between 2 and 12 carbons.
This effect is suppressed when the carbon number is 12 or higher than 12.
     A new reactor was used. It is a column packed with a porous materials.
The fermentation broth is pulsed in the purpose (i) of increases the interfacial area between the liquid medium and the dodecanol, (ii) decreases
the gas hold up.
     Alcoholic fermentations were performed at 35°C, on glucose syrup by
Saccharomyces cerevisiae, with adsorbed cells as reference,with adsorbed
cells and extractive fermentation. The results are compared and fermentation is improved.
     With this strain, by this new way of fermentation, productivity of
ethanol is multiplied by 5 and, solution of 407 g/l of glucose was totaly
fermented with yeast which cannot-in normal conditions-use totaly solutions
with more than 200 g/l of glucose.
     The feasability of a new method of fermentation coupling both liquid-
liquid extraction and fermentation is demonstrated. Extension of this
method to all microbial production inhibited by his metabolite excretion
is possible.

INTRODUCTION

Ethanol fermentation is limited by the inhibitory effect of the product which decreases the rate of ethanol production, the cell biomass concentration and the range of alcohol concentration which remain between 10 and 15 % V/V. It results that the distillation is the only method used for ethanol extraction.

One way to reduce these limitations is to remove ethanol during his production by fermentation: this process is call extractive fermentation.

In alcoholic fermentation, the vacuum fermentation process is one solution. Proposed first by BOECKELER (1948), described by RAMALINGMAN and FINN (1977), CYSEWSKI and WILKE (1977) the efficiency of this process is now established; however, the energetical cost of ethanol production is the major disadvantage: TYAGI and GHOSE (1979). Recently MAIORELLA and WILKE (1980) established that the vacuum fermentation process need, as overall energy requirement for production of azeotropic ethanol $8.352 \ 10^6$ J/m^3 of ethanol. The high productivity in ethanol production is attractive.

It is well known than ethanol is extractible from water by liquid-liquid extraction: HARTLINE (1979). Alcohol dissolves readily in some liquids that do not mix with water. By exploiting this solubility difference, alcohol can be recovered by solvent extraction, but, in extractive fermentation the non toxicity of solvent toward the cells must be established. Recently, GOMA and al. (1980) described the dodecanol as ethanol extractor and proposed to use liquid-liquid extraction to remove ethanol produced during continuous fermentation with immobilized cells while HERNANDEZ MENA and al. (1980) establish the process feasability for continuous extraction by dibutyl phtalate of alcohols from fermenting systems in batch or fed batch culture.

The aim of this paper is to describe a new technology of fermentation which combine both liquid-liquid extraction and fermentation and to show the improvement due by extractive fermentation.

MATERIAL AND METHODS

The yeast used was Sacchromyces cerevisiae UG5 described in a previous work: STREHAIANO and al., 1978.

The medium contained (gl^{-1}) KH$_2$PO$_4$, 5; (NH$_4$)$_2$ SO$_4$, 2; MgSO$_4$ 7H$_2$0, 0.04;

yeast extract, 1; glucose, variable; tap water to 1 liter; pH adjusted to 3.75 with H_3PO_4 after sterilization.

Biomass concentration was determined by filtration (Millipore 0.45mm) weighing after 12 h. at 0.2 B, 60°C. Ethanol was measured by g.l.c. (Perkin Elmer F 11) with n butanol as internal standard, glucose by dinitrosalicy-late method and viability by coloration with Ponceau red.

The aliphatic alcohols used to evaluated the toxicity on the yeast are Fluka products grade purum.

The dodecanol 1, used for extractive fermentation is commercialized by Prolabo, 12 rue Pelée 75 Paris, ref 20 840 292. In fact this commercial product was established as a mixture of dodecanol 1, 60 % and Tetradecanol 1 (40 %), traces of 1 decanol.

The influence of solvents on the yeast growth was measured in erlen at 35°C in the presence of 5 % V/V of solvents. The results related here are only that relative to the serie of homologous aliphatic alcohol.

RESULTS

1.- Choice of a non toxic liquid extractor of ethanol.

We study 31 products. In fact the most interesting results when ob-tained with the serie of aliphatic alcohol. It is well known that the affi-nity of ethanol is greater for alcohol with short carbon chain length. We measure the ethanol and biomass production in the presence of every alcohol between 0 and 14 carbon number. Strong inhibitory effects are observed for alcohol between 3 and 11 carbon number (3 and 11 included). However, no inhibitory effects were observed for alcohol with 12 and non carbon number. The figure 1 describe the effects of alcohol on ethanol production.

We must observe the difference of behaviour of Saccharomyces in the presence of 1 undecanol or 1 dodecanol. This experimental observation cannot be explained. However, similar results were reported by FINN (1966) who studied the toxicity of aliphatic hydrocarbon on yeast. The works on hydrocarbon fermentation indicate from the yeast Candida lipolytica has the maximal growth rate on normal alkane with more than 12 carbons; very slow growth were observed when the hydrocarbon have less than 12 carbons; MUNK and al. (1969) GOMA and al. (1972). GILL and RATLEDGE (1972) obtained simi-lar results with alkanes, alkyl bromide, alkenes, linear aliphatic alcohol.

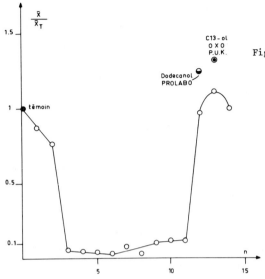

Fig. 1: Effect of carbon number of the aliphatic
l alcohol on cell
growth.

For MAHENDRA KUMARJAIN and al. (1978) short alcohols disorder more the lipid layer of cell walls than heavy alcohols.

For the following studies we use dodecanol Prolabo. The effect on batch culture, during anaerobie growth of Saccharomyces cerevisiae UG5 is descri-bed on figure 2 were is plotted the ratio µm/ µ versus the ethanol concen-tration for the growth of the yeast in the presence or not of dodecanol (5 % V/V). If we observe some lag phase in the presence of dodecanol the inhibitory effect is slight (comparison of the linear part of the curve).

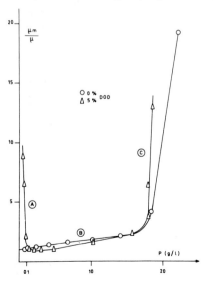

Fig. 2: µm/µ versus P during
a batch culture of
S. cerevisiae in the
presence or not of
dodecanol (DOD) initial
sugar concentration :
50 g/l, θ: 35°C.

2.- Choice of a technology of extractive fermentation

The technology of extractive fermentation coupling both liquid-liquid extraction and fermentation must have

    . a compatibility with a continuous process
    . an hydrodynamic of plug flow for reactor and extractor; this is more imperative for the dodecanol flow
    . a low carbon dioxide hold up
    . a high interfacial area between the liquid medium and the dode-canol.

We developped in the past for SCP production a new type of continuous fermentor which features an agitation aeration system hand on pulsed flow across perforated plates: SERIEYS and al. (1978). This devices was packed with porous brick in the purpose of increases the cell biomass concentration in the reactor and improve the liquid-liquid extraction by dodecanol. It is fed both by dodecanol and by fermentation medium with carbon source. The device used for extractive fermentation is described by figure 3. The reactor has 1.7 m height, an internal diameter of 0.1 m. This plant run continuously during one year and half.

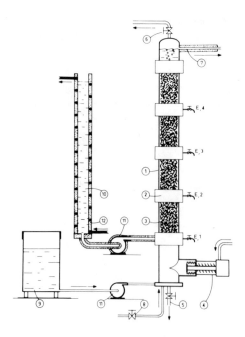

Fig. 3: Device of continuous extractive fermentation coupling liquid-liquid extraction and ethanol production by immobilized cells. 1 reactor (H=1.7m, ϕ_{int}=0.1m), 2 Heat exchange, 3 packing, 4 pulse generator (EIVS), 5,6,7 outlets, 8 air input (eventually) 9 mineral medium and carbon source reservoir, 10 reservoir of liquid for extraction (dodecanol), 11 metering pump, 12 heat (maintain dodecanol in liquid form).

3.- Improvement of alcoholic fermentation by extrative fermentation

Continuous cultures were performed with immobilized cells under conditions of extractive fermentation or not. The results are obtained in extreme conditions for Saccharomyces cerevisiae.

Tree series of experiments are performed, one of reference without dodecanol, the other, with only tiny amount of dodecanol are given in the purpose of evaluating the additive effect of dodecanol without any removal of ethanol, (surfactant effect of dodecanol, increases of the yeast clumping). The third series of experiments are extractive fermentation.

Results obtained are summarized in table I, obviously, alcoholic fermentation is improved by extractive fermentation. The residual sugar drops from 162 g/l to 2 g/l while the productivity is multiplied by 4 when So: 260 g/l and by 5 when So = 317 g/l.

Table I : Continuous alcoholic fermentation with immobilized cells without dodecanol addition (reference), with very small addition of dodecanol (experiments 1), with extractive fermentation (experiments 2,3,4).

Parameters	Experiments	Reference	1	2	3	Reference	4
Q_{DOD}	1/h	0	0.0285	1	2.55	0	1
Q_M	1/h	0.069	0.0667	0.0728	0.0698	0.068	0.068
S_{om}	g/l	260	263	260	263	315	317
S_s	g/l	162	165.9	2	1.8	242	2.9
P_{ms}	g/l	38.1	39.8	21.2	9.4	28.9	23.7
P_{DOD}	g/l		14.9	8.38	3.35		9.1
$\dfrac{Q_i P_i}{V}$	g/l.h	0.24	0.28	0.91	0.83	0.18	0.58
R	g/g	0.4	0.47	0.51	0.5	0.4	0.5

DISCUSSION CONCLUSION

The demonstration of the process feasability for continuous extraction of ethanol from continuous fermentation system is established. The improvement of the potencies of the cell indicates the efficiency of this new technique.

We must observe an increases in the yield of ethanol production, unua-
sully, the obtained yield is closed to the maximal theoretical yield. This
is not a casual observation but we cannot explain this increased yield.

Extractive fermentation occurs by reducing the product inhibitions and
we must observe than the sugar tolerance of this strain is higtly increased. In similar conditions, in a serie of eight C.S.T.R. solutions with
more than 180 g/l of glucose was not totaly fermented. Our last results
permit to use all the sugar with solutions having 407 g/l of glucose as
initial concentration.

One of the most important problem, relative to ethanol fermentation,
is to know if it is possible to lower recovery cost (energy and money needs)
of ethanol. By the utilization of membrane technique for ethanol recovery
from dodecanol and by recycling the mineral medium it seems possible to
decreases the energetical cost of ethanol fermentation but, more work is
necessary. The perspective of the development ofen isothermal ethanol recovery process is realistic.

Extraction fermentation using liquid-liquid extraction is applicable
to all microorganism culture were fermentation product can cause inhibition
and/or repression on its own synthesis or affect microbial growth.

Acknowledgements

The authors thanks the COMES, (Comissariat Energie Solaire) for his
help.
One of the authors wish to acknowledge the french C.N.R.S. for his
financial support.

REFERENCES

BOECKELER, B.C., 1948, U.S. Patent n° 2 440 925, May 4.

CYSEWSKI, G.R., C.R. WILKE, 1977, Rapid ethanol fermentations using vacuum
and cell recycle. Biotechnol. Bioeng., 19, 1125-1143.

FINN, R.K., 1966, Inhibitory cell products: Their formation and some new
metods of removal. J. Ferm. Technol., 44, 305-310.

FINN, R.K., R.A. BOYAJIAN, 1976, Preliminary economic evaluation of the
low-temperature distillation of alcohol during fermentation. In "Proc of
the Vth international fermentation Symposium". Session 3 Process Design
and product recovery. Page 48 Berlin.

GILL, C.O., C. RATLEDGE, 1972, Toxicity of n-alkanes, n-alk-1-enes, n-alkan
-1-oles and n-alkyl-1-bomides towards yeasts. J. Gen. Microbiol., 72,
165-172.

GHOSE, T.K., R.D.TYAGI, 1979, Biotechnol. Bioeng., 21, 1387-1400.

GOMA, G., A. PAREILLEUX, G. DURAND, 1972, Cinétique de dégradation des
hydrocarbures par Candida lipolytica. Arch. Microbiol., 88, 97-109.

GOMA, G., P. STREHAIANO, M. MORENO, 1980, Mechanism of inhibition during
alcoholic fermentation in strict anaerobiosis. In "Procedings of the
II^{d}Curse cum Sumposium on Bioconversion and Bioengineering, Delhi,
March 80, under publication.

GOMA, G., M. DONDE CASTRO, M. MINIER, 1980, Fermenteurs pulsés: nouvelles
technologies d'agitation aération, nouvelles potentialités de mise en
oeuvre de fermentations. Colloque Soc. Fr. Microbiol., Toulouse, in
"Fermentation: réacteur, capteur, automatisation". ed. H. Blachère,
G. Durand.

HARTLINE, F.F., 1979, Lowering the cost of alcohol. Science, 206, 41-42.

HERNANDEZ MENA, R., J.A. RIBAUD, A.E. HUMPHREY, 1980, Demonstration of
process feasibility for continuous extraction of alcohols from fermenting
systems, in "Proc. of the VIth international fermentation symposium".
Session Novel Bioreactors. F 7.3.8. (P). London-(Canada).

MAHENDRA KUMAR JAIN, J. GLEESON, A. UPRETI, G.C. UPRETI, 1978, Intrinsic
perturbing ability of alkanols in lipids bilayers. Biochimica Biophysica
acta, 509, 1-8.

MAIORELLA, B., C.R. WILKE, 1980, Energy requirements for the Vacuferm
process. Biotechnol. Bioeng., 22, 1749-1751.

MUNK, V., O. VOLFOVA, M. DOSTALEK, J. MOSTECKY, K. PECKA, 1969, Cultivation
of the yeast Candida lipolytica an hydrocarbon. III Oxidation and utilization of individual pure hydrocarbons. Folia Microbiologia, 14, 334-344.

RAMALINGHAM, A., R.K. FINN, 1977, The Vacuferm process: A new approach to
fermentation alcohol. Biotechnol. Bioeng., 19, 583-589.

SERIEYS, M., G. GOMA, G. DURAND, 1978, Design and oxygen: Transfer potential
of a pulsed continuous tubular fermentor. Biotechnol. Bioeng., 20,
1393-1406.

STREHAIANO, P., M. MORENO, G. GOMA, 1978, Fermentation alcoolique: influence
de la concentration en glucose sur le taux de production d'éthanol et le
taux de croissance. C.R.Acad. Sc. Press, 286, Serie D, 225-258.

AGRICULTURAL WASTE TREATMENT BY MEANS OF ULTRAFILTRATION

MEMBRANE ENZYMATIC REACTORS

L. GIANFREDA and G. GRECO jr°

Istituto di Farmacologia Sperimentale, Facoltà di Farmacia, University of Naples, Italy.

° Istituto di Principi di Ingegneria Chimica, Facoltà di Ingegneria, University of Naples, Italy.

Summary

In recent years, the enzymatic hydrolysis of cellulosic agricultural waste for the production of sugary solutions, that can be either employed directly or can undergo alcoholic fermentation, is becoming an increasingly studied process. As an alternative to existing layouts, and mainly in order to solve the problems related to enzymes stability and recovery, new schemes are being studied that imply the use of ultrafiltration membrane reactors.

Main purpose of the present communication is to discuss some prelimi nary experimental results obtained on cellobiose hydrolysis by β-gluco sidase (E.C. 3.2.1.21, from sweet almonds). This reaction step has to be regarded as a somewhat critical stage in cellulosic waste treatment since cellobiose inhibits the exoglucanases and hence its hydrolysis can become the overall rate-controlling step.

Cellobiase performance in an unstirred ultrafiltration membrane reactor has been analyzed at the high concentration levels attained as a consequence of the polarization phenomena taking place within this system. Comparisons have been performed with the behaviour of the same enzyme when operating in a high macromolecular (CM-cellulose) concentration environment simulating the conditions to be found in the actual system immediately upstream from the membrane surface. A considerable and encouraging increase in enzyme stability has been observed.

1. INTRODUCTION

In the near future, glucose production from cellulosic waste could become an extremely interesting industrial process. However, some fundamental
problems are still unsolved that do not allow laboratory and pilot plant re
sults to be satisfactorily translated into industrial practice. This is par
tly due to the fact that most research efforts have been mainly directed to
wards somewhat preliminary problems (cellulosic waste pretreatments, develop
ment of better cellulase complexes etc..) and little, if any, attention has
been devoted to the development of alternative enzymatic reactor layouts. In
deed, in cellulose hydrolysis by means of soluble cellulase, enzymes adsorption onto the insoluble fraction takes place. This, in turn, implies the
continuous loss of enzymes (which amount to 35% of the total process costs)
even if a recirculation system is adopted. On the other hand, the tradition
al immobilization techniques do not seem suitable, partly because of unavoid
able increases in overall costs, and mainly because of the difficulties that
arise in immobilizing different enzymes onto the same support and in dealing
with macromolecular substrates (severe mass transfer limitations).

In the last few years, our research group has devoted considerable attention to the development of ultrafiltration membrane enzymatic reac
tors, to the related dynamical immobilization techniques and to enzymes sta
bilization methods in high macromolecular concentration environments (1-9).
Purpose of the present paper is to discuss some preliminary experimental re
sults that appear to be quite encouraging for the application of the above
techniques to cellulose hydrolysis.

2. EXPERIMENTAL TECHNIQUE

Use has been made of an unstirred ultrafiltration cell of the plane mem
brane type (4cm I.D., 20 ml overall internal volume) equipped with a DDS600
ultrafiltration membrane (The Danish Sugar Co., Nakskov, Denmark; molecular
weight cut-off of 20,000). The system is fed by means of a Nitrogen-pressu-
rized vessel. A gas-chromatographic type injection system has been inserted
upstream from the reactor. This enables predetermined amounts of enzyme and

or of other macromolecular species to be injected with no perturbation of the cell pressure and permeate regimes. The reacting system is cellobiose (Merck, Darmstadt, FRG) hydrolysis by β-glucosidase (E.C. 3.2.1.21, from sweet almonds, BDH ltd., Poole, UK) and has been chosen for these prelimi nary tests since cellobiose inhibits the exoglucanases and hence its hydro lysis can become the overall rate-determining step. Glucose concentration in the permeate has been measured by means of a standard GOD-Perid kit (Bo heringer Biochemia, Mannheim, FRG). The membrane cut-off is such as to com pletely reject the enzyme. Therefore, the protein, once injected, undergoes concentration polarization phenomena that eventually result in its confine ment at fairly high concentration levels within a very narrow region immedia tely upstream from the U.F. membrane (see Fig.1). A dynamical immobilization has been thus achieved of the enzyme which obviously lasts as long as per meation takes place. Indeed, once the permeating flux is interrupted, enzy me back-diffusion occurs together with re-mixing within the whole reactor volume. Typical β-glucosidase performance under these quite unusual condi tions is reported in Fig.2 in terms of specific reaction rate vs. reaction time. It has to be outlined that the reaction temperature (55 °C) is quite high as compared to the usual operating conditions for this enzyme (37 °C) (10); accordingly, its stability is poor as shown by the fast decay in spe cific rate.

If a macromolecular solution is injected into the system, once the enzy me steady-state has been achieved, and if the molecular weight of the macro molecule exceeds the membrane cut-off, its accumulation takes place within the very reactor region where the enzyme had been previously confined dyna mically. If a linear-chain polymer is employed in sufficient amounts, the build-up in its concentration gives rise to the formation of a gel-like re gion where the enzyme is included. In a previous work (9), it has been shown that for different reacting systems this procedure results in a considera ble enhancement in enzyme stability if use is made of synthetic, water so luble polyacrylamides. Indeed, although fairly labile (it dissolves quite easily upon interruption of the permeating flux) the macromolecular gel structure is tight enough as to reduce the enzyme mobility to virtually nil and therefore to prevent, to some extent, the unfolding of the proteic macro molecule.

It is quite obvious that, in an ultrafiltration membrane reactor performing the hydrolysis of cellulose, a somewhat similar situation is bound to occur because of the accumulation of the high molecular weight fraction at the membrane surface. By the way, the corresponding concentration levels can be, in principle, controlled through the feed composition, the permeation rate and the enzymes amounts fed.

Obviously, if also the high molecular weight soluble cellulose could produce the same stabilizing effects, the application of an ultrafiltration membrane hydrolysis reactor could become extremely interesting since one could obtain the dynamical immobilization of the enzymes and their stabilization at zero additional cost. Experimental runs have been therefore performed in which limited amounts of CM-Cellulose have been injected into the system, once the steady-state β -glucosidase concentration profile had been attained. Typical results have been reported in the same Fig.2, for comparison purposes with the dynamically immobilized enzyme with no stabilization. A considerable increase in the enzyme half-life has been obtained and the enzyme activity level is virtually unaffected. This indicates that the procedure does not inactivate the enzyme and, furthermore, that extremely limited mass tranfer resistances take plece within the gel-like region.

An unusually high reaction temperature can be safely adopted, which obviously increases the overall reaction rate, thus decreasing the enzyme amounts to be used for a given overall yield.

That CM-Cellulose shows the same stabilizing effect as the one produced by a substantially different macromolecule (polyacrylamide) is hardly surprising since the chemical nature of the stabilizing species should be virtually irrelevant, the stabilization mainly depending on the reduction in enzyme mobility within the gel.

It has to be outlined that further beneficial effects could be, in principle, obtained by means of an ultrafiltration membrane reactor, namely :

i) the high molecular weight substrate accumulation at the membrane surface
gives rise to local increases in substrate concentration that could re
sult in an increase in the overall rate as compared to the usual perfect
ly stirred reactor situation;

ii) by reducing the membrane cut-off, one could simultaneously perform the
enzymatic process and a concentration stage of the outlet sugary solu
tions.

ACKNOWLEDGMENTS

This work has been supported by the EC Solar Energy Program, Project E
"Energy from Biomass", contract number ESE/060/I.

LITERATURE CITED

1 - Cantarella M., Gianfreda L., Palescandolo R., Scardi V. and Greco G. jr,
Alfani F., Iorio G., J. Solid Phase Biochem. 2, 163, (1977).
2 - Cantarella M., Rémy M.H., Scardi V. and Alfani F., Iorio G., Greco G.jr,
Biochem J. 179, 15, (1979).
3 - Greco G.jr, Alfani F., Iorio G. and Cantarella M., Formisano A., Gianfre
da L., Palescandolo R., Scardi V., Biotechnol. Bioeng. 21, 1421, (1979).
4 - Alfani F., Iorio G., Greco G.jr and Cantarella M., Rémy M.H., Scardi V.,
Chem. Eng. Sci. 34, 1213, (1979).
5 - Greco G.jr and Albanesi D., Cantarella M., Gianfreda L., Palescandolo R.
Scardi V., Eur. J. Appl. Microbiol. Biotechnol. 8, 249, (1979).
6 - Greco G.jr and Albanesi D., Cantarella M., Scardi V., Biotechnol. Bioeng.
22, 215, (1980).
7 - Greco G.jr, Alfani F. and Cantarella M., Gianfreda L., Palescandolo R.,
Scardi V., Chem. Eng. Commun. 6, 168, (1980).
8 - Greco G.jr and Gianfreda L., Albanesi D., Cantarella M., J. Appl. Bio
chem (1980) submitted.
9 - Greco G.jr,Gianfreda L., Biotechol. Bioeng. (1980) submitted.
10 - Venardos D., Klei H. E., Sundstrom D. W., Enzyme Microb. Technol. 2,112,
(1980).

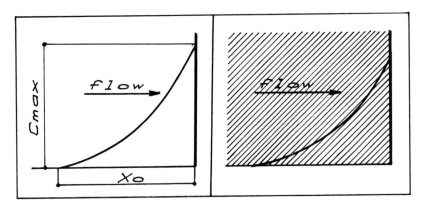

FIG. 1 - DYNAMICALLY IMMOBILIZED ENZYME CONCENTRATION PROFILES
Experimental conditions : enzyme molecular weight = 200,000
permeate rate = 0.025 ml/min sq cm, enzyme amount = 0.05 mg/sq cm
$X_o = 3.7 \ 10^{-3}$ cm, $C_{max} = 50$ mg/ml.

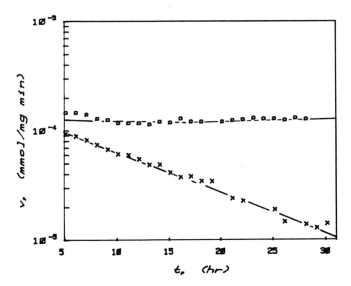

FIG. 2 - CELLOBIOSE HYDROLYSIS BY β -GLUCOSIDASE
T = 55 °C, pH = 5.0 (in 0.1 M acetate buffer), Cellobiose concen
tration = 5 mM, Enzyme amount = 0.13 mg/sq cm
x-x curve : dynamically immobilized enzyme
o-o curve : CMC-stabilized enzyme (CMC amount = 0.67 mg/sq cm)

1-THIO-OLIGOSACCHARIDES AS SUBSTRATE ANALOGS

FOR INVESTIGATIONS IN THE FIELD OF POLYSACCHARIDASES

M. BLANC-MUESSER, J. DEFAYE, H. DRIGUEZ

and E. OHLEYER

Centre de Recherches sur les Macromolécules Végétales,
CNRS, 53 X, 38041 GRENOBLE (France).

Summary

A general sequence is described for the synthesis of thio-oligosac-charides of interest as substrate analogs for the induction, purification, evaluation and elucidation of the mechanism of action of cellulases and xylanases. The sodium salt of 1-thio-β-D-glucopyranose or 1-thio-β-D-xylo-pyranose, resulting from the action of thioacetic acid in the presence of zirconium chloride, on the corresponding β-peracetates in dichloromethane, reacts smoothly respectively with a 1,6-anhydro-4-O-triflyl-D-galactose or a 4-O-triflyl-L-arabinose derivative to give after conventional treatments thiocellobiose or thioxylobiose. Further sequences involving the corres-ponding α-bromides yield either thioxylotriose or the β-nitrophenyl or nitrophenyl 1-thio derivatives of thioxylobiose and the corresponding cello compounds. From a preliminary investigation, thioxylobiose is a competitive inhibitor for Poria xylanase with a K^I similar to that of xylobiose. The β-o-nitrophenyl glycoside is a good chromogenic substrate and thiocello-biose an excellent inducer for the cellulase of Schizophillum communae.

1. INTRODUCTION

1-Thioglycosides have proved versatile as investigational tools for studies on glycosidases, serving as inducers (1), competitive inhibitors (2), or as ligands for affinity chromatography (3). Such a broad interest for this class of carbohydrate derivatives, in a field of considerable importance for biomass conversion into fuel or chemicals, has to be ascribed to a lower basicity of the sulfur atom with respect to oxygen. This results usually in a slower rate of hydrolysis of 1-thioglycosides compared to O-glycosides under chemical as well as enzymic hydrolytic conditions.

Despite of extensive research on the chemical synthesis of this class of compounds, difficulties mainly related to the control of the anomeric specificity still remained unresolved up to recently (4), which hampered their use in carbohydrate enzymology, particularly in the field of glycanases. Enzymes as amylases (5), cellulases (6) or xylanases (7) are known indeed to require oligosaccharide structures for the formation of the enzyme-substrate complex, which means that specific inducers, competitive inhibitors or ligands in this area have to be thio-oligosaccharides. Thio-maltose and thio-maltotriose have proved to be valuable compounds for studies related to α-amylase (8, 9). This paper intends to report on our results on thio-oligosaccharide synthesis as substrate analogs for xylanases and cellulases, i.e. β-1,4-thio-oligomers of D-xylopyranose and D-glucopyranose.

2. RESULTS AND DISCUSSION

β-D-1,4-linked 1-thio-disaccharides may be obtained starting either from a C-4 glycose thiolate and an α-D-glycopyranosyl halide (4, pathway A scheme 1) or alternatively from a β-D-1-thio-glycopyranose reacting with a convenient electrophilic center at C-4 of another glycosyl unit (pathway B, scheme 1). Contrary to the usual approach in oligosaccharide synthesis where pathway A is the classical one, the enhanced nucleophilicity of sulfur compared to oxygen allows the use of pathway B for thio-oligosaccharide synthesis if a convenient leaving group is located at C-4 of the alternate molecule. A trifluoromethanesulfonyl activating group was introduced for this purpose in conjunction with the use of a strong dipolar aprotic solvent, hexamethylphosphoramide (HMPA).

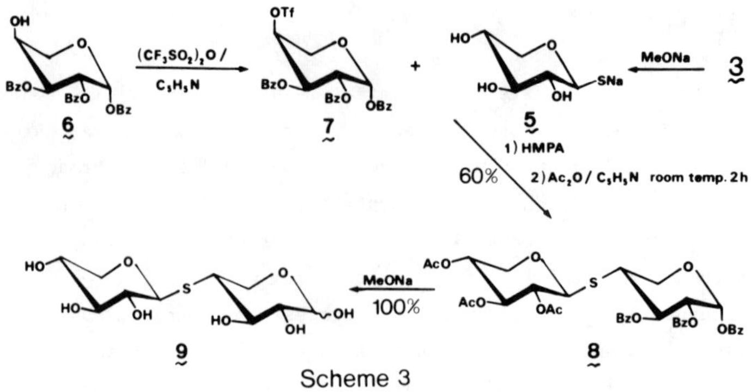

Scheme 3

The acyl disaccharide 8 is further easily transformed into the key bromo-glycosyl disaccharide 10 which leads to the β-thioacetate 11 through reaction with potassium thioacetate in acetone and gives an easy access to thioxylotriose 13 after successive thiolate activation 12, reaction with the 4-0-triflyl-L-arabinose 7 followed by conventional isolation and de-0-acylation (scheme 4). Thio-disaccharide 9 and -trisaccharide 13 are potential inducers for xylanases.

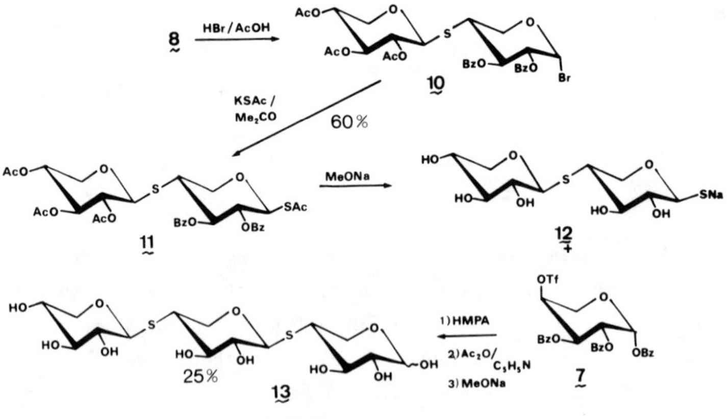

Scheme 4

The protected halo-glycosyl-disaccharide 10 is converted, in another step, to the p-nitrophenyl 1-thio-thioxylobioside 15 (p.f. 159-162°, $[\alpha]_D$ - 8.2, c 0.18, pyridine) or o-nitrophenyl thioxylobioside 17 (p.f. 177-179°,

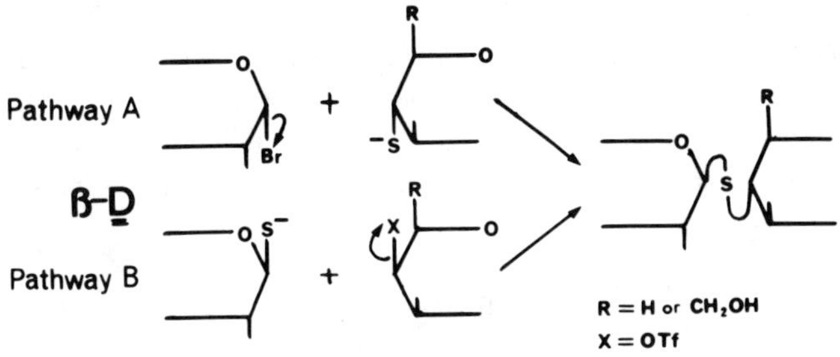

Scheme 1 ALTERNATIVE PATHWAYS FOR THE SYNTHESIS OF
β-1,4 LINKED D-GLUCO- AND D-XYLO-THIOOLIGOSACCHARIDES.

Previous results (4, 10) from this laboratory have led to a complete
stereocontrol for the introduction of thioaryl or thioacyl substituents at
anomeric centers starting either from 1,2-cis or 1,2-trans glycosyl-halides.
An even easier method for the preparation of 1-thio-β-D-glycopyranoses in
the xylo or gluco series involves the action of thioacetic acid on a β-D-
per-O-acetylated sugar (scheme 2). 2,3,4-Tri-O-acetyl-1-S-acetyl-1-thio-
β-D-xylopyranose 3 (11) and the corresponding D-glucopyranose derivative
4 (12) were prepared according to this scheme.

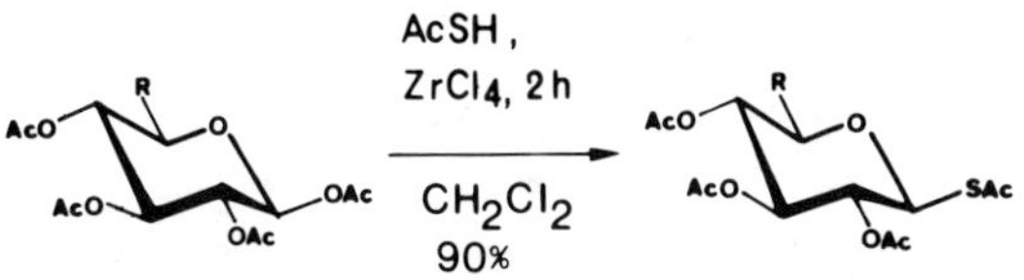

Scheme 2

The O-acyl protected thioxylobiose derivative 8 ($[\alpha]_D$ + 65°, c 0.7,
chloroform) was obtained in good yield through pathway B (scheme 1) starting from the 4-O-triflate 7, resulting from the action of trifluoromethane-
sulfonic anhydride in pyridine on 1,2,3-tri-O-benzoyl-β-L-arabinopyranose
6 (13), and the thiolate 5 obtained from 3. Conventional extraction and de-
O-acylation afforded thioxylobiose ($[\alpha]_D$ - 23°, c 2.6, methanol) in quantitative yield (scheme 3).

$[\alpha]_D$ -75.9, c 0.18, pyridine) by the respective action of p-nitrothiophenol or o-nitrophenol in acetone in the presence of potassium carbonate (scheme 5). The aryl bis-thio-disaccharide 15 is a potential ligand for the affinity chromatography of xylanases and the corresponding aryl thio-disaccharide 17 a potential chromogenic substrate for the evaluation of the activity of these enzymes.

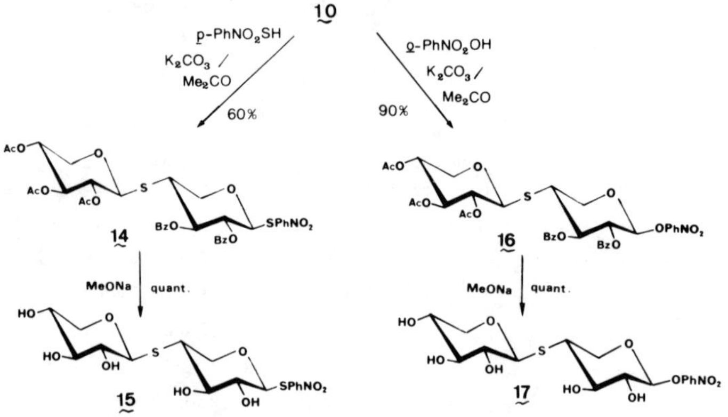

Scheme 5

A similar sequence was devised for the preparation of thiocellobiose 24. Starting from 2-O-acetyl-1,6-anhydro-3,4-O-isopropylidene-β-D-galacto-pyranose 18 (14), use of the intermediate ortho-ester 19 allows selective protection at C-3 20 through acidic opening. The corresponding triflate 21 reacts then smoothly with the activated 1-thio-β-D-glycopyranose 22, result-ing from the action of sodium methylate on 4, in HMPA at room temperature to give the 1,6-anhydro-thiocellobiose derivative 23 and then, through ace-tolysis and O-acyl deprotection, the expected thiodisaccharide 24 ([α]_D -10.2, c 0.46, water) (scheme 6). Further sequences as for xylosyl derivati-ves 13, 15 and 17 allow the preparation of cellotriose, the p-nitrophenyl 1-thiocellobioside and o-nitrophenyl thiocellobioside, homologues of 15 and 17.

3. CONCLUSION AND PROSPECT

As in the field of α-amylase where thiomaltose, thiomaltotriose and their o-nitrophenyl and p-nitrophenyl 1-thio derivatives have proved (8, 9) to be valuable substrate analogs acting either as competitive inhibitors

for the cristallographic visualisation of the active site of porcine pan-
creatic α-amylase or as defined chromogenic substrates for its enzymic activity evaluation, thio analogs in the xylo- and cello-dextrines series
appear to be potent tools for investigations in the field of xylanases or
cellulases.

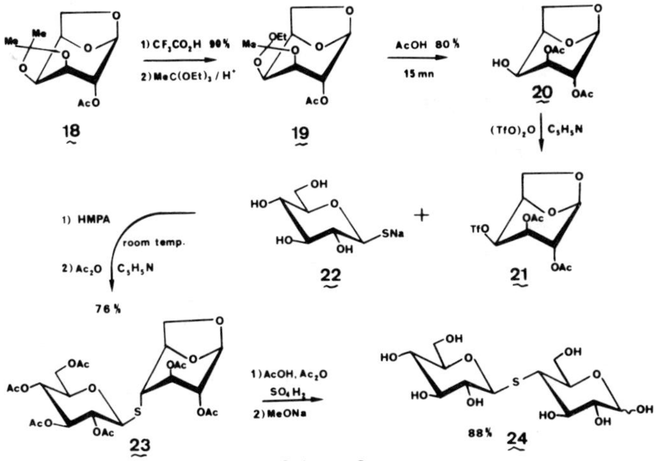

Scheme 6

Preliminary results (15) indicate that thioxylobiose $\underline{9}$ is an inhibitor
for Poria xylanase with a K^I of 8.10^{-4} M, a magnitude similar to that obtained with xylobiose. On the other hand, the corresponding nitrophenyl glycoside $\underline{17}$ is a good chromogenic substrate for the same enzyme (K_m 6.10^{-3}M)
and shall prove useful for the quantitative evaluation of its enzymic activity. Furthermore, thiocellobiose $\underline{24}$ is a potent inducer for the cellulase
of Schizophillum communae with an inductive effect twenty time that of cellobiose (16). Further investigations are in progress to evaluate the nitrophenyl 1-thio-glycoside $\underline{17}$ and its cello-oligosaccharide counterpart as
ligands for the purification of xylanases and cellulases by affinity chromatography. Up to now, thio-oligosaccharides appear to hold their promise
as potent tools for glycanase production as well as investigation at the
basic level.

REFERENCES

1. J. Monod, Enzymes, Units of Biological Structure and Function, Academic
   Press, New-York (1956) S. 7 ; ibid., Angew. Chem., 71 (1959) 685.

2. K. Wallenfels and O. P. Malhotra, Adv. Carbohydr. Chem., 16 (1961) 239 ;
   S. Tomino and K. Paigen, in J. R. Beckwith (Ed.), Lactose Operon, Cold
   Spring Harbor Laboratory : Cold Spring Harbor, New-York, 1970, p. 233 ;
   E. Steers, Jr., P. Cuatrecasas and H. B. Pollard, J. Biol. Chem., 246
   (1971) 196 ; K. L. Matta and O. P. Bahl, J. Biol. Chem., 247 (1972) 1780.

3. M. Claeyssens, H. Kersters-Hilderson, J. P. Van Wauwe and C. K. De Bruy-
   ne, FEBS Lett., 11 (1970) 336 ; M. E. Rafestin, A. Obrenovitch, A. Oblin
   and M. Monsigny, FEBS Lett., 40 (1974) 62 ; L. Kiss and E. Laszlo, Proc.
   Hung. Annu. Meet. Biochem., 18 (1978) 217.

4. M. Blanc-Muesser, J. Defaye and H. Driguez, Carbohydr. Res., 67 (1978)
   305.

5. D. French, in W. J. Whelan (Ed.), Biochemistry of Carbohydrates, Butter-
   worths, London, 1975, p. 267.

6. D. R. Whitaker, in P. D. Boyer (Ed.), The Enzymes, Academic Press, New-
   York, 3rd edit., vol. 5, 1971, p. 273.

7. R. F. H. Dekker and G. N. Richards, Adv. Carbohydr. Chem. Biochem., 32
   (1976) 277.

8. M. Blanc-Muesser, J. Defaye, H. Driguez and E. Ohleyer, Xth Intern. Symp.
   Carbohydr. Chem., Sydney, Australie, July 7-11 1980, abstract IL. 3.

9. M. Blanc-Muesser, D. Sc. Thesis, University of Grenoble, Dec. 1980.

10. M. Apparu, M. Blanc-Muesser, J. Defaye and H. Driguez, Can. J. Chem.,
    in press.

11. M. Gehrke and W. Kohler, Ber., 64B (1931) 2696.

12. F. Wrede, Z. Physiol. Chem., 119 (1922) 46 ; Methods Carbohydr. Chem.,
    2 (1963) 433.

13. J. F. Batey, C. Bullock, E. O'Brien and J. M. Williams, Carbohydr. Res.,
    43 (1975) 43.

14. H. Masamune and S. Kamiyama, Tôhoku J. Exptl. Med., 66 (1957) 43.

15. In collaboration with J. Comtat, Grenoble.

16. In collaboration with L. Jurasek, Montréal.

DEGRADATION OF CELLULOSE WITH HYDROGEN FLUORIDE

J. Defaye and A. Gadelle

Centre de Recherches sur les Macromolécules Vegetales,
CNRS, 53X, 38041 Grenoble, Cédex, France

C. Pedersen

The Technical University of Denmark, Department of
Organic Chemistry, 2800 Lyngby, Denmark

Summary

Cellulose is readily soluble in anhydrous hydrogen fluoride (HF) and is rapidly broken down by this solvent. When cellulose is treated with HF at -5° a water soluble, partially degraded cellulose can be obtained. At 20° cellulose is completely cleaved by HF in 1 h to give α-D-glucopyranosyl fluoride which subsequently undergoes condensation to a mixture of glucose-oligomers. The glucosyl fluoride and the oligomers are in an equilibrium; at low cellulose concentration the fluoride is the main product, at high concentration the oligomers dominate.

1. INTRODUCTION

Chemical degradation of cellulose and other insoluble polysaccharides to water-soluble mono- or oligo-saccharides has been performed with hydrochloric or sulfuric acid at a temperature of 80-180° (1). This procedure is energy-consuming and it causes some decomposition of the mono-saccharides resulting in reduced yields and undesirable byproducts. Trifluoroacetic acid gives better yields of degradation products (2), but it is expensive. Helferich et al. showed that cellulose is readily soluble in anhydrous hydrogen fluoride (HF) and that it is completely converted into water soluble products by this treatment (3). Similar results were obtained by Fredenhagen who also found that polysaccharides could be extracted from wood by treatment with HF (4). Later studies by Russian chemists confirmed these results (5). The water soluble product obtained by treatment of cellulose with HF was found to be a mixture of oligosaccharides and the same product was obtained when amylose or glucose was treated with HF (3,4,5). Cellulose may also be dissolved in aqueous hydrogen fluoride and carbohydrates can be separated from lignin in wood by this reagent. However, cellulose will only dissolve rapidly in highly concentrated (80-90 %) aqueous hydrogen fluoride (4,5,6).

2. RESULTS

We have now reinvestigated the reaction of HF with cellulose and through the use of ^{13}C NMR spectroscopy we have obtained more information about the products obtained.

Cellulose is readily soluble in HF as described above and solutions containing 40 % cellulose can be prepared within a few minutes at temperatures ranging from −10 to 20° (the boiling point of HF). Below −10° solubilization is slow. The products formed from cellulose and HF can be isolated from the solution either by evaporating the HF or by precipitation with diethyl ether (see experimental).

When cellulose (10 g) was dissolved in HF (15 ml) at ca. −5° and kept at that temperature for 1 h precipitation with ether gave a product which was soluble in water. A ^{13}C NMR spectrum (Fig 1A) showed that this product is almost exclusively β-1,4-linked oligosaccharides, i.e. partially degraded cellulose. Chromatographic analysis showed that it is a mixture of products with D.P. ranging from 1 to 10. When cellulose was treated with HF in the same manner, but at 20°, a different product was obtained as seen from a ^{13}C NMR spectrum (Fig 1B). This product is a mixture of oligomers having mostly α-linked glucose units. The same product was obtained when glucose or 1, 6-anhydro-β-D-glucopyranose, was treated with HF under the same conditions as also observed by previous workers.

When more dilute solutions of cellulose in HF are prepared α-D-glucopyranosyl fluoride is formed. Thus, when 10 g of cellulose was kept in 80 ml of HF for 1 h at 20° followed by precipitation with ether the main product obtained was glucosyl fluoride. When, on the other hand, the HF was evaporated from the dilute solution the residue consisted of the mixture of oligosaccharides described above (Fig 1B) and only traces of glucosyl fluoride were present. ^{13}C NMR spectra measured directly on HF solutions show that both glucosyl fluoride (formed at low concentration of cellulose) and the oligomers (high concentration) are stable in HF solution for several days.

Hence, these and previous experiments indicate that cellulose is completely degraded to α-D-glucopyranosyl fluoride in HF solution. The glucosyl fluoride is in equilibrium with a mixture of oligomers; at low concentration the fluoride is the main product, at high concentration it condenses to the oligomers. The latter probably have structures similar to the products obtained by acid catalyzed "reversion" of glucose (7).

Finally, pine wood was extracted with HF as described previously (4, 5, 8). The water soluble product (see experimental) gave the ^{13}C NMR spectrum shown in Fig 1C which shows that the product is a mixture of oligomers and glucose.

3. EXPERIMENTAL

Solutions in anhydrous hydrogen fluoride (HF) were prepared in polyethylene bottles. ^{13}C NMR spectra were measured in D_2O solution on Bruker WH-90 or HX-270 MHz instruments using dioxane as internal reference. ^{13}C NMR spectra on HF solutions were measured in Teflon sample tubes.

Treatment of cellulose with HF

Cellulose powder (Whatmann) (10 g) in a 250 ml polyethylene bottle was cooled in ice-salt while 15 ml of HF was added in portions during ca. 1 min (caution! exothermic reaction). The mixture was shaken for 3-4 min until a clear solution was obtained and it was then kept for 1 h, either at ca. -5° or at room temperature. The solution was then cooled in dry ice-acetone and cold ether (100 ml) was added precipitating a thick, syrupy product. The ether was decanted off and the product was stirred several times with ether using a polyethylene rod. This gave an amorphous powder which was filtered off and dried in vacuum. Yield 9.5 g.

Alternatively, the HF solution was evaporated in a stream of air leaving a syrupy residue. When this was stirred 3-4 times with ether a solid product was obtained which could be filtered off.

Treatment of pine wood with HF

Pine wood (25 g) was cooled in ice while HF (30 ml) was added. After ca. 15 min the HF was completely absorbed and more HF (30 ml) was added. After 1 h at room temperature the HF was decanted off and the wood was extracted with two additional portions of HF (30 ml) followed by extraction with water (2 x 30 ml). The combined extracts were evaporated in a stream of air and the residue was stirred with ether to give a brown, amorphous powder which was readily soluble in water. Yield 9.5 g.

The insoluble part of the wood was washed with methanol and dried. Yield 9.5 g.

4. REFERENCES

1.   A. E. Humphrey, Adv. Chem. Ser. 181 (1979) 25; J. P. Sachetto, Actual Chim. (1978) 65; E. E. Harris, Adv. Carbohydr. Chem. 4 (1979) 154.
2.   D. Fengel and G. Wegener, Adv. Chem. Ser. 181 (1979) 145.
3.   B. Helferich and S. Böttger, Ann. 476 (1929) 150; B. Helferich, A. Stärker and O. Peters, Ann. 482 (1930) 183; B. Helferich and O. Peters, Ann. 494 (1932) 101.
4.   K. Fredenhagen and G. Cadenbach, Angewandte Chemie 46 (1933) 133.
5.   Z. A. Rogovin and Y. L. Pogosov, Nauch. Doklady Vysshei Shkoly, Khim i Khim. Technol. (1959) 368, Chem. Abstracts 53 (1959) 22912; Y. L. Pogosov and Z. A. Rogovin, Uzbek. Khim. Zhur. (1960) 58, Chem. Abstracts 55 (1961) 24100, V. I. Sharkov, A. K. Bolotova and T. A. Boiko, Komplehs. Pererab. Rast. Syrya (1972) 39, Chem. Abstracts 80 (1974) 72199.

6.   R. Willstätter and L. Zechmeister, <u>Ber.</u> 46 (1913) 2401.

7.   I. J. Goldstein and T. L. Huller, <u>Adv. Carbohydr. Chem.</u> 21 (1966) 431.

8.   I. B. Sachs, I. T. Clark and J. C. Pew, <u>J. Polymer Science,</u> Part C (1963) 203.

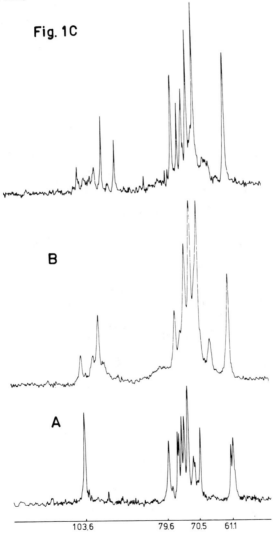

Fig. 1C

CELLULASE AND β-GLUCOSIDASE FROM CLOSTRIDIUM THERMOCELLUM

EXCRETION, LOCALIZATION AND PURIFICATION

N. CREUZET, N. AIT, P. FORGET and J. CATTANEO

Laboratoire de Chimie Bactérienne, C.N.R.S., 31, chemin
Joseph Aiguier, 13277 Marseille Cedex 9 (France)

Summary

Clostridium thermocellum is an anaerobic thermophile that produces endo and exo-β-glucanase when grown on cellulose or cellobiose as carbon source. The location of the cellulase components has been determined. Polyacrylamide gel electrophoresis of concentrated supernatant from a culture grown on cellulose resolved several peaks of cellulolytic activity corresponding to cellulases differing in the patterns of their attack on CMC and cellulose.

Partial purification of cellulase has been obtained by affinity chromatography on cellulose powder ; the results show that the preparation obtained is a multi-enzyme complex.

A β-glucosidase was isolated, purified and its properties studied. The enzyme is specific for the β-configuration ; it acts on β-glucosides and on aryl-β-glucosides but its affinity for cellobiose is less than for the synthetic substrate PNPG. β-glucosidase of C. thermocellum has been shown to be highly heat stable and could be of great interest in the industrial saccharification of cellulose.

1. INTRODUCTION

Clostridium thermocellum offers great interest for the conversion of cellulosic materials into sugars, chemicals and liquid fuels. This thermophilic
bacterium grows on cellulose, first producing sugars and then converting
them into ethanol, acetic acid, lactic acid, carbon dioxide and hydrogen (1).

In order to use C. thermocellum effectively, much has to be learned concerning the enzymes involved in cellulolysis.Cristalline cellulose is attacked by endo-1,4-β-glucanase (EC 3.2.1.4) and exo-1,4-β-glucanase (EC 3.2.1.
91) whose concerted action releases cellodextrins and cellobiose which may
be further metabolized either by a β-D-glucoside glucohydrolase EC 3.2.1.21
(β-glucosidase) or by cellobiose or cellodextrin phosphorylase (2,3).Cel-
lulase and β-glucosidase of C. thermocellum were studied.

2. CELLULASES

C. thermocellum (NCIB 10682) has been grown anaerobically at 60° C on CM3
medium as described by Weimer and Zeikus (4) with cellulose 10 g l^{-1} or cellobiose 2 g l^{-1}. Cellulase and carboxymethylcellulase were determined as previously described (5). The capacity to use different sugars as growth substrates was tested. C.thermocellum was able to grow on cellulose,cellobiose,
carboxymethylcellulose, glucose and fructose ; the specific activity of carboxymethylcellulase (μ per mg of bacterial proteins) was the same whatever
the substrate used. The cellulase of C. thermocellum appears to be constitutive. This result agrees with those of Hammerstrom et al (6) and those of
Zeikus (1) but differs from that obtained by Lee and Blackburn for a thermophilic clostridium species (7).

The localization of the cellulase has been determined. When grown on
cellobiose, cellulase seems to be truly extracellular ; about 90 % of the
total activity determined on centrifuged sample was recovered in the supernatant. When grown on cellulose the cellulase seems also to be excreted
into the culture medium but first bound to cellulose. After 40 hours of
growth only 24.8 % of the cellulase activity was recovered in the cellulose-
free supernatant but after 112 hours more than 90 % was recovered in the
supernatant. Acrylamide gel electrophoresis of concentrated supernatant
from a culture grown on cellulose resolved several peaks of cellulolytic
activity (figure 1).

In order to distinguish the mode of action of these enzymes reducing
sugar production was measured concomitantly with the decrease in viscosity
after incubation with CM-cellulose. Enzyme acting in a random manner will

produce a greater decrease in the viscosity of CMC per unit increase in reducing power than will an enzyme attacking from the end of the chain ; this may be used to distinguish "cellulases" that differ in the pattern of their attack. Peak V produced reducing power relatively strongly compared with its ability to decrease the viscosity of CMC and peaks VI and VII produced a greater decrease in viscosity for the same amount of reducing equivalents. The enzyme of peak V seemed to be the less random in its attack on CMC while the enzymes of peak VI and VII were the most ramdon.

TABLE I . RELATIVE ACTIVITIES OF C. THERMOCELLUM CELLULOLYTIC COMPONENTS SEPARATED BY GEL ELECTROPHORESIS.

Peaks	Ratio of CM-cellulase (Viscometric method[b]) to cellulase activity[a]	Changes in specific fluidity of CMC[b] for 100 μmoles of reducing equivalents liberated
I	0.25	14.4
V	0.10	9.2
VI	0.65	30.7
VII	2.12	27.2

a) Cellulase activity was determined as previously described (5) with Whatman CC41 cellulose. 1 unit was defined as the amount of enzyme releasing 1 μmole reducing glucose equivalent h^{-1}.
b) CM-cellulase (viscometric) was determined as follows : reaction mixture consisting of 0.8 % W/V CMC in 20 mM KH_2PO_4/K_2HPO_4 pH 7.0 was incubated at 60° with an appropriate amount of enzyme. The viscosity was determined at room temperature using a Ubbelhöhde viscometer. One unit of enzyme was defined as the amount of enzyme increasing the reciprocal of specific viscosity of 0.01 per minute.

From the ratio of CM-cellulase activity/cellulase activity shown in Table I, it seems also that the enzyme of peak V looks like an exoglucanase, splitting off cellobiosyl residues from the non-reducing end of cellulose whereas the enzymes of peaks VI and VII may be considered as endoglucanases attacking the glucosyl bounds of cellulose randomly.

The cellulase was partially purified from culture filtrate by affinity chromatography on cellulose powder. After having been precipitated successively by zinc chloride and ammonium sulfate (60 % S) the proteins of the supernatant were applied on a cellulose column on which the cellulases were adsorbed. The release of cellulase was obtained by decreasing the ionic strength. The preparation obtained was purified 5-6 fold with a recovery of 70 % of the initial enzyme activity.

Analytical polyacrylamide gel electrophoresis of the cellulase obtained after this step shows that the proteins of this preparation have a high molecular weight. In fact no protein seems to penetrate into 7 % gels whereas some of the proteins migrate into 5 % gels and form one band. However the electrophoresis in presence of SDS shows the presence of several bands of proteins, two of which having cellulolytic activity. These results show that the preparation obtained by affinity chromatography constitutes a multi-enzyme complex.

3. β-GLUCOSIDASE

Cellodextrin and cellobiose are generally produced by concerted action of endo and exoglucanases ; the pathways by which these products are further metabolized in C. thermocellum are not yet known. Cellodextrins and cellobiose may be phosphorolyzed by intracellular cellodextrin and cellobiose phosphorylase present in C. thermocellum (2,3) or they may be hydrolyzed by β-glucosidase. C. thermocellum was shown (8) to produce β-glucosidase. This enzyme is associated with cells and it is constitutive.

C. thermocellum β-glucosidase has been purified to apparent homogeneity (8) and some of its properties have been studied. It is active on para nitrophenyl glucoside (PNPG) and on cellobiose ; the ratio of these two activities was constant throughout the different steps of purification and maximum activity with both substrates was observed at pH 6.0 and at 65° C. These results can be taken as an indication that the same enzyme is responsible for both activities.

Based on the crude starting enzyme preparation the overall increase in specific activity of β-glucosidase was 800 fold with 8 % yield of activity. The enzyme obtained migrated as a single band in polyacrylamide gel electrophoresis showing activity on both PNPG and cellobiose.

The molecular weight estimated by gel filtration and by gel electrophoresis in presence of SDS was estimated to be 43 000 daltons and C. thermocellum β-glucosidase probably consists of a single polypeptide chain.

328

The substrate specificity of the β-glucosidase was studied. Laminaribiose (1 →3), sophorose (1→ 2), cellobiose (1→4) and gentiobiose (1→ 6) in order of decreasing susceptibility were utilized as substrates by the enzyme. The enzyme is not active on α-likage and it does not act on methyl-β -glucoside which is an alkyl-β-glucoside. It acts on aryl-β-glucoside ; PNPG is hydrolyzed very easily, salicin is also a substrate but arbutin does not seem to be hydrolyzed. Hydrolysis of p-nitrophenyl-D-galactoside is also observed whereas p-nitrophenylxyloside is a poor substrate. β-glucosidase acts on cellodextrins as shown by liberation of glucose.

Determinations of Michaelis constants show a Km value 2.6 mM and 83 mM for PNPG and cellobiose respectively ; the affinity of the β-glucosidase of C. thermocellum is much greater for PNPG than for cellobiose. Similar results have also been shown for some fungal β-glucosidases (9,10, 11).

β-glucosidase of C. thermocellum has been shown to be highly heat stable. The enzyme incubated in citrate-phosphate buffer with both substrates (PNPG and cellobiose) is quite stable at 55° C during the first 30 h and after 53 h only 26 % of its activity was lost. At 60° C the β-glucosidase loses about 40 % of its activity after 7 hours. The β-glucosidase of C. thermocellum is considerably more stable than analogous enzymes from bacteria and fungi. This could be of great interest in the industrial saccharification of cellulose. The fact of transforming cellulose into glucose in these thermal conditions may prevent some microbial contaminations.

REFERENCES

1 - NG, T.K., WEIMER, P.J., and ZEIKUS, J.G. (1977) Arch. Microbiol. 114, 1-7.

2 - SHETH, K. and ALEXANDER, J.K. (1967) Biochem. Biophys. Acta, 148, 808-810.

3 - ALEXANDER, J.K. (1968) J. Biol. Chem., 243, 2899-2904.

4 - WEIMER, P.J. and ZEIKUS, J.G. (1977) Appl. and Environ. Microbiol., 33, 289-297.

5 - AIT, N., CREUZET, N. and FORGET, P. (1979) J. Gen. Microbiol. 113, 399-402.

6 - HAMMERSTROM, R.A., CLAUS, K.D., COGHLAN, J.W. and McBEE, R.H. (1955) Arch. Biochem. Biophys. 56, 123-129.

7 - LEE, B.H. and BLACKBURN, T.H. (1975) Applied Microbiology,30,346-353.

8 - AIT, N., CREUZET, N. and CATTANEO, J. (1979) Biochem. Biophys. Res.
   Comm.90, 537-546.

9 - DESHPANDE, V., ERIKSSON, K.E. and PETTERSSON, B. (1978) Eur. J. Biochem.
   90, 191-198.

10 - BERGHEM, L.E.R. and PETTERSSON, L.G. (1974) Eur. J. Biochem. 46, 295-
    305.

11 - UMEZURIKE, G.M. (1975) Biochem.J. 145, 361-368.

FIGURE 1

Distribution of CM-cellulase and cellulase activities in gel slices
after polyacrylamide-gel electrophoresis of a concentrated supernatant
from a 2 days-old culture of C. thermocellum. Slices (2.2mm) of unstai-
ned gel were extracted with 1 ml of PO_4 buffer pH 7.0 and portions of the
extract were assayed for CM-cellulase activity by the determination of
reducing sugars from CMC (•—•—•) and cellulase activity (-▵—▵—▵-).

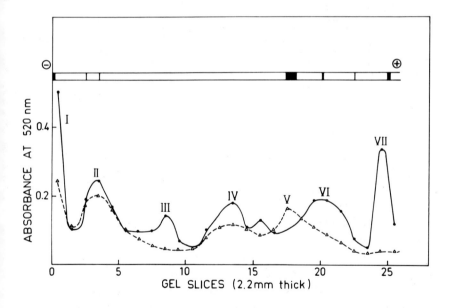

CONTINUOUS AUTOHYDROLYSIS, A KEY STEP
IN THE ECONOMIC CONVERSION OF FOREST
AND CROP RESIDUES INTO ETHANOL

JOHN D. TAYLOR

Stake Technology Ltd., 20A Enterprise Ave.,
Ottawa, Canada K2G 0A6

Summary
    A scheme for the conversion of forest and crop residues
into ethanol is described. Novel separation steps result in
maximum utilization of the three major components of lignocellulosic materials: cellulose, hemicellulose and lignin.
Technology for the continuous autohydrolysis of lignocellulosic materials has been developed. Raw materials are fed,
on a continuous basis, via a plug forming feeder, into a
steam pressurized cylindrical vessel containing a helical
screw conveyer. Autohydrolyzed materials are discharged intermittently to atmospheric pressure through an orifice.
Continuous operation effects an efficient steam consumption
and precise control of processing conditions. The autohydrolyzed product is subjected to an aqueous extraction
followed by an alkaline extraction resulting in the isolation of the three major components. The cellulose fraction is saccharified to glucose which is subsequently fermented to produce ethanol. The pentose-rich hemicellulose
fraction can be converted to furfural, xylitol or can undergo
a specific fermentation to ethanol. The lignin fraction,
which has a high value as fuel, has been shown to possess
some unique properties which should lead to other uses for
this fraction. Alcohol yields greater than 80% of theoretical have been achieved. Economic viability of the system is discussed.

1. INTRODUCTION

    World wide agricultural and forestry practices result
in the generation of billions of tons of waste lignocellulosic materials annually. These waste materials are a
potentially rich source of carbohydrates and any consideration
of their underutilization must examine the chemical composition lignocellulosic materials.

    The composition of higher plants can be grossly categor-
ized into three main fractions; cellulose, hemicellulose and
lignin. Cellulose is a linear polymer of glucose and can be
hydrolyzed either chemically or enzymatically. Hemicellulose
is a carbohydrate polymer composed primarily of five carbon
sugars. Lignin is a complex phenolic polymer which exists

in close association with the cellulose. This lignin-cellulose complex is responsible for the structural integrity of
plants but also restricts the accessibility to the cellulose
by hydrolytic agents necessitating pretreatments.

The global supplies of non-renewable energy resources
are rapidly being depleted. The conversion of renewable
waste biomass into a liquid fuel such as ethanol is an attractive prospect. High oil prices have led to more serious
consideration of biomass to ethanol schemes. For a conversion process to be economically viable, it must include an
effective low cost pretreatment and it must result in the
maximal utilization of all three fractions of lignocellulosic materials.

The necessity of utilizing hemicellulose and lignin as
well as cellulose cannot be overemphasized. The waste products of forestry and agriculture are approximately 50% cellulose. A system dealing only with the hydrolysis of cellulose to glucose for subsequent fermentation to ethanol
would be faced with utilizing only half of its raw material.
The economics of such a process are doubtful.

2. AUTOHYDROLYSIS

Lignocellulosic materials such as hardwoods and crop residues when subjected to high pressure steam for a specific period of time undergo what has been called an autohydrolysis.
Acetic acid generated from acetyl groups located in the hemicellulose fraction catalyze the hydrolysis of this fraction.
This has the effect of breaking down the lignin-cellulose complex rendering the cellulose susceptible to subsequent hydrolysis.

Stake Technology Ltd. began the development of the technology of autohydrolysis toward the goal of producing high
energy ruminant animal feeds from waste biomass. The initial
work involved high pressure batch systems, however it soon
became apparent that continuous processing would have several
advantages over the batch system, most notably: lower steam
consumption, lower capital cost and more precise control of
processing parameters, resulting in a more uniform treatment.

3. STAKE II SYSTEM

Stake subsequently developed the patented Stake II Continuous Autohydrolysis System (Figure I). Raw material is fed via a plug-forming feeder into a reaction vessel which is constantly maintained at a specific saturated steam pressure by an external boiler. The material is conveyed through the digester to the discharge zone by a rotating helical screw. The discharge system densifies the material and the product is throttled intermittently (approximately every 4 seconds) to atmospheric pressure via a ball valve.

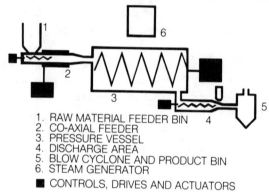

1. RAW MATERIAL FEEDER BIN
2. CO-AXIAL FEEDER
3. PRESSURE VESSEL
4. DISCHARGE AREA
5. BLOW CYCLONE AND PRODUCT BIN
6. STEAM GENERATOR
■ CONTROLS, DRIVES AND ACTUATORS

Figure I : STAKE II System

The Stake II System is now in commercial operation at two locations, producing high energy cattle feed from whole tree Aspen chips in Maine and from sugarcane bagasse in Florida. Three additional plants are now under construction.

Stake processing has been proven, in independant feeding trials, to be a highly effective method of rendering the carbohydrate fraction of several types of waste biomass digestible by ruminants (1). As the digestion of cellulose by ruminants is due to an enzymatic hydrolysis conducted by microorganisms in the rumen, it follows that these materials can be hydrolyzed by enzymes in vitro. Independent investigations conducted in 1978 revealed that the cellulose of unextracted Stake autohydrolyzed Aspen could be saccharified to

greater than 90% of theory by cellulase from Trichoderma res-
eei (2).

4. AUTOHYDROLYSIS-EXTRACTION

The concept of autohydrolysis-extraction was originated
by Lora and Wayman (3) in their investigation of the delignification of hardwoods by high pressure steaming. The hemicellulose fraction of autohydrolyzed lignocellulosic materials
is rendered highly soluble in water and processing conditions
must be carefully controlled to minimize destruction of sugars
generated. Autohydrolysis renders the lignin fraction soluble in dilute alkali. Lower molecular weight lignins soluble in alcohol can be generated by increasing the severity
of the processing conditions, that is by increasing the processing time and/or increasing pressure.

Autohydrolyzed biomass when subjected to an aqueous extraction followed by an alkali extraction yields a high alpha-
cellulose fraction which can be subsequently hydrolyzed, via
acid or enzymes, to glucose for subsequent fermentation to
ethanol. Figure II illustrates the Stake II system in a biomass to ethanol scheme. Counter-current extractions provide
efficiency with minimal dilution. Both enzymatic hydrolysis
and acid hydrolysis routes are illustrated. The possibilities
of using a modified Stake II System for the acid hydrolysis
are currently being evaluated. An economic analysis of biomass conversion to ethanol conducted by the Georgia Institute
of Technology concluded that the most favourable economics
could be achieved by performing a dilute acid hydrolysis on
a delignified substrate with employment of cellulose recycle (4).

A system employing an enzymatic hydrolysis of the high
alpha cellulose substrate resulting from autohydrolysis-
extraction would appear to have several advantages over a
system employing direct enzymatic hydrolysis of the pretreated biomass. The extraction process would remove the dilution effect of the other components resulting in higher sugar
concentrations being achievable, thus reducing ethanol distillation costs. An added advantage would be the removal

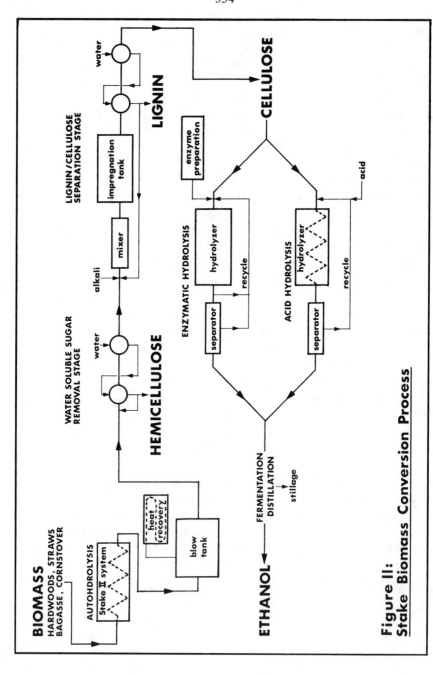

Figure II:
Stake Biomass Conversion Process

of the possibility of inhibition by products generated in the pretreatment stage.

Stake has conducted scaled up trials of the authohydroly-sis-extraction scheme. Several tonnes of Stake autohydrolyzed Aspen was subjected to an aqueous extraction followed by an alkali extraction. Scaling up was successfully accomplished by employing rotary press washers in the extraction stages.

5. CO-PRODUCTS

The hemicellulose hydrolysis products could undergo a fermentation to butanol by an organism such as _Clostridium butylicum_. It could also be converted into furfural or xylitol. Advances have been made in the fermentation of pentosans to ethanol (5) however, this has not yet devel-oped to a practical stage. Commercially, furfural is being produced from a variety of crop residues. Yields of 50% of theoretical are typical and the severe reaction conditions em-ployed result in the destruction of the cellulose and lignin fractions. The autohydrolysis-extraction system has great potential for improving the economics of furfural production.

The lignins generated from autohydrolysis could be used in the production of adhesives (phenol-formaldehyde resins) or as a chemical feedstock for production of phenol and re-lated compounds. Medical research being conducted at the Toronto Western Hospital indicates that Stake lignin might someday be used to prevent gallstone formation in humans (6).

6. ECONOMICS OF STAKE BIOMASS TO ETHANOL

The capital cost of a plant to produce 37.9 million lit-res of 95% ethanol from 114,000 dry tonnes Aspen wood via Stake autohydrolysis-extraction and enzymatic hydrolysis has been estimated at $28 million (Canadian). With enzyme costs of 9¢ per litre, variable costs and capital repayment costs would total $18 million. A raw material price of $30 per dry tonne with no allowance for co-products would result in an ethanol cost of 56¢ per litre. The high purity lignin pro-duced is expected to have a value of $500 per tonne, as an in-gredient in resin production and this would reduce the cost of ethanol to 25¢ per litre. These estimates include no allowance for dehydration of xylose to furfural and the production of crude furfural through distillation. A modest assumption would be that a profit of $100 per tonne of crude furfural could be realized and this would further reduce the cost of the ethanol to 21¢ per litre. Although the employment of acid hydrolysis results in lower ethanol yields, such a system would result in similar alcohol costs due to greater production of co-products, assuming the values noted above.

7. CONCLUSION

A biomass to ethanol scheme employing Stake autohydrolysis-extraction is presently a viable prospect. The separation process outlined successfully fractionates these lignocellulosic materials into their components generating valuable co-products and resulting in higher efficiency of hydrolysis.

References

1. Taylor, J.D. and Esdale, W.J. Increased Utilization of Crop Residues as Animal Feed Through Autohydrolysis. Proceedings of Bio-Energy World Congress, Atlanta, Georgia, April 21-24, 1980 pp 285-286

2. Wayman, M., Lora, J.H., and Gulbinas, E. Material and Energy Balances in the Production of Ethanol from Wood. American Chemical Society Symposium Series 90 Chemistry for Energy 183-201 (1979)

3. Lora, J.H., and Wayman, M. Delignification of Hardwoods by Autohydrolysis and Extraction. Tappi 61 No. 6: 47-50 (1978)

4. O'Neil, D.J. Presentation at Organization for Economic Co-operation and Development Workshop on Cellulose held in Amersfoort, The Netherlands, October, 1980. In print.

5. Wang, P.Y. et al. Fermentation of a Pentose by Yeasts. Biochemical and Biophysical Research Communications 94 No. 1: 248-254 (1980)

6. Rotstein, O.D. et al, Prevention of Cholesterol Gallstones by Lignin and Lactulose. American Gastroenterological Association Meeting Salt Lake City, Utah, May, 1980.

PRODUCTION OF ETHANOL FROM WASTES VIA ENZYMATIC HYDROLYSIS OF
CELLULOSE.

P. ANDREONI, R. AVELLA, G. DI GIORGIO, M. LOPOPOLO, A. MANCINI

Divisione Applicazioni Radiazioni, Centro Studi Nucleari Casaccia, C.N.E.N., Roma, Italy

Abstract

            In the context of the Energy from Biomass Research
and Development project of the CNEN, the objectives of this
program are to develop advanced biological technologies for
the conversion of cellulosic materials to fermentable sugars
and for optimizing the sugar-to-ethanol fermentation process.
Many types of cellulosic wastes are under study: textile wastes,
paper pulp, wheat straw, corn cobs and olive shoots.
A physico-chemical pretreatment of raw material has been tested
which allows the enzymatic conversion of cellulose to reducing
sugars with an efficiency as high as 80% in 48 hrs. Immobilization of β -glucosidase enzyme in bead-shaped alginate gels,
has been studied to optimize the enzymatic hydrolysis of cellulosic materials. Preliminary data revealed no significant
loss of activity in 30 days in enzyme-working-conditions.
Immobilized living cells of S. cerevisiae have been used for
the fermentation of glucose to ethanol. The preliminary results
have shown that the immobilized cells are able to retain their
activity for ten days at least. The yield of the sugar alcool
conversion process is greater 90% of theoretical maximum.

INTRODUCTION

The cellulose is the most abundant organic material, which can be used as source of fuels and chemicals; therefore it can also be considered as the major potential feedstock for ethanol production (1,2,3). As far as the utilization of cellulosic wastes is concerned two kinds of materials may be considered: by products arising from the agricultural activities and residues generated from industrial and community activities. In every case the raw material has some peculiar characteristics which make it hard to manage for conversion: it is insoluble, it has highly ordered crystalline fraction and it is invariabily associated in more or less degree with other molecules such as lignin.

For the above reasons three principal steps are needed for enzymatic conversion of cellulosic materials into alcohol:

1) a pretreatment process for increasing the susceptibility of cellulose to the enzyme attack;

2) the biochemical hydrolysis by which the hemicellulose and cellulose portions are converted to monomeric pentoses and hexoses;

3) the fermentation of hydrolyzate to alcohol.

1. PRETREATMENT

On an highly crystalline textile cotton waste a pretreatment has been tested using NaOH at low temperature. The X-rays analysis (fig. 1) before and after the treatment shows a significant decrease of the crystallinity order of the raw material. The effects of the above pretreatment on the enzymatic saccharification rate has been tested (fig. 2) , using ONOZUKA R-10 "cellulase" at different concentrations.

An high increase of saccharification of the low temperature treated material with respect to the control and also to the room-temperature treated has been observed. The yield of reducing sugars (D.S.A. method) is 80% in 48 hrs.

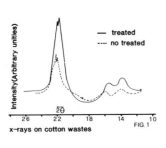

x-rays on cotton wastes   FIG. 1

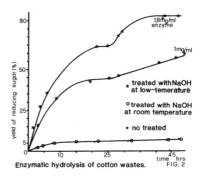

Enzymatic hydrolysis of cotton wastes.   FIG. 2

The same pretreatment was applied to other cellulosic waste materials which are more abundant in Italy: (4) wheat straw, corn cobs, urban paper waste and olive shoots, using a 6% w/v substrate concentration and 1.3 mg/ml enzyme (fig. 3, 4).

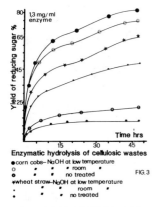

Enzymatic hydrolysis of cellulosic wastes
● corn cobs–NaOH at low temperature
o   "   " room   "
● "   " no treated   FIG. 3
✶ wheat straw–NaOH at low temperature
✶   "   " room   "
● "   " no treated

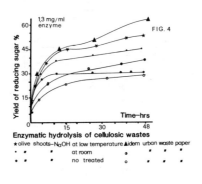

Enzymatic hydrolysis of cellulosic wastes
✶ olive shoots–NaOH at low temperature ▲ idem urban waste paper
· " " at room   o " " " "
● " " no treated   o " " " "

2. ENZYMATIC HYDROLYSIS

It has been demonstrated that the rate of saccharification of cellulase by "Trichoderma" cellulase can be significantly increased by supplementation with β-glucosidase (1,2,5). The stimulatory effect of β-glucosidase appears to

due to the hydrolysis of cellobiose which is an inhibitor
agent of the cellulase enzymes.

Therefore the immobilization of β-glucosidase seems
an important step in the whole process. β-glucosidase prepared
from a selected strain of <u>Aspergillus niger</u> ("NOVO 250 L" containing a declared activity of 250 cellobiase Units per gram)
was immobilized in bead-shaped alginate gels. Beads were formed by spraying solution of Na-Alginate, Bovine Albumin Serum,
Glutaraldehyde and Enzyme (tab. I) into a calcium chloride
solution after 2 hrs of standing. The beads, which have, in
the average, 1 mm in diameter (phot. A), show a residual enzymatic activity of 65%.

TAB. 1

Na–Alginate %W/V	CaCl 2 M	S.A.B. % W/V	Glutaral–dehyde % W/V	Residual Activity %
2	0.1	—	—	—
3	0.1	—	—	—
4	0.1	—	—	—
4	0.4	—	—	—
4	0.5	—	—	—
4	0.4	0.4	0.25	64

**Investigations of beads formation conditions
for enzyme entrapment : * Determinated after
washing of "beads" and at no leakage conditions**

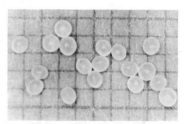

Phot.A - Gel beads (18X)
Medium diameter 1 mm.

For detecting the enzymatic activity, cellobiase was used as
substrate at 40°C and pH 5. The glucose produced was measured
with Beckman automatic glucoanalyzer (glucoxidase method).
The entrapped enzyme maintained in continuous working-conditi-
ons, retained about 90% of their maximum activity after 32
days (fig. 5) and a temperature activity profile as shown
in fig. 6.

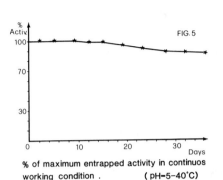

% of maximum entrapped activity in continuos
working condition .   (pH=5-40°C)

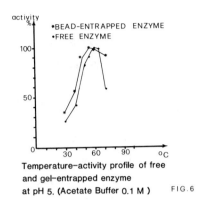

Temperature-activity profile of free
and gel-entrapped enzyme
at pH 5. (Acetate Buffer 0.1 M)   FIG.6

The beads were also stored at room temperature in calcium
chloride 0,02 M; an activity loss of 6% after 30 days was
measured. A microbial contamination was noted after 10 days,
although toluene was used as preservative.

3. ETHANOL FERMENTATION

        The considerable current interest for ethanol as
a potential liquid fueld or fuel supplement has stimulated
several kinds of research for improving the traditional fermentation process.

        Have been undertaking experiments to evaluate the
possibility of continuous production of ethanol using immobilized yeast living cells packed in a column reactor (6,7).

        Since ethanol is produced from glucose by multistep
enzyme reactions, immobilized cells must be kept in living
state; the preparation of calcium or aluminium alginate gels
can be carried out under very mild conditions to maintain a high
fraction of cells in a living state after immobilization (8,
9). S. cerevisiae (var. Elipsoideus) cells, cultivated in
liquid medium (malt broth and yeast extract), were immobilized
using 10% (w/v) suspension in 3% sodium alginate. The algina-
te-cell mixture was added to 0,1 M $CaCl_2$ solution as droplets
formed by means of a needle, where a concentric air stream

around the needle leads to decreasing particle size with increasing air velocity. After reduction of particle radius
by the gel drying, the alginate beads were hardened in 0.1 M
$Al_2(SO_4)_3$ solution (10, 11). The average bead diameter was in
the range of 0.5-0.9 mm. The number of living cells in the
beads was determined, after solubilization of gel in 0.28 M
sodium phosphate solution, by plate counting technique (nutrient algar). Micrographs B and C shows the location of S. cerevisiae cells immobilized homogeneously in gel bead slice.

Phot. B   **125 X**   **Phot. C**   **1250 X**

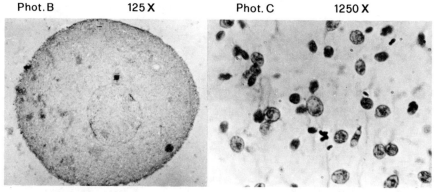

Micrographs showing the location of S. cerevisiae cells in
gel bead slice

When the immobilized living cells were packed into
a column reactor and a medium containing 150 mg/ml glucose
and nutrients was passed upward through the column (flow rate
27 ml/h, gel 70 ml, T = 30°C), ethanol was produced continuous
ly for more than 10 days (fig. 7). The efficiency of glucose
to ethanol conversion was greater than 90% of theoretical
yield (fig. 7). The reactor performance showed a significant
loss of ethanol-producing activity, probably due to a decrease of living cells in the gels (fig. 8).

343

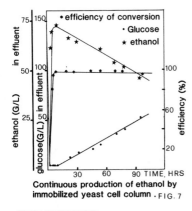

Continuous production of ethanol by immobilized yeast cell column . FIG. 7

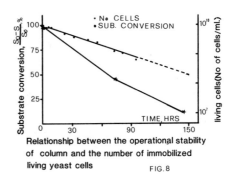

Relationship between the operational stability of column and the number of immobilized living yeast cells FIG.8

BIBLIOGRAPHY

1. Cellulase as a chemical and energy resource. - Biotechn. and Bioengin. Symp. N° 5 (1975)

2. Enzymatic Conversion of Cellulosic Materials: Technology and Applications, Biotechnology and Bioengineering Symp. N° 6 (1976)

3. FLICKINGER M.C. ; Biotech. and Bioengin. XXII, suppl.1, 27 (1980)

4. AVELLA R., DI GIORGIO G., A. MARIANI   - RT/CNEN (1980) (in press)

5. BISSETT F. and D. STERNBERG - Applied and Environmental Microbiology 35, 4, p. 750 (1978).

6. WADA M., J. KATO and J. CHIBATA - European J. Applied Microbiol. Biotechn. 8, 241 (1979).

7. GHOSE T.K. and K.K. BANDYOPADHYAY - Biotech. and Bioengin. XXII, 1489 (1980).

8. KIERSTAN M.K. and C.BUCKE - Biotech. and Bioengin. XIX, 387 (1977).

9. CHEETHAM P.J., K.W. BLUNT and C. BUCKER - Biotech. and Bioengin. XXI, 2155 (1979).

10. KLEIN J., U. HACKEL and F. WAGNER - ACS Symp. Series N°106 101 (1979).

11. KLEIN J., U. HACKEL, P. SCHARA and H. ENG - Ang. Macrom. Chemie 76/77, 329 (1979).

ENERGY PRODUCTION FROM WHEY

G. MOULIN and P. GALZY

Chaire de Génétique et Microbiologie
ENSAM - INRA
34060 MONTPELLIER Cedex France

SUMMARY

The whey, a product of cheese manufacture, has become a serious pollution problem. In France, 8.10^9 litres are produced every years. Deprotéinization, is a good process to valorize the whey. The major component of permeate whey solids is lactose. If all lactose, were fermented, it is possible to produce 2.10^9 litre of pure alcohol. In France this production is twice higher than the alcohol production coming from petroleum. A survey was made of 40 lactose assimilating yeasts, with respect to their capacity for converting lactose in alcohol. Four strains of yeast got a high alcohol yield (12 % v/v). The conversion of lactose into ethanol is assured with good efficiency (85 % of the theorical yield), for a lactose concentration in 5 - 30 % range. The fermentation time in batch process with no concentrated whey permeate is 6 hrs. With concentrated whey permeate (200 gr x 1^{-1} lactose) this time is 20 hrs. In all case the cells yeast are reusing. Energy evaluation, established for an alcohol plant and a depollution post situated in the proximity of a dairy industry, shows that the production of 100 litres of pure alcohol gives off about 150 000 kcal. If we consider the economic aspect, per litre alcohol can be paid at 2.20 FF (of pure alcohol) thus the whey permeate at 0.02 or 0.03 FF per litre.
    Consequently, whey valorisation permets:
        - a protein recuperation
        - an alcohol production by fermentation
        - and finally to reduce the pollution
            (the DCO reaches 90 % of its initial value).

INTRODUCTION

A lot of work has been done on whey treatment and its use for yeast biomass,drinks and alcohol production (Rogosa 1947; Wasserman 1960; Yoo 1975; Holsinger 1974; Moulin 1976; O'Leary 1977). A major problem met by different authors, for the production of alcohol, has been the fact that relatively few yeast are able to ferment lactose. This study present, the production of alcohol by fermentation of whey.

RESULTS AND DISCUSSION

In a previous study, (Guillaume et coll. 1979) we have shown that four out of fourthy strains of differents species of yeasts ferment lactose of whey. Further experiments were made with two strains: Candida pseudo-tropicalis I. P 513 and Kluyveromyces fragilis CBS 397 (Moulin 1980). A study of yiels as a function of initial dry-weight content of the whey (5 - 20 %) shows that observed yield with respect to theorical yield is always above 85 %.

Among the parameters studied were pH, aeration and the possibility to reusing the cell after separation.
- Fermentation at pH 3 is always slow and incomplete. When pH values varied from 4 to 5.5, fermentation can be effected under good conditions.
- Experiments conducted in 2 liters fermentor with only agitation but without aeration gave result comparable to those obtained under strict anaerobic conditions. It is therefore possible, that for Kluyveromyces fragilis and Candida pseudotropicalis, the whey contains sufficient quantities of sterol and unsaturated fatty acids to assure growth and cell activity.
- A study was made on yeast recycling, for whey permeate at two lactose concentration (50 g/1 and 200 g/1). In these two case ten successive recycles were made without any loss in cell viability or yield.

From these results, we have try, to made energy evaluation and an economic study (Moulin and coll. 1980). Energy evaluation study (table I) show that the production of 100 1 of pure alcohol gives off about 151 thermies. Economic study (table II), has been made for a unit plant treating 800 000 1 per day of whey permeate, 250 days per year. Considering only alcohol, the selling price can be fixed at 2 FF/1 of pure alcohol, the raw material can be paid for on the base of 0.02 FF/1. Product valorization

346

by recuperation of the proteins from whey and yeast biomass have not been take into account.

REFERENCES

Guillaume Maguy, G. Moulin and P. Galzy. Sélection de souches de levures en vue de la production d'alcool sur lactosérum. Le Lait n° 588 489-496, 1979.

Holsinger V.H., L.P. Posati and E.D. Devilbiss. Whey beverages: A review. J. Dairy Sciences 57, 849-859, 1974.

Moulin G., R. Ratomahenina and P. Galzy. Sélection de souche de levure en vue de la culture sur lactosérum. Le Lait, n° 553-554, 135-142, 1976.

Moulin G., Maguy Guillaume and P. Galzy. Alcohol production by yeast in whey ultrafiltrate. Biotechnol. Bioeng. 22, 1277-1281, 1980.

Moulin G., Maguy Guillaume et P. Galzy. Etude de la production d'alcool sur lactosérum. Ind. Alim. Agric. 97, n° 5, 471-474, 1980.

O'Leary V.S., R. Green, B.C. Sullivan and V.H. Holsinger. Alcohol production by selected yeast strains in lactase hydrolysed whey. Biotechnol. Bioeng. 19, 1019-1035, 1977.

Rogosa M., H.H. Browne and E.O. Whittier. Ethyl alcohol from whey. J. Dairy Science 30, 263-269, 1947.

Wasserman Aaron E. The rapid conversion of whey to yeast. Dairy Engineering. 77, 374-379, 1960.

Yoo B.W. Diss Abstr. Int. 36, 641 B 1975.

TABLE I - Energy balance

Energy required	Energy produced (in alcohol)
Fermentation and distillation processes Steam production: 364 x 0.75 .... 273 thermal units[*] (lowest calorific value)	1 hl P.A. has the equivalent of 506 thermal units (l.c.v.)
Electrical power pumps and miscellaneous: 10 kW/hl P.A., i.e.: 10 x 2.5 .... 25 thermal units/hl	
Final depollution (e.g. by forced aeration) 1 kW eliminates 1.5 kg of DCO Quantity to be eliminated: $4000 \times 4 \times 10^{-3} = 16$ kg DCO/hl thus: $\dfrac{16 \times 2.5}{1.5}$ .... 27 thermal units/hl	
Calorific equivalent of the distilling investment Depreciation over 8 years ... 30 thermal units/hl	
TOTAL .... 355 thermal units/hl	506 thermal units/hl (l.c.v.)

NOTE: If the whey is to be transported, the amount of energy required must be increased by the extra consumption: estimated at 70 thermal units for a 30 km one-way trip.

 *: 1 thermal unit = 1000 cals

TABLE II - Prospective economic balance

Depreciation
Fermentation-distillation-heating installations including final purification.
Investment: 15×10^6 FF
Cost of depreciation for a duration of 8 years:
$$\frac{15 \times 10^6}{62500 \times 8} \quad \dotfill \quad 30 \text{ FF/hl[*] P.A.}$$

Financial costs .... 12 "   "

Operating costs:
  Fuel: $\dfrac{273}{9.4}$ x 10.75 FF .... 22 "   "
  Electricity: 21 kW x 0.22 FF/kW .... 5 "   "
  Labour: $\dfrac{6 \text{ employees} \times 50 \text{ FF/kW}}{10.4 \text{ hl/hour}}$ .... 29 "   "
  Miscellaneous: reagents .... 2 "   "

      TOTAL .... 100 FF/hl P.A.

 * : 1 hl = 100 l

SEPARATION OF LIGNOCELLULOSES INTO HIGHLY ACCESSIBLE FIBRE MATERIALS
AND HEMICELLULOSE FRACTION BY THE STEAMING-EXTRACTION PROCESS

J. PULS and H.H. DIETRICHS

Federal Research Center of Forestry and Forest Products
Institute of Wood Chemistry and Chemical Technology of Wood
2050 Hamburg 80, Federal Republic of Germany

Summary

Lignocelluloses of lower lignin content like hardwoods and most agricultural residues can be treated with saturated steam at elevated temperatures
for some minutes. After additional defibration in a hot stage the fibre
material has the following properties:

> The hemicelluloses become soluble and can be washed out with
> hot water if wanted.

> The cellulose becomes accessible for enzymatic degradation in
> spite of the presence of lignin.

> The lignin can be extracted with organic solvents.

Until now only the hemicelluloses were extracted, so that the lignocelluloses are separated into a fibre fraction with the cellulose and lignin
part and into a hemicellulose fraction. The condensate contains furfural
and acetic acid in amounts, depending on the steaming conditions. Under
mild conditions 10 - 25 % hemicelluloses (based on raw material) can be
extracted as xylans and xylan fragments. After hydrolysis with acid or
immobilized xylanolytic enzymes, the solutions have xylose contents up to
80 % (total carbohydrates = 100 %).

The fibre material is digested by ruminants up to 70 - 80 % and degraded
by cellulases without further treatment to 50 - 60 %. Therefore this substrate can favourably be used for microbial and biochemical means. In the
case of enzymatic saccharification the solution has a glucose content up
to 90 % (total carbohydrates = 100 %).

In spring 1978 a pilot plant was set up in Munich. 100 to 400 kg lignocellulosic material per hour can be steamed in this installation in a
quasi continuous manner. The fiber material produced are as well fermented
by the rumen microorganisms and hydrolyzed by fungal cellulase preparation
as those obtained so far in batch cooks in the laboratory.

1. COMPOSITION OF LIGNOCELLULOSES

The qualification of lignocelluloses as a resource for energy and chemicals
is often only examined by their availability, not by their chemical constitution. At our institute of wood chemistry investigations on lignocellulose
utilisation usually begin with the determination of carbohydrate portion of
raw materials. In spite of the time consuming standard analyses for cellulose, hemicelluloses and lignin for first orientating investigations on unknown materials lignocelluloses are decomposed into their monomeric sugar
by acid hydrolysis. The acidic hydrolysates are directly injected into the
sugar chromatographic system, based on borate-ion exchange chromatography
with detection of the sugars by the noncorrosive reagent 2'2-bichinchoni-
nate. Table I shows a selection of lignocelluloses, which had been
investigated in this way.

Because the idea of steaming-extraction process is not only to produce a
highly accessible fibre material but also to make use of the extracted hemicelluloses such lignocelluloses with a low lignin content and high values
for xylose should be most favourable. Actually oat hulls with 10 % lignin
are fully digested by ruminants after steaming. Because of their high   value for xylose they are wellknown as a starting material for furfural and
xylitol. Among the straws, maize straw had the highest xylose content. A
very high lignin content (as hydrolysis residue) was detected for coconut
fibres (46 %). In this case the accessibility could not be increased by
steaming. However, the steaming-extraction process can be applied for most
other agricultural residues and hardwoods.

2. STEAMING-EXTRACTION PROCESS

Wood chips from birch, beech, and oak as well as chopped annual plants like
cereal straw, bagasse and napir grass are treated with saturated steam at
temperatures between 170° and 200°C and fiberized in the hot stage by expansion or in a refiner. The obtained fiber materials are extracted with
water (Fig. 1). Three different fractions are obtained from these treatments:

     1. condensates of volatiles
     2. fibre material
     3. water extract

When wheat straw is treated with saturated steam at 187°C the amount of xylose, which can be extracted with water from the fiberized material in form

of xylan fragments, increases with reaction time up to about 20 min and decreases when steaming is carried out for a longer period (Fig. 2). The total amount of carbohydrates calculated as monosaccharides and the total amount of solid matter in the extract follow the same tendency. The maximum yield of xylose obtained from wheat straw is about 15 % related to the dry starting material and about 75 % relative to the total amount of solubilized carbohydrates.

The yield of extracted fibre material as well as the absolute amounts of carbohydrates and residual xylose in the fibre material decrease with reaction time, those of glucose keeps constant. This shows that the cellulose is not dissolved by the steaming and extraction treatment. However, the properties of the fibre material are changed. The most striking phenomenon is that the cellulose and other residual carbohydrates in the fibre material become accessible for enzymes. In principle the same tendencies were found for all hardwoods and lignocellulosic materials so far studied. However, the optimum conditions of steaming differ from one raw material to the other.

Rumen digestibilities of some untreated lignocelluloses and of the fibre materials derived from these raw materials are given in Table II. The digestibilities of the fibre materials, that means the rates of fermentation in the rumen, are as good or better than that of good hay quality, which is 60 - 65 %.

The cellulose part of the fibre material can be converted into sugars by enzymatic saccharification. The amounts of reducing sugars obtained by enzymatic hydrolysis with commercial cellulase preparations of Trichoderma viride are listed in Table III. The yields of glucose in solution depend normally upon the cellulose content of the starting material. Wheat straw has comparatively high amounts of water-soluble glucans, which are extracted after steaming; therefore the yield of glucose based on the raw material is lower.

The steaming-extraction process may be a favourable pretreatment not only for enzymatic hydrolysis but also for acidic hydrolysis. An additional extraction of the lignin part in the fibres by alkali or organic solvents could be advantageously for reactor size, acid consumption and glucose concentration. The water extracts from the steaming process can be used for different purposes. The optimal use depends first of all on the purity of the xylan fragments relative to other carbohydrates and on their molecular distribution. We reached so far purities of 70 - 90 % and concentrations of 3 - 8 %. The xylan fragments can be hydrolyzed by conventional methods or by means of immobilized enzymes into xylose. Table IV shows the xylose yields so far reached from different lignocelluloses.

In spring 1978 a pilot plant was set up in Munich. 100 to 400 kg lignocellulosic material per hour can be steamed in this installation in a quasi continuous manner. The fibre material produced is as well fermented by the rumen microorganism and hydrolyzed by fungal cellulases as those obtained so far in batch cooks in the laboratory.

Fig. 1 Products from lignocelluloses by steaming,
extraction and further processing.

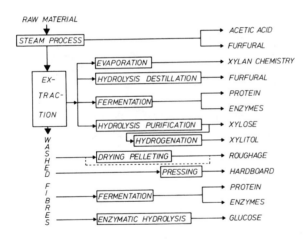

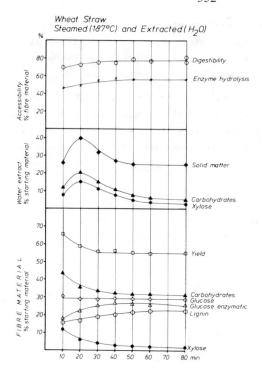

Fig.2

Composition of water-extract, fibre material and the rumen digestibility of the fibre material of wheat straw

Material	Lignin % (as hydrolysis residue)	Carbohydrates % (after total hydrolysis)			Acetic Acid % (after Klingstedt)	Ash % (TAPPI T15m-58)	Hot Water Extract % (TAPPI T1m-59)
		Total	Glucose	Xylose			
1 Birchwood	20*	68	43	22	7	0,3	2,7
2 Oakwood	20*	69	44	19	5	0,2	6,0
3 Eucalyptus	26	69	49	16	4	0,4	4,4
4 Black Wattle	20	75	51	21	5	0,3	4,3
5 Utile	41*	53	40	8	1,6	0,5	3,1
6 Wheat Straw	21	68	40	20	4,5	2,8	5,8
7 Barley Straw	6*	66	40	21	3	6,2	9,0
8 Maize Straw	17	72	41	26	-	-	-
9 Rice Straw	17*	60	38	19	2	16	12,2
10 Rice Hulls	23*	55	36	16	2,5	17	2,8
11 Oat Hulls	10	79	35	39	3	3,6	6,0
12 Sugar Cane Bagasse	22*	59	38	19	3,5	3,3	3,6
13 Napir Grass	23*	67	44	19	3	4,3	5,1
14 Cottonseed Hulls	22*	61	37	22	2	2,5	5,1
15 Cotton Stalks	25*	52	35	12	4,6	5,5	16,2
16 Residues from Pineapple Cult.	9*	46	34	6	3,4	8,1	37,5
17 Coconut Fibres	46	40	25	9	3,5	3,3	9,9

*corrected for ash

Table I   Composition of Lignocelluloses

Table II Increase of Rumen Digestibility by Steaming
and Extraction

Starting Material	Digestibility (untreated) %	Digestibility (steamed and extracted) %
Wheat Straw	45	78
Oat Hulls	45	99
Corn stoves	45	80
Sugar Cane Bagasse	28	70
Napir Grass	20	65
Aspen	6	65
Oak	7	74

Table III Glucose from Steam-treated Lignocelluloses

	Degradation rate based on fibre material %	Yield of glucose in solution based on cellulose in the fibre material %	starting material %
Wheat Straw	57	90	28
Oak Wood	52	92	31
Birch Wood	61	89	33

Enzymatic saccharification with 75 FPU Cellulase/g, 24 h, 46°C.

Table IV Xylose in Water-Extract after Steaming and Acid Hydrolysis

	Xylose % based on total solubilized sugars	Xylose % based on starting material
Wheat straw	75	15
Rice straw	69	10
Bagasse	73	14
Maize straw	70	16
Barley straw	70	12
Elephant grass	74	11
Eucalyptus	80	12
Black wattlw	81	15
Aspen wood	72	11
Birch wood	87	15

ETHYL ALCOHOL PRODUCTION FOR FUEL: ENERGY BALANCE

ROBERT Y. OFOLI and BILL A. STOUT

Agricultural Engineering Department
Michigan State University
East Lansing, Michigan 48824

Summary

A net energy analysis model is used to assess the energy balance of fuel ethanol production. The model is based on corn as the feedstock and includes energy inputs for: a) producing corn on the farm; b) transporting corn from the farm to the alcohol plant; and c) the commercial production process.

By a total energy accounting procedure, alcohol production yields 4,600 kJ/L and 2,400 kJ/L more energy than required to produce it using dryland corn and irrigated corn, respectively. A premium energy analysis shows that it takes 4,900 kJ/L and 7,100 kJ/L more premium energy to manufacture alcohol than the premium energy yielded.

The third method assumes that, of the energy inputs on the farm, off-farm transportation and the alcohol plant, 35%, 0%, and 85%, respectively, could, conceivably, be replaced by non-premium sources of energy. With this analysis, ethanol is a net premium energy producer; 15,200 kJ/L and 13,800 kJ/L more premium fuel is produced than is required for dryland corn and irrigated corn, respectively. This results in a net premium fuel gain of 2.8 for dryland corn, and 2.4 for irrigated corn.

1. INTRODUCTION

Ethanol has often been presented as a fuel that takes more energy to produce it than it yields. While this, in itself, is not a new phenomenon (e.g., it takes 3 kJ of primary energy to produce 1 kJ of electricity), the wisdom of utilizing scarce fossil fuel resources to produce ethanol at an energy loss has been questioned, especially since ethanol has been promoted as a possible solution to the liquid fuel crisis. Much attention has therefore been focused on the energy balance of producing ethanol.

Unfortunately, the data used by most previous researchers have come from the beverage alcohol industry, giving a poor picture of the energetics of producing ethanol for fuel. Reilly (1978), Kendrick and Murray (1978), Stroup and Miller (1978), David, et al. (1978), Thimsen, et al. (1978), Hofman (1979), and Krochta (1979) all reached the conclusion that when corn is used as the feedstock, ethanol production loses energy. The losses reported by these researchers ranged from 64,000 kJ/L from Reilly (1978) to 2,500 kJ/L from Krochta.

2. ENERGY ANALYSIS MODEL

The Boundary

For this model, a boundary was selected that encloses the farm, transportation of the feedstock to an alcohol processing plant, and the alcohol plant (Fig. 1). Characteristics of the model are:

1. Corn is used as a feedstock. It is grown specifically to make fuel ethanol and to produce distillers grain for animal feed.
2. Distillers grain and solubles (DDGS) are considered the only valuable by-product of the operation.
3. The solar energy driving the photosynthetic process is regarded as a free energy source.
4. The energy content of the feedstock (its heat of combustion) is not used in the model because:
   a) it does not derive solely from agricultural or cultural energy but mostly from free solar energy; and
   b) corn is not normally burned to produce energy.
5. The heat of combustion of the DDGS by-product is also ignored in the output analysis because DDGS is not normally combusted to produce energy.
6. The study is based on an industrial scale alcohol plant producing anhydrous (200 proof) alcohol and DDGS. It is assumed that the production level is at least 3.8×10^6 L of ethanol per year. At this level, maximum use can be made of heat recovery and recycling.

3. METHODOLOGY

The methodology utilizes a control volume approach and net energy analysis criteria. All definable direct and indirect energy inputs are evaluated for the farm, off-farm transportation, and the alcohol processing plant.

4. RESULTS AND ANALYSIS

Input

The energy input needed on the farm to produce corn for processing to ethanol was calculated to be 10,500 kJ/L for irrigated corn, and 8,300 kJ/L for non-irrigated corn (Table 1). Energy requirement at the alcohol plant is 19,100 kJ/L. The components are shown in Table 2.

The total energy input of the boundary is 30,700 kJ/L for irrigated corn and 28,500 kJ/L for dryland corn. The total energy input obtained here is 9.5% and 16.0% less for irrigated corn and dryland corn, respectively, than the total reported by Chambers, et al. (1979). While Chambers, et al. (1979) reported a lower plant energy requirement than this study, their farm energy input is 90% higher than the farm energy input here. However, their farm data was based on 1974 corn yields of 175 bu/ha--an unusually low yield.

Output

One way to analyze the energy value of DDGS will be to compare its protein content to that of soybean meal (SBM)--28% to 45%--and use the resulting ratio of 0.62 to determine the energy required to produce and process enough soybeans to provide 0.62 kg of SBM.

However, this scheme is not very satisfactory, because while DDGS has a lower protein content than SBM, ruminant feeding trials at the University of Nebraska have shown that the protein in DDGS is 1.37 times better utilized by ruminants than that of SMB (Klopfenstein, et al., undated; Poos and Klopfenstein, 1979; Waller, et al., 1980). Thus, in spite of its lower protein content, only 1,066 kg of DDGS is required to provide the same amount of protein obtained from 909 kg of SBM (Klopfenstein, et al., undated). Therefore, 1.0 kg of DDGS replaces 0.85 kg of SBM.

Based on Myers, et al., (1979), the farm energy input for soybean production is estimated at 4,700 kJ/kg of soybean meal. Based on information from the American Feed Manufacturers' Association, the energy required to process soybeans to SBM is estimated at 9,200 kJ/kg. The total value of distillers grain resulting from ethanol production is 9,500 kJ/L.

The energy balance of the ethanol production process is summarized in Table 3. Ethanol production yields slightly more energy than it requires-- 4,600 kJ/L for non-irrigated corn, and 2,400 kJ/L for irrigated corn. While this is not a large energy yield, the outcome is significant.

An important aspect of alcohol energy analysis is the premium fuel analysis. The term premium fuel, as used here, signifies highly versatile and portable fuels; for example, natural gas and petroleum. While all inputs currently come from premium fuels, the only premium fuel output is ethanol, since DDGS does not constitute a premium fuel by the definition here. By premium fuel analysis, 30,700 kJ/L of ethanol is used up to produce 23,600 kJ/L, an energy loss of 7,100 kJ/L for irrigated corn. Dryland corn has a better energy balance, but it still shows a loss of 4,900 kJ/L.

However, this problem can be corrected by substituting non-premium fuels for some farm and alcohol plant processes. A premium fuel analysis was done in this study based on the assumption that at least 35% and 85%, respectively, of the premium fuels used on the farm and at the alcohol plant (currently) can be replaced by non-premium sources of energy. This analysis shows that ethanol can be a net premium fuel producer (Table 4).

5. CONCLUSIONS

Three energy accounting procedures are used here. The first one examines total energy inputs and outputs. Ethanol yields 4,600 kJ/L more energy than it requires to produce it, when produced from dryland corn. If irrigated corn is the feedstock the energy balance is 2,400 kJ/L. The second method compares the premium fuel energy inputs against the outputs. It takes 4,900 kJ/L and 7,100 kJ/L more premium energy to manufacture ethanol than the premium fuel yielded from dryland corn, and irrigated corn, respectively. But this deficit can be overcome by replacing some premium energy inputs with non-premium fuels (coal, wood, biomass, etc.). If this is done, ethanol could provide 15,200 kJ/L and 13,800 kJ/L more premium energy for dryland corn and irrigated corn, respectively, than it takes to produce it. This is a gain of 2.8 and 2.4, respectively, in premium fuels.

6. REFERENCES

1. CAST (1977). Energy use in agriculture; now and for the future. Council of Agr. Sci. and Tech., report #68.

2. Chambers, et al. (1979). Gasohol: Does it or doesn't it produce positive net energy? Science, 206(4420), pp. 789-795.

3. David, M. L., et al. (1978). Gasohol: economic feasibility study.

Prepared for the Energy Res. and Dev. Center, Lincoln, NE., by the Dev. Plan. and Res. Assoc., Inc., Manhattan, KS.

4. Doering, III, O. C., T. J. Considine, and C. E. Harling (1977). Accounting for tillage equipment and other machinery in agricultural energy analysis. Agric. Exp. Stat., Purdue Univ., W. Lafayette, IN.

5. Hofman, V. (1979). Gasohol. Coop. Ext. Serv., N. Dakota State Univ., Fargo, ND.

6. Kendrick, J. G., P. J. Murray (1978). Grain alcohol in motor fuels--an evaluation. Agric. Exp. Stat., Univ. of Nebraska, Lincoln, NE.

7. Klopfenstein, T., et al. (undated). Distillers grains as a naturally protected protein for ruminants. University of Nebraska, Lincoln, NE.

8. Korff, J. H. (1979). Bohler Bros. of America (Vogelbusch Power Alcohol Division, Inc. Personal communication.

9. Krochta, J. M. (1979). Energy analysis for ethanol from biomass. W. Reg. Res. Center, Sci. and Ed. Adm., USDA, Berkeley, CA. From: Proceedings of the 2nd Int. Conf. on Energy Use Management. Fazzolare and Smith (eds.), Pergamon Press, N.Y., NY.

10. Myers, C. A., et al. (1979). Michigan farm energy audit and education program, phase I. Agric. Eng. Dept., Michigan State Univ. Final report to the Mich. Energy Adm., Mich. Dept. of Comm., Lansing, MI.

11. Pementel, D., et al. (1974). Workshop on research methodologies for studies of energy, food, man and environment, phase I. Cornell Univ., Ithaca, NY.

12. Poos, M. I., T. Klopfenstein (1979). Nutritional value of by-products of alcohol production for livestock feed. Coop. Ext. Service, Univ. of Nebraska, Lincoln, NE.

13. Raphael Katzen and Associates (1979). Grain motor fuel alcohol: technical and economic assessment study. Prepared for the Dept. of Energy by Raphael Katzen and Associates, Cincinnati, OH.

14. Reilly, P. J. (1978). Economics and energy requirements of ethanol production. Iowa State Univ.

15. Stroup, R. and T. Miller (1978). Feasibility of ethanol from grain in Montana. Montana Agric. Exp. Sta., Bozeman, MT.

16. Thimsen, D. P., et al. (1979). Production and use of fuel ethanol from corn or wheat. Agric. Ext. Serv., Univ. of Minnesota.

17. Waller, J. C., et al. (1980). Effective use of distillers dried grains in feedlot rations with emphasis on protein considerations. Dept. of Animal Science, Michigan State Univ., East Lansing, MI.

Table 1. Farm Energy Input
(Irrigated Corn)

	kJ/L
Field operations[1]	
Tillage and seeding	500
Herbicide and insecticides	100
Irrigation	3,900
Harvest	100
On-farm transportation[1]	400
Fertilizer[1]	3,700
Crop drying[1]	1,700
Labor[2]	negligible
Machinery[3]	100
Repair energy	negligible
TOTAL:	10,500[4]

[1]CAST (1977)
[2]Pimentel, et al. (1974)
[3]Doering, III, et al. (1977)
[4]Total energy input for non-irrigated corn is 8,300 kJ/L

Table 2. Plant Energy Inputs

	kJ/L
Mashing, cooking and conversion[1]	3,400
Distillation[1]	6,300
Fermentation[1]	0
Recovery of distillers by-product[1]	600
Electricity[1]	3,800
Drying of distillers grain[2]	5,000
TOTAL:	19,100

[1]Raphael Katzen Associates (1979)
[2]Bohler Bros. of America, Inc. (1979)

Table 3. Total Energy Analysis

	Irrigated Corn kJ/L	Dryland Corn kJ/L
Inputs:		
Farm energy		
input	10,500	8,300
Off-farm		
transpor-		
tation	1,100	1,100
Plant energy		
input	19,100	19,100
TOTAL:	30,700	28,500
Outputs:		
Ethanol	23,600	23,600
DDGS	9,500	9,500
TOTAL:	33,100	33,100
Energy balance	2,400	4,600
Gain (total		
outputs/total		
inputs)	1.08	1.16

Table 4. Premium Energy Analysis, Assuming Most Premium Fuels Can Be Replaced by Non-premium Sources*

	Irrigated Corn kJ/L	Dryland Corn kJ/L
Inputs:		
Farm energy		
input	6,800	5,400
Off-farm		
transpor-		
tation	1,100	1,100
Plant energy		
input	1,900	1,900
TOTAL:	9,800	8,400
Outputs:		
Ethanol	23,600	23,600
DDGS	--	--
TOTAL:	23,600	23,600
Energy balance	13,800	15,200
Gain (total		
outputs/total		
inputs)	2.41	2.81

*This analysis assumes that non-premium energy sources can replace 35%, 0%, and 85%, respectively, of the energy required on the farm, in off-farm transportation, and at the alcohol plant.

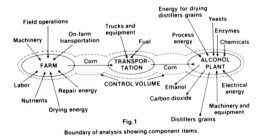

Fig. 1
Boundary of analysis showing component items.

A LOW ENERGY SYSTEM FOR THE PRODUCTION OF BIOMASS ETHANOL

H. J. GROUT and M. ENGLISH

W. S. Atkins Group Consultants
Woodcote Grove, Ashley Road, Epsom, Surrey.

SUMMARY

A development programme for the production of biomass ethanol carried out
by the W. S. Atkins Group has demonstrated a viable process with:-

- low energy requirements,
- concentrated sugar feedstocks, including molasses,
- minimum liquid effluents.

The Atkins Power Alcohol Process (ATPAL) features a novel integrated fermentation and distillation system that operates continuously. An ATPAL
pilot plant has been operated for three months, under the supervision of
W. S. Atkins Group process engineers, by the Department of Chemical Engineering at the University of Manchester Institute of Science and Technology.
The results confirm that only about one quarter of the energy equivalent,
based upon the fuel value of the alcohol, need be consumed in the total
process.

Alcohol will probably continue to be used as a gasoline extender rather
than as a fuel in its own right. It is current practice to add anhydrous
99.5% w/w (99.7% v/v)* alcohol because of fears that too much water would
prevent the alcohol mixing with the gasoline. Extensive tests have demonstrated that alcohol containing up to 3% v/v water will satisfactorily
blend in the proportion 20% v/v ethanol to gasoline.

Ethanol concentrations of up to 98.1% w/w (99% v/v) have been obtained
from the ATPAL pilot plant. This concentration can only be achieved in
practice on a full scale production plant at the expense of increased
operating and capital costs.

Conventional batch processes generate large volumes of effluent arising
from the relatively low tolerance of yeasts to alcohol (8 to 12% v/v).
This necessitates dilution of feedstocks to obtain a tolerable
concentration.

In the ATPAL process, alcohol (ethanol) is continuously removed to limit
the alcohol concentration, thus permitting the use of concentrated sugar
feedstocks.

---

* w/w means weight for weight.   Similarly, v/v means volume for volume.

1. INTRODUCTION

 The growing scarcity of oil, its consequently spiralling price, and
pressure from conservationists have inevitably led to a reappraisal of
renewable sources of energy such as crops and other vegetable matter
(biomass). The use of these sources to produce ethanol as a fuel has
received particular attention, and a feasibility study in this field was
carried out in 1978 by the W. S. Atkins Group for Rolls-Royce. It showed
considerable promise, for it indicated that an ethanol process could be
evolved which produced a net energy balance - only about one quarter of
the energy equivalent contained in the alcohol would be consumed in the
total process.

 Since then, the W. S. Atkins Group have carried out further studies
involving a wide range of crops - e.g. sugar cane, molasses, sugar beet
and cassava (a tropical plant that yields a starchy food).

 At the ATPAL pilot plant established at the University of Manchester
Institute of Science and Technology the feasibility of combining fermentation and distillation in one integrated unit has been proved. The
design and operation of the pilot plant achieved:-

- high yeast concentrations without the use of centrifuges, thereby
  resulting in much higher productivity per unit volume of fermenter,

- negligible contamination from bacterial infection,

- operating experience necessary for the design of future plants,

- high level of stability with few automatic controls.

 The ATPAL process is the subject of a patent application jointly filed
by Atkins and Rolls-Royce.

2. PILOT PLANT

 The 20 litre fermenter was operated at atmospheric pressure and vented
carbon dioxide to the atmosphere via a water-cooled condenser. The prepared
sterile feedstock, either beet molasses or glucose plus nutrients, was continuously pumped into the fermenter, and a portion of the fermenting liquor
was allowed to overflow in order to maintain the fermenter level. (On a
production plant the yeast would be removed from the overflow and the
ethanol in the liquor would be stripped in a stripping column.)

The bulk of the ethanol was removed by circulating the liquor via the circulation heat exchanger to the flash vessel where ethanol, some water and dissolved carbon dioxide were evaporated. The remaining liquor drained down a barometric leg back into the fermenter.

The ethanol-containing vapour was drawn to and through the rectifying column by the main compressor. As the ethanol vapour rises up the rectifying column, the ethanol distills and the water content reduces from initially 70% w/w to 4% w/w. In this way, the volume and power of the compressor were reduced because only highly concentrated ethanol and a smaller quantity of carbon dioxide need be compressed. The ethanol was condensed in the circulation heat exchanger and the carbon dioxide and other non-condensibles raised to atmospheric pressure by a vacuum pump. The condensed ethanol was then returned to the top of the distillation column where a reflux divider proportioned reflux and product ethanol. The product ethanol was removed via the alcohol outlet.

3. PILOT PLANT RESULTS

From practical operating experience, the following results were obtained:-

● High yeast concentrations in excess of 20 grammes per litre dry weight were achieved without the use of centrifuges. This was possible because a portion of the water in the liquor is continuously vaporised with the alcohol. Batch fermentations normally operate at about 1½ g/litre dry weight of yeast, and so they take longer to ferment the sugars to alcohol. The best commercial semi-continuous process only achieves 10 g/litre dry weight of yeast by the use of centrifuges and recycling the yeast.

● Contamination of the continuous fermentation did not prove to be a problem. This is probably due to the greater temperature tolerance of the strain of yeast used compared with that of most bacteria.

● A useful operating technique developed was to isolate the fermenter and the flash vessel from the rectifying and stripping columns when necessary for maintenance purposes. Only the fermenter, circulation pump and flash vessel need be kept aseptic.

● Operating with beet molasses as the feedstock proved to be difficult

until it was diluted to about 30% w/w sugars, due to the high concentration of soluble non-sugar components in the molasses, which become
even more concentrated in the fermenter. The soluble non-sugar components reduce the availability of water in the fermenter, so depriving
the yeast of what is usually a plentiful nutrient.

● The experimental results are still being analysed but they fall within
the ranges predicted by a computer model written by Atkins process
engineers.

4. ATPAL PROCESS ADVANTAGES

The most significant advantage the ATPAL process has is the low process
energy requirements, achieved by energy savings resulting from mechanical
vapour recompression, recycling the heat and production of a low volume of
effluent. Process energy requirements vary with feedstock concentration,
process configuration and specification of the ethanol product. Table I
illustrates the potential energy savings relative to conventional batch
processing. 96% w/w (97.5% v/v) ethanol is suitable for use as a gasoline
dilutant.

TABLE I - TOTAL PROCESS ENERGY FOR PRODUCTION OF 96% w/w ETHANOL

FEEDSTOCK	ATPAL		Conventional Batch	
	Process Energy (kcal/litre)	Process Energy Demand %	Process Energy (kcal/litre)	Process Energy Demand %
Molasses (50% glucose equivalent)	810	16	1,900	38
Cassava (30% glucose equivalent)	1,180	23	5,240	103
Cane Juice (16% glucose equivalent)	1,290	26	2,200	44
Beet (16% glucose equivalent)	1,365	27	3,439	67

Notes: ATPAL for cassava and beet includes enzyme pretreatment. By-Product
drying is excluded. Process Energy Demand is the percentage of energy
equivalent in product alcohol used to fuel process. Process energy basis:
34% electricity generation efficiency and 80% steam generation efficiency,
using fuel oil as prime energy source.

A further development involves enzyme treatment of the fibre constituents of root feedstocks such as beet and cassava. Initial laboratory

tests, using a commercially available cellulase enzyme, indicate that the fibre constituents can be made completely soluble without the addition of further water, approximately one half of the fibre being converted to fermentable sugar. The exceptionally large energy savings with cassava feedstock result from enzyme treatment of the root fibres, instead of mechanical separation of fibre as assumed in the conventional system.

In addition to the very low process energy requirements, the ATPAL process exhibits the following advantages:-

● Direct production of up to 97% w/w (98% v/v) ethanol without using the conventional benzene type entrainer distillation process.

● Reduction in effluent volumes by avoiding the use of dilution water prior to fermentation, thus reducing the size and cost of effluent treatment facilities. The dilution water required with conventional processes, expressed as a percentage of the volume of the undiluted feedstock, would be of the order of 100 to 300% for molasses, 100% for cassava and 0 to 30% for beet and cane juice.

● Less sugars lost in effluent due to reduction in effluent volume.

● Increased yield from root crops by fibre treatment, with enzymes, resulting in up to 10% lower feedstock costs.

● Lower cooling water requirements.

● Fewer centrifuges required for yeast separation.

● Higher fermenter productivity.

● Compact plant, since few items of equipment are required.

The required process equipment is within current manufacturing and operating experience. The process also offers the quality control features inherent in any continuous operating process.

The high throughput of the fermentation system and the overall low energy consumption provide the possibility of introducing ATPAL as a modification to an existing ethanol distillery as an economic means of increasing throughput. The lower energy consumption results in reduced operating costs and requires minimum expansion of existing power supply facilities. The compactness of the main process elements lends the process to application as a small mobile unit.

The process is adaptable, offering the opportunity to balance capital and operating costs to match the needs of particular feedstocks, local resources and market situations.

5. ECONOMICS

The economics of the ATPAL process can be considered by looking at its application to sugar beet. This has been chosen as a study has recently been completed and the costs are up to date. Table II demonstrates that although the total capital costs are comparable, there is a substantial reduction in energy demand. For an oil fired system, this is currently worth a saving of 2.5p per litre of alcohol, which is around 11% of the total alcohol cost (including capital charges). Thus, for a 100,000 litre /day plant, this is equivalent to £412,500 over a 165 day year, a figure which will inevitably increase.

TABLE II - TOTAL CAPITAL COSTS: BEET (thousand £)

	ATPAL	Conventional
Beet Reception	636	636
Juice Preparation	1,098	1,095
Fermentation/Distillation	2,077	1,950
Product Storage	203	203
Effluent Treatment	110	110
Pulp Drying and Bagging	911	1,139
Services	29	158
	5,064	5,291
Civils and Structures	2,364	2,468
Engineering and Project Management	1,083	1,187
	8,511	8,946
Energy Demand kcal/litre (inclusive of Pulp Drying)	3,199	5,273

Molasses presents an even more attractive proposition, as the capital costs are lower due to less feed pretreatment and reduced fermenter, distillation and effluent capacities. Energy costs are also better as there is no energy intensive pulp drying stage and less water to strip in the distillation column and treat as effluent. Several studies have been performed on beet, molasses, cane and cassava. We have concluded that the ATPAL process has significant advantages over conventional technology and is commercially viable at present day fuel prices.

THE CELLULOLYTIC COMMUNITY OF AN ANAEROBIC ESTUARINE SEDIMENT

R.H. MADDEN, M.J. BRYDER and N.J. POOLE

Department of Microbiology, Aberdeen University,

Marischal College, Aberdeen, Scotland

Summary

An estuarine mud bank situated at the mouth of a river receiving paper mill and domestic effluent was the subject of a study lasting six years. Both the microbiology and the physical nature of the sediment were investigated in an effort to elucidate the nature the bacterial communities capable of mineralising cellulose under anaerobic conditions. Field studies showed that at the depth investigated, 5 cm from the sediment/water interface, a very stable environment existed and this was only perturbed when winter storms physically disrupted the sediment. Several pure cultures of cellulolytic bacteria were isolated from the sediment and characterised; Co-cultures of these bacteria with sulphate-reducers were then prepared and studied. The results showed that sulphate-reducing bacteria could metabolise the fermentation end-products of the cellulolytics. Increased cellulolysis and higher yields of cellulase were noted. Thus both organisms appear to benefit in such co-cultures. Based on these results, and observations on microbial interactions published elsewhere models of the simplest communities required to mineralise cellulose, in the sediments studied, are presented.

1. INTRODUCTION

The continuous input of domestic and paper-mill effluent into the River Don Aberdeenshire, Scotland, has led to the lower reaches of the river being severely polluted. At the river estuary several large mud banks are present, shielded from erosion by a sand-bar which forms a natural break-water. The offensive odours released by the mud-banks, especially in late summer, led to the establishment of a project to investigate the microbiology of the mud-banks with a view to elucidating possible links between the input of cellulose, as paper fibres, and the release of hydrogen sulphide. Thus the project became an investigation of cellulose fermentation occurring at ambient temperatures, at a site which was readily accessible to seawater.

Although the project investigated the microbial ecology of a selected mud-bank the current resurgence of interest in the fermentation of plant material to provide chemical feedstocks or energy sources allows the work to be viewed in another light. The mud-bank can be seen as a natural enrichment system for cellulolytic organisms. Since this system operates at the ambient temperature of the north-east of Scotland it is expected to be markedly different, in terms of its microbiology, from the rumen - the best characterised natural system for cellulose fermentation. It is therefore possible for this ecology-based study to provide useful insights into possible biotechnological processes where ambient-temperature fermentations are envisaged in cooler climates.

2. METHODOLOGY

The methods used for mud sampling and analysis were based on those described previously (1). However all media were prepared using the Hungate technique in an unmodified form (2) Dilution blanks were also prepared by the Hungate technique, using the same reducing agent as the medium into which the diluted sample was injected. The media used for culturing and enumeration of cellulolytics were those described by Madden et al (3), while lactate-utilising sulphate-reducing bacteria were cultured in the medium of Postgate (4). Enumeration of sulphate-reducing bacteria utilised media to be described in a future publication. During co-culture studies the liquid medium of Madden et al was used with $(NH_4)_2 SO_4$ (1.2 g/l) replacing

NH_4Cl. Fermentation end-products were determined as described by Madden et al. In situ rates of cellulolysis were estimated using the litter-bag technique of Hofsten and Edberg (5) using Whatman's N°1 filter paper as substrate.

3. RESULTS/DISCUSSION

Unless otherwise stated all results relating to field studies were obtained from sediment samples taken 5 cm below the sediment/water interface.

To discover whether the input of paper-mill waste had resulted in enhanced cellulolysis in the sediments of the Don the in situ rate of cellulolysis of the mud-bank under study was measured (5) and composed with results obtained from similar sites in three other estuaries in the north-east of Scotland. It was found that the rate of cellulose degradation in the Don was about three times greater than that measured in the other estuaries, showing that more active cellulolytic communities had evolved.

A year-long microbiological and physical study was then performed to provide information on the seasonal variations occurring in the sediments of the Don. A comparison of media used for the enumeration of anaerobic cellulolytics showed the simpler medium of Madden et al (3) to be significantly superior to that previously used (1). The numbers of cellulolytics were seen to vary little during the year with recoveries being between 2 and 4 x 10^3 cellulolytics/ml of wet sediment on most sampling trips. Several other parameters measured, such as pH, Eh and salinity, confirmed that the sediment site sampled provided a relatively stable environment. For comparison it was noted that the pH of the water column fluctuated by about 2.0 pH units over a tidal cycle (12 hours) while the annual variation in the sediment was less than 0.8 pH units. This well-buffered system was only disrupted in winter when severe storms flattened the sand-bar sheilding the estuary and consequently large waves could disturb the sediment and allow a mixing with the overlying water column. Recovery of anaerobiosis after such events was very rapid, probably due to the combined oxygen demands of facultative anaerobes and precipitated metal sulphi-

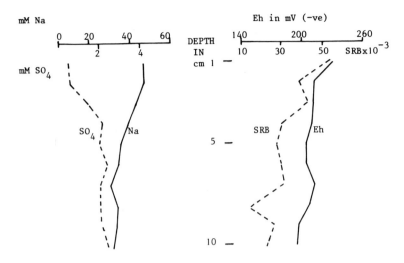

Fig.I.Sediment profile results.

des. Although Na^+ and SO_4^{--} in the sediment arise mainly from
the input of seawater their concentration profiles are not parallel, fig. I. The depletion of sulphate can readily be explained by the activity of sulphate-reducing bacteria (S.R.B.) whose
numbers are highest close to the surface. The very low Eh noted
1cm below the surface water, whose Eh was approximately 400 mV,
also testifies to high microbial activity close to the sediment
surface. The consequences of the high microbial activity close
to the sediment surface will be discussed later but it should
be noted that the sulphate concentration was reduced below
3 mM ; at which concentration S.R.B. cease the reduction of
sulphate in some sediments (6).Laboratory studies showed that
approximately 80 % of the cellulolytics enumerated were clostridia. Eight different cellulolytics were isolated in pure culture and characterised. All were clostridia and purely saccharo-
lytic. The fermentation end-products were CO_2, H_2, ethanol and
acetate. No other volatile products were detected. Lactate was
also noted. Co-cultures of one isolate, Clostridium papyrosolvens (N.C.I.B. 11394) and Desulfovibrio africanus (N.C.I.B.
8401) were prepared and gave similar results to co-cultures
with Desulfovibrio vulgaris (N.C.I.B. 8303) Fig. II.

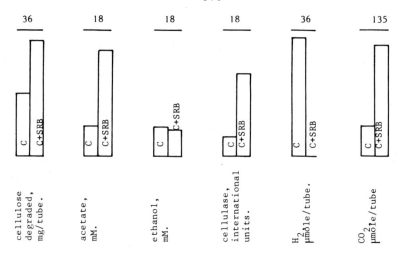

Fig.II.Product comparison of pure(C) and mixed(C+SRB) cultures.

Based on the cellulolytics end-products and the co-culture results a simple community capable of mineralising cellulose near the sediment surface can be postulated, fig.IIIa. Where sulphate is lacking methanogens and S.R.B. can co-operate in the mineralising (7) giving the model shown on fig. IIIb.

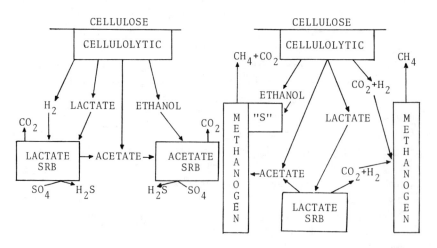

Fig.IIIa High sulphate model.   Fig.IIIb Low sulphate model.

4. ACKNOWLEDGEMENTS

The mixed culture studies were performed during the tenure of a Scientific and Technical Education grant issued by the Commission of the European Communities.

5. BIBLIOGRAPHY

1. Parkes, R.J., M.J. Bryder, R.H. Madden and N.J.Poole. 1979. Techniques for investigating the role of anaerobes in estuarine sediments.p.107-119. In Methodology for biomass determinations and microbial activities in estuarine sediments.C.D. Litchfield and P.L. Seyfried (ed.) American Society for Testing and Materials.Philedelphia.

2. Hungate, R.E. 1969. A roll-tube method for the cultivation of strict anaerobes. p. 117-132. In Methods in Microbiology 3B. J.R. Norris and D.W. Ribbons (ed.) Academic Press, London.

3. Madden, R.H., M.J. Bryder and N.J. Poole. 1980. The isolation and characterisation of Clostridium papyrosolvens (sp.nov.) p. 41-54. In Colloque cellulolyse microbienne. A.T.P. et P.I.R.D.E.S. Marseilles.

4. Postgate, J.R., 1963. Versatile medium for the enumeration of sulphate-reducing bacteria. Appl. Microbiol. 11 p. 265-267.

5. Hofsten, B.V. and N. Edberg. 1972. Estimating the rate of degradation of cellulose fibers in water. Oikos 23, p. 29-34.

6. Nedwell, D.B. and I.M. Banat. 1980. Some aspects of the regulation of sulphate reduction and methanogenesis in a saltmarsh sediment. p.123. In Abstracts of the second international symposium on microbial ecology. University of Warwick, Coventry, England.

7. Bryant, M.P., L.L. Campbell, C.A. Reddy, and M.R. Crabill. 1977. Growth of Desulfovibrio in lactate or ethanol media low in sulfate in association with H_2- utilising methanogenic bacteria. Appl. Environ. Microbiol. 33, p. 1162-1169.

THE USE OF PHASE-PLANE ANALYSIS IN THE MODELLING OF BIO-METHANIZATION PROCESSES IN ORDER TO CONTROL THEIR EVOLUTION.

S.D. ANTUNES[°] and M. INSTALLE[*]

[°] Instituto Superior Técnico, Av. Rovisco Pais, Lisboa (PORTUGAL) ; temporarily with

[*] Laboratoire d'Automatique et d'Analyse des Systèmes and Laboratoire du Génie Biologique, Université Catholique de Louvain, Place du Levant 3 (Bâtiment Maxwell) B-1348 Louvain-la-Neuve (BELGIUM).

Summary

The use of phase-plane analysis in modelling biological and physical phases in digesters give us information about on how to control them.

Study of models based on Hill and Barth's model |1| gives us some initial ideas on how to develop control strategies. Then identification of real-life digesters existing in the bio-engineering laboratory of the University of Louvain-la-Neuve let us compare model with reality.

Various types of regulators have been tested on the model and numerical results are presented and discussed.

1. INTRODUCTION

The anaerobic digestion process has been modelled in 1972 by Graef, S.P. |2| considering one bacterial population in the biological phase. Later on, in 1977 Hill,D.T. and Barth,C.L. |1| considered two types of populations with differentiated functions. Actually Bryant,M.P. |3|, Pfeffer, J.T. |4| and Zeikus,J.T. |5| are considering the existence of other types of populations, for which parameters are not yet available.

Phase-plane analysis and simulation techniques has been used by Antunes S.D. and Installé,M. |6| in order to control the biomethanization.

In this paper, we present results of testing various types of controllers on the model. Sequential logic controllers (based on biomass activity) fulfils well the advantages of automation techniques, as :

i) Sequential logic controllers are low cost regulator, easily implemented by classical automation techniques (pneumatic, hydraulic, relays,..) so as by the newest ones (microprocessors, programmable logic controllers, ..) ;

ii) Automatical maintenance of the process at a production rate close to the optimal one despite of various perturbations continuously disturbing the process. In the biomethanization process we consider the optimal state as the state which maximizes the amount of biogas produced by the unit of time, and the perturbations are the variation of the characteristics of the nutrient entering the methanizer. It is easy to see that such an automation may result in saving a great amount of money.

iii) To implement an automatic starting procedure of the process. More precisely, the task of automation is to drive automatically the process from an initial, "non-working" state towards a state of high production rate.

2. HYPOTHESES

Let us consider as in |6| an anaerobic reactor for which the characteristics of influent, effluent and reactor are assumed to be :

i) INFLUENT : the influent in liquid phase (homogeneous) is introduced in the reactor at a rate $F(1.d^{-1})$ with concentrations $L_{o,j}$ in substrate and $S_{o,j}$ in the active population where j=1 (resp.2) refers to the acidogenic subprocess (resp. the methanigenic subprocess). Concentrations are expressed in $mg.1^{-1}$.

ii) REACTOR : the liquid phase has a volume $V_D(1)$ and its mean concentrations (spatial average) in substrate and biological population are respectively L_j and S_j, $j=1,2$. The active biological phase is characterized by
- the specific growth rate $\mu_j(d^{-1})$, defined as in the Haldane-Monod model, augmented with one inhibition function $f_j(L)$:

$$\mu_j = \frac{\hat{\mu}_j}{1+K_{s,j}/L + f_j(L)/K_{i,j}} \qquad (2.1)$$

where $K_{s,j}$ (resp. $K_{i,j}$) is the saturation constant (resp. the inhibition constant)
- the specific death rates $b_j(d^{-1})$ considered as constant ;
- the yield coefficients for the transformation of one product B into C by biological action

$$R_B^C = Y_B^C \cdot \mu_j S_j \qquad (2.2)$$

so that $A_j = \mu_j S_j$ may be defined as the biological activity of every sub-process.

iii) EFFLUENT : the effluent in liquid phase leaves the reactor at the rate F and has concentrations $L'_j = k_{i,j} L_j$ in substrate and $S'_j = k_{2,j} S_j$ in active biomass where $k_{1,j}$ and $k_{2,j}$ are respectively retentions coefficient for the substrate and retention coefficient for the active biomass. The values for $k_{i,j}$ depend on the reactor geometry and possible heterogeneity.

iv) CONTROL VARIABLE : let us consider as control variable the rate of influent (F), controlled by an "on-off" regulator. The action "on" corresponds to the working-state of the feeding-pump. And for low cost, we suppose a constant flow pump : $F = F_0(1.d^{-1})$. And the action "off" corresponds to the non-working state of the pump : $F = 0(1.d^{-1})$ (which corresponds to the batch evolution of the digester).

3. PROCESS DYNAMICS

The process dynamics model considered here is described as in 6 by balance equations for substrate utilisation and by population dynamics for the biomass :

$$\frac{dS_j}{dt} = \frac{F}{V_D} (S_{o,j} - k_{2,j} S_j) + (\mu_j - b_j) S_j \qquad j=1,2 \qquad (3.1)$$

$$\frac{dL_1}{dt} = \frac{F}{V_D} (L_{o,1} - k_{1,1} L_1) - \frac{1}{Y_{L_1}^{S_1}} \mu_1 S_1 \qquad (3.2)$$

$$\frac{dL_2}{dt} = \frac{F}{V_D} (L_{0,2} - k_{1,2}L_2) - \frac{1}{Y_{L_2}^{S_2}} \mu_2 S_2 + \frac{1}{Y_{L_2}^{S_1}} \mu_1 S_1 \qquad (3.3)$$

The numerical values of the parameters are those used by Hill et al., 1977.

4. PHASE-PLANE ANALYSIS OF THE BIOMETHANIZATION PROCESS

According to the equations (3.1), (3.2) and (3.3), the state variables of the process are : L_j, S_j, j=1,2.

In order to use the phase-plane analysis, we consider separately the acidogenic subprocess (state variables L_1 and S_1) and the methanigenic subprocess (state variables L_2 and S_2).

Phase-plane analysis have been studied by Antunes,S.D. and Installé,M. |6|. The following conclusions are important :

i) there exists one QUANTITY CONDITION for influent rate, conditioning the existence of working digesters : "the mean retention time for biomass needs to be greater or equal to the growth rate minus death rate". Note that retention time for biomass is not equal to the hydraulic retention time).

ii) there exists one QUALITY CONDITION for influent, conditioning stationary state in the phase-plane.

iii) transient evolutions, such as "start-up",... depend on the amount of biomass at initial conditions in the digester. And the best condition seems to be a great concentration on biomass and the lower concentration on substrate, as verified by simulation.

Figure 1 show us phase-plane analysis of batch evolution for methagenic subprocess after different initial conditions. And figure 2, the transient evolution at a constant feed rate.

5. CONTROL OF THE BIOMETHANIZATION PROCESS

The control of a biomethanization process consists in developing a strategy which is able i) to drive quickly and safely the state of the process from an initial one to a high productivity stationary one ; ii) to maintain the state of the process close to the desired one despite of random perturbations.

6.1. MANUAL CONTROL STRATEGIES

Those control strategie s have been studied through simulation on a PDP 11/34-EAI680 hybrid computer. Figure 3 represents in the phase-planes the evolution of the state of the process when the residence time θ is decreased from 10 to 4 days. Trajectories are graduated in days. In case A, θ is decreased from 10 to 4 days in one step. One can see that this causes progressive inhibition of the process and ultimately failure of the reactor. In case B, θ is decreased in small steps such that it takes 12 days to reach 4 days. Despite this smoother action, this causes also progressive inhibition of the process. However, if θ is decreased still more smoothly, it is possible to reach a well-working stationary state.

6.2. AUTOMATIC CONTROL STRATEGIES

Two types of controllers have been tested by simulation :

i) Feed-back of the substrate concentration L_2 ("volatile acids")

Two types of controllers have been tested. The first has an "on-off" functioning mode : when $L_2 > L_{lim}$ the controller stops the influent flow into the reactor, and when $L_2 < L_{lim}$, this controller adjusts the influent flow to a predetermined level (in the examples shown, a flow corresponding to the residence time of 10 days and after 4 days). Figure 4 represents the evolution of the state of the process with such a controller, when L_{lim} was made equal to $\sqrt{K_{s,2}K_{i,2}}$, which corresponds to the maximum growth coefficient μ_2 for the methanigenic bacteria. One can see that, despite of small oscillations the desired state is reached after $\pm$ 10 days. With only 5 interventions of the controller. The second has been an "on-off" controller with functioning mode by "hysteresis" : if L_2 becomes greater than L_{lim2} controller goes "off", and when L_2 becomes smaller than L_{lim1} (with $L_{lim1} < L_{lim2}$) controller goes "on". This controller has the advantage of reducing the number of interventions if $L_{lim1} < \sqrt{K_{s,2} \cdot K_{i,2}} \quad L_{lim2}$.

However everybody knows how expensive, and how difficult are the measurements "on line" of substrate concentration L_2 (volatile acids measurement). So, such a technique is not practicable.

ii) Feed-back of the methane production rate

This controller has also an "on-off" functioning mode based on the state diagram for sequential logic controller given in Figure 7.
- If methane production rate decreases and $F = F_o$, we decide to stop the influent flow $F = 0$
- If methane production rate decreases and $F = 0$, we decide to adjust the influent flow to a predetermined value : $F = F_o$

- If methane production rate increases and $F = F_0$, we decide to maintain the influent flow at its predetermined value : $F = F_0$.

Figure 5 represents the evolution of the state of the process with such a controller. One can see that the desired state is reached after $\pm$ 8 days, with only 3 interventions of the controller (with a sampling time of 1/10 day).

7. CONCLUSIONS

The study of the methanization process through the phase-plane representation enables us to visualize the evolution of the state of the process for various input conditions. It allowed us to propose a simple but efficient regulator for a digester.

Such a methodology enabled us to define limiting conditions for a well-working mode of the reactor and to get the required insight in order to imagine control schemes of this biochemical process. Sequential logic controllers are not only easy in implementation, but they give also information on how to obtain good manual control strategies on starting up so as on transient evolution or perturbation recuperation.

8. BIBLIOGRAPHY

|1| Hill, D.T.: Barth, C.L. - *A dynamic model for simulation of animal waste digestion.* J.W.P.C.F. pp. 2129-2143 (1977).

|2| Graef; S.P. -*Dynamics and control strategies for the anaerobic digester.* Ph.D. dissertation at Clemson Univ. (1972).

|3| McInerney, M.J.; Bryant, M.P. - *Metabolic stages and energetics of microbial anaerobic digestion.* 1st Int. Symp. on Anaerobic Digestion (Cardiff - 1979).

|4| Pfeffer, J.T. - *Anaerobic digestion processes.* 1st Int. Symp. on Anaerobic Digestion (Cardiff - 1979).

|5| Zeikus, J.G. - *Microbial populations in anaerobic digesters.* 1st Int. Symp. on Anaerobic Digestion (Cardiff - 1979).

|6| Antunes, S.D.; Installé, M. - *Anaerobic digestion control using the phase-plane analysis.* Revista Técnica de AEIST, 459, pp. 383-460 (Lisboa -1980)

|7| Antunes, S.D.; Installé, M. - *The use of phase-plane analysis in the modeling and the control of a biomethanization process.* Paper submitted to IFAC - 1981.

378

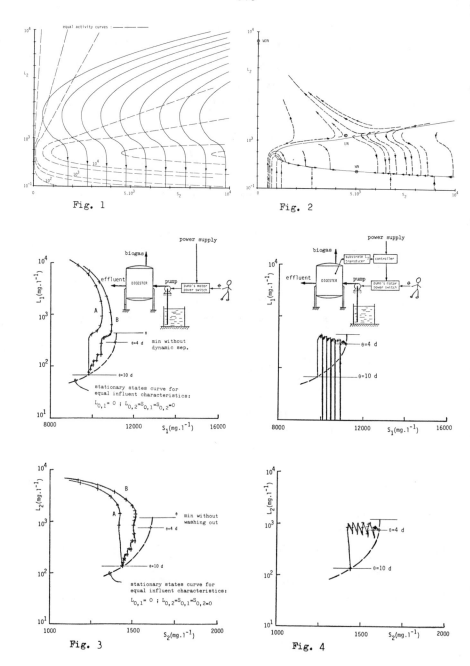

Fig. 1

Fig. 2

Fig. 3

Fig. 4

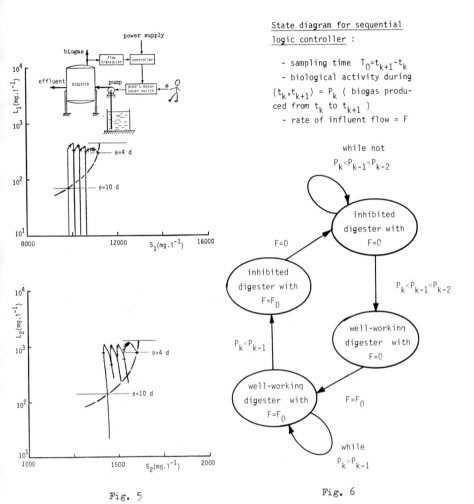

State diagram for sequential logic controller :

- sampling time $T_0 = t_{k+1} - t_k$
- biological activity during $(t_k, t_{k+1}) = P_k$ (biogas produced from t_k to t_{k+1})
- rate of influent flow = F

while not
$P_k < P_{k-1} < P_{k-2}$

inhibited digester with F=0

F=0

inhibited digester with $F=F_0$

$P_k < P_{k-1} < P_{k-2}$

well-working digester with F=0

$P_k < P_{k-1}$

F=F_0

well-working digester with $F=F_0$

while
$P_k > P_{k-1}$

Fig. 5

Fig. 6

APPLICATION OF MICROCALORIMETRY TO THE STUDY OF THE FERMENTATION

OF LIGNOCELLULOSIC COMPOUNDS

M.L. FARDEAU, F. PLASSE, J. PARTOS, J.P. BELAICH

Laboratoire de Chimie Bactérienne, C.N.R.S., BP 71

13277 Marseille Cedex 9 (France)

Summary

The fermentation of native cellulose and hemicelluloses bound to lignine is performed by complex mixed cultures which are able to convert the polysaccharides to methane and CO_2. This conversion can be separated in acidogenesis and methanogenesis. The acidogenesis is a rapid phase, of about one week ; the methanogenesis is very slow, up to two months. Study of cellulose fermentation by naturally-occuring mixed cultures is very difficult using classical techniques of microbiology, because of the heterogeneity, insolubility of the substrate and also of the cultures capable of degrading it. For our studies, cultures were isolated from marshes and soils, where microbial communities capable of degrading the substrate (straw) would be expected to exist. We describe a microcalorimetric technique which allows the rapid evaluation of the biodegradability of natural cellulose or a screening method for the evaluation of the ability of mixed cultures to degrade solid substrates. The effects of substrate pretreatments on biodegradability can also be followed. The thermograms of straw fermentation by several mixed cultures were obtained and it was shown that these figures can be considered as references for the evaluation of various pretreatments (enzymic hydrolysis, alkali and γ-irradiations. In conclusion, microcalorimetry can serve as a very useful tool for the study of the fermentation of complexes substrates by microbial communities.

INTRODUCTION

The enthalpic analysis of microbial growth were developed by numerous authors using differential microcalorimeters. Complete reviews on this subject were recently published (4). It was shown that modern microcalorimeters were a very convenient apparatus for the studies on growth occuring in liquid phase (1,2,3) in anaerobiosis or in aerobiosis (6). More recently the enthalpic analysis of the fermentation of lignocellulosic compounds was undertaken (7,11) using a Calvet microcalorimeter.

The purpose of this paper is to show how it is possible to use microcalorimetry for the rapid determination of the fermentescibility of natural lignocellulosic compounds.

MATERIAL AND METHODS

Barley straw finely ground with a knife mill was used as lignocellulosic substrate. Table 1 shows the composition of this straw obtained by the Van Soerst method in g of product for 100 grams of straw.

TABLE 1

Hemicellulose	Lignin	Cellulose	Dry matter
27.3	9.1	37.7	91.8

Mixed culture, called MCH_1 was taken from the center of a decomposing pile of weeds. The mixed culture called MCR_2 came from a rice-field of Camargue. These cultures were maintained routinely on the fermentation medium at 30° C under anaerobiosis.

The fermentation medium contained phosphate buffer (10^{-1} M pH 7) and NH_4Cl 0.5 10^{-1} M.

Treatment by cellulase. In some experiments, Trichoderma reseii cellulase (commercially available from BDH biochemicals) was added at suitable concentration (50 mg for 1 g of straw). Addition of cellulase has already been studied (9).

Treatment with alkali. The straw was boiled in 1 % NaOH for 1 h. Then the medium was cooled and neutralized by HCl (0.5 N). The inoculum was added at the end of the operation. Alkali treatment has also been considered (5,8,10) by other authors.

Treatment by γ-irradiation. The straw was irradiated with cobalt 60 (5 Mrad, 7h43'). The straw was sterilized by heating at 170° C for 1 hour.

Microcalorimetric method. A differential isothermic Calvet apparatus was used (sensitivity 600 vw^{-1}). The technique used has already been described by Belaich et al. (2,3,6,7). The pyrex microcalorimeter vessel was filled with a suitable quantity of straw contained in 10 ml of the fermentation medium. 1 ml of inoculum was then added and mixed into the medium. The calorimetric vessel was then sealed by a rubber stopper and the air was replaced by pure argon. The vessel was prestabilized in a liquid paraffin bath then transfered into the calorimeter. The heat quantities evolved were calculated by manual integration of the power-time curves. The products evolved during the fermentation were determined by gas chromatography (thermal conductivity detector or hydrogen flame ionization detector).

RESULTS AND DISCUSSION

Figure 1 shows a typical powertime curve obtained for straw fermentation by MCH$_1$. The thermogenic activity occuring during the seven first days corresponds to the acidogenic phase which is not limited by the overall cellulase activity of the culture.

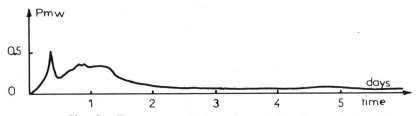

Fig. 1 - Thermograms of straw fermentation by the MCH$_1$ culture

After this period the cellulase activity becomes the limiting factor of the fermentation and the rate of fermentation is considerably decreased. Consequently the power evolved by the culture is decreased and the recording comes back almost to the baseline. It was shown (7) that the heat evolved during this acidogenic phase and the carbon recovered under fermentation products at the end of this phase were proportional to the initial straw quantities contained in the calorimetric vessel.

Pretreatments of the straw markedly increased the acidogenic phase. The result of treatment by cellulase is shown on figure 2 and Table 2. Both quantities of the heat and recovered carbon are increased by about 4 times. Approximatively the same result was obtained by alkali pretreatment (Fig. 3, Table 2).

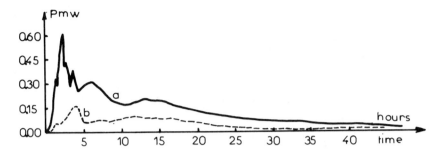

Fig. 2 - Thermograms of straw fermentation. a) with cellulase ;
b) without cellulase.

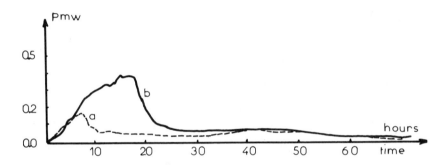

Fig. 3 - Thermograms of straw fermentation. a) without alkali pretreatment, b) with alkali pretreatment.

The pretreatment by γ irradiation produces a complete sterilization of
the straw. So the reference was prepared using straw sterilized by heat
(fig. 4).

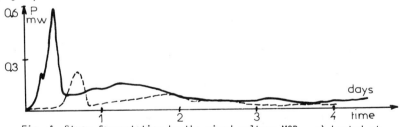

Fig. 4. Straw fermentation by the mixed culture MCR_2. a) heated at
170° C ; b) irradiated.

TABLE 2 - Straw fermentation after 7 days

Inoculum	Treatment	Total recovered carbon in mg	Degraded straw %	Heat quantity in joules	Percentage of the main products obtained					
					Acetate	Butanoate	Propanoate	Ethanol	CO_2	CH_4
MCH_1	without	26,4	6,2	47	59,1	10,8	17,1	10,1	1,1	1,4
	+ cellulase	101	23,7	172	44,6	32,3	16,3	5,3	1,1	0,4
	+ 1 % NaOH	65	15,2	171	44,9	18,9	28,3	3,0	4,1	0,65
MCR_2	Heat 170° C	14,3	3,4	27	51,0	20,3	27,3	0	0,7	0
	γ 5 Mrad	46,4	10,9	146	41,8	23,3	22,0	1,6	5,6	1,3

Table 2 shows the results of the γ pretreatment. The quantity of recovered carbon as fermentation products is increased by about 3,3 times.

Table 2 shows that the fermentation products and more precisely the proportions of the different products obtained are roughly the same in all cases. All the different pretreatments used increased markedly the fermentation by the two mixed cultures used. The more efficient pretreatment seems to be that using cellulase.

In general after seven days of fermentation only traces of methane were found in the fermentation products. It was observed in parallel experiments that methane formation occurs after this period for about one month. So the thermogenic phases recorded correspond to the acidogenic phase which is not limited by cellulolytic activity. The heat evolved during this period is proportional to the quantities of fermented sugars.

In conclusion it can be emphasized that such enthalpic analysis can routinely indicate the efficiency of a pretreatment or the catabolic potential of a mixed culture.

REFERENCES

(1) BATTLEY, E.H. (1960) Physiol. Plant, 13 : 628-639.

(2) BELAICH, A., BELAICH, J.P. (1976) J. Bacteriol., 125 : 14.

(3) BELAICH, A., BELAICH, J.P. (1976) J. Bacteriol., 125 : 19

(4) BELAICH, J.P. (1979) Biological calorimetry. Ed. Beezer, A.E., Academic Press, London.

(5) CHABAL, D.S., SWAN, J.E., MOO YOUNG, M. (1977) Developments in industrial microbiology, 18 : 433-442.

(6) DERMOUN, Z., BELAICH, J.P. (1980) J. Bacteriol., 140 : 377.

(7) FARDEAU, M.L., PLASSE, F., BELAICH, J.P. (1980) Europ. J. Appl. Microbiol. Biotechnol., 10 : 133-143.

(8) JACKSON, M.G. (1977) Animal Feed Science and Technology, 2 : 105-130.

(9) LATHAM, M.J. (1979) Pretreatment of barley straw with rot fungi to improve digestion in the rumen in straw decay and its effect on disposal and utilization. Ed. E. Grossbard, John Wiley and Sons, 131-137.

(10) PAMMENT, N., MOO YOUNG, M., HSIEH, F.H., ROBINSON, C.W. (1978) Appl. and Env. Microbiol., 36 : 284-290.

(11) PLASSE, F., FARDEAU, M.L., BELAICH, J.P. (1980) Biotechnol. Letters, 2 : 11-16.

OCCURRENCE OF CELLULOLYSIS AND METHANOGENESIS

IN VARIOUS ECOSYSTEMS

D. MARTY, J. GARCIN and A. BIANCHI

ECOLOGIE ET BIOCHIMIE MICROBIENNES DU MILIEU MARIN

C.N.R.S.   E.R. 223

UNIVERSITE DE PROVENCE

PLACE VICTOR HUGO 13331   MARSEILLE   FRANCE

SUMMARY

    Using Hungate technique, anaerobic cellulolytic and methane-producing bacteria have been detected in numerous samples from diverse origins : marine sediments (lagoons, coastal and abyssal areas), digestive tracts contents of marine Invertebrates, muds and agricultural soils, and agricultural wastes.
    For ones of these samples, a more thoroughly study of interrelations between bacterial communities present in methanogenic environments has been carried out. To the detection of mesophilic or thermophilic cellulolysis and methanogenesis, we added the study of populations densities of cellulolytic anaerobes, methane-producers, heterotrophic aerobes, heterotrophic anaerobes and sulfate-reducers.
    Cellulolytic anaerobes have been detected in all studied samples. Except agricultural wastes that exhibited thermophilic cellulolysis, other studied cellulolytic bacteria were mesophilic.
    Among all studied ecosystems; bovine and swine wastes were the best methane-producers ; they exhibited a mesophilic (30-37°C) and a thermophilic (45-75°C) methanogenesis. In the most of marine and limnic biotopes, methanogenesis has been detected. Except sediments collected in Hao Atoll that exhibited methanogenesis at 45°C, all other samples exhibited a mesophilic methanogenesis.
    Bacterial anaerobic cellulolysis and methanogenesis are ubiquitous processes in anoxic environments. These experiments showed that methane-producers which are the most strictly anaerobic bacteria known, were present in the same sediment layer with heterotrophic aerobes. This coexistence could be explained by the juxtaposition of different microniches where environmental conditions allow growth of one specific bacterial community.

* This work is supported by P.I.R.D.E.S.

1. INTRODUCTION

La production de gaz biocombustibles, et en particulier de methane, par dégradation microbienne de matières organiques riches en cellulose en milieu anoxique, est connue de très longue date. Toutefois, les espèces ou les communautés bactériennes participant simultanément ou successivement à ces processus sont très mal connues, tant du point de vue physiologique que du point de vue écologique.

Il nous a donc paru intéressant d'étudier une grande diversité de com munautés bactériennes cellulolytiques et méthanoformatrices dans le but de mieux comprendre les relations existant entre ces deux groupes et de sélec tionner les souches ou les communautés microbiennes les plus performantes dans la transformation de diverses biomasses en combustibles. Nous avons traité toute une gamme d'échantillons d'origine très diversifiée, provenant de milieux océaniques (lagunaires, littoraux et abyssaux), limniques et entériques, ainsi que de déchets agricoles.

Dans ces divers échantillons, nous avons recherché la présence de bac téries cellulolytiques anaérobies et méthanoformatrices, puis tenté de purifier ces deux types bactériens afin d'en effectuer l'étude taxonomique (morphologie, physiologie, potentialités cataboliques, génétique ...). Pour la majorité des échantillons, une étude plus approfondie a été réalisée, comprenant la distribution des bactéries cellulolytiques anaérobies et méthanogènes, suivie par leur purification et l'étude taxonomique, mais aussi la distribution d'autres populations microbiennes présentes dans l'écosystème (microflores aérobies et anaérobies dites totales ; bactéries sulfato-réductrices) et qui, par des phénomènes d'inhibition ou de synergie peuvent jouer un rôle primordial dans la méthanogénèse en influant sur telle ou telle étape intermédiaire.

2. MATERIEL ET METHODES

2.1. ORIGINE DES ECHANTILLONS ETUDIES

2.1.1. Origine océanique

- les lagunes   . Méditerranée   :   lagune du Brusc
                                          étang de Sijean
                                          étang de Prévost
                                          salines de Santa Pola
                  . Atlantique   :   bassin d'Arcachon
                  . Pacifique   :   atoll de Hao

```
- le domaine littoral      . Méditerranée   :   Baie de Calvi
                           . Manche         :   Baie de Morlaix
                           . Atlantique     :   Baie de Concarneau
                           . Mer d'Arabie   :   Golfe de Tadjoura
- le domaine abyssal       . Mer d'Arabie   :   Golfe d'Aden
                                                Mer d'Oman
```

2.1.2. Origine entérique

Tractus digestifs d'invertébrés marins provenant de la lagune du Brusc. Holothuries et Oursins.

2.1.3. Origine limnique

```
Sols agricoles   :   litière végétale
                     rizière de Camargue
```

2.1.4. Déchets agricoles

```
Lisiers   :   bovins
              porcins
              volailles
```

2.2. DETECTION ET NUMERATION

Pour chaque échantillon, deux séries de dilutions sont effectuées, l'une en aérobiose et l'autre en anaérobiose. Pour les bactéries anaérobies les milieux sont préparés selon la technique de Hungate : milieu pré-réduit réparti et stérilisé en tube en absence complète d'oxygène.

2.2.1. Etude de la distribution des bactéries hétérotrophes aérobies BHA

(aérobies strictes et facultatives) sur boîte de Petri : base minérale, extraits de levures, peptone, agar.

2.2.2. Etude de la distribution des bactéries hétérotrophes anaérobies BHN

(anaérobies strictes et facultatives) sur tube de Hungate gélosé : base minérale, extraits de levures, peptone, réducteurs, indicateur d'anaérobiose, sous atmosphère N_2-CO_2.

2.2.3. Etude de la distribution des bactéries sulfato-réductrices BSR

en gélose profonde : base minérale, extraits de levures, lactate, sulfate de fer, réducteurs.

2.2.5. Etude de la distribution des bactéries cellulolytiques anaérobies

BDC sur tube de Hungate liquide : base minérale, facteurs de croissance, bande de papier, réducteurs, indicateurs d'anaérobiose, sous N_2-CO_2.

2.2.5. Etude de la distribution des bactéries productrices de méthane BPM

sur tube de Hungate liquide : base minérale, facteurs de croissance, réducteurs, indicateur d'anaérobiose, sous atmosphère H_2-CO_2. La présence de méthane dans les tubes est détectée par chromatographie en phase gazeuse

2.3. PURIFICATION

L'isolement des cultures pures cellulolytiques ou méthanoformatrices est réalisé par repiquages successifs sur des milieux d'enrichissement sélectifs liquides et solides selon la technique de Hungate. Une fois purifiées, les souches sont conservées par cryoconservation en azote liquide.

3. RESULTATS

- LAGUNE DU BRUSC   VAR - FRANCE
5 échantillons dont deux carottes de 10 cm riches en BHA, BHN et BSR. Présence de BPM et de BDC mésophiles dans tous les échantillons.
Purification de souches cellulolytiques.

- ETANG DE SIJEAN   PYRENEES ORIENTALES - FRANCE
1 échantillon ; présence de BDC et BPM dont la purification est en cours.

- ETANG DE PREVOST   HERAULT - FRANCE
3 carottes de 10 cm riches en BHA, BHN et BSR. Présence de BDC et BPM mésophiles dont la purification est en cours.

- BASSIN D'ARCACHON   GIRONDE - FRANCE
5 séries d'échantillons très riches en BHA, BNH et BSR ; présence de BDC et BPM mésophiles dont la purification est en cours.

- SALINES DE SANTA POLA   PROVINCE D'ALICANTE - ESPAGNE
5 carottes de 10 à 25 cm riches en BHA et BHN, mais pauvres en BSR ; présence constante de BDC mésophiles en nombre parfois important ; un seul échantillon méthanogène.
Plusieurs souches cellulolytiques ont été purifiées.

- LAGON DE HAO   ARCHIPEL DES TOUAMOTOU - POLYNESIE FRANCAISE
4 échantillons riches en BHA ; présence de BDC mésophiles dans un seul échantillon ; BPM mésophiles (30-37°C) dans les 4 échantillons et BPM thermophiles (45°C) dans trois d'entre eux. Purification en cours.

- BAIE DE CALVI   CORSE - FRANCE
1 carotte de 50 cm de longueur assez pauvre en BHA et BHN ; très peu de BSR ; présence constante de BDC sur les 50 cm ; absence de BPM. Purification en cours.

- BAIE DE MORLAIX   FINISTERE - FRANCE

1 carotte de 140 cm ; BHA, BHN et BSR bien représentées jusqu'à 110 cm ; présence de BDC et BPM jusqu'à ce niveau ; purification en cours.

- BAIE DE CONCARNEAU   FINISTERE - FRANCE

1 carotte de 120 cm dont seulement les 30 premiers cm sont riches en BHA, BHN et BSR ; présence de BDC et de nombreuses BPM sur la même longueur ; purification en cours.

- GOLFE DE TADJOURA   MER D'ARABIE

Carotte de 460 cm riche en BHA, BHN et BSR dans les sédiments superficiels ; présence de BDC et BPM dans les mêmes sédiments.

- GOLFE D'ADEN ET MER D'OMAN   MER D'ARABIE

3 carottes de 560, 100 et 250 cm prélevées respectivement à 2900, 4010 et 4727 m de profondeur ; populations de BHA, BHN et BSR variables ; présence de BDC ; absence de BPM.
Certaines souches cellulolytiques ont été isolées.

- TRACTUS DIGESTIFS D'HOLOTHURIES   LAGUNE DU BRUSC VAR - FRANCE

Populations de BHA et BHN beaucoup plus importantes dans la portion postérieure ; BSR peu représentées ; présence de BDC ; absence de BPM.
Purification de souches cellulolytiques.

- TRACTUS DIGESTIFS D'OURSINS   LAGUNE DU BRUSC   VAR - FRANCE

Microflores BHA et BHN très abondantes ; BSR bien représentées ; présence de BDC ; absence de BPM.
Purification de souches cellulolytiques.

- LITIERE VEGETALE   BOUCHES-DU-RHONE - FRANCE

Présence de BPM et BDC mésophiles.
Purification d'une souche cellulolytique.

- RIZIERE DE CAMARGUE   BOUCHES-DU-RHONE - FRANCE

Très forte concentration en BHA ; présence de BDC et BPM mésophiles.
Purification d'une souche cellulolytique.

- FUMIER DE VOLAILLE ELEVAGE INDUSTRIEL HAUTE-SAVOIE - FRANCE

Présence de BDC et BPM dont la purification est en cours.

- LISIERS BOVINS   ELEVAGE INDUSTRIEL   HAUTE-SAVOIE - FRANCE

Présence de BDC et BPM mésophiles et thermophiles.
Purification d'une souche méthanoformatrice thermophile (45-75°C).

- LISIERS PORCINS   ELEVAGE INDUSTRIEL   HAUTE-SAVOIE   -   FRANCE

Présence de BDC et BPM mésophiles et thermophiles.

Purification d'une souche méthanoformatrice thermophile (45-75°C).

4. CONCLUSIONS

La cellulolyse et la méthanogénèse bactériennes sont des phénomènes largement répandus dans la nature. Ces expériences nous ont permis de mettre en évidence des bactéries cellulolytiques anaérobies dans tous les échantillons étudiés et des bactéries méthanogènes dans la plupart d'entre eux. Les études de numération ont montré que les cellulolytiques anaérobies et surtout les méthanogènes qui sont considérées comme les bactéries les plus rigoureusement anaérobies, sont généralement présentes dans les sédiments superficiels, là où les populations aérobies sont les plus importantes. D'autre part, parmi les biotopes étudiés, les méthanogènes ne sont généralement présentes que dans ceux renfermant des populations de sulfato-réductrices relativement importantes. La coexistence de populations possédant des processus métaboliques distincts et même opposés (aérobies et anaérobies ; BSR et BPM) dans une même niche écologique peut s'expliquer si l'on considère que les couches sédimentaires ou les lisiers ne sont pas homogènes microscopiquement, mais sont constitués par la juxtaposition de différentes micronicnes dans lesquelles les conditions environnementales permettent le développement de telle ou telle communauté bactérienne.

Ces études ont montré que les numérations pouvaient donner une idée des populations bactériennes présentes dans un biotope méthanogène, mais non de leur activité au sein de celui-ci. L'étape suivante de nos recherches sera donc d'essayer de définir la structure complète d'une association bactérienne participant à la méthanogénèse, c'est à dire de déterminer l'ensemble des constituants bactériens tant du point de vue taxonomique que du point de vue des exigences et des potentialités cataboliques. Il convient donc d'obtenir en culture pure tous les types bactériens présents simultanément dans une communauté méthanogène. La collection des souches isolées d'un même biotope, en l'occurrence le Bassin d'Arcachon, servira de base à des essais de reconstitution de différentes associations plus ou moins complexes. L'efficacité de ces dernières sera définie par dosage du CH_4 produit, l'utilisation de traceurs radioactifs permettant de suivre les relations interspécifiques à différentes étapes du métabolisme intermédiaire.

PRODUCTION OF METHANE FROM FRESHWATER MACRO-ALGAE BY AN

ANAEROBIC TWO STEP DIGESTION SYSTEM

C.-M. ASINARI di SAN MARZANO, H.P. NAVEAU and E.-J. NYNS

Unit of Bioengineering, University of Louvain,
1, Place Croix du Sud, B-1348 Louvain-La-Neuve, Belgium

Summary

Algae are one of the most efficient tools to trap solar energy into biomass. Biomethanation is an anaerobic microbiological process by which biomass can be bioconverted into methane. Hence, the biomethanation of algae appears as an elegant way to convert solar energy into a storable and transportable form of energy. Up to now, biomethanation of algae has only been applied to few algal species with limited success. The aim of the present work has been to assess the possibility to enhance the methane productivities by using a two step digestion process. The filamentous macro-algae *Hydrodictyon reticulatum* and *Cladophora glomerata* were grown in the cooling waters of a nuclear power plant. The algae were first pretreated in a continuous, completely-mixed, anoxic microbiological step. The pretreated algae were then biomethanized in a continuous, completely-mixed, anaerobic step. The pretreatment step, which will be used as a controlled storage of wet algae, results in a decrease in viscosity (liquefaction) of the algae, an apparent loss of around 30 % volatile solids and the steady appearance of $g \times l^{-1}$ amounts volatile fatty acids (mainly acetate and propionate). These fatty acids are used up during the biomethanation step and do not accumulate in the digestion mixed liquor. The methane production rates obtained under these conditions were three to four-fold higher than the best volume efficiencies so far reported in the literature and could be maintained for over 100 days consecutively.

1. INTRODUCTION

Algae are an efficient tool to trap solar energy into biomass. Biomethanation is an anaerobic microbiological process by which biomass, however wet, can be bioconverted into methane. Hence, the biomethanation of algae appears as an elegant way to convert solar energy into a storable and transportable form of energy (1). More particularly, mariculture on land, followed by biomethanation of the algae produced and subsequent use of the digestion mixed liquor as soil fertilizer and/or nutrient for the growing algae looks attractive (2).

Up to now, biomethanation of algae has only been applied to few algal species with limited success (3, 4). The unavoidable spontaneous liquefaction of fresh algae during storage lead to study their biomethanation in a two step digestion system, the concept of which was first introduced by Ghosh and Klass (5).

In the first step, called liquefaction, liquefying fermentative bacteria transform the complex organic material into simpler molecules. The reduced products of the first step, often consisting of volatile fatty acids from two to five C atoms, are utilised as substrate in the second step by a syntrophic community of acetogenic and methanogenic bacteria (6).

2. EXPERIMENTAL

The filamentous macro-algae, mainly belonging to the species *Cladophora glomerata* and *Hydrodictyon reticulatum*, were grown in the cooling waters of the nuclear power plant at Tihange (Belgium) by Professor Sironval (University of Liège, Belgium). They were harvested by raking and sun-dried. These algae contain between 35 and 50 g VS x 100 g^{-1} TS and, hence, from 50 to 65 g mineral matter x 100 g^{-1} TS among which 34 to 51 g carbonates expressed as $CaCO_3$.

The algae were first liquefied in a completely-mixed, anoxic microbial step. The liquefied algae were then biomethanized in a completely-mixed, anaerobic step. Both reactors were operated on a semi-continuous basis, i. e. one daily load, 5 days a week. The running conditions and the results of the two step digestion are summarized in Table I.

The methane digestion system proved very reliable with time as appears from Fig. 1, which shows the average weekly biogas production rate over a one year period.

Fig. 1. Reliability of biogas production with time in a two step methane digestion system using freshwater macro-algae

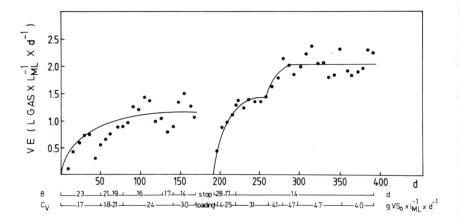

Experimental : see Table 1. The running conditions of the first step were progressively changed from A to B between day 267 and day 294. Symbols and abreviations : see Table III.

3. DISCUSSION

One of the drawbacks of the single step methane digestion is the accumulation of volatile fatty acids in the digestion mixed liquor. It is know that the accumulation of volatile fatty acids often goes together with the unreliability of the methane production rate with time (7). Table II compares the volatile fatty acid content of the digestion mixed liquor in a single step biomethanation and in a two step methane digestion system. It can be seen that, when the methanogenic substrate consists of untreated algae, which do not contain volatile fatty acids, volatile fatty acids tend to accumulate in the digestion mixed liquor. On the contrary, when the methanogenic substrate consists of liquefied algae, which consistently contain g x 1^{-1} amounts of volatile fatty acids, the anaerobic digestion tends to eliminate the volatile fatty acids from the digestion mixed liquor.

TABLE I. Two step biomethanation of freshwater macro-algae at increasing loads

	A		B	
	Liquefaction	Methanogenesis	Liquefaction	Methanogenesis
C_V(g VS$_o$ x l$_{ML}^{-1}$ x d^{-1})	22	3.1[1]	28	4.0[1]
pH	5.5	7.7	5.5	7.7
T (°C)	20	35	20	35
θ (d)	2	14	2	14
L$_o$ (g VS$_o$ x l^{-1})	44	44[1]	57	57[1]
L$_r$ (% of L$_o$)	70	34[1]	88	47[1]
VE (% CH$_4$)(l gas x l$_{ML}^{-1}$ x d^{-1})		1.4 (60)[2]		2.2 (60)[3]
v$_{CH4}$ (l x l$_{ML}^{-1}$ x d^{-1})		0.84		1.31
Y$_{ec}$ (l CH$_4$ x g^{-1} VS$_o$)		0.26		0.33
Y$_{conv}$ (g VS$_e$ x g^{-1} VS$_o$)		0.67		0.52
Y$_{eff}$ (l CH$_4$ x g^{-1} VS$_e$)		0.40		0.63
VA2 (g x l$_{ML}^{-1}$)	1.2-2.1	0.1-1.4	1.1-1.5	0.5-2.9
VA3 (g x l$_{ML}^{-1}$)	0.8-1.8	0.0-0.5	0.8-1.4	0.5-0.6
VA4 (g x l$_{ML}^{-1}$)	0.0-0.2	< 0.1	0.1-1.4	< 0.1

Full experimental details and analytical methods will be published elsewhere.
(1) Based on values at entrance of liquefactor.   (3) Id. between day 375 and day 389.
(2) Mean values between day 239 and day 266 (Fig. 1).   Symbols and abreviations : see Table III.

In the absence of volatile fatty acids in the load, the bacterial community of the single step biomethanation process does not necessarily contain appreciable amounts of acetogenic bacteria. On the contrary, steady amounts of volatile fatty acids in the load appear to stabilize the level of acetogenic bacteria in the syntrophic methanogenic community. Acetogenic bacteria are known to be slow growing. Hence, the two step digestion process will withstand environmental changes more easily than the single step process and therefore appears more reliable with time.

Table II. Volatile fatty acid content of digestion mixed liquors in the two step methane digestion system as compared to the single step biomethanation.

Digestion system	C_V $\dfrac{g\ VS_o}{1_{ML} \times d}$	θ d	v_{CH4} $\dfrac{1\ CH_4}{1_{ML} \times d}$	$[VA3]_o$ $\dfrac{g}{1}$	$[VA3]_{ML}$ $\dfrac{g}{1}$
One step	4	20	0.9	0	1.4
	4	10	0.8	0	1.3
Two step	4	14	1.3	1.4 [1]	0.6

(1) At the outlet of the liquefying step.
Two step digestion system : running conditions as in B (Table I).
One step digestion system : running conditions similar to B (Table I).
Abreviations and Symbols : see Table III.

Although there is an apparent loss in organic matter during the first step, as results from the 25 % decrease in VS, the overall methane productivities increase, both as yields and volume efficiencies in the two step digestion system as compared to the single step process.

Furthermore, observed methane production rates in the two step digestion process of 1.3 vol. CH_4 per vol. digester per day compare favorably with the values of 0.33 reported by Troiano et al. for Laminaria saccharina and of 0.44 by Chinoweth et al. for Macrocystis pyrifera.

In conclusion, the two step digestion system appears as a reliable and efficient process for the biomethanation of algae.

Table III. Abreviations and symbols

1. Concentrations (g x 1^{-1})

VS : Volatile solids

TS : Total solids

L : Load or substrate

VA : Volatile fatty acids

VA2, 3, 4 : With 2, 3, 4 C atoms

2. Mean retention time

θ : Mean hydraulic retention time (d)

3. Yields

Y_{ec} : Yield (1 CH_4 x g^{-1} VS_o)

Y_{eff} : Efficiency (1 CH_4 x g^{-1} VS_e)

Y_{conv} : Conversion (g VS_e x g^{-1} VS_o)

4. Rates

VE( ) : Volume efficiency (% methane in gas)

C_v : Volumetric load (g VS_o x 1_{ML}^{-1} x d^{-1})

v_{CH4} : Methane production rate (1 CH_4 x 1_{ML}^{-1} x d^{-1})

5. Indices

o : Influent

ML : In digestion mixed liquor

r : Effluent

e : Eliminated

4. <u>REFERENCES</u>

1. Benemann, J.R., Weissman, J.C., Koopman, B.L. and Oswald, W.J. (1977). Energy production by microbial photosynthesis. Nature (London),<i>268</i> , 19-23.

2. Wagener, K. and Inden, P. (1977). Methane production by mariculture on land. Eur. Seminar Biol. Solar Energy Conversion Systems, Grenoble-Autrans, May, 16 pp.

3. Chynoweth, D.P., Klass, D.L. and Ghosh, S. (1978). Biomethanation of giant brown kelp <i>Macrosystis pyrifera</i>. In "Energy from Biomass and Wastes", Inst. Gas Technol. ed., Chicago (Ill.), 229-251.

4. Troiano, R.A., Wise, D.L., Augenstein, D.C., Kispert, R.B. and Cooney, C.L. (1976). Fuel gas production by anaerobic digestion of kelp. Resource Recovery Conserv., <i>2</i>, 171-176.

5. Ghosh, S. and Klass, D.L. (1978). Two-phase anaerobic digestion. Process Biochem. <i>13</i>, 15-24.

6. Bryant, M.P. (1979). Microbial methane production - Theoretical aspects. J. Animal Sci. <i>48</i>, 193-201.

7. Asinari di San Marzano, C.-M., Binot, R., Bol, T., Fripiat, J.-L., Melchior, J.-L., Naveau, H., Perez, I. and Nyns, E.-J. (1980). Control of environmental factors in the biomethanation at the pilot level of economic substrates. 6th Int. Fermentation Symp., London (Ont., Canada), July, Paper F-9.1.3 (L), Abstracts, 72.

OPERATION OF A LABORATORY-SCALE PLUG FLOW TYPE OF DIGESTER ON PIG MANURE

FREDA R. HAWKES,* J.R.S. FLOYD,* D.L. HAWKES
Department of Science* and Department of Mechanical and Production Engineering, The Polytechnic of Wales, Pontypridd, Mid-Glamorgan, CF37 1DL, Wales, U.K.

Abstract

A novel plug flow anaerobic digester, designed to minimise the problems of scum formation, was successfully tested on a laboratory scale using low solids pig manure as a substrate at temperatures between 16^0 and 29^0. A Mg^{2+} tracer experiment showed that the liquid contents were fully mixed, and solids settled and accumulated rather than being displaced; the term plug flow is therefore inappropriate in this instance. Cellulose accumulated at a substantially higher rate than lipids.

Gas production responded rapidly (less than 24 hours) to changes in operating temperature and a rise from 17^0 to 25^0 increased daily gas production by a factor of 3. The CO_2 content of the digester gas also increased with temperature. The overall gas yield of 0.492 m^3CH_4/kgVS added was greater than has been reported for most conventional digesters operated on similar wastes at higher temperatures. Ammonia concentration within the digester averaged 3280 ppm, a level often considered inhibitory.

1. INTRODUCTION

Plug flow systems theoretically may have advantages over conventional stirred reactors. A true plug flow reactor is more efficient than a stirred tank for reactions where rate increases with substrate concentration, since this concentration is maximised. Anaerobic digestion is generally assumed to follow such first order kinetics. Also, in an idealised plug flow reactor, all elements of the feed have the same residence time, in contrast to the stirred tank. For digestion to proceed if true plug flow occurred, inoculation of the feed with digester contents would be necessary.

In practice, true plug flow is unlikely to occur during the digestion of heterogeneous wastes such as animal slurries, since solid settling and gas evolution will introduce longitudinal mixing. An attempt has been made in this study to establish the actual nature of the liquid flow through a laboratory-scale plug flow type digester, operating on a relatively low strength piggery waste.

Although digesters of this type will depart from true plug flow, they retain advantages of simplicity and reduced energy input, since mechanical mixing is not required. However, previous reports (1,2) have shown that blockage by scum can be a problem in unmixed digesters. To overcome this, a novel design of plug flow digester in which the surface where scum can accumulate is minimised has been proposed by D.L. Hawkes and H.R. Horton of The Polytechnic of Wales. To date, two laboratory-scale digesters have been built to this design, and data has been collected during several months' successful operation.

If digesters could be operated successfully at the same temperature as their environment, heat exchangers and insulation could be omitted. Pig houses where temperatures are in the range $16^{\circ}-20^{\circ}$ might provide such a location. The rate of gas production increases with temperature (3,4) and digesters are normally operated at about 35°. The relationship between gas production and temperature has been studied here in the range $16^{\circ}-29^{\circ}$. It is expected that gas composition will vary with temperature because of the increased solubility of CO_2 in the effluent at lower temperatures (5), and this variation is also reported on here.

2. MATERIALS AND METHODS

The experiments were carried out using a 15.3 l capacity perspex digester fitted with sampling ports along its length (Fig.1). The digester was operated inside an insulated cabinet and temperatures were controlled using an electric heater and an aquarium thermostat. Gas volumes were measured using a 0.25 l wet gas meter (Wright & Co., Tooting, London). The digester was seeded with contents from a conventional high rate digester operating at 35° on pig manure, and started up at 15°, feeding with pig manure (1.49% TS, 56.90% VS, LR = 0.25gVS l $digester^{-1}day^{-1}$, RT = 34 days) for 14 weeks. During the 116-day experimental run which followed start-up, the feed, of composition 1.01% TS, 46.86% VS, was stored refrigerated, 0.75 l being added daily from Mondays to Thursdays and 1.50 l on Fridays. Gas volume and composition and median temperature were recorded on these days.

To follow the passage of liquids through the digester, $MgCl_2$ was used as a soluble tracer at a concentration less than that reported inhibitory (6). After 53 days' operation, 0.75 l of feed was dosed with 500 ppm Mg^{2+} as $MgCl_2$ and the effluent was assayed for Mg^{2+} daily until 20 days later when 250 cm^3 samples were taken from the feed, effluent and each of the horizontal sample ports 1-3 (Fig.1). The samples were assayed for solids, lipid, cellulose, ammonia-N and total Kjeldahl-N.

Assays: CH_4 and CO_2 compositions were assayed as previously reported (7). Total cellulose was determined gravimetrically following delignification with Na chlorite and 0.3% v/v acetic acid at 75° (8). Total lipids were determined by extraction with chloroform:methanol:water, 1:2:0.8 v/v (8). Ammonia-N was measured by titration following steam distillation into boric acid-indicator solution. Total Kjeldahl-N was determined as for ammonia-N after 20 hours' digestion in 50:50 v/v H_2SO_4:water plus selenium dioxide. Total and volatile solids (TS and VS) were determined as previously reported (7). Mg^{2+} levels were measured using a Varian Techtron 1100 atomic absorption spectrometer; 10,000 ppm lanthanum as $LaCl_3$ was needed to suppress interference. $LaCl_3$ caused precipitation so samples were precipitated fully with 9000 ppm La^{3+} and filtered before adding 10,000 ppm La^{3+} and assaying.

3. RESULTS

During the start-up period gas production increased slowly to a fairly stable level of 0.5 l day $^{-1}$. Table I shows the operating parameters and gas yield treating the 116-day experimental run as a whole. The solids composition of pig manure from different sources can vary widely; that used here was particularly dilute. Variations in temperature, gas volume and gas composition throughout the run were marked, and it is important to note that

Table I presents average figures. The extent of daily temperature and gas volume fluctuations is shown in Fig.2 which demonstrates the rapid response of gas production to temperature changes. Mean gas volumes have been calculated for each median temperature and are plotted in Fig.3. The increased spread of data at temperatures above 22^0 reflects the small number of replicates available in this range. When mean values of % CO_2 for each median temperature are plotted (Fig.4), an increase in % CO_2 with temperature is apparent but experimental variation is large. Unlike gas volume, % CO_2 did not respond to daily temperature fluctuations in a simple way.

Fig.5 shows the Mg^{2+} concentrations recorded in the effluent plotted against time after the addition of 0.75 l of feed containing 500 ppm Mg^{2+} as $MgCl_2$. The background concentration of 5.2 ppm has been subtracted from the data and the times refer only to days when the digester was fed; on Fridays the digester was fed twice, 2 samples were collected and treated as if from 2 days. The theoretical curves on the graph were calculated for the following conditions:-
1. Complete mixing of liquid throughout the whole volume within 24 hours.
2. Complete mixing in 24 hours in the upper 5 l of the tube, the remainder being occupied by unmixed sediment.
3. Partial mixing in the whole volume assuming plug flow-like movement. Half the Mg^{2+} in each 0.75 l cylindrical segment dividing equally between the two adjacent segments each 24 hours.
4. Partial mixing as in 3, but liquid flow restricted by sediment to the upper 5.25 l.
The calculations have allowed for dilution in the 2.715 l effluent tank. The results of assays on feed, contents along the digester and effluent are presented in Table II.

4. DISCUSSION
Comparison of the results of the Mg^{2+} tracer experiment with theoretical curves shows that the aqueous phase in the digester is fully mixed throughout the whole volume despite accumulated sediment which appeared to restrict freely circulating supernatant to the upper third. The compositions of samples taken along the length of the digester do not reflect the progression of digestion in a simple way; the assay results show that solids accumulated in the digester rather than being displaced in plug flow. The samples were taken from the centre of the digester tube (see Fig.1) which was below the surface of the sediment so the solids concen-

trations in them were higher than those in the digester as a whole, but since sediment filled more than half of the digester the sample concentrations are probably less than twice the overall ones.

Cellulose and lipids accumulated differently, the results showing about 36 times more cellulose in the digester than in the feed but only about 7.5 times more lipid. This reflects the rapid sedimentation and slow digestion of cellulose because of its relatively large particles. Sediment was deepest at the lower end of the digester tube and this probably accounts for the changes in solids content along its length. The samples were taken late in the experimental run when temperatures and gas volumes were relativly high and solids accumulation had ceased since gas bubbles carried particles to the outlet faster than they were added in the feed. This had not been happening long enough to have greatly affected the composition of the digester contents.

The higher ammonia-N levels in the contents and effluent than in the feed arise from the degradation of nitrogen containing organic compounds. The high levels in the digester (mean=3283 ppm) did not prevent good gas yields, in agreement with van Velsen (9) who recorded a decrease of only 20% in gas production rate when ammonia nitrogen in pig waste was increased from 605 to 3075 ppm at pH7. The difference between total Kjeldahl-N and ammonia-N is also greater in the digester contents than in the feed, probably indicating some bacterial assimilation and accumulation of e.g. protein.

The overall gas yield of 0.492 m^3CH_4 kgVS added^{-1} was high in comparison with published data for pig manure at similar temperatures. van Velsen (4) reported yields of 0.226 m^3CH_4 kgVS added^{-1} at 20^0 and 0.258 at 27^0 and a retention time of 20 days, and Summers and Bousfield (3) 0.179 m^3CH_4 kgTS added^{-1} at 25^0 and a 10-day retention time. The high yield obtained here can be attributed to sedimentation which produces very long solids retention times.

Daily gas volumes responded rapidly to temperature fluctuations. Since temperature changes were short term the gas volumes recorded probably represent changes in metabolic rate rather than the size or composition of the bacterial population. The long solids retention time probably permitted the rate-limiting population to persist at low temperatures when it would have been washed out in fully mixed digesters. An increase in % CO_2 with temperature was recorded. The solubility of CO_2 decreases as temperature increases so the varying amount of CO_2 dissolved in the effluent may account for the observed relationship. CO_2 entering and leaving solution in

the digester contents during temperature fluctuations probably contributed
to the large experimental variation in the data. Summers and Bousfield (3)
however found no significant variation in % CO_2 between 25° and 35° and
van Velsen et al (4) no temperature-related change in % CO_2 between $20-55^O$.

Under the conditions of the experimental run reported here the digester showed no problems due to scum formation, but subsequent experience with
this and another digester of similar design has shown that at higher
temperatures and solids contents blockages can occur because gas production
is much more rapid. Modification to include a simple manual agitator at
the digester exit relieved these problems.

The results reported here on low solids piggery waste show that this
type of digester can give high gas yields due to its ability to accumulate
solids. Accumulation is more effective at low temperatures so operation in
this range may be an advantage.

REFERENCES

1. L.F.Fry. Practical building of methane power plants for rural energy
   independence. Standard Printing, Santa Barbara (1974).
2. W.J.Jewell in Fuel gas production from animal & agricultural residues
   and biomass. Dynatech Report No. 1845 C00-5099-4, p. 41-57 (1978).
3. R.Summers and S.Bousfield. A detailed study of piggery-waste anaerobic
   digestion. Agricultural Wastes 2, 61-78 (1980).
4. A.F.M. van Velsen, G. Lettinga, D.den Ottelander. Anaerobic digestion
   of piggery waste. 3. Influence of temperature. Neth. J. Agric.Sci.27,
   255-267 (1979).
5. J.T.Pfeffer. Anaerobic Digestion Processes in D.A.Stafford,B.I.Wheatley
   and D.E.Hughes (eds). Anaerobic Digestion, Applied Science Publishers,
   London (1980).
6. P.L.McCarty. Anaerobic waste treatment fundamentsls III.Public Works
   95, 91-94 (1964).
7. F.R.Hawkes and B.V.Young. Design and operation of laboratory-scale
   anaerobic digesters: Operating experience with poultry litter.
   Agricultural Wastes 2, 119-133 (1980).
8. S.E.Allen. Chemical analysis of ecological materials. Blackwell
   Scientific Publications, London (1974).
9. A.F.M. van Velsen. Adaptation of methanogenic sludge to high ammonia-
   nitrogen concentrations. Water Research 13, 995-999 (1979).

TABLES:

Table I. overall operating parameters and gas yield.

Duration of run	116 Days
Mean temperature	20.3^0
Feed	Pig Manure 1.01%TS, 46.86%VS
Total volume fed	67.5 l
Hydraulic retention time	34.4 Days
Loading rate	0.184 g VS l digester^{-1} day^{-1}
Mean gas volume produced	1.38 l day^{-1}
Mean % methane	85.2
Mean gas yield	0.492 m^3 CH$_4$ kg VS added^{-1}

Table II   Composition of Feed, Contents and Effluent

	Feed	Port 1	Port 2	Port 3	Effluent
Volatile Solids, gl^{-1}	4.74	38.72	28.53	23.98	5.58
Cellulose, gl^{-1}	0.57	25.40	19.61	16.38	0.56
Cellulose as % of VS	12.0	65.6	68.7	68.3	10.0
Lipids, gl^{-1}	0.54	4.72	3.80	3.53	0.73
Lipids as % of VS	11.4	12.2	13.3	14.7	13.1
Ammonia-N,ppm (AN)	2500	3400	3250	3200	3280
Total Kjeldahl-N,ppm(TKN)	3050	4100	3950	3800	3450
TKN - AN, ppm	550	700	700	600	170

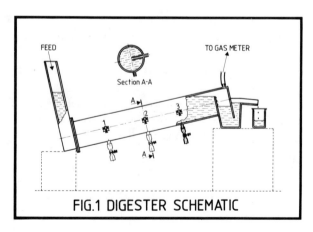

FIG.1 DIGESTER SCHEMATIC

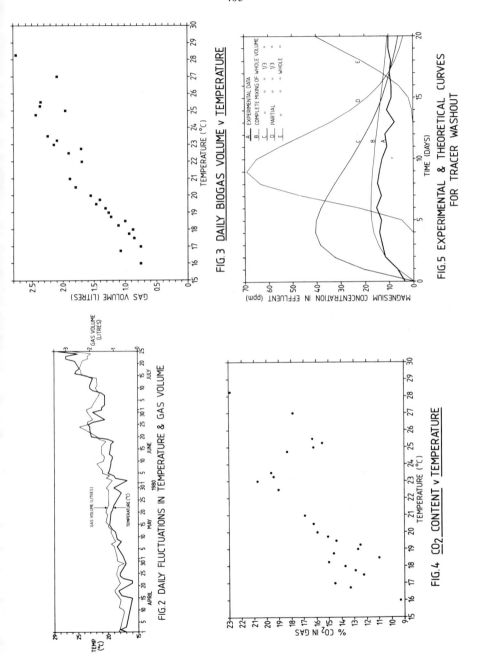

FIG.3 DAILY BIOGAS VOLUME v TEMPERATURE

FIG.5 EXPERIMENTAL & THEORETICAL CURVES FOR TRACER WASHOUT

FIG.2 DAILY FLUCTUATIONS IN TEMPERATURE & GAS VOLUME

FIG.4 CO₂ CONTENT v TEMPERATURE

FERMENTATION TO BIOGAS USING AGRICULTURAL RESIDUES

AND ENERGY CROPS

D A Stafford and D E Hughes
Department of Microbiology University College Cardiff

Introduction

There is in the UK about 24 mtce available annually from organic wastes of
which farm wastes are a 4.9 mtce. In many parts of Europe pig and cattle
farms present a local environmental problem and waste disposal difficulties.
This pollution problem may be turned into a residue asset by digesting
anaerobically to produce methane gas, fertilizer and animal feedstuffs.
Similarly energy crops may be grown for bioconversion to methane using
digester units.

Potential Gas Production

Loading rates in digesters, in terms of VS added, can vary from 0.7 – 5 Kg VS/m^3/day[5]. Higher loadings may be possible, but this will depend on whether toxic build up of volatile fatty acids occurs and on the water content of the manure. One of the major problems in the anaerobic digestion of animal wastes is the 'build up' of toxic concentrations of ammonia, produced from the high nitrogen content of the manures. They may be 0.4 g Nitrogen/Kg animal for cattle, 0.9 for pigs and 0.15 for humans. Toxic concentrations of ammonia on anaerobic digestion are difficult to determine and this will depend upon the hydraulic retention time and water content of the feed. One digester system was inhibited at unionized ammonia concentrations of 150 mg/l, whilst another showed successful digestion of swine waste at concentration of about 5000 mg/l. Our systems on pig wastes have been operating successfully at ammonia concentrations of around 4000 – 5000 mg/l.

Conventional digesters have been improved upon in recent years in that automated systems are available to farmers to produce a treatment procedure to reduce the pollution load, to produce methane energy and fertilizer for the farms. Such a system has been developed by the author in collaboration with Hamworthy Engineering Ltd, Poole, Dorset, UK (the manufacturers) and Chediston Agri-Systems Chediston, Suffolk UK (Licensee). A 350m^3 digester is at present treating the waste from 4,500 pigs and it is intended also to feed the waste, to the digester, from 120 cattle. The methane gas is converted to electrical energy using a modified Ford engine and generator set producing 20 kw.

When considering a choice of digester system the newer types of units developed for smaller farms (up to 600 pigs or 60 cattle), include the plug-flow digester type. In this laboratory (Cardiff) we have been treating pig waste in conventional mixed systems and also in plug-flow anaerobic digesters. The relative performances show that gas yields per Kg of volatile solids added are about one third of the conventional digester, but the temperature of operation is however much lower. It is thus possible that the net gas yields may be comparable since less energy is required to heat the digester and to mix the contents.

BRACKEN   Hydraulic Retention Time = 100 days
(Pteridium aqualinum)

Total Solids   = 30% (Fresh weight)
Volatile Solids = 90%

GAS YIELDS m^3 BIOGAS/Kg Volatile Solids (added)

Week No   1.   0.296
           2.   0.383
           3.   0.447
           4.   0.433

ASSUME   20 t/Ha growth yield (Fresh weight)
AND 0.4 m^3 BIOGAS/Kg V S added
1 Ha could produce 8,000 m^3 BIOGAS

Area of Bracken in U K   = 3470 Km^2
Total potential gas production = 2776 x 10^6 m^3 BIOGAS
At 25 MJ/m^3 Calorific Value   = 69.4 x 10^{15} J
                         = 2.57 mtce

JAPANESE KNOTWEED
(Polygonum cuspidatum)

Hydraulic Retention Time = 100 days

Total Solids   = 28% (Fresh Weight)
Volatile Solids = 93%

GAS YIELDS

m^3 BIOGAS/Kg Volatile Solids (added)

Week No		
	1.	0.238
	2.	0.469
	3.	0.458
	4.	0.923
	5.	0.673

ASSUME

25 t/Ha growth yield (Fresh Weight) AND 0.55 m^3
BIOGAS/Kg V S added
1 Ha could produce: 13,750 m^3 BIOGAS

CORD-GRASS
(Spartina anglica)

Hydraulic Retention Time = 100 days

Total Solids   = 30% (Fresh Weight)
Volatile Solids = 83%

GAS YIELDS

m^3 BIOGAS/Kg Volatile Solids (added)

Week No		
	1.	0.314
	2.	0.435

ASSUME

1 ot/Ha growth yield (Fresh Weight)
AND 0.38 m^3 BIOGAS/Kg V S added
1 Ha could produce 3,800 m^3 BIOGAS

410

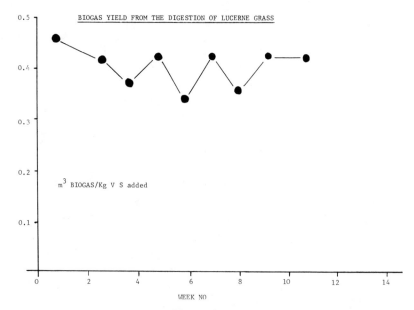

Figure 1

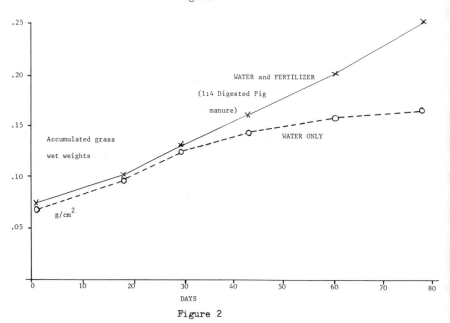

Figure 2

HORIZONTAL-ROTATING-DRUM CONTINUOUS FERMENTOR

V. A. VASEEN, P. E.

President, AVASCO

Summary

The Horizontal-Rotating-Continuous Fermentor is applicable to processing many kinds of fermentables which are the result of bio-organisms reacting with organic solutions. The specific description of monosaccharides converted to "beer" is illustrative of the operation of the apparatus for use of wort to beer, molasses or syrup to beer, must to wine, milk to cheese, waste organics anaerobically to methane or aerobically to carbon dioxide. The apparatus is conducive to production of industrial ethanol wherein, taste, color, and odor, are secondary to quantity and efficiency of ethanol conversion from monosaccharides. The 6 to 15% ethanol solution provided is then distilled for concentrated ethanol production, or filtered and used to generate electric power by fuel cells.

FERMENTATION → BEER

VIA THE HORIZONTAL ROTATING DRUM CONTINUOUS FERMENTOR

Continuous fermentation, that is, the molasses, syrup or other monosacchrides feed to the fermentor at a constant rate per unit of time, and the
beer, CO_2 and yeast removal also being at a constant rate per unit of time;
has long been the object of the fermentation industry, especially the industrial alcohol group.

The HORIZONTAL ROTATING FERMENTOR is designed to retain the sterility or
aseptic environment from monosacchrides supply to beer discharge. This is
accomplished by a proprietary inlet manifold which provides for the slow
rotation of the horizontal drum while adding monosacchrides solution at a
constant rate to the entry or head end of the fermentor. The yeast and
particulates produced in the fermentation process is by the use of "paddles"
mounted on the internal circumference of the drum in an approaching of
spiral shape, or rifling, move the yeast and particulates as produced, to
the head end of the drum. The yeast and particulates is moved countercurrent to the monosacchrides solution flow and thus innoculates the new
monosacchrides solution which, with control of removal rate, assures
proper yeast to monosacchrides balance. The yeast and particulates as
collected at the head end of the drum is picked up in scoops or cups
mounted on the head end of the drum for removal. Near the top of the drum
retained in a vertical position by the specially designed inlet manifold-
ing, is a yeast and particulates slurry collection hopper which collects
the slurry and removes it as it drains out of the vessel through the specially designed inlet manifold system to slurry dewatering, processing, and
salvage.

This counterflow moving of yeast and particulates solids and returning, through the <u>influent monosacchrides solution,</u> along with mixing of the fermenting <u>beer</u> with the new <u>monosacchrides solution,</u> assures acceleration and greater completion of fermentation process under maximum controlled conditions.

The internal wall paddles are heated and/or cooled by a heat exchanger fluid also introduced through the specially designed manifold at the drum head end.

The combination of counterflow, physical mixing and temperature control reduce the time of fermentation to the least time possible, as well as assure the greatest efficiency of fermentation conversion of fermentable sugars and monosacchrides materials, to ethanol.

An additional efficiency factor is the removal of carbon dioxide formed in the fermentation process. The carbon dioxide is more efficiently stripped from the beer by the mixing. The removal of the carbon dioxide is made possible by means of a gas removal orifice or nozzle located above the high liquid level, where it is then removed through the specially designed discharge manifold at process completion, or tail end of the drum.

The <u>beer</u> is removed from the drum by the liquid level overflow "Weir" at the "tail" end of the drum, and flows aseptically to collection vats and/or other processing.

The heat exchanger fluid introduced at the head of the drum and circulated along the drum walls, transferring heat to or from the internal paddles, is removed in a similar manner from the tail of the drum.

The drum is a self-supporting shell resting on rollers which, when rotated by prime mover, cause the drum to rotate. The structural design of the drum indicates the drum have a structural ring around its circumference at two or more locations which not only receive the transfer of horizontal bending movement from the drum skin, but transmit the rotary motion to the drum as a vehicle.

The drum head and tail influent/effluent manifolds systems are able, for the most part, be made up of standard commercially available, ASA type flange fittings, or welded pipe fittings. Each end connection is fitted with a special designed rotary coupling, with "O" ring or other self-sealing, easily replaceable, sliding type separators between the monosacchrides solution, yeast and particulates, and heat exchange fluid on the head end; and the beer, carbon dioxide and heat exchange fluid on the tail end.

Continuous fermentation by vertical design fermentors, as previously used for production of "wine" or "beer" by most various apparatus designs has not been generally, commercially acceptable, or successful.

*DESIGN PARAMETERS are as follows for (10) ten-hour fermentation with 12% ethanol by weight:

Production for						
Anhydrous ethanol - GPD	120	240	480	1200	2400	4800
**Monosacchrides Solution (GPM)	10	20	40	100	200	400
(Barrels/day)	44	53	59	83	107	126
***Diameter Drum - Ft.	3.5	4.5	6.0	8.0	10.0	13.0
***Length (minimum) of drum, Ft.	44	53	59	83	107	126

*** = Subject to specific fermentation design schedules.

** = Based on drum liquid depth equals 80% of diameter.

* = Patented apparatus.

The HORIZONTAL ROTATING DRUM FERMENTOR eliminates most of the problems
found to exist with these earlier designs; and now provides a means of
using any <u>monosacchride solution</u> not suitable for human consumption by
producing <u>beer</u> or <u>wine</u>, and is specially usable for such alcohol products
as ethanol for industrial uses, at a rate and efficiency as well as
quality, which can be made competitive in the industrial alcohol industry.

FERMENTATION OF MONOSACCHRIDES WITH THE HORIZONTAL-ROTATING
FERMENTOR FOLLOWED BY: MECHANICAL CONCENTRATION OF ETHANOL -
THEN ELECTRIC POWER FROM THE CONSTANT STRENGTH FUEL CELL
OR POLISHING DISTILLATION AND DEHYDRATION.

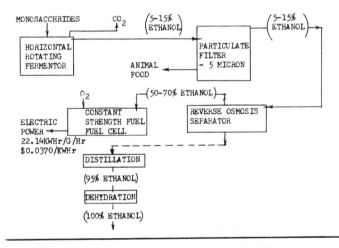

CALCULATED OPERATIONAL COST OF ETHANOL AND POWER PER GALLON
(Does not include Manpower or Capital investment costs)

	DISTILLATION	FUEL CELL
MONOSACCHRIDES	$0.55	$0.55
HORIZONTAL ROTATING - FERMENTATION	0.05	0.05
PARTICULATE FILTERING	0.01	0.01
REVERSE OSMOSIS - ETHANOL SEPARATION	0.01	0.01
POLISHING DISTILLATION	0.01	0.00
DEHYDRATION	0.02	0.00
COST PER GALLON ETHANOL	$0.65	$0.62
	100%	70%
OXYGEN		0.20
COST TO OPERATE FUEL CELL		* $0.82

* PER 22.14 KWHr = $0.0370/KWHr

BIOCONVERSION OF AGRICULTURAL WASTES INTO
FUEL GAS AND ANIMAL FEED

E. COLLERAN, A. BOOTH, M. BARRY, A. WILKIE,
P.J. NEWELL AND L.K. DUNICAN

Department of Microbiology, University College, Galway, Ireland.

SUMMARY

A two-stage anaerobic digestion system consisting of a liquefaction
stage followed by a methanogenic stage utilising a fixed bed reactor
design, was developed for biogas production from pig slurry. COD
removal rates of 80 - 90% were obtained at a liquid retention time of
3 days at 30°C with gas yields close to the theoretical maximum. Initial
results indicate that the process is also suitable for the digestion of
solid agricultural residues such as straw. Liquefaction of straw-slurry
mixtures, using a system of batch replacement of the supernatant every
5 - 6 days, indicated that 50-60% of the solids fraction was
solubilised within 18 days. The cellulose content of the straw
decreased by 63% during the liquefaction period. Passage of the
solubilised fraction to a fixed bed reactor resulted in COD removal
rates of 90% with a yield of 0.36L CH_4 per gramme COD removed.

Alternatively, it was demonstrated that straw can be upgraded
to a protein rich animal feed by a dual fungal fermentation utilising
the lignolytic fungus, Sporotrichum pulverulentum and the cellulolytic
species, Chaetomium cellulolyticum. The utilisation of this novel
mixed culture system obviated the need for costly pretreatment processes
and yielded a product containing 30-32% protein (% d.w.) when chopped
straw was used as substrate. The operation of an integrated system
for fuel gas and animal feed production from agricultural wastes is
discussed.

1. INTRODUCTION

It is estimated that of the total 1,800 million tonnes of waste produced within the E.E.C. countries in 1979, over half (950 million tonnes) was contributed by the agricultural sector(1). Agricultural wastes are derived from a variety of different inputs, ranging from animal manures to a diverse selection of crop residues. The high organic content of pig, poultry and cattle manures from intensive husbandry units ensures their suitability as substrates for anaerobic digestion and the treatment process combines the dual advantages of fuel production and pollution control. Sugar-cane bagasse and cereal straw are, quantitatively, the most important of the agricultural non-wood plant residues (2). The annual production of cereal straw in the U.K. currently exceeds 12 million tonnes (3). Worldwide, annual production of the five common slender stemmed cereals is in excess of 750 million tonnes of dry matter, of which an increasing proportion is disposed of by burning (4). The high carbohydrate content of cereal straws (60 - 90% D.M.) suggests that surplus world straw represents an enormous substrate for the production of biogas or animal feed through microbial fermentation.

Anaerobic digestion research in Galway has focussed on the application of the anaerobic fixed bed reactor design to the digestion of agricultural wastes (5). The fixed bed reactor is a modification of the anaerobic filter pioneered by Coulter and co-workers (6) and by Young and McCarthy (7). The flow of the input feed to the anaerobic filter is upwards through a bed of gravel, plastic or ceramic chips or rings which support and entrap the active microbial floc within the digester. The fixed bed reactor can be utilised in a one-step system for the digestion of liquid wastes of low solids content (1%). For wastes containing higher proportions of solid material, a two-stage system which includes a predigestion or liquefaction phase must be employed.

2. ANAEROBIC DIGESTION OF PIG SLURRY

A two-stage system was developed for the anaerobic digestion of pig slurry. The first stage was essentially hydrolytic and fermentative

and occurred naturally at ambient temperatures in a holding tank or slurry
lagoon. Gravity settlement, after 12 days liquefaction, yielded a solids
fraction of c.20% dry matter and a liquid fraction which contained up to
80% of the total COD. The liquid fraction was initially pumped to a
laboratory scale fixed bed reactor with the results shown in Table I .

Table I. Anaerobic digestion of Pig Slurry by a two-stage fixed bed
reactor process.

	COD mg/L	BOD	V.F.A.	Solids content g/L	pH
Raw Untreated Slurry	73,500	28,300	9,200	40	7.0
Hydrolysed and Separated Slurry	58,800	22,600	11,700	8	6.8
Digested Effluent	7,050	2,800	1,212	8	7.4

Characteristics of Fixed Bed Reactor: - Liquid capacity of 9.2
litre, operating temperature, $28^{\circ}C$; liquid retention time, 3 days;
average COD removal, 88%. Organic Loading rate, $19.6kg/m^3$.day.

It is apparent from table I that a 3-day liquid retention time
in the fixed bed reactor is sufficient to achieve efficient digestion of
the liquefied fraction of pig slurry. The percentage methane in the
biogas produced varied between 80 and 85% and an average yield of
$0.33 L CH_4$ per gramme was obtained. A pilot scale reactor of $9m^3$ capacity
was constructed and installed in a pig fattening unit and initial
results parallel those obtained at laboratory scale. Following laboratory
trials on milk washing wastes, a second pilot scale reactor of similar
capacity was installed in a local milk processing plant. During a 6
months trial period, the input waste had an average COD of 2,500 mg/L

and removal rates of 80 to 85% input COD were routinely obtained at a
liquid retention time of 12 hours. The milk was fed directly to the
reactor without pretreatment and shock loadings and wide fluctuation
in the pH of the input waste (pH 2-10) did not affect digester
performance. Passage of the digester effluent through a small aerobic
trickling filter yielded a final effluent which conformed to Royal
Commission standards.

Table II. Anaerobic Digestion of Milk Wastes by a Pilot-Scale fixed
Red Reactor

	COD mg/L	BOD mg/L	Solids content g/L	pH
Dairy Influent	2,456	819	1	5.8-10.2
Effluent	450	150	<1	7.2- 7.3

Characteristics of Fixed Bed Reactor: Liquid capacity, $9m^3$;
operating temperature, $28^{\circ}C$; liquid retention time, 12 hours; average
COD removal, 82%; Organic Loading rate, $4.9kg/m^3$. day.

3. ANAEROBIC DIGESTION OF SOLID AGRICULTURAL RESIDUES

Studies on the anaerobic digestion of straw residues were
focussed initially on the development of an efficient first stage
system for the liquefaction of the digestible components of the straw
substrate. Incubation of straw with pig slurry (40g straw / L pig slurry)
in an unstirred reactor revealed a rapid solubilisation and fermentation
of cellulose and hemicellulose during an initial 6-8 day period.
Thereafter, hydrolysis rates and volatile fatty acid production
declined rapidly and no further solubilisation occurred until the
onset of methanogenesis at day 16-20. The initial rapid rate of
solubilisation of the digestible fraction of straw was shown to be

maintained by a liquid displacement regime in which the supernatant in
the stage 1 reactor was withdrawn every 5-6 days and replaced by an equal
volume of effluent from a fixed bed reactor treating liquefied pig slurry.
Solubilisation of 50% of the input solids was achieved in 18 days during
which period the cellulose content decreased to 37% of the initial level.
Pretreatment of the straw substrate with alkali in order to increase
digestibility markedly increased both the rate and the extent of cellulose
solubilisation in this system. Over 90% of the initial cellulose content
was solubilised by a 15 day cycle in the reactor. The liquefied material
from the stage 1 reactor was pumped directly to a fixed bed reactor for
methane generation. Using a hydraulic retention time of 3 days, COD
removal rates of 80-90% were achieved with gas yields of 0.37L CH_4 per
gramme COD removed.

Hydrolysis of cellulose and hemicellulose is apparently the major
rate-limiting factor in the anaerobic digestion of straw residues.
The intimate association between lignin and the cellulose and
hemicellulose of straw also presents problems in the aerobic upgrading
of straw residues to animal feed by fungal or bacterial conversion (8).
Single cell protein (SCP) processes reported to date have relied heavily
on costly chemical and/or physical pretreatment procedures to enhance
the accessibility of straw cellulose and disrupt the lignin seal (9).
An alternative approach involving the co-culture of lignolytic and a
cellulolytic fungal species on chopped non-pretreated straw yielded
the results shown in Table 111.

Table 111. Typical properties of the mixed fungal SCP product obtained
by co-culture of _Sporotrichum pulverulentum_ and _Chaetomium
cellulolyticum_ on chopped wheat straw.

Colour:	Brown
Texture:	Granular
Fungal Biomass:	60% D.M.
Cellulose Content:	18% D.M.
Lignin Content:	7% D.M.
Crude Protein Content:	30-32% D.M.
Ash Content:	4% D.M.

Chopped straw was included in the medium at 2% w/v and the pH was initially set at 5.5 and thereafter uncontrolled. The flasks were innoculated with a 10% mixed mycelial innoculum and cultivated at $37^{\circ}C$ for 9 days.

The inclusion of the lignolytic species greatly enhanced the rate of cellulose and hemicellulose utilisation of chopped straw. The protein content of the final product exceeded that obtained by growth of Chaetomium cellulolyticum in mono-culture on chopped straw which had been physically pretreated by ball-milling to pass through a 40 mesh sieve.

Preliminary studies indicate that the non-carbon nutrient requirements of the fungi may be largely satisfied by the inclusion of digested effluent from a fixed bed anaerobic reactor treating liquefied pig slurry. An integrated process is envisaged in which animal manures and/or agricultural residues or energy crops are subjected to the two-stage anaerobic digestion system described above. The biogas produced may be used for space-heating or for electricity generation and/or to supply the operating energy for the aerobic dual fermentation of straw residues to animal feed, the main non-carbon nutrient supplements being provided to the aerobic system by the NPK rich effluent from the fixed bed anaerobic reactor.

References

1. Newell, P.J. (1980). In "Proceedings of Symposium on Energy Conservation and the use of solar and Other Renewable Energies in Agriculture, Horticulture and Fish Culture", 15 - 19 September, London, Pergamon Press, London.

2. Virkola, N.E. (1975). In "Symposium on Enzymatic Hydrolysis of Cellulose", p 23-40. Edited by M. Bailey, T.M. Enari and M. Linko. Finnish National Fund for Research and Development, Helsinki.

3. Hughes, R.G. (1979) in "Straw Decay and its Effect on Disposal and utilization", p 3-10. Edited by E. Grossbard. John Wiley, Chichester.

4. Staniforth, A.R. (1979) in "Cereal Straw" p 1-5, Clarendon Press, Oxford.

5. Newell, P.J., Colleran, E. and Dunican L.K. (1980) The application of the anaerobic filter to energy production from strong agricultural and industrial effluents; p 17-21 in "Anaerobic Digestion Poster Papers", Edited by D.A. Stafford and B.I. Wheatley; A.D. Scientific Press, Cardiff.

6. Coulter, J.B., Soneda, S. and Ettinger, M.B. (1957) Anaerobic contact process for sewage disposal. Sewage and Industrial Wastes 29, 468-477.

7. Young, J.C. and McCarthy, P.L. (1969) The anaerobic filter for waste treatment. J. Water Pollution Control Federation 41, R 160- R173.

8. Worgan, J.T. (1976) Wastes from crop plants as raw materials for conversion by fungi to food or livestock feed. In "Food from Waste" Edited by G.G. Birch, K.J. Parker and J.T. Worgan. Applied Science, London.

9. Han, H.W. (1978) Microbial Utilisation of straw. Adv. Appl. Microbiol, 23, 119-153.

PROPOSALS FOR THE ANALYTICAL MODELLING OF A DIGESTER

E. LAVAGNO, P. RAVETTO and B. RUGGERI

Istituto di Fisica Tecnica ed Impianti Nucleari

Politecnico di Torino

Summary

The present work is devoted to define and discuss an analytical model to describe the performance of a digester to be used for the anaerobic processing of agricultural and animal wastes to produce biogas. The lack of such a model is particularly evident whenever the engineer undertakes the task of designing a medium or large scale digester. The plant is characterized by well defined input and output variables. The main biological, chemical and physical parameters influencing the phenomena involved in the digestion process are indentified. Afterwards, mathematical correlations among them are studied, justified and discussed. The implications for the solution of some engineering problems is then investigated. This research stands only as a first step in the development of a complete and satisfactory model of a digester, due to some restrictions and simplifying assumptions which had to be imposed and accepted. Of course, it is not a concluding work, but has to be regarded as an effort to fill the lack of a mathematical means to be used by the engineer for the design of a digester and to be set aside to all the significant experimental works which have been performed in the field of biomass energy recovery.

1. INTRODUCTION TO THE ENGINEERING PROBLEM

When facing the task of designing a digester of organic material, the
engineer needs an analytic instrument through which from the input parame
ters he may compute a certain number of significant variables and choose,
among the many alternatives, the most convenient configuration. This work is
a first stage approach to define a biochemical and thermal model of the di
gester.

The input fundamental data are essentially the volume to be digested
available per unit time, which is defined by the rate Q_1 and its organic con
tent x_1 which may be translated in BOD_5 units, which seem more significant
than the TS ones, for instance, from the biochemical point of view.

What is expected from the model shall be the amount of the gas produ
ction, upon which consideration on its use, the eventual accumulation and
its economical turnout are based, and the output BOD_5, which characterizes
the pollution impact of the activity to which the digester is connected. All
the above outlined calculations shall be volume dependent, and, as a conse
quence, an optimization procedure could be started to find its most conve
nient value.

2. SCHEMATIZATION OF THE BIOCHEMICAL PHENOMENON AND MATHEMATICAL FORMALIZA-
TION. INTERPRETATION OF THE RESULTS.

In studying the methane production through the fermentation of organic
substances, typically agricultural products or wastes, such as manures and
straw, for instance, three phases may be identified (1):
1 - A first phase, during which an enzymatic hydrolisis process takes place
and the organic substance undergoes transformations which produce a substra
te for the bacteria that determine the second phase;
2 - A second acid phase, during which specific acidogenic bacteria produce
acids to be fed to the methanogenic ones;
3 - A third final phase, when methanogenic bacteria produce biogas (CH_4+CO_2)
that is liberated.

Obviously in a steady state situation, where fresh material is conti
nuously (or almost continuously) fed into the digester and the digested pro
duct, together with the generated biogas, is extracted, the three phases are
to be intended as contemporary.

Complex phenomena take place in each phase. We have attempted an analy
tical schematization supposing that the equations regulating the multicompo
nent actual system take the same form as the classic ones for a single com
ponent system.

Let us introduce a little nomenclature:

M,G   gas mass and volume rates

Q_1, Q_2 affluent and effluent volume rates

V,S   volume and surface of the digester

x_1, x, w, z organic material concentrations (1= inlet)   /1/

y   acidogenic bacteria concentration

y'   methanogenic bacteria concentration

J   mass gas production per unit volume of the digester $(M = dJ/dt.V)$

Keeping in mind (a) the equations for the enzymatic process of Michae
lis-Menten (2), (b) the equations for the bacterium growth of Monod (3) and
(c) the equations for the substrate evolution (4), we may write the follo
wing six equations differential non-linear system which expresses the biomass
balance for the digester:

$$
\begin{cases}
dx/dt = -ax/(k_M + x) + Q_1 \, (x_1 - x)/V \\
dy/dt = \mu zy/(k_S + z) - ky - Q_1 y/V \\
dy'/dt = \mu' wy'/(k_S' + w) - k'y - Q_1 y'/V \\
dz/dt = ax/(k_M + x) - (A_z \, dy/dt + B_z \, y) - Q_1 z/V \\
dw/dt = A_z \, dy/dt + B_z \, y - Q_1 \, y/V - dy/dt - (A_w \, dy'/dt + B_w \, y') - Q_1 w/V \\
dJ/dt = A_w \, dy'/dt + B_w \, y' - dy'/dt - Q_1 y'/V
\end{cases}
\qquad /2/
$$

A good agitation is supposed, so that the whole mass is perfectly homo
genized.

A few comments are needed for the correct interpretation of /2/. In the
r.h.s. of each equation, except the one for the biogas concentration J, a

term connected to the input volume rate Q_1 (in all pratical calculations $Q_1 \simeq Q_2$) appears as a contribution to the time variation of each function describing the system. For instance, in the first equation which is a balance for x, a source term $Q_1(x_1 - x)/V$ appears, which represents the net introduction (input minus output) per unit time of organic matter. In all other equations obviously only a subtractive term appears.

The system /2/ is non linear and its solution looks rather difficult. It may be noticed however, that the first equation is not connected to the remaining five, and, as a consequence, it may be solved separately.

Whenever the digester feeding may be supposed continuous, the steady state situation may be looked for by setting all time derivatives to zero, except dJ/dt, since it represents the biogas production rate that is extracted. We are than about to solve a non-linear algebraic system, which is an easy task.

We have performed some calculations, assigning trial values to the system coefficients, since there was no available data in the required form, and the results are summarized in fig. 1. It is a picture of x and of the gas production G vs. the volume of the digester V. The cost trend is also qualitatively plotted, as suggested in (6). A well defined saturation phenomenon may be observed, which is well known to be present in all biological processes and that has been detected in direct experiments upon digesters (5).

The importance of the "retention time" $RT = V/Q_1$ for engineering evaluations is immediate. Since Q_1 is a designing input parameter, from it the volume of the digester may be directly established.

3. THE THERMAL MODEL

The system performance depends upon its thermal behaviour, since all its coefficients are more or less temperature dependent.

To calculate the heat power which is necessary to keep a certain temperature, the following relation may be used:

$$P = qV + h_t S \, \Delta T \qquad\qquad /3/$$

where h_t is the overall heat exchange coefficient and may be found in the literature (7), ΔT is the temperature difference between the interior and the outside environment and is season dependent, q is the global reaction volumic heat. The P curve is qualitatively plotted in fig. 1, too.

4. THE SYSTEM COEFFICIENTS.PROPOSALS FOR EXPERIMENTS

We herewith propose a systematic series of experiments which should allow calculating the parameters of system /2/ and disclosing their dependence from the physical and biochemical variables, particular care being devoted to the temperature. The reaction heat q should be measured too, to permit a satisfactory thermal calculation.

5. THE OPTIMIZATION PROBLEM. CONCLUSIONS AND FURTHER DEVELOPMENTS OF THE RESEARCH

The optimization problem may be set down taking into account the following considerations:

- System /2/ is temperature dependent and different solutions may be found at different temperatures. Eq. /3/ must be kept in mind to calculate the fraction of the produced biogas which must eventually be burnt to yield the chosen temperature.

- Going to working points beyond the saturation threshold $aV/(Q_1 x_1) = 1$ will not improve the gas production per unit volume of the digester and it will increase the output polluting content; costs and heating needs will however decrease.

- Some costraints such as a limiting value of the polluting content in the discharge must be allowed. Besides, going to small values of V,i.e. to small values of the retention time, the model is going to fail.

- Some further parameters should not be forgotten, such as the land fertilizing properties of the discharge, if the plant is inserted in the agricultural context, and the gas use and storage, which could possibly modify the

428

costs or introduce specific constraints.

To well define the optimization problem, more investigation is needed, which shall be the further development of the present work. Closer attention is also to be devoted to the cost curves, to improve the trend that is plot ted in fig. 1. An analysis of experimental works is finally needed, to have a better test of the performance of the model, and more realistic calcula tions upon prototype plants should be welcome to draw some definite conclu sions.

REFERENCES

(1) J.LeGall, "Biochimie et Microbiologie de la Methanogenese", APRIA, Jour nees d'etudes, Paris, 13,14 March (1979).
(2) A.L.Lehninger, "Biochemistry",Worth Publishers, Inc. (1972).
(3) H.J.Gold, "Mathematical Model of Biological Systems", Wiley ISP (1977).
(4) W.W.Eckenfelder et al., "Biological Water Treatment", Pergamon Press, London (1961).
(5) R.Summers et al., "A Detailed Study of Piggery-Waste Anaerobic Digestion", Agricultural Wastes, 2, 61 (1980).
(6) G.Grignaschi, ENECO S.p.A., Private Communication (1980).
(7) B.Lagrange, "Biomethane", Edisud, Aix-en-Provence (1979).

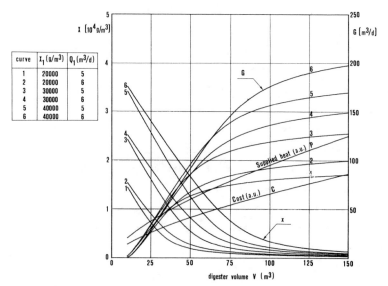

Fig. 1 – Significant characteristics of the digester as functions of its volume

COMPUTER AIDED DESIGN OF ANAEROBIC DIGESTERS

FOR ENERGY PRODUCTION

D.L. HAWKES

Senior Lecturer

Department of Mechanical and Production Engineering

The Polytechnic of Wales

Pontypridd, Mid Glamorgan, CF37 1DL

Summary

Anaerobic digestion of biomass could provide up to 4% of the United Kingdom energy consumption. The factors which influence the net energy production from digesters are many and the relationships between some of them are known to be complex. A Relational Data Base has been developed at the Polytechnic of Wales and is a suitable way of storing data generated by experimentation or operating experience and of examining the relationships between sets of data. Modelling of digester systems has the advantage that the effects of changes in one or more variables can be examined quickly and likely performance can be predicted, thus obviating the need for lengthy and often costly 'trial and error' operation of a full scale digester. A computer model has been developed at the Polytechnic for predicting net energy production from a mesophilic high rate digester.

It has been estimated that the energy potentially available from bio-mass by anaerobic digestion could be approximately 4% of the United Kingdom inland energy consumption (1978), a figure similar to the present nuclear electricity contribution (1).

1. FACTORS INFLUENCING NET ENERGY PRODUCTION

There are many factors which influence the net energy production from a digester (2) and the relationship between these can be complex. The energy that can be obtained from a digester is the difference between the gross energy in the gas produced by the bacteria and that used to maintain the process. The former is largely dependent upon the material input to the digester, especially the volume, the solids content and the biodegrad-ability. It also depends upon the operating temperature as well as the presence of materials such as inhibitors. The energy required to maintain the process depends upon the temperature, both of the digester and the in-coming feed, together with the ambient conditions. Other factors are the surface area of the digester, the efficiency of the heat exchangers and the degree of insulation provided. The mixing and feeding mechanisms also require an imput of energy.

A quantitative assessment of the effect of various parameters shows that net energy production from a conventional stirred tank digester is most sensitive to changes in the total solids content (% dry matter) in the feed and the gas yield (3). Gas yield, usually expressed in terms of m^3 of biogas per kg of volatile solids added, is particularly affected by temperature of operation, retention time and solids concentration. In order to obtain accurate predictions for the performance of a digester it is thus necessary to store many experimental results, to analyse them statistically and to examine the relationships between various sets of data (4).

2. DATA ACCUMULATION, STORAGE AND RETRIEVAL

At the Polytechnic of Wales, work is progressing on the development of a 'Relational Data Base' with a 'Query' langauge, so that data may be retrieved from the system in various combinations of parameters and ranges of values. The use of such a Relational Data Base facilitates the examin-ation of sets of data and their possible relationships to others. One example of this is in the work carried out at the Polytechnic of Wales under SRC grant number GRB13893, 'Factors affecting the rate of methane production during the anaerobic digestion of chicken litter'. This work includes an investigation of the relationship between gas yield, retention time and solids concentration.

3. COMPUTER MODELLING

As well as for the storage and retrieval of data another use of computers to aid digester design is in the field of modelling. By this means the possible effects of changes in one or more variables can be examined using a mathematical model of the process, thus obviating the need for the lengthy and usually costly 'trial and error' operation of the digester.

4. MODELLING METHODS

The usual method for determining gas production in mathematical models of anaerobic digesters involves reaction kinetics (5), (6), (7). Kinetic equations are incorporated in the models, the constants being found from plots of experimental data and forming some of the inputs. One method involves plotting the volume of gas produced against a function of the substrate used and the retention time to obtain the proportionality constant. The substrate affinity constant and the maximum specific growth rate can be determined from a graph of retention time against the reciprocal of the available subsgrate concentration. Another method involves the use of a single rate of reaction constant found from a plot of retention time and gas yield. Both of these methods have at times been used by, for example, Athonisen and Cassell (8) and Pfeffer and Quindry (9). It was however decided to develop a third method using parameters such as retention time and gas yield directly, rather than the kinetic constants derived from them. It is felt that this method is likely to be more accurate since less manipulation of the original experimental results is involved.

5. THE POLYTECHNIC OF WALES MODEL

The Polytechnic of Wales Anaerobic Digester Research and Development Unit has developed a program for determining net energy production from a full size digester, the program being run on the Polytechnic's DEC system 20 computer. It comprises the main program PFADP2 which performs the calculations together with data files for a number of different wastes and a curve fitting program which computes gas yields from experimental data stored in the files. Since gas yield depends upon many factors, for

example physical condition of the substrate, pH, inhibitors, etc., these
are thus accounted for in the model, providing the experimental data were
obtained on similar waste to the system for which the predictions are being
made. This proviso applies to any of the methods that are used. The temperature of digestion, if other than $35^{\circ}C$, can be taken into account by
means of a relative gas production curve arrived at by using a curve fitting subroutine with experimental results. As further results become available of the effects of temperature on gas yield, they can be added to the
data store from which the relative gas production curve is obtained.

No attempt has been made to predict effluent quality or to include
information on any system components before or after digestion.

The model also includes cost calculations and has a subroutine for
determining the digester gas mixing requirements.

5.1 DATA INPUT

The data for the model includes a number of parameters known to
affect net energy production (2). The daily volume of waste available
for digestion is included as input data. A retention time is specified although the optimum retention time is usually a factor which is
to be determined. This is done by a series of computer program runs
at different retention times. Net energy production can then be
plotted against retention time to determine the optimum.

A digester 'excess capacity factor' is included as a safety
factor where this is needed. Digester insulation 'U' factors are
included to take account of different constructional methods which
may be employed in the digester manufacture.

The temperature of digestion is an input since this can vary at
the design stage. A relative gas production curve based on experimental data is used to compute the gas yield that can be expected if
the digester operating temperature is not $35^{\circ}C$. Minimum expected
ambient temperature and incoming slurry temperature together with mean
annual temperatures are input.

The gas yield can either be computed from an existing data file
where this is available or input separately where not enough data
exists for computation.

The design of digester for which this program is suited assumes that some of the gas produced may be used to heat the digester. The efficiency of the heat exchanger is therefore input to the program.

5.2 COMPUTATIONS

The program PFADP 2 computes the digester volume required for the conditions outlined and then the geometry. Since relatively little energy is lost through the digester walls if well lagged, no attempt was made to compute digester dimensions to minimise surface area. Instead the dimensions are computed to give a digester diameter:height ratio of 1:0.44, which was substantiated by experiment to be the best for mixing by gas recirculation. A similar program exists which allows for various aspect ratios. The area of the digester walls in contact with the contents is calculated as is the area of roof in contact with the gas. This is used to compute the heat loss through the walls and roof. The heat required to raise the feed to the digester operating temperature is also calculated.

The gross gas production calculation requires knowledge of the gas yield. This is either input as a single value or in the form of a set of experimental data for each type of waste. Where this is done the data is in the form of arrays of gas yield for particular retention times and percentage solids.

5.3 OUTPUT

The output includes a full listing of the input data together with the calculated values such as gas production and net energy available after heating and mixing the digester.

ACKNOWLEDGEMENTS

The assistance of the Polytechnic of Wales Computer Centre is gratefully acknowledged, in particular that of two senior programmers, J.E. McBride and Dr. B.L. Rosser.

REFERENCES

1. HAWKES, D.L, HORTON, R., 'Anaerobic Digester Design Fundamentals, Part III, Process Biochemistry, In press.

2. HAWKES, D.L., 'Factors affecting net energy production from mesophilic anaerobic digestion', pp 131-149, in Anaerobic Digestion, Ed. Stafford, D.A. et al, Applied Science Publishers, London (1980)

3. STAFFORD, D.A., HAWKES, D.L., HORTON, R., 'Methane Production From
    Waste Organic Matter', C.R.C. Press, Boca Raton, Florida (1980)

4. HAWKES, D.L., HORTON, R., 'Optimisation of anaerobic digesters for
    maximum energy production', Proc. International Colloquium
    Energetics and Technology of Biological Elimination of Wastes,
    Rome (17-19 Oct. 1979).

5. QUINDRY, G.E., LIEBMAN, J.C., PFEFFER, J.T., 'Biological Conversion
    of Organic Refuse to Methane', Vol. II. Final report UILU-ENG-
    76-2021, Department of Civil Engineering, University of
    Illinois at Urbana, Illinois (1976).

6. ASHARE, E., WISE, D.L., WENTWORTH, R.L., 'Fuel gas production from
    animal residues', Dynatech report No. 1551, Camb. Mass.
    (Jan. 1977).

7. GRAEF, S.P., 'Dynamics and control strategies for the anaerobic
    digester', Ph.D. dissertation, Clemson University (May 1972).

8. ATHONISEN, A.C., CASSELL, E.A., 'Methane recovery from poultry waste',
    North Atlantic Region of ASAE, Paper No. NA74-108, 28, (1974).

9. PFEFFER, J.T., QUINDRY, G.E., 'Biological conversion of biomass to
    methane', Beef lot manure studies, Report No. UILU-ENG-78-2011,
    University of Illinois, Urbana (May 1978).

A NOVEL PROCESS FOR THE ANAEROBIC DIGESTION OF SOLID WASTES

LEADING TO BIOGAS AND A COMPOST-LIKE MATERIAL

B.A. RIJKENS

Institute for Storage and Processing of Agricultural Produce (IBVL)

Wageningen - The Netherlands

Summary

The complete anaerobic digestion of solid organic matter is achieved in a batch reactor (R_I) by percolating water through the decaying solid mass. The dissolved organic matter and volatile fatty acids (v.f.a.) are leached out continuously, thus preventing the preservation by a low pH, that would otherwise occur. The effluent of the reactor R_I, containing these dissolved organic compounds, is treated in a conventional continuous anaerobic digester (R_{II}) with a sludge blanket of bacterial mass, under the evolution of biogas. The treated water is recirculated to the batch reactor R_I.
Of the total digestible matter in the reactor R_I about 70 % is leached out within 14 days and converted into biogas in the digester R_{II}. After 2 to 10 days the R_I also starts to produce biogas, the R_{II} can be disconnected while the breakdown of the solid mass proceeds in the R_I that has become self-sustaining and takes over completely.

1. INTRODUCTION

It is well known that solid organic matter, stored in a vessel under exclusion of air, will be broken down by anaerobic bacteria to volatile fatty acids (v.f.a.) which lower the pH, so the activity of the micro-organisms is stifled and the organic matter will be preserved (ensiled). Under these conditions the development of methane bacteria will be prevented, since they are strongly inhibited by the low pH and also by the relatively high redox potential. The redox potential in a freshly decaying organic matter will readily be low enough for all kinds of micro-organisms performing the first step of the decay, mainly leading to the production of v.f.a., but not sufficiently low for the more sensitive methane bacteria which can transform v.f.a. into biogas. The inhibition is probably caused by prior organic intermediates like sugars etc. which, at higher concentrations, will act as toxins for the methane bacteria.

The idea of our novel process is based upon the assumption that the anaerobic breakdown of solid organic matter in a batch digester will continue when the inhibiting substances are leached out by the percolation of water through the solid mass.
When the rapidly decaying compounds have disappeared, the concentrations of dissolved intermediates will decrease and the conditions will become favourable for the development of the methane bacteria. As soon as their capacity to break down the v.f.a. will meet its production rate, the process will be self-sustaining, so the percolation can be ceased and the decay will proceed under the development of biogas.

The percolate from the batch reactor can be treated in a conventional continuous anaerobic digester with a sludge blanket of bacterial mass, under the evolution of biogas. The treated water will be recycled to the batch reactor.

2. THE MAIN RESULTS FROM THE EXPERIMENTS

Experiments were carried out in 12 L batch reactors, filled with the solid material, seeded with the residue from a previous batch and percolated with recycled water. The connected continuous anaerobic digester was started up with bacterial sludge from a waste water treatment plant and adapted to the percolate from the batch reactor.

The s y s t e m , consisting of the batch reactor (R_I) and the digester (R_{II}), connected with tubing and provided with a pump, gasometers etc.,was placed in a thermostated cupboard (fig. 1).

Various raw materials were tested, like the organic fraction of municipal solid waste, obtained after shredding and air classification, sugarbeet pulp, poppy-seed hulls and wheatstraw, "enriched" with some pig manure. It was found that all these waste materials were readily broken down in the batch reactor (R_I) during the percolation. The R_I produces a clear effluent, containing v.f.a. and dissolved organic matter. The R_{II} begins to produce biogas immediately after the circulation in the system is started (fig. 2). After 2 to 10 days, depending on the kind of raw material, the R_I starts to produce biogas as well, while the concentrations of dissolved organic matter in the effluent of R_I is decreasing. So eventually R_I fully takes over and no more gas is developed in the R_{II}; the R_{II} can be disconnected.

These results prove that the hypothesis was right, the inhibition of the methane production can be prevented by percolation and the s y s t e m works.

3. FURTHER DETAILS ABOUT THE EXPERIMENTS

For further optimisation and for the assessment of technological parameters and concepts it was necessary to have a standardised raw material available during the whole year. After the preliminary experiments we choose for a mixture of 50 % by weight dry matter sugarbeet pulp, 35 % wheat straw and 15 % cow manure (P.S.M.). The figures shown are obtained from runs with P.S.M. at mesophylic temperature (35°C).

We assume that all the organic matter will be digestible anaerobically, except for the lignin and an equal percentage of the cellulose; this is in agreement with the fact that peat consists of roughly 50 % cellulose and 50 % lignin.
Postulating this, the maximal digestible organic matter, the $COD_{digestible}$ ($COD_{dig.}$) can be defined and can be calculated from the analysis. For PSM we calculated a total COD of 1.15 g/g d.m. and a $COD_{dig.}$ of 0.80 g/g d.m.

In fig. 2 the methane production, calculated als percentage of the $COD_{dig.}$ is depicted. The disconnection of the R_{II} does not show an influence on the total gas production.

The $COD_{dig.}$ decay as calculated from the COD in the effluent of R_I is very similar to the $COD_{dig.}$ decay calculated from the total methane production, as is shown in fig. 3.

The COD in the effluent of R_I is depicted in fig. 4. The curve shows the rapid decline of the concentration of dissolved organic matter in the time, decreasing to about zero in 18 days. Besides the v.f.a. there is always some other dissolved organic matter present.

The pH curves in fig. 5 show that the R_{II} is functioning well and can treat the effluent of R_I effectively, even in the beginning when the pH is low and the concentration of dissolved COD is high.

4. SCOPE OF THE PROCESS

This process might be a solution for the treatment of solid wastes like municipal solid waste, agricultural residues etc. The biogas is the most valuable end product, the compost also has some value for agriculture and furthermore is less bulky, does not smell and is easily to dispose of. Much more development work must be carried out before this principle is translated into a process that suits the purpose and the raw material. The economics of processes of this kind have to be established.

5. ACKNOWLEDGMENT

This project is funded by the Commission of the European Communities, Directory General XII in Brussels by a contract under the Solar Energy Programme.

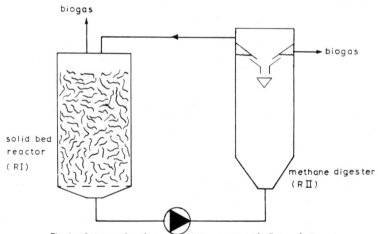

Fig.1 Schematic diagram of the anaerobic "plug-in" system.

Fig 2 – CH₄-production of mesophilic digested PSM by the solid bed reactor (RI) and the methane digester (RII) as a percentage of the calculated maximum possible CH₄-production (COD_{dig})

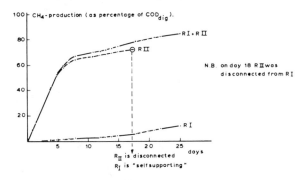

Fig 3 – The cumulative proportional COD_{dig}-decay calculated from the effluent of RI and from the CH₄-production by RII and RI (NB: on day 18 RI was disconnected from RII)

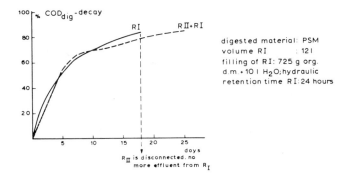

Fig. 4 – Concentrations of total COD and fatty acid COD in the effluent of RI

Fig. 5 – pH of the effluent of RI and RII

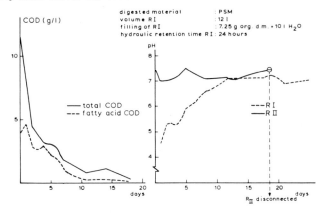

APPLICATION OF RECENT DEVELOPMENTS IN ANAEROBIC
WASTE WATER TREATMENT TO SLAUGHTERHOUSE WASTES
(GERMANY)

G.D.LINDAUER and H.SIXT

AGRAR- UND HYDROTECHNIK GMBH, ESSEN

O.SCHULZE GMBH & CO. KG, GLADBECK

Summary

    A slaughterhouse in Germany produces 10 400 kg waste per
day. This waste substrate has been filled into a digester
of about 300 cbm. The biogas, 600 cbm/d, is first stored in a
tank and subsequently used to drive the cold-store compressor.
The heat given off from this process is used for heating the
substrate for a pasteurisation process.
After all requirements of the plant have been covered, the
heat from the biogas available for other uses amounts to:

min:	1,290 KW/d
max:	1,435 KW/d

Once the substrate has been dehydrated, it can be spread on
agricultural land as a fertilizer.

1. INTRODUCTION

Under the effect of the increasing energy prices, the application of anaerobic processes has lately regeined interest. According to recent results, the process of anaerobic decomposition is more complicated than that of the well-known two-phase-model. In compliance with the present level of research (1) the decomposition can be divided into a hydrolysis-, acidification-, acetate- and methane phase(see Fig.1). The objective of most anaerobic plants in operation was to obtain decomposition to a large extent with a long period of putrefaction. The latest attempts aim at the application of the anaerobic prepurification as an economic solution in the case of high loads and short periods of putrefaction providing the microbiological boundary conditions.

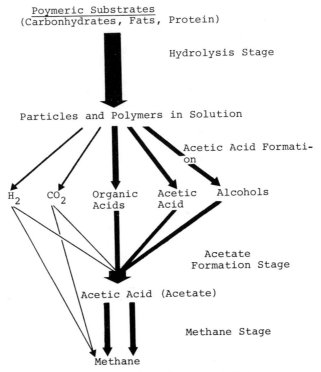

Fig. 1 Multi-phase model of anaerobic decomposition

It is the acetic acid formation phase which is the "bottleneck" for the decomposition of polymers that are biologically
easily convertible. However, in the case of materials contained in sewage that are difficult to break down biologically,
the hydrolysis stage can be also the speed-limiting factor
because only when the fermentative bacteria have broken down
the polymers into soluble components will it be possible to
achieve a total decomposition.

Implications of these new research findings have been applied
to the planning and construction of a biogas digester at a
German slaughterhouse.

2. The problems

2.1 Slaughterhouse waste

The slaughterhouse produces considerable quantities of
waste. Approximately 85 per cent are made up of contents of
giblets and pig intestines and contents. At present these waste products are transported to a carcass disposal centre.
However, the quantities of waste involved are so great
(10 400 kg/day) that the centre has great difficulty in coping.
In addition, there is the problem of smell. For these reasons it was necessary to find an alternative method of disposal.

2.2 Energy situation

At present, the pig bristles are burnt off by means of
a natural gas flame. This process requires 150 cbm of natural
gas per day. This could be replaced by approximately 200 cbm
of biogas.

The cooling plant for the cold-store is driven by natural gas;
this could also be at least partly replaced by biogas
(approximate 60% of the energy requirement).

2.3 Aspects of hygiene

The solid and liquid wastes from slaughterhouses contain
animal remains which, according to the law for the disposal
of animal carcasses of 02.09.1975 (2) and the regulation

governing animal carcass disposal centres of 01.09.1976 (3),
have to be purified accordingly. (t=20 min; T= 133°C;
p= 3 bar; the material is to be agitated continuously
throughout the process). In order to use waste sludge from
municipal sewage works as a fertilizer on agricultural land,
hygienic conditions are having to be met which correspond to
at least a pasteurisation of the sludge.
This requirement is also to be applied to the waste matter
from slauhterhouses.
These two conditions pose certain complications to the simple
process of extracting biogas, in that they require expensive
pre-treatment processes,These processes should however facilitate the decomposition process by breaking down the organic
substances.

3. Choice of process

3.1 Selection criteria

The first criterion is to be seen in a balanced energy
policy. This necessitated an anaerobic process involving the
reclaiming of waste heat and the utilization of gas in gas
engines which make use of a coupled heat/power system.
The second criterion is determined by the aspect of hygiene.
In order to satisfy this condition, the substrate should be
subject to a preliminary pasteurisation process and to a long
period of decomposition. Alternatively, a comparison with
autoclave treatment in order to fulfil legal requirements
will have to be made.
The third criterion is concerned with the aspect of plant
safety, process stability and the transporting of substrate
during the pasteurisation process and in the decomposition
tank.

3.2 Process

Half the daily amount of substrate is first stored,
heated and agitated in one of the pasteurising tanks. Once
the pasteurisation temperature has been reached, the substrate is stored until the other half of the substrate has

been filled into the other pasteurisation tank. The pasteurised substrate is cooled by circulation, and the heat which it gives off is used to heat the second amount of substrate. The pasteurised substrate is cooled until it reaches the temperature required in the decomposition tank (+37°C). It is then injected into the agitation process.

In the decomposition tank, the substrate is agitated by external pumps. These produce two circulations, thus preventing a layer floating to the top or sinking to the bottom. The decomposed substrate then passes to a tank where the sludge is stored before proceeding to a dehydration process.

It is also possible to return substrate from the storage tank into the decomposition chamber in order to increase the aspect of hygiene. Once the substrate has been dehydrated, it can be spread on agricultural land as a fertilizer. The biogas which is extracted is first stored in a tank and subsequently used to drive the cold-store compressor. The heat given off from this process is used for heating the substrate in the pasteurisation process.

3.3 Energy aspects

The substrate is only produced for 8 hrs/day, which means that the heat which is applied to the pasteurisation process can only be reclaimed fully if the substrate is stored in the pasteurisation tank until the following morning. Because of the danger posed by fermentation, it will not always be possible to apply this procedure, with the result that only half the possible heat can be reclaimed. Figure 2 shows that a large part of the energy required for driving the cold-store compressor can be provided by the biogas. Due to the pasteurisation process, higher figures than those found in the literature (4,5,6,7) can be expected. After all requirements of the plant have been covered, the heat from the biogas available for other uses amounts to:

min: 1,290 kW/d
max: 1,435 kW/d

4. Research aspects of the slaughterhouse biogas plant

As this biogas plant is one of the first to envisage the use of slaughterhouse waste, it is still strongly research orientated. In particular the following parameters are to be investigated:

- optimum process conditions for the substrate (temperature, agitation, duration of decomposition process)

- composition of the gas

- influence of the composition of the substrate on gas yield, composition of the decomposed sludge and the possibilities of further use

- formation of a floating layer on the substrate

- tendency to foam.

Notes

1) SIXT, H.,WERNECKE,S. and K.MUDRACK: Neuere Entwicklungen auf dem Gebiet der anaeroben Abwasserbehandlung. In: Korrespondenz Abwasser, vol.27/I (1980),p.22-27

2) -   Tierkörperbeseitigungsgesetz 9/75

3) -   Tierkörperbeseitigungsverordnung 9/76

4) SIXT,H   Reinigung organisch hochverschmutzter Abwässer mit dem anaeroben Belebungsverfahren am Beispiel von Abwässern der Nahrungsmittelherstellung
Veröffentlichungen des Institutes für Siedlungswasserwirtschaft der Universität Hannover, Heft 50/1980

5) Reinhold,F   Neue Weg zur Beseitigung und Verwertung von Schlachthofabfällen Fleischwirtschaft 6 (1954)

6) BAADER,W   Biogas in Theorie und Praxis
   DOHNE,E   KTBL-Schrift 22 9
   BRENNDÖRFER,M   Landwirtschaftsverlag Münster-Hiltrup

7) HEGEMANN,W   Untersuchung über den Einfluß der
   WECHS,F   Pasteurisierung von Rohschlamm auf Faulung
                    Vortrag BMFT-Statusseminar
                    ATV-Jahrestagung Mainz Juni 1980

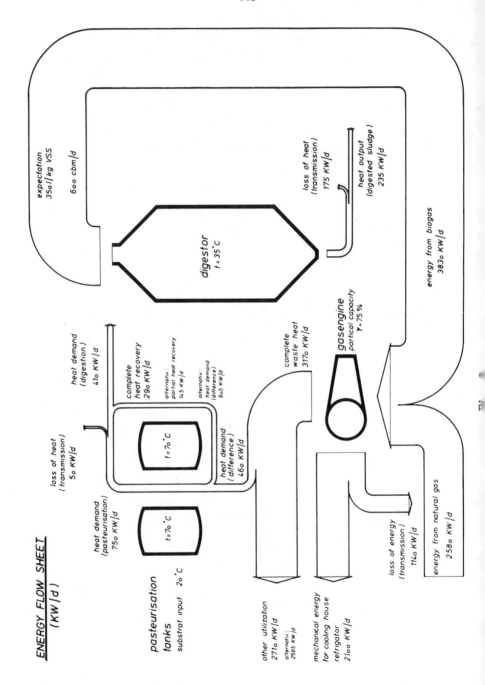

ENERGY FLOW SHEET
(KW/d)

447

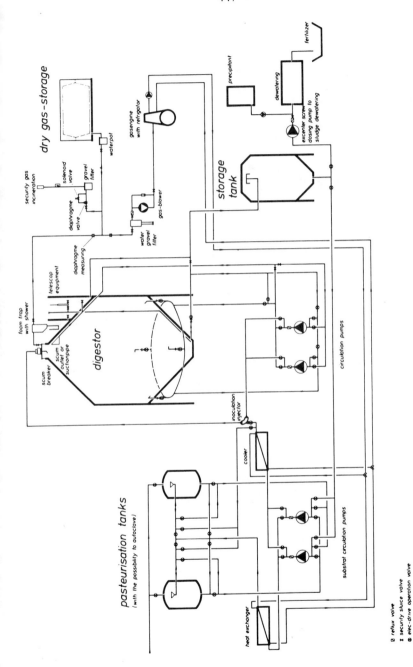

Fig. 3

"FINDINGS OF B.A.B.A.'S SUBCOMMITTEE TO PRODUCE A CODE OF PRACTICE ON
SAFETY IN AND AROUND ANAEROBIC DIGESTERS".

P.J. Meynell, Secretary
British Anaerobic and Biomass Association Ltd.,
The White House,
Little Bedwyn,
Marlborough, Wilts. U.K.

Summary

The British Anaerobic and Biomass Association Ltd., hereafter referred to
as BABA, formed a sub-committee of engineers, chemists and microbiologists
to produce the framework for a draft "Code of Practice for Installation
and Maintenance of Anaerobic Digestion Plant". This draft was intended to
highlight the safety aspects of such installations. This is not a
definitive document but the discussion document circulated among U.K.
members for constructive criticism.

GENERAL

The general topics which were considered were as follows:

1. Standards for installation procedures

2. Safety requirements in installation design

3. Safety standards during operation or maintenance

4. Performance criteria

1. STANDARDS FOR INSTALLATION PROCEDURE

The following is intended as an indication of the preliminary conditions under which a U.K. company would install a digester plant in the U.K.

SECTION A   GENERAL CONDITIONS

SCOPE OF WORK

The works shall be to build a digester tank etc.

SECTION B

STATUTORY OBLIGATIONS

The contractor shall give all notices required by any local statutory authority, and where required shall submit any component for test or approval before incorporating it in the works, properly certified. The contractor shall pay all fees involved in complying with these requirements.

All materials shall comply with the requirements of:

a.   The Health and Safety Executive

b.   The Local Authority and the Statutory Undertakers

c.   The current British Standards and Codes of Practice

d.   The Shops, Offices and Railways Act

CONTRACTUAL OBLIGATIONS

Materials are to comply with relevant British Standards irrespective of particular mention in any specificatiion.

Samples of all materials are to be submitted on request.

TOOLS, SPARES AND OPERATING INSTRUCTIONS

The contractor shall provide the client with the following:

A complete set of tools and spares for the proper operation and maintenance of the plant for 12 months. This to include:

    a. Spanners wrenches and all tools special to the equipment as recommended by the manufacturers.

    b. Spare washers, discs, seats, diaphragms, packing materials needed for general maintenance.

A set, in duplicate, of all manufacturers guarantees, operating and maintenance manuals.

A set of general instructions written in layman's language for starting, running, shut down, protection and emergency procedures. These shall indicate normal daily maintenance and shall be explained verbally to the client or his staff to his satisfaction.

NAMEPLATES AND LABELS

Label all plant and apparatus with the maker's name, reference number, size, type, speed or any relevant particulars on engraved plastic laminate.

Label all valves and switches with appropriate number on engraved plastic laminate.

Provide operating instructions and a valve identification chart mounted on a glazed panel fixed to the wall, together with a plant schematic.

TESTING

Each section of pipework is to be tested on completion to the scheduled pressure, for a 30 minute period, with the test pump disconnected, with no detectable drop in pressure.

Make good test failures to the satisfaction of the Client.

All services shall be retested on completion.

Pressure test required schedule.

    a. Water supply circuit   900 kn/m(2)

    b. Gas supply and connection circuits   150 kn/m(2)

    c. Digester heating circuits 200 kn/m(2)

    d. Gas reinjection circuits   200 kn/m(2)

Test underground pipework before backfilling.

Test all instruments installed for correct performance, against certified gauges, adjust or replace incorrect gauges.

The Client is to be informed when final testing is to be done and facilities to be made available for witness or inspection.

Where available, Codes of Practice issued by the Chartered Institute of Building Services and manufacturers instructions are to be used.

The Contractor will provide all instruments required for testing and commissioning.

On completion, three copies of all test certificates are to be provided for the Client.

COMMISSIONING

The sequence for commissioning shall be as follows:

Certify that all plant and systems have been tested. Present copies in an approved form to the Client with notice that the commissioning is to commence.

Flush out all systems to remove dirt from all pipes, filters, strainers, drain pans, etc.

Lubricate all moving parts of plant and controls to the manufacturers recommendations.

Fill the boiler and heat exchange circuits with water.

THIS MUST BE DONE BEFORE FILLING THE DIGESTER TO PREVENT THE HEAT EXCHANGERS FROM FLOATING.

Fill the digester tank to 80% capacity with water.

Start up boiler and bring digester to normal working temperature.

Check all supports and fittings.

Operate system for at least 6 hours before regulating circuits for correct flow and performance.

Start gas reinjection circuit. Test compression and adjust water flow. Circuit is to be run on nitrogen fed or CO2 in upstream of the pump to provide stirring at the rate of 20 minutes in the hour. Maintain stirring from the start of the loading process.

Load digester according to an agreed schedule, via the loading pump which is to be adjusted to provide the correct loading rate. Monitor the pH and in the event of a serious fall in the PH lower the loading rate to compensate. If this proves impractical neutralise with bicarbonate of soda. pH value to be neutral.

Remove nitrogen supply as soon as gas performance is maintained at correct level. The gas mixture is not important for this purpose as long as air is excluded.

Switch over boiler to methane supply when supply available.

Final adjustment of system to take place at this point, and certificates provided for the Client stating the final performance figures.

MAINTENANCE

The contractor shall be responsible for the maintenance for a period of 12 months from the date of the certificate, as follows:

Maintain the plant during normal working hours.

Lubricate plant as required.

Be responsible in the event of shut down for draining down protection, reloading and recommissioning.

Answer all emergency calls on a 24 hour basis providing a competent engineer on a 24 hour basis.

Replace all parts damaged or defective or showing undue wear.

Carrying out two comprehensive examinations (at times agreed with the Client) and making all necessary adjustments.

Keeping equipment clean and free from dirt.

PIPEWORK AND FITTINGS

The contractor will install all pipework as follows:

    a. Agree all pipe runs with the Client.

    b. Fix all pipeworks at levels and gradients to facilitate draining down, provide all necessary drain down cocks.

    c. Fix all pipework to the manufacturers recommendations.

    d. Use materials for each service as scheduled.

        Generally: exposed external pipework in ABS,
                   buried external pipework in Alkathene,
                   internal pipework in black steel.

Fix in all pipework in a neat and workmanlike manner.

All pipes are to be even bore, clean and smooth throughout and free from injurious external blemishes and rust. Remove all burred ends to maintain the full bore of the pipe.

All vertical pipes are to be truly plumb.

Protect all open ends of pipework during installation.

No joints are to be within the thickness of any structure.

No pipes or lagging shall be within 50mm of any other service, except where connected to that service.

Slope horizontal pipes and provide vents at high points to facilitate removal of air from all systems and condensate traps must be avoided.

Make joints between dissimilar materials with suitable adaptors to prevent interaction.

Install all fittings, valves, strainers, instruments such that they are accessible.

Install all pipes or lagged pipes so that there is 50mm between pipe or lagging and wall and 100mm between pipe or lagging and floor.

Install all pipework, joints, branches and supports to accommodate the linear expansion between cooling and heating, without distortion. Where possible this is to be done by diversion.

Form swan-necks, offsets and springs by an approved cold bending process.

Support all adjustable valves and fittings on either side of the fitting.

Connect gas supply lines out of top of distribution pipe, with the exception of line to level tank which shall be from the invert of the distribution pipe - to act as drain.

2. SAFETY REQUIREMENTS IN INSTALLATION DESIGN

The following if intended as a guide to U.K. requirements of the U.K. Health and Safety Executive relating to fire/explosion risks only.

INTEGRITY OF PLANT

The digester, gas collection facilities and associated plant and pipework should be designed, constructed and installed to avoid, so far as is reasonably practicable, the risk of leakage of gas.

The plant, including pipework and fittings, should be suitable for the pressure and temperatures involved. The pressure relief facilities (i.e. valves and vents) should be capable of preventing the pressure inside any part of the system ever exceeding that which it can safely withstand. Effective steps should be taken to ensure that the operation of relief devices is not impaired when the plant is in use (e.g. by deposition of solidified slurry).

Safety relief devices should be arranged to discharge at a safe position in the open air, remote from any building opening, so as to prevent impingement of escaping gas on any structure or personnel, and well away from any ignition source.

The number of joints in pipework should be kept to the minimum.

INSTALLATION

The plant should be installed by a competent person who is aware of the
risks associated with the unit and the precautions required. Suitable
checks and tests should be carried out to ensure the integrity of seals,
joints etc. and the safe operation of pressure relief facilities before
the plant is taken into use. The installers should fully implement advice
given by the plant designers in this matter.

Plant Siting

The boiler/engine house should be sited as far as is reasonably
practicable from the digester. A separation distance of at least 6 metres
is normally recommended, but allowance could be made for the following
factors:

    a.   The wall of any plant building facing the digester tank is
        imperforate and of at least half hour "fire resisting"
        construction (see British Standard 476) and,

    b.   effective steps are taken to prevent any gas leakage at the
        digester entering the building. The ventilation inlets should be
        at least 5 metres from the tank. A screen (e.g. sheet steel)
        should be provided between the air inlet and the digester
        extending to a height above the top of the air inlet and
        horizontally to a point at least parallel to the edge of the
        digester.

Any gas compressor should preferably be located in a fully open air
location, or:

    a.   The partition separating the compressor housing and the boiler
        house is imperforate, forms a gas tight seal and is of at least
        half hour "fire resisting" construction, and additional screening
        is provided where necessary to prevent any gas leakage entering
        the boiler/engine room (e.g. via ventilation opening).

    b.   Adequate permanent natural ventilation is provided throughout the
        housing to safely disperse any likely leakage. There should be
        no "dead pockets" where high concentrations of gas may reside.

    c.   The plant housing is provided with adequate and suitable
        explosion relief. The relief area should be as large as is
        reasonably practicable. An area not less than that of the
        largest side of the housing would be acceptable.

In order to achieve conditions (b) and (c) it is recommended that the
sides other than those forming the partition with the boiler house are
left completely open, or if security is required, covered with course weld

mesh. If cladding must be provided on the external walls, adequate ventilation openings at high and low level must be incorporated. Panels should be of light weight material and only loosely held in position so that they will easily be displaced in the event of an explosion, before the pressure in the housing rises to a dangerous level. It may be possible to use light weight aluminium louvred panels for cladding. Suitable restrainer chains should be provided for the panels to prevent them becoming dangerous missiles in the event of an explosion.

The digester should be located well away from any other feature which may create a hazard. The separation distance required may need individual consideration, but in general a minimum horizontal separation distance of 6 metres should be provided between the plant and any naked flames or other ignition sources, including electrical equipment which is not specially protected or effectively shielded. A minimum horizontal separation distance of at least 8 metres should be provided between the installation and any flammable liquid or gas storage (e.g. small stocks of liquefied petroleum gas, fuel oil or stocks of combustible materials such as hay or timber).

The area around the plant should be concreted and kept free from weeds and other growth. Chlorate based weed-killers should not be used in this area.

SOURCES OF IGNITION

Sources of ignition should be eliminated from any point where a dangerous concentration of flammable gas may reasonably be expected to occur (e.g. in event of leakage).

a. Naked flames and smoking materials (e.g. sparks from vehicles etc.), should not be permitted within 6 metres (horizontal distance) of the digester or any part of the plant where gas leakage may reasonably be expected to occur (e.g. the compressor). Notices should be posted to this effect.

b. Electrical equipment should be selected, installed and maintained in accordance with British Standard 5345 Part I: 1976. In order to minimise expense, electrical equipment should, where practicable, be sited in a safe position remote from any source of flammable gas. Electrical equipment necessarily sited in a hazard area should be suitably protected.

c. If the compressor is sited in a fully open air position, the area within 1.5 metres of the compressor in all directions should be considered a Zone 1. The area between 1.5 metres and 3 metres of the compressor in all directions should be considered to be Zone 2.

d. If the compressor is installed in the proposed housing, the area inside the housing should be considered to be Zone 1. The area outside the housing and within 2 metres of any opening in the housing, should be considered to be Zone 2. It is understood that the only electrical equipment installed in the vicinity of the digester is a pump and float switch. These components should be checked to ensure that they are suitably protected or, alternatively, they should be sited in a safe position.

e. Assuming that the plant/pipework etc. contained in the boiler
    house is (like the remainder of the installation) designed,
    constructed and installed to avoid, so far as is reasonably
    practicable, the risk of gas leakage, provision of explosion
    protected electrical equipment would not be insisted upon here.

PRECAUTIONS AGAINST INGRESS OF AIR

All necessary steps should be taken to prevent the introduction of air
into the plant when it is being used to generate and/or hold flammable
gas. The matters requiring attention in this respect include:

1. Low Level Cut-off
    To prevent the formation of a negative pressure inside any gas
    collection cap (e.g. in the event of excessive demand), suitable
    facilities should be provided to ensure that the gas outlet of
    the digester is automatically shut-off when the level of gas
    falls to the minimum safe level.

2. High Level Cut-off
    Suitable facilities should be provided to prevent over filling
    any of the gas collection.

3. Gas Compression
    A suitable low pressure detection device should be provided at
    the inlet to the gas compressor. This should be suitably
    interlocked to cut off the compressor if the pressure of inlet
    gas falls to the minimum safe level and prevent its re-start
    other than by means of manual intervention.

4. Gas Scrubbers
    The design of the gas scrubbers should be checked to ensure that
    air cannot be introduced into the gas by this plant (e.g. if the
    water supply is lost).

PURGING

Whenever a dangerous air/gas mixture could occur within the plant (e.g. on
initial start-up or service of the digester etc.), an effective inert gas
purging procedure should be adopted to prevent the formation of such a
mixture. Before the flammable gas is introduced into plant containing
air, the oxygen content inside the plant should be reduced to a safe level
(below 5% v/v). Before air is introduced into plant containing flammable
gas, the concentration of flammable gas inside the plant should be reduced
to 25% Lower Flammable Limit. (The lower flammable limit for methane =
approx. 5% v/v).

Nitrogen or inert gas from a suitable generator may be used for purging.
A volume of inert gas equivalent to at least five times the internal
volume of the plant to be purged, is generally adequate, but instrument
checks are recommended in the first instance. Suitably calibrated oxygen
meters and flammable gas detectors may be used for this purpose. For the

precautions to be adopted in the use of flammable gas detectors, reference should be made to Guidance Note Chemical Safety /1 entitled "The Industrial Use of Flammable Gas Detectors". This document is available from the Area Office of the Health and Safety Executive.

STARTER FUEL

If a temporary supply of Liquefied Petroleum Gas in cylinders will be used to fuel the boiler at start-up, the cylinders should be located in a safe, adequately ventilated position in the open air, remote from any source of ignition or building opening. A position against the imperforate "fire resisting" wall of the plant house would be acceptable on the understanding that:

1. The cylinders will be removed immediately after use and will not be stored in this position.

2. The number and size of cylinders used should be as small as is reasonably practicable.

3. The appropriate separation distances are maintained from building openings, sources of ignition etc. as specified in the Health and Safety Executive Guidance Note entitled "Code of Practice for the Keeping of Liquefied Petroleum Gas in Cylinders and Similar Containers". The booklet is available from H.M. Stationery Office.

GAS PURIFICATION

In view of the impurities in the gas it is understood that it will not be used to fuel domestic ovens etc. Combustion products from all heaters, engines etc. should be vented to a safe position in the open air. The composition and supply pressure of the gas should always be such as to afford safe and stable burning conditions.

FIRE PRECAUTIONS

The advice of the Fire Brigade should be sought in this matter. It is recommended that a sufficient number of portable fire extinguishers are installed in easily accessible positions by the installation for "first aid" fire fighting purposes.

SECURITY

A security fence should be designed such as to prevent unauthorised access to the installation. It should incorporate at least two means of escape which should not be adjacent to one another. Gates should open outwards, should not be self-locking and should at all times provide easy means of escape from within. The gates should be kept locked when the compound is not occupied.

SERVICING AND MAINTENANCE

The plant should be regularly and thoroughly checked and maintained by a competent person. Particular attention should be given to relief valves, seals, venting facilities, flame safeguards, safety shut-off valves and other safeguards.

INFORMATION AND INSTRUCTIONS

Adequate written instructions should be given to operators on the safe use of the plant. These should include emergency procedures, purging methods etc. The advice of the plant suppliers should be sought in this matter.

3. SAFETY STANDARDS DURING OPERATION OR MAINTENANCE

Most accidents have occurred after the commissioning of plant when the owner or one of his employees neglects commonsense practice. The subcommittee is presently drawing up the toxicity level information and do's and don'ts for the maintenance of digestion and ancillary plant.

4. PERFORMANCE CRITERIA

As with many technologies where there is a rapid awakening in the commercial sphere of its potential, how does a prospective purchaser know what he is buying and indeed what plant performance should be expected of it in terms of energy output, pollution reduction etc. BABA is currently drafting a standard contract procedure and documentation to alleviate this problem.

AN ASSESSMENT OF SOME COMMERCIAL DIGESTERS IN THE U.K.

P.J.MEYNELL

(SECRETARY)
British Anaerobic and Biomass Association Ltd.

SUMMARY

Commercial digesters now being built in the UK have appreciable differences in design, although performances remain within a well-known range. These differences result from the differences in previous experience of the different companies and in the marketing approaches. In arriving at a design which is both technically and economically viable, various cost/effective compromises have been made each in line with the particular companies approach. The broad base of the British industry is demonstrated, springing as it does from three standpoints - water pollution control, farming and farm machinery and energy conservation. The marketing approaches vary between the 'ready-to-wear' unit and the tailor-made plant. Various cost ranges for different sizes of plant are given. British manufacturers are beginning to adapt the technology developed for farm waste treatment to other organic matter.

1. INTRODUCTION

In recent years, a number of British companies have designed and erected commercial anaerobic digesters for the treatment of farm waste and sewage sludges. All of these companies have a different approach to the problems of digester design and this affects the emphasis, performance and costs of their products. It is appropriate at this stage of development to see where these approaches have led. Whilst a critical comparison is impossible in the absence of comparable performance data, it is useful to demonstrate the strengths of these different approaches and the resulting designs. The British Anaerobic and Biomass Association Ltd. is in a position to provide an overview of the British high-rate digesters, but no attempt is made to pick out any one design above the rest; in different situations, some designs will be more appropriate than others, and it is for the client to make the final choice.

2. COMMON FEATURES OF BRITISH DIGESTERS

All the commercial digesters in the UK designed for treatment of farm wastes are based on the high-rate, mesophilic type with complete slurry heating and mixing. In design they follow the research leads provided in particular by the Rowett Research Institute at Aberdeen, University College, Cardiff and the Polytechnic of Wales at Pontypridd. The latter two are closely connected with the development of a digester design with Hamworthy Engineering Ltd.

Thus all digesters share the common feature of the digestion tank, completely covered to ensure anaerobic conditions. By and large retention times are fairly uniform varying mainly with the type of waste. Typical retention times are as little as 10 days for pigs and up to 20 days for cattle and poultry wastes. The digesting slurry, usually heated to between 30-35°C, is kept completely mixed by regular stirring. Other features in common include gas storage and extensive insulation to prevent heat loss. Given these common features, the actual choice of equipment and arrangement within the design shows much variation. However, the actual quantities of gas given off from similar wastes does not differ significantly with different designs, ranging from 0.25-0.4 m³ per kg.TS.

The conclusion from this is that, with these common features, different designs will alter the efficiency of net energy recovery rather than the gross gas production. Indeed the cost and efficiency of component parts of digesters will make or break the viability of the system, so that cost effective compromises have to be made in every commercial design.

3. BACKGROUNDS OF COMPANIES IN DIGESTION

The hypothesis underlying this paper is that differences in design arise from the different approaches taken by different companies based upon their previous experience in other associated fields. Often the decision to move into the field of anaerobic digestion has been taken because it offers a diversification and an outlet for existing products.

Thus the commonest way into the digester field has perhaps been through waste water treatment and pollution control. Until the comparatively recent interest in energy recovery, pollution control of animal wastes has been the main impetus to development of digestion. In particular the firms of Hamworthy Engineering Ltd, Denis Evers and Associates Ltd. and IMES Ltd. (or ADBAC Ltd) have been working in water pollution control for many years, and their experience has proved valuable in adapting this technology to the changed requirement upon the farm. Denis Evers and Associates Ltd. for example offer a complete treatment system - the ANOX process - to enable discharge to a water course. Hamworthy Engineering Ltd. have another interest in their gas boilers, so they have a source of experience to build upon.

A second approach has come from the farmers themselves and the manufacturers of farming equipment. Kilbees Slurry Digesters Ltd. built one of the earliest farm scale digesters in recent times and have now linked up with Farrow Irrigation Ltd for the UK and N. Europe marketing. Farrow Irrigation Ltd. have come into the field of digestion as a diversification for their slurry handling equipment. Similarly A.O.Smith Harvestore Ltd. have used their experience in the manufacture of silos and storage tanks in their digester design. Indeed the storage tanks have been used extensively by other digester designers. Similarly the UK subsidiary of the Dutch company, Paques (UK) Ltd. uses their own vitreous enamel tanks as the basis for their designs.

Farm Gas Ltd, one of the companies which has been in the digester business longer than many, lies in between this and the third approach. Their emphasis is on producing gas on the farm rather than on pollution control; energy is the prime concern. This emphasis characterises the third approach illustrated by Helix Multiprofessional Services, a firm of architects and engineers with a common interest in energy conservation and alternative enrgy sources. Their approach to design has led to some very novel developments.

4. <u>DIFFERENCES IN DIGESTER DESIGN</u>

The most obvious design differences lie in the digester tanks and gas storage facilities; from these, other differences may well result. The commonest form of digester seems to be the above ground sealed unit with a fixed roof and separate gas store. The digester tanks are similar to the basic slurry store, and such units are produced by Farm Gas Ltd., A.O.Smith Harvestore Ltd., Paques (UK) Ltd. and Denis Evers and Associates Ltd. The gas store usually consists of a floating 'bell-over-water'.

In contrast to this, the Kilbees/Farrow digester has a flexible membrane covering the digester tank, in which the gas collects. The membrane inflates at constant gas pressure with gas production and collapses as it is used; it obviates the need for a separate and often costly gas holder. Similarly the Helix/ADBAC system has integral gas storage under modular gas caps which float on the surface of the digesting slurry. Because of their design, these caps can be fitted together to cover any area or shape of tank, so that the digester can be as simple as a hole in the ground, or as complicated and sophisticated as many other designs. As can be seen in the well-publicised Bore Place digester, the tank is below ground, offering savings in both insulation and pumping costs. The adaptability of these gas caps is a real positive feature giving flexibility of maintenance etc., since they can be removed individually.

Starting with these basic differences in tank design, others follow. For example in mixing, the commonest method is gas re-circulation, but draft tubes and mechanical stirrers are also found. Stirring to minimise dead spaces and maintain the solids in suspension is constantly being improved, but differences arise more from specific tank configurations, rather than other considerations. The Farrow/

Kilbees system combines mixing with external heat exchangers, through which the slurry is pumped. Most other systems use an internal heat exchanger through which hot water is pumped to heat the slurry whilst it is inside the digester. Waste heat from a generator and heat reclaimed from the outgoing slurry can be supplemented by gas boilers running on the gas produced. Insulation requirements will be largely dictated by the design and manufacturers have the choice between internal or external insulation - often a sprayed polyurethane foam.

Apart from the digester itself, the most important decision which affects the design and the cost, is whether to install a generator or not. This depends upon the adequacy of the gas production and the alternative uses for the gas. All manufacturers offer a generating option, and this characteristically adds 20-30% to the cost in terms of the generating sets and additional automatic controls required.

5. MARKETING APPROACHES

As well as having different designs, the various companies offer different packages which fit in with their marketing styles etc. These too alter the cost structures as much as the quality of the design, although high costs and quality are not necessarily synonymous. There are two marketing approaches which can be identified as the 'ready-to-wear' and 'tailor made'designs, although most sales would fall somewhere between the two.

Thus Farm Gas Ltd. operating at the lower end of the market offer two types of package, an $11m^3$ modular fibreglass unit suitable for use in a series of six units, and a range of steel digesters from 40-830 m^3 capacity. Their largestplant to date is $390m^3$ near Scarborough in Yorkshire. Offering standard units and a short on-site construction time (4-6 weeks) their prices are competitive.

The Farrow/Kilbees unit consists of one size of tank only - 450 m^3. Judging this to be the most economically marketable size which can act as a module for larger units, they are able tostandardise ancillary equipment so that the unit fits alongside other farm machinery in terms of both spares and the farmers knolwedge of his equipment.

The Paques (UK) Ltd. package - or Biopaq - also comes in a number of standard sizes, based on their vitreous enamel tanks. Production runs of equipment for these standard units make them a good example of the 'ready-to-wear' approach. A.O.Smith Harvestore are in a similar position with their steel tanks.

Hamworthy Engineering Ltd. possibly lie in between the two approaches , as shown by their recently opened digester at Chediston in Suffolk. Although all manufacturers make use of standard equipment wherever possible, it is the fitting and adaptation of designs to a given situation which charaterises the 'tailor-made' approach. It is the extent of adaptation required which differentiates the supplier from the consultant.

Thus whilst a sizeable portion of the market can be supplied with standard units, some situations require a tailor-made unit and it is for this portion of the market that companies such as Denis Evers and Associates Ltd., IMES/ADBAC and Helix Multiproffessional Services operate as consultants/manufacturers. Obviously the larger plants are more attractive, because the added design costs can be more easily accounted in the overall costs, but the approach is applicable to smaller plants as well where the material to be digested is not straightforward or where the client requirements are more specialised.

6. COSTS OF DIGESTERS

In the absehce of a particular situation, it is difficult to present cost comparisons. Ranges of costs per metre cubed capacity can be given as an indication; the larger the digester the smaller the volume cost. Thus smallest units might cost over £300/m^3, whereas a unit of 150 m^3 will have a price range of £150-220/m^3. A digester of about 400 m^3 will cost about £100-125/m^3 and one of 800m^3 will have a range of about £80-90/m^3: For units of about 400m^3 the payback period on energy costs alone is often quoted as about 5 years.

7. CONCLUSIONS

It is impossible in the short space available here to provide a full technical assessment of the various digesters currently available – it would not do them justice even if comparable information was available. Large scale digesters have been operating now for over three years and all manufacturers have working demonstration plants. Design improvements are being made as practical knolwedge increases. This paper attempts to outline the design and marketing approaches of the

of the different companies and to demonstrate the broad base of the
British Industry. Whilst all companies have started by designing digesters for the treatment of animal wastes for both technical and marketing reasons, all recognise the growing importance of ther forms of
biomass and are expanding the technology developed for farm wastes to
deal with other organic material.

N.B. All companies mentioned above can be contacted through the Secretary, BABA Ltd., The White House, Little Bedwyn, Marlborough, Wilts.

ECONOMIC ASPECTS OF BIOGAS-PRODUCTION FROM ANIMAL WASTES

Kleinhanss, W.

Institute of Farm Economics

Federal Research Center of Agriculture

Braunschweig, Germany (F.R.)

Summary

The importance of biogas production and its contribution to the substitution of fossil energy is determined inter alia by economic competition. In this paper we have investigated the range in which biogas is competitive with conventional sources by analysing farm structures in Lower-Saxony (Germany) and consequences for the adjustment of technology.

As an example for biogas production the following results are related to pig farms in the size of 90 - 750 livestock units:

1. Biogas plants, which are sized for the maximum heating requirements of a farm in the winter season, are economically competitive in piggeries beginning with the size of 90 livestock units for a given price of 1,25 DM per litre of petrol.
2. Farm with more than 200 livestock units can sell surplus biogas to their neighbours. In the size range of 750 livestock units the biogas production can compete at a price of 1,00 DM per litre petrol.
3. The transformation of biogas to electricity would cause 0,30 DM per kWh in piggeries with 90 livestock units resp. 0,13 DM per kWh for 750 livestock units. This exeeds the price of 0,035 to 0,055 DM per kWh actually paid by the electric power station for surplus energy.

To cover requirements of heating energy for a farm 50 livestock units are necessary. Therefore the use of this technology will be limited to almost 5 percent of all farms in Germany, i.e. only a small part of the potential energy from animal wastes can be used by biogas under the given structural conditions of the German agriculture.

1. <u>INTRODUCTION</u>

The role of biogas-production as energy-supply for farms is determined by economic competition to other energy-sources. In this analysis (2) we evaluated the economic competition of biogas-production on pig farms in Lower-Saxony. The influence of biotechnical efficiency, gas-utilisation and farm-structure to economic of biogas-production is discussed. The utilisation of biogas plants with respect to pollution control and amelioration of the efficiency of fertilizer of animal-wastes will not be discussed here.

2. <u>TECHNICAL PROBLEMS OF BIOGAS-UTILISATION AS ENERGY-SOURCE</u>

Biogas-production in plough flow digersters spend continuous quantity of gas (1). The potential of biogas-production per farm depends of wastes availability and gas-production efficiency. Compared with this, the energy-requirement of farms is characterized by extremely seasonal changes and a demand for heat-energy, which is almost equal in different cattle farms.

Only the requirements for electrical energy increase with livestock. Therefore the following consequences results:

1. Farms with more than 50 to 70 livestock units can cover heat-energy-needs in winter season (gas-production rate = 0,3 m^3 per kg organic dry matter (ODM)). This technical limitation restricts the biogas-production to big cattle farms. In Lower-Saxony more than 80 % of cattle farms have less than 50 livestock units. Hence, only a small rate of biogas-production-potential would be usable by biogas-plants on single farm level.

2. Farms with more than 50 livestock units can produce biogas from their total potential of animal wastes. If biogas-utilisation is possible on farm level only, it is favourable to dimension the biogas plant to max. requirement of heat-energy in the winter season. Nevertheless, only 50 % of biogas can be used, 30 % are necessary for the biogas plant itself and the rest is surplus. The seasonal biogas-excess is greater, if biogas plants are dimensioned to serve all energy needed for drying-processes.

3. The usage of all biogas potentials is meaningful, if it is possible and economically to sell gas to neighbours and to produce electrical energy.
This is especially advantageous when increasing size of biogas plants correspond with economies of scale.

The amount of investment for industrial biogas plants

depends on its size and on gas utilisation. For biogas plants which are dimensioned to heat-energy-requirements in the winter season on farm level, the investment is about 90 000 - 110 000 DM (2). In relation to other energy-utilisation systems the investment is very high. For this situation the reduction of investment for biogas plants by technical amelioration should be a very important question for research in this area in the future.

3. ECONOMIC COMPETITION OF BIOGAS-PRODUCTION FOR FARM-HEAT-ENERGY-REQUIREMENT

If biogas is used for heat-energy-requirements of farms only, the dimension of biogas plant depends on this energy-demand. In large cattle farms, only a small part of energy potential from animal wastes can be used for biogas-production. In the top of figure 1 you see some results of economic evaluation of biogas-production in pig farms with more than 50 livestock units in Lower-Saxony.

Fig.1: Maximum amount of investment at gas-production-rate of 0,3 m^3/kg ODM and break even prices of biogas-plants, dimensioned of heat-energy-need in the winter season

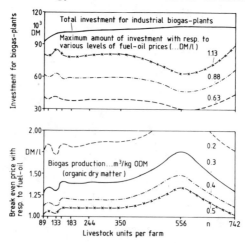

At actual fuel-oil-prices (0,63 DM/l) the maximum amount of investment for biogas plants may be up to 30 000 - 45 000 DM. At future fuel-oil-prices of 1,13 DM/l the maximum amount of investment is 65 000 to 90 000 DM. The investment for industrial biogas plants is higher. Hence, biogas-production can be economically, if: a) fuel-oil prices are higher than 1,13 DM/l, or b) investment for biogas plants can be reduced drastically.

At the bottom of figure 1, break even prices (with respect to fuel-oil) for biogas plants are given by the current costs of biogas plants. At a biogas-production rate of 0,3 kg per kg ODM (organic dry matter) economic competition of biogas-production is given by fuel-oil-prices higher than 1,25 DM/l. In contrary to farms with 89 livestock units the economic competition of biogas-production in larger farms is less, because: a) the utilisation rate of biogas for hay drying decreases, b) in farms with 556 livestock units biogas utilisation for hay and grain drying is not possible.
The important effect of biogas-conversion rate to economic realisation is demonstrated by the curves of biogas-production of 0,4 and 0,5 m^3/kg ODM. A biogas-production rate of at least 0,3 m^3/kg ODM, (better 0,5 m^3/kg ODM) is necessary for biogas-production being valuable energy source at farm level in the future.

4. ECONOMIC COMPETITION OF BIOGAS PRODUCTION FOR GAS SALES

The bio-energy potential from animal wastes in large cattle farms can be utilised only, if farm-extern gas utilisation is possible. In German-village structure biogas selling is possibel to households surrounding the biogas-producer-farm, as regional gas distribution systems are not present. In figure 2, break-even-prices for biogas-production are given with premisse, that gas can cover the heat-energy-needs of one or more households. On this condition gas sale is possible in farms with more than 244 livestock units at biogas-production rate of 0,3 m^3/kg ODM. Dependent of gas-production rates the results are as follows:

Fig.2: Break-even-prices for
biogas production in pig-
farms with gas sales to
neighbours

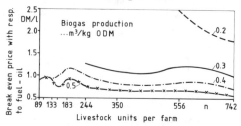

Livestock units per farm

1. With a biogas-production of 0,2 m³/kg ODM the profit is less than by biogas-utilisation for heat-energy on farm level only.

2. At biogas-production rate of 0,3 m³/kg ODM the profit of biogas-production with gas sales is substantially higher in large cattle farms. Nevertheless, biogas-production is competitive at fuel-oil-prices higher than 1,- DM/l.

3. In large cattle-farms and at biogas-production-rate of more than 0,4 m³/kg ODM, biogas-production can be economic at fuel-oil-prices, which are to be expected in the next future.

In view of Lower-Saxonie's cattle farm structure, this favourable economic competition exists only in less than 5 % of all cattle farms.

5. POSSIBILITIES AND LIMITS FOR ELECTRICAL-GENERATION BY BIOGAS

The utilisation of biogas to electrical-energy-generation is favourable in view of using seasonal biogas-excess and exhausting biogas-production potential from animal wastes in large cattle farms. Two types of technical installations are suitable:

1. Electrical energy-generation from biogas, which is not needed as heat-energy by the farm. Seasonal fluctuations of electrical energy supply cause problems, because harmonisation with demand of electrical power is very difficult.

2. If all biogas is used for electrical generation, the heat-energy-requirements of farms must be satisfied by heat-energy-recuperation (Totem-system). With this type of installations it is possible to adapt electrical generation to the demand.

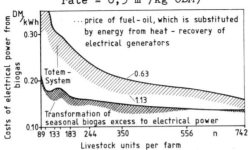

Fig.3: Costs of electrical generation
from biogas (gas-production-
rate = 0,3 m³/kg ODM)

In figure 3 costs of electrical-energy-generation from
biogas are given. In the most favourable condition, the production costs of electrical power are 0,13 DM/kWh. This is
about 30 % higher than the price demanded by electrical power
stations. The return of electrical energy produced by biogas
plants is only 0,03 to 0,05 DM/kWh. Hence, electrical generation by biogas for sales is not favourabel in this time. Electrical generation by biogas may be possible and economic in
the future: a) in large cattle farms, which can utilize the
most of generated electrical power itself and, b) if the price for electrical power is rising.

As the production costs for electrical power are less when
only seasonal biogas-excess is used, this types of installations should be favourised. In view of electrical utilization
the electrical generation by Totem-system is favourable. It
can be expected that higher prices for electrical power will
be paid, when electrical energy is generated by Totem-system.

Summarising, the utilisation of biogas for electrical
power has advantages in view of total utilization of biogas,
but the economic competition is less than direct biogas-uti-
lization for heat-energy-requirements.

References:

1. BAADER, W.; DOHNE, E.; BRENNDÖRFER, M.: Biogas in Theorie
   und Praxis, KTBL-Schrift 229, Münster-Hiltrup 1978

2. KLEINHANSS, W.: Wirtschaftlichkeit der Biogaserzeugung aus
   tierischen Exkrementen. Berichte über Landwirtschaft,
   H.4. Hamburg, Berlin 1980.

METHANE FERMENTATION IN THE THERMOPHILIC RANGE

O.KANDLER, J.WINTER and U.TEMPER

Botanical Institute, University of Munich, Menzinger Strasse
67, D-8000 München 19

Summary The effect of different agitation rates and increasing loading rates combined with decreasing detention
times on methane production, degradation of volatile solids
and dewatering properties of the fermented sludge were
studied in fermentations of municipial sewage sludge at meso-
philic (35°C) and thermophilic (50°, 56°, 60°C) temperatures.
Stable fermentations were achieved at loading rates up to
7.3 g v.s./l.d and a detention time of 3 days with an optimum
at detention times of 4-5 days corresponding to loading rates
of 4.4-5.4 g v.s./l·d. 50°C was superior to either 35° or
60°C with respect to gas production per fermentor volume and
volatile solids degraded.

Introduction Methane fermentation in the range of thermophilic temperatures ($50-65^{\circ}$C) is supposed to offer several
advantages over that at traditionally applied mesophilic
temperatures (1,2,3,4,5), such as shorter holding times,
higher yield of gas per reactor volume per day, better dewatering properties and last not least a complete hygienisation of the sludge. The obvious disadvantages of thermophilic
systems include higher energy requirement to heat the influent and to maintain the fermentor temperature; poor process stability, a strong odour and a higher COD of the
effluent. As yet, most studies on the effect of thermophilic
temperatures on methane production have been carried out with
manure or artificial substrates. This paper reports preliminary results of the comparative fermentation of municipial
sludge at various temperatures.

Substrate: A homogenised mixture of equal portions of primary
sludge of municipial sewage and surplus sludge from the activated sludge process was dispensed into plastic bottles, each
to provide the desired daily input of sludge for the fermentors. The bottles were stored at -20°C until use. The sludge
contained: Dry weight: 31 g/l, volatile solids: 22 g/l,
ammonia: 0.3 g N/l, Kjeldahl N: 1.7 g/l.

Digesters: Cylindrical, complete-mix 2 l fermentors(Braun,
Melsungen, BRD) with a filling volume 1.5 l were used. Mixing
was performed 5 cm from the digester bottom with 3 propeller

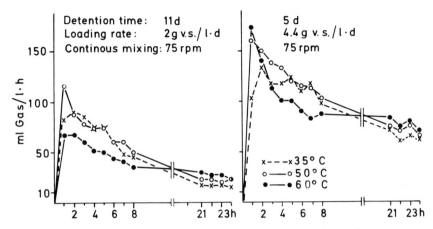

Fig. 1. Kinetics of gas production at mesophilic and thermo-
    philic temperatures.

type stirrers of 6 cm diameter ca. 5 cm apart from each other
on a central shaft.

Experimental design: Fermentation was carried out in a semi-
continous mode. New feed was added every 24 h subsequent to
the removal of the desired volume of fermented sludge. Samples
for the analytical procedures were withdrawn before the new
substrate was added. To evaluate the influence of different
loading and agitation rates, it was necessary to establish
baseline performance data for each of the 3 temperatures
applied. Fermentations conducted at the respective temperatures with a constant detention time of 11 days and a loading
rate of 2 g v.s./l·d under continous mixing at 75 rpm served
as baseline runs. All data given in the following Figures and
Tables were determined after the fermentation had reached a
steady state at the particular condition.

RESULTS

Kinetics of gas production within the feeding interval

As shown in Fig. 1, gas production reaches its maximum within
the first hour at thermophilic and within the second hour at
mesophilic temperatures at both loading rates. It declines
within 8 hours and reaches a constant level at least after
20 hours. The final level of gas production is slightly
higher at 60° and slightly lower at 35°C than at 50°.

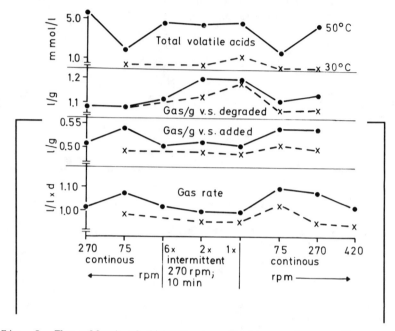

Fig. 2. The effect of different agitation rates on the fermentation at a loading rate of 2 g v.s./l·d.

The effect of agitation on the fermentation at 30 and 50°C

Although methane fermentation of sewage sludge does usually not require continous agitation, the latter may be of advantage at higher temperatures. Therefore the agitation rate was changed every 3-4 weeks in fermentations which were performed at 30, 50 and 56°C, respectively, at a constant loading rate of 2 g v.s./l·d. Since the curves for 50 and 56°C were virtually identical, only the curve for 50°C is shown in Fig. 2. It is obvious, that the effect of different agitation rates on gas production is very small, but intermittent stirring tends to be less favorable than continous mixing at low speed. There is, however, a very distinct increase of total volatile acids at 50°C at either fast stirring or at intermittent stirring whereas slow continous stirring results in a lower

	h	Detention time 11d; loading rate 2 g v.s./l.d.							Detention time 5d; loading rate 4.4 g v.s./l.d.						
		COD per liter sludge	Sediment		Supernatant				COD per liter sludge	Sediment		Supernatant			
			ml/l	COD total	ml/l	COD/l	COD total	COD in %		ml/l	COD total	ml/l	COD/l	COD total	COD in %
35°	2	25.5	900	25.0	100	5.4	0.5	2.0	28.5	470	23.8	530	9.2	4.7	16
	4		750	24.4	250	4.7	1.1	4.5		370	22.8	630	9.0	5.6	20
	24		330	23.2	777	3.5	2.3	9		250	22.0	750	8.7	6.4	23
50°	2	25.0	480	21.0	520	7.9	4.0	16	27.5	600	24.6	400	7.3	2.9	11
	4		450	20.7	550	7.7	4.2	17		510	24.2	490	7.0	3.3	12
	24		380	21.3	620	5.8	3.7	15		300	23.2	700	6.3	4.3	16
60°	2	28.0	420	21.3	580	11.8	6.7	24	29.4	475	23.6	525	11.1	5.7	20
	4		410	21.3	540	11.6	6.7	24		420	23.3	580	11.0	6.1	21
	24		370	22.2	630	9.1	5.8	21		260	22.2	740	10.0	7.2	24

Table 1. Dewatering properties of sludge fermentation at different temperatures. Sedimentation time 2.4 and 24 h, respectively.

level of acids. The experiment shows that fermentation in the thermophilic range does not require any stronger agitation than in the mesophilic range.

Dewatering properties of mesophilic and thermophilic fermented sludge

As shown in table 1 the dewatering properties of the mesophilic and thermophilic fermented sludge are fairly similar with respect to the final volume and the total COD of the sediment at least after 24 h sedimentation. However, the mesophilic sample tends to be superior to the thermophilic samples after shorter sedimentation times at the fairly high loading rate of 4.4 g v.s./l·d. At the lower loading rate, the COD of the supernatant of the two thermophilically fermented samples exhibits a significantly higher COD than that of the mesophilically fermented sample, but is almost the same at the higher loading rate. Similarily, the pronounced odour found in the thermophilic samples is absent in the mesophilic samples at the low loading rate.

The effect of increasing loading rates at different temperatures

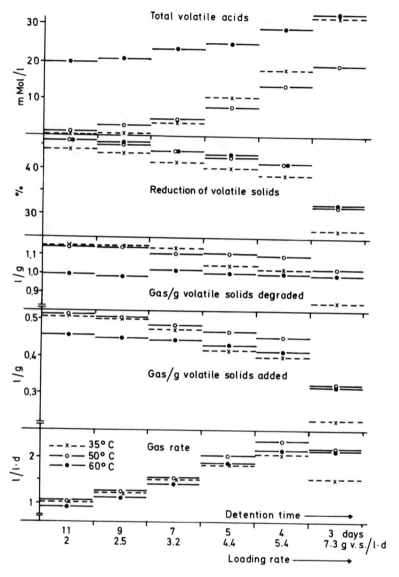

Fig. 3. The effect of loading rates and detention times on gas production, degradation of volatile solides and the level of organic acids at different temperatures.

Table 2. The effect of different temperatures on pH, destruction of protein, accumulation of acetic acid and propionic acid and on the methane content of the produced gas.

t	pH		NH$_3$ mg N/l	org. N in %	% CH$_4$		Acetic acid mMol/l		Propionic acid mMol/l	
	a	b	a+b	a+b	a	b	a	b	a	b
35°C	7.30	7.20	950	47	76	72.5	0.6	4	-	13
50°C	7.50	7.35	1050	55	74	72	1.0	5	0.3	9
60°C	7.60	7.45	1130	59	73	70	5.0	12.5	13	15

a = Detention time 11d; b = Detention time 5d

Increasing loading rates coupled with decreasing detention times were achieved by adding increasing amounts of the same batch of stored sludge to the fermentors. The loading was increased at intervals of at least 4 times the detention time. The data shown in Fig. 3 and Table 2 represent the steady state levels established at the end of each period. The maximum rate of gas production per fermentor volume is reached at a detention time of 4 days at all temperatures tested whereas a detention time of 3 days is inferior. Surprisingly, in spite of the high loading rate and short detention time gas production did not break down completely at any temperature. The percentage of the degradation of volatile solids decreases steadily with increasing loading rates whereas the level of volatile acids increases, as expected.

Final remarks: The differences in the degradation of volatile solids and gas production caused by the different fermentation temperatures are surprisingly small. Only at the higher loading rates is fermentation at 50°C superior to that at 35° by about 10%, whereas fermentation at 60°C is less favourable. These results are in agreement with those of the most recent experiments (4) employing beef cattle waste. They indicate that fermentation of sludge at thermophilic temperatures does not result in a drastic improvement of gas production or degradation rates. The application of temperatures between 50° and 56°C is most likely of advantage, however, when strict hygienic rules require a complete hygienisation of the fermented sludge, which is achieved at these temperatures without additional treatment. More detailed studies are required to evaluate the potential economic feasibility of fermenting sludge at thermophilic temperatures at a large scale.

Literature: 1) Buhr,H.O.; J.F.Andrews. Water Res.11, 129-143 (1977); 2) Therkelsen,H.H., D.A.Carlson. J.WPCF51, 1949-1964 (1949); 3) Varel, V.H., A.G.Hashimoto, Chen,Y.R. Appl.Envir. Microbiol. 40, 217-222 (1980); 4) Varel,V.H., H.R.Isaacson, Bryant,M.P. Appl.Environ.Microbiol.33, 298-307 (1977); 5) Garber,W.F. et al. J.Water Poll.Control Fed.47, 950 (1975)

Acknowledgment: This work was supported by the Government of the Federal Republic of Germany under contract PTB 8151.

SESSION IV: THERMOCHEMICAL ROUTES TO GASEOUS AND LIQUID FUELS

Chairman: K. ROBINSON
Energy Division, National Board for
Science and Technology, Ireland

Summary of the discussions

Invited papers

Gasification - The process and the technology

The combustion, pyrolysis, gasification and liquefaction
of biomass

Poster papers

Energy from straw and woodwaste

Farm straw as fuel for power

Pyrolysis and combustion of wood in relation with its chemical
composition

On the influence of the different parts of wood on the production
of gaseous combustible products by pyrolysis

The pyrolysis of tropical woods: the influence of their chemical
composition on the end products

Gaseous fuel from biomass by flash pyrolysis

Fast pyrolysis/gasification of lignocellulosic materials at short
residence time

The conversion of biomass to fuel raw material by
hydrothermal treatment

A process for thermo-chemical conversion of wet biomass into heat
energy

Mass and energy balances for a two fluidised-bed pilot plant
which operates on wood fast pyrolysis

Two new types of biomass gasifiers developed at C.N.E.E.M.A.

Kinetic studies of pyrolysis and gasification of wood under
pressure with steam and oxygen

Methanol catalytic synthesis from carbon monoxide and hydrogen
obtained from combustion of cellulose waste

480

Process and equipment for the fluid bed oxygen gasification of wood

Techno-economic evaluation of thermal routes for processing biomass to methanol, methane and liquid hydrocarbons

Thermal processing of biomass to synthesis gas - a programme of experimental investigations and design studies

Economics of combustion energy from crop residue

SUMMARY OF THE DISCUSSIONS

Session IV   THERMOCHEMICAL ROUTES TO GASEOUS AND LIQUID FUELS

Rapporteur :   R. OVEREND, Canada

Speakers :   W.P.M. VAN SWAAIJ, Netherlands
                  T. REED, USA

Chairman :   K. ROBINSON, Ireland

Poster Session : 16 papers presented

Biomass energy can only be realistically viewed as a holistic system
encompassing the chain from land/water resource management to end use
application. To separate out one set of conversion technologies from
the context of the system (resource, straw/wood/animal residues, transport, conversion, delivery and end use application) results in the loss
of perspective of both social costs and benefits to be gained from
bioenergy utilisation. As a result of this distortion the session was
rather inconclusive on the issue of thermochemical technologies for
the EC or North America while being extremely definitive about the
application of combustion and gasification or electricity generation
in developing countries. The paradox resulting from this was disturbing;
many audience members (their memories of world war II technologies
somewhat blurred by the passage of time) suggested that these technologies, while not meeting the environmental "niceties" of the developed
world, were quite adequate for the developing world !

The general impression gained was that the application of the combustion of difficult materials such as straw for on-farm applications or
as a supplementary fuel with high sulphur coal and the coupling of wood
fired gasifiers to boilers previously fueled on premium fuels, should
be treated as demonstration projects to accelarate their introduction
into the free market system.

Because not many discussed the current activities of national governments and the EC in accelerating the development of biomass derived
syngas to methanol synthesis, it is evident that the role of land use
strategies and scale effects on the economics of liquid fuels are not

receiving significant public consideration. The poster paper by Ader
et al showed that at plant scales approaching those of modern pulp
mills, an energy product (methanol) from either the yet to be proven
technologies of steam or oxygen/steam gasification, is very close to
commercial viability. This study is comparable to others in Canada and
Sweden which draw similar conclusions, thus reinforcing the need for
the construction of process development units of around 50-100 tons per
day dry feedstock capacity to finally confirm cost and yield data.

While both session speakers stressed the antiquity of gasification
technology Van Swaaij managed to convey clearly the lack of refinement
of the technology that during WW II produced more than 1/2 million
vehicle gasifiers. A concentrated program at Twente University on the
co-current gasifier has resulted in the development of both a satisfactory technology along with an adequate chemical engineering design
basis, thus enabling calculation of scale up/down parameters. The dark
horse development in gasification is the fluidised bed which offers
feedstock throughputs as high as 2 tons/m2/hr using air at atmospheric
pressure. While not commercially viable below about 10 MW the evidence
from experiments in Canada suggests that fluidised bed reactors with
wood feedstock give high yields of saturated and unsaturated hydrocarbons (CH_4, C_2H_4, C_2H_6) with little tar. Poster papers 6,7 and 10 concerned with experiments using rapid pyrolysis rates of more than 200 $^\circ$C/sec
and brief residence times followed by rapid quenching, indicate high
yields of hydrocarbons at the expense of CO & H_2 yields. In the limit
at heat fluxes more than 100 W/cm^2, Reed reported that more than 20%
mass conversion to olefins are possible. Fast pyrolysis in fluidised
beds evidently could lead to processes producing ethanol for example
by the hydration of ethylene which would be far superior to current
hydrolysis and enzymatic routes from wood.

Slow or traditional pyrolysis is still the subject of scientific interest despite the well-known difficulties in applying the data obtained
to "fixed bed gasifier design". The general knowledge that slow heating
rates result in less gas production with greater proportions of CO & H_2
as well as char are verified by papers 3 and 4. The influence on pyro-

lysis products of differing feedstock composition obtained by fractionating the plant material was demonstrated in papers by Richard and Caubet.
Much of the current thermochemical laboratory work and technology development is specific to dry feedstocks. To obviate the requirement to
evaporate water with its attendant loss of energy before pyrolysis a
paper by Bobleter discussed a hydrothermal treatment which offers the
opportunity of fractionating the feedstock with the recovery of chemical
feedstocks as well as energy products. A paper by Have outlined a combustion scheme that would effectively recover the latent heat of evaporation in drying feedstocks such as animal residue - the restriction
being that only low grade (less than 80°C heat) would be available as
a product though at relatively high efficiency.

Liquefaction by indirect methods such as syn-gas to methanol are seen as
the near term opportunity while direct liquefaction is recognised as
being some time away from even pilot plant demonstration. Syn-gas to
methanol synthesis over selective catalysts is seen to be far more viable than Fischer-Tropsch type synthesis with its accompanying diverse
product slate. Current technology is operating at high pressure requirement and even into less selectivity, thus producing a methyl fuel with
higher alcohols will give significant gains in process efficiency. The
paper by Masson suggests a promising route to lower pressure (and therefore process energy requirements).

The energy requirements for gasification to MJV (medium joule value) or
LJV gas are less than 10% of that contained in the wood feedstock. For
countires with significant non-fossil fueled electricity generation the
use of electricity to provide this energy in gasification appears promising as a route to syn-gas or MJV gas, as discussed in the paper by
Divry.

As a worker in the field I would like to make a plea concerning the
gathering of experimental data. Biomass quantities should if at all possible be given on a dry mass basis. Where moisture is a significant parameter, the moisture content should if possible be stated as a total
mass basis i.e. $\%H_2O$ = mass H_2O/(mass H_2O + mass Biomass).

As far as possible SI units should be used rather than Btu or k Cal.
The reporters of experimental data should take care to analyse their
chemical products and close their mass balances – particularly with
respect to the char produced. Deglise noted that the molecular weight
of a 700 C pyrolysis char at 16.7 amu is considerable in excess of the
theoretical value of 12 amu assumed in most studies. Likewise weight loss
data presented in terms of Ahrenius factors $\left(A \exp(-E/T)\right)$ presumes a
mechanistic understanding which is just not available at this time. In
fact the poster papers at this meeting clearly show our lack of understanding of fundamental factors in the process that we casually call pyrolysis. Feedstock variation, heat transfer and heating rates, moisture contents, thermal history, the immediate gaseous environment, etc., all are
significant factors in this puzzle. Nevertheless, the meeting did show
significant advances in the technologies of gasification despite the
lack of fundamental understanding.

"GASIFICATION - THE PROCESS AND THE TECHNOLOGY"

W.P.M. VAN SWAAIJ

Twente University of Technology
The Netherlands

Summary and Conclusions

Thermochemical gasification of biomass can produce low, medium and high joule value gases. The characteristics, applications and potential of the different processes and reactor types are discussed. The introduction of biomass gasification on a large or intermediate scale for the production of power, SNG, methanol etc. will depend on developments in coal and municipal solid waste gasification and on the prices of biomass. Biomass and especially wood is a clean fuel and therefore its direct combustion using modern equipment will be a strong competitor for energy generation via gasification. Gasification is also attractive for small scale, power and power heat generation and developments nessecary for its widespread acceptance are discussed. It is further concluded that, on the small and intermediate scale, new processes which require minimum feedstock preparation and preferentially producing M.J.V. gas without the use of oxygen, should be developed.

INTRODUCTION

Thermochemical gasification of solid fuels such as biomass, peat and coals has been studied and applied for about 140 years (1). A complete review would easily fill several textbooks. Therefore the discussion will be restricted to some of the fundamentals and the different types of technology used in the field of biomass gasification and will then concentrate on some applications which are expected to penetrate the energy market in the near future.

The aim of a gasification process is to transfer the combustion value of the solid fuel to a gaseous energy carrier, preferably in the form of chemical energy and not in the form of sensible heat. Gasification is performed because of the advantages of a gas over a solid fuel: gases are easy to clean, to transport and to combust with a low excess of air and there is little resulting pollution. Further, gases can be burned in an internal combustion engine (gasturbine, reciproking engines) and can be easily applied in combined cycles.

Biomass gasification can also be carried out via biological gas generation. Thermochemical routes have the advantages of compact equipment due to the relatively short residence times required (1 - 10.000s), easy start-up and stable operation and there are no requirements on the nutrient value of the feedstock. Its disadvantages are that relatively dry feedstocks are required and that the ash produced has little value as fertilizer. We will only discuss the thermochemical routes further.

In the gasification process the biomass is succesively heated up, dried and pyrolysed to produce gases and char. These products then react further in a complex way with a gasification agent which can be air, oxygen, CO_2, steam, mixtures of these gases of hydrogen. The reactions that take place between the char and the gasification agents can be described broadly speaking by the following equilibrium reactions:

$$C + O_2 \rightleftharpoons CO_2 \qquad \text{exothermic} \quad (1)$$
$$C + CO_2 \rightleftharpoons 2CO \qquad \text{endothermic} \quad (2)$$
$$C + H_2O \rightleftharpoons CO+H_2 \qquad \text{endothermic} \quad (3)$$
$$C + 2H_2 \rightleftharpoons CH_4 \qquad \text{exothermic} \quad (4)$$

The composition of the productgas is determined by the biomass feedstock and gasification agent used, and by process conditions such as pressure, temperature, residence time and heat loss or external heat input. The external heat source can be a nuclear plant (2), concentrated solar radiation or another chemical reaction (e.g. external combustion of part of the fuel, CO_2

acceptor process (3) etc.). Most gasification processes are authothermic however and we will consider only those cases. Gasification produces several types of gases and these can be roughly divided into three catagories, according to their heat of combustion per m^3 at ambient conditions (see table 1.)

	CO	H_2	CO_2	CH_4	C_nH_m	N_2	Joule value MJ/m^3
L.J.V. gas %	17.0	18.3	14.2	1.8	-	48.7	4.75
M.J.V. gas %	61.0	28.0	2.0	8.0	-	1.0	13.61
H.J.V. gas %	-	-	0.9	81.2	3.4	14.5	31.65

Table I. Typical compositions of dry, clean product gases (vol.%).

Low Joule Value gas. $(3.5 - 7 MJ/m^3)$
A typical production and application scheme is given in fig. 1.

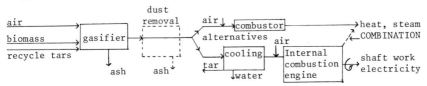

fig. 1., L.J.V. gas production and application.

Air is the gasification agent. In most cases it is simply a two stage combustion process but sometimes the gasifier is retrofitted to an existing gas/oil boiler, kiln or motor. The lower scheme is specially attractive if electricity generation can be combined with a usefull application of the sensible heat of the exhaust gases (drying, heating, etc.). Due to the low specific energy content the product gas cannot be transported or stored economically.

Medium Joule Value gas. $(9 - 15 MJ/m^3)$
Two typical production schemes are given in fig. 2 together with possible applications.

A.

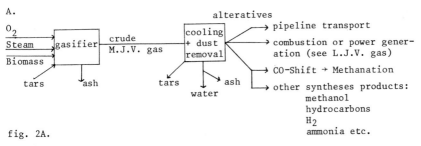

fig. 2A.

B.

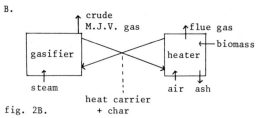

fig. 2B.

fig. 2., production and application of M.J.V. gas.

Scheme A is well known. However, Oxygen is usually expensive and scheme B
avoids its use. The endothermic char-steam reaction is separated in location from the exothermic reaction (combustion with air). Solid reactants
and/or heat carriers are circulated between the gasifier and the heater in
the same way as in fluid bed catalytic cracking. To some extent it can be
considered as a continuous variant of the old intermittend water gas process.
This scheme B (4,5) has not yet been proven commercially. The M.J.V. gas can
be economically transported over large distances and used for (combined)
power/heat purposes. Furthermore the product gas (also called synthesis gas)
can be used to produce a wide variety of chemicals and energy carriers such
as methanol, H_2, gasoline, SNG, etc.

High Joule Value gas. (20 - 36 MJ/m^3)

Mostly these gases are used to substitute or supplement natural gas. They
are usually produced from M.J.V. gas (see fig. 2) but in a few schemes direct production by gasification is being developed (see fig. 3)

A.

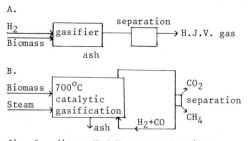

fig. 3., direct H.J.V. gas production by gasification.

Apart from gasification at high pressure with hydrogen (A) the EXXON (6)
process for coal looks promising. The almost thermally neutral reaction:
$2 H_2O + C \rightarrow CH_4 + CO_2$ is realized at approximately 600-700°C utilizing
$K_2 CO_3$ as a catalyst. Methane is continuously separated from the recycle
system by cryogenic destillation. However such processes are not yet commer-

cially available.

Most of these type of processes have been developed for coal or peat but some have also been tested for biomass or municipal waste. Biomass contains much more oxygen and hydrogen than coal, reducing the amount of steam nessecary to effect gasification. Furthermore, its sulphur content is usually low. On the other hand, as produced, biomass often contains large amounts of water and after drying it will attract an equilibrium amount of water if not properly stored. Apart from energy losses in drying, this adds to the complexity of the process. The biomass is often in an inconvenient shape and unless the gasifier is specially adapted, extensive feedstock preparation may be required such as grinding or pelletizing. The cost of these operations can be considerable (up to \$ 25,-- per ton (7)) and consume up to 10% of the heat of combustion of the biomass. Another important property of biomass is the dispersed nature of its production giving rise to important collection and transportation costs specially for large scale conversion processes. The different technical solutions proposed for the gasification reactors will now be discussed in relation with the properties of the biomass feedstock.

REACTOR TYPES

Most of the reactor types have already a long history. Fig. 4 gives the different principles together with temperature and conversion profiles. It should be realized that reactor properties do not necessarily reflect the process performance. Tars and heat can be recovered and recycled to the reactor. However this complicates the overall process.

The countercurrent moving bed reactor has the longest history and is widely used (1, 8, 9, 10) both with solid and liquid (10, 11) ash removal. Its advantages are its simplicity of operation, that no solids flowrate control is nessecary and that there is internal heat exchange of the product gas with the biomass feedstock. (T_{out} = 400°C). Its disadvantages are that large amounts of tars are produced, and that channeling due to sticking tarry particles may occur, nescessitating the use of rotating grids. Pelletizing may be nessecary depending on the feedstock.

The co-current moving bed reactor (1, 12, 13, 14) is also simple to operate, produces an almost tarfree productgas but is more difficult to scale-up (see later). It has a higher productgas temperature (700°C) and can also require feedstock pelletization.

The fluid bed (15, 16, 17) reactor can handle a wide range of feedstocks but has a high product gas temperature (e.g. 900°C), and important

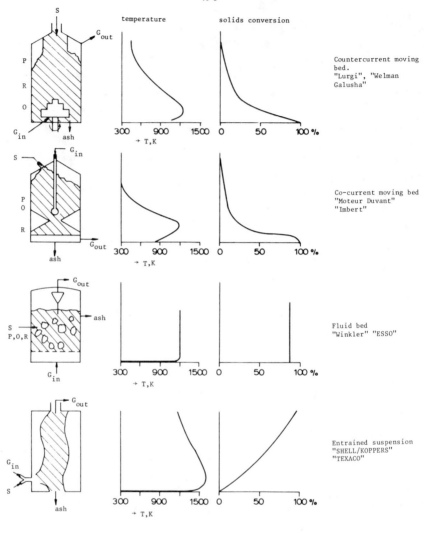

s = solids feedstock input, G_{in} =
gasification agent input, G_{out} =
product gas output, P = pyrolysis
zone, O = oxydation zone, R =
reduction zone

fig. 4 - Gasification reactor types with temperature and solid
conversion profiles

tar production, limited solidsconversion, severe particle entrainment and
is more difficult to operate (solids flow rate control etc.).

Also, the entrained bed reactor (1, 18, 19, 20) is omnivorous, operates tar free and with molten ash due to the prevailing high temperatures.
However the process is complex in operation, requires powdered biomass and
extensive heat exchange. We shall not consider here the molten salt (21)
and molten iron processes (22) as the conversions and temperatures of these
processes are somewhat simular to the fluid bed process. Also rotary kilns
are used for co- and counter current operation. Furthermore intermediates
between co- en counter current operation existed: double shaft and double
fire gasifiers (see (1), (12)).

PROSPECTS FOR BIOMASS GASIFICATION

The important factors for all gasification processes are the gasification agent, the operational pressure and the unit capacity. Fig. 5 gives a
general picture of the situation as seen by the author.

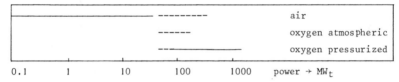

fig. 5., application of biomass gasification processes.

With regard to large scale biomass units (>100 MW_t) if feasible the criteria
for process selection are not unlike those of the coal gasifiers and pressurized gasification with oxygen seems logical. Process pressures will be
30 Bar or higher in relation to the subsequent syntheses process (methanol,
SNG, etc.). For coal gasification only the LURGI process (8) has reached the
commercial stage here but many others (Texaco (19), SHELL/KOPPERS (18)) are
in an advanced stage of development. For electricity production from coal,
air gasification at e.g. 30 Bar is often proposed (23). With biomass gasification the problem of sulphur removal is far less important (or absent)
and therefore direct combustion (e.g. in powder flames) seems more likely.
The situation may change if high efficiency power units (combined cycles)
based on gasification of coal can attract widespread use.

For the intermediate scale (10-100 MW_t) the situation is not yet clear.
Direct combustion in powder flames or (fast) fluid beds (24) are strong
competitors to gasification. The preference for gasification depends much on
the existing site facillities, feed preparation requirements and heat/work
demands of the applications. It is very difficult to give general rules but

sure many more technical/economical data are required. Gasifiers for municipal waste based on oxygen (25) or air (26)(27) gasification are slowly finding
application and similar biomass gasification may become more attractive.

Retrofitting existing boilers, kilns etc. previously fired by coal,
gas or oil is widely discussed in the literature (28, 29, 30) especially in
relation to wood. No derating is nessecary if M.J.V. gas is produced. However the system requires expensive oxygen or processes still to be made
commercially available. Derating becomes important with air gasification if
the heating value of the product gas is below 7.4 MJ/m^3. (28). An important
factor in air blown gasifiers can be the conservation of the sensible heat
of the product gas by fitting the gasifier directly to the boiler. Increasing the joule value or duel fuel firing may also counteract derating.

For small and medium scale power generation 0,1 - 20 MW$_t$ the set up
given in fig. 1, is very attractive and it is likely that gasifiers in connection with dualfuel diesel engines or gas engines for combined power/heat
generation will capture a important slice of the market in the near future.
These plants have a relatively high efficiency at different loads and are
basically simple. Several manufactorers are marketing such systems (31, 32)
and some units are already operating or are in the construction phase. Some
units operate in the countercurrent mode but for smaller plants co-current
operation seems to be particularly attractive because of its low tar production. We will discuss these units in more detail.

CO-CURRENT MOVING BED GASIFIERS

Although these gasifiers have been known for more than 100 years only
approximative describtions of the reactor and only purely empirical design
rules can be found in literature. Groeneveld et al (13, 33) recently studied
the reactor mechanics in more detail. (see fig. 6). The solid feedstock
enters the top of the gasifier and the pyrolysis zone is situated above the
hottest zone i.e. the oxidation zone. A critical requirement is that tars
and other pyrolysis products should pass the hottest zone and not escape
via the reduction zone into the product gas. In the reduction zone oxygen
is absent and due to the endothermic reactions (2) and (3) the temperature
is relatively low. Therefore the reaction time for the gasphase is not
sufficient for tar conversion in this zone. It was found that for tarfree
operation a double vortex induced by the incoming airflow should fit
into the throat geometry. With this in mind and knowing the time required for complete pyrolyses of the particles it is possible to understand to some extent the empirical rules for throat design found in the

litterature (13)

The product gas composition and the temperature can be estimated by using simple heat and mass balances, the estimated methane formation (mainly in pyrolyses zone (34)) and assumptions on heat loss and "equilibrium" temperature (see e.g. Schläpfer model (35)). Although the concept of equilibrium cannot strictly be applied to this process, these simple models give good results once fitted for a specific unit. Groeneveld et al (33, 13) have made an approximate kinitic/transport model for the processes occuring in the reduction zone. From this model the relationship between reactor volume conversion, particle size, carbon conversion profile in the particle, solidsflow, gascomposition and temperature can be understood (see fig. 7). This model can be used as and additional guide for design and gives some background information about the limits of the simple "thermodynamic" models.

A very critical design problem concerns the scaling up of the throat without increasing tar production. A possible solution might be the use of a cylindrical anular throat (34). The problem might also be solved by recycling of pyrolyses gas over the top of the gasifier, possibly via an external combustion chamber to the airinlet. However this increases the complexity of the system. Such systems have been in operation for some years (1, 12) and is now being used in designs of "Moteur Duvant" (32). An additional advantage of this system could be a lowering of the highest oxidation zone temperature and thus reducing the risk of ash fusion. However the exact describtion of the highest solid temperature in the oxidation zone specially in relation to pyrolyses gas recycle needs more investigation.

Although the future for the small scale units appears bright, new demands concerning automation, safty, flexibillity and environmental protection have been imposed on these units as compared to former applications. In designing modern units we must consider the following: a. automation of solid preparation and feeding, start up, stand by, turn down, control of dual fuel ratio and speed control. b. problem free heatexchangers and residual tar soot and ash removal. c. safety with respect to CO and to the danger of explosions of gas/air mixtures. d. evironmentally acceptable solutions for the condenswater containing compounds such as sufides, cyanides ash, traces of organic compounds (phenolic compounds, etc.).

Another important factor is the selling/buying policies of electricity companies which will influence the economics of all decentralized electricity generation units. Furthermore feed preparation steps may be too

expensive in some cases. (drying, pelletizing). Drying should be integrated
as far as possible with exhaust heat recovery and for extremely wet feedstocks compression drying or other special upgrading processes may be required. Because feedpreparation is so costly (7) more complicated techniques (fluid bed reactors in combination with new type heat exchangers)
may gain preference over the simple moving bed system. The application area
can be extended to much larger capacities if intergration with gas turbines
can be realized. Also small scale units (30 - 50 kW_t) might become economically attractive for induvidual energy provision for farms, homes and for
villages in developing countries. In a Dutch-Tanzanian development
cooperation program (12, 36) 40 kW_t units running on maize cob spills and
used for maize mulling in the villages are being tested for the technical
economical and social vialibility of this concept.

On the long term even small scale production of pure H_2 via shifting
and separation could become attractive depending on possible developments
in e.g. fuel cell and hydrogen storage technology. Generally the production
of syntheses gas from L.J.V. gas deserves more attention in research programs.

References

(1) Meunier, J. Vergasung fester Brenstoffen Verlag Chemie GmbH, Weinheim (1962).

(2) Peters, W., Jüntgen, H. and Van Heek, K.H. Chemical Engineering in a Changing World. Elsevier Amsterdam (1976) 285-299.

(3) Dravo Corpor; Handbook of gasifiers and treatment systems; NTIS, FE 1772-11 Feb. (1976).

(4) Jackson, F.R. Energy from Solid Waste Noyes data Corp. New Yersey U.S.A. (1974) 106-120.

(5) Feldman H.F. Energy from Biomass and Wastes. I.G.T. symp. Washington D.C. (1978) 537-557.

(6) Gallagher, J.E. and Marshall, H.A. Coal Process Techn. V (1979) 199-204.

(7) Jones, D. and Jones, J. Energy from Biomass and Wastes. I.G.T. symp. Florida (1980) 223-249.

(8) Pollaert, T.J. C.E.P. aug. (1978) 95-98.

(9) Hahn, O.J. Fuel Process. Techn. 2. (1979) 1-16.

(10) Davidson, P.E. and Lucas, T.W. A.S.C. symp. series 76 (1978) (ISSN0097-6156).

(11) Aubo, J.C. and Glaser, D.P. Energy from Biomass and Wastes IV. I.G.T. symp. Florida (1980) 387-401.

(12) Groeneveld, M.J. and Van Swaaij, W.P.M. Applied Energy 5 (1979) 165-179.

(13) Groeneveld, M.J. Thesis Twente University of Technology (1979)

(14) Hos, J.J., Groeneveld, M.J., Van Swaaij, W.P.M. Energy from Biomass and Wastes IV I.G.T. symp. Florida (1980) 333-351.

(15) Franke, F.H. V.G.B. Kraftwerktechnik 59, H.9 sep. (1979) 697-702.

(16) Keairns, D.L. AIChE Symp. Series No. 156 vol 72 (1977) 103-116.

(17) Beck, S.R. and Wang, M.J. Ind. Eng. Chem. P.D.D. 19 No. 2 (1980) 312-317.

(18) Vogt, E.V. and V.d. Burgt, M.J. C.E.P. March (1980) 65-72.

(19) Cornils, B. Chem. Ing. Tek. 52 No. 1 (1980) 12-19.

(20) Levy. S.J. Proc. CRE conf. (1975) Montreux, Swiss (1975) 226-231.

(21) Kohl, L. C.E.P. Aug. (1978) 73-79.

(22) La Rosa, P.J. Symp. Clean Fuels from Coal. Coal. I.G.T. Chicago (1973).

(23) McCallister, R.A. and Ashley, G.C. Proc. Am. Power Conference, 36 (1974) 292-299.

(24) Engstrom, F. and Ahlström, O. Energy from Biomass and Wastes IV I.G.T. symp. Florida (1980) 555-566.

(25) Cheremisinoff, P.N. and Morresi, A.C. Energy from Solid Wastes; M Dekker Inc. New York (1976) 168-177.

(26) Mark, S.D. Energy from Biomass and Wastes IV I.G.T. symp. Florida (1980) 577-589.

495

(27) Helmstetter, A.J. and Sussman, B.D.
Proc. National Waste Processing Conf.
(1978) Chicago U.S.A. May (1978) 465-476.
(28) Bailie, R.C., Energy from Biomass and
Waste IV I.G.T. symp. Florida (1980)
423-443
(29) Reed, T.B. Paper presented at React 78
Biomass Energy Inst. Winnipeg, Canada
October 3-4 (1978).
(30) Overend, O. proc. "Hardware for Energy
generation in forest products ind.
seattle Washington. Feb. (1979) 107-119.
(31) "Imbert Energietechnik" 5760 Arnsberg 2
Steinweg 11 F.D.R.
(32) "Moteur Duvant" BP 599
59308 Valenciennes Frances
(33) Groeneveld, M.J. and Van Swaaij, W.P.M.
Chem. Eng. Sci. 35 (1980) 307-314.
(34) Groeneveld, M.J. and Van Swaaij, W.P.M.
A.S.C. symp. Thermal conversion of
Biomass and Wastes, Washington D.C. (1979)
(35) Schläpfer, P. Tobler, J. Theoretische
und praktisch untersuchungen über den
Betrieb van Motorfahrzeugen mit Holzgas,
Bern (1937).
(36) Stassen, H.E.M. Twente University of
Technology, Research report (1980).

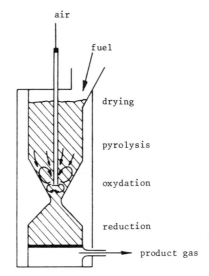

fig. 6.
Typical Co-current moving bed
gasifier.

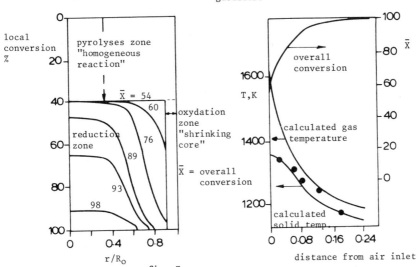

fig. 7.
Co-current moving bed gasifier

typical local conversion inside
a wood particle of initial radius
$R_0 = 12,4 * 10^{-3}$ m (Groeneveld (13)).

typical profiles
• observed by thermocouple
readings

THE COMBUSTION, PYROLYSIS, GASIFICATION AND LIQUEFACTION OF BIOMASS

T. B. Reed

The Solar Energy Research Institute (SERI), Golden, Colorado

Summary

We can provide all the products now obtained from oil by thermal conversion of the solid fuels biomass and coal. As a feedstock, biomass has many advantages over coal and has the potential to supply up to 20% of U.S. energy by the year 2000 and significant amounts of energy for other countries. However, it is imperative that in producing biomass for energy we practice careful land use.

Combustion is the simplest method of producing heat from biomass, using either the traditional fixed-bed combustion on a grate or the fluidized-bed and suspended combustion techniques now being developed. Pyrolysis of biomass is a particularly attractive process if all three products-- gas, wood tars, and charcoal--can be used. Gasification of biomass with air is perhaps the most flexible and best-developed process for conversion of biomass to fuel today, yielding a low energy gas that can be burned in existing gas/oil boilers or in engines. Oxygen gasification yields a gas with higher energy content that can be used in pipelines or to fire turbines. In addition, this gas can be used for producing methanol, ammonia, or gasoline by indirect liquefaction. Fast pyrolysis of biomass produces a gas rich in ethylene that can be used to make alcohols or gasoline. Finally, treatment of biomass with high pressure hydrogen can yield liquid fuels through direct liquefaction.

1. COAL VS. BIOMASS - THE PAST AND FUTURE ROLE OF SOLID FUELS

Until a little over a century ago, mankind relied on the solid fuels-- biomass (mostly wood) and coal--to supply space and industrial heat, energy for transportation, and power. With the discovery of plentiful, convenient, low cost oil and gas, the use of solid fuels greatly diminished. Now that shrinking gas and oil reserves necessitate a return to the traditional solid fuels, several new factors which favor the use of biomass must be considered in choosing between biomass and coal for various uses.

Wood was the principal fuel of mankind until about 1800, when many nations began to use coal for a variety of reasons. In some cases (England) the wood supply was exhausted by overharvesting. In other countries (the United States) coal was plentiful and could be obtained more cheaply than wood. Coal also has the advantage of being a denser fuel than wood, and it can be stored and transported more conveniently.

Now that we are turning back to solid fuels (reluctantly), this past experience suggests that where possible we will choose coal rather than biomass. However, a number of technical and social factors have changed and can tilt this decision toward wood and other forms of biomass. Coal is now less desirable because we are less willing to tolerate the despolia- tion of the land associated with mining. We are concerned about contamina- tion of our air and water by the sulfur in coal. We also recognize the possibility that continued use of fossil fuels may upset the CO_2 balance of the atmosphere and our climate. Finally, the race of men willing to toil and moil in the bowels of the earth for pennies a day has disappeared dur- ing the era of low cost oil energy.

Biomass, on the other hand, can be obtained on a renewable basis with actual improvement of the land. The techniques for planting and harvesting biomass for fuel are similar to those used for agriculture and forestry; biomass can be produced continuously on a renewable basis as we now produce food, lumber, and paper. Biomass production can be a source of revenues for land improvement. Table I lists six areas where biomass production has resulted in the improvement of the environment.

In the past biomass production has often resulted in land destruction, and Table II is a cautionary list of examples of the destruction caused by unregulated use of biomass for food, lumber, paper, or fuel production. Thus, biomass is a two-edged sword for mankind, and we must learn to produce it only on a renewable basis (1,2).

TABLE I.   EXAMPLES OF LAND RESURRECTION

REGION	DATE BEGUN	ACTIVITY	RESULT
Jordan, Syria, Iraq	1970	Dam construction, irrigation, sprinkling	Agriculture being reestablished
Israel	1900	Drainage, irrigation, terracing	Productive farms, orchards and forests
China	1950	7% reforestation, shelterbelts halt "sanddragon"	Decreased silting, dust storms reduced, improved forestry
England, Holland	1300	Dyking, pumping	Productive farmland
France (Provence)	1910	Reforestation	Forestry, farming, industry, towns
United States (Kans., Nebr., Colo., Okla.)	1935	Contour plowing, crop rotation, windbreaks	Productive farmland

Biomass is a desirable fuel ecologically because it contains very little sulfur and generally much less ash than coal. With the new agricultural and forestry machines now available, its production is no longer as labor intensive as it was a century ago and the work is certainly less hazardous than mining.

In the past, the principal form of biomass burned for energy was wood because of its relatively high density and availability. The process of densification, in which biomass residues (sawdust, straw, food processing wastes) are compressed into pellets, briquettes, or logs, promises to make other biomass forms, now wasted, equally attractive (3). These processes produce a fuel that has approximately three-fourths of the energy of coal, both on a mass and volume basis, and thus densified biomass could be called "instant coal". A number of plants are now operating in the United States to produce densified biomass.

Thus, though coal will no doubt be the staple solid fuel in many countries, recent technical and social developments favor biomass, and ultimately, when the supply of coal is depleted, mankind will have to learn to

TABLE II.   EXAMPLES OF LAND DESTRUCTION

REGION	DATES	ACTIVITY	RESULT
Mesopotamia (Iraq, Syria)	6000 BC to 1200 AD	Canal irrigation and intensive agriculture	Silting and salting - 20 M population at peak, now 3.5 M.
Israel	3000 BC to 100 AD	Agriculture, terracing, grazing	Soil loss through erosion, overgrazing, created hardpan
Phoenecia	4000 BC to 400 AD	Lumbering	Deforestation, erosion
China	3000 BC to 1950 AD	Deforestation, intensive agriculture	Silting, flooding, dust storms
North Africa (Tunis, Algeria)	0 to 600 AD	Corn, wheat, olive production	Erosion, desertification
England	1500 AD to 1700 AD	Lumbering, charcoal manufacturing	Hardpan formation leading to moors, few forests
France (Provence)	1800 AD to 1900 AD	Charcoal manufacture to support steel industry	Hardpan formation, high moors
United States (Kans., Nebr., Colo., Okla.)	1900 AD to 1934 AD	Wheat farming	Dust bowl, wind erosion, soil depletion, overgrazing

live on its "energy income" because its "energy capital" will be exhausted. A recent analysis of possible energy futures for Sweden suggests that she could rely solely on renewable energy sources, primarily various forms of biomass, and maintain her current standard of living (4). A recent evaluation of U.S. biomass resources suggests that 12 to 17 x 10^{15} kJ (12-17 quads) could be obtained from biomass by the year 2000 (2). The potential for biomass production in many other countries is also enormous.

2. COMBUSTION OF BIOMASS

In the past, energy has been produced from wood by simple combustion to produce heat. Combustion now provides about 1.5×10^{15} kJ (1.5 quads) of energy in the United States, primarily from waste liquors from the Kraft paper process, but increasingly from combustion of wood wastes and in home stoves. This combustion is most familiar in the form of a fixed bed of fuel on a grate (or andirons), but we are increasingly turning to fluidized-bed combustion or suspended combustion, which permits cleaner, more automatic combustion in a smaller volume and allows the use of fuel particles as small as sawdust (5). Figure 1 shows a variety of such combustion devices.

3. PYROLYSIS OF BIOMASS

The combustion of wood produces only heat directly (and power indirectly through steam). However, pyrolysis and gasification make possible the production of more refined fuels, which can be used to produce power directly in engines and turbines or for the synthesis of the liquid fuels such as methanol or gasoline on which we have become so dependent.

A familiar experiment in the chemistry laboratory is the heating of a wood splint in a test tube to pyrolyze (rearrange by heating) the wood. A clean combustible gas issues from the mouth of the tube, a brown oil (pyrolysis oil) collects near the cool mouth of the tube, and charcoal remains in the lower portion. In situations where all of these products are needed, pyrolysis processes such as this, on a larger scale, are highly recommended because they are very simple and have a high thruput. A number of pyrolytic processes are being developed (6). Unfortunately, the oil that is produced is corrosive and presents handling problems. Yet ultimately, when we learn how to separate the valuable components of this oil, we may look on it as the most valuable product obtainable from biomass.

The amount of charcoal produced by pyrolysis depends strongly on the rate of heating. If the heat is supplied fast enough, little or no char results, greatly simplifying the subsequent processing. "Fast pyrolysis" processes are discussed in Section 5.

4. GASIFICATION OF BIOMASS

Gasification is the process of treating biomass so that only gas (and sometimes oil vapors) are produced. The charcoal resulting from pyrolysis is converted to gas by partial combustion with oxygen or steam:

501

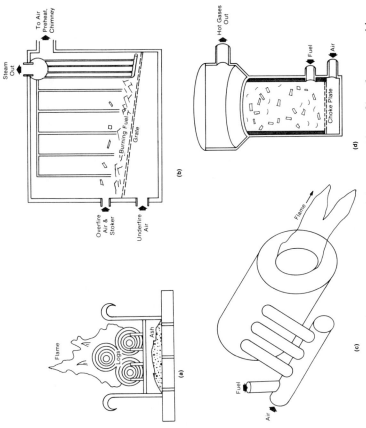

Figure 1. Biomass Combustion Devices. (a) The familiar fireplace provides unlimited air access and poor regulation, control, and heat transfer. (b) Large commercial biomass combustors provide accurate control of air flow and efficient heat transfer to steampipes. Small particles (e.g., sawdust) can be burned in (c) suspension or (d) fluidized-bed combustion units.

$$C + \tfrac{1}{2} O_2 \longrightarrow CO$$

$$C + H_2O \longrightarrow CO + H_2$$

This can be accomplished in the updraft, downdraft, fluidized-bed, or suspension gasifiers shown schematically in Figure 2 (6).

In an updraft gasifier for biomass or coal, air or oxygen is injected under a grate supporting charcoal causing combustion and reduction of the gases. The resulting hot gases then rise through the incoming biomass at the top of the shaft furnace of Figure 2, producing oils and water by pyrolysis and drying. The resulting gas must be burned directly (close coupled) because the tars are difficult to remove. Updraft gasifiers are especially appropriate for retrofitting existing gas- or oil-fired boilers. There are now several dozen manufacturers of updraft and downdraft gasifiers in the United States (7).

In the downdraft gasifier air or oxygen is injected above the char mass, causing pyrolysis of the incoming biomass and producing char and oils. These oils then pass over the hot char and are cracked to gases; as a result, very little oil is produced. For this reason, downdraft gasifiers are particularly suitable for running internal combustion engines and over a million were used in Europe during World War II (8).

Although not as well developed, fluidized beds for biomass gasification offer a number of theoretical advantages over the fixed-bed gasifiers discussed above. Because of their very high recirculation rates, fluidized beds have a high heat transfer rate and high thruputs. They are also able to process a wide range of biomass sizes. However, because the contact time is short, they are not as efficient as downdraft gasifiers in consuming char or cracking the oils and tars. There is also a tendency for the light biomass fraction and lighter char fractions to separate from the bed prematurely. Even less well developed is suspended gasification, though again the suspension gasifier offers high thruputs and the processing of small particles as potential advantages.

When air is used for biomass gasification, a low energy gas [typically 5200 kJ/nm^3 (150 Btu/scf)] results because of the nitrogen dilution. Though suitable for operating engines or close coupling to boilers, it is not economical to distribute this gas in pipelines. When oxygen is used for biomass gasification a "medium energy gas" [typically 10,400 kJ/nm^3 (300 Btu/

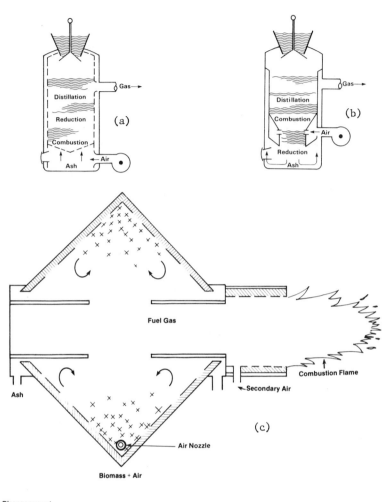

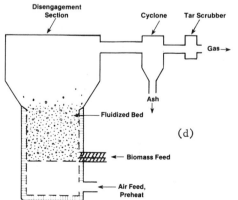

Figure 2. Schematic Diagrams of (a) Updraft Gasifier; (b) Down-draft Gasifier; (c) Suspended-Fuel Gasifier; and (d) Fluidized-Bed Gasifier.

scf)] is produced that can be distributed in pipelines and can be used to power turbines or to synthesize chemicals. We are presently testing the high pressure, downdraft gasifier shown in Figure 3, using oxygen to produce a very clean gas composed of only hydrogen and carbon monoxide.

5. INDIRECT LIQUEFACTION

In many situations gaseous fuels are desirable because they can be distributed by pipeline and are most easily burned with minimum emissions and maximum efficiency; however, gaseous fuels are difficult to store. We will probably require liquid fuels at least for transportation fuels and possibly (depending on cost) for turbine and power plant operation in isolated locations where pipeline costs make gas transport too costly.

Methanol (wood alcohol) was originally made by the destructive distillation of wood, but yields were typically several percent and the major product was charcoal and other chemicals. Now the following reaction is used to make methanol from synthesis gas:

$$CO + 2 H_2 \longrightarrow CH_3OH$$

The synthesis gas is compressed to 50-200 atm and passed over a chromium or copper oxide catalyst at 250°-350°C. Synthesis gas is likely to be the key industrial chemical of the post-oil era because, once obtained by gasification of biomass (or coal), it can be used to make methanol, ammonia, gasoline, and many other chemicals and fuels. The gasifier shown in Figure 3 is being developed at SERI especially for the production of fuels and chemicals from biomass.

Other important gases for chemical synthesis are ethylene and other olefins (primarily propylene and butylene). The slow pyrolysis discussed in Section 3 produces oxygenated tars and methane gas. However, it has recently been recognized (7,8,9) that sufficiently high heat transfer rates (typically in excess of 100 W/cm^2) induce fast pyrolysis, producing high yields of olefins and often no charcoal at all. Such a process was initiated at the Naval Weapons Center, China Lake, and now is being developed at the Solar Energy Research Institute. High yields of olefins are produced from biomass (typically over 14%). These olefins can in turn be made into plastics (by polymerization), gasoline (by thermal polymerization), or alcohol fuels (by hydration).

505

SERI Oxygen Biomass Gasifier

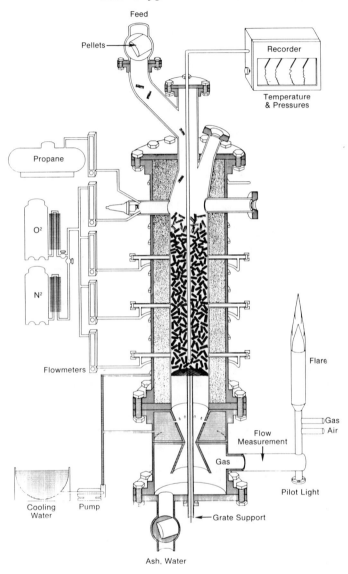

Figure 3. SERI Oxygen Biomass Gasifier

6. DIRECT LIQUEFACTION

It is desirable to be able to convert biomass directly to a liquid fuel, bypassing the gaseous intermediate. In work undertaken during World War II it was demonstrated that alcohols and hydrocarbons could be produced in significant yields by the direct, high pressure hydrogenation of wood or lignin over a catalyst (see Table III)(10). Work along these lines is being pursued now in Canada (11,12). Up to 50 (weight)% of biomass has been converted to oxygenated oils using nickel carbonate catalysts.

TABLE III. EARLY WOOD LIQUEFACTION RESULTS

PROCESS	FEEDSTOCK	PRODUCTS
Hydrogenation	Wood	5% methanol, 15% propanol, 40% pulp
Hydrogenation	Lignin	8% methanol, 13% alcohols, 23% hydrocarbons

More recently, an attempt has been made to apply the methods developed for coal liquefaction to biomass in the "Albany, Oregon" biomass liquefaction process (13). In this process, 30 wt% of biomass is slurried with anthracene oil and heated to 370^0C in high pressure (160 atm.) hydrogen or carbon monoxide. A number of difficulties have plagued this pilot plant because the oil does not allow high concentrations of wood to be injected into the reactor. An aqueous pretreatment has permitted higher concentrations of more reactive wood to be used in the reactor and several barrels of oil have been produced with a relatively low oxygen content (13).

Thus it can be seen that the thermal conversion of biomass can follow the various routes outlined here to provide the heat, power, fuels, and chemicals currently produced from petroleum.

REFERENCES

1. Reed, T. 1978. "Biomass Energy: A Two-Edged Sword." Paper presented at the International Solar Energy Society (ISES); Denver, CO; 28 Aug. 1978. SERI DDS-011. Golden, CO: Solar Energy Research Institute. T.E. Bull, Project Director. 1980.

2. Energy from Biological Processes. Office of Technology Assessment.

3. Reed, T.; Bryant, B. 1978. Densified Biomass: A New Form of Solid Fuel. SERI-35. Golden, CO: Solar Energy Research Institute. 1980.

4. Lonnroth, M.; Johansson, T.B.; Steen, P. "Sweden Beyond Oil: Nuclear Commitments and Solar Options." Science. Vol. 208: p. 557.

5. See, for example: (a) O'Grady, M.J. 1979. "Grate, Pile, Suspension and Fluid Bed Burning." Wood Fuel for Small Industrial Energy Users. Conference Proceedings; North Carolina State University, Raleigh, NC; 2-3 Oct. 1979. (b) Zerbe, J.I. 1979. "Wood Processing, Forestry and Agricultural Wastes." New Fuels and Advances in Combustion Technologies. Symposium sponsored by the Institute of Gas Technology; New Orleans, LA; 26-30 Mar. 1979. (c) Bio-Energy Council. Bio-Energy Directory. Washington, DC. pp. 515-526.

6. Reed, T. 1979, 1980. A Survey of Biomass Gasification. SERI/TR-33-239. 3 Vols. Golden, CO: Solar Energy Research Institute.

7. Solar Energy Research Institute. 1979. Retrofit '79: Proceedings of a Workshop on Air Gasification. Seattle, WA; 2 Feb. 1979. SERI/TP-49-183. Golden, CO: Solar Energy Research Institute. 1979.

8. Solar Energy Research Institute. 1979. Generator Gas: The Swedish Experience from 1939-1945. SERI/SP-33-140. Translation of Gengas, published by the Swedish Academy of Engineering; 1950. Golden, CO: Solar Energy Research Institute.

9. See, for example: (a) Diebold, J. 1980. Research into the Pyrolysis of Pure Cellulose, Lignin, and Birch Wood Flour in the China Lake Entrained-Flow Reactor. SERI/TR-332-586. Golden, CO: Solar Energy Research Institute. (b) Solar Energy Research Institute. Forthcoming. Specialists' Workshop on Fast Pyrolysis of Biomass. Copper Mountain, CO; 20-22 Oct. 1980. Golden, CO: Solar Energy Research Institute.

10. Stern, A.J.; Haris, E.E. 1953. The Chemical Processing of Wood. New York: Chemical Publishing Co.

11. FMC Inc. Engineering Analysis. 1979. "A Study on the Design and Optimization of Biomass Liquefaction Processing Units." 2 Vols. For Project C-69, Dept. of Canadian Forestry Service.

12. Boocock, D.G.B.; Mackay, D. 1980. "The Procession of Liquid Hydrocarbons by Wet Hydrogenation." Energy From Biomass and Wastes.

IGT Symposium; Lake Buena Vista, FL; 21 Jan. 1980. p. 765. See also Boocock, D.G.B.; Mackay, D.; Franco, H. 1980. "The Direct Liquefaction of Wood Using Nickel Catalysts." Thermal Conversion of Solid Wastes and Biomass. Jones and Radding, eds. ACS Symposium Series #130. Washington, DC: American Chemical Society. p. 363.

13. See, for example: (1) Ergun, S. "An Overview of Biomass Liquefaction" and following papers in 3rd Annual Biomass Energy Systems Conference Proceedings. 5-7 June 1979. SERI/TP-33-285. Golden, CO: Solar Energy Research Institute. (2) Ergun, S.; Schaleger, L.; Seth, M. 1980. "Albany Biomass-to-Oil Project." Design and Management for Resource Recovery. Vol. I. T.C. Frankiewicz, ed. Ann Arbor, MI: Ann Arbor Science.

ENERGY FROM STRAW AND WOODWASTE

A. Strehler

Bayerische Landesanstalt für Landtechnik, TU München

(Technical University of Munich)

D 8050 Freising

Summary

Since 1974 research has been carried out in Weihenstephan on the production of energy from cereal straw. First the fuel straw was characterised by determining its calorific value, taking into consideration moisture content, storage conditions, species, varieties, growth conditions and fertilisation. Other characteristics of the straw - such as percentage of volatiles and chemical elements, demand for combustion air and specific fuel gas volume - were determined.
Measurements were made with serial-furnaces (mainly the through-burning type). These furnaces were developed in order to increase efficiency and decrease emission. Automatic stoking systems had to be tested. Test runs were also carried out on prototype furnaces with both under-burner systems and hand stoking. A special under-burning furnace with a low capacity automatic charging system (for heating a domestic dwelling) had been developed in Weihenstephan.
Different types of air heater, using both high pressure bales and roto bales, were constructed. Economic calculations show that straw is a cheap fuel under certain conditions. An EEC study shows the energy demand and energy production potential of straw and woodwaste in various rural areas.

1. INTRODUCTION

Heating oil has become very expensive in the last few years and its availability is uncertain. In the agricultural sector there is the possibility
of using straw and woodwaste to produce energy. The characteristics of
straw as a fuel were investigated. Subsequently, the first serial-furnaces
were tested and a number of furnaces constructed in order to give farmers
the opportunity of using their home-produced fuel straw with high efficiency at low cost.

This work, which began in 1974, was financed by the BMFT (Bundesministerium
für Forschung und Technologie or Federal Ministry for Research and Technology) and the EEC. The most important results are outlined in this paper.

2. MEASUREMENTS AND RESULTS WITH DIFFERENT FURNACES:

2.1 Through-burning boilers:

Fifteen factories are meanwhile constructing through-burning boilers. These
consist of a vertical cylindrical combustion chamber which can accomodate
between 2 and 8 small bales or long woodpieces up to a diameter of 30 cm.
The chamber has a length of 0,8 - 1,5 m and a diameter of 0,5 - 1,2 m.
Charging is carried out manually through the big fire door at the front
of the furnace. About 50 000 of these furnaces have already been sold,
mainly for the combustion of big sized woodpieces. Only about 10 % of
these furnaces are used for straw in form of high pressure bales.

Measurements on these plants revealed that their combustion efficiency is
not yet satisfactory when using straw and wet wood. In this case emission
is rising up from 300 to 1500 mg dust and tar in 1 cubicmeter smoke. In
Germany, for example, the emission has to be below 300 mg/m^3; other countries don't mind about high emission in straw and wood combustion in rural
areas. In order to diminish emission we developed a special combustion
chamber which can be fitted in the usual chamber afterwards (fig. 1).

A second cylinder leads to higher temperatures in the primary combustion
chamber, additional tubes as secondary combustion chambers lead to lower
emission and higher efficiency.

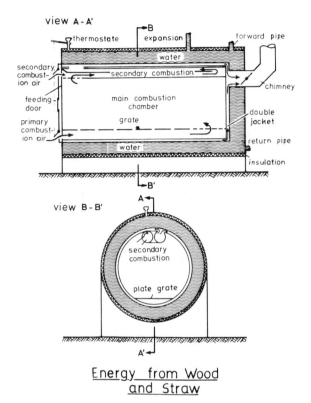

throughburning boiler for high pressure bales
with double jacket and secondary combustion

view A-A'

thermostate expansion forward pipe

water

secondary combustion air secondary combustion chimney

feeding door main combustion chamber

primary combustion air grate double jacket

water return pipe

insulation

view B-B'

secondary combustion

plate grate

Energy from Wood
and Straw

Fig. 1

2.2 Underburner furnaces

Tests on various prototypes with underburner system and secondary combustion chamber were carried out. In this type of furnace only the lower part of fuel is in combustion. The combustion quality of this system is satisfactory, even when using straw. But there are some problems to solve with regard to selection high temperature resistant material and the optimal way of charging.

High power furnaces (over 300 kW) for roll bales from PSW, a German manu-
facturer, are operating satisfactory from the emission (below 300 mg/m³)
and efficiency (over 73 %) points of view. The prices are, however, very
high at about 30 000 DM for one bale furnace with front loader charging
and 300 kW and at 60 000,- DM for a furnace with two bales and crane charging with a heating power of 500 kW (fig. 2).

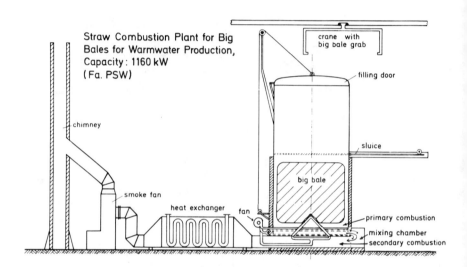

Straw Combustion Plant for Big
Bales for Warmwater Production,
Capacity : 1160 kW
(Fa. PSW)

Energy from Wood and Straw

Fig. 2 :   Roto bale furnace with crane charging 500 kW PSW

2.3 Airheaters

A low-price solution for drying purposes was built in Weihenstephan with
charging by front loader (fig. 3). The heating power is 500 kW. Power
regulation by hand is satisfactory. Ash is conveyed by two angers in a
dustproof box. The movable air-cooled grate gives a high power even at the
end of combustion.

Meanwhile, even a hand charged airheater for drying purposes using high
pressure straw bales and big woodpieces is produced by a German factory.

The price is 12 000 DM by a heating capacity with big sized woodpieces. A under-burning plant is just in construction; it shall bring less emission in straw combustion.

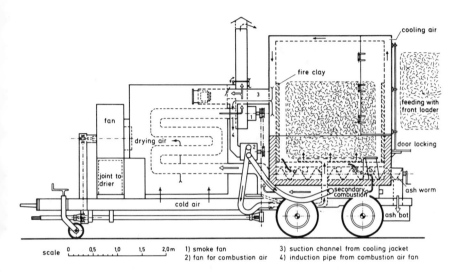

scale

1) smoke fan
2) fan for combustion air
3) suction channel from cooling jacket
4) induction pipe from combustion air fan

Energy from Straw
warm air producer with big bale furnace

Fig. 3 : Movable heater with a one bale furnace

The manually stoked furnaces take 10 - 30 minutes work-time a day. Continuous stoking systems were constructed to reduce this work load. Such plants have been built first in Denmark by Passat (fig. 4).

2.4 Continuous stoking systems

Automatically stoked furnaces with high heating power are used a long time for the combustion of saw-dust and shaving. Some furnace systems of this type were adapted to the use of chopped straw as fuel. There are some problems in plants with blowing systems. The charging by angers seems to be the better way.

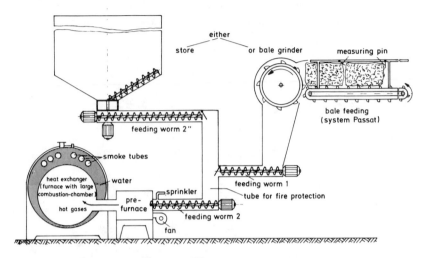

Energy from Straw and Wood

Fig. 4 : Straw furnace with automatic feeding system - Passat , ∅ mm, DK

On the right a chain conveyor of any length automatically supplies the bale grinder with small bales. Chopped straw is stoked by angers, automatically regulated by the heating power required. The combustion quality is satis- factory. Meanwhile 5 other producers build furnaces with automatic charging systems for high pressure bales.

From Sweden came furnaces with automatic charging systems for wood-chips, which can be fitted to oil boilers after putting away the oil burner.

The efficiency in combustion is satisfactory. The price of 6000,- DM for one day fuel container, conveyor, combustion chamber and electrical equip- ment with a heating power of 40 kW is very low. No change of chimney and in the heat distribution system is neccessary. The collection of wood- chips is nearly solved, from cutting to drying, storage and conveying.

There was no cheap and satisfactory low power furnace with automatic char- ging for straw and wood-chips on the market. Therefore Weihenstephan tried to build such a furnace. It was constructed as an under-burner type with

feedback of a part of the smoke in order to reduce the combustion temperature in the main combustion chamber.

Construction work on 5 air heaters for drying plants was continued. Plans
were made for the possible connection of straw boilers, oil boilers and solar collectors with heat stores and various consumers of heat such as domestic dwellings, stables, workshops and driers.

3. ECONOMY

Some calculations were carried out concerning the costs of heating domestic
dwellings. Fuel oil has become very expensive recently. Therefore 2000,- DM
can be saved each year by using straw as fuel oil, when the conditions on
the farm are appropriate. (Fig. 5)

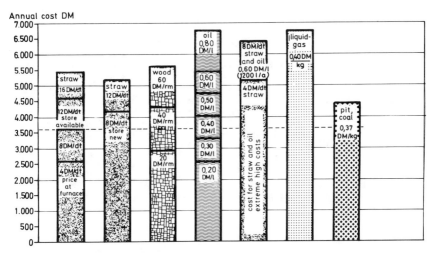

Comparison of annual costs for domestic dwelling
heating using different fuels
(Costs for fuel, labour, capital; 7·3 t/a fuel oil)

Fig. 5 : Comparison of annual costs for domestic dwelling heating using
different fuels under consideration of different straw-prieces,
including costs for fuel, labour, capital; 7 t fuel oil/year

Our work for the next years should be done mainly on:

    Point 1   Building up further pilot plants
    Point 2   Research on wood and straw gasification for power supply.

FARM STRAW AS FUEL FOR POWER

M.W. Thring and R.J. Crookes

Department of Mechanical Engineering

Queen Mary College (University of London)

Summary

The most feasible use of solar energy in a temperate climate is through the production of high pressure steam by the combustion of the biomass by-products of food production. Cereal straw burnt in the fields of Britain every year is equivalent to well over 10^6 tonnes of fuel oil and causes fire hazards and pollution. The straw must be stored and burnt within a few km of the fields. Balers already in regular use on farms produce bales of density 110 kg/m^3 but this density could be doubled by modifications to existing machines. Downjet System developed in the forties by M W Thring for coke could be applied to give regular efficient combustion of straw bales with wide flexibility. This could be used in farming areas to burn all the baled straw in:

1.   (100 - 500 kW) Local electricity generation
2.   Cooperative production of N-fertiliser
3.   Mobile power production (tractors)

1.   THE USE OF SOLAR ENERGY IN TEMPERATE CLIMATES

In temperate climates there is very frequently light cloud which makes a focussing solar heater useless.   For example we are developing at Queen Mary College a mechanical-power 'tower' type of solar heater for use in the tropics and we are very rarely able to test it in this country.   Flat solar heaters can produce reasonable amounts of hot water in the summer and one of the authors (MWT) has installed a 5m^2 commercial one in Suffolk but the summer water heating saved each year represents only a few per cent of the capital cost.   It is probable that this will not become a common method of saving energy in Britain until -

a)   some fiscal encouragement such as low interest loans are available.

and b)   a well designed flat solar heater with large water storage capacity is combined with a heat pump for winter whole-house heating.   Ultimately this will be combined also with a windmill to operate the heat pump since the wind is most available in winter.

The other use of flat plate solar heaters which may become economical as fossil fuel costs rise is for heating air for grain drying and preheating it before combustion for grass drying.   A 20 m^2 solar air heater for drying 40 t of grain in a bin on his farm is planned by M W Thring.   This will require about 60 h of sunshine to remove the 0.8 t of water corresponding to 2% of moisture in the grain.   Solar grass dryers would also be desirable if they could be produced but the firing rates used in drying correspond to something like a hectare of sunshine collecting area.   This is why hay has been produced for thousands of years while the much more efficient use of land for growing frequently cropped grass has been available only while the fossil fuels used for drying have been cheap.

The most likely use of solar energy in a country like Britain is to use the sunshine to grow fuel crops.

Clearly the growth of crops purely as a fuel will be totally uneconomic as long as coal is available at a reasonable price, both because of the cost of handling the crops and because of the absorption of land area which could otherwise be used for food production.   There is one exception to this which is already being used in the country and this is coppicing or pollarding broad leaved trees to produce regular crops of firewood.   Possibly also the extended collection of seaweed, mainly as an economical raw material could be used but the total tonnages are relatively small. The most important use of solar energy in Britain is likely thus to be the production of a combustible fuel as a by-product of food production. The crops of wheat in this country have been increased (primarily by the use of chemical fertilisers (PKN) and sprays) from 2.5 t/ha to occasionally more than 10 t/ha now. This is, of course, at the expense of a considerable consumption of fossil fuel, primarily oil, especially for the fixation of nitrogen from the atmosphere, but also for manufacturing and operating machines. The amount of straw produced has remained fairly constant at about 2.5 t/ha because one of the main ways of increasing the yield has been to develop wheats with a short straw and carrying more grains to each straw. Barley straw is at present baled because it can be used either directly or after treatment with alkali to break down the cellulose as a ruminant feed. In the past days of mixed farming, farmers used their wheat straw as animal bedding so that it could be fed back to the land enriched with nitrogen and partially rotted. The development of modern farming machinery has led to mono-cropping or at best to a rotation of cereals. On his farm M W Thring grows wheat, barley and oil seed rape. Wheat straw is, wherever possible, burned on the field in the row as it comes out behind the combine, for two reasons: one is that the combustion helps to destroy pests in the ground, particularly those which build up when the same crop is grown year after year (this is especially successful if most of the stubble is burned) and secondly because it is generally accepted among farmers that if one breaks up and ploughs in the straw one will require more nitrogen the following year to break down the cellulose. For example in an 8 ha field on which the burning of straw can only take place when the wind blows from the East, if the straw cannot be burned it costs an additional £300 to break up and plough in the straw and add the extra nitrogen. The cost of burning the straw on this field is only a few pounds. The rape straw similarly has to be burned. However, the burnt straw has the very serious disadvantage to the farmer that he has to take precautions to ensure he does not cause damage to the surroundings and, of course, is a very unpleasant and often dangerous form of atmospheric pollution. It is much more convenient to get rid of a field of straw (as with barley straw) by baling; thus we have a very cheap fuel already available as are the machines for baling it into conventional bales.

In the book 'Cereal Straw' by A R Staniforth (1) the figure is given that in 1972 there was a yield of 9.5 Mt of cereal straw and that 36.6% of this was burned.   This corresponds therefore to 3.48 Mt of straw which could be baled and used as a fuel for power.   The amount burnt is probably greater now.

2.   BALED STRAW AS A FUEL

When it is first harvested the straw contains about the same moisture content as the crop, that is 16% and its gross calorific value is 15 MJ/kg. This is almost exactly one-third of that of fuel oil so that the 3.5 Mt of straw which was available in 1972 is equivalent to 1.15 Mt of fuel oil or 5.2 x 10^{16} J.   The cost of baling straw is given by Staniforth (2) at £7-10/t if done by contractor and the costs of carting and stacking would increase the total to £12-15/t. If one multiplies these figures by three to make them equivalent to oil one still obtains a price very much lower than that of oil but, of course, bales are a very bulky fuel and it is expensive to carry them any great distance from the farm.   We must therefore look for means of burning it within a few miles of the farm where it is produced.

There are, of course, already on the market a number of straw burning domestic heating boilers, mostly made in Denmark and several hundred have been installed in this country.   However, these are concerned with burning perhaps three bales in 24 hours, a bale weighs some 18 kg and is thus equivalent in total heating capacity to 6 kg of oil.   An 8 ha field of wheat will produce some 20 t of straw or 1100 bales.   If one farm uses three bales a day for six months it will consume 500 or 600 bales.   Even a small cereal farm of 40 ha will produce ten times as many bales as is needed for the farmhouse boiler.

The standard small bale has approximate dimensions 0.36 m x 0.46 m x 1.0 m and weighs approximately 18 kg, with an approximate density of 110 kg/m^3.   Conventional bales can be adapted to give bales of up to twice this density.   If they are to be handled manually then the best way is to make them shorter to keep the weight about the same.   There has been research done on making brickettes or cubes from straw which would make it as convenient as a conventional solid fuel but it is found that the power required is a high proportion of the total energy of the fuel.   The fuel we have to consider therefore is a series of rectangular bales weighing 18 kg and with dimensions the same as those given above except that the 1 m dimension might be reduced to 0.5 m with higher compaction.   It has been suggested that the rise in the cost of oil would increase the cost of baling but in fact fuel costs are a very small fraction of the total baling cost.

One other factor which must be mentioned is the cost of storage. It
is necessary to cover the bales in the stack otherwise they absorb too much
moisture. Staniforth gives the cost of storing in the range £2.5 - 6/t
but this would be roughly halved if the density could be doubled.

3. DOWNJET COMBUSTION

One of the authors of this paper developed a method of combustion in
the 1940s (3) in which the combustion of a bed of non-caking coal or coke
could be adjusted to give almost exactly complete combustion, without excess
air, and without using separately controlled secondary air. In this downjet combustion the whole of the combustion air was blown into a surface of
the fuel bed through a nozzle of variable aperture so that the velocity of
the air jet could be varied while the quantity was kept constant. The air
entered the fuel bed through a free surface and the combustion gases left
the long side through the same free surface after describing a horseshoe
path in the fuel bed. By adjusting the velocity of the air the depth of
penetration of the gases in the fuel bed could be varied and thus the contact between gases and the fuel adjusted to the optimum, at which there was
neither excess free oxygen on the one hand nor unburned carbon monoxide or
volatiles on the other. The fuel had to be fed in such a way that this
free surface remained approximately constant in relation to the jet of air.
It is also possible with this method of combustion to vary the rate of air
supply over a wide range by keeping the velocity constant and reducing the
area of the nozzle, or varying the number of nozzles used. In the present
research it is possible to apply this method of combustion to the burning
of baled straw. The bales could either be pushed horizontally against a
barrier of one or two water cooled tubes between which the combustion zones
would be arranged or the bales could be fed by gravity down a 45° slope.
In the former case the combustion surface would be vertical and in the
latter it would overhang by 45°. When bales of straw are burned in
conventional fashion (Staniforth (2)) the loose outer layers burn rapidly
leaving a dense mass and it is difficult to control the air flow to give
efficient combustion. Once the necessary characteristics of a downjet
system have been developed this would enable the bale of straw to burn
steadily through like a candle and the volatiles will be burned at the same
time as the residual charcoal. Research would be required to determine
the optimum ratio of turndown, positions of the water cooled stop tubes,

air velocity, distance of nozzle and so on.

Another problem which requires research is the behaviour of the ash in the straw which corresponds to about 6% and is mainly silica. Again a conventional combustion tends to form a clinker which blocks the grate whereas in the downjet system it can be removed continuously as has been successfully done in coke burning. The best mechanism for this would have to be developed but a small continuous belt grate could be used to draw it out and dump it in a well.

4. STATIONARY POWER

As has been said it would be to the advantage of farmers to find an economic use for wheat straw which would cover the costs of baling, storing and carting it and there is far too much to be burned for domestic heating. There is therefore a strong case for having a small local power station for a village or a group of farms producing a few 100 kW of electricity. In this case the bales would have to be fed automatically to the boiler with a minimum of human handling as a 200 kW(e)plant would be consuming 8-9 bales per hour (30%). Clearly no one will do this at the moment but as the costs of fuel rise and the pressure against field burning becomes more strenuous this will become a worthwhile development.

5. NITROGEN FIXATION

To fix 1 kg of nitrogen from the atmosphere requires about 67×10^6 J in terms of the actual fossil fuel that has to be extracted from the ground (4). To produce 7.5 t/ha wheat requires 0.5 t/ha of 34% nitrogen fertiliser so the energy required to fix this is 11.5×10^9 J/ha. This hectare of wheat produces 2.5 t of straw with a fuel calorific value of 40×10^9 J/ha, thus the straw from this field could be burned to produce over three times as much nitrogen as is required for the field. Moreover, it is true that the chemical companies in the developed countries are finding it increasingly uneconomic to fix nitrogen for farmers owing to the high cost of fuel required. Here again therefore there will be a very strong case in the not too distant future for farmers to form co-operatives and set up their own factory in which bales of straw are burned to produce steam for power to run nitrogen fixation plant. In this way high output farming can become independent of the rising cost of fuels.

6. MOBILE POWER

There is a long history of tractors run on straw in the old days of steam tractors but, of course, the development of the diesel tractor running on liquid fuel has enabled one man to plough or do other operations over an enormously increased area every day. Tractors up to 75 kW are now being used but if diesel fuel becomes really unavailable it would probably be quite sufficient to have tractors of 25 or 30 kW, in which case two or three bales per hour, depending on the thermal efficiency, could be operated by having a horizontal shelf with a reservoir capacity of three or four bales. It will only be necessary to load these on once every hour or so from a trailer and the whole system becomes very much more convenient than returning to horses which is the other way of operating ploughs and machinery off the land without the elaboration of gas producers or producing liquid fuel from the crops. It would be necessary to have a condenser and use high pressure superheated steam to obtain a thermal efficiency (net CV of fuel to shaft power) of the order of 15%. A steam engine with variable cut off is much more suitable for traction than a diesel engine.

Alex Moulton has built a compact modern 3 cylinder single acting 15 kW steam engine weighing 45 kg to operate from a once through boiler producing steam at over 60 bar and running on small coal. Such a boiler could readily be rearranged to take the combustion gases from a downjet straw combustion chamber as it consists of a single tube which can be appropriately costed.

7. REFERENCES

(1) A.R. Staniforth, 'Cereal Straw', Oxford Science Publication, 1979.
(2) A.R. Staniforth, Power Farming, September 1979, p.16.
(3) M.W. Thring, Transactions, Faraday Society, No.286, Vol.XLII, March 1966, p.16.
(4) M. Slessor, The Energy Requirements of Agriculture, 'Food, Agriculture and the Environment', Blackie, 1975.

PYROLYSIS AND COMBUSTION OF WOOD IN RELATION

WITH ITS CHEMICAL COMPOSITION

H. MELLOTTEE and J.R. RICHARD

C.N.R.S., Centre de Recherches sur la Chimie de la Combustion

et des Hautes Températures,

Orléans, France

and

B. MONTIES

I.N.R.A., Laboratoire de Chimie Biologique et de Photophysiologie,

Grignon, France

Summary

Pyrolysis and combustion of wood has been investigated by thermo-gravimetric analysis (TGA) and by thermodifferential analysis (TDA) in view to correlate the pyrolysis kinetic and thermal data to the physico-chemical characteristics of wood. Different parts of the wood such as the bark, the sapwood and the core of different species (e.g. poplar, pine, oak and beech) were compared. The study was also extended to the different constituents comprising wood : lignin, cellulose, holocellulose and parietal residues.

The behaviour of the different samples which were studied (on the form of sawdust or fine powder) was characterized by two heating rates (10 and 20°C/min) and under four different atmospheres ranging from air to pure nitrogen. One of the main results obtained points to distinguish two stages in the degradation process, one at temperatures ranging between 180 and 350°C and another one at temperatures higher than 400°C. In the first stage, the oxygen content in the oxidizer does not affect percepti-bly the behaviour whereas the influence of oxygen becomes significant at high temperature.

The differences which were observed on the various samples studied should contribute to making a better choice in the application of the py-rolysis of biomass materials as an alternative source of energy.

1. INTRODUCTION

La valorisation énergétique de la biomasse passe par le développement des études sur l'augmentation des quantités disponibles et mobilisables de bois et sur le comportement en combustion et pyrolyse des bois dans des conditions appropriées à la valorisation recherchée. Dans ce domaine, il importe en particulier de caractériser la nature et la quantité des gaz dégagés au cours des traitements thermiques, la cinétique des dégradations, les effets thermiques. C'est à ces deux derniers aspects qu'est consacrée la présente étude.

Les corrélations entre cinétique de dégradation, effets thermiques et propriétés physico-chimiques du bois ont été recherchées. Les échantillons étudiés ont été non seulement diverses espèces, telles que peuplier, sapin, chêne et hêtre, mais aussi des fractions telles que l'écorce,l'aubier ou le coeur. En vue de permettre l'interprétation des résultats, le rôle de la constitution chimique des échantillons de bois peut être évoqué. Aussi les mêmes études cinétiques et thermiques ont-elles été appliquées aux fractions constitutives principales des bois, lignines (extraites au dioxanne), cellulose, holocellulose, résidu pariétal.

2. PARTIE EXPERIMENTALE

Tous les essais ont été réalisés dans une thermobalance,équipée d'un analyseur thermique différentiel simultané. La montée en température est programmée, à la vitesse de 10 ou 20°C/mn. Quatre compositions de l'atmosphère ont été utilisées : air, azote, mélanges N_2/O_2 à 5% ou 10% d'oxygène. Le système permet l'analyse des gaz émanents, mais les résultats obtenus ne sont pas rapportés ici.

Les échantillons ont été employés sous forme de sciures pulvérulentes de 0,5 mm de diamètre environ, ou dans le cas des lignines, cellulose, hollocellulose, résidus pariétaux de poudres fines de granulométrie moyenne 20 _ 50 µm. Les bois n'ont pas subi de traitement préalable particulier avant leur emploi.

3. RESULTATS

Les résultats donnés ci-dessous concernent le pourcentage de perte de masse des échantillons, les effets thermiques et les paramètres cinétiques de la dégradation en fonction de la température.Quelques exemples significatifs de l'influence des divers paramètres expérimentaux sont proposés.

Dans tous les cas, les bois, fractions ou constituants des bois s'oxydent en deux étapes. Une première étape se situe vers 180 - 350°C. A la fin de cette étape, la perte de masse est comprise, selon les cas, entre 50 et 70% et s'élève pour la cellulose à plus de 80% de la masse initiale. Une seconde étape se situe au dessus de 400°C. Si la dégradation a lieu sous azote, seule la première étape apparaît. Les paramètres cinétiques de cette première étape sont voisins, quel que soit le gaz utilisé. Il s'agit donc d'une étape de pyrolyse essentiellement. L'étape de haute température, plus sensible à la nature du gaz de traitement, est une étape d'oxydation.

3.1. Influence du gaz de traitement

Les courbes ci-dessous montrent cette influence dans le cas de l'aubier de sapin, à la vitesse de montée en température 10°C/mn.

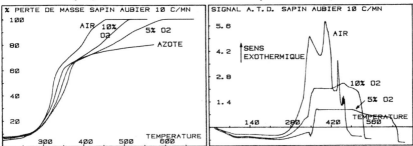

L'appauvrissement en oxygène retarde la dégradation, surtout l'étape d'oxydation, et reporte vers les hautes températures l'effet thermique, dilaté dans l'échelle des températures et donc moins intense. Les maximums de vitesse des deux étapes sont aussi déplacés vers les hautes températures (basse température : 325° pour l'air, 365° pour l'azote; haute température: 406° pour l'air, 507° pour 5% d'oxygène).

Paramètres cinétiques d'ordre 1	Air		10 % O_2		5 % O_2		Azote
	B.T.	H.T.	B.T.	H.T.	B.T.	H.T.	B.T.
Energies d'activation kcal/mol	30,6	34,5	23,3	23,8	21,3	18,3	20,9
log A (A en s^{-1})	9,1	8,9	8,1	4,7	7,1	2,5	6,8

3.2. Influence du prélèvement sur une même espèce (coeur, aubier, écorce)

Les courbes ci-dessous montrent une telle influence dans le cas du sapin, sous air, à la vitesse de montée en température 10°C/mn.

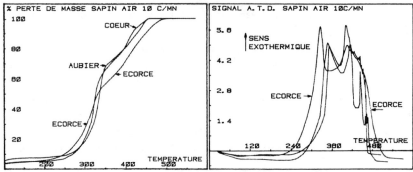

Aubier et coeur se comportent sensiblement de la même façon, tandis que l'écorce se distingue en se dégradant à plus basse température et plus lentement à haute température. Les maximums de vitesse sont situés aux mêmes températures pour l'écorce et l'aubier (323°), décalés vers les hautes températures pour le coeur (337°).

Paramètres cinétiques d'ordre 1	Coeur		Aubier		Ecorce	
	B.T.	H.T.	B.T.	H.T.	B.T.	H.T.
Energies d'activation kcal/mol	31,6	28,8	30,6	34,5	20,8	13,2
log A (A en s^{-1})	9,2	6,8	9,1	8,9	5,3	1,7

L'écart entre l'écorce et les autres fractions du bois apparaît nettement sur le tableau des paramètres cinétiques.

3.3. Influence de l'espèce du bois (étude sur l'aubier à 10°C/mn)

Sous air, les résultats sur les quatre espèces étudiées (chêne, sapin, hêtre, peuplier) montrent une grande similitude entre le hêtre et peuplier, un comportement voisin du sapin, un comportement différent du chêne (à basse température, sa dégradation commence plus tôt et est plus rapide jusque vers 320°, plus lente ensuite).

Paramètres cinétiques d'ordre 1	Hêtre		Peuplier		Sapin		Chêne	
	B.T.	H.T.	B.T.	H.T.	B.T.	H.T.	B.T.	H.T.
Energies d'activation kcal/mol	26,9	31,5	26,9	26,5	30,6	34,5	30,3	36,4
log A (A en s^{-1})	7,7	8,1	7,7	6,4	9,1	8,9	9,3	9,5

Le chêne a effectivement des énergies d'activation élevées, en particulier à haute température.

3.4. Influence des parties constitutives du bois

Une comparaison entre les dégradations de l'aubier de peuplier et de la cellulose sous air à 10°C/mn montre la différence des comportements. On constatera des différences encore plus grandes entre cellulose, lignine, holocellulose et résidu pariétal extraits du peuplier.

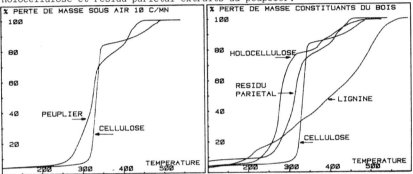

Les mécanismes de basse température sont simples et conduisent à des grandeurs cinétiques du premier ordre nettes, sauf pour la lignine. Les mécanismes de haute température sont généralement plus complexes et sont une succession et superposition de plusieurs réactions. Seule la réaction de haute température de la lignine apparaît avoir un mécanisme simple.Pour les autres corps étudiés, la cinétique du 1er ordre n'est, à haute température, qu'une représentation approchée.

Paramètres cinétiques d'ordre 1	Lignine		Cellulose		Holocellulose		Rés. pariétal	
	B.T.	H.T.	B.T.	H.T.	B.T.	H.T.	B.T.	H.T.
Energies d'activation kcal/mol	4,9	22,1	93,6	33,5	25,9	20,5	32,2	36,3
log A (A en s^{-1})	0,1	2,8	19,5	6,4	9,0	3,5	11,0	10,5

4. DISCUSSION ET CONCLUSION

Il existe de nombreuses méthodes permettant d'obtenir les paramètres cinétiques. Dans ce travail, ils ont été calculés à partir de l'équation $dm/dt = -km$ où m est la masse actuelle et k est de la forme $A.\exp(-E/RT)$. La figure ci-dessous donne un exemple de l'exploitation cinétique des résultats, où le logarithme de $(1/m)(dm/dt)$ est exprimé en fonction de $1/T$ (diagramme d'Arrhénius). Il s'agit dans le cas présent de l'aubier de sapin à différentes compositions de l'atmosphère de traitement.

528

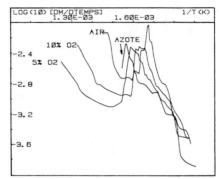

On aura remarqué les écarts sensibles entre les deux mécanismes de basse et haute température, celui-ci correspondant à un phénomène d'oxydation.

L'étude a mis en évidence le très grand rôle de la composition de l'atmosphère de traitement, le comportement particulier de l'écorce par rapport à l'aubier, les différences sensibles selon l'espèce du bois. Elle a montré en outre les écarts importants de comportement entre le bois et ses constituants, cellulose et lignine en particulier.

Deux orientations se dégagent pour l'avenir. Le couplage des mesures d'ATG et ATD avec les analyses de gaz doit compléter les critères de choix de matériaux les plus aptes à une valorisation de la biomasse. On peut en outre espérer modéliser le comportement des bois en les considérant comme des associations en proportions variables de leurs constituants principaux.

ON THE INFLUENCE OF THE DIFFERENT PARTS OF WOOD ON THE PRODUCTION

OF GASEOUS COMBUSTIBLE PRODUCTS BY PYROLYSIS

J.R. RICHARD and M. CATHONNET

Centre de Recherches sur la Chimie de la Combustion
et des Hautes Températures
C.N.R.S. - 45045 ORLEANS CEDEX - France

The aim of the present work has been to determine the optimum conditions leading to gaseous combustible mixtures by pyrolysis of different varieties of wood and its different parts.

The following parameters were especially studied
- Wood sample : cellulose, fir tree, poplar and beech - Differents parts of the wood, i.e. sapwood, bark.
- Temperature : ranging from 400°C to 900°C.
- Pyrolysis method :
    - Slow, at atmospheric pressure,
    - Rapid, at low pressure.

The main results point to the following :
A - Overall yield in gaseous products
    1. The percentage of solid residue is lower in the case of sapwood compared to bark. Cellulose leads to an even smaller percentage of residue. Concurrently gaseous products yields are higher for sapwood than for barks.

    2. At higher temperature the percentage of solid residue decreases and the yield in gaseous product increases.

    3. Higher gas phase yields are favoured by rapid pyrolysis.

B - In so far as the distribution of gaseous species is concerned (CO, CO_2, H_2, CH_4, C_2H_4, C_2H_6) the following is noted :
    1. Whatever type of pyrolysis is used, sapwood produces more CO than bark wood. Yields in C_2H_4 and C_2H_6 are the same.

    Low pyrolysis at 700°C yields more CH_4 in the case of sapwood than in that of bark wood.
    Rapid pyrolysis shows that sapwood produces more hydrogen and less CO_2 than bark wood.

    2. All yields tend to increase with temperature except in rapid pyrolysis for that of CO_2.

    3. Rapid pyrolysis increases the yield of CO but decreases those of CO_2 and CH_4.

In conclusion the production of gaseous species is favoured by an increase in temperature. Their distribution can be controlled by the pyrolysis regime. In all cases sapwood and especially poplar sap appears to be the best generator of combustible gaseous mixtures.

I - <u>INTRODUCTION</u>

Le présent travail vise à déterminer les conditions expérimentales de la pyrolyse de matériaux cellulosiques susceptibles de conduire à un mélange gazeux combustible.

On considère généralement que durant une pyrolyse, les matériaux cellulosiques se comportent de façon à peu près identiques quelle que soit l'origine du produit.

Cette étude a pour but de rechercher les différences qualitatives et quantitatives éventuelles des produits obtenus par pyrolyse de différentes variétés de bois et de ses diverses parties constitutives.

Les méthodes expérimentales utilisées ont permis de déterminer :

1. <u>la stabilité thermique des différents échantillons</u> ;

2. <u>la nature des produits gazeux obtenus</u> et leur proportion respective.

Les paramètres suivants ont été étudiés.

. nature de l'échantillon: cellulose, sapin, hêtre, peuplier
. température variant de 400 à 900°C
. méthode de pyrolyse : lente à pression atmosphérique
           rapide à basse pression.

II - <u>METHODES EXPERIMENTALES</u>

Diverses techniques expérimentales ont été mises au point pour déterminer la résistance à la dégradation thermique d'une part et la nature et la proportion des espèces gaseuses libérées au cours de la pyrolyse d'autre part.

A) <u>Analyse thermogravimétrique isotherme</u>

Il s'agit de porter l'échantillon dans un délai minimum de la température ambiante à la température choisie dans une atmosphère contrôlée.

Un four mobile, maintenu en permanence à une température choisie, recouvre une nacelle porte-échantillon reliée à une balance monoplateau. L'évolution de la perte de masse en fonction des paramètres imposés est suivie en continu.

B) <u>Analyse de la phase vapeur</u>

    a - <u>Pyrolyse lente à pression atmosphérique</u>

Il s'agit, après avoir déterminé le bilan massique de chacune des fraction (résidu, phase liquide, phase gazeuse) après pyrolyse, d'analyser la nature et la composition de la phase gazeuse libérée.

Une nacelle porte-échantillon placée au bas d'un four vertical est portée à une température choisie, les produits pyrolysés sont entraînés par un écoulement de gaz et traversent dans la partie supérieure du four une zone de température isotherme où ils subissent une post-pyrolyse sur environ 50 cm.

La composition de la phase gazeuse est déterminée au moyen d'une sonde de prélèvement mobile dans le four. Les échantillons sont analysés par chromatographie en phase gazeuse.

Les fragments de bois utilisés en pyrolyse lente ont une taille moyenne de 0,5 mm. Le poids de l'échantillon est de 2g. La vitesse de montée en température de l'échantillon est voisine de 250°C/mn. Les produits gazeux de pyrolyse sont entraînés dans un courant d'azote ayant une vitesse voisine de 1cm/s.

b - Pyrolyse rapide à basse pression

Un tube en quartz fermé à une extrémité est placé dans un four et relié à un ballon par l'intermédiaire d'un piège à azote liquide ; le vide ayant été fait dans l'appareil, l'échantillon est porté à la température désirée. Les produits de pyrolyse sont recueillies dans le piège et dans le ballon. Après réchauffage du piège les produits gazeux sont analysés par Chromatographie en Phase Gazeuse.

La masse des échantillons utilisés, en pyrolyse rapide est de 100 mg. Ils sont placés dans une zone du four maintenu à 500°C environ, les produits de pyrolyse traversent une zone à 800°C ou 900°C sous une pression de 5 à 20 torrs avant d'être recueillis.

III - RESULTATS

A) Analyse thermogravimétrique isotherme

Par analyse thermogravimétrique nous avons déterminé les pourcentages de résidu recueilli après pyrolyse.

On note que lorsque la température augmente les pourcentages de résidu diminuent.

Ainsi, le pourcentage de résidu pour l'aubier de peuplier passe de 30% à 400°C à 15% à 700°C.

Dans tous les cas étudiés, les écorces sont plus stables que les au-

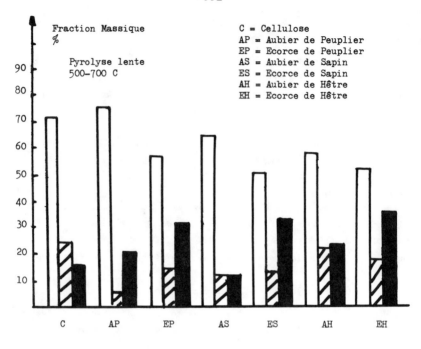

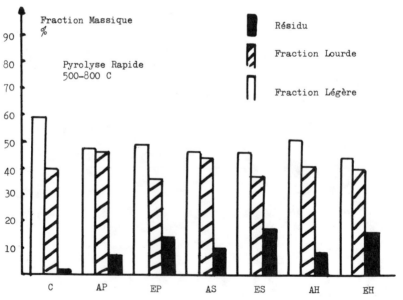

biers. Les expériences effectuées avec la cellulose conduisent à une stabilité plus faible que n'importe quelle autre partie de bois.

B) Répartition des produits de pyrolyse lente

Pour chaque échantillon et pour des températures de post-pyrolyse variant de 450 à 700°C nous avons caractérisé les trois phases.

1) fraction légère ; 2) phase condensable ; 3) résidu de pyrolyse

Comme l'indique les schémas précédents, les produits gazeux légers
sont plus abondants pour les aubiers que pour les écorces en pyrolyse lente.
La différence est moins marquée pour aubiers et écorces dans le cas de la
pyrolyse rapide où les produits lourds sont nettement plus abondants.

Dans tous les cas les pourcentages de résidu sont plus faibles en pyrolyse rapide.

C) Analyse de la phase gazeuse

L'analyse par chromatographie de la phase légère permet de déterminer
l'évolution de chacune des espèces analysées, CO_2, H_2O, CO, H_2, CH_4, C_2H_4
C_3H_6, C_3H_8, CH_3OH, CH_3CHO.

Une méthode de calcul a été mise au point ; elle permet de déterminer
la fraction massique de chaque espèce à tout instant de la pyrolyse connais-
sant les fractions volumiques de chaque espèce recueillie.

Masse de produits gazeux libérés
pour 100 g. de produit initial

pyrolyse lente

	Peuplier				Sapin				Hêtre				Cellulose	
	aubier		écorce		aubier		écorce		aubier		écorce			
T(c)	500	700	500	700	500	700	500	700	500	700	500	700	500	700
CH_4	1,05	7,85	1,20	4,14	0,95	5,70	0,75	5,45	1,00	6,40	1,05	5,30	0,35	4,00
CO	10,90	26,70	5,20	15,90	10,75	19,90	6,50	14,60	9,25	19,80	6,40	13,50	11,30	23,00
CO_2	27,20	35,15	25,70	27,50	26,50	34,70	15,50	31,30	27,00	28,50	19,00	28,50	21,50	28,70
C_2	1,12	4,10	1,10	3,15	1,15	3,05	1,25	3,30	1,10	2,80	0,75	2,70	0,60	3,60
H_2	0,15	0,65	0,05	1,00	0,15	0,60	0,10	2,05	0,15	0,60	0,15	1,75	----	0,40

A partir des tableaux et schémas, on peut faire les remarques
suivantes :

| | Peuplier | | | | Sapin | | | | Hêtre | | | | cellulose | |
	aubier		écorce		aubier		écorce		aubier		écorce			
$T(c)$	800	900	800	900	800	900	800	900	800	900	800	900	800	900
CH_4	2,50	3,60	2,40	3,10	2,30	3,40	2,20	3,40	2,50	3,90	2,50	2,60	1,90	4,40
CO	31,00	39,50	23,00	25,40	27,20	35,80	21,20	28,20	29,50	39,50	21,80	22,80	34,00	54,70
CO_2	6,40	4,30	14,00	8,10	6,20	4,30	9,50	7,60	5,50	4,80	10,50	6,80	2,60	2,20
C_2	3,80	4,70	5,30	5,60	3,20	7,40	3,90	5,30	3,40	4,70	3,80	4,20	3,30	4,80
H_2	0,40	0,70	0,34	0,50	0,35	0,60	0,30	0,50	0,40	0,50	0,30	0,70	0,60	1,10

pyrolyse rapide

Pour les deux méthodes, la température de pyrolyse de l'échantillon est de 500°C. On constate que le résidu solide est plus faible en pyrolyse rapide qu'en pyrolyse lente. Les rendements en produits gazeux augmentent avec la température de traitement des produits pyrolysés. Dans la fraction pyrolysée, les produits gazeux légers sont plus importants en pyrolyse lente qu'en pyrolyse rapide bien que la température de traitement de cette fraction soit moins élevée. Ceci s'explique par un temps de séjour beaucoup plus long. Quelle que soit la méthode de pyrolyse utilisée, l'aubier produit plus de CO que les écorces. Les fractions de C_2H_4 et C_2H_6 sont très comparables. En pyrolyse lente le CH_4 est plus abondant pour l'aubier que pour l'écorce. En pyrolyse rapide, l'aubier produit plus d'hydrogène et moins de CO_2 que l'écorce. Les rendements en CO augmentent alors que ceux de CO_2 et C_2H_4 diminuent.

IV - CONCLUSION

Les résultats obtenus ont mis en évidence les différences de comportement de chaque échantillon sous l'action d'une élévation de température en fonction de la méthode de pyrolyse étudiée et des parties où ont été prélevées ces échantillons. On a noté que la production d'espèces gazeuses, dans la gamme de températures utilisées, est favorisée par une élévation de température et leur distribution peut être contrôlée par le régime de pyrolyse.

Dans tous les cas l'aubier et plus particulièrement l'aubier de peuplier apparaît comme le meilleur générateur de mélange gazeux combustible.

THE PYROLYSIS OF TROPICAL WOODS : THE INFLUENCE

OF THEIR CHEMICAL COMPOSITION ON THE END PRODUCTS

by G. PETROFF and J. DOAT

Centre Technique Forestier Tropical

Summary

Six samples of wood were pyrolysed at the laboratory scale :
- Three tropical mixtures consisting of woods :
  . rich in lignin
  . rich in extracts
  . rich in carbohydrates
- Two reafforestation species :
  . Gmelina arborea
  . Eucalyptus
- A mixture of French hardwoods as a reference

It was observed that tropical woods gave slightly less pyroligneous products than temperate zone woods. Furthermore, the products obtained seem linked with the chemical composition of the wood :
  . a high lignin content is reflected in a higher charcoal yield.
  . a high percentage of pentosanes gives more acetic acid and acetates.
  . a high cellulose content results in more methanol.
  . woods containing large amounts of alcohol benzene extractives may cause tarry deposits in the circuits.

From a pratical point of view, it seems possible, without too great a risk, to replace temperate hardwoods by tropical species to supply an industrial pyrolysis line.

1. <u>INTRODUCTION</u>.

On connaît mal les conditions d'emploi énergétique des bois tropicaux dont la nature et les constituants diffèrent quelque peu de ceux des bois tempérés. Aussi, pour parvenir à une exploitation plus rationnelle de ces essences, on a procédé dans un premier stade de recherche, à des essais de carbonisation en laboratoire sur des bois ou mélanges de bois.

2. <u>CONDUITE DE LA RECHERCHE</u>.

21 <u>Matières premières utilisées</u>.

On s'est approvisionné en échantillons de bois tropicaux offrant des caractéristiques aussi variées que possible. L'échantillonnage était constitué :
- d'Eucalyptus saligna x tereticornis (essence de reboisement à croissance rapide):(E)
- de Gmelina arborea (essence de reboisement à croissance rapide):(G)
- d'un mélange de bois tropicaux riches en lignine:(T1)
- d'un mélange de bois tropicaux riches en extrait à l'alcool benzène :(T2)
- d'un mélange de bois tropicaux pauvres en lignine et en extrait à l'alcool benzène (c'est-à-dire riches en cellulose et hemicelluloses):(T3)
- d'un mélange de feuillus français : Bouleau, Peuplier, Hêtre, Chêne à titre de référence:(F.F.)

On a donc au total testé six échantillonnages de bois de composition et provenances diverses.

22 <u>Appareillage et mode opératoire</u>.

La pyrolyse des bois a été effectuée en cornues placées dans des fours électriques régulés automatiquement. Les résultats de carbonisations correspondent à la moyenne de plusieurs essais (3 à 7 selon le cas).

Les carbonisations ont été conduites à 500° pendant 4h, avec une montée en palier de 4h soit environ 2 degrés à la minute. Les produits dégagés ont été refroidis dans des réfrigérants à eau et les condensats recueillis dans des flacons barboteurs refroidis à la glace. Les produits gazeux non condensables ont été récupérés dans des ampoules ou des ballons de plastique selon les quantités. Les volumes ont été mesurés à l'aide d'un compteur à gaz.

23 <u>Essais effectués et résultats obtenus</u>.

<u>Analyse des bois</u>.

On a procédé tout d'abord à l'analyse physicochimique des bois.

On a déterminé la densité, le pouvoir calorifique supérieur, l'extrait alcool benzène, l'extrait à l'eau, la lignine, les pentosanes, la cellulose, les cendres, la silice. Les résultats sont donnés au tableau I.

Carbonisation des bois.

Les bois ont été fractionnés en plaquettes de quelques centimètres sur quelques millimètres puis pyrolysés comme indiqué précédemment. Les rendements en charbon, en pyroligneux et en gaz sont donnés au tableau II.

Analyse des charbons.

Les pourcentages de matières volatiles, cendres, phosphore, soufre et carbone fixe sont donnés au tableau III.

Analyse des pyroligneux.

Une analyse des pyroligneux a été effectuée par chromatographie gazeuse pour chacun des six échantillons. Les résultats sont donnés au tableau IV.

Analyse des gaz.

Les gaz recueillis au cours des tests définitifs de pyrolyse ont été analysés. On a utilisé l'appareil d'Orsat pour le dosage du gaz carbonique, de l'oxygène, de l'oxyde de carbone et de l'hydrogène formés. Les hydrocarbures (méthane, éthane, éthylène et autres composés de C_3 à C_5) ont été dosés par chromatographie gazeuse. Les résultats sont donnés au tableau V.

3. ANALYSE DES RESULTATS.

31 Conduite de la carbonisation.

Le processus de carbonisation avec un palier à 500° n'a pas mis en évidence de difficultés particulières d'emploi des bois tropicaux par rapport aux feuillus français. Toutefois, dans le cas du mélange T2 riche en extraits et, dans une moindre mesure, dans le cas du Gmelina, des dépôts plus ou moins importants sont apparus sur les tuyaux d'évacuation du pyroligneux et des gaz. Dans la pratique industrielle, un encrassement des circuits serait donc possible au cours du traitement de certains bois tropicaux riches en produits extractibles.

32 Rendements à la carbonisation.

Pour les bois ou mélanges de bois tropicaux, le rendement moyen en charbon de bois varie de 30,6 à 36,6 % alors que le mélange de feuillus français donne 30,1 % de charbon.

Les quantités moyennes de pyroligneux obtenues à partir des essences tropicales vont de 42,9 à 50,6 %; les feuillus tempérés ont donné 53,1 % de pyroligneux. Il semble donc que dans l'ensemble, les feuillus tempérés donnent moins de charbon et plus de produits condensables que les bois tropicaux.

En ce qui concerne les gaz formés, on ne note pas de grandes différences entre les bois. On a environ 15 litres de gaz totaux pour 100 g de bois sec, à l'exception du Gmelina qui en a fourni 17,7 litres.

On a remarqué que ce sont les bois riches en carbohydrates qui donnent le moins de charbon et le plus de pyroligneux. A l'inverse, les bois ou mélanges de bois contenant beaucoup de lignine et d'extraits ont un meilleur rendement en charbon. Le pyroligneux varie aussi comme les teneurs en pentosanes des bois.

Tableau n° I : Caractéristiques physicochimiques des bois

	E Euca- lyptus	F.F. Feuillus Français	G Gmelina	T1 Mélange riche en lignine	T2 Mélange riche en Ext.A.B.	T3 Mélange pauvre en lignine + Ext. A.B.
Caractéristiques physiques						
Densité	0,68	0,57	0,36	0,75	0,73	0,51
p.c.s.	4820	4740	4875	4800	5015	4590
Caractéristiques chimiques (% bois sec)						
Ext. A.B.	1,5	2,15	2,95	0,95	16,6	2,5
Ext. eau	0,9	1,2	0,4	2,85	0,5	2,8
Lignine	31,6	22,7	25,05	35,3	26,55	23,5
Cellulose	44,5	42,05	48,15	42,8	37,0	40,7
Pentosanes	16,5	24,9	16,15	13,5	13,7	22,0
Cendres totales	0,3	0,5	0,9	1,0	0,95	2,7
SiO_2	0,018	0,007	0,05	0,26	0,009	0,01
Lignine + Ext. A.B.	33,1	24,85	28,0	36,25	43,15	26,0
Cellulose + Pentosanes	61,0	66,95	64,65	56,3	50,7	62,7

Tableau n° II : Résultats moyens de carbonisation

	E	F.F.	G	T1	T2	T3
Rendement en charbon (1)	32,4	30,1	30,6	33,4	36,6	32,9
Rendement en pyroligneux (1)	49,7	53,1	50,1	45,5	42,9	50,6
Gaz totaux (2)	14,7	15,1	17,7	15,4	14,9	15,6

(1) : en % du bois sec - (2) : litres % sur bois sec

Tableau n° III : Analyse physicochimique des charbons

	E	F.F.	G	T1	T2	T3
Densité charbon	0,27	0,22	0,16	0,33	0,29	0,26
p.c.s.	8170	8070	8185	8000	7840	7710
Matières volatiles (1)	12,95	14,0	12,9	13,1	16,0	11,8
Cendres (1)	1,05	1,45	2,85	2,9	2,35	6,6
Phosphore (1)	0,105	0,031	0,034	0,128	0,009	0,016
Soufre (1)	0,014	0,024	0,024	0,035	0,020	0,063
Carbone fixe (1)	86,0	84,55	84,25	84,0	81,65	81,6

(1) : En % du charbon sec

Tableau n° IV : Analyse du pyroligneux (% du bois sec)

	E Euca- lyptus	F.F. Feuillus Français	G Gmelina	T1 Mélange riche en lignine	T2 Mélange riche en Ext.A.B.	T3 Mélange pauvre en lignine + Ext. A.B.
Pyroligneux total	50,5	52,1	47,4	46,5	44,8	51,2
Produits déposés	1,9	0,5	7,2	0,9	12,0	2,4
Jus de pyro-ligneux :						
Phase aqueuse	38,8	42,3	33,9	35,1	30,9	40,7
Phase organique	8,3	6,8	5,2	8,7	1,7	5,6
Non récupérés	1,5	2,5	1,1	1,8	0,2	2,5

Principaux constituants (dans phase aqueuse et organique) % bois sec

	E	F.F.	G	T1	T2	T3
Méthanol	1,84	1,40	1,94	1,45	1,02	1,43
Ethanol + acide formique	0,10	0,09	0,04	0,02	0,03	0,03
Acétone	0,11	0,15	0,11	0,13	0,13	0,11
Acétate de méthyle	0,32	0,76	0,56	0,24	0,12	0,64
Acide acétique	2,35	3,62	2,44	1,89	1,96	3,82
Méthyléthylcétone +acétate d'éthyle	0,13	0,17	0,15	0,15	0,15	0,15
Gaïacol	0,61	0,46	0,29	0,75	0,21	0,37
Phénol+0. crésol	0,37	0,58	0,07	0,47	0,17	0,23
m. + p. crésols	0,19	0,13	0,05	0,28	0,07	0,13

Tableau n° V : Analyse des gaz - Composition pondérale (% bois sec)

	E	F.F.	G	T1	T2	T3
H_2	0,02	0,03	0,03	0,01	0,02	0,03
O_2	0,42	0,10	-	0,33	0,40	0,56
CO_2	14,1	14,0	15,7	15,1	14,0	13,0
CO	5,55	5,7	7,35	6,05	5,4	5,7
CH_4	0,98	1,21	1,36	0,76	1,01	1,65
C_2H_4	0,09	0,07	0,10	0,08	0,08	0,11
$C_3H_6+C_4H_8$	0,19	-	0,21	0,22	0,29	0,41
C_2H_6	0,20	0,21	0,27	0,19	0,20	0,33
Total hydro-carbures	1,46	1,49	1,94	1,25	1,63	2,64

Ces remarques ont d'ailleurs été confirmées par un essai ultérieur de carbonisation d'une pâte cellulosique blanchie, ne contenant pratiquement que de la cellulose et des pentosanes, qui a donné un rendement en charbon inférieur à celui des bois, soit 28 % et sur de la lignine qui a donné 59 % de charbon.

33 Qualité des charbons.

Les caractéristiques physicochimiques des charbons diffèrent assez nettement selon l'espèce : densité variant de 0,16 à 0,33, p.c.s. de 7710 à 8185, taux de matières volatiles de 11,8 à 16 %, cendres de 1,05 à 6,6 %. Les charbons de bois obtenus à partir de feuillus français se classent à l'intérieur de ces fourchettes et ne se distinguent pas de l'ensemble des bois tropicaux.

Le mélange tropical T3 constitué de bois pauvres en lignine et en extraits accuse un taux de cendres très élevé (6,6 %). Ceci pourrait entraîner un refus pour certaines utilisations industrielles.

On a confirmé, au cours de cette étude, qu'une forte corrélation existait d'une part entre la densité des bois et la densité des charbons résultants, d'autre part entre les teneurs en cendres des bois et celle des charbons. Inversement, on n'a pas observé de corrélations entre les p.c.s. des bois et ceux des charbons. En fait les p.c.s. des charbons dépendent de leur constitution chimique finale (produits volatils, taux de cendres et de carbone fixe).

34 Composition des pyroligneux.

La composition qualitative des pyroligneux est la même. Quel que soit le bois ou le mélange de bois pyrolisé, on retrouve les mêmes constituants.

Cependant, l'analyse quantitative révèle quelques différences (on rappelle que les rendements totaux en pyroligneux n'étaient pas identiques pour toutes les essences). Ainsi, pour les bois tropicaux, on a recueilli 1,02 à 1,94 % de méthanol; 1,89 à 3,82 % d'acide acétique; 0,21 à 0,75 % de gaïacol; 0,12 à 0,75 % de phénols + crésols, soit, selon les cas, deux à six fois plus de produits. Seule l'acétone ne semble pas dépendre de la nature du bois puisque l'on n'a observé que peu de variations (0,11 à 0,15 %).

Il faut noter qu'ici aussi les feuillus tempérés ne se distinguent pas particulièrement des bois tropicaux.

On a cherché à établir l'influence de la composition chimique des bois sur la formation de certains composés de pyrolyse. On a noté que des quantités plus importantes de méthanol sont produites par les bois riches en cellulose.

D'autre part, l'acide acétique et l'acétate de méthyl augmentent quand les bois contiennent beaucoup de pentosanes.

De plus, les dépôts condensables semblent suivre les teneurs en extraits alcool-benzène des bois.

Enfin, les bois riches en lignine donneraient moins d'acide acétique et les bois riches en extraits (alcool-benzène et eau) moins de méthanol.

On n'a pas observé de tendance très nette en ce qui concerne la formation de gaïacol, de phénols ou de crésols. Toutefois, d'un point de vue théorique, ces constituants devraient être issus de la lignine car on retrouve des structures chimiques analogues. C'est effectivement pour le mélange T1, riche en lignine, que l'on observe un rendement maximum en gaïacol, phénols, crésols.

35 Composition des gaz.

Quel que soit le bois ou le mélange de bois, les gaz de pyrolyse sont composés essentiellement de gaz carbonique, d'oxyde de carbone et de méthane. On trouve en plus petites quantités, de l'oxygène et de l'hydrogène, de l'éthane et de l'éthylène. La composition des gaz varie selon l'essence carbonisée. Le gaz carbonique est le constituant principal (de 45,7 à 53,8 % en volume ou de 13 à 15,7 % par rapport au bois initial); vient ensuite l'oxyde de carbone (de 31,2 à 35,7 % en volume ou de 5,4 à 7,35 % par rapport au bois initial). Les quantités de méthane varient de 7,4 à 16 % en volume (ou 0,76 à 1,65 % par rapport au bois) et les taux d'hydrocarbures totaux de 9,5 à 20,1 % en volume (ou 1,25 à 2,64 % par rapport au bois) soit du simple au double. On remarque que les résultats trouvés pour les feuillus français, se situent dans la fourchette moyenne des résultats globaux.

Le p.c.s. des gaz varie en fonction des teneurs en hydrocarbures totaux de 2170 à 3380 calories par litre, avec un chiffre de 2450 pour les bois français. Si l'on calcule le nombre de calories pouvant être fournies par 100 g de bois sec, des différences apparaissent également, les résultats s'étageant entre 33,4 à 52,7 kilo calories. Le mélange T3 et le Gmelina se détachent nettement et donnent les chiffres les plus favorables.

La recherche de corrélation entre la composition chimique du bois et celle des gaz s'est soldée par un résultat négatif. Il est donc très difficile de formuler des hypothèses sur l'origine des composés gazeux.

4. CONCLUSION.

Les essais réalisés montrent que l'on peut envisager sans trop de risques le remplacement de bois feuillus français par des bois ou mélanges de bois tropicaux pour l'approvisionnement d'une chaîne de carbonisation avec production de charbon, de pyroligneux et de gaz. Toutefois, certaines différences pourront apparaître dans les rendements en produits de carbonisation, tant en ce qui concerne le charbon que les composants liquides ou gazeux. Il peut en résulter de légères modifications du réglage des appareils. La valeur énergétique des produits gazeux en particulier peut varier d'un mélange de bois à l'autre. Enfin, des risques accrus d'encrassement peuvent apparaître dans le cas de quelques bois tropicaux riches en extraits aux solvants organiques.

Il n'en reste pas moins démontré que l'utilisation énergétique ou chimique des bois tropicaux est envisageable.

D'un point de vue théorique, l'étude fournit une première série d'informations sur l'origine des corps obtenus à la pyrolyse. On pourrait accroître la production d'acide acétique et d'acétates par l'emploi de produits riches en pentosanes comme la paille ou augmenter le rendement en charbon en utilisant des bois riches en lignine.

Bibliographie.
- Colloque sur la valorisation énergétique des sous-produits agricoles 13-14 mars 1979 (APRIA, 35 rue Gal Foy, 75008 PARIS)
- Bois et Forêts des Tropiques : n° 159 (Janv.-Fév. 1975, p. 55) et n° 177 (Janv.-Fév. 1978, p. 51)
- Carbonisation des bois et carburants forestiers, par Ch. Mariller, Dunod 1941
- Le Coke, par Loison Foch et Boyer, Dunod 1970

GASEOUS FUEL FROM BIOMASS BY FLASH PYROLYSIS

S. CAUBET, P. CORTE, C. FAHIM, J.P. TRAVERSE

Laboratoire de Recherche sur l'Energie. Université Paul Sabatier,
31062 TOULOUSE Cédex - FRANCE

Summary

Thermal decomposition of biomass in an inert atmosphere is studied through basic components : cellulose, lignin, wood.

Flash pyrolysis (heating rate 250 °C/s allows the production of a medium heating value synthetic gas with gas phase thermal efficiency of up to 95%.

Light hydrocarbons (CH_4 , C_2H_4 , C_2H_2 , C_2H_6) provide about 50% of the energy recovered in the gas. Ethylene represents 5% (Vol.) of the pyrolysis gas.

Résumé

On étudie la décomposition thermique de la biomasse en atmosphère neutre à travers ses composants de base : cellulose, lignine, bois.

La pyrolyse flash avec des vitesses d'échauffement d'environ 250 °C/s, permet de produire un gaz de synthèse à pouvoir calorifique moyen avec un rendement énergétique de conversion atteignant 95%.

Les hydrocarbures légers (CH_4 , C_2H_4 , C_2H_2 , C_2H_6) représentent jusqu'à 50% du contenu énergétique du gaz. L'éthylène représente 5% en volume du gaz de pyrolyse.

1. INTRODUCTION

Large scale use of the biomass as a source of non-fossil renewable carbon is the only long-term solution to fossil fuel depletion. Flash pyrolysis allows potentially useful conversion of lignocellulosic materials to gaseous mixtures, which can be used either as gaseous fuel or as raw materials for chemical synthesis. The purpose of this work is to complete available data on this chemical process.

2. EXPERIMENTAL

Materials

The experiments were carried out separately with the two basic components of the biomass : cellulose and lignin, and with beech wood, chosen as a typical example.

All samples were dried for 24h at 95 °C.

Wood and cellulose were shredded to 1-2mm particule size. Lignin was in the form of a fine powder (100 μm). Composition of the materials is given in Table I.

TABLE I

	Weight composition (%) dry ash-free basis					Elementary composition				
	WATER	ASHES	C	H	O	C	H	O	H/C	O/C
CELLULOSE	2,0	0,3	44,4	6,2	49,4	6	I0	5	I,67	0,83
LIGNIN	0,0	0,0	63,8	6,4	29,8	6	7,2	2,I	I,20	0,35
BEECH WOOD	2,5	0,5	48,3	6,0	45,I	6	8,9	4,2	I,49	0,70

Apparatus and method.

Figure I shows the laboratory reactor used for the experiments. After heating the reactor to the desired temperature, weighted amples of 1.30g were introduced into the hot zone with a feed rate of about 0.3g/mn for cellulose, 0.4g/mn for beech wood and 0.9g/mn for lignin. An original feeder was designed.

An inert gas flowed through the reactor.

The particles fell on the top of a porous alumina bed where they are pyrolysed. The pyrolysis gases were extracted at the bottom of the reactor. The total volume of gas produced in one run was measured. Gaseous products were then analysed with a gas chromatograph.

The average heating rate of the particules was measured. For an 850 °C reactor temperature, the heating rate reached 250 °C/s.

The residence time of the gases in the hot zone was roughly calculated.

After the pyrolysis stage, air was blown through the reactor instead of inert gas ; the char then burns. Analysis of the combustion gases gives the amount of carbon in the char.

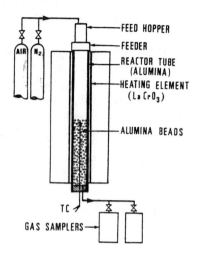

FEED HOPPER
FEEDER
REACTOR TUBE (ALUMINA)
HEATING ELEMENT ($La CrO_3$)
ALUMINA BEADS
TC
GAS SAMPLERS

Fig. I : *Laboratory apparatus used for flash pyrolysis.*

3. RESULTS

Experiments show that the most useful temperature range for gasification is 600-1000 °C. The gas product is primarily composed of carbon monoxide, carbon dioxide and hydrogen. Substantial amounts of hydrocarbons especially methane and ethylene are present in the gas. The gas volume always increases with temperature. For example, at 850 °C, 1g of cellulose yields 0.91 of pyrolytic gas witch contains 12.2% (Vol) of methane and 5% (Vol) of ethylene. Table II presents the detailed results of the experiments.

The carbon, hydrogen and oxygen balances are given. The mass balance is also given assuming that char is almost pure carbon. Mass losses are due to non-recovered products : deposits of carbon black and tar outside the reactor hot zone ; water and other condensibles.

Figures II and III represent variations of gas yield (Y) and gas phase thermal efficiency (e) versus reactor temperature. The gas yield is defined as :

$$y = \frac{\text{gaseous product weight}}{\text{weight of organic matter introduced}}$$

and the gas phase thermal efficiency as :

$$e = \frac{\text{available energy in pyrolysis gas}}{\text{available energy in biomass sample}}$$

TABLE II

Gas analysis , % by volume

	T°c	t	H₂	CO	CO₂	CH₄	C₂H₄	C₂H₆	C₂H₂	hv	V	y	e	Hg/H	Og/O	Cg/C	Cs/C
CELLULOSE	600	7.5	13.85	42.50	32.35	6.40	3.70	1.10	/	3065	0.274	35.25	20.75	12.4	42.5	30.8	23.5
	700	5.2	22.15	42.55	18.45	11.75	3.75	1.15	0.20	3865	0.564	59.8	53.9	47.5	64.75	57.8	19.3
	800	4.1	24.60	42.80	13.60	13.30	4.75	0.70	0.25	4260	0.806	79.9	83.3	75.6	81.7	80.8	13.6
	850	3.9	26.90	44.20	11.10	12.20	5.00	0.40	0.20	4180	0.883	84.0	91.1	82.4	84.7	85.9	8.0
	900	3.7	28.60	43.40	11.25	12.00	4.35	0.35	0.20	4045	0.939	87.6	93.8	87.1	89.4	88.3	8.1
	1000	3.4	34.00	46.00	8.00	10.40	1.60	0.04	/	3675	1.063	91.1	96.6	91.9	95.3	88.9	7.9
LIGNIN	700	3.7	17.00	41.20	13.55	25.10	1.95	1.30	/	4665	0.317	32.2	21.9	20.1	51.8	23.0	/
	800	3.0	23.40	39.25	11.35	21.60	2.90	0.83	0.15	4555	0.425	39.8	28.7	44.6	63.1	27.3	/
	900	2.8	29.90	37.80	9.00	19.60	3.20	0.25	0.27	4485	0.507	43.3	33.8	54.3	66.8	30.1	/
	1000	2.2	36.80	37.80	8.25	15.60	1.23	0.02	0.29	3990	0.623	49.7	36.9	61.7	81.1	32.4	/
BEECH WOOD	700	5.0	18.75	43.24	17.11	16.26	3.54	0.94	0.15	4140	0,445	47.5	41.9	40.7	54.5	43.7	24.8
	800	3.8	21.10	43.10	15.60	14.80	4.34	0.72	0.31	4175	0.680	70.5	64.5	62.5	79.9	65.5	19.3
	850	3.7	21.50	42.80	13.30	16.70	4.80	0.46	0.43	4400	0.713	71.6	71.3	70.3	78.3	68.6	15.0
	900	3.5	24.85	41.80	13.20	15.35	4.37	0.23	0.20	4210	0.767	74.6	73.4	74.3	82.8	70.1	12.4
	1000	3.0	37.00	42.80	7.60	11.70	0.70	/	/	3655	0.954	77.2	79.2	87.7	87.5	69.2	12.0

t : Residence time of gases (s)
hv : Heating value of pyrolysis gas (kcal/Nm³)
V : Total volume of pyrolysis gas (Nl)
y : Gasification yield
e : Gas phase thermal efficiency

Hg/H : Hydrogen balance
Og/O : Oxygen balance
Cg/C : Carbon balance (gasification)
Cs/C : % Initial carbon in the char

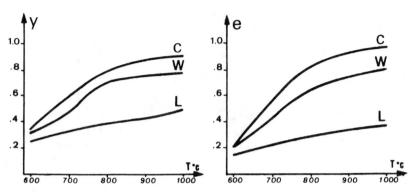

Fig.II:*Gas yield (y) vs temperature* Fig.III:*Gas phase thermal efficiency (e) vs temperature*

4. DISCUSSION

The results of the present work show that biomass materials can be converted to gaseous product with high gas yield and gas thermal efficiency.

Operating conditions : residence time of gases (4s), atmospheric pressure in an isothermal reactor and continuous feeding can be easily obtained in an industrial process.

Our results are consistent with other processes described in the literature, considering the difference of raw materials and procedures (Ref. 1 to 4).

Pyrolysis gas can be burned directly as gaseous fuel without problem, because of its relatively high calorific value : (about 40% of that of natural gas).

Light hydrocarbons of high calorific value represent more than 20% of the gas volume i.e. 50% or more of the energy available in the gas : see curve on Fig. IV. The most interesting chemicals products are unsaturated hydrocarbons. Ethylene production reaches 5% of gas volume. Better ethylene yields could be obtained with shorter residence time and faster quenching of the gases.

The reaction enthalpy can be estimated ; assuming an average calorific value of 8000 kcal/Kg for unknown products. We find :

H = 630 kcal/Kg for cellulose

H = 475 kcal/Kg for beech wood.

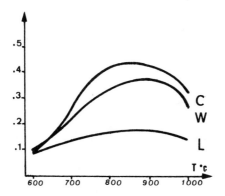

Fig.IV:*Thermal hydrocarbons yield :*
available energy in hydrocarbon gas
available energy in pyrolysis gas

These values represent less than 15% of the initial calorific value of the biomass.

From an industrial point of view, fluidized bed will be the most appropriate reactor. Fluidized bed studies will shortlhy be undertaken to optimize the conversion.

References :

1 - S. PRAHACS, H.G. BARCLAY, S.P. BHATIA.
   "Pulp and paper magazine of canada". 72, 6. (1971), 69-83.

2 - J. LEDE, P.BERTHELOT,J.VILLERMAUX,A.ROLIN,H.FRANCOIS,X.DEGLISE
   "Revue Phys. Appl." 15 (1980), 545-552.

3 - R.S. BURTON, R.C. BAILIE.
   "Combustion". Feb. (1974), 13-19.

4 - E. RENSFELT and Al.
   "Symp. pap. Energy from Biomass and Wastes". Aug. 14-18 (1978) -
   Washington D.C., 465-594.

FAST PYROLYSIS/GASIFICATION OF LIGNOCELLULOSIC
MATERIALS AT SHORT RESIDENCE TIME

X. DEGLISE, C. RICHARD, A. ROLIN and H. FRANÇOIS

Laboratoire de Photochimie Appliquée
(Groupe de Recherches de Chimie du Bois)
Université de Nancy I - 54000 NANCY FRANCE

Work is funded by the Délégation Générale à la
Recherche Scientifique et Technique (DGRST)
and Commissariat à l'Energie Solaire (COMES)

Summary

The results discussed in this paper point to the advantages of Fast
Pyrolysis over air and oxygen gasification as a source of medium energy
gas. Both rapid heating and short residence time of gases are employed in
a single step pyrolysis-gasification of sawdust samples in a free-falling
system. Gas yields are increasing with temperature and moisture content
(for temperatures upper than 700°C). Tables of products, H_2, CO, CO_2, CH_4,
C_2H_4, C_2H_6 and C_2H_2 are shown with volume balance and overall formulae of
char for different temperatures and moisture contents. Experimental results
provide indirect information on the heat of reaction which shows that fast
pyrolysis is slightly endothermic for temperatures upper than 600-700°C,
and exothermic under 700°C.

1. INTRODUCTION

    Two main types of gasifiers have to be distinguished ([1]) :

    - The production of a low energy (air gasification : 5000 KJ/m^3) or a medium energy (oxygen gasification : 12500 KJ/m^3) gas requires near-equilibrium product distribution, large particles, slow heating rates and long residence times.

    - The increasing in energy content of the gas (medium energy gas : 19000 KJ/m^3) with non equilibrium product distribution requires small particles, fast heating rates and short residence times.
Fast pyrolysis, corresponds to the second type of gasifier and is able to produce hydrocarbons such as C_2H_4 and C_2H_2 giving a higher heating value to the gas and valuable products for chemical industry.

2. EXPERIMENTAL ([2]) ([3])

    Pyrolysis of sawdust samples (100 to 300 mg, particle size : 25 to 400 μm) was performed in a free falling system : vertical quartz tube in an electrical furnace (from 500 to 1000°C). Heating rate of 1000°C/s and residence times of less than a second were estimated. After an experiment the gaseous products were expanded into calibrated volumes and analyzed by gas chromatography (flame ionization detector for hydrocarbons, and thermal conductivity detector for H_2, CO and CO_2). Water formation during the pyrolysis of dry wood was determined by the difference between the total pressure measured on the gauge and the pressure of the analyzed gas. (The monitored gases : H_2, CO, CO_2, CH_4, C_2H_4, C_2H_6, C_2H_2 and H_2O represent more than 99 % of the products, only traces of other hydrocarbons were detected). From carbon, hydrogen and oxygen balances the overall formula of char is obtained at different temperatures.
Materials : softwood (Douglas fir) and hardwood (Beech) were studied at different moisture contents (from 0 to 100 %).

3. RESULTS AND DISCUSSION

    The yields of the major pyrolytic products versus temperature are presented in Table 1 for the fast pyrolysis of dry material.

    The char yield decreases with increasing temperature and levels out around 800°C. The yield of the gas increases with temperature, and the water yield remains almost constant. The overall formulae of char are given in Table 2. The overall composition of char appears not to be very different for Douglas fir and beech from 500 to 1000°C. These results are in good agreement with literature data ([3]).

TABLE 1 - Mass yields (%) of the major pyrolytic products of dry imput feed versus temperature.

a) Dry softwood

$T°C$	500	600	700	800	900	1000
Gases*	13.85	26.8	52.25	59.6	60.3	61.3
Char	79.4	61	33.5	25.6	27	26
Water	6.75	12.2	14.25	14.8	12.7	12.7

b) Dry hardwood

$T°C$	500	600	700	800	900	1000
Gases*	16.7	29.7	48.8	58	60.4	67.6
Char	75.8	57.5	37.8	29.6	26.4	21
Water	7.5	12.8	13.4	12.4	13.2	11.4

(*) H_2, CO, CO_2, CH_4, C_2H_4, C_2H_6, C_2H_2

TABLE 2 - Overall formulae of char versus temperature.

	Douglas fir	Beech
500°C	$CH_{1.33} O_{0.51}$	$CH_{1.33} O_{0.48}$
600°C	$CH_{1.15} O_{0.39}$	$CH_{1.14} O_{0.35}$
700°C	$CH_{0.76} O_{0.22}$	$CH_{0.90} O_{0.24}$
800°C	$CH_{0.52} O_{0.12}$	$CH_{0.73} O_{0.19}$
900°C	$CH_{0.48} O_{0.18}$	$CH_{0.58} O_{0.14}$
1000°C	$CH_{0.28} O_{0.16}$	$CH_{0.43} O_{0.07}$

The variations of the yields of gases are shown in Table 3. The nature of

TABLE 3 - Yields of the major gases (% volume a : dry beech, b : dry Douglas fir).

$T°C$		500	600	700	800	900	1000
CO	a	55.5	59	54	51	50	48.5
	b	59	62	56	53.5	51.5	46.9
CO_2	a	26.5	13	9	8	7	6.5
	b	22	11.5	5	5	4.5	4.5
H_2	a	2	8	14	18	21	26
	b	2.5	8	15	19	24	30
CH_4	a	11.5	14	15	15	14	13
	b	12.2	13	16	14.5	13.5	13
C_2H_4	a	3.5	4.7	6.5	6.5	6.0	3.8
	b	3.5	4.2	6.5	6.5	5.6	3
C_2H_6	a	0.7	1.1	1	0.5	0.5	0.2
	b	0.7	1	1	0.5	0.2	0.1
C_2H_2	a	0.3	0.2	0.5	1.0	1.5	2.0
	b	0.1	0.3	0.5	1.0	1.7	2.5

the wood studied has apparently no influence on the composition of gases
and residues and yields of products.

The yields of CO and CO_2 show significant decreases when increasing
pyrolysis temperature. The hydrogen yield shows a dramatic increase in our
temperature range. The CH_4 and C_2H_4 yields increase, present a maximum
around 700 and 750°C respectively and gradually decrease for higher temperatures. The C_2H_2 yield increases with temperature. The heating values of
the pyrolytic gas increase to a maximum of 20000 KJ/m^3 around 750°C and
then show a slight decrease until 18000 KJ/m^3 at 1000°C.

For fast pyrolysis of Douglas fir and beech, the moisture content of
wood has practically no influence for temperature below 700°C. The yields
of gases versus temperature and moisture content of wood (Figure 1, Douglas
fir) are increasing with temperature over 700°C and moisture content.

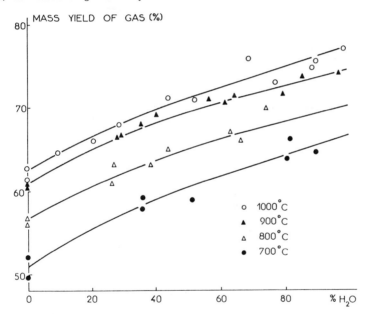

Figure 1 - Yields of gases versus temperature and moisture content of
wood (Douglas fir).

For 1 kg of wood, pyrolysed at 1000°C with a 100 % moisture content, we
obtain 1 m^3 of gas and a heating value of 16000 KJ/m^3. The influence of
moisture content on the evolved gas composition is shown in figure 2.

At constant temperature the hydrogen yield shows a dramatic increase

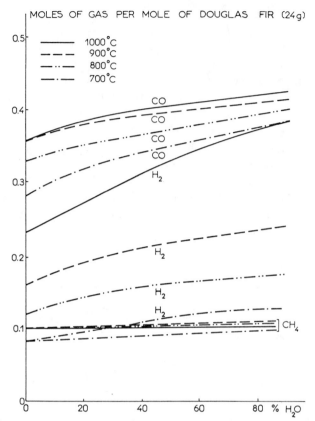

MOLES OF GAS PER MOLE OF DOUGLAS FIR (24g)

Figure 2 - Influence of moisture content on the evolved gas composition.

with the moisture content. For short residence time the influence of water does not result entirely from steam-gasification as the increase of H_2 does not correspond to the increase of CO.

The heating values of the products (gas + char) are very close to the heat content of the dry input feed material. At 600°C the char represent 76 % of the total heating value of the input feed, whereas at 900°C it represent only 40 %.

Calculated data from experimental results show that fast pyrolysis reaction is slightly endothermic only above 700°C. Heats of reaction are shown in Table 4.

As the heats of combustion of dry wood are : Douglas fir : 19980 KJ/kg, beech : 19540 KJ/kg ; the heat of reaction for gasification to a medium energy gas (17000 KJ/m^3) is less than 10 % of the wood heat content.

TABLE 4 - Heat of reaction for fast pyrolysis (KJ/kg of dry wood).

1) Softwood

T°C	500	600	700	800	900	1000
0 % H$_2$O	-500	-410	40	210	460	580
50 % H$_2$O	-625	-540	200	540	1040	1400

2) Hardwood

T°C	500	600	700	800	900	1000
0 % H$_2$O	-40	16	375	630	740	1030
50 % H$_2$O	-210	-160	430	1080	1460	1970

4. CONCLUSION

These bench scale fast-pyrolysis experiments over the temperature range 500°C to 1000°C and for moisture contents from 0 to 100 %, provide data on the yields of the pyrolytic products, the composition of the gaseous phase, the heating value of the products, and the heat of reaction. These experiments shows that moisture content enhance the gasification but with higher heats of reaction. These data are useful as guidelines in the operation of future continuous flash-pyrolytic pilot plants we plan to develop ([4]).

References

[1]   T. MILNE - A survey of Biomass Gasification, Vol II, chapter 5 SERI July 1979.

[2]   J. LEDE, P. BERTHELOT, J. VILLERMAUX, A. ROLIN, H. FRANÇOIS and X. DEGLISE - Revue Phys. Appl. 15, 545 (1980).

[3]   X. DEGLISE, C. RICHARD, A. ROLIN and H. FRANÇOIS - Revue Gén. Thermique, on press. November 1980.

[4]   X. DEGLISE, J. MORLIERE and Ph. SCHLICKLIN - Energy from Biomass Conference - Brighton G.B. November 1980.

THE CONVERSION OF BIOMASS TO FUEL RAW MATERIAL BY HYDROTHERMAL TREATMENT

O. BOBLETER, H. BINDER, R. CONCIN
and E. BURTSCHER

Institute for Radiochemistry, University of Innsbruck
Innrain 52 a, A-6020 Innsbruck

Summary

The present paper deals with the hydrothermolysis of wood at temperatures between 180° and 200°C. Under these conditions a degradation of the hemicelluloses and of parts of the lignin takes place. The yields of the different reaction products (xylose, furfurals and low molecular lignin compounds) varies according to the flow rate and the temperature in the reaction vessel.

In addition to the almost complete conversion of the hemicelluloses to water soluble substances, the lignin is also degraded to an average trimeric aromatic units. For the conversion of the cellulose remaining from the first temperature step at 180° and 200°C a hydrothermal treatment at 270°C is necessary.

1. INTRODUCTION

The increasing costs and the coming shortages of liquid fuels and chemical feedstocks have renewed the interest in the utilization of biomass.

Besides the nonselective thermochemical processes such as gasification and pyrolysis mainly hydrolytic treatments are used for the degradation of carbohydrate components of the wood.

Thereby, the cellulose (which constitutes 50 % of the wood) is converted to glucose, whereas hemicelluloses (20 - 35 % of the wood) mainly yield mannose from softwoods and xylose from hardwoods. One of the greatest disadvantages of these hydrolytic processes (catalysis by acids or enzymes) is the fact that the lignin remains as a solid residue.

In recent years we have developed a new degradation process for plant matter (1-3). Through this method wood is converted to water soluble substances with pure water at elevated temperatures without nearly any residue. For filter paper conversion rates of cellulose to glucose of about 70 % were found (4).

With wood the behaviour of the hemicelluloses and the lignin during the degradation is of major importance. In the present work this conversion characteristics were studied with poplar wood which is frequently used in short rotation forestry with regard to its utilization as fuel raw material.

2. EXPERIMENTAL

The hydrothermolysis was carried out in the apparatus shown in Fig. 1. The wood material was filled into the reaction vessel (6 cm^3) and then water was pressed through safety valves and a preheating system to the autoclave (flow rates from 3.0 to 11 ml/min). The precise temperature in the reaction cell is guaranteed by an electric heating unit. The hot water (180° or 200°C; 270°C) degraded the plant material and the solutions left the system by passing a cooler, a metering valve and a UV detector. The UV-signal was used to control the progress of the reaction. The eluent was collected in frac-

tions and then the dry weight, the contents of sugars, furfurals, lignin degradation products as well as the molecular
weights of the different higher molecular products were determined by means of chromatographic methods (5-8).

The composition of the residues from the first degradation step (180° and 200°C) was analysed according to Klauditz
(9) and Kürschner (10).

The combined eluates of the main fractions and the cellulose residue were subjected to an acid hydrolysis with 0.01 N
H_2SO_4 in stainless steel autoclaves.

The degradation and analyses scheme is shown in Fig. 2.

3. RESULTS AND DISCUSSION

A characteristic elution profile for poplar wood (dry
weight versus time) is depicted in Fig. 3. In a first step the
hemicelluloses were degraded at 180°C and in a second step
rising the temperature to 270°C the cellulose was converted.
The remaining material (7.1 %) was degraded at temperatures
higher than 300°C.

In order to determine the degree of the hemicellulose
conversion and the behaviour of the lignin, the hydrothermolysis was stopped after the first degradation step. These experiments were carried out at two different temperatures
(180° and 200°C) and with different flow rates of the eluting
water (3 - 11 ml/min).

At a temperature of 180°C 41 to 43 % of the wood were
converted to water soluble substances whereas at 200°C the
soluble part of the wood amounted to 49 to 55 % (Fig. 4A).
The results of the wood residue after the hydrothermolysis at
180° and 200°C are given in Fig. 4B.

The hemicelluloses were almost completely degraded
whereas about 50 % (at 180°C) and 70 % (at 200°C) of the original lignin became soluble.

The cellulose was only slightly degraded (about 8 %) at
200°C, which is due to the amorphous parts of the cellulose.
The composition of the residues showed only little dependence
on the flow rate of the eluting water. The reaction time,

however, was reduced from 60 minutes with a flow rate of 3 ml/min to 10 minutes with a flow rate of 11 ml/min. The character of the water soluble products of the first reaction step (hemicelluloses and lignins) was determined by chromatographic methods. For the determination of the molecular weights of saccharides and lignins calibrated Sephadex gels were used. The products of the hemicellulose showed molecular weights above 5000 D, whereas the lignin degradation products lay on an average at 500 D (Fig. 5).

The sugar analyses (HPLC) revealed that the solutions only contain little amounts of xylose (max. 0.8 mg/ml). These xylose concentrations decrease with increasing flow rates.

The low molecular lignin degradation fraction consists of a few monomeric compounds (Fig. 6). The quantity of these compounds increase with higher flow rates. This is due to the characteristic behaviour of the lignin (11).

In order to investigate the utilization of the hemicellulose products, the combined fractions were hydrolysed with 0.01 N H_2SO_4 at 200°C. The xylose formation curve is shown in Fig. 7A, a chromatogram of the sugar analysis is given in Fig. 8B. The recent works on fermentation of xylose to ethanol (12) open new possibilities for the utilization of the hemicellulose fraction of the wood.

The lignin degradation products are of small molecular size and are soluble in organic solvents (e.g. ethanol). These products could become a new feedstock for aromatics.

To assess the quality of the cellulose in the residue, the remaining hemicelluloses, lignins and extractives were removed as discribed above. Afterwards the purified cellulose was subjected to acid hydrolysis (Fig. 7B and 8A). The kinetic data of the acid hydrolysis of this purified cellulose almost correspond to the data found with filter paper. Therefore, it can be assumed that the structure of the cellulose is not affected in the first step of the hydrothermolysis.

ACKNOWLEDGEMENT

The authors are indebted to the Fonds zur Förderung der Wissenschaftlichen Forschung (Wien) for the financial support.

REFERENCES

(1) Bobleter, O., R. Niesner, M. Röhr, J. Appl. Polymer. Sci.
    20 (1976) 2083
(2) Bobleter, O., H. Binder, Holzforschung 34 (1980) 48
(3) Bobleter, O., H. Binder, R. Concin, Chem. Rundschau
    32 (1979) 1
(4) Bobleter, O., G. Bonn, R. Concin, in press
(5) Niesner, R., W. Brüller, O. Bobleter, Chromatographia
    11 (1978) 400
(6) Binder, H., J. Chromatogr. 189 (1980) 414
(7) Concin, R., E. Burtscher, O. Bobleter, J. Chromatogr.
    198 (1980) 131
(8) Burtscher, E., R. Concin, H. Binder, in press
(9) Klauditz, W., Holzforschung 11 (1957) 110
(10) Kürschner, K., St. Karácsonyi, Holzforschung 15 (1961) 107
(11) Bobleter, O., R. Concin, Cellul. Chem. Technol. 13
    (1979) 583
(12) Wang, D.I.C., C.L. Cooney, R.F. Gomez, A.L. DeMain,
    A.J. Sinskey, in "Proc. 3rd Annual Biomass Energy Systems
    Conf.", Solar Energy Research Institute, Golden, Colo.,
    June 5-7, 1979

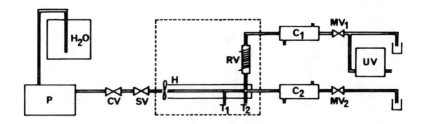

Fig. 1 Scheme of the apparatus for hydrothermolysis
P pump, CV check valve, SV safety valve, T thermocouples,
RV reaction vessel, H preheating unit, C cooler, MV metering
valve, UV detector.

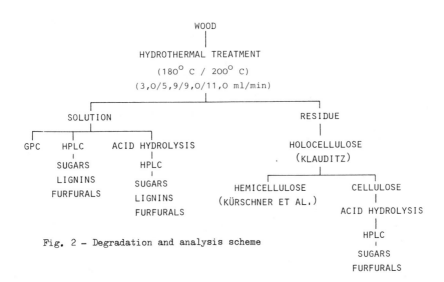

Fig. 2 – Degradation and analysis scheme

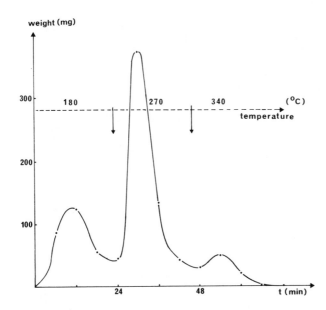

Fig. 3 – Elution profile of a hydrothermal degradation of poplar
wood. The dry weight of the fractions is depicted versus
the reaction time. Sample weight: 1245 mg; flow rate:
12.3 ml/min; three temperature steps: 180–270–340° C.

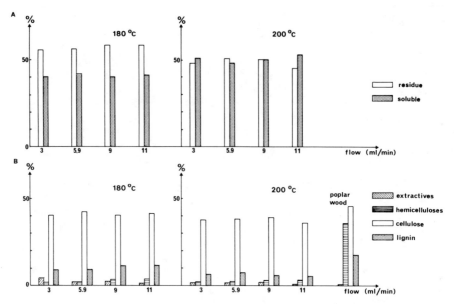

<u>Fig. 4</u> – A: Soluble part and residue of the original wood amount after
hydrothermolysis at 180° and 200°C with different flow rates
B: Composition of the residue

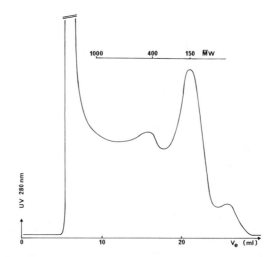

<u>Fig. 5</u> – Gelchromatographic separation of the lignin degradation products
on Sephadex LH 20; 38 x 1.0 cm I.D.; eluent dioxane-water 7 : 3;
flow rate: 0.28 ml/min.

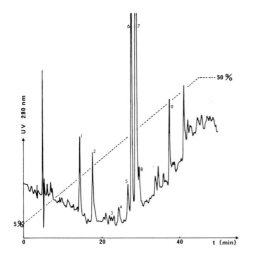

Fig. 6 - HPLC-separation of the
monomeric lignin degradation
compounds and furfurals on a
RP C-18 column; 250 x 4.6 mm
I.D.; mobile phase: gradient
5-50 % acetonitrile in phosphate buffer (0.01 M, pH 2.2);
flow rate: 0.8 ml/min; temperature 50°C.
1 furfural, 2-9 monomeric
lignin degradation products.

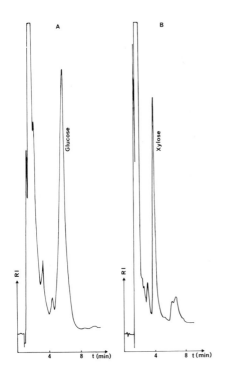

Fig. 7 - Acid hydrolysis of the
converted hemicelluloses and
the cellulose residue with
0.01 N H_2SO_4 at 200°C.
A : Formation curve for
    xylose
B : Formation curve for
    glucose

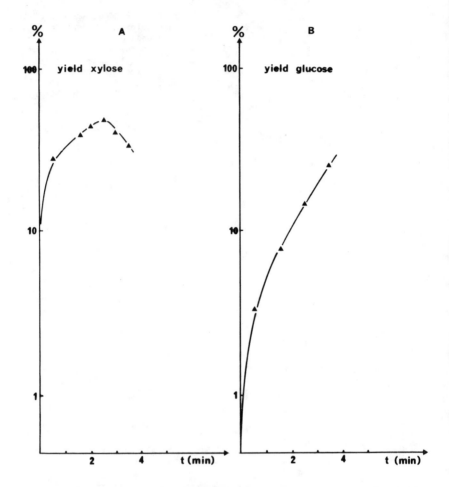

Fig. 8 – HPLC-separation of monomeric sugars after the acid hydrolysis of hemicelluloses and cellulose on a NH_2-column;
250 x 4.6 mm I.D.; solvent acetonitrile-water 75:25, flow
rate: 2.0 ml/Min; RI detection.

A: glucose

B: xylose

A PROCESS FOR THERMO CHEMICAL CONVERSION OF WET BIOMASS INTO HEAT ENERGY

H. HAVE

Jordbrugsteknisk Institut
The Royal Veterinary and Agricultural University
Denmark

Summary. The described system provides possibilities for conversion of
wet biomass (e.g. animal manure and catch crops) into low grade heat
energy suitable for space and water heating.
The system consists of a drier, a burner and a gas scrubber. The heat
from combustion of already dried biomass is used for drying of wet biomass on the way to combustion and the heat of the humid flue gas from the
drier is regained by means of the gas scrubber.

Prediction of the performance of a suggested design shows that the thermal efficiency after inclusion of anticipated losses could be around 65%
and that the maximum acceptable moisture content in the biomass is around
81% on wet basis when no auxillary fuel is used. The maximum temperature
level of supplied heat is around $80^{\circ}C$.

1. INTRODUCTION.

Most sources of biomass available or produced for energy production have a high content of moisture which make utilization through direct combustion inefficient or impossible.

For sources as animal manure and catch crop anaerobic digestion is a possible method of conversion which leads to the high grade energy of methane. However this method has disadvantages of long retension time and low efficiency.

Composting is another method of conversion that can be applied to these sources of biomass (1). The produced energy is here low grade heat suitable for space heating etc. This method seems to have a similar efficiency as anaerobic digestion. But the retension time is only one third of that for anaerobic digestion.

Most countries in the temperate zone of the world have a considerable need for low temperature heat for space heating purposes. For these purposes it is possible to use wet biomass by means of a new conversion system which seems to be more economic and more efficient than the biological methods.

The system, which comprises drying, combustion and regaining of heat, is described in the following.

2. DESCRIPTION OF THE SYSTEM.

The system is shown schematically in Fig. 1. Wet biomass is led to a drier by a feeding mechanism. When drying has taken place it passes on to combustion in a furnace. The flue gas produced by combustion is mixed up with a suitable amount of cooling air and used for the drying. The humid air from the drier is led to a gas scrubber which works as combined heat exchanger and gas cleaner i.e., heat energy (including latent heat) and impurities are transmitted to the washing water while the cooled and cleaned flue gas is led to the open.

The hot water from the scrubber is led to a heat exchanger connected to a space heating system or similar. After cooling here it is led into a clarification container where impurities are settled before the water is recirculated to the scrubber. The condensate is led away through an over-

565

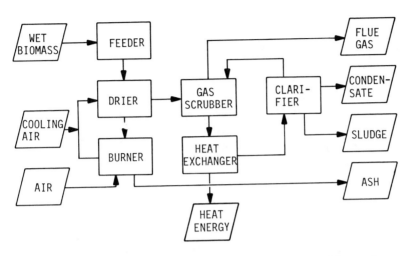

Fig. 1 Flow chart showing system for heat extraction from biomass by combustion. (Patent pending.)

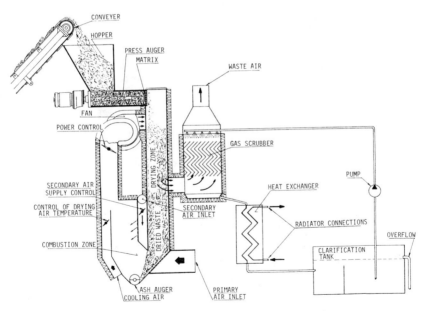

Fig. 2 Suggested design of plant for combustion of animal manure.

flow opening in the clarification container.

In Fig. 2 a sketch is shown of the first attempt to a practical design of a conversion unit for animal manure. The metering mechanism is here an auger with knife which presses the manure through a matrix. In this way it is given an aggregate structure, which facilitates efficient drying and combustion.

In the drier flue gas and manure flow in the same direction. This allows high drying air temperature without ignition hazards and provides a fast surface drying of aggregates just formed.

The process may be started up by an oil burner or simply by filling up the drying column with some dry straw or wood.

3. PREDICTION OF PERFORMANCE.

The conditions assumed for the plant considered are:

Water temperature in radiator supply line: $60^{\circ}C$
Water temperature in radiator return line: $30^{\circ}C$
Temperature difference in heat exchanger : $7^{\circ}C$
Temperature difference in gas scrubber ..: $3^{\circ}C$
Ambient temperature ....: $5^{\circ}C$
Moisture content of ambient air ....: 0,0047 kg/kg

Under these conditions the flue gas flowing from the drying zone to the scrubber must have a temperature of at least $70^{\circ}C$ in order to maintain the $60^{\circ}C$ in the radiator supply line.

Saturated air at $70^{\circ}C$ has a moisture content of 0.276 kg per kg dry air and an enthalpy of 795 kJ/kg. It is assumed that 10% of the energy is lost in the drier. Therefore the enthalpy of air entering it must be at least 795 1.1 = 874 kJ/kg to avoid temperature drop below $70^{\circ}C$ at the outlet.

Taking this into consideration and taking notice of moisture entering the drier with the flue gas, it can be calculated that the temperature of flue gas flowing to the drier should be at least $700^{\circ}C$ to hold the above enthalpy. This rather high temperature seems acceptable as the flue gas is brought in contact with wet and cold manure in the drier.

If it is now assumed that the flue gas leaving the scrubber is saturated and has a temperature of $40^{\circ}C$ the enthalpy of it is 166 kJ/kg compared to 795 kJ/kg at the entrance. Thus the thermal efficiency of the scrubber is 79%. Inclusion of a 15% loss in combustion in addition to the

10% that was assumed lost in the drier yields a total thermal efficiency
of 0.85 0.9 0.79 = 0.60.

Since the exhaust from the scrubber has a much higher enthalpy than
ambient air it would be an advantage to recirculate part of it to the inlet of the combustion unit. Such an arrangement would reduce the loss and
increase the efficiency to about 0.65.

The maximum content of moisture that can be accepted in the waste depends on the amount of energy available for drying. If it is assumed that
65% of the energy content can be made usefull, the highest allowable content would be

$$\frac{17 \cdot 0.65}{2.5} = 4.4 \text{ kg moisture/kg dry matter}$$

since the calorific value is about 17 MJ/kg dry matter (2) and the heat of
evaporation is 2.5 MJ/kg water. This moisture content corresponds to 81.5%
(w.b.).

In practice animal manure is found with water contents in the range
75 to 85% with an average of about 80%. Catch crops and peat are nearly in
the same range. Thus the process should make it possible to utilize a
great deal of the available wet biomass directly.

4. RETAINMENT OF PLANT NUTRIENTS.

Usually biomass has a considerable content of K, P and N. Of these
potassium and phosphorus remain in the ash by combustion at moderate temperature. Potassium in the ash is in a form easily accessible for plants,
while phosphorus is fixed in a compound of low solubility which is accessible for plants only after treatment with sulphuric acid (3).

Nitrogen of organic compounds is expected mostly to disappear as nitrogen or nitrogenoxides during combustion. However, in animal manure a
considerable part of nitrogen is present in the form of ammonium carbonate
which has been formed by decomposition of organic nitrogen compounds. This
ammonium carbonate will most likely be driven off during drying and afterwards be dissolved in the washing water of the scrubber.

5. EMISSION.

It is expected that the scrubber can be designed to separate most particles as well as NH_3 and SO_2 of the flue gas. In addition some separation
of odour is expected (4).

Therefore, since the exhaust from the scrubber has a low temperature
and is cleaned, a chimney may not be necessary.

6. INVESTIGATIONS OF DRYING.

Some preliminary experiments with structurizing and drying of pig manure have been carried out in cooperation with the manufacturing company Vølund A/S. The equipment used was a feeding and drying unit similar to the one shown in Fig. 2, but fitted with an electric heater. So far successful drying has been accomplished by mixing up the wet manure with some of the dried and by use of a small drum drier. The trials have shown that manure adhere very strongly to any solid surface it is in contact with during drying.

The investigations also showed that aggregates formed by the feeding mechanism became sufficient stable for gentle handling at moisture content of 3 kg/kg dry matter or less. At moisture contents below 0.5 kg/kg the aggregates were stable enough for relatively rough mechanical handling.

7. POTENTIAL SOURCES OF BIOMASS.

The process is applicable for any type of biomass with sufficient content of energy for evaporation of the content of moisture. Sources as solid animal manure, green plants and peat may be used directly as the moisture content usually is in the range 80-85% w.b. Other sources as slurry and sewage sludge with moisture contents above this level need preliminary dehydration or addition of supplementary energy.

+++++++++

REFERENCES:

1. Vemmelund, N. and Berthelsen, L. (1977): Udnyttelse af komposterings-varme fra staldgødning. Research Report no. 28, Jordbrugsteknisk Institut, The Royal Veterinary and Agricultural University, Copenhagen.

2. Traulsen, H. (1971): Verfahren zur Beseitigung tierischer Exkremente. - KTL-Berichte über Landtechnik no. 147, 1971.

3. Willems, M., Pedersen, B. and Jørgensen, S.S. (1976): Compostition and reactivity of ash from sewage sludge. - AMBID, Vol. 5, no. 1.

4. Olsen, H.J. and Berthelsen, L. (1975): Rensning af staldluft for varmegenvinding. - Research Report no. 26, Jordbrugsteknisk Institut, The Royal Veterinary and Agricultural University, Copenhagen.

MASS AND ENERGY BALANCES FOR A TWO FLUIDISED-BED
PILOT PLANT WHICH OPERATES ON WOOD FAST PYROLYSIS

X. DEGLISE   : Prof. Université de Nancy 1
               C.O. 140 - 54037 NANCY CEDEX,FRANCE
P. MORLIERE   : R and D Director TNEE (Tunzini-Nessi Entreprises d'Equipe-
               ments) - 1, Place H. de Balzac - 95100 ARGENTEUIL,FRANCE
Ph. SCHLICKLIN : Dr Ing. Centre de Recherches de Pont-à-Mousson
               B.P. 28 - 54700 PONT A MOUSSON,FRANCE

Abstract

Laboratory results provide for the drawing and calculation of the flow-sheet of a possible industrial facility, which can produce a medium BTU gas (17-19 MJ/Nm^3) not diluted with nitrogen and without need for oxygen plant. It operates on small size particles of wood residues.

With the two main input parameters (pyrolysis temperature and moisture content of the feed), the mass and energy balances were computed and the operating conditions are available for self sufficient energy operation. The efficiency of the system in term of energy was previously fixed.

The results of calculations for the case of pyrolysis of beech at 750°C and 10 % moisture content are given.

The pilot facility (1 t/day) will be demonstrated at the Research Center of Pont-à-Mousson, a subsidiary of Saint-Gobain - Pont-à-Mousson Company, in Pont-à-Mousson (France).

A partir de travaux menés au Laboratoire de Photochimie Appliquée de l'Université de Nancy 1 (Professeur X. Deglise), nous avons cherché à concevoir un système industriel permettant la transformation des déchets lignocellulosiques, et plus spécialement de bois, en un gaz combustible suffisamment riche pour pouvoir être transporté et distribué. Les expériences de laboratoire ont montré en effet que l'on pouvait par pyrolyse rapide dans la gamme de température de 700 à 1000°C obtenir 50 à 80 % du contenu énergétique du bois sous la forme d'un gaz possédant un PCI compris entre 17 et 19 MJ/Nm^3 (4100 à 4600 Kcal/Nm^3) tout-à-fait comparable au gaz de four à coke ou au gaz dit "de ville".

L'installation décrite ici -qui a fait l'objet d'une demande de brevet français- constitue un des meilleurs schémas utilisant au mieux l'énergie thermique générée pour assurer la bonne marche du système en couvrant les besoins thermiques de la pyrolyse proprement dite et de la préparation de l'alimentation (séchage des matières au degré de siccité requis).

Le schéma de l'installation, dont un pilote est prévu au Centre de Recherches de Pont-à-Mousson, est donné par la figure 1, qui montre en outre les résultats du calcul des bilans matière et énergie effectués pour des conditions réalistes de fonctionnement qui seront justifiées plus loin.

Le bois entrant à 50 % d'eau (1 kg de bois anhydre + 0,5 kg d'eau) est séché jusqu'à une valeur comprise entre 10 et 20 % d'eau. Le séchage est effectué dans une tour à entraînement (licence Air Industrie du Groupe SGPM) qui utilise, d'une part, de l'air préchauffé par les gaz sortant du séchoir lui-même et, d'autre part, par les gaz de combustion du résidu carboné solide ou "char" provenant de la pyrolyse.

Le bois séché et légèrement préchauffé est envoyé dans le réacteur de pyrolyse où il est mélangé, dans un lit fluidisé par une recirculation des gaz de pyrolyse, à des particules réfractaires portées à la température adéquate par la combustion du "char" qui est effectuée dans un réacteur séparé travaillant lui aussi en lit fluidisé.

Les gaz produits servent à un premier préchauffage de l'air de combustion du "char", cette récupération assurant la production d'un gaz sec par la condensation de la vapeur d'eau existant dans ces gaz.

L'air de combustion du "char" reçoit encore deux préchauffages avant son admission dans le réacteur de combustion :
- par les gaz de combustion sortant de la jaquette du réacteur de pyrolyse qu'ils contribuent à maintenir à la température désirée,
- par les gaz de combustion lors de leur passage dans le cyclone à la sortie du réacteur de combustion destiné à les débarasser des cendres du "char".

Les gaz de combustion sortant du réacteur de combustion du "char" sont utilisés en 4 endroits :
- pour le 3ème préchauffage de l'air de combustion dans le cyclone de décendrage,
- pour maintenir le réacteur de pyrolyse à la température de réaction,
- pour le 2ème préchauffage de l'air de combustion,
- enfin pour le séchage du bois d'alimentation.

Nous avons établi les bilans matière et énergie pour l'installation décrite ci-dessus pour différents cas caractérisés par un couple de valeurs: température de pyrolyse - humidité du bois. Cette première approche comporte des approximations car un certain nombre de paramètres ne pourront être déterminés que par l'expérience sur le pilote projeté. Néanmoins ces calculs permettent de cerner la zone de travail possible, c'est-à-dire les conditions d'équilibre du système supposé en régime établi. Un modèle plus complet de l'installation est en cours d'élaboration.

Nous donnons ci-après un exemple de ce calcul effectué pour la pyrolyse effectuée à 750°C de bois de hêtre amené à 10 % d'humidité.

Le tableau de la page suivante donne toutes les valeurs nécessaires aux bilans calculées à partir des données expérimentales. Ce tableau présente les valeurs pour les températures de pyrolyse comprises entre 700 et 900°C par pas de 50°C.

L'ensemble montre la cohérence des résultats expérimentaux sur lesquels il s'appuie. On notera en particulier que la somme des énergie contenues dans le gaz et le "char" conduit à une valeur très proche de celle obtenue pour le bois tant expérimentalement qu'en utilisant la formule de Dulong et Boile. Nous avons calculé par cette formule les pouvoirs calorifiques du bois et du char.

Pour le bois dont la formule réduite (ramenée à 1 atome-gramme de carbone) est :
$$CH_{1,44} \, O_{0,66}$$
et la formule pondérale :
$$C_{50} \, H_6 \, O_{44} \quad (50 \text{ \% en masse de C, } 6 \text{ \% H et } 44 \text{ \% O})$$
la formule de Dulong et Boile :
$$PCS = 0,352 \, (C) + 1,162 \, (H) - 0,111 \, (O) \quad MJ/Kg$$
donne
$$PCS = 19,775 \, MJ/Kg$$
d'où l'on tire :
$$PCI = 18,457 \, MJ/kg$$
Pour le char de hêtre pyrolysé à 750°C on a la formule :
$$CH_{0,815} O_{0,215} \text{ ou } C_{73,8} \, H_{5,0} \, O_{21,2}$$
qui donne un $PCI = 28,389 \, MJ/Kg$.

Les calculs de bilans sont menés de la manière suivante :
1) On calcule d'abord les débit et température du mélange d'air préchauffé et de fumées de combustion permettant de sécher le bois à la valeur désirée. Pour ne pas commencer la décomposition du bois dans le séchoir, il faut que le gaz chaud ne dépasse pas 250°C. Dans le cas d'un séchage de 50 à 10 % le gaz est admis à 210°C.

HETRE à 10 % d'eau

Soit 1 kg bois sec + 0.1 kg d'eau
1 mole bois + 0.133 mole d'eau
(24 g)   (2.4 g)
(= 1 at-g de C)

	Température de pyrolyse (en °C)				
	700	750	800	850	900
G A Z					
- Mole de gaz par mole de bois					
CO	0.275	0.300	0.325	0.337	0.349
CO_2	0.047	0.051	0.055	0.0535	0.052
H_2	0.077	0.1015	0.126	0.144	0.162
CH_4	0.079	0.088	0.097	0.0985	0.100
C_2H_4	0.033	0.0375	0.042	0.040	0.038
C_2H_6	0.004	0.0035	0.003	0.0025	0.002
C_2H_2	0.003	0.0045	0.006	0.0085	0.011
H_2O produite	0.123	0.112	0.100	0.105	0.105
- Gaz sec					
Mole	0.518	0.586	0.654	0.684	0.714
Masse g	12.308	13.527	14.746	15.07	15.394
Volume $\frac{Nm^3}{kg\ bois}$	0.483	0.547	0.610	0.638	0.666
- Taux de gazéification					
en mole	0.518	0.586	0.654	0.684	0.714
en masse	0.513	0.564	0.614	0.628	0.641
sur le carbone	0.481	0.530	0.579	0.591	0.603
- Energie dans le gaz MJ					
pour 1 mole de bois (24 g)	0.213	0.240	0.268	0.276	0.285
pour 1 kg de bois sec	8.874	10.017	11.156	11.520	11.884
PCI MJ/Nm³	18.376	18.309	18.289	18.042	17.841
C H A R					
- Mole	0.519	0.470	0.421	0.409	0.397
- Formule	$CH_{0.90}\ O_{0.24}$	$CH_{0.815}\ O_{0.215}$	$CH_{0.73}\ O_{0.19}$	$CH_{0.655}\ O_{0.165}$	$CH_{0.58}\ O_{0.14}$
- Masse moléculaire	16.74	16.255	15.77	15.255	14.82
- Masse					
par mole de bois (g)	8.688	7.640	6.639	6.255	5.883
par kg de bois sec (kg)	0.362	0.318	0.277	0.261	0.245
- PCI calculé MJ/kg	27.854	28.389	29.021	29.779	30.545
ENERGIE RECUPEREE (MJ/kg bois)					
- sur gaz	8.874	10.017	11.156	11.520	11.884
(%)	(46.8)	(52.6)	(58.1)	(59.8)	(61.3)
- sur char	10.084	9.029	8.037	7.761	7.489
- Energie totale du bois (PCI)	18.958	19.046	19.193	19.281	19.373

2) On calcule ensuite toutes les valeurs intéressant le réacteur de pyrolyse :
   - humidité du gaz produit en tenant compte de l'eau de l'alimentation,
   - chaleur de réaction pour la pyrolyse du bois(par interpolation des
     valeurs expérimentales)
   -chaleur totale de réaction en tenant compte des contenus calorifiques
     des réactifs et des produits ainsi que du réchauffage des gaz de flui-
     disation
   -débit de caloporteur pour couvrir les besoins calorifiques du réac-
     teur de pyrolyse en admettant une chute de température de 100°C pour
     le caloporteur.
3)Puis on calcule les différentes étapes du préchauffage de l'air stoechio-
   métrique de combustion du char.
4) Enfin on calcule le bilan de la combustion du char en déterminant
   d'abord la température théorique de combustion de celui-ci. Puis on vé-
   rifie que l'énergie contenue dans le char permet, une fois les besoins
   calorifiques nécessaires au réchauffage du caloporteur satisfaits, et
   compte tenu du préchauffage de l'air de combustion, de porter les gaz à
   une température au moins égale à la température théorique de combustion
   calculée plus haut. Dans le cas de l'exemple : hêtre à 10 % d'eau, température de pyrolyse 750°C, on arrive aux chiffres suivants :
           température théorique de combustion : 2 200°C
           température possible pour les fumées : 2 255°C

En effectuant ces calculs pour différentes conditions de température
de pyrolyse et d'humidité du bois, on peut tracer une courbe frontière
dans la représentation : % d'humidité - température de pyrolyse, au-delà de
laquelle le système n'est plus autosuffisant en énergie avec les seuls
résidus solides de l'opération.

Cette représentation est donnée à la figure 2 ci-après où figurent
également la limite pratique de dessication du bois (5 %) et une courbe
correspondant à une fraction déterminée de l'énergie contenue dans le bois
et récupérée dans le gaz. Sur la figure 2 nous avons adopté les valeurs de
50 % en trait plein et 55 % en trait pointillé.

On définit ainsi un domaine de travail où le système trouvera un
point d'équilibre, la température de pyrolyse s'ajustant à la teneur en
humidité de l'alimentation. Mais il faudra tenir compte des effets rétro-
actifs des différentes parties du système et c'est ce que nous permettront
de vérifier le pilote et le modèle mathématique prévus.

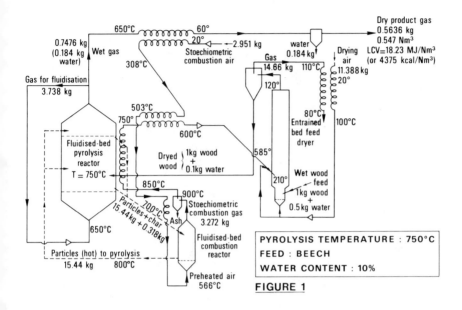

PYROLYSIS TEMPERATURE : 750°C

FEED : BEECH

WATER CONTENT : 10%

FIGURE 1

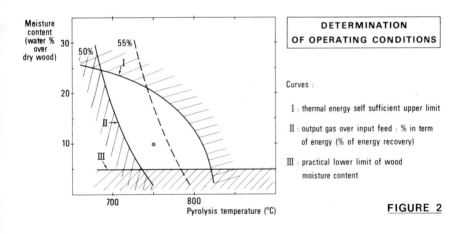

DETERMINATION
OF OPERATING CONDITIONS

Curves :

I : thermal energy self sufficient upper limit

II : output gas over input feed : % in term
of energy (% of energy recovery)

III : practical lower limit of wood
moisture content

FIGURE 2

TWO NEW TYPES OF BIOMASS GASIFIERS
DEVELOPED AT C.N.E.E.M.A.

J.F. MOLLE

Centre National d'Etudes et d'Expérimentations du Machinisme Agricole
(C.N.E.E.M.A.) FRANCE

Summary

In FRANCE, wood gasification technology was and is still used on an industrial scale since a long time (De Lacotte devices). Its further development was thwarted for economical difficulties. This technology is suited for granular materials. In order to take advantage from other non granular products (straw, sawdust, rice hulls, peanut husks, etc ...) an agglomerating process was necessary but too expensive to be implemented. So, such products are to be used as such. The problem was to be solved from two standpoints :

1. Modernization of packed bed gasifiers for rough products,
2. Development of a suspension bed gasifier for fine products.

These devices are now available on the french market.

The present packed bed gasifier is two times smaller than the older types. Consequently investment costs are reduced. This gasifier is designed so that ashes are carbon free and gases tar free, thus its output is increased and its polluting effect eliminated.

For the suspension bed gasifier (horizontal shaft cyclone) those same problems are solved. The high turbulence generated by gases causing the product to be in suspension, gives rise to very fast mass and heat exchanges, thus gasification reactions are also very fast.

I. INTRODUCTION

In FRANCE, gasifiers and the associated problems are known from a long time and equipment manufactured on an industrial scale have proven their worth (✗). The both prevailing De LACOTTE and DUVANT down draft gasifiers are still sold and allow several gasification systems in the world to be presently running. These systems are fixed bed gasifiers in wich wood is not entrained by gas but is fixed. Down draft gasifiers are also sold presently by the IMBERT company in the Federal Republic of GERMANY. Both De LACOTTE and Down draft gasifiers are expensive devices, and cannot process non granular products (straw, sawdust, rice hulls, ..), Problems C.N.E.E.M.A. has been working on for some years.

II. GASIFICATION TECHNOLOGIES

The De LACOTTE gasifier (Cf. figure)

This gasifier which consists of two vessels (A and B) and an important combustor (C) is top fed.

Pyrolisis of wood is made in the upper vessel. Products driven off from pyrolitic process pass to the combustor C in which they burn with oxygen (when synthesis gas is required) or air (when only lean gas is required).

Hot gases (1 000 - 1 100°C) released from combustion flows through the constricted zone between the two vessels, practically following the axis of the gasifier. Consequently gases are well distributed within the wood pile. One quarter of gas released flows up to vessel A and ensures wood pyrolisis, the three quarters left flowing through vessel B. Charcoal produced in vessel A flows down through vessel B where it is gasified according to the following reactions by water vapour and carbon dioxide contained in hot gases.

$$C + CO_2 \quad 2\ CO\ (BOUDOUART)\ \Delta H = + 41\ Kcal$$
$$C + H_2O \quad CO + H_2\ (water\ gas)\ \Delta H = + 31\ Kcal$$

✗ *Gasifiers used on cars during World War II which were not all properly designed are not referred to here.*

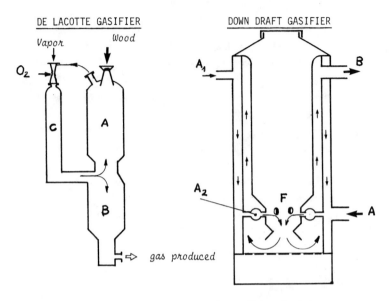

DE LACOTTE GASIFIER

DOWN DRAFT GASIFIER

A = Main air inlet
A_1 = Additional air inlet
A_2 = Heating chambers
B = Gas outlet
F = fire box

It can be drawn from the above-mentioned information that the "prime mover" of gasification comes from heat produced by combustion of part of the wood product ; combustion gas heat content is used in endothermic reactions of charcoal gasification (BOUDOUARD and water gas). When gasification begins if the temperature is 0°C (or temperature at which charcoal is gasified by combustion gases) combustion gases give up their sensible heat between 0°C and 600°C, temperature under which gasification process is practically ineffective. This sensible heat quantity reduces the charcoal gasification rate but two solutions which are not incompatible can be implemented for improving this rate.

1) Increase temperature of combustion gases.
2) Increase the volume of combustion gases passing through charcoal.

The first solution cannot be applied owing to the peculiar characteristics of Wood fuel. At temperatures higher than those above mentioned (1 000 - 1 100°C) i.e. about 1 300°C, ashes become very fluid entailing slag production which hinders running of the unit.

Down draft gasifier (Cf. figure) ✗

This gasifier is less sophisticated and consist of only one vessel. Gases can only be withdrawn from the base of the unit. They are forced to pass through a constriction in which air supplies create a combustion area. In this area pyrolisis of wood is made as well as cracking or combustion of tars produced. Then charcoal is gasified by hot combustion gases in the conical area formed after the constriction.

The main advantage of this gasifier is its simple design. But for high powers (higher than 500 KW), the diameter of the vessel is important so that air supplied do not reach to produce a flame zone up to the constriction centre. In this way, cracking or combustion of tars is not made and tars can be found in gases produced.

Air is blown at a very high speed in order that a combustion zone could be generated up to the bed centre. Temperature at the air intake level is then very high ("torch effect") reaching about 1 300°C. Matters and gas flowing is hindered under these conditions by ash changing into slag just like the De LACOTTE gasifier when a total gasification is aimed at.

This gasifier is not well suited to an oxygen-blown gasification with a view to produce synthesis gas. On the one hand, temperature reached with an axygen combustion is very high and to bring down it to 1 300°C, which is the degree reached with air, oxygen has to be diluted with a great deal of water vapour. On the other hand, gasifier sizes required for chemical synthesis units are huge. For example, in the case of a "modest" plant for methanol sunthesis producing 500 T/day (plants presently running produce from 1 000 up to

✗ It's a gasifier diagram of a down draft gasifier for use with trucks during World War II and not a DUVANT gasifier. However the running principle is similar

1 500 T/day). 2.5 Kg of dried wood are required to produce one kilo of methanol. Consequently 500 × 2,5 = 1 250 T i.e. 52 T/h should be gasified.

It's impossible to foresee more than 10 gasifiers which implies that they would have an output of 5 T/h, but presently a down draft gasifier (bigger than those wich produce 700 Kg/h of methanol) releasing gases without tars has not yet been developed.

III.C.N.E.E.M.A. RESEARCH WORKS

If the gasification principle is not new, improvements can be made concerning size of gasifiers (and consequently investement) their output (without carbon in ashes and without tars in gases) and their easy operation (problems raised by flowing of solid products - bridging effect- or gases).

Moreover, other products lighter than wood can be processed for example, straw, sawdust, rice hulls, ground nut or coffee husks, etc ... Such residues cannot be processed in a fixed bed gasifier. When they are in deep beds, gases travel along preferential flue shaped passages on account of their low density and resistance to gas flow. As this phenomenon increases, the gasifier cannot go on running. If granulated these residues for use with fixed bed gasifier are too expensive. Two gasifier types taking account of the crop fuel initial granulometry (because modification of granulometry would be too costly) are required as follows :

- Fixed bed gasifiers for wood, coconut husks, maize cobs and more generally for all heavy products with a high granulometry.
- Fluidized bed gasifiers for light and fine products. Such gasifiers are not yet available on the market.

1) Fixed bed gasifier

a) The De LACOTTE principle consists in burning pyrolisis products in a separate chamber. Gases produced are free from tars. This process is used again.

b) A special heat resisting steel fan has been designed to activate the hot gas circulation in the bed in order to maintain the top column temperature at 600°C. Consequently the new fuel is immediatly dip into this atmosphere at 600°C where it is subjected to a thermal shock. Thermal transfers are more rapid with the previous technology (top column temperature at 100°C). Drying and pyrolisis of the product are carried out in a very little sized zone. The total size of the unit is divided by a factor equal to 2 for a same power.

Uncompletely burnt gases from combustion of volatile matters is recycled by a fan on the reacting charcoal. During the gasification process, a quantity of gases, two or three times higher than that of a gasifier without recycling, passes through charcoal. The temperature at wich charcoal is gasified by such gases, with a same quantity of calories can be brought down from 1 300°C to 900°C maintaining a complete gasification. At 900°C ashes are not under a melting slag state. Ash fertilizer content (almost all phosphorus and potassium initially contained in the fuel) can be assimilated by crop roots if ashes are recycled.

A horizontal gasifier in which a load of wood is pushed on by a piston has been choosen. Very hot gases can uniformly pass through the entire section of the pile of wood standing at a certain natural angle of rest. There is no problem of material mechanical strength because gases can directly pass through charcoal. Hot gas distribution is easy on the wood pile which remains identical whichever the gasifier width. Very big units can be foreseen. As the De LACOTTE process allows an oxygen blown gasification, the unit is matched to synthesis gas production. With the horizontal design, savings can be made as far as structure, skip hoists, ladders and observation passageways are concerned ... Beside these specifications, this unit runs just like the De LACOTTE gasifier.

A prototype of 200 Kwe allows to prove that this process is fully justified. Running of such prototype has been tested with wood, coconut husks and maize cobs.

The following results were obtained :
- 1Kg of wood - 10 % moisture content - 3 800 Kcal/Kg
- Gases produced :
  - mean composition $CO = 16$

       % in volume $CO_2 = 12$

                                       $H_2 = 20$

                                       $N_2 = 50$

                                       $CH_4 = 2$
  - LHV   = 1 170 Kcal/Nm3
  - Volume   = 2,5 Nm3
- Output $\dfrac{1\ 170 \times 2,5}{3\ 800} = 0,77$

If such gas production is burnt in a dual-fuel engine the following energy equivalence is :
- 1 Kg of wood with a 10 % moisture content = 1 Kwh

A second prototype of 700KW is presently under investigation (march 1978). Tests will be carried out with urban wastes.

2) PILLARD fluidized bed gasifier

a) Light materials concerned such as chopped straw, sawdust, coffee husks, rice hulls, etc... have a wide range of granulometry. A cyclone type furnace with horizontal axis was used. Whichever the particle size may be, particles are uniformly laid on the wall by centrifugal force, remaining nearly the same time can be processed under the same conditions.

b) Two zones have to be foreseen in this unit, i. e. :
- A zone operating at a mean temperature (500°C) where the crop material is fed. In this zone, a rapid pyrolisis of the product is made.
- At the other end of the unit another zone operates at very high temperatures (1 000°C) owing to air supplies and ensures charcoal gasification as well as tar cracking. The gas withdrawn from the two zones at 600°C is recycled. It allows :
  - The product fluidization (since gasification air would

be insufficient).

. The cold gas supply around flames at the level of air
inlet to maintain steel sheets at a low temperature, so that too
important stresses are not applied to sheets and ashes do not adhere
to them.

Tests were carried out with a prototype of 400KW
(Ø 1.2 m l = 2.5 m) by Ets PILLARD at MARSEILLE (FRANCE) since
January 1978. Satisfactory results were achieved with straw, rice
hulls and coffee husks ; performances look like those above-mentioned
for wood material.

Charcoal dust obtained by pyrolisis can be uncompletely gasified. If such dust is pelleted charcoal can be produced
from annual crops such as straw or from wastes which can be of
interest for many African countries.

c) In fluidized bed gasifiers wood (or other crops) fuels
are fluidized by gases, a reaction which entails high speed heat and
mass exchanges between fuels. Chemical reactions are faster than
with fixed bed gasifiers. These units are more compact.

3) Future of these technologies

It seems particularly attractive for the future to produce
from biomass a liquid fuel which can be stored and transported such
as methanol. Fertilizers (synthesis of ammoniac) and basic carbonaceous products for chemical industry (hydroliquefaction) can also
be derived from biomass.

In the two first cases only, gasification of plant
materials with a view to produce synthesis gas may require a new
approach since further synthesis imply techniques which proved their
worth.

Hydroliquefaction consists in changing directly at 250 bars,
300°C, biomass into oil comparable with fuel oil in the presence of
a synthesis gas produced from biomass gasification.

A study carried out by TECHNIP, for C.N.E.E.M.A., and concerned with methanol and ammoniac synthesis from plant material shows that raw material account for half the final cost of methanol (✗) or ammoniac.

The importance of a good output of gasifiers which can take advantage of plant materials under the best conditions is underlined.

Such processes were not necessarily cheap with old technologies. Il would be interesting presently to have gasifiers, matched to products to be processed, available since there remains the key of many future theories dealing with biomass utilization.

C.N.E.E.M.A.-PILLARD Fluidized bed gasifier

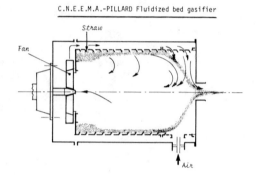

Fixed bed gasifier - C.N.E.E.M.A.

A = Air
F = Fan for recycling pyrolitic gases
G = Gasifigation

P = Pyrolisis
R = Recycling

✗ 100 $/t in MALI, i. e. the present world rate

KINETIC STUDIES OF PYROLYSIS AND GASIFICATION
OF WOOD UNDER PRESSURE WITH STEAM AND OXYGEN.

A. DIVRY, P. DUBOIS and J.C. RENARD

Laboratoires de Marcoussis, Centre de Recherche de la
Compagnie Générale d'Electricité, 91460-Marcoussis-France

Summary

These kinetic studies are part of the wood gasification programme of
the "Compagnie Générale d'Electricité". The general purpose of this programme is the building of plants for methanol synthesis from wood. This
programme is conducted by NOVELERG and is carried out by "Les Laboratoires
de Marcoussis, Centre de Recherches de la Compagnie Générale d'Electricité",
"Les Ateliers et Chantiers de Bretagne" and by "CGEE-Alsthom".
    The aim of the present studies is to determine wood pyrolysis and
gasification parameters which are needed for the design and the operation
of a fixed bed gasifier working under pressure with steam, oxygen and
partial or total electrical heating.
    In this paper we describle an experimental apparatus which has been
built to carry out these kinetic studies. This apparatus comprises main
four parts : a test vessel fitted with an electrical heater, a steam
generator and an oxygen feed, a treatment and measurement device for the
output gas and a quench device for gas sampling from the test vessel.
    The main parameters which will be measured are the optimal residence
times of wood and charcoal in the pyrolysis and gasification zones, the
gasification rate and yield and the electrical properties of charcoal
pyrolysed at different temperatures.
    Every part of the apparatus has been tested and experimental studies
are now under progress.

1. INTRODUCTION

These kinetic studies are part of the wood gasification programme of the "Compagnie Générale d'Electricité". The general purpose of this programme is the building of plants for methanol synthesis from wood. This programme is conducted by NOVELERG and is carried out by "Les Laboratoires de Marcoussis, Centre de Recherches de la Compagnie Générale d'Electricité", "Les Ateliers et Chantiers de Bretagne" and by "CGEE-Alsthom".

The aim of the present studies is to determine wood pyrolysis and gasification parameters which are needed for the design and the operation of a fixed bed gasifier working under pressure with steam, oxygen and partial or total electrical heating. The results of these studies will be used to design and to analyse working conditions of a wood gasifier, operating in the 1-20bars range and which will be in operation at Marcoussis by the and of the first term of 1981.

2. EXPERIMENTS

2.1. Apparatus

Figure 1 gives the flow diagram of the apparatus which comprises four parts :
- a test vessel fitted with an inner electrical heater.
- a steam generator and an oxygen feed
- a treatment and measurement device for the output gas.
- a quench device for gas sampling from the test vessel.

A - Test vessel

The test vessel is a room temperature pressure shell fitted with an inner thermal insulation and an inner electrical heater (see figure 2). Wood or charcoal are placed in an inconel basket lined with alumina. Produced gas is collected at the top of the basket and for gasification studies, steam and oxygen are introduced underneath the basket. This vessel is equipped with a savety valve, a gaz sampling tube (see-D) and with several thermocouples for gas and wood or charcoal temperature measurements.

B - Steam generator and oxygen feed

Water stored under nitrogen pressure in a stainless-steel vessel is supplied to the evaporator through an electrical flow-meter (see figure 1). The evaporator is held at a temperature higher than the saturation temperature so that the steam flow-rate is equal to the measured flow-rate of the water.

Oxygen under pressure is heated at a temperature close to the steam temperature and mixed with the steam at the test vessel inlet. Oxygen flow-rate is measured with an electrical flow-meter.

C - Treatment and measurement device for the output gas

At the outlet of the test vessel, gas is cooled down through a water heat-exchanger and flow through a decanter in order to condense non-volatile products. The clean gas flows through an upper pressure regulator and leaves the apparatus at the atmospheric pressure. Pressure and flow-rate of the clean gas are measured, upstream the pressure regulator, with electrical pressure-gauge and flow-meter. In addition, below the pressure regulator, there is a sampling gaz device for chromatography composition analysis.

D - Quench device for gaz sampling from the test vessel

The gas sampling tube of the test vessel is connected to a quench device joined it self to a vacuum pump (see figure 1). This device comprises a sampling vessel, in which gas samples are taken at low pressure (a few tens of Torr) and a draining vessel to clean the piping before sampling. A sampling cycle is automatically achieved through electromagnetic-values driven by logical circuit.

2.2. Measured parameters

The following parameters are measured :

A) - Residence times of wood and charcoal in the pyrolysis and gasification zones as a function of :
   . the sizes of wood and charcoal logs
   . the operating pressure
   . steam and oxygen partial pressures
   - Gasification rate and gasification yield as a function of the state of pyrolysis of the wood.

Theses parameters are determined through the measurements of the flow rate and of the composition of the gas produced at different temperatures between 200°C and 900°C.

B) Electrical properties of charcoal produced from wood pyrolysed at different temperatures and pressures. The table 1 gives room temperature electrical resistivity of charcoal from different kinds of wood pyrolysed at atmospheric pressure.

Table 1 - Electrical properties of charcoal

Kind of wood	Pyrolysis temperature (°C)	Resistivity at 20°C (Ωcm)		$\dfrac{\rho_\perp}{\rho_{//}}$
		Parallel to the fibre direction, $\rho_{//}$	Perpendicular to the fibre direction, $\rho_\perp$	
Balata rouge	500	2,86	3,36	1,2
	700	$5,2 \ 10^{-2}$	1,82	19,2
AKO	500	$1,35 \ 10^{5}$	$5,95 \ 10^{5}$	4,4
	700	$1,16 \ 10^{-1}$	$1,15 \ 10^{-1}$	1
Bilinga (Gabon)	500	$1,45 \ 10^{-1}$	$4,1 \ 10^{-1}$	2,8
Jaboty (Guyanne)	500	$1,34 \ 10^{-1}$	$6,5 \ 10^{-1}$	4,8
Beech	500	$1,6 \ 10^{5}$	$1,35 \ 10^{6}$	8,4

3. CONCLUSION

The apparatus for kinetic studies of pyrolysis and gasification of wood has been built and every part of it has been tested. Experimental studies are now under progress.

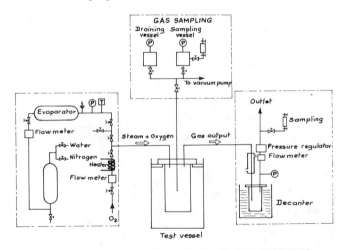

Figure 1 : **FLOW DIAGRAM OF THE APPARATUS FOR PYROLYSIS AND GASIFICATION KINETIC STUDIES**

587

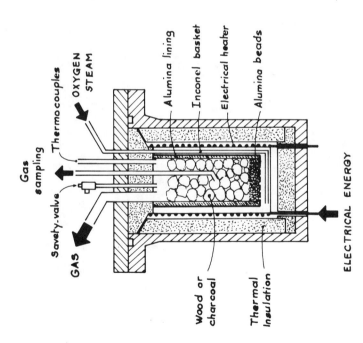

Figure 2 : TEST VESSEL

METHANOL CATALYTIC SYNTHESIS FROM CARBON MONOXIDE
AND HYDROGEN OBTAINED FROM COMBUSTION OF CELLULOSE WASTE

C. MASSON, A. BOURREAU, M. LALLEMAND

F. SOUIL and J.C. GOUDEAU

Groupe de Recherches de Chimie Physique de la Combustion
Université de Poitiers
Domaine du Deffend
MIGNALOUX-BEAUVOIR
86800 St JULIEN L'ARS (FRANCE)

Summary

    Certain industrial catalysts for methanol synthesis have been tested at low pressure (20 to 50 bars) in an apparatus under continuous rating on pure carbon monoxide and hydrogen gases. Several factors other than working pressure, e.g. temperature, stoichiometry, mass of catalyst and flow rates have been studied. Interesting results have been obtained mainly concerning selectivity and yield.

    Taking into account the results obtained, we carried out trial tests on home made catalysts in view of a low pressure operation (50 bars) yielding methanol with medium selectivity as efficient as industrial processes from diluted gases. We shall study the effects of nitrogen and other gases like ammonia or sulphur dioxide on the activity of the catalyst.

1. INTRODUCTION.

   Nous avons entrepris l'étude de la synthèse catalytique du méthanol
à partir de gaz de synthèse ($CO + 4 H_2$) dans un domaine de pression compris entre 20 et 50 bars. Dans une première phase, nous avons réalisé une
installation de synthèse fonctionnant en régime dynamique et testé les
catalyseurs industriels de synthèse du méthanol par une exploration sys-
tématique de différents paramètres tels que : temps de contact, débits de
gaz, température et pression. Dans une deuxième étape, nous avons préparé
des catalyseurs de synthèse à partir d'oxydes métalliques et les avons
également testés dans notre installation. Ces catalyseurs sont destinés à
être utilisés directement à partir des gaz issus de la combustion de matière végétale, c'est-à-dire des gaz contenant diverses impuretés telles
que l'azote ou l'ammoniac notamment.

2. DESCRIPTION DE L'APPAREILLAGE ET FONCTIONNEMENT.

   L'appareillage représenté sur la figure (1) comporte trois arrivées
de gaz : monoxyde de carbone, hydrogène et azote. Le circuit d'azote peut
être utilisé lors de la réduction des précurseurs ou pour les études sur
gaz dilués. Les débits gazeux sont mesurés par débitmètres thermiques et
réglés par vannes aiguilles, ces débits sont maintenus constants dans le
temps par introduction de volumes tampon placés en amont et par des capil-
laires placés en aval des débitmètres. La pression est maintenue constante
dans l'installation à l'aide d'un déverseur à dome.

   Le chauffage du réacteur tubulaire en acier inoxydable est assuré
par trois fours à régulations indépendantes. Le lit catalytique est placé
au milieu du réacteur et sa température est maintenue constante malgré
l'important dégagement de chaleur issu de la réaction de synthèse du
méthanol. Ceci a pu être obtenu par la réalisation d'un lit catalytique
composé de 15 cm^3 de catalyseur et de 15 cm^3 de carborundum, la dilution
du catalyseur permettant une plus grande surface d'échange thermique.

   Les produits formés sont condensés et séparés des gaz n'ayant pas
réagi qui sont détendus dans le déverseur et dirigés vers un chromato-
graphe en phase gazeuse pour analyse.

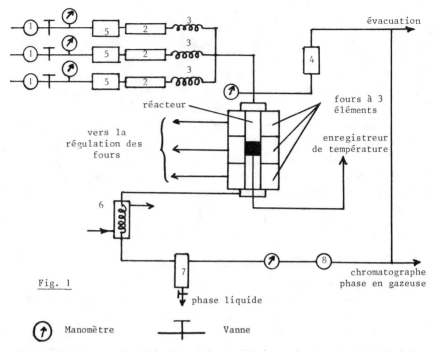

Fig. 1

⊕ Manomètre ⊢ Vanne

1. Manodétendeur ; 2. Débitmètre ; 3. Capillaire ; 4. Soupape de sécurité ;
5. Volume tampon ; 6. Condenseur ; 7. Séparateur de phases ; 8. Régulateur
de pression amont.

3. RESULTATS EXPERIMENTAUX.

Dans un premier temps, nous avons étudié la synthèse du méthanol dans
les conditions stoéchiométriques, c'est-à-dire selon la réaction :

$$CO + 2 H_2 \rightarrow CH_3OH + 21,7 \text{ Kcal}$$

Sur de tels mélanges, nous avons étudié l'influence de la température du catalyseur et du temps de contact du mélange gazeux avec certains
catalyseurs industriels et avec des catalyseurs préparés au laboratoire
sur le taux de conversion du monoxyde de carbone et la sélectivité en
méthanol.

a/ Influence de la température du catalyseur.

Cette étude a été menée sur deux catalyseurs industriels (catalyseurs à base d'oxydes métalliques : Cu, Zn, Cr, Al en proportions variables) et sur deux catalyseurs de notre fabrication (catalyseurs également à base d'oxydes métalliques : Cu, Zn, Cr, Al, Th, Ce en proportions variables).

Les conditions opératoires sont les mêmes pour les quatre échantillons c'est-à-dire : pression de 50 bars et temps de contact de 0,8 seconde.

Les catalyseurs contenant du chrome voient leur activité diminuer très rapidement avec la température (au-dessus de 340°C). Par contre, les catalyseurs contenant de l'aluminium présentent une excellente stabilité à température élevée, l'alumine stabilise donc le catalyseur et limite sa désactivation. Un séjour sous atmosphère d'hydrogène lui permet en outre de retrouver son activité initiale.

Ces résultats indiquent que l'alumine peut avantageusement remplacer l'oxyde de chrome dans la composition des catalyseurs.

b/ Influence du temps de contact.

La figure n° 2 représente les variations du taux de conversion du monoxyde de carbone en méthanol en fonction de la température du catalyseur pour différents temps de contact du mélange gazeux avec ce même catalyseur. Les conditions opératoires sont les suivantes : pression 50 bars, catalyseur à base d'oxyde de thorium.

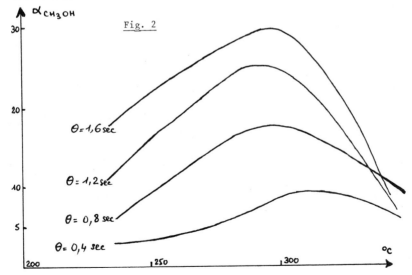

Le taux de conversion maximum augmente rapidement avec le temps/de contact mais la quantité de méthanol obtenue par unité de masse de catalyseur et par unité de temps diminue. Inversement, lorsque le temps de contact est faible, la quantité de méthanol obtenue augmente et le taux de conversion diminue. On est donc amené à déterminer des conditions optimum qui correspondent à un temps de contact voisin de 0,8 seconde. Signalons que ce temps de contact correspond à un VVH (volume de mélange gazeux à l'entrée du réacteur par unité de volume de catalyseur et par heure) de 4 500 voisin des VVH industriels.

La figure n° 3 représente les variations du taux de conversion du monoxyde de carbone en fonction de la teneur des catalyseurs mis au point dans notre laboratoire en oxyde de thorium. Les conditions opératoires correspondent à une pression de 50 bars et à un temps de contact de 0,8 seconde.

Fig. 3

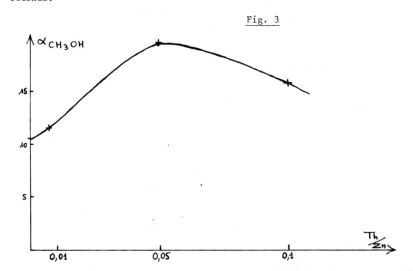

La courbe obtenue montre l'influence importante de la teneur en oxyde de thorium. On constate en effet la présence d'un maximum de taux de conversion pour un rapport de concentration $\frac{[Th]}{[Zn]}$ de 0,05. Dans tous les cas, la sélectivité en méthanol est supérieure à 95 % et dépasse 99 % pour certains catalyseurs de notre fabrication.

4. CONCLUSION ET PERSPECTIVES.

L'installation réalisée permet de tester directement les catalyseurs de synthèse sur des gaz de provenance et de composition variées dans des conditions identiques à la synthèse industrielle.

Les travaux effectués à ce jour ont utilisé deux gaz de synthèse purs CO et H_2. Les produits de départ pouvant être d'origine et de composition diverses, des études sont actuellement en cours avec des mélanges gazeux contenant de l'azote, de l'ammoniac, de l'anhydride sulfureux et du dioxyde de carbone. Nous étudions plus particulièrement l'influence de ces gaz sur les paramètres de la synthèse : taux de conversion et sélectivité et sur l'activité du catalyseur : vieillissement et empoisonnement.

Dans ces études nous avons utilisé des catalyseurs de type classique : à base d'oxydes métalliques, nous nous proposons de développer notre recherche de catalyseur sur ce même type de composition : influence de la teneur en certains oxydes et de faire appel à des supports tels que l'alumine et les tamis moléculaires. Nous porterons notre effort sur une limitation de la pression de synthèse.

PROCESS AND EQUIPMENT FOR THE FLUID BED

OXYGEN GASIFICATION OF WOOD

G. CHRYSOSTOME

Engineer - Energy division

CREUSOT-LOIRE

B.P. 31

F - 71208 LE CREUSOT

Summary

The aim of the research is to develop a fluidized bed oxygen gasification of wood. A new process as well as the equipment allowing its implementation are to be developed. The combination of fluidized bed and oxygen allows to obtain a medium B.T.U gas free of any nitrogen. The use of fluidized bed allows to reduce wood preparation and feeding. Sawdust can also be gasified. Due to fluidized bed temperature homogeneity and to the fact that counter flow is avoided the medium B.T.U gas is free of any tar. The research project involves three main stages : 1 - getting kinetic and thermodynamic data with the help of a low thermal inertia research reactor ; 2 - developing and implementing a fluidized bed gasogene that may be fed with wood, bark and sawdust ; 3 - operating a pilot reactor under various operating conditions.

For CREUSOT-LOIRE research on wood gasification is the object of a very large programme. Only a part of it will be funded by the European Communities. All the programme will be described in this paper. The part which is to be sponsored by the European Communities and makes the object of the final report will be described later.

The programme first presented to the European Communities includes four main stages :

Stage 1 : <u>Getting kinetic and thermodynamic data on wood gasification</u> :

This stage includes :

- construction of a laboratory test rig for wood gasification, equipped with oxygen, steam and carbon dioxyde feed systems.

- the a priori computation of gaseous balance in terms of the reagent analyses and of the operating conditions;

- the experimental study of wood gasification in terms of temperature and pressure conditions, of the reactive medium analyses and of the size of biomass pieces to be gasified.

Stage 2 : <u>Design and implementation of a fluidized bed gasifier</u> :

The equipment mostly includes a fluid bed gas generator of 400 mm inside diameter.

In order to reach a swift steady load operation, a low thermal inertia unit is to be provided. The gasifier will thus include a steel casing insulated from the outside, equipped at its base with a perforated grate supporting the fluid bed.

The reactor is to be equipped with a sufficient number of temperature probes to allow the establishment of the unit thermal chart.

Pressure probes will allow to check the pressure under and above the fluid bed. By holding the pressure slightly below atmospheric pressure above the fluid bed emission of harmful gas (CO) into the station atmosphere can be avoided.

The fluid bed will be fed with pure oxygen and steam ; injection of CO_2 may be tested.

A part of the gas produced by the gasifier will be sent to gas analysers. The most important part will be fired from a special burner into an ancillary combustion chamber in order to avoid any discharge of polluting gas into the atmosphere.

The reactor will be completed with a cyclone and equipped with wood injection systems. The larger wood pieces are to be fed onto the fluid bed by an endless screw system. Sawdust will be sent through a pneumatic conveying system within the fluid bed itself.

All fluid and solid flows are to be metered with appropriate devices.

Stage 3 : Gasifier operation

The main parameters having consequences on the resulting gas analysis which are to be systematically tested are :

- fluidization gas analysis : $\dfrac{O_2}{H_2O}$ ratio
- fluid bed temperature
- fluidizing velocity
- influence of the stoechiometric ratio
- dimensions of the pieces of wood

Besides their influence on the resulting gas analysis the effect of such parameters on the rate of elutriation and on the possible tar content is to be examined.

Periodical samplings taken in the bed while the unit is in operation will allow to determine the analysis of the fluid bed. The main purpose of the sampling is to detect any possible enrichment of the fluid bed by alcaline elements.

These elements are liable to entail the sintering of the fluidized bed while in operation on a continuous industrial basis.

Stage 4 : Preliminary project for an industrial unit :

The preliminary project of an industrial unit, the size of which is still to be determined, will allow to estimate the corresponding investment. The cost of the unit plus that of utilities (O_2, H_2O, etc...) will permit to calculate the price of the resulting gas.

Several hypotheses as to the cost of raw materials, will be considered.

At the request of the E.C. we have shortened the programme to be sponsored :

This programme non includes :

- erection of the fluid bed gasifier.
- experimental study of the ways to introduce the wood into the gasifier.
- influence of the dimensions of wood pieces on gasification.

At the end of the research the following results are expected to be available :

- ways to introduce the wood into the gasifier : it means out of the bed or into the bed.
- best size of wood pieces : it means pieces of wood or sawdust.

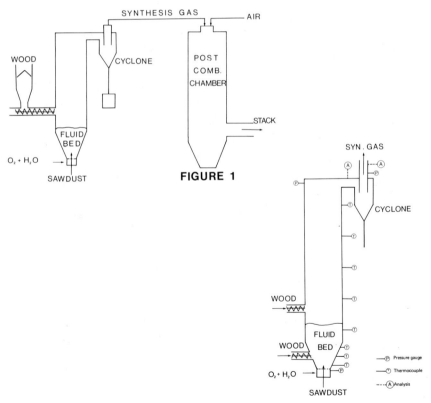

FIGURE 1

FIGURE 2

TECHNO-ECONOMIC EVALUATION OF THERMAL ROUTES
FOR PROCESSING BIOMASS TO METHANOL, METHANE
AND LIQUID HYDROCARBONS

G. ADER

Ader Associates Ltd., P.O. Box 69 West Wickham, Kent U.K.

A.V. BRIDGWATER and B.W. HATT

Department of Chemical Engineering
University of Aston in Birmingham,
Gosta Green, Birmingham B4 7ET. U.K.

Summary

    Conceptual studies have been carried out on a number of specific
case studies to define realistic process flow-sheets, develop mass
and energy balances and derive capital costs and total production
costs for conversion of biomass to useful fuels such as methanol,
methane (SNG) or liquid hydrocarbons (gasoline and syncrude). The
thermal process case studies covered low and high temperature pyrolysis,
oxygen gasification and steam gasification. In all cases the gaseous
products were assumed to be processed through a series of conversion
steps (such as shift reaction, steam reforming, carbon-dioxide
removal) to give suitable material for methanol synthesis, methanation,
or Fischer-Tropsch synthesis. Plant capacities were initially set
at relatively low but realistic levels of production. Further
calculations are included on what is considered to be a maximum
achievable biomass availability at any one point in order to explore
the ultimate effect of economies of scale. The results indicate that
steam gasification might be the most promising thermal process route
although it is not yet fully developed; they also indicate methanol
as being the most economically promising fuel product. Oxygen
gasification also shows significant promise, although it does not
appear to be as attractive as steam gasification: it is in principle,
however, available commercially.

1. INTRODUCTION

It is known that biomass materials can be thermally processed to gaseous products that may be converted to "synthesis gas", a well established raw material for the production of a range of fuels such as methanol, methane and liquid hydrocarbons.

The composition of products from thermal processing of biomass is highly variable and depends on processing conditions employed. It cannot be predicted on theoretical grounds as sufficient relevant data are not available due to the complexity of the process. A sophisticated approach to process optimisation cannot be realised and the main objectives of this study have been restricted, therefore, to; (a) developing outline flowsheets; (b) evaluating selected case studies in terms of overall mass and energy balances; (c) deriving approximate capital and production costs. Evaluation of mass and energy flow-sheets for processes starting with pyrolysis have been based on published compositions of products for medium and high temperature pyrolysis of wood; for oxygen gasification; and for the steam gasification. These were chosen for the wide range of conditions encountered in thermal processing. The pyrolysis and steam gasification case studies were based on laboratory scale experimentation, while oxygen gasification was based on substantial demonstration scale work. Many flow sheets for the several products have been examined in order to select the more favourable for detailed study.

The data base for the conceptual conversion of the thermal decomposition products to synthesis gas, operations to purify and adjust the composition of the raw gas and the final conversion to useful products were all derived largely from well established information.

Hydrogenation to produce liquid and gaseous hydrocarbons directly were not included in this study.

Capital cost estimates were developed for each process on a consistent capacity basis. Calculations were based on specified cost models for process units where historical costs are available; for other items the functional unit step counting method was adopted. All costs are quoted on a U.K., mid-1980 basis. As the main purpose of this study was comparison of costs for different rountes, sensitivity analyses for different capacity plants were not undertaken.

Production costs were computed from operating costs and financial charges based on ten year amortisation at 10% interest. Raw material costs are assumed at £20 per dry tonne. All energy requirements are assumed to be met "in-house" and there are therefore no fuel and power costs.

2. METHANOL

Mass and energy balance calculations for production of methanol by a low pressure process for each of three thermal routes investigated, showed steam gasification to have the highest overall yield, both in terms of the weight of methanol produced per unit weight of dry biomass feedstock, and in terms of net energy recovered. Table I summarises the results.

Capital cost and production cost estimates are summarised in Table II.

Pyrolysis routes show the highest costs, even the more economical high temperature process has been shown to give costs as high as £209 per tonne methanol.

TABLE I   Methanol from Biomass - Summary of Yields

	Net Energy Yield	Weight Yield (dry weight basis)
Pyrolysis (500°C)	35%	29%
Pyrolysis (900°C)	41%	34%
Oxygen gasification	48%	35% (50% on a d.a.f.basis)
Steam gasification	55%	51%

N.B. Net energy yield is defined as the energy content of the product divided by the energy content of the feed plus other energy inputs. The latter is zero for all these processes as they are self-sufficient in energy by increasing feedstock and/or using byproducts.

TABLE II   Methanol from Biomass — Summary of Costs

	Pyrolysis		Gasification	
	Medium Temp.	High Temp.	Oxygen	Steam
Raw material input, t/d as recived @ 30% water	1250	1060	1000	700
t/d, d.a.f. basis	870	742	500	490
Methanol Output, t/d	250	250	250	250
Capital Cost Estimates £ million	39	31	28	23
Production Cost Breakdown				
Raw Materials (%)	31	29	33	29
Labour (%)	4	4	4	5
Capital related (%)	65	67	63	66
Methanol Production Cost £/t	253	209	193*	149

Current Methanol cost is around £120–£130/t

* becomes £174/t if feedstock is costed on the same basis as the other processes.

For the oxygen gasification route it should be noted that the cost of £193 per tonne methanol was derived from data on processing of urban refuse, as no reliable data were available on oxygen gasification of wood.

The steam gasification route indicates the lowest methanol cost at £149 per tonne.   This applies to a 250 t/d methanol plant; if capacity was doubled, a capital cost increase of about 50% can be anticipated leading to a methanol production cost of around £120 per tonne which is comparable to current costs.   The uncertaintly in scaling up laboratory-derived process data to full scale commercial operation should not be underestimated.

It must be concluded that both in terms of yield and costs, thermal processing of biomass by steam gasification may prove the optimum route for producing methanol from biomass if the process can be satisfactorily scaled up.   However, this is at such a relatively early stage of development that direct comparison with alternatives might be misleading. On current available technology, methanol production via oxygen gasification may give comparable costs at a reasonable throughput.

3. METHANE

Analogous calculations have been carried out on production of methane by the same three processes. The methanation stage has been assumed to proceed according to the steam-moderated multi-bed process that is currently being developed for coal-to-methane processing.

Mass and energy balance calculations for production of methane by each of the three thermal routes showed weight yields to be low, but net energy yields to be comparable to methanol apart from the relatively high figure for steam gasification. The following Table III summarises the results:

TABLE III Methane from Biomass - Summary of Yields

	Net Energy Yield	*Weight Yield (dry weight basis)*
Pyrolysis (900°C)	*39%*	*13%*
Oxygen gasification	*48%*	*13% (19% on d.a.f.basis)*
Steam gasification	*77%*	*29%*

Estimates of capital costs and production costs were computed for plants assumed to have a capacity of 250 t/d methane. The relative importance of the various cost elements are highlighted in Table IV.

TABLE IV Methane from Biomass - Summary of Costs

	Pyrolysis (900°C)	*Oxygen Gasn.*	*Steam Gasn.*
Raw material input			
t/d as received basis - 30% water	*2660*	*2620*	*1235*
t/d d.a.f. basis	*1860*	*1310*	*865*
Methane output, t/d	*250*	*250*	*250*
Capital cost estimates, £ million	*58*	*46*	*29*
Methane production costs			
- £ per tonne	*428*	*378*	*209*
- £ per GJ	*7.7*	*6.8*	*3.8*
Production cost breakdown			
Raw materials (%)	*38*	*44*	*37*
Labour (%)	*2*	*2*	*4*
Capital related (%)	*60*	*54*	*59*

4.   LIQUID HYDROCARBONS
      Current technology indicates the following alternative approaches to
the production of liquid hydrocarbons from biomass:
(a) Thermal processing to synthesis gas that is converted into a crude
      liquid hydrocarbon by Fischer-Tropsch synthesis
(b) Thermal processing to synthesis gas, converting to crude methanol
      and subsequently converting to a motor fuel by the M-gasoline
      process.

      The process of direct liquefaction of biomass to crude oil has been
excluded from the studies.

Fischer-Tropsch
      The Fischer-Tropsch reaction has been assumed to proceed according
to data available for the SASOL coal-to-oil plant.   The information
available on mass and energy balances and on capital costs for the
Fischer-Tropsch plant is very scant and the level of reliability of
data for this process is much lower than in the other processes.   Moreover, the liquid hydrocarbons from a Fischer-Tropsch process are a crude
product requiring further refining, and therefore strictly comparable
only to a high grade petroleum crude.   Recent developments however claim
a greater specificity to particular fractions.

      Mass and energy balances and estimates of capital costs and prod-
uction costs were computed for plants assumed to have an output of 100t/d
liquid hydrocarbon.   The selected process sequences result in mass and
energy yields and in production costs summarised in Table V.

      It is evident that compared to high grade petroleum crudes (current
price around £100 per tonne), liquid hydrocarbons produced from biomass
via Fischer-Tropsch reaction are not likely to be economically viable.
Even at the largest conceivable scale of operation, and provided further
improvements will achieve the ultimate yields that are theoretically
feasible, costs would still be around three times the price petroleum
crudes are likely to reach by the end of the decade.

M-gasoline process
      Conversion of biomass to liquid hydrocarbons via the M-gasoline
process has been visualised as an add-on process.   Biomass is first
converted to methanol by any of the thermal routes outlined, followed by
conversion to gasoline by the process recently developed by the Mobil

Oil Company and currently being scaled up.

TABLE V   Liquid Hydrocarbons from Biomass via Fischer-Tropsch
Synthesis - Summary of Yields and Costs

	Oxygen Gasn.	Steam Gasn.
Raw materials, t/d, as received 30% water	3040	1160
t/d, d.a.f. basis	1520	812
Liquid hydrocarbon output t/d	100	100
Other hydrocarbon output t/d	350	110
Net weight yield (dry basis)(%)		
- liquid hydrocarbons	5	12
(7 on d.a.f. basis)		
- total hydrocarbons (%)	20	26
(29 on d.a.f. basis)		
Net energy yield (%)		
- liquid hydrocarbons	14	32
- total hydrocarbons (ignoring in-plant requirements)	65	72
Capital costs £ million	86	55
Production costs - per tonne of liquid hydrocarbons, £	1400	770

The data published so far indicates that the M-gasoline process will yield about 38 tonnes of stabilised gasoline for every 100 tonnes of methanol processed.   Estimates of capital and production costs for a plant with methanol input of 250 t/d have shown incremental production costs to be equivalent to £22 per tonne methanol.   Adding this cost to those previously estimated for melthanol gives the results summarised in Table VI.

Compared to current price of gasoline of around £210 per tonne, it is evident that the estimated cost of £450 per tonne applicable to the steam gasification route may have some prospect for becoming economically viable at a higher production rate as the price of fossil fuels continues to escalate in the future.   Based on currently available technology, it appears that conversion of biomass to gasoline via the M-gasoline process shows better prospects of viability than conversion to mixed liquid hydrocarbons via the Fischer-Tropsch reaction.

5. MAXIMUM CAPACITY AND MINIMUM COST

It is difficult to envisage any biomass conversion process operating with a feedrate on a dry basis in excess of 500000 t/y or 1500 t/d. The transport and handling costs and logistics preclude consideration of a larger plant.

All of the processes examined have been reassessed on this maximum feed rate basis of 1500 t/d dry and relatively ash free material. For the oxygen gasification alternative the results will probably approximate to a wood feed. The results are shown in Table VII. The production costs are probably the minimum achievable by exploiting the economies of scale.

TABLE VI   Liquid Hydrocarbons from Biomass via M-Gasoline Process
Summary of Yields and Costs

	Pyrolysis (900°C)	Oxygen Gasn.	Steam Gasn.
Feedstock, t/d, as recieved 30% water	*1000*	*1000*	*700*
t/d, d.a.f. basis	*742*	*500*	*490*
Methanol production, t/d	*250*	*250*	*250*
Gasoline output, t/d	*95*	*95*	*95*
Overall yields, %			
(gasoline on feedstock)			
- weight	*13*	*13*	*19*
- net energy	*38*	*46*	*52*
Total capital cost, £ million	*36*	*33*	*28*
Production costs, £			
per tonne methanol			
- for biomass/methanol	*209*	*193*	*149*
- for methane/gasoline	*22*	*22*	*22*
Total production costs, £			
per tonne gasoline	*608*	*566*	*450*

N.B.   L.P.G. is produced concurrently at the rate of 12% by weight of the gasoline produced.   This is not included in any way in the above figures.

TABLE 7 Maximum Yields and Minimum Costs

FEED : 1500 t/d dry biomass (2143 t/d at 30% water biomass).

	High Temperature Pyrolysis	Oxygen Gasification	Steam Gasification
METHANOL			
Yield t/d	505.5	753.0	760.3
Capital cost £ million	49.5	57.0	47.2
Production cost .£/t	179.9	131.3	115.8
METHANE			
Yield t/d	191.4	286.0	433.4
Capital cost £ million	49.2	54.0	40.9
Production cost £/6	474.9	336.4	188.9
£/GJ	8.6	6.1	3.4
GASOLINE			
Yield t/d	192.1	286.1	288.9
Capital cost £ million	53.6	62.3	52.6
Production cost £/t	509.5	375.6	334.9

ACKNOWLEDGEMENTS

The work outlined in this report is the result of a three year project sponsored by the Energy Technology Support Unit of the Department of Energy.

BIBIOGRAPHY

Ader G., Bridgwater A.V. and Hatt, B.W.,
Conversion of Biomass to Fuels by Thermal Processes - Phase 1 Review and Preliminary Assessment.   Report to Energy Technology Support Unit, Department of Energy, December 1978.

Ader G., Bridgwater A.V. and Hatt, B.W.
Conversion of Biomass to Fuels by Thermal Processes - Optimisation Studies for Conversion to (a) Methanol, April 1980,
                              (b) Methane, July 1980,
                              (c) Liquid Hydrocarbons, September 1980.
Reports to Energy Technology Support Unit, Department of Energy.

THERMAL PROCESSING OF BIOMASS TO SYNTHESIS GAS -
A PROGRAMME OF EXPERIMENTAL INVESTIGATIONS AND
DESIGN STUDIES

E.L. SMITH, B.W. HATT, G.A. IRLAM and A.V. BRIDGWATER

Department of Chemical Engineering,
University of Aston in Birmingham,
Gosta Green, Birmingham B4 7ET U.K.

Summary

Extensive technical and economic evaluation of thermal processes
to produce synthesis gas from Biomass for subsequent conversion to
high value fuels, has revealed many areas of uncertainty and lack
of knowledge. An experimental programme has been initiated to
obtain a greater understanding of reaction systems and develop
design procedures for thermal gasifiers.

1. BACKGROUND

The conversion of biomass to mixtures of carbon oxides and
hydrogen is attractive, as this "synthesis gas" can be converted subsequently to a range of relatively high grade fuels, such as methanol
and methane, by conventional chemical engineering operations.   Much
work has been reported throughout the world, particularly in the last
decade, on the gasification of wood, straw, refuse, manure, peat and
other forms of biomass.   A report in which this work is comprehensively
reviewed has been prepared for the Energy Technology Support Unit of
the Department of Energy (1).   The following were among the conclusions
drawn from that study:

(a) The effects of reaction system parameters on the composition of
     the products and their rate of formation is highly complex and
     not well understood.

(b) There are apparent inconsistencies between some of the published
     data, possibly due to incomplete monitoring and reporting of
     reaction conditions.

(c) While many types of reactor system have been investigated by
     different workers, very few attempts have been made to compare
     them under controlled conditions.

(d) Reliable design procedures for biomass gasifiers have yet to be
     developed.

Subsequent discussions led to the initiation of a programme of
experimental investigations and design studies: this is briefly
described below.

2. THE RESEARCH PROGRAMME

The programme is being carried out in the Department of Chemical
Engineering, the University of Aston in Birmingham, and is funded by
the Department of Energy.

The experimental work will be aimed at finding the relationships between various reaction parameters and the formation of products from a range of feedstocks. In addition to thermogravimetric and differential thermal analysis of individual particles, purpose-built equipment will be used to examine and model the interactive effects occurring in multi-particle systems. The reaction parameters to be investigated will include feedstock size and preparation, heating rate and temperature, and the composition of the gaseous environment. The data obtained in this part of the overall programme and that reported elsewhere will be used to select and design a number of gasification reactors for the optimal production of synthesis gas.

The research programme at Aston is to be run in parallel with and be complementary to a related programme of research and development being carried out by Foster Wheeler Power Products (FWPP) Limited of London. A 40 kg/h multi-purpose test facility is being constructed by FWPP that will enable several reactor types and process configurations to be investigated at pressures up to 30 atm. and temperatures up to 1200°C. Initially, runs will be done with the Cross-Flow reactor already developed by FWPP. Subsequently, reactors designed at the University of Aston will be installed and evaluated in the test facility (2).

3. REFERENCES

(1) Ader, G., Bridgwater, A.V. and Hatt, B.W.
     "Conversion of biomass to fuels by thermal processes -
     Phase 1", Energy Technology Support Unit, Department of Energy,
     December, 1978.
see also Hatt, B.W. and Bridgwater A.V. "Conversion of Biomass to
     Fuels by Thermal Processes", in "Energy from the Biomass"
     the Watt Committee on Energy Ltd., Report No. 5, 62-66,
     June 1979.

(2) Smith, E.L. and Wilson, H.T., These proceedings.

ECONOMICS OF COMBUSTION ENERGY FROM CROP RESIDUE

F.J. HITZHUSEN and M. ABDALLAH

Associate Professor of Resource Economics, Department of Agricultural
Economics, The Ohio State University, Columbus, Ohio - U.S.A.
and Assistant Professor of Economics, Faculty of Administrative
Sciences, Ryiadh University, Ryiadh, Saudi Arabia

Summary

    This research evaluated the economic feasibility of corn stover
(remaining residue after grain harvest) as a coal supplement in coal
burning steam-electric power plants in the North Central United States.
This is consistent with the worldwide search for renewable sources of
energy accentuated by the rising prices of nonrenewable energy sources
(e.g., gas and oil) and their declining availability. When burned with
high sulfur coal, corn stover (or other residues) can also reduce sulfur
emissions and may result in liquid fuel savings from reduced transport of
low sulfur coal for blending.

    Two case steam-electric plants were analyzed utilizing three harvest
and collection systems, two boiler types and alternative values for
several key technical and economic parameters. The stack harvest system
was least costly. Sensitivity analysis found the price of coal, the BTU
content of stover and the harvest cost significant determinants of feasibility. The optimization of boiler size and transport costs occurs
between 100 and 168 tons of stover per day. These results combined with
data on a sample of 53 steam-electric plants show considerable potential
in the North Central United States. Crop residues may also have potential
in other countries with coal burning installations in proximity.

Note: A copy of the cited references can be secured from the authors.

1. INTRODUCTION

The energy content of crops and crop residues results from solar energy captured by plants in the photosynthesis process. Part of that energy is digestible and used as food. The major indigestible part of the plant energy is contained in agricultural crop residues and is a major subset of total biomass. Among crop residues in the North Central United States, corn stover has the highest BTU value per unit of weight because of its relative efficiency in utilizing atmospheric carbon dioxide (CO_2) (7). The estimated BTU value of corn stover dry matter ranges between 6,500 BTU/lb. (3) to 8,000 BTU/lb. (2,8). The average heat value for coal is 12,200 BTU/lb.

Corn stover contains only 0.017% sulfur (0.053 lbs. SO_2/MBTU). When burned with high-sulfur coal, it can potentially function as a sulfur emission control material to meet air pollution standards of the U.S. Environmental Protection Agency (EPA). The EPA air quality standards differ from one state to another but, generally speaking, coal of more than 1% sulfur (1.639 lbs. SO_2/MBTU, on the average) is considered "low quality coal" (4). Further, 62% of the higher sulfur coal reserves in the United States are found east of the Mississippi, where 90% of the coal-fired power generation occurs. To meet the EPA sulfur emission standards, it is necessary for coal burning power plants east of the Mississippi to import western coal, or use stack scrubbers, fluidized beds, or some other technology which may be more expensive than supplemental corn stover combustion. Corn stover is also a soil erosion control material. It provides soil nutrients and it can also be used for livestock feed and bedding.

In determining the feasibility of corn stover as a coal supplement, the benefit of reducing the sulfur emissions externality and the external and private costs (soil and nutrient loss) associated with the removal of stover from the soil should be considered. Other costs include the real and opportunity cost of harvest, storage, and transport of the stover. Thus, for corn stover to be economically feasible as a coal supplement, the cost of producing a certain amount of energy (one million BTU is the conventional unit) from corn stover should be less than or equal to the cost of producing an equivalent amount of energy using coal. The major task of this research is the determination of the cost of a unit of heat generated using corn stover for a variety of situations and assumptions.

2. METHODOLOGY

Two case studies of coal burning steam-electric power plants chosen from the population of power plants in the North Central United States are evaluated including sensitivity analysis. The Peru, Indiana power plant represents the numerous small plants (50 tons of stover/day) which may have limited potential for resource recovery from solid waste or garbage (6). The Ames, Iowa, power plant, converted in 1975 to use solid waste, represents slightly larger power plants (150-200 tons of stover/day) and its operators have experience in the combustion of solid waste with coal. The selection of these two power plants also represents the two most common types (pulverized coal and stoker) of boilers.

Due to the lack of a market price for corn stover, break-even point and linear programming analysis are found to be the most appropriate methodologies to determine the feasibility of corn stover as a fuel supplement for coal. The market for corn stover is conceptualized as having a supply (farm sector), a demand (power plant sector), and a transportation sector to link supply and demand. The quantitative relationships that determine the break-even points of each of the three sectors are formulated. The interaction between the three sectors makes up the complete economic model of stover as a supplemental fuel. For complete formulation of the individual sector models including the linear programming model for optimizing fuel mix and data collection procedure, see Hitzhusen and Abdallah (5).

The break-even point of the system is realized when the summation of the farm sector cost/MBTU and transportation cost/MBTU is equal to the maximum price the power plant can pay for stover per MBTU. The complete model is as follows:

(1.a) $CH + CS + NVS + LUC + CLM \times D = Pc + PRc + Ec - CUS$, or
(1.b) $CH + CS + NVS + LUC + CLM \times D - Pc - PRc - Ec + CUS = 0$

where,

CH = harvest cost for stover generating one MBTU of heat
CS = storage cost for stover generating one MBTU of heat
NVS = savings of foregone chopping costs minus fertility loss for amount of stover generating one MBTU of heat
LUC = loading and unloading costs for amount of stover generating one MBTU of heat
CLM = cost of hauling stover generating one MBTU of heat
D = hauling distance required to meet plant capacity
Pc = price of coal per MBTU
PRc = processing costs of coal per MBTU

Ec = sulfur emission control costs per MBTU

CUS = cost of using stover in power plant per MBTU.

If equation (1.b) is less than or equal to zero, stover is assumed to be feasible as a coal supplement at 1977 prices. If it is greater than zero, stover is infeasible.

To attempt to apply the results of the two case studies to other power plants and associated stover sheds in the region, fifty-three (53) power plants were randomly chosen from the power plant population (a 22% sample) in the North Central United States. Data on coal prices, custom rates, power plant stover capacity and availability were collected for each of the fifty-three power plants and associated stover sheds and are presented in Abdallah, pages 170-73 (1).

3. RESULTS

The Ames power plant is already converted to burn solid waste. It has two stokers and one pulverized coal boiler. To reduce the sulfur emissions of the high sulfur Iowa coal, Ames is using low sulfur Colorado coal to conform to the EPA sulfur standards. The stover combustion capacity of one of the stoker boilers is 150 tons/day. To compare the results of the analysis of the stoker boiler with the pulverized coal boiler, a 150 ton/day level of throughput is also used for the pulverized coal boiler.

Four scenarios are analyzed for this power plant: stoker boiler without emission control costs, stoker boiler with emission control costs (Ec), pulverized coal boiler without emission control costs and pulverized coal boiler with emission control costs (Ec). The farm, transportation, and power plant sector costs are estimated for each of the four scenarios. The cost of sulfur emission control is imputed using the linear programming model mentioned earlier. The capital cost of modifying the power plant is amortized using 5%, 9%, and 13% interest rates as part of the sensitivity analysis. However, the 9% rate is assumed to be the most likely or appropriate rate of interest over the life of the project.

By adding the farm and transportation sector costs and subtracting the maximum price the power plant can pay, the feasibility of stover is determined for the four scenarios at the Ames power plant. At 150 tons/day throughput, a 9% interest rate on capital, and 1977 prices of coal, positive values are obtained for both boiler types including the sulfur emission control values and utilizing the stack harvest system. A positive value reflects an economically feasible or "better than break-even" alternative. Excluding the sulfur emissions control costs from the cost of using coal at the Ames power plant, corn stover is not feasible under any of the four scenarios. More detailed development of the Ames case results is presented in Abdallah, pages 81-106 (1).

The Peru case results are generally less promising due to lower custom rates and the lower per unit modification costs at Ames. Per unit modification costs are one and one-half times higher at Peru because of the size economies of boiler conversion. The higher level of throughput at Ames yields a lower modification cost per unit of corn stover burned ($0.295/MBTU) compared to Peru ($0.476/MBTU) for the same type of boiler (pulverized coal) and the same rate of interest (9%).

Sensitivity analysis is done on the major variables that affect the feasibility of stover as a fuel. Specific independent variable changes required for the system to break-even are determined holding all others constant. The major variables are: price of coal, custom rates, BTU content of stover, plant modification costs, level of throughput, changes in harvest technology, and hauling distance.

Independence of the major variables can be assumed in the sensitivity analysis with the exception of level of throughput which may affect plant modification cost, hauling distance and/or transport costs. Plant modification costs might also increase as the result of such things as rapid escalation of wages and older plant and equipment. Hauling distance may increase or decrease at a constant level of throughput due to variation in stover density. The price of coal, BTU content of stover, and harvest cost decreases from new technology can all be assumed independent. A separate sensitivity analysis is done on the interrelated variables or parameters. At 9% interest on capital for the stack system of harvest at both the Ames and Peru plants, the system is generally most sensitive to the independent variables of coal price, stover BTU content, and harvest cost.

Sensitivity analysis tests the impact of the size of boiler converted or level of throughput on the per unit cost of firing and transporting the stover (the scale effect). Five boiler sizes or full capacity levels of throughput at 50, 75, 100, 125, and 150 tons/day of stover are employed to show the scale effect for the Ames, Iowa stoker boiler. Extrapolated estimates are made for two additional levels of throughput (175 and 200 tons/day). It is found that as the boiler size increases, the per unit modification cost decreases. However, as boiler size increases, the per unit transportation costs also increase. The optimum level of throughput is

100, 137, and 168 tons/day respectively at 5%, 9%, and 13% interest.

The results of the sensitivity analysis and data from the sample of power plants (n=53) can be utilized to get some notion of the potential of corn stover combustion with coal in steam-electric boilers in the North Central States. Information for the sample plants on stover density, harvest and transport, custom rates, coal prices, sulfur emission standards, boiler type, number and size, combined with the assumption that large multiple boiler plants can convert only part of their capacity to corn stover combustion provides some basis for extending the case study results. It would appear that from one-third to one-half of the coal burning power plants in the North Central United States have more favorable or comparable conditions for feasibility of stover as a fuel to those of Ames, Iowa.

4. CONCLUSIONS

If the price of coal increases at a faster rate than the harvesting, transport, and firing costs of stover, the feasibility of corn stover as a fuel at small power plants similar to the Peru, Indiana plant is foreseeable in the near future. Assuming sulfur emission standards are required to be met, the Ames, Iowa case is economically feasible at current prices.

The sulfur emission control costs proved to be important for the economic feasibility of stover. If the sulfur standards in the United States become more stringent and are more rigorously enforced, stover feasibility would improve. If the proposed requirement of installing stack scrubbers in all the coal-using power plants is imposed, the value of corn stover as a sulfur emission reducing material would probably decline.

More research is needed on the potential for combustion of corn and other crop and forest residues. This is particularly true with coal burning industrial and institutional boilers which are probably more numerous than steam-electric plants in the United States. This work should also attempt to determine any potential net liquid fuel savings from reduced transport of low sulfur coal to meet emission standards. Preliminary assessment of these savings looks promising. Further research is also needed on the economic feasibility as well as the energy balance of alternative uses of corn stover and other crop and forest biomass for methane, ethanol, and other thermochemical and biological conversion products. The recent discovery at Purdue University of a more efficient solvent process for the conversion of cellulosic material to ethanol (a liquid fuel) may be promising in this regard (10).

PRODUCTION OF A LOW-BTU FUEL GAS BY COCURRENT GASIFICATION OF SOLID WASTES

J.J. Hos, A.A.C.M. Beenackers,

F.G. van den Aarsen and W.P.M. van Swaaij

Twente University of Technology

P.O. Box 217,

Enschede, The Netherlands.

Summary

At Twente University of Technology a research programme is being
carried out on the conversion of biomass into a low-Btu fuel gas in co-
current moving bed gasifiers. The purpose of this programme is to study
the suitability of this reactor for the conversion of organic solid wastes
into a clean fuel gas.
Important features of the cocurrent moving bed gasifier are:
- a fuel gas free of tar and acids can be produced;
- gasification together with firing of the fuel gas requires
  low excess air rates;
- the fuel gas can be supplied either to a boiler or to an
  internal combustion engine;
- the solid waste is exposed to temperatures in between
  1000-1600 °C.
Experimental results have been obtained on gasification of wood wastes,
maize cobs and municipal solid waste.
Wood wastes such as chips, shredded crates, shavings and sawdust were
gasified in reactors with a capacity of 20 - 50 kg of wood waste per hour.
The lower heating value of the gas produced was 3000 - 6000 kJ/Nm^3,
depending on the moisture content of the feedstock.
With respect to wood wastes the stage of commercialization has been reached.
Shelled maize cobs were gasified in reactors with a capacity of 10 kg per
hour. The gas was fired succesfully in a diesel engine.
Muncipal solid waste was gasified to yield a fuel gas with an L.H.V. of
4000 kJ/Nm^3. A problem area is the removal of cindering or slagging residues
from the gasifier. Further research will be carried out concerning the
application of this gasification process to feedstocks containing components
with melting points below 1000 °C.
Existing thermodynamic models were verified and a kinetic model was developed. A new research subject is the development of an omnivorous fluid bed
gasifier producing a clean fuel gas.

1. INTRODUCTION

Due to scarcity of energy resources a good deal of attention is paid
to the conversion of solid wastes into energy. The most wide spread method
to attain this conventional incineration, remains unsatisfactory in many
respects. Not only high excess air rates are usual, but also corrosion
prevention and meeting the emissivity standards are often problematic.
Gasification, defined here as the complete conversion of carbonaceous
solids into a combustible gas, is a way out for dealing with these problems,
since a fuel gas can be cleaned easily before burning and no high excess
air rates are required.

Taking this into consideration and noticing the ever increasing
quantities of solid wastes being produced, a research programme on gasification was set up at Twente University of Technology. The cocurrent moving
bed reactor was selected, mainly because of its simple construction. In
addition the fuel gas produced with this reactor type contains only small
amounts of condensible organic matter and also in situ gascleaning was in
principle possible.

Initially only wood waste and maize cobs were gasified in reactors
having a capacity of 2 - 5 kg of solid waste per hour. About 2.5 Nm^3 of
dry gas per kg of solid waste with a lower heating value of 5000 kJ/Nm^3 was
obtained. Later on fundamental research was carried out to develop reaction
engineering models for different parts of the gasifier (1). Separate projects were set up to evaluate the utilization of the cocurrent moving bed
gasifier with respect to various types of solid waste.
The projects concern gasification of municipal solid waste, coal and the
implementation of gasifiers for agricultural wastes in developing countries.

In this contribution attention will be paid to the principles of gasification in cocurrent moving bed reactors and to the results achieved with it
at Twente University of Technology. More over, new research programmes,
recently started at THT will be described briefly.

2. THE COCURRENT MOVING BED GASIFIER

The main part of the reactor is a vertical shaft, in which air and
solid waste enter at the top (Fig. 1).

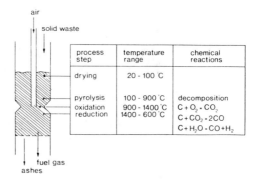

Fig. 1 Principles of cocurrent gasification.

Gas and ashes leave the reactor at the bottom. The solid waste is ignited
in the narrowest cross section of the reactor very near the place where
the air is introduced. In this section partial combustion of the solid
matter takes place. The incoming oxygen is rapidly consumed to form char
and a steam and carbondioxide containing gasmixture. This is called the
oxidation zone of the gasifier and temperatures are as high as 1500 $^{\circ}$C.
The heat produced in this zone has two functions. Firstly it brings about
drying and thermal decomposition of the solid matter in the pyrolysis zone.
Secondly the char below the oxidation zone reacts endothermically with
carbondioxide and steam to form hydrogen and carbonmonoxide. This zone is
called the reduction zone. As a result the temperature of the gas leaving
the gasifier has dropped to 700 $^{\circ}$C. The flowrate of the solids through the
gasifier is determined by the rate of char conversion and the rate of ash
removal. Consequently the air flow rate is an important control variable.
The reduction of the cross sectional area of the gasifier and the position
of the air inlet tube both are very important factors for the production of
a fuel gas which is free of organic condensables produced in the pyrolyses
zone (tars). In a well designed gasifier both factors contribute to the
formation of flowpatterns that prevent pyrolyses products from flowing
directly into the reduction zone along the relatively cold wall of the
gasifier. Instead all pyrolyses products are forced to pass the area where
the highest temperatures prevail and they are completely decomposed. Rules
for designing and upscaling the required reactor geometry were developed
by Groeneveld (2). The design of a cocurrent moving bed gasifier was found

to be strongly dependent on the shape and size of the solid waste to be gasified, as well as its moisture content and inert content (3).

3. EXPERIMENTAL RESULTS

Experiments were carried out with gasifiers ranging in capacity from 2 - 50 kg of solid waste per hour. For detailed descriptions see (3). Woodblocks, chips, maize cobs, shavings and saw dust were gasified in experimental runs from 6 to 8 hours. Shredded M.S.W. was used as M.S.W. as received contains objects too large to be gasified in the present reactors. Data concerning typical experiments are summarized in table I. The efficiency mentioned in this table is defined as the percentage of the lower heat of combustion of the solid waste that is recovered as lower heat of combustion in the fuel gas.

Table I: experimental results.

Solid Waste	Wood, chips	Wood, shavings sawdust	Maize cobs	M.S.W. shredded
Dimensions				
average (cm)	3	0.5	10	5
Moisture (weight %)	8	9	7.5	5
Exp. time (hr)	6.5	8.0	6.0	2.0
Materials				
IN:				
Solid waste	21.4	17.3	13.0	12.5
Air	42.9	35.0	25.6	29.8
OUT:				
Dry gas	58.9	46.8	37.7	36.6
Water	5.0	4.3	0.6	4.3
Ash	0.2	0.9	0.2	4.2
Dry gas composition				
CO (vol %)	25.4	21.7	18.6	16.7
H_2 "	14.2	15.8	14.0	12.4
CO_2 "	7.8	9.6	13.9	10.3
N_2 "	49.4	50.4	51.5	58.0
CH_4 "	1.8	1.7	1.4	1.0
Ar, O_2 "	1.2	0.8	0.6	1.5
Energy				
Solid waste (MJ/kg)	17.15	16.87	13.40	12.60
Dry gas (MJ/kg)	4.75	4.50	3.42	3.22
Efficiency (%)	77	73	74	75

The ashes together with charfines were recovered from the ashchamber and a cyclone, the total weight being 1-5 % of the solid wastes gasified. The amount of tars, char and ashes recovered in the scrubber following the cyclone was about 500 mg/Nm^3 fuel gas. After a total of 100 hours of operation it was not yet necessary to clean the piping following the scrubber.

Special subjects under investigation are the influence of moisture content

620

of the feedstock and air preheating on the properties of the fuel gas,
removal of slagging or cindering residues from the reactor and in situ
removal of acid gaseous components. Until so far the following conclusions
are drawn:

- Solid wastes can be gasified without additional measures if average
  dimensions are in between 10^{-3} - 5.10^{-2} m., if the moisture content
  is below 30 wt. % and if ashes and inert do not form cindering or
  slagging residues.
- Drying of the feedstock and preheating of the gasification air
  considerably increase the heating value of the fuel gas produced.
- Preliminairy tests indicate that in situ removal of acid gaseous
  components should be possible (3).
- Producing a tar free gas is highly dependent on reactor geometry.
  An interdependence exists between tar production, solids flow rate
  and solids diameter (1).
- The main problem area in gasifying M.S.W. is the removal of cinders
  from the reactor.
- Fuel gas produced from maize cobs can be fired in a diesel engine.
  The lower heating value of the fuel gas was 4000 KJ/Nm^3 and it was
  operated dual fuel (4).

4. FUTURE DEVELOPMENTS

A second generation gasifiers will be developed aiming at better
efficiency, less tar, larger feedstock flexibility and higher production
capacity per unit reactor volume.

Concerning the cocurrent moving bed gasifier, further research
concentrates on the effects of recycling of pyrolysis gases to the feed air,
followed by external combustion to realize higher capacities at a similar
low level of tar content of the producer gas.

Because some biomass feedstocks are difficult to gasify in moving bed
reactors, new research has started in the field of fluidized bed gasification which is expected to be suitable for gasifying any biomass material.
Additionally it is expected that with this reactor type very large
capacities per unit volume of reactor can be realized

Special attention will be paid to the utilization of waste heat and
tars within the process to obtain an omnivorous gasification system that

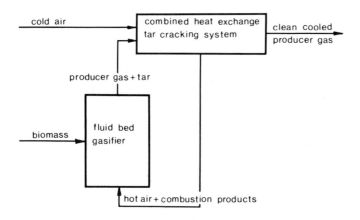

operates tar free and with a high overall thermal efficiency. The kinetics of gasification in fluidized beds, the mechanism of tar cracking and the modelling of the gasifier will receive special attention.

ACKNOWLEDGEMENTS

This research is sponsored by the Dutch Ministry of Public Health and Environmental Hygiene and by the commission of the European Communities, Directorate General for Research Science and Education.
The authors wish to express their gratefullness towards the members of the committees for stimulating our research and towards all co-workers at Twente University.

REFERENCES CITED

1. Groeneveld, M.J. and Van Swaaij, W.P.M. The Design of Co-current Moving Bed Gasifiers fueled by Biomass.
Symposium papers ACS "Thermal Conversion of Solid Wastes and Biomass", Washington D.C. (1979), pp. 463-479.
2. Groeneveld, M.J. The cocurrent moving bed gasifier, PhD-Thesis, Twente University of Technology, Netherlands (May 1980).
3. Hos, J.J.; Groeneveld, M.J.; and Van Swaaij, W.P.M. Gasification of organic solid wastes in cocurrent moving bed reactors.
Symposium Papers IGT "Energy from biomass and wastes IV", Orlando, (Fla) U.S.A. (1980), pp. 333-351.
4. Stassen, H.E.M. Gasification by partial combustion project in Tanzania, Progress report, Twente University of Technology, Netherlands (July 1980)

SESSION V : NEW CONCEPTS IN FUELS BY BIOLOGICAL ROUTES

Chairman: Dr. R. RABSON
Department of Energy, USA

Summary of the discussions

Invited papers

Prospects for abiological synthesis of biomass

Algae as solar energy converters

Poster papers

Model system for the continuous photochemical production of
hydrogen peroxide mediated by flavins

Renewable hydrocarbon production from the alga botryococcus
braunii

New concepts in solar biotechnology

H_2 production by the photosynthetic bacteria, rhodopseudomonas
capsulata entrapped in alginate gels

First experiments of production of macrophytes with waste water
and methanization of biomass

Valorization of aquatic macrophytic biomass - Methane production -
Depollution and use of various by-products

Growth of marine biomass on artificial structures as a renewable
source of energy

Mariculture on land - A system for biofuel farming in coastal
deserts

Algal biomass from farm waste - a pilot plant study

Photobiological production of fuels by microalgae

Increasing biomass for fuel production by using waste luke-warm
water from industries

Algal fermentation, a promising step in biomass conversion

SUMMARY OF THE DISCUSSIONS

Session V NEW CONCEPTS IN FUELS BY BIOLOGICAL ROUTES

Rapporteur :	K. WAGENER, Germany
Speakers :	G. PORTER, U.K.
	W.J. OSWALD, U.S.A.
Chairman :	R. RABSON, U.S.A.
Poster Session :	12 papers presented

The photochemical and photobiological routes of solar energy conversion and storage are potentially of great interest and practical importance since they may lead to higher overall efficiencies than the photosynthetic yields of whole plants. The status of this new concept is still in the phase of basic, directed research. The main problems are presently the low efficiency and short life time of the catalysts. However the field is developing very quickly and has progressed rapidly since the initial discoveries over the last 6-7 years.

Much attention is presently paid to algae as energy crops, because they are clearly high producers. Under eutrophic outdoor conditions, productivities of 20-30 g/m^2/day are obtainable, equal to 70-100 t/ha/yr. The main constraint for still higher productivities is given by the rate at which carbon can be supplied to the cultures. Under laboratory conditions higher productivities have been reached demonstrating that it is not the plant itself which sets the limit. Algal ponds as developed in California for waste disposal are proving very interesting in other parts of the world, also for possible energy and animal feed production.

Algae grow in seawater as well as in fresh or brackish water. Therefore, using the marine environment for growing algae avoids the competition between such energy farming and food production on arable land. Two approaches are presently being investigated, the one concerning open sea mariculture on floating structures, the other with mariculture on land on coastal areas. Open sea mariculture is still in the planning and small pilot phase but should continue to be carefully scrutinized. Mariculture on land work by the EC has now 3 years of practical experience with continuous outdoor cultures in southern Italy. Productivities between 60 and 90 t/ha/yr are obtained as average values over more than two years, and the key operations such as a cheap pond construction, pond management, harvesting and the possibility of anaerobic fermentation to methane with recycling of the mineral nutrients, have successfully been demonstrated; presently 1000 m^2 of pond area are under construction.

As the large-scale production of algal biomass seems to be possible, it is important to assess possible ways of its use. Various routes are under discussion :

(1)   Fermentation to methane. In this approach natural gas can be substituted and such a unit could possibly serve in isolated situations, since it is entirely self contained. The biogas can be used for heat and power.

(2)   Using the living algae for the production of hydrogen or hydrocarbons. The hydrocarbons are produced extracellularly ; in this case they can be separated without damaging the cells, and thus the cells could be used several times before they finally are fermented. With a hydrocarbon content of about 30% or more, and a doubling time of 2-3 days, this system could operate faster than so-called "gasoline trees".

PROSPECTS FOR ABIOLOGICAL SYNTHESIS OF "BIOMASS"

Sir George Porter, FRS

Director

The Royal Institution, 21 Albemarle Street

London W1X 4BS

Summary

Natural photosynthesis is an elegant and impressively versatile process but, as a source of fuel, it has some disadvantages. It is not very efficient as a solar energy store, it requires large quantities of water, fertiliser and chemical controls, its production is labour intensive and, since it does not yield directly the product required, further expensive processing is necessary. The situation may be improved by using plants which have been modified by selective breeding or genetic manipulation. Alternatively, we may seek to construct an abiological photosynthetic system.

Much progress has been made with the latter approach over the last year. Most attempts are based on the two photon/electron system of the natural Z scheme and use sacrificial donors or acceptors .. it has not yet been possible to put together the two photosystems, although a single photon/electron system liberating both hydrogen and oxygen has been reported. Nearly all the work is at present concerned only with the splitting of water to hydrogen and oxygen which is energetically similar to the overall production of oxygen and carbohydrate from carbon dioxide. The present state of the art will be reviewed.

The somewhat contradictory title of this lecture is a consequence of my presence here, as a photochemist, at a meeting which is discussing energy from biomass. Nevertheless, since the photochemical approach to solar energy utilisation is derived almost entirely from the natural biological process of photosynthesis, what I have to say is, I hope, quite close to the main themes of this conference.

Owing to the intermittent character of sunlight and also to the prime need for transportable forms of energy, methods of solar energy collection, which include the possibility of storage over periods of at least one year, have a great advantage. At present the biological route to storable solar energy is the only one available and has served man very well, if one includes fossil fuels, up to the present time.

There are, however, some important disadvantages to the natural photosynthesis of biomass :

1. The efficiency of solar energy collection is low; it rarely exceeds 1% on a year-round basis.

2. Large quantities of water are needed for irrigation, thereby excluding much of the land area with the highest insolation.

3. Efficient agriculture is energy intensive since it requires fertiliser, pest control and heavy machinery.

4. The products of photosynthesis require further processing before they are suitable for use as gaseous or liquid fuels.

Much of the discussion at this conference is concerned with overcoming these difficulties by improving agricultural products and the processing of them, accepting always the natural process of photosynthesis which we have inherited. Very recently, as the mechanism of the natural process has become better understood, photochemists have begun to consider the possibility of constructing an entirely artificial system which, although it derives its inspiration from the methods used by the living plant, is modified and simplified so as to be more suitable to man's fuel requirements.

In reviewing the prospects for a successful development of such an artificial system I will consider in turn the fundamental theoretical restraints on efficiency, the present stage of experimental development and the future feasibility of economic development of such systems.

Theoretical efficiency considerations

The theoretical limit to which we may hope to improve the efficiency of conversion of incident solar energy into chemical free energy is set principally by three factors. An exact calculation requires that all three factors be considered together for each wavelength of solar radiation and the result depends on ambient conditions as well as on what assumptions are made about theoretical kinetic efficiencies (the third factor below). A sufficiently close estimate of what is possible can however be derived as follows :

(a) Thermodynamic limit determined by the second law. Owing to the small solid angle which the sun subtends at the earth's surface its temperature is reduced from 5800 K to an effective thermodynamic temperature of 1350 K which dictates that the maximum thermodynamic efficiency for conversion to work is 78%.

(b) Spectral distribution limitations. These arise because not all wavelengths of the sun's radiation are absorbed and of those that are absorbed the energy is not fully used. In the case where the absorbing substance has a threshold wavelength below which all wavelengths are absorbed but energy in excess of that of the threshold wavelength is wasted (a common and fairly readily realised situation), the maximum fraction of the sun's energy which can be utilised is about 47% (at air mass 1) and this occurs for a threshold at 1100 nm. If the threshold is at 700 nm (as it is for green plant photosynthesis) the maximum efficiency falls to 33%.

(c) Kinetic limitations. These arise because a sequence of reactions must proceed very rapidly, and therefore must not involve an activation energy and, furthermore, in order to prevent the occurrence of back reactions the forward reactions must actually expend energy. The extent to which this is necessary depends on how the reaction is carried out and is a matter of some debate at the present time but it seems probable that at least 0.3 ev and perhaps nearer to 0.6 ev will have to be expended in this way out of the total energy of the photon (which is 1.76 ev for a threshold of 700 nm).

Taking all of these limitations into account leads to a maximum efficiency between 20% and 30% depending principally on the kinetic

requirement. With photovoltaic cells, which have somewhat similar
theoretical limitations, actual efficiencies approaching 20% have already
been achieved.

Present state of experimental progress

Natural photosynthesis occurs according to the reaction :

$$H_2O \; + \; CO_2 \; \xrightarrow{\text{light}} \; (CH_2O) \; + \; O_2 \qquad \Delta G = 500 \text{ kJ}$$

where (CH_2O) represents carbohydrate in the form of glucose and ΔG refers
to the free energy change under ambient conditions. In fact, this simple
equation describes the overall process of an exceedingly complex multistep
sequence of reactions. The light is absorbed principally by chlorophyll
with a threshold at 700 nm and, although four photons are energetically
sufficient, at least eight photons are used in practice for the liberation
of one oxygen molecule. It seems to be well accepted that the overall
reaction proceeds in two steps, each step involving four photons, as
follows :

Photosystem 2. Water is oxidised to oxygen and an intermediate
substance Q is reduced to Q^- (or $4Q^-$ for each O_2 liberated).

Photosystem 1. The four electrons of $4Q^-$ are raised by a second
photochemical reaction to a higher reducing level and, via several other
intermediates, are eventually used to reduce carbon dioxide.

Although oxygen liberation and Photosystem 2 are an essential part of
green plant photosynthesis the final product of Photosystem 1 may, instead
of reduced carbon dioxide, be reduced nitrogen or reduced protons. In the
latter case Photosystem 1 reduces water and two molecules of hydrogen are
liberated. The energy stored for each oxygen molecule liberated is very
similar whether the reduced products are carbohydrate, ammonia (reduced
nitrogen) or hydrogen. Since the latter process, which amounts to the
dissociation of water into hydrogen and oxygen, is chemically the simplest,
and since hydrogen is a potentially valuable fuel (to replace natural gas,
or perhaps to reduce coal to liquid hydrocarbons), most attention is at
present directed to the problem of the photodissociation of water using
visible, and preferably red, light.

The experimental work has mainly been concerned with the easier of
the two steps, Photosystem 1, using in place of Q^- a sacrificial donor,

that is a reducing agent which is used up in the process. The principal difficulty is the need to accumulate two electrons for the liberation of one hydrogen molecule and this has now been solved by the use of colloidal metal catalysts, particularly platinum. As an example of several systems which have been used successfully we may quote the following :

$$Sens \xrightarrow{\text{light}} Sens^*$$

$$Sens^* + MV^{2+} \longrightarrow Sens^+ + MV^+$$

$$Sens^+ + D \longrightarrow Sens + D^+$$

$$2 MV^+ + 2H^+ \longrightarrow 2MV^{2+} + H_2$$

Here, MV^{2+} is methyl viologen (paraquat) and the sensitiser (sens) may be chlorophyll, or other metal-organic complexes. Using, as sensitiser, the water soluble zinc tetramethylpyridyl porphyrin, a molecule closely related to chlorophyll, Harriman and Richoux in our laboratory have measured a quantum yield of hydrogen production, with light of wavelength 560 nm, of 60%.

The modelling of Photosystem 2 has proved to be more difficult because of the requirement to accumulate four positive charges for the elimination of one oxygen molecule. However, it has recently been dis-covered that certain metal oxide powders in suspension, and ruthenium oxide in particular, are effective as accumulators of positive charge with low overpotential and the overall process of water oxidation, using sacri-ficial electron acceptors in this case, can now be achieved, albeit with poor efficiency.

There remains the problem of linking the two systems so that the acceptor of system 2 becomes the donor of system 1, neither of them being sacrificed so that the overall process is cyclic. This has not been solved using the eight photon, two photosystem scheme of natural photosynthesis but, rather surprisingly, the whole process has recently been reported in a simple four photon system by Kalyanasundaram and Gratzel. In their experiment only a sensitiser, a carrier (methyl viologen) and the two catalysts, platinum and ruthenium dioxide were used in a single solution and both hydrogen and oxygen were eliminated. Although the yield was very low and the reaction terminated after a short time, this result has caused considerable excitement and a hope that it may be possible to carry out the water splitting process using only half as many photons as are used in nature.

In addition to these essentially homogeneous systems, much attention has also been given to semiconductor electrodes as light absorbers for water splitting, a method used by Honda and colleagues in the early seventies, using titanium dioxide crystals as the irradiated cathode. At present these electrode systems are only able to dissociate water by using ultra-violet light or when assisted by added potential.

Economic feasibility

Typical insolation in a desert region, averaged throughout the year, day and night is 250 w/m^2. From what has been said it would appear reasonable to expect an efficiency of solar energy storage of 10% or 25 w/m^2. The capital cost of electrical power generating station today is about $1/installed watt so that, to be competitive, the capital cost of installation of a solar collector of free energy should not exceed $25/m^2$. If the fuel has then to be converted into electricity, further costs would be incurred but if, conversely, the electrical power of a generating station has to be used to generate storable chemical fuels the photochemical fuel generator would have the advantage. This latter situation is the more likely if nuclear power is further developed whilst fossil fuels are inevitably depleted, in which case electricity will be plentiful but transportable storable fuels will be in short supply.

The cost of the chemical materials mentioned above, in a photochemical reactor of the type envisaged, are unlikely to exceed one or two $/m^2$ of reactor and the principal cost will be the construction of the reactor itself. It seems probable that this could be mass produced within the economic limits given.

In conclusion, it must be stressed that a complete photochemical solar fuel generator has not yet been developed even on the laboratory scale. Nevertheless, when such a reactor is developed on a laboratory scale, scaling up to a large-scale collector, covering many square miles, should not involve any new problems and could probably be brought about very rapidly.

ALGAE AS SOLAR ENERGY CONVERTERS

WILLIAM J. OSWALD

Professor of Sanitary Engineering and Public Health

University of California

Berkeley, California

USA

Summary

Three types of algae systems have been proposed as photosynthetic energy converters. Macro marine algae grown, harvested and fermented to produce methane micro, blue green algae for hydrogen and biomass, and micro green algae to produce oxygen, animal feed and biomass. This paper deals with the energy and oxygen producing potential of the micro green algae. In long term experiments real productivity was found to be directly proportional to the quantity of solar energy above a threshold of 125 cal cm^{-2} day^{-1} in Richmond and 225 cal cm^{-2}day^{-1} in Manila. The difference may be due to temperature effects on respiration of the algae. Productivity in the Philippines varied from 3.5 gm meter^{-2}day^{-1} on cloudy days to 28 gm meter^{-2}day^{-1} on clearest days. Productivity in Richmond varied with season from 6 to 35 gm meter^{-2}day^{-1}. There peak productivity exceeded 125 metric tons ha^{-1}yr^{-1} but declined at highest solar energy input, probably due to excessive light for the experimental conditions. Flow mixing velocities of 5 to 15 cm sec induced in the pond channels with paddle wheels increased settleability of algae from virtually zero without paddle wheels to 60 to 90 percent with paddle wheels. Higher flow velocities disrupted settling and used excess energy. Waste oxygenation with micro green algae is now economical in sunny locations saving 65% of system first cost and 90% of energy required for conventional organic waste treatment.

I. INTRODUCTION

Several kinds of algae have been proposed as solar energy converters including the giant kelp and other large marine algae, (1) the microscopic nitrogen fixing blue green algae which, under nitrogen starved conditions can be induced to produce hydrogen, (2) and the green micro algae that grow well in domestic sewage and most liquid organic wastes, (3) and produce oxygen, animal feed and fermentable biomass.

This presentation will be limited to the green microalgal systems that we have studied for various engineering applications over a number of years most recently under sponsorship of the U.S. Department of Energy, the World Health Organization and the Laguna Lake Development authority in Manila. In this paper I will describe our most recent findings but before doing this it is well to review some of the established facts concerning large scale micro algae cultures and applications (4). Algae growing on freshly decomposing sewage in which ammonium is the principle source of nitrogen synthesize cell material and release oxygen as shown in equation (1).

$$1.0 \ NH_3 + 7.62 \ CO_2 + 6.34 \ H_2O \ \xrightarrow[\text{Chlorophyll in algae cells}]{\text{Sunlight}}$$

$$(C_{7.62} \ H_{8.08} \ O_{2.53}N) + 7.6 \ O_2 + 3.81 \ H_2O \qquad (1)$$

As is well known all of the oxygen evolved on the right of equation (1) comes from water on the left; thus algae catalyze the photochemical decomposition of water to produce oxygen and hydrogen. The oxygen, released into the water as dissolved molecular oxygen (O_2) is immediately available for aerobic bacterial oxidation of organics in the wastes. The hydrogen is retained by the micro algae to reduce CO_2 and incorporate it with NH_4^+ into their rapidly growing cell material. Production of one unit dry weight of algae is accompanied by release to the wastes of about 1.6 units of free molecular oxygen. Accordingly, the rate of organic oxidation is determined by the rate of algal growth and oxygen production in the system; that is by the algal productivity.

As with other biomass systems productivity in microalgal cultures is defined as the dry weight of algal organic matter produced in the pond per unit of area per unit of time. We are concerned with both O_2 productivity and biomass productivity. The most commonly used units of productivity are grams dry wt meter^{-2}day^{-1} and metric tons hectare^{-1}year^{-1}. A theoretical expression relating productivity to pond area and depth, solar energy flux,

algal concentration and algal heat content is useful for engineering design
of algal systems as well as needed to determine productivity.

In continuous cultures a pond with volume V and daily input or output
liquid volume, Q, has a hydraulic detention period Θ,

$$\Theta = V/Q \qquad (2)$$

Also for a pond with a surface area A and depth d the volume is:

$$V = Ad \qquad (3)$$

which by substitution for V in (2) yields:

$$\Theta = Ad/Q$$

and by rearranging one has:

$$d/\Theta = Q/A \qquad (4)$$

Both d/Θ and Q/A are important parameters since they give the hydraulic
loading   velocity on the pond and therefore, when multiplied by the area,
A, the volume of culture discharged.

If A is selected to be 1.0 m^2 the daily volume, Q, discharged per
$meter^2$ will be

$$Q = d/\Theta \times \frac{10,000}{1,000} \text{ liters meter}^{-2}\text{day}^{-1} = 10 \frac{d}{\Theta}$$

and if C_c is the concentration of algae in the discharged water in mg
$liter^{-1}$, the weight of algae discharged $meter^{-2}day^{-1}$ will be the productivity, P thus:

$$P = 10 \frac{d}{\Theta} C_c \text{ mg meter}^{-2}\text{day}^{-1}$$

Since P is usually expressed in grams $meter^{-2}day^{-1}$ we must divide P^1 by
1000 thus:

$$P = \frac{10 \, d \, C_c}{1000 \, \Theta}$$

or

$$P_{algae} = 0,01 \, (d/\Theta)C_c \qquad (5a)$$

$$P_{O_2} = .015 \, (d/\Theta)C_c \qquad (5b)$$

To calculate the efficiency of utilizing solar energy, one measures
the solar energy flux, S, in cal $cm^{-2}day^{-1}$. The input solar energy in
calories $m^{-2}day^{-1}$ is then 10,000 S and, to express productivity in the
same units of gm cal $m^{-2}day^{-1}$, the algal concentration must be multiplied
by its heat of combustion, h in calories mg^{-1} giving

$$\text{output energy} = 10 \, (d/\Theta)C_c h$$

The efficiency is then output/input or

$$F = \frac{10(d/\Theta)C_c h}{10,000 \, S} = \frac{dC_c h}{1000 \, S\Theta} \qquad (6)$$

in which all terms were previously defined. It should be emphasized that equations 5 and 6 apply only to homogeneous completely mixed systems.

One might assume from equations 5 and 6 that by making a pond very deep he would attain high productivity and efficiency but unfortunately, beyond a certain depth the concentration of chlorophyll synthesized by each algal cell increases and the number and concentration of cells declines more rapidly than the depth is increased. On the other hand, if a pond is too shallow it becomes extremely difficult to mix, and the solar energy may overheat the algae causing a large increase in their respiration and sometimes even bleaching of their chlorophyll, and hence a decline in their growth rate and concentration. A result is that for each location and set of conditions there is an optimum depth and both efficiency and productivity decline above and below that depth.

Temperature is also an important factor that although it does not appear in equations 5 and 6, influences the concentration of algae and chlorophyll attained and in turn as will be shown later affects the relationship between productivity and the daily input of solar energy.

One concludes that pilot experimental work should be done at each location to optimize the use of algae as solar energy converters.

EXPERIMENTAL WORK

In 1978 we completed 3 months of intensive experimental work with two 100 M^2 pilot ponds in the Philippines (see figure 1) and in 1979 we completed two years of operations with two 1000 M^2 ponds at the University of California, Berkeley, Sanitary Engineering Research Laboratory in Richmond. In each case we measured solar energy and algal productivity and harvestability among other parameters of water quality and waste treatment. This work has been covered in detailed technical reports Oswald (5), Benemann, et. al. (6), and therefore I will only repeat here essential details and pertinent results of the experiments.

In both Manila and Richmond we used domestic sewage for nutrient for algal growth. In both cases the sewage was quite weak having Biochemical Oxygen Demand (BOD) on the order of 70 to 150 mg/l, total Nitrogen about 10 to 30 mg/l, and Phosphorus from 1 to 4 mg/l. The sewage in Manila was screened to remove course settleable and floatable particles whereas in Richmond the sewage was passed through a 30 minute detention period sedimentation and flotation tank and then passed over a cascade screen to remove settleables and floatables. The sewage was then introduced continu-

ously to the algae growth ponds at precisely metered rates. Pond depths
were determined by overflow conduit elevations and detention periods were
determined by the quantity of waste introduced daily. Mixing in both systems was provided by slowly turning paddle wheels. In the Manila experiments the linear flow velocity induced in the continuous channels was about
5 cm/sec^{-1} with a faster mix at 30 cm/sec^{-1} for 6 hours beginning at 2 A.M.
daily and for 15 minutes at noon each day. In Richmond the linear flow
velocity was 10 cm/sec^{-1} continuously with no fast mix.

The paddle wheel mixing of algal growth ponds does not accomplish any
net oxygen aeration. Although some oxygen is absorbed from the atmosphere
at night when the water is below O_2 saturation this is more than offset
during the day when, due to rapid growth and O_2 production by the algae,
losses to the atmosphere far exceed the night time gains. One major advantage of continuous mixing is prevention of thermal stratification which if
permitted causes the surface algae to be overheated. Another advantage is
that it prevents anoxic conditions from occurring on the bottom where most
bacterial activity occurs in unmixed ponds. It also greatly improves harvestability of the algae.

PRODUCTIVITY

The results of determinations of productivity as a function of solar
energy input for Manila are shown in figure 2. Each point represents the
mean value for both productivity and solar energy over a two to three week
period. According to the figure productivity of algae varied from about
3.5 grams $M^{-2}day^{-1}$ when the solar energy input was 275 langleys day^{-1} to
about 28 gm $M^{-2}day^{-1}$ when the solar energy flux was about 540 langleys
day^{-1}. The mean was about 14 gm $M^{-2}day^{-1}$ at a solar energy input of about
380 langleys day^{-1}. Related to larger land areas the mean productivity
was about 51 metric tons hectare$^{-1}yr^{-1}$. During this period temperature
varied little in the pond water ranging from a low of 27°C to a high of
30°C. Because of this it can be inferred that in these experiments productivity was not greatly influenced by temperature. It is interesting to
note, however, that the regression line of best fit for productivity vs.
solar energy had an intercept at 225 langleys day^{-1} on the solar energy
axis probably resulting from the high respiration rate of algae at the
mean temperature of 28.6°C.

Recalling that the ratio of O_2 to algae is about 1.6, O_2 production
varied from about 20 met ton ha$^{-1}yr^{-1}$ to about 150 met ton ha$^{-1}yr^{-1}$ with a
mean of about 75 met ton ha$^{-1}yr^{-1}$. The data indicate that for algae pro-

duction:

$$P_{algae} = 0.09(S_T - 225) \tag{7}$$

and

$$P_{oxygen} = 0.14 (S_T - 225) \tag{8}$$

in which P_{algae} and P_{oxygen} are productivities in grams $M^{-2}day^{-1}$, and S_T is the total solar energy in cal $cm^{-2}day^{-1}$ (langley day^{-1}).

The results of determinations of productivity in the 0.1 ha ponds in Richmond, California, are portrayed in figure 3. Each point represents mean values for the parameters over a period of one month for two years. Highest monthly productivities within the two year period are shown. As indicated in the legend the regression line of best fit for the data is

$$P_{algae} = 0.108(S_T - 125) \tag{9}$$

and

$$P_{oxygen} = 0.162(S_T - 125) \tag{10}$$

For the minimum S value of about 195 langleys day^{-1} was 7.56 gm M^{-2} day^{-1} and for 400 langleys day^{-1} the productivity was 29.7 gm $M^{-2}d^{-1}$. In terms of larger land areas the productivity varied from 27.6 to 108 met tons ha^{-1}yr^{-1}. The drop in productivity above 460 langleys at Richmond is attributed to excessive light and heat for Microactinium sp., the micro algae that predominated in the ponds. The lower solar energy intercept of 125 langleys day^{-1} is attributed to the much lower temperatures in Richmond corresponding to the lower solar energy inputs. These lower temperatures were on the order of 7°C to 8°C. The lower temperatures were accompanied by lower nocternal respiration rates for the algae.

It is somewhat unexpected to find lower peak productivities in the tropics than in a cool coastal temperate climate such as that at Richmond. However, much of the Manila work was done during the monsoon season (permitting the large variations in S observed) and it is possible that a full year's data would give a modified picture. Our work in Manila is being resumed now and hopefully a longer data base will become available. We are also initiating pilot plant studies in Jamaica near Kingston to determine productivity where there is less cloud cover than in Manila.

SEPARATION

When we were designing the pilot plant in Manila we used paddle wheels so that they could be constructed and repaired locally rather than depend on international orders for equipment and parts which is often a problem.

We did not use the paddle wheel velocities of 30 cm per second that had been reported for installations for Clorella growth in Formosa because theoretical calculations indicated the energy demand would be too high. In open channels the required input energy increases as the cube of the velocity, and more than 200 times as much energy is required to mix at 30 cm/sec than at 5 cm/sec the velocity selected.

The mixing regime selected was 5 cm/sec continuous and 30 cm per second for several hours per day. The results of the Manila work were unexpected in that effluents from the ponds showed removals of micro algae by simple sedimentation which ranged from 60% to 95%. Highest removals occurred when the pond was operated at 30 cm depth, a hydraulic loading of 7.5 cm per day and when the duration of fast mixing was only 2 hrs per day.

The conclusion tentatively reached was that fast mixing probably was not needed and that gentle paddle wheel mixing of the cultures resulted in the development of flocs of sufficient size to settle readily when removed from the mixing environment. Observations of the Manila ponds also indicated that some algae were settling out in the pond and that a more rapid mix of 10 cm per second should have been used. Based on the Manila data we modified our 1000 M^2 Richmond ponds to be paddle wheel mixed at 10-15 cm/sec. These had previously been mixed with propellar pumps intermittently at flow velocities ranging from 20 to 30 cm per second. The algae had never tended to settle in the ponds except on days when the pH was sufficiently high (10-11) to cause autoflocculation (Oswald and Golueke, 1968).

The results from modifying the Richmond ponds for paddle wheel mixing have been even more impressive than the Manila results. The cultures were stable with fewer predator problems and in effluents subjected to simple sedimentation removals always exceed 70% and often exceed 95% virtually independent of other operating conditions in the ponds such as depth or hydraulic loading. The exception being that at very low hydraulic loadings, predators tended to take over, decreasing the concentration of algae. For hydraulic loadings in the range of 8 to 12 cm day^{-1} this has not occurred. In our most recent experiments, when paddle wheel mixing is suspended simple sedimentation of the algae stops almost completely.

With plain sedimentation, as described by Eisenberg (7), algal concentrations of about one percent solids are obtained. This concentration can be increased by gravity thickening for 1 day to about 3 to 5 percent solids a concentration suitable for introduction to a methane fermentation chamber.

ENERGY TRANSFORMATIONS

Methane fermentation of naturally settled and thickened micro algae has been described recently by Eisenberg (7). His results are similar to those found by Golueke et. al. (8) and by Uziel (9). Under optimum conditions at least 50% of the solar energy fixed in the micro algae can be recovered in the form of methane.

ECONOMICS

We are hastening to repeat the paddle wheel mixing experiments in other locations to assure ourselves that the simple sedimentation phenomenon is not a peculiarity of Richmond and Manila sewage algae or climate. If the sedimentation occurs everywhere that it is properly designed - and this will take several more years to determine - then there will be an extremely strong case for using micro algae as primary producers of biomass for large solar energy fixing systems.

Our work on productivity has already convinced us that in spite of the expressed doubts of many (10) micro algae simply have no peer in productivity in the biotic world. The Manila and Richmond data strongly indicate that 60 metric tons hectare^{-1}yr^{-1} is a conservative basis for design in sunny locations and that with the established energy conversion of 50% from algae to methane a net energy of 500 million BTU hectare^{-1}yr^{-1} may be realized. At the current price of $5.00 per million BTU this would gross $2,500 hectare^{-1}yr^{-1}. How much capital expenditure and operations and maintenance this would pay for at today's prices is speculative. If ponds could be built for $10,000 to $20,000 per hectare the prospects appear good. A federally subsidized program such as has been used for large dams in the USA would be ideal from the standpoint of financing since it would permit optimum systems to be built with long term low interest loans. By amortizing these systems over fifty to 100 years the cost of energy might actually be decreased!

The question of micro algae as energy converters no longer appears to be solely technical, but now involves engineering financing and the question of governmental subsidies. We are now taking on a new project in California that may resolve some of these questions.

ENERGY CONSERVATION AND CAPITAL SAVINGS

As noted in the introduction there is another important aspect of micro algae - namely their energy conservation through oxygen generation in waste treatment. Domestic sewage and organic industrial and agricultural wastes require large amounts of O_2 for their stabilization. This oxygen must be supplied using energy through chemical, physical or biological systems. In figure 4A we show a convention physical system for liquid waste oxidation using electrical energy to power mechanical aerators. In sewered communities complete oxidation of the liquid - born wastes of one person - or person equivalent (P.E.) - requires about 0.25 pounds of O_2 day^{-1}. Mechanical aerators will transfer about 2.5 pounds of O_2 from the atmosphere to the water for each kilowatt hour (KWhr) expended at the aerator. Output energy at the aerator is therefore about 0.1 KWhr per person per day. As shown in figure 4A by assuming a petroleum refinery efficiency of 80%, a generator efficiency of 35%, and an electrical transmission and transforming efficiency of 80%, the fuel input energy must be about 0.5 KWhr $person^{-1}day^{-1}$. Assuming the energy is generated using No. 6 sulfur free oil (11) the fuel cost will be about \$0.025 per KWhr, or about \$0.0125 per person per day. By contrast in the micro algae aeration system shown in figure 4B oxygen is released from water during photosynthesis and the only mechanical energy input is that required to induce flow mixing of the culture. Studies of this energy requirement indicate it to be about 20 KWhr $hectare^{-1}day^{-1}$. Since sufficient oxygen is produced by each hectare of culture to oxygenate the wastes of 2000 P.E. the per capita energy is about 0.01 KWhr. Assuming, as in the case of mechanical aeration, a 20% overall efficiency the input energy is about 0.05 KWhr and the net saving is 0.45 KWhr per person per day. Assuming the micro algae aeration process could be applied to 25% of the U.S. population (due to climatological limitations) the savings would be about \$225,000,000 per year. This is sufficient to purchase 7.5 million barrels of oil at current prices, but it is far from the only saving. The capital investment for micro algae systems is approximately 1/3 that for the mechanical systems shown in figure 4A. Mechanically aerated systems of 2 to 5 million gallons daily capacity have a first cost of about \$5.00 per U.S. gallon processed daily. Since U.S. individuals produce about 100 gallons of liquid waste per day the capital cost is about \$500.00 per capita which assuming a 11% per annum for interest and amortization works out to about\$ 0,015per person per day. A plant which we recently designed for Hollister, California, (12), (figure 5)cost only \$1.50 per gallon with a resultant capital saving of \$0.10 per person day. Since the population served is 10,000 persons, the savings in interest and capital together with energy amounts to \$1,100 per day or more than $400,000 per year. Applied to 25% of the U.S. population the saving could be over two billion dollars per year, and applied to agricultural and industrial wastes in the same areas of the U.S. the savings could approach $5.6 billion per year. Micro algae systems also eliminate much of the cost of operations and maintenance of conventional systems giving further savings.

CONCLUSION

To summarize, the micro algae together with bacterial systems have through extensive experimental and practical work proven to have great potential as solar energy converters, capable of utilizing most types of liquid oxygen wastes and transforming more than 500,000,000 BTU per hectare yr^{-1} of solar energy into methane. The fixation rate into crude biomass is about 1 billion BTU per hectare yr^{-1}. As importantly, at this time, micro algae have the potential of immediately contributing significantly to the energy conservation effort by reducing both the first cost and the energy cost of liquid waste treatment in sunny locations the world around.

ACKNOWLEDGEMENTS

This work was supported by the U.S. Department of Energy, the Sanitary Engineering Research Laboratory of the University of California, the World Health Organization and the Laguna Lake Development Authority. I wish to acknowledge the continued support and efforts of Mr. Don M. Eisenberg, Graduate Research Engineer, Dr. Joseph Weissman, Research Biophysicist and Dr. John Benemann, Research Biochemist, Sanitary Engineering Research Laboratory, University of California, Berkeley.

I also wish to acknowledge the support of Dr. E. W. Lee, Environmental Health Officer, the World Health Organization, Manila, and Mr. Ben Adam, Executive Engineer, Laguna Lake Development Administration, Rizal Province, Metro Manila.

I am especially grateful to Ms. Annette Mora, Administrative Assistant University of California, for preparing this manuscript.

REFERENCES

1. Lapointe, B. E., L. D. Williams, J. C. Goldman, and J. H. Ryther, (1976),"The Mass Outdoor Culture of Macroscopic Marine Algae," Aquaculture, 8:9-20.

2. Benemann, J. R. et. al. (1977), Solar Energy Conversion with Hydrogen-Producing Algae, Final Report, Sanitary Engineering Research Laboratory, University of California, Berkeley.

3. Oswald, W. J., and C. G. Golueke (1960), "Biological Transformation of Solar Energy," Advances in Applied Microbiology II, Academic Press, New York, New York.

4. Oswald, W. J., (1977), "The Engineering Aspects of Micro algae," Handbook of Microbiology, 2nd Ed., Vol. II, C.R.C. Press, 18901 Cranwood Parkway, Cleveland, Ohio 44128.

5. Oswald, W. J., (1978), "Pilot Plant High Rate Pond Study for Waste Treatment and Algal Production," Final Report. World Health Organization, Regional Office for the Western Pacific, Manila, Philippines.

6. Benemann, J. R., J. C. Weismann, B. L. Koopman, D. M. Eisenberg, R. P. Geobel and W. J. Oswald, (1978), "Large Scale Fresh Water Microalgal Biomass Production for Fuel and Fertilizer," Final Report, DOE Contract EY-76-S-03-0034. Sanitary Engineering Research Laboratory, University of California, Berkeley.

7. Eisenberg, D. M., W. J. Oswald, J. R. Benemann, R. P. Goebel and T. T. Tiburzi, (1979), "Methane Fermentation of Micro algae." Proceedings of the First International Symposium on Anaerobic Digestion, University College, Cardiff, U.K.

8. Golueke, C. G., W. J. Oswald and H. B. Gotaas, (1957), "Anaerobic Digestion of Algae," Appl. Microbiol.

9. Uziel, M., W. J. Oswald and C. G. Golueke (1975), "Integrated Algal Bacterial Systems for Fixation and Conversion of Solar Energy," Presented before the Annual Meeting of the American Association for the Advancement of Science Section on "Energy-- New Sources," New York City, New York.

10. Anon., "The Economics and Engineering of Large-Scale Algae Biomass Energy Systems." MIT Sea Grant Program MITSG 78-11. Massachusetts Institute of Technology, Cambridge, MA 02139.

11. Williams, S., (1980), Pacific Gas and Electric Company, San Francisco, California, Personal Communication.

12. Swanson, Oswald Associates (1978), "Integrated Waste Treatment Ponds for Hollister, California," Design Report SOA Inc., 594 Howard Street, San Francisco, California 94105.

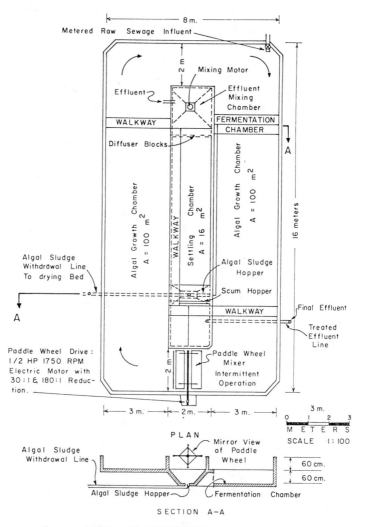

Fig. 1 DETAIL OF EXPERIMENTAL PONDS (2 UNITS)

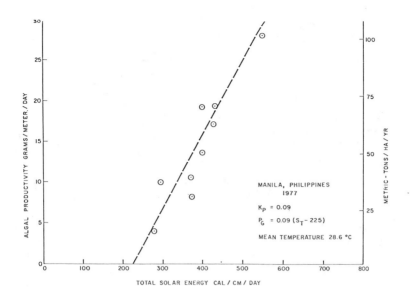

Fig. 2 - Productivity of algae in high rate ponds

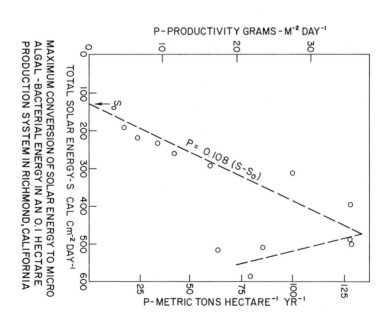

Fig. 3 - Maximum conversion of solar energy to micro
         algal-bacterial energy in an 0,1 hectare
         production system in Richmond, California

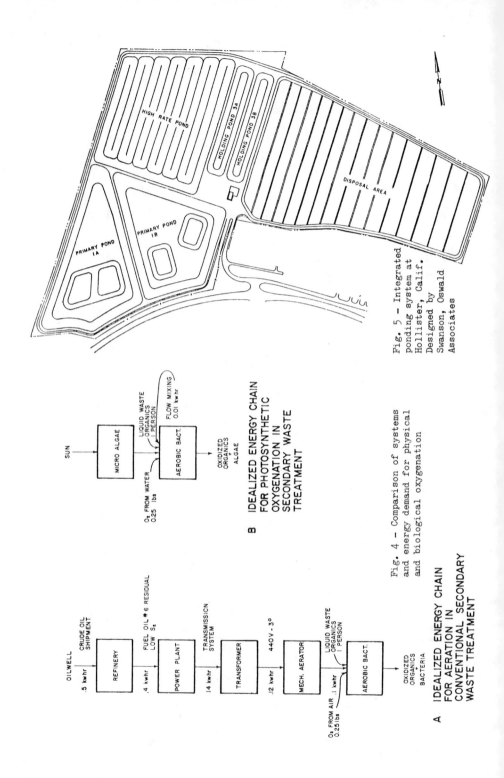

Fig. 5 – Integrated ponding system at Hollister, Calif. Designed by Swanson, Oswald Associates

HIGH RATE POND

HOLDING POND 3A

HOLDING POND 3B

PRIMARY POND 1A

PRIMARY POND 1B

DISPOSAL AREA

Fig. 4 – Comparison of systems and energy demand for physical and biological oxygenation

B IDEALIZED ENERGY CHAIN FOR PHOTOSYNTHETIC OXYGENATION IN SECONDARY WASTE TREATMENT

SUN → MICRO ALGAE → AEROBIC BACT. → OXIDIZED ORGANICS ALGAE

LIQUID WASTE ORGANICS 1 PERSON

FLOW MIXING 0.01 kwhr

O_2 FROM WATER 0.25 lbs

A IDEALIZED ENERGY CHAIN FOR AERATION IN CONVENTIONAL SECONDARY WASTE TREATMENT

OILWELL

CRUDE OIL SHIPMENT .5 kwhr → REFINERY → FUEL OIL # 6 RESIDUAL LOW S_t .4 kwhr → POWER PLANT → TRANSMISSION SYSTEM .14 kwhr → TRANSFORMER → 440V - 3° .12 kwhr → MECH. AERATOR → AEROBIC BACT. → OXIDIZED ORGANICS BACTERIA

LIQUID WASTE ORGANICS 1 PERSON

O_2 FROM AIR .1 kwhr 0.25 lbs

MODEL SYSTEM FOR THE CONTINUOUS PHOTOCHEMICAL PRODUCTION OF HYDROGEN PEROXIDE MEDIATED BY FLAVINS

C. Gómez-Moreno, A.G. Fontes and F.F. de la Rosa

Departamento de Bioquímica, Facultad de Biología, Universidad de Sevilla and CSIC, Sevilla, Spain

Summary

Irradiation of flavin compounds in the presence of different molecules that act as electron donor, and in the presence of oxygen, generates hydrogen peroxide. We have shown that significant amounts of this compound are formed under continuous illumination, providing a model system for the conversion and storage of solar energy.

The rate of production of hydrogen peroxide depends on variables such as the nature and concentration of the photosensitizer, intensity of light and concentration of the electron donor. A linear increase in hydrogen peroxide formation when changing the flavin concentration was observed only if the ratio EDTA/flavin was kept constant, the optimal being 200. Under these conditions, a yield of 6 mM H_2O_2 after 60 minutes of illumination was observed.

Besides EDTA, semicarbazide, tricine, and different thiol compounds, act as efficient electron donors for the photochemical reaction. Higher rates are obtained, as well as protection against the photosensitizer degradation, at high EDTA concentrations.

1. INTRODUCTION

Hydrogen peroxide is an intermediate compound in the reduction of oxygen to water that can loose 2 electrons to produce O_2 or gain 2 electrons and become H_2O. In thus tends to decompose into these compounds, specially in the presence of catalysts such as metal ions or catalase, according to the following exergonic reaction:

$$2\ H_2O_2 \longrightarrow 2\ H_2O + O_2 \qquad (\Delta G_o' = -48\ kcal)$$

It can then be considered as a potent fuel material, although the high cost of industrial production by conventional methods makes its use unfeasible for such purpose. There are, then, powerful reasons for searching new and more economic methods for the production of hydrogen peroxide as an energy rich compound.

In this context the reaction by which oxygen is photochemically reduced into hydrogen peroxide provides an interesting way of transforming solar into chemical energy. Although different photochemical systems using solid catalysts have been described (1), the most attractive ones are those involving homogeneous solutions. The biologically important flavins, as well as some organic compounds (2-5), get excited in the light becoming more easily reducible, so that very weak reductants (such as ascorbate, thiourea and even EDTA) act as efficient electron donors.

The reduced flavin (FlH_2) so obtained can reduce a variety of compounds among which is oxygen. This reaction with oxygen generates hydrogen peroxide directly, as well as through the formation of the intermediate superoxide ion

$$FlH_2 + O_2 \longrightarrow Fl + H_2O_2$$

In this paper we describe the conditions required for the photochemical production of hydrogen peroxide using flavin compounds as photosensitizers. This method can be considered as a model system for capturing and storing solar energy introducing a new kind of fuel material.

2. MATERIALS AND METHODS

Irradiation was performed in a lucite (2 x 5 x 10 cm) cuvette containing initially 60 ml of solution, from which aliquots were withdrawn at the desired times to measure the H_2O_2 content and spectral characteristics of the solution. For the actinic illumination a slide projector with a 150 W white light lamp was employed. The light was perpendicular to the walls and vigorous aireation was achieved (1300 cc/min) with a small air pump connected through a flowmeter.

<u>Determination of H$_2$O$_2$ concentration</u>

A modification of the enzymatic method described by Bernt and Bergmeyer (6) was used. The reaction mixture contained 200 μmoles of phosphate buffer pH 7.5, 300 nmoles o-dianisidine and 0.5 PG units of horseradish peroxidase in a final volume of 4 ml. After addition of the corresponding H$_2$O$_2$ containing aliquot, the mixture was incubated at 30°C for 15 min and the absorbance measured at 440 nm.

3. RESULTS AND DISCUSSION

Irradiation of a flavin containing solution under vigorous aireation and in the presence of EDTA promotes the continuous formation of hydrogen peroxide. Fig. 1 shows that the reaction is maintained linearly for the first 60 min reaching a final concentration of 6 mM. At this point the reaction stops due to the desappearance of EDTA since addition of more of this compound restablishes the reaction. The process under the same intensity of sunlight proceeds at somewhat slower rate but

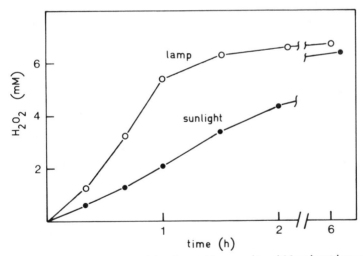

Fig. 1. Hydrogen peroxide formation under illumination with incandescent lamp and sunlight. The 60 ml solutions contained 100 μM FMN, and 10 mM EDTA in 0.2 M phosphate buffer pH 7.5. The illumination with both the lamp and sunlight had an intensity of 400 W/m^2.

Table I. EFFICIENCY OF DIFFERENT FLAVIN COMPOUNDS
AS PHOTOCATALYSTS FOR H_2O_2 PRODUCTION

Flavin compound	H_2O_2 (μmol/min)
FMN	1.85
FAD	0.38
Lumiflavin	0.88
Riboflavin	1.71
5-deazariboflavin	0.00
Proflavin	0.00

60 ml solutions of 10 mM EDTA in 0.2 M phosphate buffer pH 7.5, containing the different flavin compounds at a 50 μM concentration were irradiated at 200 W/m^2 and the H_2O_2 content determined at different times.

reaches, after a longer period, the same 6 mM level. Photodegradation of FMN, which is observed after prolonged illumination is prevented by increasing the EDTA concentration which causes also higher rates of formation of hydrogen peroxide. 10 mM was found the optimal EDTA concentration (data not shown) since higher ammounts of this compound show an inhibitory effect on the overall photochemical reaction.

Among the different flavins and flavin analogs (see Table I) those containing the isoalloxazine ring (except FAD) were the most efficient. The flavin related compounds proflavin and 5-deazariboflavin, which are characterized for their very negative redox potentials, show a very slow rate of reoxidation by oxygen as compared with other flavins. EDTA can be substituted as electron donor in the photochemical reaction for the production of hydrogen peroxide by many other compounds of very different nature (Table II). In fact, other compounds such as semicarbazide or tricine are even more efficient than EDTA. Thiol compounds reported as good electron donors for the proflavin mediated hydrogen production are also acceptable donors for H_2O_2 production.

By increasing the amount of photosensitizer higher rates of H_2O_2 production are obtained reaching the saturation at 80-100 μM FMN concentration. Keeping a constant EDTA/flavin concentration ratio a linear correspondence between the rate of reaction and FMN concentration was obtained. The highest rates were obtained with a ratio of 200.

The reaction by which the photochemically reduced flavins react with oxygen to produce hydrogen peroxide shows a marked dependency on pH but mostly on the nature of the buffer. We have found phosphate and pyrophosphate buffers pH 7.5-8.5 among the most efficient of the non electron donating buffers.

The results summarized here stablish the optimal conditions required for the production of hydrogen peroxide catalyzed by flavins and driven by light energy. Although hydrogen peroxide is not, at present days, considered as a fuel material, if sufficiently economical methods for large scale production are found it would contribute to make the energetic per-

Table II. EFFICIENCY OF THE ELECTRON DONOR IN THE PHOTOCHEMICAL H_2O_2 FORMATION REACTION

Electron donor	H_2O_2 (μmol/min)
EDTA	2.25
DTE	2.62
Hydroxylamine	1.00
Tricine	2.40
Semicarbazide	7.50
Imidazole	0.67
Triethanolamine	2.50
2-mercaptoethanol	3.46

60 ml solutions of 50 μM FMN in 0.2 M phosphate buffer pH 7.5, containing the electron donor at a 10 mM concentration were irradiated at 200 W/m² and the H_2O_2 content determined at different times.

spective more optimistic in the future.

In spite of the limitations of the photochemical reaction presented here for the production of hydrogen peroxide -i.e. photodegradation of the pigment and requirement of an expensive compound as electron donor- it constitutes a highly promising model system for capturing and storing solar energy.

Other photochemical reactions in which flavin derivates have been used to produce other fuel materials such as H_2 have been described (7,8). Those systems require, besides flavin and electron donor, a catalyst (platinum-asbestos or the enzyme hydrogenase) as well as the mediator methyl viologen. Such systems show several folds lower rates of reduction than the obtained with the EDTA-FMN system, probably due to the requirement of very negative potentials for the reduction of methyl viologen as compared with oxygen.

Acknowledgements: This work was supported by grants from Philips Research Laboratories (The Netherlands) and Centro de Estudios de la Energía (Spain). The authors wish to thank Prof. M. Losada for his encouragement and stimulating discussion and Mrs. M.J. Pérez de León for her secretarial assistance.

REFERENCES

(1) Dixon, R.G. and T.W. Healy (1971) Aust. J. Chem. 24, 1193--1197.

(2) Oster, G. and Wotherspoon (1957) J. Am. Chem. Soc. 79, 4836--4838.

(3) Bonneau, R., J. Joussot-Dubien and J. Faure (1973) Photochem. Photobiol. 17, 313-319.

(4) Mauzerall, D. (1962) J. Phys. Chem. 66, 2531-2533.

(5) Losada, M. (1979) Bioelectrochem. Bioeng. 6, 205-225.

(6) Bernt, E. and H.U. Bergmeyer, in Methods of Enzymatic Analysis (H.U. Bergmeyer, ed.) (1963) Verlag Chemie, Academic Press, New York, pp. 633-635.

(7) Krashna, A. (1979) Photochem. Photobiol. 29, 267-276.

(8) Adams, M.W.W., K.K. Rao and D.O. Hall (1979) Photobiochem. Photobiophys. 1, 33-41.

RENEWABLE HYDROCARBON PRODUCTION FROM THE ALGA BOTRYOCOCCUS BRAUNII

C. LARGEAU, E. CASADEVALL and D. DIF

Laboratoire de Chimie Bioorganique et Organique Physique,
ERA CNRS 685, E.N.S.C.P., 11 rue Pierre et Marie Curie
75231 PARIS Cedex 05, France

C. BERKALOFF

Laboratoire de Botanique-Cytophysiologie Végétale,
LA CNRS 311, E.N.S., 24 Rue Lhomond 75005 PARIS - France

Summary

The green alga Botryococcus braunii is characterized by unusually high hydrocarbon levels and could supply directly an alternative energy source to fossil fuels. In the present work we will consider, first, the location of the sites of hydrocarbon accumulation and formation: It appears that the bulk of B. braunii hydrocarbons is not only located but also produced in the outer walls. The effect of growth conditions on productivity is also examined ; it is shown that, in the case of B. braunii hydrocarbon accumulation does not necessarily require that the cells are starved and in a resting stage. So it is possible to achieve simultaneously high biomass yield and high hydrocarbon level. The features which are favourable to a large-scale hydrocarbon production via B.braunii are finally examined and it is concluded that such a production may be feasible.

1- Introduction

It is known that photosynthetic organisms can produce hydrocarbons, i.e. compounds exhibiting a particularly high energy content. Unfortunately, in nearly all species, hydrocarbon levels are low. In fact, it seems that only one major exception is known at this time : the green unicellular alga Botryococcus braunii. In nature this colonial alga shows hydrocarbon content ranging from 15 % to 75 % of dry weight (1), instead of ca 0.1 % in other algae. Moreover geochemical studies on algal coal formation and the occurrence, nowadays, of some important "blooms" indicate that B.braunii can yield large amount of biomass (2). Therefore it appears, at first sight, as an interesting species for a renewable production of hydrocarbons which could supply directly an alternative energy source to fossil fuels. Contrary to usual biomass, which contains mainly products having fairly low heat of combustion, no transformation via pyrolysis or fermentation is required in the case of B.braunii. The aim of the present work was to get information on the main features of hydrocarbon production from this alga and to test the effect of culture conditions on its productivity.

2- Location of the sites of hydrocarbon accumulation and production in B. braunii.

The main feature of B.braunii, observed on light microscopy, is the presence of refringent droplets (G) associated with most of the colonies, and frequently as large, or larger, than the cells. Electron microscopy on ultrathin sections (3) shows that these droplets are in fact located in the lumen of the trilaminar sheaths (TLS) which constitute the outer wall of the cells (each cell, besides its cellulosic wall, being surrounded by successive TLS). In addition some small oily droplets are also observed within the cells ; these cytoplasmic inclusions (g) show the same staining reactions as those of the lipidic globules (G) associated with outer walls. The above observations were consistent with the assumption that B.braunii hydrocarbons accumulate in two distinct sites. In vivo studies, using a Raman microprobe, allowed us to confirm this assumption. Moreover in vivo Raman spectra show that these sites of accumulation are specific (no other class of lipids but hydrocarbons being observed).

These two hydrocarbon pools (internal one in cytoplasmic inclusions and external one in globules associated with outer walls) were separated

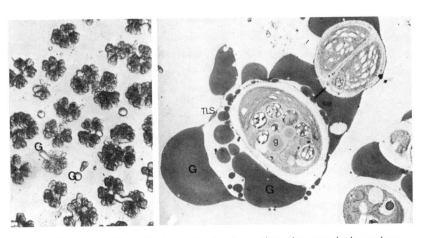

by selective extraction. Their analysis shows that the same hydrocarbons accumulate in the two sites (series of unbranched dienic compounds with odd carbon number ranging from C_{25} to C_{31} and a C_{29} triene). However the relative percentage of these long-chain hydrocarbons depends on the location (shorter compound more abundant in the internal pool). It is also to be noticed that the bulk of B.braunii hydrocarbons (ca 95 %) is located in the external pool. Owing to this extracellular location, the cultures can exhibit both an unusually high hydrocarbon content and, within the cells, low levels close to these usually observed in algae. Such a finding suggested that some excretory process was occurring : hydrocarbons first stored in cytoplasmic inclusions would be afterwards excreted in outer walls in order to keep the internal level at a normal value. Nevertheless some results (relative abundance of hydrocarbons in the two pools, absence of special structures related to excretion in electron micrographs) seemed not in keeping with such a process.

Therefore various feeding experiments were carried out (3) with tritiated palmitic acid as precursor. The effect of feeding duration (time-course experiments) and also of subsequent transfer in a cold medium (pulse -chase experiments) allowed us to show that no hydrocarbon transfer takes place from the internal pool to the external one. Accordingly the bulk of B.braunii hydrocarbons is not only located but also produced in the outer walls ; it arises from a biosynthetic pathway, the terminal steps of which occur in these walls. Feeding experiments indicate also that B.braunii,under normal culture conditions, cannot catabolize its own hydrocarbons.

3- Effect of growth conditions on hydrocarbon productivity.

The determination of O_2 evolution against chlorophyl content shows that B. braunii photosynthetic activity is similar to that of fast growing algae. Nevertheless, with standard conditions (temperature 20°C, unshaked and unaerated batch cultures), we observed a slow growth with a mean generation time of about one week. Then, the amount of biomass is fairly low (Table I) and the hydrocarbon level close to the minimal value observed in nature.(However, extrapolation with a water depth of 0.2 m (note 4) shows that such conditions would yet provide about 8.5 T. of hydrocarbons/ ha. /year).

On the other hand, with the same mineral medium but with some changes in other culture parameters (temperature 26°C, batch culture shaked and aerated by air lift 1 % CO_2), an important improvement is obtained and a growth curve is observed with a mean generation time of 2.5 days during the exponential phase (5).

Table I.

growth conditions	dry biomass (g/l)	Hydrocarbon concentration (g/l)	Hydrocarbon level (% of dry wt)	Hydrocarbon productivity (g/l/day)
1	2	0.34	17	0.011
2	6	1.77	29.5	0.084
3	1.33	0.24	18	0.080
4	1.33	0.21	16	0.053
5	0.71	0.10	14	0.050
6	5.3	1.50	29	0.088
7	7	2.52	36	0.148

1- batch culture, "standard" conditions, harvested after 30 days.
2- batch culture, aerated and shaked by air lift, t=26°C, harvested after 21 days (resting phase). 3- as 2, but harvested after 3 days (beginning of the exponential phase). 4- continuous culture, same conditions as 2, generation time 4 days. 5- as 4, but generation time 2 days.
6- batch culture as 2 but harvested after 17 days.
7- as 6, but initial nitrate concentration x 2.

The study of hydrocarbon content was carried out on samples corresponding to the beginning of the exponential phase (culture 3) and to the stationnary phase (culture 2) on the curve. It appears (Table I) that hydrocarbon level is, as expected, higher in 2 (it corresponds to an extrapolated (note 4) hydrocarbon production of c.a 60T./ha./year) ; however hydrocarbon-rich cells are also observed in the fast growing stage.

It is known that many algae accumulate lipids (mostly fatty acid esters) during the resting stage and, especially, when growth stops as a result of nitrogen starvation. Hydrocarbon accumulation in B. braunii might appear, at first sight, as possibly related to such a behaviour : the starvation resulting from the presence of the successive outer walls which may decrease the rate of nutrient uptake. The slow growth reported, so far, in the case of B. braunii laboratory cultures was consistent with such an assumption. But the above results show that fast growing B. braunii cells, although less rich than resting cells, exhibit yet unusually high hydrocarbon levels. (Consequently, batch cultures harvested during exponential growth, such as 3, show nearly the same hydrocarbon productivity as cultures grown until they enter the resting phase like culture 2). Therefore it appears that the abundant hydrocarbon production typical of this alga does not necessarily require starved and resting cells. This was confirmed by continuous cultures (5) with a decreasing generation time (Table I). Thus in the case of culture 5, which cells are maintained in the exponential stage, the hydrocarbon content is still 14 % of dry weight. The study of culture 7 (initial nitrate concentration X 2)indicates also that nitrogen starvation does not seem to play an important role in hydrocarbon accumulation by B. braunii. In fact, the nitrate-rich culture exhibits not only higher cell density but also higher hydrocarbon level.

Therefore, in the case of B. braunii, it is not essential to carry out slow growth in order to obtain an important hydrocarbon content and it is possible to improve both biomass yield and hydrocarbon productivity. It appears also that the variations in growth conditions tested here do not give way to changes in the nature and in the relative abundance of B. braunii hydrocarbons.

4. CONCLUSION

Bearing in mind that unicellular organisms are particularly efficient to obtain large amounts of biomass , if one considers that :

- B. braunii hydrocarbons can be readily transformed, via cracking, into fuels (2) and feedstocks for petrochemicals -

- this alga is ubiquitous and grows in fresh and brachish water - and if one takes into account some of the features observed in the present work :

- due to the low density of hydrocarbons no centrifugation is required to harvest the biomass -

- owing to their extracellular location most of B. braunii hydrocarbons are recovered (without extraction) just after a mild mechanical treatment -

- the hydrocarbon-freed biomass can be again brought into culture -

- preliminary results on growth conditions show that high biomass production and high hydrocarbon levels can be achieved simultaneously with B. braunii -

" It appears that renewable and direct large-scale hydrocarbon production may be feasible via this alga ".

REFERENCES AND NOTES

(1) BROWN,A.C., KNIGHTS,B.A. and CONWAY, E. (1969) Phytochemistry 8, 543.
(2) HILLEN,L.W. and WAKE,L.U. (1979) A.I.E. National Conference, Newcastle.
(3) LARGEAU,C. CASADEVALL,E. and BERKALOFF,C. (1980) Phytochemistry 19, 1043 and 1081.

(4) There are extrapolated values, they are only indicative of the actual production which could be obtained from large-scale cultures.
(5) These growth experiments were carried out by C. GUDIN and D. CHAUMONT (Laboratoire d'Héliosynthèse, S.F.B.P., LAVERA)

NEW CONCEPTS IN SOLAR BIOTECHNOLOGY

C. GUDIN, D. CHAUMONT and E. BERRA

Laboratory of Heliosynthesis BP. COMES and CEE Contractors

Summary

A controlled continuous culture of microalgae (Scenedesmus acutus) in a tubular cultivator having an area of 1m2 exposed to sunlight (near Marseilles, 1976-1980) resulted in an average productivity of dried biomass of 21 g/m2/day, corresponding to an average yield of 4% sunlight conversion. The possible maximum is 34 g/m2/day corresponding to a yield of 6.6% of the available sunlight in Marseilles. Compared with results from plants, those obtained with microalgae show that photosynthetic cell in controlled cultures are more efficient bioconvertors. However, the energy balance is $\leq$ 1 and has to be improved, since the present cost of making biomass is high approximatively 5-8 FF/kg, and consequently only a high value product, $\geq$ 10 FF/kg could be economically produced. Different means of improving the energy balance in controlled cultures are reviewed. We also present a concept of cellular and extra-cellular biomass in photosynthetic conditions (polysaccharides, hydrocarbons). A photoreactor containing immobilized microalgae on a translucent polyurethane matrix has been operational in the absence of nitrogen for 6 months producing an extracellular sulphonated polysaccharide. This type of photoreactor offers a way to improve the energy balance by exploiting photosynthesis directly to the end product.

1. INTRODUCTION

From the point of view of photosynthetic biomass production, using sunlight as an energy source and CO_2 as a carbon source, microalgae give the best yield per unit of time and surface illuminated (10-100 t/year/ha of dried biomass) and can therefore be regarded as the most efficient system of converting sunlight to biomass (1-5% of yield depending upon the sophistication of the biotechnology involved).

If a culture is well managed and optimised the photoconversion can be close to the maximum expected at the cell level i.e. 6.6%, accepting the assumption of 8 quanta required to evolve 1 mole O_2. However to get a maximum of biomass productivity a high energy input is necessary to grow the microalgae under the best conditions leading to an energy balance close to 1.0.

Simplified technology for growing microalgae can reduce the energy input but at the same time reduces the biomass productivity. However in certain geographical and economic situations, simplified technologies are acceptable eg. "Mariculture on Land" Project.

If we want to use the selective advantage of microalgae as the most efficient converter of sunlight, the biomass productivity must be maximised and this involves sophisticated biotechnology. However, there appear to be two ways in which the energy output of the biomass can be improved :

1. Increase the productivity per unit of time and illuminated surface, which, in a certain way, overcomes the "theoretical maximum conversion yield" without significantly increasing the energy.
2. Reduce the energy input achieved with a lower energy. Consuming bio-technology while maintaining the present realistic photoconversion yield of approximatively 4%.

2. CONTROLLED CONTINUOUS CULTIVATION OF MICROALGAE AND PHOTOTROPHIC BACTE-RIA.

Since 1975, in Mediterranean conditions (Martigues, near Marseilles) we have developed a tubular cultivator of 30 l with a sunlight trapping surface area of 1 m2. This cultivator can operate continuously day and night throughout the year and all parameters (temperature, CO_2 input, pH, nitrogen, nutrients) can be controlled making sunlight the only limiting factor of biomass production.

In this tubular system we have grown <u>Scenedesmus acutus</u> (fresh water green alga) in optimised conditions and have produced a realistic estimates of monthly production throughout the year (table 1).

The feasibility of this biotechnology has been established by growing different types of microalgae :

<u>Spirulina maxima</u> (blue-green alga),

<u>Porphyridium cruentum</u> (marine red alga ; produces in limiting nitrogen, 50-100% of the dried cellular biomass in extra-cellular sulphonated polysaccharide).

<u>Botryococcus braunii</u> (green alga ; produces extra-cellular hydrocarbons).

<u>Rhodospirillum rubrum</u> (phototrophic <u>athiorhodobacterium</u> ; fixes CO_2 in anaerobic photosynthesis with an organic source of protons).

HALL D. et al have isolated an active hydrogenase from R. rubrum biomass (1).

The last, bacterial, species has shown that continuous anaerobic photosynthesis is technically feasible over a long period (3 months) in a tubular cultivator. However, the choice of an athiorhodobacterium is, at present, uneconomic since the organic source of protons is insufficiently cheap from an energetic and economic point of view. A thiorhodobacterium, using SH_2 as a proton source, is a better choice.

The potential role of phototrophic bacteria in biomass production

Phototrophic bacteria use the green and near infra red bands of the sunlight spectrum, which are not used by microalgae. Also, these bacteria do not photolyse water and hence, produce biomass at a lower light intensity than microalgae. The biomass production characteristics of algae and phototrophic bacteria can therefore be combined to overcome the "theoretical maximum conversion yield", using a double layer tubular structure made of sunlight-transparent polythene, with algae in the upper layer and bacteria beneath (2).

However, if the feasibility of such a biotechnology is to be realised the first target must be to study how to avoid doubling the energy input in growing two types of photosynthetic microorganisms.

3. ENERGETIC AND ECONOMIC EVALUATION OF A ONE LAYER TUBULAR CULTIVATOR PRODUCING 1,000 TONNES/YEAR OF DRIED ALGAE BIOMASS.

3.1. Energétic cost : Nutrients   1681 kcal/kg dried biomass
             Electric energy 1782   ""   ""
                Polythene 2222   ""   ""

           Total   5685 kcal/kg dried biomass.

This analysis gives an energy balance of 0.8. This value is approximate and probably pessimistic, but indicates where improvements in the technology can be made. Firstly, the short life span of plastic material (3 years) is the main conbributor to the low energy balance. Secondly, renewable energy (as in the "Mariculture on Land" project) could be used instead of electric energy.

3.2. Economic cost : Nutrients   51.4 c/kg dried biomass
              Electric energy   12.1   ""   ""
                  Polythene   33.6   ""   ""
              Thermal energy   6.9   ""   ""

                        104.0 c/kg dried biomass
                  Labour   82.0   ""   ""
          Capital investment 318.0   ""   ""

Total without a drying process =   520 c/kg dried biomass
Total with a drying process   =   720   ""   ""

    i.e. a production cost of 5.2-7.2 FF/kg dried biomass using present technology.

The economic analysis shows that with this type of biotechnology capital investment is high. Accepting these values as rough estimates it appears that the present economic target should be a high value product i.e. $\geqslant$ 10 FF/kg. Algal products, such as starch, glycerol are less expensive than 10 FF/kg, however, other products eg. sulphonated polysaccharide produced extra-cellularly, cost in the order of 10 FF/kg.

4. CONTROLLED CONTINUOUS PRODUCTION OF A CHEMICAL OR FUEL IN A BIOPHOTO-REACTOR USING IMMOBILISED PHOTOSYNTHETIC CELLS.

Energetic and economic evaluation of a controlled process based on the continuous cultivation of photosynthetic cell biomass leads to two key ideas :

1. The energy input can be reduced by eliminating the circulation of the biomass (924 kcal/kg dried biomass), and by passing the separation of

of cellular biomass from the liquid medium (418 kcal/kg dried biomass).

2. Producing an extracellular chemical of economic interest ($\geq$ 10FF/kg) which is easy to harvest.

An immobilised bed of continuously fed photosynthetic cells, producing the target chemical in the effluent liquid opens up interesting possibilities in solar biotechnology.

During the past 6 months in cooperation with D. Thomas (U.T.C., Compiègne, France), we have set up such a biophotoreactor in which cells of Porphyridium cruentum are immobilised in a light translucent bed of polyurethane, forming a real biosponge structure.
This photoreactor is continuously fed with artificial light (as a first step), a stream of CO_2 and a nutrient solution. For the last 4 months sulphonated polysaccharides have been continuously produced under conditions of nitrogen starvation. Harvesting the polysaccharides in the effluent is easy, since the viscosity of the effluent depends upon the dilution and can be controlled. The polysaccharides can then be precipitated with ethanol. However the economical evaluation of such a route has to be done.

Much work remains to improve this biotechnology but the concept appears promising and will be applied to hydrocarbon production using Botryococcus braunii (in cooperation with E. Casadevall et C. Langeau, Paris University).

Such a system could be applied to plant cell as we are able to grow them fully autotrophically (3, 4).

REFERENCES

(1) ADAMS M.W.W. and HALL D.O. Bioch. and Biophys.Res.Com. vol 77 n° 2 ;
     p. 730-732, (1977).

(2) GUDIN C., US Patent 9,955,317 (1976).

(3) GUDIN C. and PEEL E., U.K. Patent 1,401,681 (1975).

(4) DALTON C., Journ.of Exp.Bot. vol 31 n° 122 pp. 791-804 (1980).

PREVIOUS WORK

GUDIN C. - Solar energy in agriculture. Reading U.K. - ISES p. 48-51,
          Sept. 1976

GUDIN C. and CHAUMONT D. in "Héliosynthèse et Aquaculture" Martigues,
    CNRS (available at AFEDES - 48 rue de la Procession Paris 15°)
    pp. 59-62; pp. 71-84; 117-118; 149; Sept. 1978

GUDIN C. and CHAUMONT D.; Bedford Colloquim in Biochemical Journal
    (London Bioch. Soc.) under press - April 1980

GUDIN C., CHAUMONT D., DESANTI O. and PIOLINE D.; 99ème Congrès de
    l'AFAS - Amiens, in "Revue du Palais de la Découverte", under
    press, Sept. 1980

GUDIN C. and CHAUMONT D., EC Solar Energy Programme, Project E - Brussels -
    internal report (1978)
    Taormina - internal report (1979)
    Final report - under press (1978-1979)

GUDIN C., French Patent 76 07901 (1976)

GUDIN C., French Patent 77 24365 (1978)

Table 1 - PRODUCTIVITY OF S. ACUTUS in MARTIGUES - LAVERA

Lat N 43-22-48 Long E 4-58-56	Solar Energy dayly average kcal/m2/day	Biomass Productivities (dried matter) g/m2/day		Photoconversion yield %
Months	measured under glass	Calculated maximum based on 6.6% yield	Experimental results	1 g = 5 kcal
J	1100	14.5	11	5
F	1550	20.5	13.4	4.3
M	2450	32.3	14.8	3
A	2850	37.6	21.2	3.7
M	4350	57.4	30	3.4
J	4450	58.7	33.6	3.8
J	4375	57.8	30	3.4
A	3750	49.5	30	4
S	2500	33	21.9	4.4
O	1700	22.4	21.5	6.3
N	1300	17.2	11.8	4.5
D	950	12.5	11	5.8
Daily average	2610 kcal/m2/day	34.5 g/m2/day	20.9 g/m2/day	4%
	10,868 KJ/m2/day 126 W/m2/day	126 t/ha/ year	76 t/ha/year	

H_2 PRODUCTION BY THE PHOTOSYNTHETIC BACTERIA, RHODOPSEUDOMONAS CAPSULATA

ENTRAPPED IN ALGINATE GELS

F. PAUL and P.M. VIGNAIS

DRF/Biochimie (INSERM U. 191 / CNRS-ER 235), Centre d'Etudes Nucléaires,
85X, 38041 Grenoble cedex, France

Summary :

We are presenting preliminary results obtained with Rhodopseudomonas capsulata cells entrapped in barium alginate beads. Optimized conditions for cell entrapment (cell and alginate concentration, viscosity of the alginate solution), rates of H_2 production by free and entrapped cells, stability of gel beads upon storage or during functioning in a reactor are described. Yield of entrapment measured just after fixation varied between 30% and 80% depending on the concentration of cells in alginate solution, the type and viscosity of the alginate solution, the nature of the divalent cation used. Storage stability of entrapped cells was the same as that of free cells. After 20 days at 4°C, under anaerobiosis, the capacity for H_2 production had decreased to 25% of its original value in both types of cells.

1. INTRODUCTION

Rhodopseudomonas capsulata, strain B10, evolves hydrogen gas at high rates (130 µl H_2 x h^{-1} x mg^{-1} protein) under 100% H_2 atmosphere, through the activity of its nitrogenase (1, 2). Other attributes such as easy growth on a variety of simple organic substrates, resistance to harsh environments, make it suitable for further biotechnological studies in view of its use in H_2 production (see (3) for review).

The immobilization of R. capsulata chromatophores by entrapment in alginate gels for continuous ATP regeneration has demonstrated the suitability of this entrapment method for photosynthetic material (4). Preliminary studies of H_2 production of R. capsulata entrapped in alginate gel are presented in this paper.

2. MATERIALS AND METHODS

2.1 Bacterial strain and culture

R. capsulata, strain B10, was obtained from the photosynthetic bacteria group, Department of Microbiology, Indiana University, Bloomington, USA, and was cultured photosynthetically at 30°C for 16 h in a mineral salt medium (1, 5) supplemented with 30 mM malate and 7 mM glutamate as carbon and nitrogen sources, respectively. Cultures were illuminated by 100 W incandescent lamps (ca. 10 000 lux).

2.2 Preparation of resting cell suspensions

Cultures were harvested at late exponential phase and centrifuged (25 000 x g, 20 min, 4°C). The pellets were washed once with 10 mM Tricine-NaOH buffer, pH 7, and recentrifuged as above before resuspending in mineral salt medium (diluted to 1/10th) supplemented with 10 mM Tricine-NaOH buffer, pH 7. The resting cell suspension was stored under argon at 4°C in the dark until required.

2.3 Preparation of immobilized cells

Sodium alginate was obtained from CECA S.A., Département "Algues et Colloïdes", Usine de Baupte, 50500 Carentan, France. The entrapment method using calcium or barium as gel-inducing agent was as described in details in (4) except that a 10 mM Tricine-NaOH buffer pH 7, supplemented with mineral salt medium diluted to 1/10th the normal concentration was used in all steps of the preparation. The method is briefly outlined in Table I.

2.4 Incubation conditions

Free or entrapped cells (ca. 1 mg protein) were incubated in a medium made of the growth mineral salt medium (without C or N source) diluted to 1/10th and 10 mM Tricine-NaOH buffer. Assays were carried out in either 4-ml vials containing 1 ml of cell suspension or 12-ml vials containing 2 ml of the suspension. Vials were capped with rubber septa and gassed with argon. The reaction was started by addition of 3.75 mM malate. H_2 produced in the gas phase by either free or entrapped cells was monitored by gas chromatography. Reaction vessels were shaken at 30°C in Warburg bath uniformly illuminated by 40 W incandescent lamps from below (ca. 5000 lux). Aliquots (50 µl) were withdrawn from the gas phase of the vials and injected into an Intersmat IGC 120 gas chromatograph. Protein concentration was measured by the Folin method using bovine serum albumin as standard.

3. RESULTS

3.1 Kinetics of H_2 evolution from free and entrapped cells in barium alginate gel

Initial rates of H_2 evolution were measured in illuminated flasks and H_2 evolution was monitored in the gas phase. For free cells a lag time of about 5 min was observed before the rate of the reaction appeared linear. This effect was increased with cells in alginate beads (lag time of 30 min) (Fig. 1). For entrapped cells, diffusion of carboneous substrate and gazeous product can be rate-limiting (6).

3.2 Effect of the sodium alginate concentration in the entrapment solution

In sodium alginate gels (0.4 to 2%, w/v) of relatively low viscosity (satialgine S300, CECA S.A., France), the activity of entrapped cells remained the same. Above 2% concentration, diffusional limitations (for malate, H_2 and light) increased and the activity of entrapped cells decreased by a factor of 40%. To obtain beads with sufficient mechanical stability a concentration of sodium alginate of 1% (w/v) was necessary. (Fig. 2).

3.3 Effect of cell concentration in the entrapment solution

Cell concentration in the entrapment solution varied from 0.75 mg protein/ml to 6.75 mg protein/ml. From about 4 mg protein/ml, a decrease in H_2 production was observed (about 25%) probably due to limitation in light diffusion (Fig. 3).

3.4 Storage stability of R. capsulata cells

Storage stability of entrapped cells was the same as that of free cells. After 20 days at 4°C, under argon in the dark, the capacity for H_2 production had decreased to 25% of its original value in both types of cells. (Fig. 4).

3.5 Yield of entrapment. Effect of incubation in the light

Yield of entrapment measured just after fixation varied between 30% and 80% depending on factors such as the concentration of cells in sodium alginate solution, the viscosity of the alginate solution and the nature of divalent cation used (calcium or barium). (Table II). The specific activity of entrapped cells increased during the first hours of incubation in the light and reached values as high as in free cells incubated in the same conditions (Fig. 5). The phenomenon is due to limited microbial growth in alginate gel, as already observed by Ohlson et al. (7) and to additional synthesis of the nitrogenase enzyme upon illumination of resting cells (8).

CONCLUDING REMARKS

These preliminary studies indicate that alginate beads containing entrapped cells of R. capsulata can be used in fluidized-bed bioreactors for continuous H_2 evolution. The method affords mild, non denaturing conditions for the cells which can divide after embedding. Gel beads can be prepared in anaerobic and sterile conditions ; they form homogeneous particles, stable over long periods of time allowing a good diffusion of light and little diffusional constraints for the gazeous product (H_2). In addition, by affording mild conditions of entrapment and excellent physical properties of the gel beads, the barium alginate method appears to be a suitable tool for the construction of bioreactors with living cells and organelles.

REFERENCES

(1) Hillmer, P. and Gest, H. (1977) J. Bacteriol. 129, 724-731.

(2) Hillmer, P. and Gest, H. (1977) J. Bacteriol. 129, 732-739.

(3) Weaver, P.F., Lien, S. and Seibert, M. (1980) Solar Energy 24, 3-45.

(4) Paul, F. and Vignais, P.M. (1980) Enzyme Microbiol. Technol. 2, 281-287.

(5) Weaver, P.F., Wall, J.D. and Gest, H. (1975) Arch. Microbiol. 105, 207-216.

(6) Kronwel, P.G. and Kossen, N.W. (1980) Biotechnol. Bioeng. 22, 681-687.

(7) Ohlson, S., Larsson, P.O. and Mosback, K. (1979) Eur. J. Appl. Microbiol. Biotechnol. 7, 103-110.

(8) Meyer, J., Kelley, B.C. and Vignais, P.M. (1978) J. Bacteriol. 136, 201-208.

TABLE I

ENTRAPMENT OF CELLS IN ALGINATE BEADS

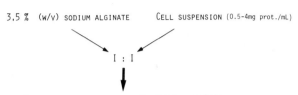

3.5 % (W/V) SODIUM ALGINATE   CELL SUSPENSION (0.5-4mg prot./mL)

I : I

ADDITION AS DROPLETS INTO 50 MM CaCl$_2$ OR BaCl$_2$
BUFFERED WITH I0 MM TRICINE, PH 7.

SHAKING FOR I H. IN THE DARK UNDER ARGON

WASHING WITH COLD DISTILLED WATER THEN WITH COLD
2.75 MM TRICINE, PH 7.

STORAGE IN 2.75 MM TRICINE, PH 7 + MINERAL SALT
MEDIUM DILUTED TO $\dfrac{I}{I0}$

TABLE II
H$_2$ EVOLUTION BY CELLS OF R. CAPSULATA ENTRAPPED IN ALGINATE BEADS

	CALCIUM ALGINATE GEL		BARIUM ALGINATE GEL		
EXPT. NB. :	I	2	I	2	3
H$_2$ EVOLVED (µMOL.H.$^{-I}$.MG^{-I} PROTEIN)	2.8	I.4	2.4	3.0	4.4
ENTRAPMENT YIELD (%)	56	6I	3I	59	82
CELL CONCENTRATION IN ALGINATE SOLUTION (MG PROTEIN.ML^{-I})	3.35	3.75	0.72	0.72	0.47

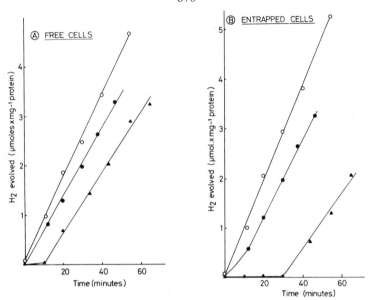

Figure 1 - Kinetics of H_2 evolution by free cells (A) and cells entrapped in barium alginate beads (Satialgine S300) (B).
H_2 evolved upon illumination of cells under argon was measured either immediately (—▲—) or after preillumination (30°C ; 5000 lux) of resting cells for 6 h. (—●—) or 17 h. (—O—). Preilluminated cells were again sparged with pure argon before measuring H_2 evolution at time zero.

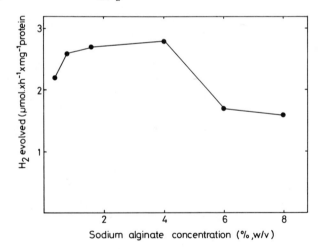

Figure 2 - Effect of alginate concentration on H_2 evolution by cells entrapped in calcium alginate beads.

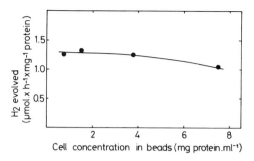

Figure 3 - Effect of cell concentration in alginate beads on H_2 evolution by entrapped cells.

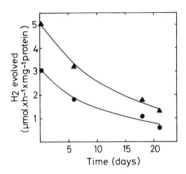

Figure 4 - Storage stability of free and entrapped cells. Free cells (—▲—) ; cells in calcium alginate gel (—●—).

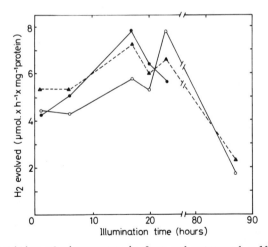

Figure 5 - Stability of nitrogenase in free and entrapped cells (barium alginate).
Free cells (--▲--) ; cells entrapped in sodium alginate S550 (—O—) ; cells entrapped in sodium alginate S800 (—●—).

FIRST EXPERIMENTS OF PRODUCTION OF MACROPHYTES WITH
WASTE WATER AND METHANIZATION OF BIOMASS.

by Marie-Luce CHASSANY-DE CASABIANCA[°] and François SAUZE[°°]

[°] Ecology and macrophytic production. C.N.R.S. Route de Mende -
34060 Montpellier FRANCE.

[°°] Méthanization. Station d'Oenologie I.N.R.A. Narbonne FRANCE.

Summary

        The basic principle is to obtain energy production by methaniza-
tion of macrophytes cultivated on waste water which is simultaneously de-
polluted.
        Two types of culture have been experimented in semi artificial
 basins of 20 square meters, in the open, strictly fed with waste water.
It concerns firstly Lemna minor species of soft water being able to adapt
themselves to waste water and secondly Enteromorpha intestinalis, typical
brackish water green alga and being able to adapt itself to a high percentage of organic matter.
        The first experiments show :
a) How to make macrophytes grow on this waste water in which phytoplankton
spontaneously develops. So it allows to precise the optimal conditions of
growth and the conditions of competition with phytoplankton.
b) The experiment allows also to precise from the analysis of the natural
ponds the conditions for optimising and maintaining artificial macrophytes
production by acting on the different parameters : continuous or disconti-
nuous feeding, length of the stay of waste water in the basins, depth, temperature, enlightment, quantity or quality of organic matter.
c) The efficiency of depollution is established.
d) Experiments of methanization of Enteromorpha at laboratory scale has
given yields of about 0,3 liter of biogas per gramme of dry organic matter
treated, with a rate of more of 60 % methanic digestion of plankton and of
phragmites.

Ces premiers essais se situent dans le cadre de l'utilisation de macrophytes aquatiques (en l'occurence, pour la production d'énergie par méthanisation) en épurant des eaux résiduaires.

1 - Cultures de macrophytes aquatiques sur eaux résiduaires : premières données écologiques. Par M-L. CHASSANY-DE CASABIANCA.

1.1. Introduction.

Cette démarche résulte d'une double question :

- est-ce que l'on peut obtenir une production de biomasse macrophyte (facilement récoltable) intéressante ou pas sur eaux résiduaires, ou en principe se forme préférentiellement du phytoplancton, et est-ce que cette production peut lui être supérieure ?

- est-ce que l'on peut trouver des espèces dont l'adaptation puisse s'inscrire dans les limites des contraintes les plus diffiles qu'offrent les stations d'épuration, même si la production optimale nécessite une marge de variabilité écologique plus restreinte ? Les contraintes étant:bassins à l'air libre à fortes variations, forte eutrophisation autour des moyennes (pour eau lagunée : NH4 : 4,2 mg/l ; NO3 : 0,1 ; PO4 : 1,3 mg/l ; C : 8000 mg (10^{cc} et N : 550). Eau brute : NH4 : 7,6 ; NO3 : 0 ; PO4 : 5. Compétition éventuelle avec le phytoplancton, et difficulté de compenser l'évaporation par de l'eau douce.

Les premières tentatives de cultures ont été réalisées sur des espèces à haut pouvoir d'adaptation et/ou pouvant supporter une certaine eutrophisation. Il s'agit de Lemna minor, espèce d'eau douce (teneur moyenne en milieu naturel de : 0,45 mg/l NH4, 0,05 NO3 ; 1 PO4 ; 1300 mg/10c de C et 770 mg/10 cc N.) et d'Enteromorpha intestinalis, algue d'eau saumâtre (5-20 S%° environ, 0,2 mg NH4 ; 0,2 NO3 ; 0,06 mg/l PO4 ; 3200 C mg/10cc 200 N mg/10cc).

Ces cultures ont pour but, de déterminer les limites de leur adaptation et d'autre part, de comprendre l'impact des facteurs écologiques abiotiques (profondeur, alimentation,... type d'alimentation...) et biotiques (cycles des espèces, compétition avec le phytoplancton...) sur les variations du taux de croissance de ces macrophytes, plus que d'obtenir des rendements de production proprement dit qui ne pourraît être ensuite efficacement obtenus qu'à partir de cultures menées d'après un protocole déterminé après une série d'essais préliminaires.

1.2. Marges d'adaptation des deux espèces sur eaux résiduaires (essais en cuvettes).

. Les premières constatations sont les suivantes :

- Pour E. intestinalis, les résultats de production les plus significatifs, ont été obtenus avec de l'eau lagunée par rapport à l'eau brute où les algues ont été brûlées à 50 %. Les meilleurs résultats étant obtenus avec de l'eau lagunée complétée à l'eau, soit 3 fois supérieur à l'eau lagunée complétée en lagunée.

- Pour L. minor, des résultats comparables semblent obtenus sur les 2 substrats (brut et laguné).

- Le meilleur taux de déseutrophisation est obtenu avec E. intestinalis (soit pour les orthophosphates 0,03 g/j/1g P.S. produit).

. On peut conclure :

1°) A l'importance de la biodéposition du phytoplancton, éliminé par la faible profondeur, en eau lagunée (permettant de maintenir la croissance des macrophytes pendant 15 jours au moins) et donc du problème de sa compétition.

2°) A haut pouvoir épurateur de E. intestinalis.

3°) Que l'emploi de l'eau brute semble devoir être plus nuancée dans le cas de E. intestinalis que dans L. minor, laissant entrevoir un champ d'investigation.

1.3 Essais en bassins sur eau lagunée (Figure 1 et Poster).

★ Processus expérimental général résumé (figure 1).

Bassin de 15 m2 en plastique blanc sur 10 cm de profondeur. Essais sur 3 mois (mai-août).

- Alimentation lagunée discontinue par compensation de l'évaporation (soit en moyenne 100 l/j à raison de 2 alimentations hebdomadaires). Ensemencement faible (1 à 3 kg P.H. pour les 2 espèces). En ce qui concerne les lentilles d'eau en particulier les essais comportent 3 phases sur 2 bassins :

1°) Une phase à alimentation réduite a) avec eau lagunée décantée Bassin IV
b) avec eau lagunée non décantée, amené régulièrement. Bassin III.

2°) Une phase à alimentation régulière et récolte lente (1 fois par mois).

3°) Une phase à alimentation régulière et récolte rapide (1 / 3 jours).

- D'autres essais, en continu, ont eu pour variable la durée de séjour.

★ Résultats.

A/ Production et problèmes écologiques préliminaires posés.

Sur eau lagunée, le taux de production journalier a varié de $0,1 - 10^{-2}$ $0-20.10^{-2}$ pour les lentilles et de 2 à 19.10^{-2} pour les Enteromorpha (le maximum étant obtenu pour une faible biomasse en présence (0,2 à 0,6 kg P.H./m2).

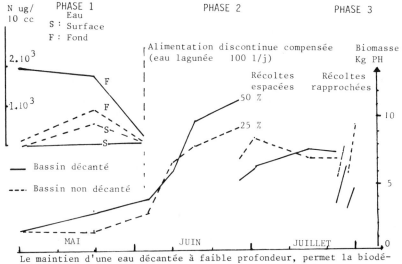

Le maintien d'une eau décantée à faible profondeur, permet la biodéposition du phytoplancton et la restitution du matériel organique, qui peut
être recyclé sans compétition par les macrophytes. Il en résulte que l'on
observe un taux de croissance 7 fois plus important sur eau lagunée décantée
que sur eau lagunée fraîchement alimentée. (phase 1 du schéma). Ou sur eau
lagunée en circulation continue lente, qu'en eau lagunée à temps de séjour
court. Dans ce dernier cas, la production de macrophytes, est comparable à
la production phytoplanctonique. Au cours de l'été, de nombreux facteurs
sont à intégrer, qui mettent en cause l'influence grandissante de l'ensoleillement et de la température sur le cycle de l'algue, mais encore, l'importance de l'évaporation qui a été compensée par un apport plus fréquent
d'eau lagunée (phase 2 et 3). Malgré cela, l'importance de la récolte apparaît.

B/ Deseutrophisation (absorption et précipitation).

Compte tenu de la complexité du recyclage de la matière détritique et
minérale par les macrophytes et le phytoplancton, nous résumerons ici cette
déseutrophisation par l'évolution comparée des orthophosphates au cours de
la période du 5 au 25/6, où les différents bassins subissent une phase de
croissance comparable, et dont les résultats montrent la déseutrophisation
préférentielle avec E. intestinalis, confirmée par ailleurs.

Bassins	Enteromorpha 1	Lentilles 3	Lentilles 4
PO4 éliminé	5 824 mg	2 600 mg	2 660 mg
Biomasse algue produite (Kg PH)	4	6	7,6

1.4 Conclusion : cette étude fait ressortir :

. La possibilité d'utilisation des eaux résiduaires en cultivant des macrophytes à haut pouvoir d'adaptation présentant un pouvoir productif et/ou eutrophisant correspondant à l'objectif souhaité.

. Par quelques exemples pris ici, l'importance des facteurs biotiques et abiotiques menant à la production de biomasse et l'intérêt d'en maîtriser les effets par l'étude du fonctionnement de systèmes simples, afin de définir pour ces études préliminaires, les modalités et la méthodologie permettant d'engager véritablement des actions de production.

2 - Méthanisation des macrophytes aquatiques. Par F. SAUZE.

2.1 Performances

2.1.1. Digestion statique.

3 végétaux expérimentés en fermentation mésophile (35°).

	Enteromorpha	Macrophytes	Phragmites
Capacité digesteur	1 l	2 l	2 l
Levain	Boues résidu-aires urbaines	Purin+boues r. urbaines	Purin+boue de méthanogénèse (enteromorpha)
Teneur initiale en MES (g./1-1)	36	25	10
Rendement volumique biogaz (1.PO^{-1})	0,45-0,50	0,3	0,1
Richesse en méthane %	65	63	62 - 64
Rendement d'épuration (DCO %)	87	85 - 90	-

Rendement "économique" (massique) : de 0,25 à 0,35 l de biogaz par g. de solve volatile introduit.

2.1.2. Fermentation semi-continue mésophile (Enteromorpha)

Capacité digesteur : 20 l.

Levain : boue extraite des fermentations discontinues + un peu de purin

Teneur initiale en MES : 25 g l^{-1} - Série de 8 fermentations à 35°c -

(Variables : - charge volumique θ en MSV - durée de séjour - recyclage ou non

Rendement volumique en gaz Y_G : max. 0,35 P l$_D^{-1}$ j^{-1}

Rendement massique ("économique") Y_{ec} : max : 0,36 l j^{-1}. g MSV^{-1}

(optimisation pour C_V = 1,12, θ = 20 jours. Richesse en CH4 : 62-65 %).

Action favorable du recyclage (réduit "lavage" de biomasse).

Taux de conversion du θ en CH4 : 40-45 %.

Taux d'épuration de l'influent (en % DCO éliminée) = 87-91 %.

Temps de séjour θ : Essais effectués avec θ = 10,15,20 et 40 jours, en vue de mettre en évidence une éventuelle corrélation avec les rendements et le taux d'épuration.

2.2 Influence et optimisation des paramètres de fermentation.

(Substrat : Enteromorpha).

Charge volumique introduite C_V (g.1_D^{-1} MSV$_0$)

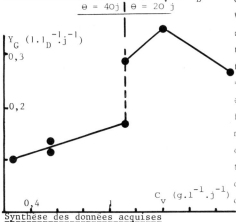

Variation sensiblement linéaire du rendement en gaz lorsque le temps de séjour θ est long (partie gauche de la courbe). Avec θ = 20 jours, et C_V 1 le rendement est élevé, mais la variation en fonction de la charge accuse un maximum (0,35) pour C_V = 1,4. La capacité de production du digesteur n'étant probablement pas encore atteinte, on peut espérer obtenir des rendements 0,5 à 0,6.

Synthèse des données acquises

* Aptitude à la fermentation mésophile de l'algue macrophyte Enteromorpha.

. Possibilités pour celle de Phragmites, sous réserve d'amélioration de la phase de biodégradation ligno-cellulosique par prétraitement.

A partir d'un levain de première fermentation le traitement du substrat d'Enteromorpha, réalisable au laboratoire en semi-continu, est assez rapide, stable et reproductible.

* Taux d'épuration très important après digestion, variant grosso modo comme le rendement en gaz avec divers paramètres.

Perspectives

* Des essais à l'échelle du pilote, actuellement en préparation, pourraient fournir des performances plus élevées, notamment le rendement volumique.

* Même avec les rendements actuels, intérêt économique et écologique des végétaux aquatiques.

. Potentialités très importantes en biomasse dans les zones littorales même en se limitant au milieu naturel.

. Possibilité de culture en eaux résiduaires, urbaines ou d'IAA, avec épuration des nutriments qu'elles contiennent (notamment N et P).

VALORIZATION OF AQUATIC MACROPHYTIC BIOMASS. METHANE PRODUCTION
DEPOLLUTION AND USE OF VARIOUS BY-PRODUCTS.

M-L. CHASSANY-DE CASABIANCA, L. CODOMIER, A. GELY, F. SAUZE.

1 - Ecologie et production de macrophytes. C.N.R.S. Montpellier.

2 - Cultures in vitro. Université de Perpignan.

3 - Mécanisme agricole. C.N.E.E.M.A. Nîmes.

4 - Méthanisation - Réalisation des pilotes - actions de développement.
I.N.R.A. Narbonne.

The aims of the group of study and research on aquatic biomass
are :
a) Evaluation of the macrophytic biomass (algae and phanerogames) in natural environnemt (brackish water ponds) in several stations of Languedoc-
Roussillon with determination of the physico-chemical conditions of their
development.

b) Exploitation of macrophytes in natural environnment and valorization of
their production by harvesting and depolluting.

c) Laboratory study designed to determine on different nutritive solutions
and for each species the physico-chemical conditions of their development
and their biocycle.

d) Production of macrophytes in a basin of waste water.

e) Development and testing of a harvester.

f) Tests of methanization of biomass from the natural environnment and
experimental tanks.

g) Possible recycling of residual organic matter in agriculture and aquaculture.

h) Economic evaluation of the operation.

The integrated project which concerns two types of operations
(the first in brackish water ponds, the second in aquaculture basins) are
presented in posters.

The results concerning macrophytes aquaculture and their methanization are detailed in other posters (M-L. CHASSANY-DE CASABIANCA,
F. SAUZE).

Il s'agit d'étudier la possibilité d'utiliser les macrophytes pour la production d'énergie, en milieu naturel ou artificiel. Ceci en respectant l'équilibre du milieu naturel ou en utilisant les possibilités de dépollution et de déseutrophisation des divers milieux :

Matière organique + soleil → Energie + dépollution + sous-produits.

Le programme général du groupe peut être schématisé de la façon suivante :

1 - RECHERCHES DE BASE .

. A/ MILIEU NATUREL (ETANGS, SALINES, CANAUX).

- Détermination en milieu naturel des conditions physiques (température et ensoleillement) et chimiques (éléments nutritifs) d'obtention optimale de la biomasse macrophytique. Les espèces étudiées sont des algues : Ulva sp., Cladophora sp., Chaetomorpha sp., Enteromorpha sp., Gracillaria sp., des phanérogames aquatiques : Zostera sp., Ruppia sp., Potamogeton sp. et des phanérogames terrestres : Phragmites sp.

- évaluation de la biomasse macrophytique dans quelques stations types des étangs du Languedoc-Roussillon

- étude du caractère renouvelable de cette biomasse : influence des modalités d'exploitation de cette biomasse (récolte).

. B/ CULTURE. - En laboratoire :

- détermination du biocycle in vitro

- contrôle de la sporulation et de la gamétogenèse

- essais de multiplication végétative

- choix des milieux de culture

- conditions optimales de développement en fonction de la température et de l'éclairement.

- confection d'un appareil de culture en continu.

. C/ AQUACULTURE. - En bassins extérieurs :

- essais de culture de différentes espèces avec divers effluents en bassin de l'ordre de 20 m2

- mêmes essais de culture en bassin pilote de quelques centaines de m2

- études des principaux paramètres de culture (débit d'effluent par unité de surface, densité de plantation, etc...).

. D/ RECOLTE ET METHANISATION (Cela concerne à la fois les opérations 1et2)

- Récolte et séchage :

        - Mise au point et essai de prototype de récolteuse-chargeuse

        - Mise au point et essai de prototype de séchoir solaire (type séchoir de fourrage en vue du stockage de la biomasse).

- Méthanisation et utilisation des sous-produits :

. Méthanisation.

        - essai successifs ou parallèles in vitro et en digesteurs extérieurs de plus grande capacité des biomasses issues du milieu naturel et des bassins de culture.

        - paramètres étudiés : température, pH, charge journalière moyenne, etc...

        - rendements en biogaz et en résidus de fermentation.

. Utilisation éventuelle des sous-produits

        - recyclage dans les cultures du CO_2 de la méthanisation et du résidu humique de la fermentation.

. E/ RECYCLAGE DES MATIERES ORGANIQUES RESIDUAIRES en tant que fertilisant au niveau de l'agriculture et de l'aquaculture.

II - RECHERCHE DE DEVELOPPEMENT : UNITE PILOTE

        Conception et réalisation d'une unité, comprenant étang naturel ou bassin de culture, machines de récolte, de séchage et de transport, digesteur à méthane.

III - CALCUL DES ELEMENTS DE COUT DE L'OPERATION.

IV SCHEMA RESUMANT LE PROGRAMME DE RECHERCHE.

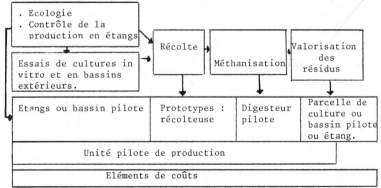

GROWTH OF MARINE BIOMASS ON ARTIFICIAL STRUCTURES AS A RENEWABLE SOURCE OF ENERGY.

J.G. MORLEY[*] and J.M. JONES[†]

[*] Wolfson Institute of Interfacial Technology,
University of Nottingham, Nottingham.

[†] Department of Marine Biology, University of
Liverpool, Port Erin, Isle of Man.

Summary.

It is estimated that the equivalent in energy terms of 1000 tonnes of
coal could be produced per square kilometre of sea surface per year through
the growth of native species of marine biomass. Studies are in hand to
confirm expected biomass growth rates on inexpensive subsurface flexible
structures. Preliminary estimates of the commercial feasibility of the
process indicate that methane could be produced from the biomass at costs
comparable with existing fuel costs. The objective of the programme is
to provide data by which firmer estimates of the technical and commercial
viability of the process can be made.

1. INTRODUCTION.

    A preliminary feasibility study of the growth of local species of
marine biomass on artificial substrates as an energy source is under
investigation for the sea areas adjacent to the United Kingdom.
Studies on similar lines utilising california giant kelp as the
preferred biological species have been carried out in the United States.
Structural problems have been encountered during the initial U.S. trials
due to the buoyant fronds and very large size of giant kelp. As a
consequence of wave action this produced large differential movements
between the fronds of the plant and its point of attachment to the
subsurface structure. It is expected that the much smaller size of
local species will reduce structural problems to manageable
proportions. Although the growth rates of individual plants of giant
kelp is very high, the amount of biomass produced per unit area of sea
surface per year by this species is of the same order of magnitude as
that known to be attainable with local species.

2. SCALE OF POTENTIAL SOURCE OF ENERGY.

    Existing experimental data indicates that biomass growth rates in
the region of 2 kg (dry weight) per square metre of sea surface per
year are achievable in United Kingdom waters having an energy content
approximately half that of the equivalent weight of coal. It follows
that a production rate of the order of 1000 tonnes of coal equivalent
per square kilometre of sea surface per year is achievable in principle.

3. THE GROWTH OF BIOMASS ON THE SUBSTRATE.

    It is proposed to use partly submerged inexpensive polymeric
ropes to provide points of attachment for the biomass. Flexible
structures can be designed which are expected to deform under wave
action without large forces being developed. It is also expected that
it will not be necessary to design the structure to meet rarely occurring very severe sea conditions, since under these circumstances, some
or all of the seaweed would be washed away from the structure allowing
its integrity to be preserved. This load limiting feature is expected
to allow the use of relatively low strength and hence inexpensive
structures. The capital cost of the structure is clearly a very
important factor controlling the commercial feasibility of the project.

Attention will be paid to Laminaria saccharina and Saccorhiza polyschides as potential energy crops. The former is the fastest growing species of the genus in the North East Atlantic. Saccorhiza polyschides is an opportunistic species, growing (under natural conditions) on freshly cleared subtidal rocks before other species can become established. A considerable quantity of gametophyte material can be produced under laboratory conditions and this will be used for initial seeding. It is hoped that by controlling the harvesting period enough plants should be fertile to ensure continued 'seeding' of the substrates. Similar techniques are used for the growth of seaweed for food elsewhere. Brown seaweeds have been cultivated on ropes in Japan for decades, and L. japonica has been grown on a large scale (1,300,000 wet tonnes/yr.) in China. L. saccharina is of similar morphology, so developments of these techniques should not encounter fundamental biological and engineering problems.

4. HARVESTING.

The mass of seaweed growing on one metre of rope (wet weight) is expected to be in the region of 15 kg. Only a very preliminary assessment of possible harvesting techniques will be feasible within the scope of the proposed investigation. Harvesting may be achieved by passing the weed covered rope through cutters arranged on each side of a boat, the weed being removed continuously and conveyed to the hold. It seems reasonable to suppose that such processes will not demand rope tensile strengths greater than a few tonnes, both during harvesting and plant growth. It follows that a structure using polypropylene sheet in the form of rope could be manufactured at a cost of about £0.1 per metre. This should grow about 1 kg of coal equivalent per year per metre having a value of about £0.03 at current prices.

Currently marine biomass is harvested using small boats with dredges and mechanical harvesters for alginate production. This method consumes about 10% the energy content of the seaweed produced. Hence, improved harvesting techniques would demand a negligible proportion of the total energy content of the crop.

5. PROCESSING THE BIOMASS.

Marine biomass has a high water content (up to 85%) which

precludes direct combustion. At present the preferred route for the utilisation of marine biomass is by anaerobic digestion to produce methane which is commercially much more valuable as a fuel than coal. Energy recovery efficiencies of 55% are currently reported for laboratory scale digesters.

6.   COMMERCIAL CONSIDERATIONS.

The primary capital costs associated with this potential renewable energy source concern the artificial structure on which the seaweed would be grown and the cost of the anaerobic digester.   The cost of the former will be governed by the forces encountered during plant growth and harvesting.   The latter largely by structural design and degree of utilisation.   Preliminary, very approximate, estimates indicate that methane could be produced at costs comparable with present day fuel costs.   Experimental data on all aspects of the process is required before firmer estimates of commercial feasibility can be made.

MARICULTURE ON LAND -

A SYSTEM FOR BIOFUEL FARMING IN COASTAL DESERTS

K. WAGENER

Dept. of Biophysics, Technical University Aachen, FRG

Summary

A system for biofuel farming in coastal deserts is presently under investigation in a cooperative European project
(1). The biomass production is based on seawater which is
pumped through shallow artificial ponds. Due to the high productivity of phytoplankton, easy fermentation to methane, and
complete mineral nutrients recycling, the net energy balance
(output/input) is expected to be 14 for methane and 7 for
methanol, respectively. Since no fertile land or fresh water
is needed, there is no competition with food production on
arable land. Such a sysem is entirely energy self contained,
what makes it attractive for remote desert areas and developing countries.

During the first two years of collaboration on this
project the feasibility of all crucial steps could be demonstrated: productivities of 60 - 90 t dry matter $\cdot$ hectare^{-1} $\cdot$
$\cdot$ yr^{-1} obtained in southern Italy, a cheap way of pond construction, harvesting, fermentation, and recycling of the
sludge (2). A pilot plant is under construction.

1. The goals and principles of Mariculture on Land (MCL) are:

- To overcome the technological difficulties of open sea mariculture

- To make use of non-arable land for biofuel farming, especially arid coastal areas

- To make use of the high productivity of phytoplancton

- To use seawater instead of the rare and valuable fresh water

- To establish highly productive cultures with mineral nutrients recycling

- To produce a feedstock easy convertable into a clean fuel.

The recognition of these objectives led to the system shown in Figure 1. Marine algae are grown in shallow ponds located in coastal deserts. Fresh seawater is continuously pumped through these ponds at a rate sufficient to provide the carbon supply for intensive cultures. Pumping seawater on coastal land needs per unit volume only 1 permil of the energy necessary for desalination. In the case of growing micro algae, the entering seawater is passed through a sand filter to keep off zoo-plancton. The harvesting procedure depends on the type of algae. In the case of micro algae, the only economic way is by straining or settling which implies the selection of filamentous or coagulating types. Macro-algae, as e.g. various types of Gracilaria, can be cut like grass. Fermentation is possible even in seawater, without problems of H_2S-formation. The sludge contains all the minerals, especially phosphorus and nitrogen, so that recycling the sludge into the ponds means closing the mineral cycle. Finally, the CO_2 content of the biogas can be stripped with seawater and also recycled into the ponds.

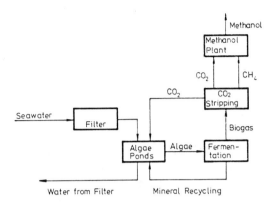

Fig. 1   Flow diagram for biofuel production by mariculture on land

2. The ponds.   The tested pond shape is a long channel. Pilot scale dimensions are 4 by 100 m, technical scale may be larger by a factor of 2.5 in linear dimensions. Preferred areas for MCL are remote deserts along the coast which normally have sandy soils with a certain fraction of clay. The pond construction includes the removal of soil to a depth of 0.1 m what can be done by a caterpillar without reloading into a truck. The solidification of the even surface obtained in this way and forming the pond is done by a well established procedure for road construction: a surface layer of 10 - 20 cm is removed, mixed with mortar or bitumen, and spread out back in the same place. The costs for this fully machine operated procedure range between 5 and 8 $/m^2, including material cost.

3. Productivity.   Outdoor experiments for testing different types of micro-algae under varied culturing conditions have been carried out in small tanks of 3 m^2 surface each located in southern Italy at the Gulf of S. Eufemia, Calabria (2). Using Aphanoteca sp. and Dunaliella sp., partly also mixtures with others, harvesting is possible by straining or settling, respectively. The average productivity over 12 months, including the strong 1978/79 winter, was 62 t dry matter/ha/yr.

During the summer months only (May - October), the productivity was 91 t/ha/yr. Thus, in places having favorable conditions all the year round, as e.g. in subtropical or tropical
regions, an annual production around 100 t/ha can be expected.

4. Seawater pumping.   Seawater is the principal carbon source
in the MCL process. 15 - 20 % of the total carbon demand can be
covered by recycling the CO_2 from the biogas. The fermentation
residue contains about 50 % of the primarily fixed carbon which
can be recycled partially directly - as far as low molecular
weight compounds are concerned - or after oxidation, respectively. The rest of about 35 % has to be supplied from the marine
carbonate system. The necessary rate of water exchange in the
ponds for this purpose depends on the degree to which the total
carbon content of seawater can be exhausted without becoming
the limiting factor for the productivity. An exhaustion of
40 to 45 % is easily obtained; additional measures allow a
recovery of up to 80 %. In this way a productivity of 100 t
dry matter/ha/yr requires a daily water turnover of 2000 t
fresh seawater/ha, or 0.2 m per day. Assuming a pumping height
of 10 m above sea level including head loss due to friction
in the tubes gives a necessary installation of 2.2 kW/hectare
of pond area. If this energy is taken from the biofuel production, then 12 % are consumed to run this entirely self contained system.

5. Fermentation and nutrient recycling.   Fermentation of
marine algae in seawater has successfully been tested, so far
in batch cultures (2). No problems of H_2S-formation have been
observed. The fertilization effect of the mineral and other
nutrients contained in the sludge is as expected. The partic-
ulate organic matter needs oxidation before recycling, because
otherwise the light shieldung effect upon the algae is too
high.

6. The   cost analysis   for MCL methane is shown in the
Figure 2. As in all biofuel processes, the biomass production
costs are dominant. The methane production costs are
8.5 US $/GJ at a productivity of 80 t/ha/yr, and 7.5 US $/GJ at

100 t/ha/yr. Conversion into methanol gives 25 t methanol/
ha/yr at a price of 14 US \$/GJ or 0.25 US \$/liter.

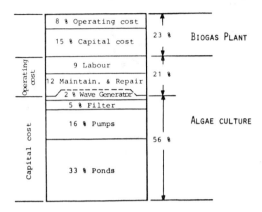

Fig. 2   Cost analysis of MCL methane

7. <u>Potential of MCL and outlook.</u>   For this purpose, the category of land has to be defined which is open for the potential
use by MCL with no competion with other types of agriculture.
Using the following characteristics:

- Desert, semi-desert, savanna
- Annual precipitation less than 500 mm (for tropical and
  subtropical regions)
- Productivity less than 5 t/ha · yr (for Europe)
- Average winter temperatures > 10 $^{\circ}$C
- Flat coastal zones

then the total length of adequate coast lines is:

Europe 1,000 km, Africa 17,000 km, total world
25,000 - 30,000 km.

Potential methanol production by MCL using a 1 km (3 km) coastal strip:

Europe	2 Mio t/yr	(6)
Africa	50 Mio t/yr	(150)
Total world	80 Mio t/yr	(240)

The world consumption of gasoline and Diesel fuel for transportation in 1977 was about 700 Mio t.

The general chances of applicability of MCL for energy purposes can be summarized in this way:

As a completely self-sufficient methane producing unit to substitute natural gas, city gas, and LPG

- for remote coastal settlements
- for towns and cities up to 300 km from the coast in the back country
- as raw material for methanol production as gasoline substitute
- as raw material for the chemical industry.

References

(1) Commission of the European Communities, Brussels, contract # 462 - 78 - 1 ESD.

(2) Final report on period 1978/79 of contract given under (1). To be published by the Commission of the European Communities, Brussels.

ALGAL BIOMASS FROM FARM WASTE - A PILOT PLANT STUDY

M.K. GARRETT and H.J. FALLOWFIELD

Department of Agricultural and Food Chemistry,

The Queen's University of Belfast

and

Department of Agriculture for Northern Ireland

Summary

Extensive laboratory work attests to the suitability of the liquid phase of pig slurry as a medium for algal biomass production. The process is attractive in that it combines the removal of potentially polluting nutrients such as phosphorus, as well as BOD, with biomass production. Moreover, if the autotrophic nutrition of the algae can be effectively exploited, the productivity of the system can exceed that of many conventional agricultural systems. A pilot plant for culture of the green alga Chlorella vulgaris has been established to assess the feasibility of algal culture in slurry as an "on farm" operation for slurry purification and biomass production. The pilot plant has been designed to process on a continuous basis the effluent from a 100 pig fattening unit, producing a liquid with a BOD of 4 mg ℓ^{-1}, SS = 130 mg ℓ^{-1} and soluble phosphorus concentration of 2.5 mg ℓ^{-1}. The system should yield approximately 4 g DM algal biomass ℓ^{-1} of slurry input with production rates in the range 30 - 75 mg ℓ^{-1} h^{-1}. The product should have a gross energy content of approximately 5 k cal g^{-1}.

1. INTRODUCTION

Many liquid organic wastes, especially those having a low Carbon to Nitrogen ratio, are ideal media for the growth of photosynthetic algae, permitting solar conversion efficiencies of approximately 3%. Such systems combine pollution reduction with biomass production, and micro-algae are especially attractive because of their autotrophic nutrition, high protein content and ability to accumulate potentially polluting nutrients such as phosphorus.

Oswald and co-workers (1) have demonstrated the feasibility of algal culture in settled domestic sewage in California with average yields of algae in excess of 100 kg (dry matter) ha^{-1} and peak production rates of several times this figure. Similar work in Germany, France, Belgium, Isreal, Canada, Mexico, Algeria, Japan, Czechoslovakia and many other countries illustrates the widespread use of this technology . Thus much of the methodology for this process has been established and provides a good basis for an evaluation of the process under UK conditions. Such an evaluation would also depend, <u>inter ailia</u>, upon the selection of algal species suited to local climatic and cultural conditions, upon a thorough knowledge of the biology of the culture system, upon the demonstration of the biological feasibility of the process under simulated local climatic conditions, and upon an evaluation of potential uses and estimated value of the algal biomass produced as a by-product of the process. Most of these aspects have been researched by Garrett and co-workers (2-10).

2. PILOT PLANT OBJECTIVE

The overall objective of the pilot plant study is to evaluate the feasibility of algal culture as a method for the treatment of the liquid phase of pig slurry. The principal aspects of the evaluation are to validate and refine projected performance data derived from controlled environment work, to optimize the physical and operational design of the pilot plant, to assess the suitability of the method for settled domestic and anaerobic digestion effluent and to analyse the performance of the system in both energy and economic terms.

3. PILOT PLANT DESIGN

The main components of the design are illustrated in Figure 1.

693

FIGURE 1 PILOT PLANT LAYOUT

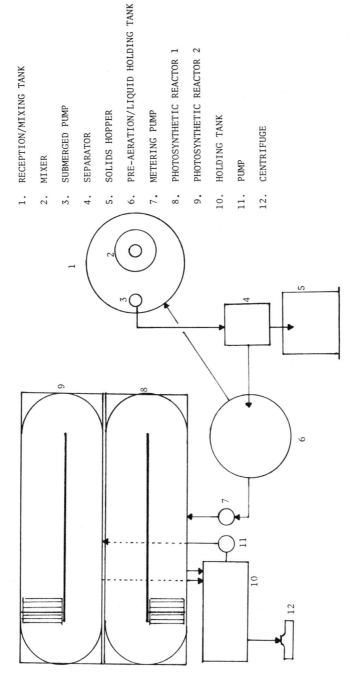

1. RECEPTION/MIXING TANK
2. MIXER
3. SUBMERGED PUMP
4. SEPARATOR
5. SOLIDS HOPPER
6. PRE-AERATION/LIQUID HOLDING TANK
7. METERING PUMP
8. PHOTOSYNTHETIC REACTOR 1
9. PHOTOSYNTHETIC REACTOR 2
10. HOLDING TANK
11. PUMP
12. CENTRIFUGE

Slurry is delivered to a 2000 ℓ reception tank, mixed and separated into crude solid and liquid fractions on a rotary screen press separator. This unit has a capacity of 0.3 m³ h⁻¹ of pig slurry at 50 g total solids ℓ⁻¹ and produces a liquid fraction containing approximately 25 g total soilds ℓ⁻¹ and stackable solids containing 250 g total solids ℓ⁻¹. The liquid fraction is held in a 2000 ℓ tank where it may receive mild aeration pre-treatment if required, any solids generated being returned to the reception tank. The liquid is delivered to the first of two photosynthetic reactors via a precision metering pump, the rate of addition determining algal growth rate in this reactor. Feed back control of liquid addition rate will be attempted at a later stage in the project. An overflow from this reactor is allowed a short settlement period to facilitate harvesting by centrifugation and return of the liquid to the second photosynthetic reactor for further algal culture.

Thus it is intended to operate the first photosynthetic reactor in a continuous mode to optimize biomass production and the second reactor in either a batch mode or continuously at a low growth rate to optimize phosphorus removal. The design and layout permits other operational combinations which will also be assessed.

The photosynthetic reactors are constructed of marine plywood sealed with resin bonded fibreglass and finished with a white epoxy paint. Each reactor tank consists of a rectangular channel with perpendicular sides, rounded ends and a surface area of 10 m². Liquid is circulated by a 12 blade paddle wheel driven by a 0.5 horse power electric motor. Convention-al gearing is being compared with thyristor control to regulate paddle wheel speed. The liquid depth in these reactors may be varied between 5 and 30 cm.

Complete recovery of the products of this unit will not be attempted until after a period of intensive monitoring and analysis of light and temperature inputs, biomass production and phosphorus removal rates, culture stability etc. Subsequently it is planned to compare various harvesting techniques for the recovery of the algal biomass.

4. PREDICTED PERFORMANCE OF THE SYSTEM

The effect of light and temperature upon various parameters of algal growth and phosphorus removal has been reported elsewhere (10). These

Studies produce data such as that in Table 1 in which cell biomass doubling
time is related to mean monthly temperature and light conditions for Northern Ireland.

Table 1 - Effect of mean monthly light and temperature conditions in
Northern Ireland on doubling time of Chlorella vulgaris in the liquid
phase of pig slurry.

Month	Jan	Feb	Mar	Apr	May	Jun	Jul	Aug	Sept	Oct	Nov	Dec
Temp (°C)	3.7	4.2	5.9	7.9	10.5	13.3	14.7	14.5	12.7	9.7	6.6	4.9
Radiation (Langley)	41	97	179	298	351	444	378	297	233	113	59	34
Doubling time (h)	78	68	48	36	27	21	19	19	22	29	43	58

Strain (3) developed regression equations defining inter alia cell number
and dry weight for a continuous culture system of the selected alga.
These predict production rates ($R = 0.92$, $P < 0.001$ for cell number,
$R = 0.97$, $P < 0.01$ for cell dry weight) and consequently enable the energy
amplification inherent in the system to be predicted. Thus, for example,
at 5×10^7 cells ml^{-1}, a likely operating concentration in the continuous
mode, the culture contains 318 mg algal dry matter ℓ^{-1}. In the second stage
of the process a batch culture would yield 1.7 g ℓ^{-1} giving a total yield
of 2.018 g ℓ^{-1}. (This assumes a 50% dilution of input slurry supernatant).
The gross energy content of the harvested material is 5.032 k cal g^{-1}
giving a total energy yield of 10.154 k cal ℓ^{-1} as biomass. This represents
an energy amplification of 1.5 - 2.0 fold as a result of algal photosynthesis. These predictions may indicate that the product will be more
valuable as a single-cell protein rather than an energy source.

5. CONCLUSIONS

This study seeks to evaluate the feasibility of algal culture as a
potential "on farm" method for the treatment of liquid organic wastes such
as pig slurry. If feasibility is established the method could contribute to
the solution of the many problems associated with disposal of the waste
products of intensive livestock production. If the study fails to establish
this possibility it will still yield new data indicating the potential of

algal biomass production from liquid organic wastes under cool temperate conditions.

6. ACKNOWLEDGEMENTS

This work is supported jointly by the Department of Agriculture for Northern Ireland and the Department of Energy of the United Kingdom.

7. REFERENCES

1. OSWALD, W.J. and GOLUEKE, C.G. (1968). Large-scale production of algae. In Single-Cell protein, ed. by R.I. Mateles & S.R. Tannenbaum, 271-305. Massachusetts Institute of Technology.

2. GARRETT, M.K. (1975). Biochemical basis of the relatively high productivity of mixed algal-bacterial systems. J. Brit. Phycol. Soc. 10, 311.

3. STRAIN, J.J. (1976). Studies on the composition and nutritive value of Chlorella. PhD. Thesis. The Queen's University of Belfast.

4. GARRETT, M.K. and ALLEN, M.D.B. (1976). Photosynthetic purification of the liquid phase of animal slurry. Envir. Pollut. 10, 127-139.

5. GARRETT, M.K., STRAIN, J.J. and ALLEN, M.D.B. (1976). Composition of the product of algal culture in the liquid phase of animal slurry. J. Sci. Food Agr. 27, 603-611.

6. ALLEN, M.D.B. and GARRETT, M.K. (1976). Effects of the culture of Chlorella upon the bacterial flora of the liquid phase of animal slurry. Proc. Soc. Gen. Microbiol. III (3), 131.

7. GARRETT, M.K., WEATHERUP, S.T.C. and ALLEN, M.D.B. (1976). Microalgal culture in the liquid phase of animal slurry; effect of light and temperature upon growth and phosphorus removal by Chlorella vulgaris. Proc. Soc. Gen. Microbiol. III (3) 131.

8. STRAIN, J.J. and GARRETT, M.K. (1976). Composition of the product of algal culture in the liquid phase of animal slurry and nutritive value of the algal cells. Proc. Soc. Gen. Microbiol. III (3) 131.

9. ALLEN, M.D.B. and GARRETT, M.K. (1977). Bacteriological changes occurring during the culture of algae in the liquid phase of animal slurry. J. appl. Bacteriol. 42, 27-43.

10. GARRETT, M.K., WEATHERUP, S.T.C. and ALLEN, M.D.B. (1977). Algal culture in the liquid phase of animal slurry. Effect of light and temperature upon growth and phosphorus removal. Envir. Pollut. 15, 141-154.

PHOTOBIOLOGICAL PRODUCTION OF FUELS BY MICROALGAE

S. LIEN

Solar Energy Research Institute (SERI)*

Golden, Colorado 80401, U.S.A.

Summary

This report discusses a long-term research program to establish the scientific and technical foundations of aquatic, photobiological fuel production systems, based on the oxygenic photosynthesis of microalgae; gaseous fuel in the form of hydrogen, and liquid fuels in the forms of algal oils or hydrocarbons, are of great interest. The technical problems of using algae for hydrogen production are presented together with new observations on the catalytic properties of the algal hydrogenase. Finally, a simple but highly efficient technique for evaluating the oleaginous potential of various algal species presented together with some preliminary but encouraging observations on algal production of oils and hydrocarbons.

1. INTRODUCTION

To meet future energy needs mankind will be increasingly more dependent upon the renewable energy sources. Photobiological production of fuels is probably one of the most important new avenues towards establishing a significant source of renewable energy supply. As part of a comprehensive long-term photobiological research program at SERI, we attempt to adapt and modify (physiologically, biochemically, and genetically) the electron tran-

*SERI, is a Division of MRI operated for the U.S. Department of Energy (DOE) under Contract EG-77-C-01-4042. The author acknowledges the vital support for portions of this research from the DOE, Office of Energy Research, under its field task proposal No. 006-80.

sport reactions and associated dark biosynthetic processes of the oxygenic photosynthetic organisms to produce fuels directly. The principle of photobiological fuel production linking to the oxygenic photosynthesis is illustrated in Fig. 1.

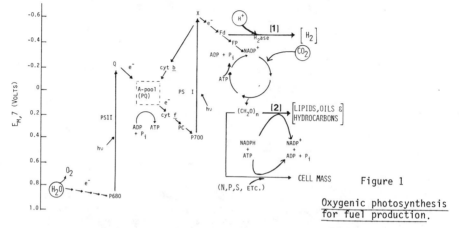

Figure 1

Oxygenic photosynthesis for fuel production.

Electrons derived from the photosystem II (PSII) catalyzed water-splitting reactions of the oxygenic photosynthetic organisms are further energized by a second light-driven reaction of photosystem I (PSI) to a redox potential nearly 0.2 V more negative, i.e., more strongly reducing, than the midpoint redox potential of the H_2/H^+ couple at neutral pH. These highly energetic electrons are stabilized as a reduced physiological low-potential redox carrier, such as the reduced ferredoxin (Fd). In the presence of an appropriate hydrogen catalyst (e.g., hydrogenase in some species of algae), the reduced Fd can donate its electron to a proton (H^+) and produce H_2, a gaseous fuel, as indicated by reaction 1 in Fig. 1. Most frequently, the reduced Fd is utilized to generate another physiological redox carrier, NADPH, which is used in the photosynthetic carbon fixation and other biosynthetic reactions yielding cell mass as the final product. Under special conditions some species of algae can be induced to channel a large fraction of their photosynthetically fixed carbon into the metabolic pathway leading to lipid biosynthesis (reaction 2 of Fig. 1). Oils and hydrocarbons may accumulate. In some cases, they constitute a large fraction of the total algal mass (1). These oleaginous algae may lend themselves to future adaptation and development as efficient solar energy to liquid fuel transducers.

2. ALGAL HYDROGENASE AND ALGAL HYDROGEN PRODUCTION

The scientific feasibility and various technical difficulties of photobiological H_2 production were reviewed in detail elsewhere (2,3). Fig. 2
illustrates the technical problem areas associated with algal H_2 production
linked to photobiological oxidation of water.

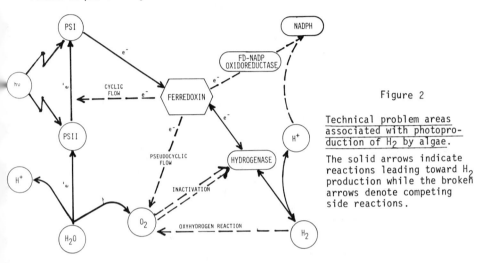

Figure 2

Technical problem areas
associated with photopro-
duction of H_2 by algae.

The solid arrows indicate
reactions leading toward H_2
production while the broken
arrows denote competing
side reactions.

Clearly, most of the technical problems arise because whenever water serves
as the ultimate substrate in the production of H_2, O_2 is liberated as an
obligatory by-product in a stoichiometric ratio of 2 H_2:1 O_2. As O_2 concentration begins to build up, various O_2-induced back reactions (including
oxy-hydrogen reaction, auto-oxidation of the photosynthetic redox carriers)
are accelerated. These back reactions tend to reduce the yield of H_2 production and, consequently, reduce the net energy efficiency of the system.
More critically, for any system using algal hydrogenase as the hydrogen
catalyst, oxygen molecules, even in very low concentrations, rapidly inactivate the catalytic activity of the enzyme. Thus, algal photoproduction of
H_2 from water is generally limited to very short durations. In most cases,
simultaneous O_2 and H_2 production by eucaryotic algae lasted only for a few
minutes. To obtain basic information needed to solve the technical problems associated with algal hydrogenase, we have undertaken a systematic
characterization of the hydrogenase system in the unicellular green alga,
C. reinhardtii, emphasizing: (a) understanding, at the molecular level,

biochemical events leading to the appearance of active algal hydrogenase during anaerobic adaptation in vivo and (b) characterizing the biochemical and catalytic properties of the enzyme in vitro. We obtained experimental evidence indicating that an energy-requiring (or ATP-consuming) step is involved in hydrogenase activation during anaerobic incubation (4). Most recently, we observed a very strong stimulatory effect by various anions on the catalytic activity of the algal enzyme (see Fig. 3) in cell-free reactions involving the non-physiological redox carrier, methyl viologen (MV).

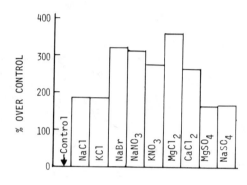

Figure 3

Anionic stimulation of hydrogenase activity.

The rate of hydrogenase catalyzed H_2 production using dithionite+MV as electron donor system was measured in the absence and presence of various salts (all at 0.5 M concentration). The control rate (with no salt added) was 100 ±20 µmoles $H_2 \cdot$ mg Chl^{-1}.hr^{-1}.

Although there is a considerable amount of literature concerning the effects of salt and anions on the enzymatic reaction in general, there are very few reports dealing specifically with hydrogenase. The first documented salt-effect on this enzyme appeared in 1979 concerning the hydrogenase isolated from a rumen bacterium (4). To the best of our knowledge our finding about the anionic stimulation on MV-mediated hydrogen production represents the first observation on an algal hydrogenase. Further analysis of the anionic effect is crucial for it may not only further our understanding on the basic catalytic mechanism of the algal hydrogenase, but may also contribute toward technical improvements to develop a cell-free photobiological H_2-production system.

3. ALGAL OIL AND HYDROCARBON PRODUCTION

The term "algal oil and hydrocarbon" is loosely defined as that class of carbonaceous, lipoidal compounds that are: (1) produced by the algal cells; (2) not covalently linked to the cellular proteins or carbohydrates; (3) characterized by a high content of reduced carbon (such as -CH$_3$, =CH$_2$, ≡CH, etc.) relative to that of the oxygenated carbon (for example,-COH,

=C=0, or -COOH, etc.). Thus, algal oil and hydrocarbons are, in principle, readily extractable, and easily separated from other cellular constituents. Because of their high reduced carbon content, algal oil and hydrocarbons have high caloric value and may be converted into fuels without a large net input of additional chemical energy. To facilitate research on algal oil and hydrocarbon production, we developed a highly efficient cytochemical staining technique suitable for screening and evaluating a large number of algal species for their oleaginous capacity. This staining technique is based on a drastic change in the solubility property as well as a large spectral shift accompanying the protonation and deprotonation of the dye molecule "CI Basic Blue 12". In an acidic aqueous solution (pH<5) the dye is protonated and absorbs strongly in the red spectral region. The blue protonated dye in acidic aqueous solution is not readily extracted into the n-nonane. At neutral to mild alkaline pH, the dye readily partitions into the lipoidal phase as the deprotonated free base that absorbs strongly in the blue spectral region, making the dye yellowish-orange. When samples of algae are pulse-stained with the dye "CI Basic Blue 12" under appropriate conditions, the intracellular oil droplets of the cells can be readily visualized as yellowish-orange structures, while other more polar cellular constituents are stained in deep blue. Using this technique, a large number of algal species grown under diverse nutritional conditions can be readily screened for their ability to produce and accumulate oils. Our initial screening identified several species of soil alga from two closely related genera,Neochloris and Chlorococcum as good potential oil or hydrocarbon producers. Fig. 4 illustrates the capricious amount of oil produced by these algae. Studies on the effect of nutritional and physiological parameters affecting relative rate of algal mass and oil content of these organisms is currently being investigated. We hope that a continuous and active research in the area of algal oil and hydrocarbon production will lead to the development of a truly renewable source of liquid fuels in the future.

Figure 4. Examples of oleaginous algae

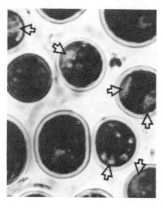

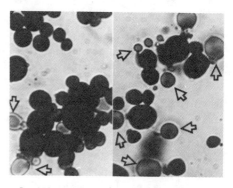

B. Oil droplets (arrows) are readily
released from nitrogen-starved
cells of N. oleoabundans

A. Old cells of C. oleofaciens
accumulate capricious amount
of oil (arrows).

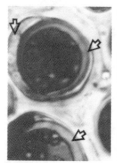

C. The cells of N. pseudostig-
mata produce oil of low
apparent viscosity. Under
moderate mechanical pressure
the intracytoplasmic oil
(arrows) accumulates in the
space between the partially
raptured cytoplasmic mem-
brane and the intact cell
wall.

REFERENCES

1. Maxwell, J.R., A.G. Douglas, G. Eglinton, and A. McCormick, Phytochemistry 7, 2157-2171 (1968).

2. Lien, S., and A. San Pietro, "An Inquiry into Biophotolysis of Water to
   Produce Hydrogen". NSF/RANN Report (1976).

3. Weaver, P.F., S. Lien, and M. Seibert, Solar Energy 24, 3-45 (1980).

4. Seibert, M., S. Lien, P.F. Weaver, and A.F. Janzen, in "Proceedings of
   the 1980 International Symposium on Solar Energy Utilization" in press.

INCREASING BIOMASS FOR FUEL PRODUCTION BY
USING WASTE LUKE-WARM WATER FROM INDUSTRIES

C. PIRON-FRAIPONT, E. DUJARDIN and C. SIRONVAL

Photobiology Laboratory; LIEGE University, B.22
4000 Sart Tilman BELGIUM

Summary

Waste luke-warm waters of many industries do not often contain enough calories to be recuperated with a good [yield/investment] ratio by techniques such as : heat pumps, heat exchangers...

On the other hand, it is known that when algae are moderately warmed in the laboratory, the rate of photosynthesis increases, and hence biomass accumulation.

So it seemed interesting in temperate zones to grow algae in waste luke-warm waters at temperatures between 20 and 35°C, in order to recuperate the calories by increasing the yield of biomass. This was shown to be possible with filamentous and microscopic algae.

1. Cultures made in lagoons utilize the polluted water of the Meuse river, the temperature of which is increased by circulating through the 3rd cooling circuit of the Tihange nuclear power plant. The temperature of the water is 12°C higher than that of the river. Cultures of filamentous algae like Hydrodictyon or Cladophora exhibit a yield increase of 50 to 200 % in the luke-warm water. In that way, it has been possible to harvest in Belgium 12 to 15 tons of dried algae/ha/year.

2. It is also possible to use polluted warm water of industrial origin to regulate the temperature of unicellular algae cultures. This can be performed by means of appropriate heat exchangers.

This new concept appears now to be the basis of a "depolluting bio-industry" to be developed in temperate industrial countries. Even after extraction of high-value substances from the biomass, the residue is still the source of fuel through methanisation, or of energy by direct combustion.

1. INTRODUCTION

The culture of algae is an interesting method of accumulating biomass. Depending on the climatic conditions and on the species 10 to 150 tons (or even more) dry biomass are harvested per year and per hectare. The organic matter produced by the algae is an excellent substrate for anaerobic fermentation. Naveau, Nyns et al have shown that 0,3 l methane are produced per one g volatile matter, which is 50 % of the maximum, theoretical yield (0,55 l/g volatile matter; 1,2).

2. CULTURE OF ALGAE ON LUKE-WARM WATERS FROM INDUSTRIES

In Belgium, the mass culture of fresh-water algae yields 10 to 30 tons dry biomass/ha/year depending on the temperature of the summer and on the species. It has been shown that the temperature of the water is an essential factor for the control of the growth rate. Fig. 1 gives the form of the dependance of photosynthesis on water temperature in Scenedesmus quadricauda. Such a curve represents also the temperature dependance of biomass accumulation as a result of photosynthetic activity. It illustrates the fact that the algae make a better use of solar energy when the temperature is raised by some degrees between 20° and 30°C.

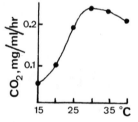

Fig.1: Temperature dependance of photosynthesis in Scenedesmus quadricauda.

Luke-warm waters are available in industrial countries. In Belgium the case of the Meuse and Sambre valleys is typical. These rivers collect warm waters from many factories of different kinds. This provokes an appreciable perturbation of the ecological equilibrium. By growing algae on the rejected waters, one increases the biomass yield, and one contributes to the river depollution by cooling and by cleaning the water before rejection in the river. This concept seems particularly adapted to electrical power plants which furnish a great amount of wasted calories and relatively clean warm water (3,4). Our experiments have shown that in this particular case, the algae biomass yield increases substantially. The harvested biomass is suitable for feeding herbivorous fishes in warm waters; very fast growth rates of fishes have been obtained in this way (5,6).

Several industries introduce mineral salts or organic matter in the warm waters (siderurgy, food industry, textile industries, etc...). The culture of algae allows to recuperate part of the salts, as well as organic matters together with the calories. The carbon dioxide from industries, especially from fermentation industries, may be used to increase the biomass yield and the depolluting activity of the algae.

3. WARMING ALGAE CULTURES

a. Macroscopic algae

Fig. 2 is a scheme of the lagoon system constructed recently in Tihange by the Photobiology Laboratory of the Liège University.

The depth of the lagoons is 30-40 cm. Luke-warm water circulates as indicated in the scheme. The water comes from the river Meuse. It is warmed by the nuclear electric power plant of Tihange I, the temperature of the water increasing by about 12°C above that of the Meuse. This allows to increase the biomass yield appreciably. Table I gives the dry weight of filamentous algae (essentially Hydrodictyon reticulatum) collected from May to July 1979 in warm water, and in August-September 1979 in non-heated water. Table II gives the composition of the collected biomass. Mineral salts are accumulated in the biomass, showing that the culture is also an excellent method for recuperating salts from the river. Thus the system

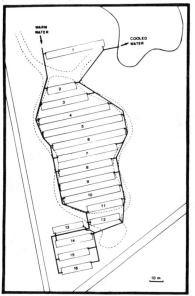

Fig. 2 : The lagoon system in Tihange (Belgium).

depollutes the water chemically and thermically. It utilizes the wasted calories for increasing the efficiency of photosynthetic conversion of solar energy, and it replaces an expensive chemical cleaning of the polluted water by a non-expensive photobiological cleaning.

It has been shown by Naveau, Nyns et Asinary that 1 g of the dry matter collected in Tihange produces about 0,1 liter methane gas (1).

TABLE I

Hydrodictyon harvested in 1979 in Tihange (Kg dry algae/400 m2 of lagoons).

Month	Dry weight	Mean daily water t°
May	85	20-20
June	79	26-30
July	85.5	25-27
August	53.6	18-20
Septem.	30.5	19-20
October	45	23-30

TABLE II

Hydrodictyon composition (% dry weight)

Carbon	28.4
Hydrogen	3.3
Nitrogen	2.6
Sulfur	1.5
Ashes	35.7
Carbon (as carbonates)	5.1

b. Microscopic algae

Microscopic algae (Scenedesmus, Chlorella) are grown either in ponds in which the suspension is more or less moved continuously, or in a circulating thin layer, as the one shown in fig. 3. In both cases part

of the suspension is removed from time to time for harvesting the algae; the clarified water is recycled. In the case of the thin-layer cultures, a heat-exchanger has been conceived in Liège for transferring heat from water warmed at a temperature between 40° to 60°C, to the culture medium.

Fig. 3 : View of 1 exposed surface for growing microscopic algae at the Photobiology Laboratory in Liège (Belgium).

The general concept is the following. The suspension of algae flows
along a surface which exposes the
algae to light. It is collected into a
reservoir at the bottom of the exposed
surface. A pump recycles the suspension
to the top after saturation with CO_2.
The heat-exchanger which has been conceived by Delvaux and Tran "Laboratoire
de Génie chimique", Liège University
(7) with the collaboration of Piron*
consists in a series of metal tubes in
which the warm water is circulated as
shown in fig. 4. The number of the tubes
and the input of warm water vary according to the temperature of the latter.
The warm water is cooled very rapidly in
such a heat-exchanger and therefore its

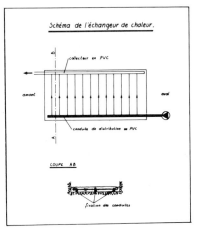

Fig. 4 : Heat exchanger for the installation described in fig. 3.

flow must be very rapid. For a very fast input, the water is still warm at
the output of the exchanger. It is then introduced into lagoons where filamentous algae are grown.

In Liège, a culture of Chlorella or Scenedesmus may produce 20 to 30
tons of dry algal biomass per effective year (6 month) if it is not warmed.
We hope to be able to increase the yield by a factor of 50 to 100 % by circulating 40° to 60°C warm water through the heat-exchanger.

The collected biomass, or even better the residue collected after
extraction of usefull chemicals, may be used as fuel. It produces about
0,3 to 0,4 liter methane per g volatile matter (8). When grown on wastes
containing a certain amount of organic molecules Scenedesmus and Chlorella
cultures assimilate organic matter from the medium. We hope to be able to
increase in this way the biomass content in volatile matter, and consequently to increase the methane yield.

4. DEPOLLUTING BIOINDUSTRIES

When one takes the costs of conventional thermal and chemical depollution into account, the real cost of the algal biomass produced as indicated
in § 3 becomes rather low. This is still more true if the biomass furnishes

* Laboratoire de Génie civil, Université de Liège.

708

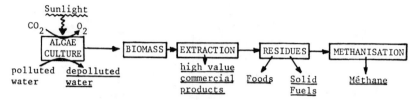

Fig. 5 : The general concept of a depolluting bioindustry.

interesting high-value commercial products. The extraction residues may be
processed into food for animals and man, or may be used to produce either
solid fuels, or methane. The ashes or the fermentation residue are good
fertilisers for traditional agriculture and for the algae cultures.
Depolluting bioindustries may be developped on this general basis (fig. 5).

(E.E.C. Contract n°ESD/021/B).

5. ACKNOWLEDGMENTS

We thank the INIEX for chemical analyses and Mr S. Sougné for
technical assistance.

6. REFERENCES

(1) H.P. Naveau, E.J. Nijns and C.M. Asinary : Methane production by anaerobic digestion of algae; 1st Coordination Meeting of Contractors E.E.C.
Solar Energy Project E, Energy from Biomass, 18-19 sept. 1980, Amsterdam.

(2) ibid., this meeting.

(3) E. Dujardin : Cultures d'algues d'eau douce et dépollution; in:Helio-
synthèse et Aquaculture, (P. Chartier and C. Gudin, edrs), Séminaire de
Martigues, sept. 1978, CNRS Edition, France.

(4) E. Dujardin : Récupération des déchets et des chaleurs résiduelles pour
augmenter le rendement d'une culture énergétique; to be published in:
Actes des Journées Européennes de Bioénergie, Conseil de l'Europe,
Dijon, 29-31 oct 1980.

(5) C.Mélard and J.C. Philippart : personal Communication.

(6) P. Fontaine-Delcambe and C. Hough : personal Communication.

(7) L. Delvaux, M. Piron and H.L. Tran; in;Etude sur la faisabilité tech-
nico-économique d'un pilote industriel pour des cultures d'algues
chauffées à partir de chaleurs à basse température, y compris l'extraction de produits commercialisables; Société de Développement Régional
pour la Wallonie (S.D.R.W.), rapport interne, 1979.

(8) Golueke, C.G. and Oswald, W.J. : Biological Conversion of Light Energy
to the Chemical Energy of Methane, Applied Microbiology, 7, 1959.

ALGAL FERMENTATION, A PROMISING STEP IN BIOMASS CONVERSION

K. KREUZBERG

Institute of Botany, University of Bonn, D-5300 Bonn 1, FRG

SUMMARY

The unicellular green alga Chlorogonium elongatum was cultivated continuously under photoheterotrophical conditions in a Phauxostat to produce biomass. With an irridiance of 37 W/m² and constant acetate concentrations at 15 mM a specific growth rate of 0.27 h⁻¹ results. Biomass production increased from μ = 0.024 h⁻¹ to 0.15 h⁻¹ and decreased in faster growing cells about 30 %, whereas starch content rised up to 32 % of cell mass. After transition to anaerobic dark conditions 80 % of this starch was fermented in Chlamydomonas moewusii, Chl. reinhardii, Chl. reinhardii Y-1, Chlorella fusca and Chlorogonium elongatum to H_2, CO_2, acetate, ethanol, glycerol, formate, lactate and butanediol within 6 hours. Two fermentation types could be distinguished among algal species on the basis of reoxidation of the anaerobic produced reduction equivalents. Whereas in Chlamydomonas moewusii glycerol and molecular hydrogen is produced for balance of reduction equivalents, in Chlamydomonas reinhardii, Chl. reinhardii Y-1, Chlorella fusca and Chlorogonium CO_2 accepted the electrons and formate was produced. The fermentation type determined the degradation rate for starch by changing the glycolytic flux. Ratio and type of fermenation products could be influenced by pH. Increasing the production of attractive compounds (e.g. H_2, ethanol, butanediol and glycerol) algal fermentation seems to be a promising procedure for biomass conversion.

INTRODUCTION

Unicellular green algae can be used for biomass production in continuous culture either photoautotrophically be reducing CO_2 (1, 2, 3) or photoheterotrophically by assimilation of organic carbon (4). This biomass can be directly fermented by the algae to energy-rich products under anaerobic dark conditions. Several algal species were investigated and compared with regard to their ability to produce fermentation products grown in batch- and continuous culture. Systematic investigations for conditions increasing the production of attractive compounds like H_2 or ethanol by modifiing fermentations activities were done.

MATERIAL AND METHODS

Chlorogonium elongatum L 12-2E, Chlamydomonas moewusii 11-5/10, Chlamydomonas reinhardii 11-32/90 and Chlorella fusca 211-8 b were obtained from the Sammlung von Algenkulturen, Göttingen, FRG. Chlamydomonas reinhardii Y-1 was a kind gift from the culture collection of the University of Texas. Stock cultures were maintained in the light and transfered monthly using liquid medium of composition similar to growth medium. Chlamydomonas strains were grown photoautotrophically on a mineral medium described by Klein and Betz (5), whereas for Chlorella the medium of Klein et al. (6) and for Chlorogonium that of Stabenau (7) was used. Ch. reinhardii 12-32/90 and Y-1, Chlorella and Chlorogonium were grown under photoheterotrophic conditions in the presence of acetate.

Chlorogonium and Chlorella were cultivated continuously under photoheterotrophic conditions in the Phauxostat (4). The culture medium was the same except that sodium acetate was increased to 12 mM. The dry weight of algal cultures and their content of RNA and DNA were estimated as described by Kreuzberg and Hempfling (4). The protein of the cells was determined as follows: 4 ml methanol (Chlorogonium, Chlorella) or acetone (Chlamydomonas) respectively, were added to 1 ml algal suspension and mixed vigorously. After centrifugation for 10 min at 3000 x g the pellet was resuspended in 2 ml of 2 % NaOH. In this solution, protein concentration could be measured after Lowry et al. (8).

The fermentation activities of batch-grown algae cells including the determination of the gases CO_2 and H_2 as well as the estimation of the produced glycerol, ethanol, acetate, formate, lactate and 2.3 butanediol were analysed by the methods described by Klein et al. (6). Fermentation products of cells from continuous culture were determined after the procedure of Kreuzberg and Hempfling (4).

RESULTS

Biomass Production

The production of biomass by continuous culture of <u>Chlorogonium</u> <u>elongatum</u> under photoheterotrophic conditions depends on the illumination rate. From 2 to 37 W/m^2 the specific growth rate increases from 0.024 h^{-1} to 0.27 h^{-1}. Several properties of <u>Chlorogonium elongatum</u> cells obtained from continuous culture are described in Fig. 1.

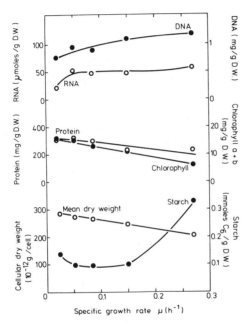

Fig. 1. Cellular parameters as function of specific
growth rate of <u>Chlorogonium elongatum</u> grown
in light limited continuous culture.

DNA content rises from 0.76 mg/g dry weight at growth rate of μ = 0.024 h^{-1} to 1,29 mg/g dry weight at μ = 0.15 h^{-1}. RNA concentration increases 22 μmol/g dry weight in slow growing cells about 1.4 fold to 52 μmol/g dry weight at μ = 0.052 h^{-1} and reaches 58 μmol/g dry weight with the highest growth rate. By contrast, protein drops from about 32 % at μ = 0.024 h^{-1} linear to about 24 % at μ = 0.27 h^{-1}. Chlorophyll a + b con-centration also decreases from 16 mg/g dry weight in slow growing cells to

6 mg/g dry weight at maximal practical growth rates. Similar to protein and Chlorophyll a + b, also cellular dry weight changes from 288×10^{-12}/g cell for about 27 % to 209×10^{-12}g/cell at $\mu = 0.27$ h^{-1}. Starch content of algae cells varies within growth rates: algae growing at $\mu = 0.024$ h^{-1} contain about 160 μmol glycosylunits starch/g dry weight. This concentration declines to 100 μmol glycosylunits during the growth from $\mu = 0.052$ h^{-1} to 0.15 h^{-1} and increases 3.3 fold to 330 μmol glycosylunits/g dry weight at the highest growth rate.

Biomass Fermentation

Starch is a common product during photoautotrophic and photoheterotrophic growth of unicellular green algae. This starch can be easily degraded to glucose, which is further fermented during anaerobic dark conditions. In evaluation of the conversion of algae biomass, not only starch concentration but also its degradation rate as well as the energy content of fermentation products are of importance. The fermentation type of starch breakdown depends highly on the strain, chosen for biomass production. This is shown in table I.

Table I. Comparison of starch breakdown and fermentation products in four different algae during 6 hours dark incubation.

μMol/mg Chl	Chlamydomonas moewusii	Chlamydomonas reinhardii	Chlorogonium elongatum	Chlorella fusca
Starch (C_3)	51.0	31.2	5.0	1.1
Glycerol	24.2	0.0	0.1	0.1
Acetate	15.1	15.4	2.6	0.6
Ethanol	8.2	9.2	2.8	0.3
Formate	0.6	23.5	4.3	1.1
D-Lactate	2.9	0.0	0.0	0.0
L-Lactate	---	0.0	0.0	0.1
Butanediol	2.2	---	---	---
CO_2	23.9	1.5	0.3	0.1
H_2	10.8	1.8	0.1	0.2

Starch breakdown differs significantly among green algae. Whereas in Chlorella cells 0.6 μmol glycosylunits/mg chl are degraded during 6 hours of dark incubation, Chlamydomonas moewusii cells and ethanol as well as the gases CO_2 and H_2, while formate, D-lactate and also butanediol are found only in low concentrations. In contrast to Chlamydomonas moewusii, cells of Chl. reinhardii, Chlorogonium and Chlorella seem to use another fermentation mode, where acetate, ethanol and formate are the main fermentation products, while no glycerol, D-lactate or butanediol are detectable.

Only low concentration of L-lactate, CO_2 and H_2 can be found in this cells. Hence, not only the amount of fermented starch but also the mode of fermentation depends on the algae strain, chosen for biomass production.

Within one single species fermentation rate and the ratio of fermentation products can be influenced by experimental conditions (e.g. pH), which is shown for Chlamydomonas reinhardii Y-1 in table II.

Table II. pH dependent starch breakdown and fermentation in Chlamydomonas reinhardii, Y-1 during 5 hours dark incubation

Product in μMol/mg Chl.	pH 4.5	%	pH 6.1	%	pH 7.3	%	pH 8.0	%
Starch (C_3)	3.0	100	5.5	100	5.1	100	8.1	100
Glycerol	0.2	7	5.5	0	0.0	0	0.0	0
Acetate	0.6	13	1.3	16	2.1	28	1.3	11
Ethanol	2.1	47	3.3	40	3.1	41	5.0	41
Formate	1.8	20	2.6	16	3.7	24	3.1	13
CO_2	0.9	10	1.9	12	1.0	7	1.4	6
H_2	0.2	-	0.7	-	0.3	-	0.2	-
H_2 Product. from 100 μMol C_6	13.3	-	26.0	-	11.8	-	4.9	-

Starting at pH 4.5 only 1.5 μmol glycosylunits of starch are fermented during 5 hours dark incubation to ethanol, formate and lower amounts of glycerol, acetate and both gases CO_2 and H_2. At pH 6.1 starch degradation rises 1.8 fold. No glycerol is produced further and ethanol drops by 7 %, while acetate concentration increases only by 3 %. At pH 7.3 much more acetate (12%) and formate (8%) are found, but the amount of ethanol did not change. With further increase of pH to 8 , starch breakdown is 1.6 fold higher, but acetate and formate decrease, while ethanol still remains the main fermentation product. The capacity of fermentative H_2 production is optimum at pH 6.1 with 26 μmol H_2 from 100 μmol glycosylunits of starch. This indicates a dependence of fermentative H_2 production from pH rather than from the degradation rate of starch.

DISCUSSION

Quantity and quality of biomass from continuous culture depend strongly on the specific growth rate. The mean cellular dry weight decreases with higher growth, whereas biomass production (dry weight/ml) like continuous culture of Chlorella (3) rises about 25 % up to μ = 0.15 h^{-1} and then drops by 30 % at μ = 0.27 h^{-1}. In contrast to Chlorella, starch content does not follow biomass yield but increases with high growth rate up to 32 % of cell mass. Optimum conditions for biomass fermentation (high starch content) coincide with maximum practical growth rates.

Conversion of biomass in unicellular green algae depends on starch content, fermentation mode and experimental fermentation conditions. Optimum conditions are those, where large amounts of starch are fermented to energy rich products, which can be directly used further like H_2, ethanol, butane-diol and glycerol. As shown in table I, all these interesting compounds can be produced by algal fermentation. The most important fact seems to be, by which way anaerobically produced reduction equivalents can be reoxidized to ensure continuous operation of glycolysis. On one hand, in cells of Chlamydomonas moewusii, reduction equivalents are balanced by production of glycerol and the evolution of H_2. On the other hand, in cells of Ch. rein-hardii, Chlorogonium or Chlorella, CO_2 accepts electrons from reduction equivalents resulting in the production of formate. Therefore in the latter cell type only low concentrations of CO_2 and H_2 and no glycerol can be detected. Thus, the first fermentation type saves less energy from one fermented glycosylunit than the latter type. Reoxidation of reduction equivalents seems to be easier in the first type and a higher glycolytic flux results although starch content is nearly equal in all algae strain investigated.

Optimizing biomass production by continuous culture for easy fermentable compounds and further influencing the kind and the rate of fermentation by choice of algae species, mutation of algae and change of reaction conditions should lead to very efficient bio-(solar) energy converison systems. In this way algal fermentations seems to be a promising step in bioconversion.

REFERENCES
1 Meyers, J. and L.B. Clark (1944). J. Gen. Physiol., 28, 103-112.
2 Oswald, W.J. (1973). Solar Energy, 15, 107-117.
3 Pirt, M.W. and S.J. Pirt (1977). J. appl. Chem. Biotechnol., 27. 643-650.
4 Kreuzberg, K. and W. Hempfling (1980). Proc. VI Intern. Fermentation
        Symp., London, Canada, in press.
5 Klein, U. and A. Betz (1978). Physiol. Plant., 42, 1-4.
6 Klein, U., K. Kreuzberg and A. Betz (1980). Proc. VI Intern. Fermentation
        Symp., London, Canada, in press.
7 Stabenau, H. (1971). Biochem. Physiol. Pflanzen, 162, 371-385.
8 Lowry, O.H , N.J. Rosebrough, A.D. Farr and R.J. Randoll (1951). J.Biol.
        Chem., 193, 265-275.

SESSION VI : IMPLEMENTATION - DEVELOPING COUNTRIES

Chairman: Dr. J. CARIOCA
Federal University of Ceara, Brazil

Summary of the discussions

Invited papers

An Indian village agricultural ecosystem - Case study of Ungra
Village
- PART I - Main observations

- PART II - Discussion

Fuelwood and charcoal in Africa

Poster papers

A strategy for rural development in the third world via biomass
resource utilisation

The present use and potential for energy from biomass in Tanzania

Fractionation of biomass (sugar cane) for animal feed and fuel

Biogas - Kenyan case

Community biogas plants - Implementation in rural India

A 4 to 6-MW wood gasification unit : scaling-up the gasification
technology into an unknown size category (case study Guyana)

A design strategy for the implementation of stove programmes in
developing countries

The United Nations system and biogas activities

<u>SUMMARY OF THE DISCUSSIONS</u>

Session VI   <u>IMPLEMENTATION</u> - DEVELOPING COUNTRIES

<u>Rapporteur</u> :   E. DA SILVA, UNESCO
<u>Speakers</u> :   A.K.N. REDDY, India
                  E.M. MNZAVA, Tanzania
<u>Chairman</u> :   J. CARIOCA, Brazil
<u>Poster Session</u> : 8 papers presented

The agricultural village of Ungra, South India, was studied as an eco-system, with special emphasis being given to the production of biomass and its utilization by human beings and livestock. A.K.N. Reddy in his presentation provided valuable dat$_a$ on land-use and cropping patterns. In addition, above-ground plant biomass productivities and their divi-sion into various components for utilization by the village human and livestock populations. The energy aspects of the ecosystem, in particu-lar draught power and cooking fuel, were discussed in terms of enhancing the productivity and/or self-sufficiency of the whole ecosystem. In considering the limitations of the study, attention was given to the timeliness and impact of many agricultural operations. Ploughing, for example, has profound effects on biomass productivity and energy requi-rements. Following on the presentation of the parameters of the agri-cultural ecosystem at the village level, a basis for rural development was discussed.

Similar parameters have been employed in the modelling of strategies for rural development by the use of biomass resources.
For instance, under simulated conditions and given a stable population, the integrated use of bio-gas plants, algal cultivation, legumes, ener-gy plantations and local woodfuel growing, etc., could result in signi-ficant provision of energy over the coming years. Likewise, the biocon-version of sugarcane and other carbohydrate crops into animal feed and fuel could make impact on rural economies and development.

Mnzava's paper focused on the fuelwood and charcoal situation in Africa. Utilizing Tanzania as a point of reference, several issues were identi-

718

fied such as efficient use and development of the existing renewable
sources of energy, the creation of new and renewable sources, the
impact of fuelwood and charcoal exploitation on the environment, production technologies especially for charcoal, priority areas of use
in different ecological situations (rural and urban area differences)
the use of wood wastes and agricultural residues, and allowing local
communities to collect fuelwood as a free wood from the forest. Future
strategies would focus on the conservation and rational utilization of
the existing resources e.g. improved stoves and boilers, increased
efforts to establish energy woodlots/plantations, increased utilization
of forest and wood industrial wastes, policy implications and research
on the use of subsitutes and alternative energy sources, the formulation
of realistic energy budgets, research and development on the use of
carbohydrate materials for transportation by extracting alcohols which
can be mixed with gasoline, and the ability to mobilize people, especially the rural communities, so that they become fully involved in
formulating and implementing short – and long – term energy plans.

In this context, a design strategy for implementing stove programmes
in developing countries was discussed. Attention was given to certain
factors responsible for the failure of previous stove programmes in
developing countries ; efforts aimed at overcoming the constraints
dealt with improved design and laboratory and field testing. Similar
programmes could also help to conserve valuable biomass reserves as
well as aiding the formulation of suitable bioenergy technologies, e.g.
alcohol production, and biogas generation.

Valuable information on biogas was provided through Tanzanian and
Kenyan experiences as well as through a case study concerning the implementation of community biogas plants in rural India.

A presentation outlining the range of biogas activities as executed
within the framework of the U.N. system was presented and discussed.

AN INDIAN VILLAGE AGRICULTURAL ECOSYSTEM - CASE STUDY OF UNGRA

VILLAGE PART I: MAIN OBSERVATIONS *

N.H. Ravindranath, S.M. Nagaraju, H.I. Somashekar,

A. Channeswarappa, M. Balakrishna, B.N. Balachandran

and Amulya Kumar N. Reddy

with the assistance of

P.N. Srinath, C.S. Prakash, C. Ramaiah and

P. Kothandaramaiah

ABSTRACT

The South Indian agricultural village of Ungra (population - 932, households - 149, total area - 360.2 hectares) has been studied as an ecosystem, with special emphasis on the production of biomass and its utilisation by human beings and livestock.

This first part of the paper reports the main results of the ecosystem study. In particular, a quantitative report is given of the land-use and cropping patterns, the above-ground plant biomass productivities, the disaggregation of the plant biomass into various components, the utilisation of these components, the food consumption by human beings and livestock, the materials and energy flow through the ecosystem, and its imports and exports.

* This paper is a condensed version of a full paper to be published in "Biomass", Volume 1, 1981, Applied Science Publishers Ltd.

INTRODUCTION

ASTRA, the Centre for the Application of Science and Technology to Rural Areas, is the instrument of an inter-disciplinary, inter-departmental programme of the Indian Institute of Science to generate technologies for rural development. ASTRA has established and operated an Extension Centre amidst a cluster of villages near Bangalore in South India. This Extension Centre has been used mainly to study village needs, particularly those pertaining to energy, water, housing, employment, etc.

It soon became obvious that the needs of food, fuel and fodder are inter-related in a number of ways and that this interplay is influenced by the resources of land, energy (particularly solar energy stored as biomass), water, livestock - and, of course, human beings. It was also realized that an understanding of these complex and intricate inter-relationships must be a pre-condition for intervention into rural life. Since the villages in the region of the Extension Centre are agricultural villages, the main focus had to be on plant biomass, its production and utilization.

The present study was initiated in 1976. The first phase of surveys, data collection and measurement was completed in April, 1980, and what follows is a brief report on some of the main observations.

UNGRA VILLAGE

The village of Ungra is located about 113 kms from Bangalore. Its height above sea level is 670.6 metres and the mean annual rainfall is 727 mms. In December, 1979, it had 149 households with a total population of 932. Ungra's total livestock population was 949 consisting of 460 cattle (111 bullocks, 143 cows, 93 calves and 113 buffaloes) and 489 goats and sheep (214 goats and 275 sheep). The area of the village is 360.2 acres. Ungra is a typical village of the dry belt in Karnataka.

METHODOLOGY

The present study has adopted a census approach in which <u>all</u> the land within the village boundary, all the households and all the domesticated animals were covered. It pertains to the agricultural year 1979-80 and includes the kharif (monsoon) and kar (summer) seasons.

Table 1: Land-use Pattern

		Hectares	%
Crop land	Cultivated (22)*	243.7	67.7
	Fallow (1)	47.0	13.0
		290.7	80.7
Grass land (3)	....	15.8	4.4
Marsh land (1)	....	15.4	4.3
Plantation (coconut, fuel) (2)		0.8	0.3
Water bodies (2)	....	0.2	-
Settlement (Houses, Roads etc.)(4)		37.3	10.3
		360.2	100.0

* Number in brackets refers to number of items aggregated.

Viewing fallow land as fodder-producing, along with grass land and marsh land, the ratio of pasture land (fallow+grass+marsh land) to cultivated land is 22:68, i.e., the area of pasture lands is approximately one-third the area of cultivated lands. This underlines the importance of pasture lands to the village ecosystem.

Table 2: Cropping Pattern - AY 1979-80

Crop	Kharif June/July-Nov/Dec.		Summer	
	Area(ha)	%	Area	%
1. (a) Paddy (local)	126.7	52.0	-	-
(b) Paddy (HYV)	33.7	13.8	-	-
	160.4	65.8		
2. Ragi	41.1	16.9	10.1	50.1
3. Sorghum	8.7	3.6	-	-
4. Sugar-cane	7.1	2.9	7.1	35.6
5. Horsegram	6.7	2.7	-	-
6. Coconut	4.6	1.9	-	-
	228.6	93.8	17.2	86.3
7. Others (20)	15.1	6.2	2.7	13.7
Total	243.7	100.0	19.9	100.0

Considering the kharif season (June/July to November/December), it is seen that paddy accounts for two-thirds of the cropped area, and paddy plus ragi (Eleusine coracana) for 82.7% of the kharif area. Further, the cropped area for the summer season is only 8.2% of the kharif area; hence, the kharif yields are equivalent to the annual yields.

Table 3: Above-ground Plant Biomass Productivity

Entity	Productivity (tonnes/ha)	Ratios	Production (tonnes)	%
1. Crops				
1(a). Paddy(local)	6.9	1.38	791.5	41.4
1(b). Paddy (HYV)	7.1	1.42	216.5	11.3
			1008.0	52.7
1.2 Ragi	3.8	0.76	111.6	5.8
1.3 Sorghum	6.1	1.22	50.1	2.6
1.4 Sugar-cane	27.3	5.46	194.0	10.1
1.5 Horsegram	1.9	0.38	13.3	0.7
1.6 Coconut	9.2	1.84	97.4	5.1
1.7 Others	-	-	24.2	1.3
			1498.6	78.3
2. Grass land	5.0	1.0	73.7	3.9
3. Fallow land	5.3	1.06	249.6	13.1
4. Marsh	5.0	1.0	76.7	4.0
5. Shrub	12.5	2.5	14.4	0.7
			414.4	21.7
Total average	6.0		1913.0	100.0

Paddy, Sorghum, Coconut and Sugarcane yield 40%, 22%, 84% and 55% greater productivity than grass land, but ragi gives 25% less productivity. In fact, the weighted average is only 28% greater than grass land - this indicates the poor return on human efforts in Ungra. All the crops taken together account for about four-fifths of the plant biomass compared to grass + marsh + fallow + shrub land which yield one-fifth of this biomass.

Table 4: Disaggregation of Above-ground Plant Biomass

Crop	Grain %	Straw (%)	Other
1. Paddy	36.68	61.58	1.74
2. Ragi	32.34	67.66	-
3. Sorghum	11.87	88.43	-
4. Horsegram	37.83	62.17	-

Crop		
5. Sugarcane	Stem	72.26%
	Leaves	27.74%
6. Coconut	Nuts (Shell+husk)	21.98%
	Leaves	61.41%
	Inflorescence	4.84%
	Copra	11.77%

The table reveals that the ratio of grain to straw for the main food crops: paddy, ragi and horsegram is 1:2. Sorghum has an approximately 1:9 ratio of grain to straw because it is of the fodder variety grown solely for cattle.

Table 5: Fodder Consumption by Livestock

	Total	Bullocks	Cows	Calves	Buffalo	Goats & Sheep
Paddy straw	571	180	207	17	167	-
Sorghum grain	1	1	-	-	-	-
Sorghum straw	49	20	23	-	6	-
Grass + Fallow + Marsh land	400	78	91	-	124	107
	1021	279	321	17	297	107

Fodder from crops = 621 (60.8%)

Fodder from grass)
Land + Fallow land + } = 400 (39.2%)
   Marsh land)

Table 6: Food Consumption of Human Beings

	Annual consumption (tonnes)	Consumption (capita/day)
Rice	143.19	0.42 kg
Ragi	89.88	0.26 kg
Pulses	52.25	0.15 kg
Oil	3.00	8.75 gms
Jaggery	15.22	44.74 gms
Sugar	2.25	6.63 gms
Milk	18710(lit)	55 ml
Meat	5.11	15.02 gms

The important observation here is that livestock derive about 60% of their fodder requirements from crop residues and 40% from grass, fallow and marsh lands. Crop residues are important, but they must be supplemented by grazing from pasture land.

Table 7: Energy Sources & Activities in Ungra

Source	Units	Agriculture	Domestic	Lighting	Industry	Total
1. Human	hours	218,941	612,133	-	132,116	963,190
1.1Man	"	(143,549)	(168,671)	-	(110,135)	(422,355)
1.2Woman	"	(75,392)	(305,304)	-	(19,426)	(400,122)
1.3Child	"	-	(138,158)	-	(2,555)	(140,713)
2. Animal	"	36,654	-	-	3,381	40,035
3. Firewood	kg	-	411,636	-	68,085	479,721
4. Agro-waste	"	-	42,808	-	-	42,808
5. Electricity	KWh	25,894	-	9,524	3,000	38,418
6. Kerosene	litres	-	144	5,383	756	6,283
7. Diesel	"	32	-	-	-	32
8. Coal	kg	-	-	-	1,500	1,500

Table 8: Ungra Energy Source-activity Matrix (x 10^6 kcals/annum)

	Source	Agriculture	Domestic	Lighting	Industry	Total
1.	Human	51.0	119.9	-	31.7	202.6
(1.1	Man)	(35.9)	(42.2)	-	(27.5)	(105.6)
(1.2	Woman)	(15.1)	(61.1)	-	(3.9)	(80.1)
(1.3	Child)	-	(16.6)	-	(0.3)	(16.9)
2.	Animal	84.3	-	-	7.8	92.1
3.	Firewood	-	1,564.2	-	258.7	1,822.9
4.	Agro-waste	-	162.7	-	-	162.7
5.	Electricty	22.3	-	8.2	2.6	33.1
6.	Kerosene	-	1.3	48.3	6.8	56.4
7.	Diesel	0.3	-	-	-	0.3
8.	Coal	-	-	-	9.9	9.9
		157.9	1,848.1	56.5	317.5	2,380.0

Total village energy = 2.766×10^6 KWht/year = 7578.5 KWht/day
= 8.13 KWht/cap/day

Table 9: Energy requirement in Agriculture

	Man hours	Woman hours	DAP hours
Paddy	690	349	163
Ragi	328	284	179
Sorghum	271	-	66
Sugarcane	2237	1180	175
Horsegram	102	59	132
Coconut	216	-	66

Table 10: Energy requirement of Critical Operations in Agriculture

1. Paddy

Man - Transplanting (35%), Harvesting + Threshing (23%)

Woman - " (34%), " + " (43%)

DAP - Ploughing (41%), Manuring (27%)

2. Ragi

Man - Ploughing (24%), Transplating (21%)

Woman - Transplanting (42%), Harvesting+Threshing (51%)

DAP - Ploughing (45%), Manuring (25%)

Table II: Self-sufficiency in food

	1 Prodn.	2 Exports	3 % Expor- ted	4 Avail- able for food	5 Im- ports	6 Consump- tion	4/6 %
Rice	370	176	47.4	124	18	143	87.4
Ragi	36	1	2.8	35	55	90	39.1
Pulses	6	-	-	6	46	52	11.5
Oil seeds	0.4	-	-	0.4	2.6	3	3.8
Jaggery	52	48	92	4	11	15	27.2
Milk(in ltrs)	25897	6935	26.8	18961	-	18961	100

Table 12: Fuel Sources

From Ecosystem		tonnes/yr
Coconut	....	61
Other crops	....	2
Trees (Twigs)	....	172
Trees (Felled)	....	20
Shrubs	....	5
		260
Imports from outside ecosystem		
FW (Bought)	....	203
FW (Gathered)	....	59
		262
Total	....	522

REFERENCES - Part I

Odend'hal, S. (1972) Human Ecology. 1 3

Seshadri, C.V. (1978) Report of the Murugappa Chettier Research Centre, Madras, India

Briscoe, J. (1979) Population and Development Review 5 615

Moctar, B.A. (1979) Environnement Africain: Serie Etudes et Recherches No.30

Odum (1971) "Environment, Power and Society".

Ravindranath, N.H. et al (Forthcoming report)

AN INDIAN VILLAGE AGRICULTURAL ECOSYSTEM - CASE STUDY
OF UNGRA VILLAGE PART II: DISCUSSION*

Amulya Kumar N. Reddy

ASTRA, Indian Institute of Science, Bangalore 560012, India.

ABSTRACT

The observations reported in Part I of this paper have been
discussed in detail here. It has been argued that the agricultural
village of Ungra is an open, dependent, land-human beings-livestock ecosystem. Simple expressions have been derived for the
carrying capacity of the ecosystem with respect to cereal crops,
the pasture land-cropland ratio, the number of draught animal pairs,
the human to draught-animal-pair ratio and the human to cattle
ratio. The agreement between the calculated and observed parameters
have been taken as evidence that there is a rationale underlying
the ecosystem. The energy aspects of the ecosystem, in particular
draught power, and cooking fuel, have been discussed in terms of
enhancing the productivity and/or self-sufficiency of the ecosystem.
Some of the important limitations of the study have been identified
in order to indicate the directions of recent work. It has been
concluded that an understanding of the logic of village agricultural ecosystems should be the basis of rural development.

* This paper is a condensed version of a full paper to be published in
"Biomass", Volume 1, 1981, Applied Science Publishers Ltd.

LAND-HUMAN BEINGS-LIVESTOCK ECOSYSTEM

One of the most important conclusions emerging from the observations reported above is that the agricultural village of Ungra is
principally a land-human beings-livestock ecosystem. Land, human
beings and livestock are in a delicate balance with energy mediating
their inter-relationships.

To procure their food requirements, the majority of Ungra's
adult human beings must engage in agriculture, which not only
requires land but also draught power. The latter comes from
draught animals. But draught animals need fodder which can be
obtained either from pasture land or from crop residues. With
land being scarce because of high population densities, the minimum possible land must be allocated for grazing - hence, as much
fodder as possible must be obtained from crop residues, and only
the fodder essential to supplement the crop residues must come
from pasture lands. Since both human beings and cattle must depend
on crops - and on different components of the crop biomass - all
components of crop biomass are vital to the functioning of the
ecosystem.

Thus, the traditional strategy has been to produce all the
various crop biomass components in precisely the ratios required
to sustain the populations of human beings and cattle. This is
in sharp contrast to the green revolution strategy which exclusively emphasises the grain in its dwarf varieties. By yielding a
lower percentage of crop residues, the fodder output of the dwarf
varieties is reduced - and this either necessitates a separate
production of fodder, or more pasture land, or a replacement of
animal draught power with mechanized equipment which then creates
a demand for fuel.

Crop varieties, therefore, must be chosen to suit the agricultural ecosystem into which they are introduced. A crop variety
may be appropriate for one ecosystem but be totally inappropriate
for other ecosystems which are basically different. This point is
borne out by the Ungra data which shows that the high-yielding
varieties of paddy give about 12% more grain at the cost of 6%

less fodder and 117% more wastage.

OPEN, DEPENDENT ECOSYSTEM

The situation with regard to food is such that only in the case of milk does the ecosystem have enough _net_ production (production minus export) to meet consumption requirements. This does not mean that the ecosystem cannot meet its requirements, the fact is that it does not - this is an "irrationality" of the ecosystem which needs comment later.

For instance, about half the total paddy production is exported - this is tantamount to dedicating about a third of the ecosystem's cropland for exports. The effect is that the ecosystem becomes highly dependent on imports for pulses, oil-seeds and jaggery (crude sweetener) - in fact, the village imports almost all its requirement of oilseeds.

About 90% of the inanimate energy requirements of Ungra come from firewood. Only half of this firewood comes from internal sources, the remaining half has to be imported from outside the ecosystem. Further. only 23% of these imports are obtained by firewood-gathering expeditions, with 77% being purchased as a commercial commodity. Also, the bulk of the firewood, about 93% of the firewood, is obtained as twigs, leaves, etc., i.e. without the felling of trees and deforestation.

In addition, Ungra imports electricity at the rate of about 38,500 KWH/year but about 67% of this is consumed by its agricultural pumpsets for water-lifting. The remaining electricity is shared between its industry (8%) and its domestic needs, which are overwhelmingly lighting (25%). However, only 20% of its households are electrified.

The Ungra ecosystem is highly dependent on the external world for several items including food and fuel.

LAND-USE PATTERN

(a) <u>Carrying Capacity with respect to Cereal Crops</u>

Simple calculation shows that, due to increases in population,
the Ungra ecosystem is currently bearing a population density
which is only slightly less than its carrying capacity with respect
to ragi. Even this carrying capacity has been calculating from the
observed ragi productivity for one season, and may be lower in bad
years. The approach of the actual population density to the
carrying capacity for ragi was clearly appreciated by local wisdom
because the process of abandoning ragi as the main crop began
several decades ago.

Since then, the change in the cropping pattern has been in
favour of paddy which has a much higher carrying capacity. The
present measurements yield an ecosystem carrying capacity for
paddy of 8.6 persons/ha which is almost <u>double</u> that for ragi.
But, the change-over from ragi which was a rain-fed crop to paddy
has made the ecosystem dependent on the import of water. This
import is through a chain of seven tanks supplied by a large
reservoir. Even this linkage of the Ungra ecosystem to the
reservoir has not assured it of a guaranteed water supply every
year. The reservoir itself collects rain water from a large
catchment area, and if the rains fail in the region, then water
is not let out to the chain of tanks and the ecosystem does not
get sufficient water for paddy. In such bad years, the Ungra
farmers have no alternative except to fall back on ragi in which
case they suffer food scarcity - perhaps, even starvation.

What has been described here is a principle which is being
revealed in almost every one of ASTRA's investigations: <u>the traditional has ceased (or is rapidly ceasing) to be adequate, but the
modern is invariably inaccessible except to a few</u>. This is a
central dilemma of development, and this dilemma also defines the
crucial role of science and technology, which is generating a
variety of adequate and accessible options.

(b) <u>Summer Crop Area</u>

The summer crop area is only about 8% of the kharif area. This is primarily due to the lack of water, but represents a tremendous under-utilization of capacity. To increase the potential of the ecosystem, it would be necessary either to decrease imports by water conservation along with an optimized utilization of the Marconhalli reservoir + tanks system or to enhance the raising of ground water with inputs of energy.

LIVESTOCK

(a) Number of Draught Animal Pairs (DAP) and Human to Draught Animal Pair Ratio

The ratio of human beings to draught animal pairs should be 7.41. The observed ratio is 16.79 if only bullocks are used for draught purposes. But, there is a belt in Southern India where cows are also used for ploughing - a common practice in the Ungra ecosystem. If cows are also counted in calculating the observed ratio of human beings to draught animal pairs, an observed value of 7.34 is obtained with which the predicted value is in agreement.

The reason why cows are used for ploughing in this part of the country, and not in other parts, has not yet been established. But, it is almost certain to be related to the hardness of soils, as modified by the soil moisture, because this hardness defines, through an inverse relationship, the power requirements for ploughing.

(b) Weight of Ungra Cattle

Cattle in Ungra are much smaller and lighter than cattle in western countries - the average weight of bulls is about 230 kg and that of cows is about 193 kg. However, observations in Ungra have shown that draught animals are used for ploughing only for about 20 days in a year. When an item of equipment has such a low number of working hours per year, the obvious strategy is to incur on it as little expenditure as possible. Thus, their lean and hungry draught animals represent their optimization of costs and benefits of feeding cattle - the farmers increase their fodder supply to draught animals before the ploughing season in order to increase their strength.

ENERGY

(a) Draught Power

The central role of animal energy in a traditional agricultural ecosystem is based on the crucial importance of the ploughing operation and its requirement of draught power. It appears that draught power is a critical constraint on Ungra's agriculture. An augumentation of draught power is essential for increasing the carrying capacity of the ecosystem. There are many options for achieving this objective.

The first option is to increase the number of draught animal pairs.

The second option is to increase the productivity of draught animal pairs, i.e., the number of hectares which can be ploughed by a draught animal pair. This productivity can be increased in three ways:- (1) by increasing the power output of draught animal pairs, i.e., by using stronger draught animals, (2) by decreasing the hardness of the soil, and (3) by increasing the mean speed of ploughing.

The third option is to replace draught animal power by mechanized equipment (power tillers and tractors) fuelled with inanimate energy.

Much deeper analysis is required before one of these options, or a combination of two or more of these options, is chosen.

(b) Human Energy

Human energy also appears to be a critical constraint in agriculture, particularly for the transplanting and harvesting operations.

(c) Domestic Fuel

The present consumption of fuel in Ungra is 442 tonnes/year, or 7.5 kg/household/day, or 0.44 tonnes/capita/year, and as pointed out above, about 50% of this has to be imported from

outside the ecosystem. The question, therefore, arises: can
the ecosystem be made self-sufficient with respect of cooking
fuel?.

The efficiency of cooking stoves, which is presently less than
10%, can be approximately doubled leading to a halving of fuel
consumption. Since fuel imports constitute approximately half
the total consumption of fuel in Ungra, the approach of doubling
stove efficiencies can lead to fuel self-sufficiency. Cattle
wastes can be anaerobically fermented to generate biogas for
cooking. It is estimated that Ungra's cooking needs can be met
with the dung from 390 cattle. This means that if the dung from
all Ungra's 460 cattle are processed in biogas plants, there will
be excess gas available for other purposes.

(d) Fuel for Brick-burning

About 20 tonnes of firewood were used in 1980 for brick-
burning. This firewood was obtained be felling 0.5% of the tree
population of the ecosystem. It appears that brick-burning is
far more responsible than cooking (which is largely done with
twigs) for rapidly reducing the tree resources of the Ungra ecosystem. Fortunately, it has been demonstrated that unburnt
compacted mud-blocks can serve the same purpose. Thus, the demand
for fuel from brick-kilns can be eliminated.

(e) Energy for Lighting, Water-lifting, etc.

Electricity can easily be generated from biogas which means
that the excess biogas available after cooking needs can be used
to provide electric lighting and eliminate kerosene imports.
And, pumpsets can be run on gas generated from wood or charcoal -
the requirement is about 3 tonnes of wood/pumpset/year. With the
replacement of firewood for cooking and of burnt-bricks for house-
construction, the demand on firewood is so drastically reduced
that it becomes available in plenty for alternative uses such as
running pumpsets.

In conclusion, therefore, the plant biomass resources of the
village ecosystem can be shown to be abundant enough to meet its

present and future energy requirements.

CONCLUSION

The study of the village agricultural ecosystem described
here has revealed some of the important inter-relationships
involved in its functioning. In fact, it has shown that there
is a great deal of rationality in the ecosystem as demonstrated
by the agreement between many calculated and observed ratios
pertaining to important parameters of the system. This shows
that these parameters are not matters of chance, but they emerge
from an inherent logic. An understanding of this logic relating
to the complex linkages, particularly between human beings,
livestock, land, energy, and water, and between food, fuel and
fodder, should be a pre-condition for intervention into village
ecosystems. Otherwise, the actual results of interventions
may be totally different from what is intended. Thus, an ecosystem approach must be the basis of rural development; for,
it necessarily leads to the so-called "holistic" and "integrated"
viewpoints.

REFERENCES - Part II

Jagadish, k.s. (1979) Proc.Ind.Acad.Sci. C2   305

Reddy Amulyn Kumar N. (1979) Contributions of Science and Techno-
    logy to National Development, (Indian National Science Academy,
    New Delhi)

Vaidyanathan, A. (1978) Ind.J.Agri.Econ. 33   1

FUELWOOD AND CHARCOAL IN AFRICA

E. M. Mnzava
Director of Forestry, Ministry of Natural Resources and Tourism, Dar es Salaam

1. INTRODUCTION

Fuelwood and charcoal are the most traditional sources of energy in Africa, providing more than 90% of energy in rural areas and 50 to 60% of energy in the urban areas. These renewable resources are primarily used for cooking and heating, though increasing amounts are being utilized for village industries such as tobacco curing.

A bulk of the fuelwood is produced and consumed within the village confines as a "free" good. In such environments monetary transactions are minimal, hence much of it goes unrecorded. Therefore, the production figures often quoted in many publications are just little more than intelligent guesses. So it is always rational to refer to potential rather than actual fuelwood production. The potential sustainable supply in Africa is estimated at more than 500 million cubic metres annually (FAO,1980). The estimated total consumption per year is over 370 million cubic metres (equivalent to about 134 million tons of coal), assuming a per capita consumption of 0.8 m^3/annum and growth rate of 1.9% from 1975 to 1980.

Unlike fuelwood a substantial amount of charcoal is normally transported and marketed in the semi-urban and urban areas. Charcoal is essentially a commercial source of energy.

Unfortunately, these renewable resources are dwindling fast and becoming progressively less available to the consumer, both physically and economically. In East Africa, for instance, fuelwood is collected from distances up to 100 kilometres away. A similar trend is also observed for charcoal. In Dar es Salaam, Tanzania charcoal is transported from distances up to 180 kilometres away. This is partly due to irrational forest exploitation, as well as some areas being naturally deficient of forest land. The Sahelian Africa is a classical example. Under such circumstances the questions of efficient distribution and utilization of the resources are vitally important. Transportation logistics, socio-economic and environmental factors need a thorough investigation. The potential use of alternative energy sources is, no doubt, a strategy also to be looked into in order to ease this situation.

In the final analysis the production and utilisation of fuelwood and charcoal should be blended into the overall national (and possibly regional) energy policy. Several issues need attention; these include:

(a) Conservation and development of the existing renewable sources of energy.

(b) Creating new and renewable sources of energy.

(c) The impact of fuelwood and charcoal exploitation on the environment.

(d) Production technologies, especially for charcoal: "traditional" versus "modern" or combinations of the two.

(e) Priority areas of use: in different ecological situations; rural and urban areas.

(f) The use of wood wastes, including agricultural residues.

(g) Allowing local communities to collect fuelwood as a free good from the forest.

This paper reviews the present fuelwood and charcoal situation in Africa and discusses future strategies. Tanzania is used as the main reference point.

2. POLICY

The most important point to remember when formulating a fuelwood and charcoal policy is the fact that it has to be integrated with the overall national energy policy (and perhaps regional energy policies as well). For example, what measures should be taken to conserve various sources of energy? What technologies should be employed? When and where should a country use say, charcoal instead of coal (or vice versa)? How about the replacement of the new and renewable sources of energy? There are so many other questions which have to be answered by policy statements. Such a policy has to fulfil the function of providing a basis upon which decisions can easily be taken. It also has to ensure a co-ordinated approach to the utilisation of the various energy sources. If care is not taken the overlaps can become a source of friction among different producers and users.

Every government has to be very clear on the energy consumption priorities both in terms of the type and where it is consumed. Should emphasis of a given energy source be in the rural areas rather than in the urban? How about areas of acute fuelwood and charcoal resources? It would appear that the majority of the people in Africa live in the rural areas; as such whatever source of energy is considered, the emphasis should be placed among the local communities.

The question of substitutes and alternative energy sources such as coal, electricity, biogas, solar energy etc. has to be thoroughly examined, as well.

3. SUPPLY AND CONSUMPTION

3.1 FUELWOOD

3.11 Production and Supply

It is estimated that between 330 and 370 million people in Tropical Africa depend primarily on fuelwood and charcoal as a source of energy for cooking and heating.

In Tanzania, for example, more than 15.2 million people or 87% of the population use almost exclusively this renewable source of energy for the above purposes. As stated before, much of the fuel-wood is normally collected as a "free" good, at least in the adequately forested areas. As such, monetary transactions are very limited, hence much removal from the forest land is unrecorded. Nonetheless, the potential sustainable supply in the wood rich and wood poor parts of Africa is 330 and 170 million cubic metres per annum respectively (FAO, 1980)*, or 500 million cubic metres per annum in total. Total figures for Latin America and Asia are respectively 365 and 560 million cubic metres.

As it will be seen in section 3.12, the total estimated consumption figure is 370 million cubic metres per year. Superficially,

* Tech. P. on F & Ch. Iss. P. UNERG/FP/1/1)

this looks like a healthy situation: an apparent surplus of 130 million cubic metres. The reverse seems to be the real situation because:

(a) the figures are average figures which include the "havenots", like the Sahelian Africa

(b) even in those areas of apparent abundance the fuelwood collection distances are increasing day in and day out, partly because of the corresponding deforestation

(c) the use of fuelwood for household needs has to compete with other demands such as timber, village industries etc.

(d) fuelwood is relatively bulky; and is both inconvenient and expensive to transport to main population centres

The net result is that fuelwood collection is becoming time-consuming, hazardous, exhausting and costly in terms of opportunity costs. In East Africa fuelwood collection distances of up to 100 km (round trip) are not uncommon. This causes a lot of fatigue to women and children who carry a headload of fuelwood which may weigh 25 to 30 kg. Moreover they are exposed to ferocious wild animals, excessive sun, rain, cold and ultimate ill-health. In Central Tanzania a family may use between 200 and 250 mandays per annum for fuelwood collection; a labour equivalent to US $ 400-500. What a benefit for a village family to forgo!

The solution to this difficulty seems to hinge on sound and practical policy on conservation as well as efficient use of the existing forest resources. And new and renewable sources such as energy plantations have to be developed. The utilisation of suitable agricultural crops, such as cashewnut, mango and coconut should also be encouraged as should the whole concept of agro-forestry.

3.12 Fuelwood Consumption and Utilisation

The use of fuelwood in Africa accounts for 65-70% of total energy requirements. This consumption is rising steadily, as the

fossil fuel supplies are getting progressively exhausted and prices
escalate. Oil prices are 80% higher today than they were in 1978.*
The average annual consumption figure of fuelwood for the period
from 1960-75 was 171 million cubic metres; the present figure is
estimated to be 375 million cubic metres. This figure may increase
to over 450 million cubic metres in 1995, assuming a steady per
capita consumption mentioned above, and a low yearly growth rate of
about 2-3%.

For convenience the use of fuelwood is categorised as follows:
household, village industries and large scale industries needs.

3.121 Household Needs

More than 90% of the fuelwood harvested is consumed in rural
areas for household requirements - for cooking and heating. The
most prevalent method for cooking is still open fire cooking. This
is a very wasteful method compared to stove cooking. In Tanzania
this simply involves three stones or clay bricks supporting a pot
with firewood underneath. The wood utilisation efficiency is hardly
more than 10-12%. A metal Dover stove on the other hand is 30-50%
effective.

Assuming a household with six people uses $10.8m^3$/year of
Eucalyptus spp wood, a cubic metre of which sells at an average
price of US $ 15, about US $ 146.0-152.6 "goes up in smoke" per year
in open cooking as compared to US $81.0-113.4 if a dover stove is
used. In addition, open cooking is unhealthy because of the smoke
produced; hazardous, as it may burn houses and children; untidy and
takes considerable space in the house or kitchen.

Although in many rural parts of Africa fuelwood is still free,
as stated above, this practice is changing. Quite a number of
communities are forced to make special budgets for purchasing fuelwood. The Sahel zone and many parts of west and north-eastern
Africa such as Somalia and Ethiopia fall in this category. Purchase
of fuelwood in the Sahel accounts for 20 to 30% of a household's

(* Tanz.Daily News 18/8/80)

total expenditure every year (Morgan and Icerman, 1979). In the
future, fuelwood may become inaccessible to the rural communities,
since their annual earnings are often meagre.

Lastly, bearing in mind the inefficient methods of fuelwood use
for household purposes, future consumption surveys should concentrate more on effective energy consumption rather than round wood
requirements. In semi-arid Tanzania, for instance, the estimated
fuelwood consumption rate per household is roughly 3100 kwh/year.

3.122 Fuelwood for Village Industries

The utilisation of fuelwood for village industries is causing
"the second fuelwood crisis". Some of these industries include
tobacco curing, burning bricks, lime and cement making, fish smoking,
baking, local beer brewing, tea drying, and village metal works.

(a) Tobacco curing is likely to become a "curse-rather-than-
blessing" activity, if revolutionary steps to replenish and replace
the forest land are not taken in a number of African countries. The
main tobacco growing regions are in east and central Africa-
Zimbabwe, Zambia and Malawi, Tanzania etc. Our experience in Tanzania indicates that a hectare of tobacco - under average growth
conditions - produces 450 - 600 kg which need 50 - 60m^3 of fuelwood
for curing. In 1978 about 16.2 million kilograms of tobacco were
produced which needed a minimum of 1.79 million cubic metres of wood
to cure. Corresponding figures for 1985 are estimated at 21 million
kilograms tobacco produced and 2.33 million cubic metres of wood
consumed. More than 98% of the wood comes from "miombo" woodlands
(with Brachystegia, Commiphora Spp) whose mean annual increment per
hectare may be as low as 1-5 m^3/yr. So where is all this wood going
to come from in the future?

(b) Brick burning is another village operation gaining momentum
in developing Africa. Countries like Tanzania have made deliberate
policies for the rural communities to use these bricks instead of
cement ones, partly because of inadequate supplies of cement. But
burning 1000 bricks needs 1.0-1.5m^3 of wood. So, the Kilimanjaro
Region, expecting to produce about two million bricks in 1980/81,
will need two to three million cubic metres of fuelwood for bricks

alone.

(c) So far there has been very little mention in the literature on the fuelwood consumption for local beer brewing. Let us consider Tanzania again. Experience has shown that to process 180 litres of "mbege" in Kilimanjaro 1 m^3 of fuelwood is utilised. It is estimated that the Region processes about 21.2 million litres of "mbege" per annum; this needs 530,000 m^3 of fuelwood. In Kisarawe District (Coast Region) there are 38 local brew shops each of which make 440 litres of "kangara" every day using 0.5 m^3 per 200 litres. In a year (250 days) 10,450 m^3 of wood are required.

(d) Tea drying is also putting heavy demands on fuelwood. In Korogwe District in Tanzania there are two tea factories, and these need about 12,000 m^3 stacked per annum. A more detailed survey for the whole country is needed.

(e) Fish smoking as a means of preservation is very common among the fishing communities. The Sahelian experience is well documented. The fuelwood quantities utilised for this purpose for the continent are not precisely known. At Nyumba ya Mungu Dam in Kilimanjaro 7,500 m^3 stacked are used to smoke the 500 tons of fish processed each year.

Lime and cement and metal works and baking also require considerable amounts of fuelwood. Here again, precise quantities are not known. Investigations are, therefore, important.

3.123 Large-Scale Industrial Fuelwood and Biofuel

Relatively small amounts of fuelwood are being used for large-scale industrial activities in Africa. In fact wood waste rather than fuelwood in the round is more common - from sawn timber, veneer, plywood, carpentry shops etc. The wastes are utilised in boilers to generate steam, which in turn can be passed through turbines to generate electricity. Using miombo woodland species one air-dry ton of wood (roundwood equivalent) can produce 900-1200 kwh.

The use of fuelwood as a source of biofuel is still on an

experimental basis in Africa as it is in many countries of the
world like Brazil and Japan. Eucalyptus and Pinus are among the
species being investigated.

Because of handling difficulties as well as high costs of fuelwood transportation over long distances, more and more is being
converted into charcoal.

3.2 CHARCOAL

Charcoal is more of a commercial fuel than fuelwood. And because
of its superior transport properties, its utilisation is more
prevalent in urban and semi-urban areas than in the rural regions.
It is easy to handle and store, and has a relatively low bulk
density of about 30-40% that of wood. With an efficient stove its
heat value per unit weight is almost twice that of fuelwood, but
smaller than that of kerosene - the respective figures are 7.4,
3.7 and 11.1 kcal/gram, dry weight basis (Morgan and Icerman 1979).

3.21 Production

Less than 25% of the total round wood harvested annually in the
world is converted to charcoal (Booth). A corresponding figure for
Africa is estimated at an average of 35-40%, though there are
considerable variations. For example, the percentages for Tanzania,
Gambia and Senegal are 10-15, 26,and 50. Good quality charcoal is
obtained from fairly high density wood, such as the miombo woodland
species of Zambia and Tanzania. But since these resources are
dwindling rapidly, more charcoal is being produced from agricultural
crops such as cashewnut (Anacardium occidentale). The use of sawmill waste, bark etc. to make briquet charcoal is still on a small
scale. Some countries have established "charcoal plantations". But
the careful choice of species is of the utmost importance. An
investigation on the acceptability and use of pines for making
charcoal was done in Tanzania 1978. The results indicated that,
though effective calorific value per unit weight of charcoal was
almost the same as that of hardwoods the market was unwilling to
accept the charcoal. This was despite the fact that the price for
pine charcoal was reduced by 50% (Mnzava, 1980). Perhaps this is a
question of promoting a new innovation rather than a technical

problem.

For the purpose of this discussion three production techniques
are considered.

3.211 Earth Kilns

More than 80% of all charcoal produced in Africa is still made
by the traditional Earth Kiln Technique. This method is very wasteful of energy as anything between 50 to 90% of the energy is lost to
the air. In Zambia 92% and in Ghana 85 to 95% of the energy is thus
lost (Brown and Morgan/Icerman). The other drawback of this method
is that there is normally no way of collecting valuable by-products
like pitch, creosote, alcohols, acids etc. There is still some room
for improving the earth kiln method, for it is cheap and requires
minimal technical knowhow, factors which are essential among the
rural communities. More research is needed.

3.212 Portable Metal Kilns

These are more sophisticated than the earth kilns. They are more
efficient, and allow recovery of a few by-products such as tar. The
most common kilns are the Mark V and VII. Using hardwoods under
favourable conditions the output is about 1.3 to 1.5 tons of charcoal per shift of several hours duration.

The main problem with this technique is that it requires some
skill, which is normally lacking in villages; and is more expensive
than the earth kiln. It is improbable that a villager could afford
it. A co-operative effort is needed such as is the case in Tanzania
where a kiln may be owned by an Ujamaa (Familyhood) village.

3.213 Permanent Kilns and Retorts

Use of this technique is still in its infant stage. And much of
the charcoal thus produced is frequently used for industrial purposes. The capital investment is relatively high and this limits
their use by the rural communities.

2.22 The Utilisation of Charcoal

The use of charcoal is increasing in not only the urban but also in semi-urban areas. And this trend is likely to continue beyond this century. In Tanzania, for example, the use of charcoal was increasing at the rate of 3% annually in 1974; the percent will be 25 in year 2000. Consumption figures for Dar es Salaam are shown below.

TABLE I

RECORDED CHARCOAL CONSUMPTION IN DAR ES SALAAM

YEAR	CONSUMPTION (TONS)
1967	28,400
1973	432,500
1974	733,300
1975	1,150,000
1976	775,000
1977	900,000
1978	2,025,000
1979	1,400,000

Notes:

- There is some charcoal consumed that goes unrecorded

- Although there have been some consumption fluctuations during 1976-77 and 1979, the trend still remains: rapid increase.

- The 1976-77 fluctuations are partly due to management problems experienced by the main organisation then supplying a bulk of charcoal to Dar es Salaam.

It would be misleading to quote an average per capita consumption figure for the whole of Africa. Here it will suffice to quote a few examples to give a general impression. In the Sahel zone the per capita consumption is 180 kg/year (Jullander & Stockman); while in Tanzania it is 200 kg/year (Mnzava, 1980). Zambia produced and consumed 250,000 tons this year; the domestic demand is 240,000 tons/year (Brown, 1980).

In the past considerable quantities of charcoal have been exported, particularly to the Far East. Some countries, like the East

African, are reconsidering the issue because of the environmental
questions, the subject of the next few paragraphs.

4. THE ENVIRONMENTAL IMPLICATIONS

"Large areas of our country have already been denuded of trees
and still people cut without planting. We are beginning to feel
the effects but not everyone has yet made the connection between
water shortage alternating with flooding and tree cutting in which
we casually indulge. We in Tanzania still have time to avoid
disaster if we take action now".
(President Nyerere in his banquet speech to the President of
Cape Verde in Dar es Salaam, January 1980).

As stated in section 3.11, Africa consumes approximately 370
million cubic metres annually of fuelwood. A bulk of this wood is
harvested from the natural forests or forest land.

Land clearing for agriculture is also claiming a large share
of the wood - especially shifting cultivation. The result of the
heavy fuelwood and charcoal demand is intense land denudation.
According to Morgan and Icerman, Algeria, Tunisia and Morocco have
less than 11% of their forest area left. Gambia has only 4%. Ivory
Coast, once densely forested with 12 million hectares, has now only
4 million hectares left.*

A more dramatic example is that of Tabora Region in Tanzania,
the most important tobacco growing area. It is estimated that at
the current rate of forest clearing for tobacco cultivation and
curing, 700,000 ha (22% of total woodland area) will have been
cleared by the year 2000 (Tem,1980)

This means clearing at the rate of 0.21 ha/minute for tobacco
alone. In total, tropical Africa has only 33% of its original
forests. And information at hand also indicates that tropical
forests in West and East Africa are the most badly hit by this
depletion. It is estimated that between 1975 and 2000 the projected losses are 13,200,000 and 6,500,000 ha respectively (IUCN,UNEP,
(* Tanz. Daily News 27/8/80)

WWF, 1980) The closed forest is barely 9%. All this warns Africa that it is marching towards desertification, unless urgent corrective measures are taken.

One of such steps is intensifying regeneration of the forests. This is relatively easy in many countries of west and parts of central Africa. But in Eastern Africa where savanna woodlands are dominant the mean annual increments are very poor, sometimes even below $1m^3$/ha/year. Similar difficulties are also prevalent in Upper Volta, Mali, Senegal etc. It is, therefore, rational to propose that afforestation and reforestation are the obvious strategies. Village afforestation should be encouraged, since it is likely to involve the rural communities directly, thereby educating them and reducing woodlot establishment costs. The provision of alternative energy sources will no doubt help to ameliorate the situation. This subject is further discussed in section 6.

The discussion of the fuelwood and charcoal situation has so far highlighted production, consumption and environmental status. All these have socio-economic considerations.

5. SOCIO-ECONOMIC CONSIDERATIONS

The consumption and benefits of fuelwood and charcoal outlined above have already elucidated on the social and economic burdens imposed on village and urban communities. Here it will suffice to re-emphasize the cost of the resources - both direct and indirect.

The best method of pricing these renewable sources of energy is through what the market is prepared to pay, since "production" costs of a natural unreserved forest or forest land are complex. An average family of six in Tanzania needs 10.8 m^3/year of fuelwood. The price of $1m^3$ of miombo species is approximately US $ 12. The basic wage is $ 60. A household, therefore uses 20% of its income for buying this energy - assuming this family lives in semi-arid central Tanzania.

The per caput consumption of charcoal is about 200 kg/annum; and this is equivalent to about 2.5 m^3 of fuelwood if produced by an

earth kiln. This per caput consumption means $ 30, assuming no
other sources of energy are used.

In the absence of fuelwood and charcoal substitutes would have to
be used. Kenya, for example, with a big proportion of its population using fuelwood, would have to spend an extra $ 240 million a
year, if paraffin were to be substituted for fuelwood (Chandler and
Spurgeon, 1979). The opportunity costs for diverting labour for
fuelwood collection, using cow dung and agricultural residues as
fuels can similarly be valued.

6. SUBSTITUTES AND ALTERNATIVES

The supply of fuelwood and charcoal will continue to be a problem,
unless readily available, cheap substitutes are found. These include
natural gas, biogas (mainly from organic residues). coal, electricity
and solar energy.

6.1 Biogas (mainly Methane): Biogas production using cow dung is
relatively new in Africa when compared with the supply in countries
like China, India and the Far East. In Tanzania the technique was
introduced in 1975. And to date there are more than 40 plants of
various capacities, the gas of which is used primarily for cooking
and lighting. A few plants are designed for industrial use and are
capable of producing 10 to $40m^3$ of gas. The main objection to this
alternative energy source is that it diverts cow dung from agriculture. Although there has been a complaint that the resultant
cow dung contains few minerals, hence unsuitable for agriculture,
investigation in Tanzania has proved the reverse: the cow dung is
richer. Moreover, transportation can be expensive for communities
living in non-livestock rearing areas, as indicated below.

TABLE 2

Cowdung Transportation (1km)

Mode of Transport	Load Capacity (Kg)	Trips/ton	Time (hrs)
Foot/head	7	142	37.75
Wheelbarrow	28	36	14.5
Handcart	350	3	2.5
Ox-cart	360	2	0.2

Source: Ministry of Agriculture,Uyole Research Station,Tanzania(1978)

Lastly, it may be inconvenient for a consumer in urban areas to keep livestock for biogas. A family of six people needs about $7m^3$/day of biogas for cooking and lighting; so it has to keep a minimum of three to four cows.

The use of natural gas is limited to semi-urban and urban areas. Like oil it is a non-renewable source of energy, and often involves foreign exchange for its importation.

6.2 Coal

It is mainly used for industrial purposes. Research has revealed that it has a high potential in the village and in small-scale industries.

6.3 Electricity

Hydroelectricity is most common. It is normally used for lighting in rural areas and its use for cooking in towns and cities is increasing.

6.4 Solar Energy

Vigorous research is being carried out in this field. Some countries have already started using it on a small scale for various purposes.

To reiterate, fuelwood and charcoal will continue to be the main sources of energy, especially in the rural areas. This is partly because they are more readily available and are cheaper, as indicated below.

TABLE 3

Prices of Alternative Energy Sources in Kenya - Per Caput/year

Fuel	Units	Joules/unit x 10^6	Appliance Efficiency %	Units Req.	Ann/Caput Price US $
F/wood	m^3	9,670	10	1.5	20
Charcoal	Kg	28	25	187	20
Paraffin	litre	38	50	77	22
L.P. Gas	Kg	46	60	52	29.5

Source: Openshaw & Moris (1979): The Socio-Economics of Agroforestry

7. CONCLUSION AND FUTURE STRATEGIES

It is obvious from the above coverage that fuelwood and charcoal will continue to be the poor man's energy source for a long time to come, especially in rural Africa. So the question of having a sound policy on the conservation and development of these renewable resources is crucial. In order to arrest the decline of forests and forest land reforestration and afforestration programmes have to be immensely expanded. Tanzania, for example, is planting about 7 to 9,000 ha annually in the villages. But according to consumption figures she has to plant some 15 to 20,000 ha annually to the year 2000.

Substitutes will no doubt augment the energy from natural forests and woodlot plantations. But their socio-economic implications have to be taken into account. All this calls for national and regional energy consumption surveys. This will necessitate working out the logistics and costs of transporting these renewable sources of energy from adequately-supplied to deficit areas. Regional co-operation which is almost non-existent today, is needed. Ultimately these issues, including the environmental considerations, must be integrated into the total national energy policy. In more practical terms the future strategies should focus on:

- (i) the conservation and rational utilisation of the existing resources
- (ii) increased efforts on establishing energy/woodlots plantations
- (iii) increased utilisation of forest and wood industrial wastes
- (iv) efforts to use wood resulting from agricultural clearings
- (v) continual assessment of the energy sources, including energy consumption surveys
- (vi) improved resource utilisation - e.g. improving cooking stoves
- (vii) establishing priority use areas for a given energy source - rural urban, deficit areas
- (viii) the policy towards, and research into, the use of substitutes and alternative energy sources.
- (ix) the formulation of realistic energy budgets
- (x) research on use of wood for transportation by extracting chemicals like alcohols which can be mixed with gasoline
- (xi) ability to mobilise people, especially the rural communities, so that they become fully involved in formulating and implementing short- and long-term plans. Bearing this in mind, the conference may wish to discuss: "What is the role of the people, particu-

larly the rural communities, in the development of the
fuelwood and charcoal energy sources?"

REFERENCES

(1) Booth, H.: Charcoal in Energy Crisis of the Developing World

(2) Brown, J.B.(1980): Charcoal Industry in Zambia.
                    Paper prepared for the Eleventh Comm.
                    Conference, Trinidad

(3) Chandler and Spurgeon (1973): International Co-operation in
           Agroforestry. Proc. of Int.Conf.Nairobi

(4) FAO(1980): Tech. Paper on Fuelwood and Charcoal Issue Paper
           UNERG/FP/1/1

(5) IUCN,UNEP,WWF (1980): The World Conservation Strategy

(6) Jullander, I. & Stockman, L.(1970): The Life Cycle of Wood.
           STU Information No.110

(7) Mnzava (1980): Village Afforestation: Lessons of Experience
           in Tanzania - a FAC/SIDA publication

(8) Morgan, P.R. & Icerman, L.J.(1979): Applied Tech. for Renewable
           Resource Utilization

(9) Temu, A.B.(1980): Fuelwood and Tobacco Production in Tabora.
           University of Dar es Salaam Record No.12

A STRATEGY FOR RURAL DEVELOPMENT IN THE THIRD WORLD VIA BIOMASS RESOURCE
UTILISATION

M.SLESSER, C.W.LEWIS and I.HOUNAM

Energy Studies Unit, University of Strathclyde, Glasgow, Scotland, U.K.

Summary

The concept of enhancing the self-reliant development of Third World
rural communities by the implementation of appropriate biotechnologies is
one worthy of serious examination.   A community increasing its indigenous
energy production, mostly via biological processes, is becoming a potentially more independent community.   The questions are : can such self-
reliance be achieved, and if so, under what conditions?   What are the
impediments;   what are the dynamics of change; and what are the side
effects?

In an attempt to answer these questions a systems simulation modelling
technique has been used to monitor the effects of biomass utilisation
within a hypothetical but plausible Third World rural community (x) through
time.   Under the conditions simulated the integrated installation of biogas plants, algal cultivation, legumes, energy tree plantations, and so
forth, can result in annual per capita increases in nutritional calorific
intake, protein, and energy availability of respectively 125%, 185% and
600% within 15 years, given a stable population.   Energy and nitrogenous
fertiliser imports are reduced to zero, while excess productivity is
exported to give a valuable annual balance of payments surplus.   Similarly
encouraging results are also presented using as the data base the profile
of an existing village (I) in southern India.

1. INTRODUCTION

This paper summarises the current state of progress of the International Federation of Institutes for Advanced Study's (IFIAS) self-reliant development programme (SRD). The programme aims to indicate which mix of integrated agricultural, solar and biotechnologies (e.g. biogas plants, algal cultivation, energy crop plantations, etc) is best suited for the maximal development of rural communities in developing countries. The correct mix will vary depending on existing conditions and the potential for change as determined by the community's physical, economic, social and political characteristics. Self-reliance in no way implies isolationism or complete self-sufficiency but is considered to represent the ability to meet basic needs locally. Such an integrated system potentially seeks to raise productivity, but the salient question is whether this novel approach to development can reach levels of productivity sufficiently high to be attractive to development ambitions, and if so, can it be fast enough? For example, if a complex system of biotechnologies raised prosperity by 10% in 20 years it would probably be disregarded, whereas 50% in five years would be meaningful. Thus four crucial issues need to be resolved : What is the best use of bio- and agricultural technologies to raise ouput? What are the dynamics of change? What level of productivity may be ultimately attainable? What social or economic changes may be needed to make such a self-reliant development possible?

To begin to answer the above questions a systems simulation technique is used which has the merits of versatility and speed, allowing for the examination of a large number of scenarios in a short space of time. The technique has thus far been used to monitor the effects through time of increased biomass utilisation, firstly within a hypothetical, but plausible, Third World rural community (X) and then in an existing southern Indian village (I).

2. SYSTEMS ANALYSIS

Briefly, the systems analysis approach involves reducing the complex entities and interactions of the real life system to a set of observations consistent with the system under study, but sufficiently simple to be understandable to the observer. This is known as a model. Systems analysis considers all facets of a situation simultaneously and thus deals in interrelationships between the component parts of a system rather than in the parts themselves.

To illustrate this point imagine the following cycle of events : cows

in a field discharge dung which is transferred to a biogas generator,
thereby evolving methane gas for cooking and light, and an effluent sludge
which can act as a nitrogen fertiliser, which is then applied to the pasture to be grazed by cows. The cycle, now complete, then repeats itself.
In systems language we say we have closed the loop. The food and energy
collecting activities of the community may be seen as a series of events,
each leading to the closing of various loops. These are not simultaneous
events, but sequential and so time-dependent. The modeller must design a
simulation model that effectively simulates the various happenings, taking
into account the dynamics of change. The village itself exists within a
fixed system boundary within which events are tracked in mass and energy
terms (gigajoules of fuel energy, kilocalories of food energy, kilograms of
protein, and so on). If a material good, like rice, or an activity passes
through the system boundary to the outside world, it is changed into money
and used for purchases such as fertilisers and transistor radios. Man-
hours can also be tracked. The name of the model, realised for the computer in DYNAMO 2 language, is MERDA : Model for Energy and Resource Development Analysis.

3. DEVELOPMENT OF HYPOTHETICAL COMMUNITY (X)

The principal features forming the data base, or initial conditions,
of community X at time zero are as follows:

Population: 600 living in 100 households.

Land area: 350 hectares (205 under cultivation, 45 grazing,
                     45 forests, 10 fallow, 45 wasteland).

Livestock: 300 cattle, 100 goats, 50 sheep, 200 poultry.

The average steady-state unimproved productivity of the rural community at
time zero is estimated as follows:

Crop productivity: rice 235 t/yr, ragi (millet) 94t/yr; minimal legumes.

Food energy/capita: 2507 kcal/day.

Protein/capita yr: 18.35 kg (all plant protein) = 50.3 g/day.

Non-food energy/capita yr: 8 GJ (firewood, kerosene).

Financial export/yr: 360,000 rupees (Indian) = U.S.$ 45,000.

Other points to note are:

1. Average rainfall is 1300 mm/yr which allows 20% of the land designated
for food crops to grow one crop of rice and one of ragi each year, the
remainder providing only one crop of rice. 50% of rainfall occurs in
October/November.

2. Average temperature is 30°C; 27°C in winter and 33°C in summer.

3. Annual insolation is 26 PJ.

4. The village is non-electrified. All energy accrues from wood (7% of which is imported) and kerosene (all 4750 litres of which are imported).

5. Animal manure is used as a fertiliser but not as an energy source.

6. Population growth is zero (though scenarios in line with the Indian population rate of increase of 2% per annum were also examined).

By monitoring the four crucial variables of food energy, protein, and non-food energy available per capita, together with the community's export, a good indication is gained as to whether the community is becoming more prosperous, healthier, and better developed, or whether it is going into a further decline. Of the scenarios tested that which appeared to be of the most all-round benefit is presented below. The intervals between the implantation of some of the technologies are conservative, stressing the need for the village to become acclimatised to major changes before the next is introduced (1). The sequence is as follows:-

Year	Implantation
5	Artificial fertilisers are imported at 21tN/yr. This provides 85.4 kgN/ha crop to realise the potential of high yielding rice strains.
10	A community biogas plant sufficient to receive the dung of 300 cattle is installed.
10	The 45 ha of natural forest are converted into a Casuarina energy tree plantation with the nitrogen-rich leaves ploughed into the paddy fields.
12.5	N-fixing blue-green algae are cultivated and applied to the fields at 10 kg/ha.
15	A second msaller biogas plant is installed, receiving human waste plus unused natural vegetation such as water hyacinths.
20	20 hectares of rice land are given over to legume cultivation.

The result is that from the 5th year, when the first changes are being made, to the 20th year, the overall annual per capita increases in nutritional calorific intake, protein and energy availability are respectively 125%, 185% and 600%. Moreover, energy and N-fertiliser imports have been reduced to zero, and there is an annual balance of payments surplus of Rs 1 million (US $ 127,000). This means that food intake is now adequate, making the villagers nutritionally, and presumably medically, healthier, while welfare in terms of energy availability has increased dramatically.

4. DEVELOPMENT OF INDIAN VILLAGE (I).

The following profile of an existing village I in south-east India is derived mainly from data made available by the Murugappa Chettiar Research

Centre, Madras (2) and supersedes all previously published data (3).

Population: 944 living in 211 households (485 males, 458 females and including 389 children under 15 years old).

Land area: 352 hectares (56 wet lands under food crop cultivation, 210 dry lands for tree growth and some grazing, 85 poramboke of which 7 are pond area and the remainder canals, cremation grounds, common land for grazing etc., 3 fallow).

Livestock: 81 cows, 65 bullocks, 47 cow calves, 76 buffalo, 24 buffalo calves, 163 goats, 127 sheep, 2606 poultry.

The average steady-state unimproved productivity of the rural community at time zero is estimated as follows:

Crop productivity: rice 76 t/yr, ragi 3 t/yr; minimal legumes.

Food energy/capita: 963 kcal/day.

Protein/capita yr: 8.46 kg (reference protein 54.5% animal; 45.5% plant) = 23.2 g/day.

Non-food energy/capita yr: 6.6 GJ (mostly firewood, some kerosene/diesel, minimal electricity and dung cakes).

Financial export/yr: Rs 1.285 million (estimated to equal village expenditure).

Much of the people's protein intake is derived from fish, made possible by the village's proximity to the Bay of Bengal. Apart from poorer agricultural yields and lower nutritional standards, village I has much in common with community X.

Of the future scenarios tested the most all-round benefit accrues from that below. The time intervals between technology implantations are reduced to a level still acceptable to the inhabitants of the village. The sequence is as follows:

Year	Implantation
1	A community biogas plant sufficient initially to receive 140 t/yr cattle dung; dung collection efficiency rises from 50-75%.
2	_Casuarina_ energy plantation set up covering 10 ha producing 44t/hayr dry wt, with N content of leaves 1.8 kgN/t wood; imported insecticides phased out over 10 year period due to utilisation of indigenous bioinsecticidal neem tree leaves.
3	Stoves twice as efficient as existing types introduced gradually over five years,; rice variety grown with net protein utilisation increased from 50% to 57%; post-harvest crop losses reduced from 20% to 10%.
4	N-fixing blue-green algae established in paddy fields over four years providing 30 kgN/ha crop.
5	Pulse (rice bean) grown as third crop-yields 1.5t/ha over 10 ha.
6	2nd community biogas plant accepting human waste + 50tdry wt water hyacinths; 35 windmills introduced over five years to

irrigate 5% paddy field area.

7   2 Fiat Totem-type engines running on biogas producing 30 kW
       electricity for lighting + some additional heat.

At present the farmers of village I only cultivate 25% of their available arable land in summer and 43% in winter. By utilising the indigen-
ously produced biogas (and windpower) for irrigation allied to the cultivation of biofertilisers, maximum use can be made of the paddy field area for
rice cultivation. The labour requirement increases, but is still well within the capabilities of the village. As a result of this scenario implementation over a period of 12 years, yields leap by 340%, N fertiliser levels
per hectare nearly double, protein and calorie intake per person rise 140%
and 245% if necessary, the village becomes self-sufficient in energy and
has a surplus of 750 GJ/yr, as well as an annual balance of payments surplus
of Rs 384,000. Solar energy capture has quadrupled in this time and the
village is now well set for the initiation of one or more agro-industries.

5. CONCLUSION

The preliminary investigations have demonstrated that the system modelling methodology has much to offer in the evaluation of the various options
available for rural Third World self-reliant development. To achieve a
higher level of development two goals must be attained : increased output
along with decreased imports. Both these goals are attainable via the
installation of an appropriate integrated system of biotechnologies allied
to improved agricultural practice and renewable energy technologies. The
national and international repercussions of such a self-reliant development
strategy remain to be ascertained.

REFERENCES

(1) Reddy, A.K.N., and Subramanian, D.K. (1979). Proc.Indian Acad. Sci.
        C2, Part 3, 395.

(2) MCRC (1980). First Annual Report of the IFIAS Self-Reliant Develop-
        ment Project, Murugappa Chettiar Research Centre, Madras, India.

(3) Slesser, M., Lewis, C.W., and Hounam, I. (1980). Paper presented at
        6th International Conferences on Global Impacts of Applied Micro-
        biology, Lagos, Nigeria, 31 Aug.- 6 Sept. 1980.

THE PRESENT USE AND POTENTIAL
FOR ENERGY FROM BIOMASS IN TANZANIA

G. Mauer

University of Dar es Salaam, Faculty of Engineering
Dar es Salaam
Tanzania

Summary

The energy supply situation in Tanzania is characterized by an ever increasing deficit of firewood in the rural
areas and by oil import costs that absorb more than half of
the country's export earnings. It is shown that energy for
cooking purposes can to a large extent be provided by
simple low-cost biogas plants which would be built in regions where large numbers of livestock are kept. When fully
developed, biogas can provide about 60 percent of the rural
population's need for cooking energy, at a cost lower than
that of any other alternative energy source. Oil imports
can be partly substituted by ethanol obtained from molasses.
A projected plant will produce 8,000 cubic metres of ethanol
and 1,000 tons of baker's yeast annually, which could lead
to import savings of 19 million U.S.-Dollars annually. Production of alcohol could be raised to 30,000 tons annually
by 1990.

1. INTRODUCTION

The Energy Situation in Tanzania

Energy in Tanzania is at present provided by three
principal sources:
- imported oil
- electricity generated from water power
- firewood and dung

Electricity from water power is relatively cheap and reliable.
After the initial investment has been made, foreign currency
expenditures are only required for the importation of spare
parts. On the other hand, oil imports are putting a severe
stress on the country's balance-of-payments situation. Although only about 700,000 tons of crude oil and petroleum
products are imported annually, at present prices more than
half of Tanzania's export earnings are absorbed by oil purchases.

Firewood is the major source of energy at the household level in rural Tanzania, where over 90 percent of the
population live. Ninety-five percent of cooking is done
with wood; charcoal and kerosene account for the remainder.
The annual demand for wood fuel stands at about $2m^3$ per
capita (1). The present level of plant regeneration does not
satisfy this demand, which is constantly increasing as the
rural population grows. According to estimates of the Department of Forestry the demand for wood will rise to 50 million
m^3 annually by 1985 (2). The growth increment from existing
forest is 17 million m^3 per year, leaving an enormous wood
deficit. As a consequence, the prices of firewood and charcoal are rising rapidly, and soil erosion renders large
areas of land infertile.

The Government of Tanzania supports activities to improve
living conditions at the village level. Activities are
coordinated by the Tanzanian National Council of Science, in
cooperation with several ministries. However, as will be
elaborated later, there have been few efforts until now to
tap biomass-energy for use in the villages.

2. PROSPECTS FOR ENERGY FROM BIOMASS IN TANZANIA

In view of the facts stated above, two areas of high urgency can easily be identified:

1. supply of energy to the rural population
2. substitution of oil imports by other energy sources.

Biomass can contribute to the solution of both problems. Considering the prevailing conditions, two approaches are most promising:

-introduction of biogas digesters in areas with a
high number of cattle

-use of sugar cane molasses to produce ethanol as an
additive to gasoline.

3. BIOGAS DIGESTERS

Although well known in India, China and other developing countries, biogas digesters have not been used yet on a large scale in Tanzania. There are about 200 small digesters in operation in the countryside, most of them being located in the Arusha region in Northern Tanzania. One obstacle to their large scale introduction is the scarcity of building materials in the country which results in high prices for the construction of the biogas plant.

Using an Indian design, a small digester for gas production of $2m^3$ per day costs between 500 to 600 U.S. - Dollars. However, by adapting the design to cheaper local materials, the Arusha Appropriate Technology Project has succeeded in reducing the price to less than 100 U.S.-Dollars for a plant with the same capacity. At this price, biogas will be cheaper than any other energy alternative, the price for cooking energy amounting to less than 0.01 U.S.-Dollars per kilowatt-hour. Livestock is kept in many parts of Tanzania, the largest cattle herds being in Arusha and Kilimanjaro regions and around Lake Victoria. In other areas, mainly the South, South West and along the Coast, few domestic animals are kept, often because of tsetse fly infestation. When fully developed, biogas could meet as much as 60 percent of the future energy needs of the villages of Tanzania, according to an estimate made by the Small

Scale Industries Development Organization (3).

4. ETHANOL FROM MOLASSES

4.1 Availability

There are four sugar factories in Tanzania with a total
annual sugar output of between 93,000 to about 130,000 tons.
These fluctuations in sugar production are mainly determined
by the rainfall in a given year, in the absence of artificial
irrigation. Molasses production varies from about 40 to
60 percent of the sugar produced, in some cases reaching 90
percent. On the average, molasses production amounts to about
50 percent of the sugar output, i.e. 47,000 to 65,000 annually.
Because of increased sugar production 87,000 tons of molasses
will be available by 1983-84, according to Sugar Development
Company estimates. It was found that for the production of
one ton of ethanol 4.22 tons of molasses are required. Thus
by 1984 approximately 22,000 tons of ethanol could be produced. The alcohol can be mixed with gasoline. Experience
in Brazil has demonstrated that "gasohol" can contain up to
15 percent ethanol without necessitating engine changes.

4.2 Economic Aspects of Ethanol Production

The Tanzania Industrial Studies and Consultancy Organisation (TISCO) conducted a feasibility study on the
possibilities of producing alcohol, baker's yeast and polyethylene from sugarcane molasses(4). TISCO considered an al-
cohol/baker's yeast plant with an annual capacity of 8,000
cubic metres of ethanol and 1,000 tons of baker's yeast.
The total cost for the construction of this plant was estimated by TISCO at 216 million Tanzanian Shillings,
of which about 132 million shillings (16million U.S.-
Dollars) would be required in form of foreign currency. At a
price of six shillings per litre of ethanol, the annual internal rate of return would be 14 percent before tax for
the first ten years of operation. If the alcohol is sold at

4 shillings per litre, the internal rate of return would be six percent.

For a country such as Tanzania the possibility of import savings is of paramount importance. TISCO's study demonstrates that once the plant is fully operational, the annual import savings for fuel and baker's yeast would be 155 million shillings (19 million U.S.-Dollars) at 1979 prices. The initial foreign currency expenditure of 16 million U.S.-Dollars would therefore already be compensated by corresponding savings in the first year of operation.

The National Development Corporation, a parastatal organisation plans to build a plant in Moshi in Northern Tanzania with an annual capacity of 10,000 cubic metres of alcohol and 1,000 tons of baker's yeast. The plant should become operational in 1984.

5. CONCLUSION

Despite obvious advantages, the country's biomass potential, except for traditional uses, is virtually untapped. Two methods to obtain energy from biomass have been discussed, biogas digesters and ethanol production from biomass. In light of their economic advantages, development projects in these two fields seem to be especially promising. Whereas the Government seems to be proceeding with an alcohol plant project, major efforts will be needed to introduce biogas digesters on a large scale in the rural areas.

REFERENCES

1. Tanzania National Scientific Research Council:
   Workshop on Solar Energy for the Villages of Tanzania,
   Dar es Salaam 1977.

2. Tanzania National Scientific Research Council:
   Solar Energy for Villages Pilot Project, Background
   Paper, Dar es Salaam 1978.

3. Small Industries Development Organisation: Biogas Development Programme Report, Dar es Salaam 1978

4. Tanzania Industrial Studies and Consulting Organisation:
   Manufacture of Polyethylene and Baker's Yeast from Molasses, feasibility study for the National Development Corporation, Dar es Salaam 1980

FRACTIONATION OF BIOMASS (SUGAR CANE) FOR ANIMAL FEED AND FUEL

T R Preston, M Sánchez and R Elliott

Facultad de Medicina Veterinaria y Zootecnia, Universidad de Yucatán, Apartado 116-D, Mérida, Yucatán, México

Two recent papers[1,2] have called for a new look at photosynthesis by plants as a means of providing renewable quantities of food, fibre, fuel and chemicals. The potential for fixing the required amounts of both energy and food certainly exists. The problem is to develop the appropriate technologies in order to use photosynthesis efficiently.

Lipinsky[2] has argued in favour of the development of what he describes as "adaptive" systems which will provide a means of integrating the production of fuel from biomass with the production of food and other raw materials. He cited as examples the use of the sugar cane plant to produce ethanol, sugar and electricity; the maize plant to provide human and animal food and ethanol; and guayuli as a potential source of natural rubber and of fuel. This approach was compared with the alternatives of energy farms and the use of agricultural residues.

In a study of potential renewable energy sources for the European Communities[3], it was proposed that "Some quite urgent attention is recommended for the selection of perennial crops that would be capable of fractionation into an animal feed and a fuel fraction". This approach is in line with Lipinsky's philosophy. It also promises to provide a potential solution to the very practical problem of how to develop intensive aniaml production systems in the tropics which can compete in biological and economic efficiency with those used in temperate countries.

Tropical regions considerably exceed the temperate climate areas in their capacity to support growth of biomass, mainly because of the high average temperature, combined with adequate solar radiation. Unfortunately, because of this very favourable condition for plant growth, the most productive species are those that grow fast and relatively tall. As a result, tropical plants are more lignified than typical plants of temperate regions and therefore so much inferior in terms of their nutritive value for animal production.

Sugar cane is an excellent example of this phenomenon. It is one of the most efficient plants known for fixing solar energy into carbohydrate. This carbohydrate is divided almost equally into two different forms: the soluble sugars which are rapidly and completely digested by mammalian and microbial enzymes; and the fibrous cell wall material which has a very slow rate of degradability and is only attacked by microbial enzymes.

It has been shown in feeding trials with cattle that when whole sugar cane is fed, it is the soluble sugar fraction which contributes the greater part of the metabolizable energy that the animal obtains from this feed[4]. The fibrous component is digested to only a minimum extent (about 20%), but of greater concern is the fact that it has a detrimental effect on the overall nutritive status of the animal, because its slow rate of degradability results in a very low rate of turn over of digesta whithin the digestive tract. Thus feed intake is low and so is the efficiency of rumen microbial synthesis[4]. To use sugar cane efficiently for animal feeding, it would therefore be highly desirable to extract and use only the soluble sugars and to find an alternative use for the fibre.

It is proposed that the conversion of this fibre into charcoal and/or producer gas (a mixture of carbon monoxide, hydrogen, nitrogen and carbon dioxide which results from the gasification of carbonaceous materials) is likely to be the most economical, the most efficient and the most appropriate in terms of satisfying an important part of the energy needs for rural areas in the tropics.

In contrast with procedures being developed for extraction of juice from temperate crops, it is comparatively easy, at low energy cost (about 3 KWh/t of cane) to extract the juice from sugar cane using a simple roller mill, as has been practised in small scale sugar manufacture for centuries.

To keep the technology simple (in line with that used by small scale sugar producers), the cane stalk is passed through the mill only once. On average this extracts some 62% of the total juice (about 31% of the total biomass). The unextracted sugars would be converted with the fibre into fuel.

A typical flow sheet for the process is set out in Figure 1. Although the model refers to a fairly large unit of some 40 ha, in fact, the technology can be set at any desired level of scale, since the two principal items of machinery — the roller mill and the gasifier — can be built and operated at levels ranging from small and unsophisticated (animal or even human power can be used for the mill) to completely integrated units which operate automatically.

The appeal of this technology for food and fuel production will depend upon the technical and economic feasibility of the two component sub-systems. The technology for gasification of organic residues was first developed by a British engineer, Mr J E Dowson, at the end of the 19th century and was closely related with the development of the internal combustion engine. Although it was soon superceded by the use of more convenient oil based liquid fuels,

the technology found ready application again in many European Countries,
during the Second World War. In Sweden, it is estimated that some 75,000
producer gas units were in commercial use, particularly in trucks and buses[5].

Recent developments with gasifiers in Sweden have favoured the use of
wood chips and as the partially pressed sugar cane stalk, when chopped into
small pieces, closely resembles wood chips, it is considered that it will
be a comparatively simple matter to adapt the gasification technology for use
with sugar cane.

The relatively unresearched area concerns the use of the sugar cane juice
as an animal feed. This is presently being investigated in laboratories of the
University of Yucatan, México and the State Sugar Council in the Dominican
Republic, where experiments are being carried out with both ruminants and
monogastric animals.

The first trial has just been completed and the results have been particularly encouraging. The fresh sugar cane juice was used as the basal diet for
growing cattle. Molasses was selected as a control treatment, since this has
been the basis of the most successful tropical feeding system developed so far,
from locally available resources[6].

For both basal diets, urea was added to provide fermentable nitrogen for
the rumen micro-organisms, restricted amounts of a tropical grass (African Star
Grass) were fed as a source of roughage and an evaluation was made of providing
a supplementary protein source. The trial lasted 75 days and involved
16 animals in a 2 X 2 factorial arrangement of treatments (cane juice vs molasses;
and with and without sunflower meal) with two replications. The principal findings
are shown in Table 1 and Figure 2.

The highlights of the results are:

(1) The very fast rate of growth (1.35 kg/d) on the supplemented cane
juice. This is comparable with the level of performance expected on cereal
grain based rations in Europe and North America.

(2) The very acceptable growth rate of 800 g/d which was obtained on cane
juice in the absence of protein supplementation (other than that in the forage).

(3) The superiority of both cane juice treatments over either of the
molasses diets.

Further trials are in progress, but it already appears that the animal
feeding system resulting from the fractionation of sugar cane, is not only
technically feasible, but promises to be comparable in every way with the best
of temperate climate production systems.

The use of high protein forages such as the legume <u>Leucaena leucocephala</u>[9] and the foliage of cassava (<u>Manihot esculente</u>)[10] has permitted the elimination of the protein supplement in molasses-based diets with no reduction in animal performance. Similar results can be expected with these forages in cane juice based diets and the model in Figure 1 is based on this assumption.

On the energy side, it has already been shown that the pressed cane stalk after partial juice extraction, can be converted easily into charcoal, using open ended 55 gallon oil drums[7]. The first tests on gasifying the pressed sugar cane stalk have also given encouraging results and a pilot unit, of Swedish design should be operating very soon.

The socio-economic perspectives for the application of this technology in developing countries in the tropics can be estimated from the model shown in Figure 1. The yearly production from the 40 ha unit is estimated to reach 100,000 litres of milk, 11,000 kg of carcass beef and 610,000 KWhr of electrical energy. Assuming that 10% of the energy output would be required to provide power to the production unit, and that an average family of 5 would consume daily some 2.7 litres of milk, 300 g of meat and 16 KWhr ofelectricity, then one square kilometre (100 ha) of cultivable land utilizing this technology would provide all the animal protein and energy needs for a community of some 1,300 people.

References:

[1] Hall D O 1979 Nature 278:114

[2] Lipinsky E S 1978 Science 199:644

[3] Anon 1979 Overview of Opportunities for Energy from Biomass in the European Communities. General Technology Systems Ltd, Brentford:UK

[4] Preston T R & Leng R A 1980 Proc. Vth Symposium Ruminant Physiology, In press

[5] Anon 1979 Generator Gas: The Swedish Experience from 1939-1945, Solar Energy Research INstitute, Colorado:USA

[6] Preston T R 1972 Molasses as a feed for cattle. <u>In</u> Wld Rev Nutrit Dietet. (Ed G H Bourne) Karger:Basle

[7] Ffoulkes D, Elliott R & Preston T R Trop Anim Prod 5:in press

[8] Sanchez M & Preston T R 1980 Trop. Anim. Prod. 5:in press

[9] Hulman B, Owen E & Preston T R 1978 Trop Anim Prod 3:1

[10] Ffoulkes D & Preston T R 1978 Trop Anim Prod 3:186

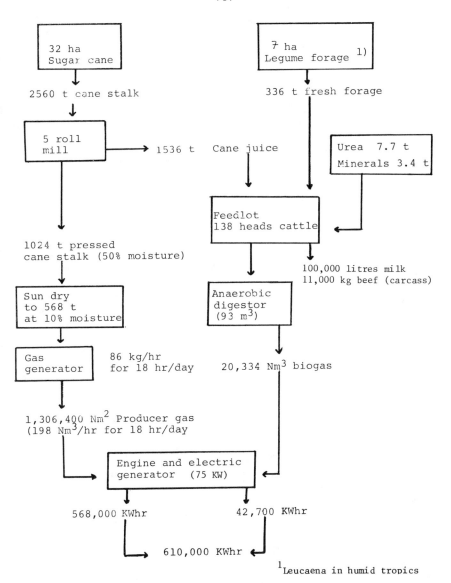

Fig. 1 : Conversion of sugar cane biomass into animal feed and fuel
(data are on annual basis unless indicated otherwise)

Table 1 : Mean values for liveweight gain, feed intake and conversiom of crossbred bulls fed basal diets of cane juice for molasses (from Sanchez & Preston 1980)

	Molasses		Cane Juice		
	No Supplement	1 kg/d Sunflower meal	No Supplement	1 kg/d Sunflower meal	$SE_{\bar{x}}$ (Prob)
Liveweight, kg					
Initial	279	?66	261	279	
Final	300	304	309	361	
Daily gain	0.252	0.595	0.795	1.315	$\pm$.17(P<.001)
Feed intake, kg/d					
Sugar cane juice	-	-	22.7	31.9	
Molasses	3.95	4.00	-	-	
Forage	8.3	8.4	8.2	9.1	
Supplement	-	1.00	-	1.00	
Total dry matter	5.44	6.42	5.85	8.43	$\pm$.22(P<.001)
Feed conversion[1]	21.54	11.78	7.42	6.44	$\pm$.57(P<.001)

[1] Dry matter intake/live weight gain

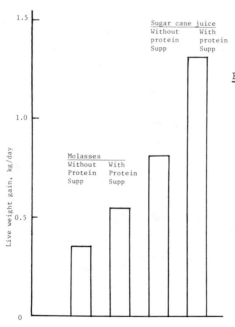

Fig. 2 : Growth rate of cattle fed sugar cane molasses or sugar cane juice with or without a supplement of 1.0 kg/day of sunflower seed meal(from Sanches & Preston 1980).

B I O G A S - K E N Y A N C A S E

P.N. KARIUKI DIC, PhD
MINISTRY OF ENERGY
P.O. BOX 30582
NAIROBI

Summary:

Due to increasing prices of oil and the subsequent dependence on indigenous Renewable resources, the increasing population in Kenya firewood and charcoal is being depleted at an alarming rate. The need for looking for alternative sources of energy has enhanced chances for successful Biogas development unlike in the last 30 years. Efforts being made by the Kenyan community have been sketched out to give a rough impression.

INTRODUCTION:

Kenya has, like many countries, faced energy problems in both the commercial and domestic sectors. Oil importation accounts for about 30% of the total foreign exchange expenditure. 80% of all the oil imported is consumed by the commercial sector leaving the rest to go to the public utility and very little to the domestic sector.

90% of Kenya's 15 million people leave in the rural areas and depend on the land for biomas energy mainly firewood and charcoal. The available land is not sufficient to continue supporting the increasing population with biomass energy.

However, while afforestation, improved techniques for utilization of firewood charcoal & briquettes (from sawdust, coffee waste etc.) are being developed, other renewable sources of Energy are being studied. Among these is BIOGAS.

Biogas appears an attractive alternative particularly for the rural sector which is agriculturally based. It was first used by Mr. Hutchinson about 30 years ago who then started manufacturing digesters around 1958. Apparently not much interest developed amongst farmers partly because of abundance of alternative sources of energy. Now the crunch has come and alternatives for firewood and charcoal, fossil fuel based sources must be found.

Rural Energy Estimates

Table 1 was developed by Muchiri using limited information available from statistical abstracts published by the Central Bureau of Statistics.

Table 1: Estimated current energy consumption and estimated demand in the year 2000 assuming no drastic measures are taken

Table 1: Estimated current energy consumption and estimated demand in the year 2000 assuming no drastic measures are taken

SOURCE OF ENERGY	ENERGY USED IN KCAL x 10^{10}								% Total
	Agriculture		Domestic		Rural Industries and transport		Total		
	Current	2000	Current	2000	Current	2000	Current	2000	
Human	58	118	38	76	10	20	106	212	3.2
Animal	13.6				3.4		17		.5
Kerosine			59.6	119			59.6	119	1.8
Firewood			3100				3100	6200	92.9
Fertilizer	54.19	108.38					54.19	108.38	1.6
Total	125.79	226.38	3198	195	13.4	20	3336.8	6639.38	100
% Total	3.8		95.8		.4		100		

Biogas and Slurry System

Since 1978 the Biogas story has received a lot of attention from farmers research institutions (mainly agricultural) and government institutions hence the creation of the Ministry of Energy in Nov. 1979.

Technical Consideration

Most of the basic science in methane production is assumed, but the diagrams below will demonstrate the cycles considered.

TECHNICAL CONSIDERATIONS:

1.   Natural Cycle.

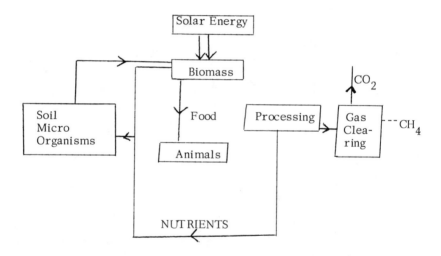

ALTERNATIVE USES OF BIOMAS

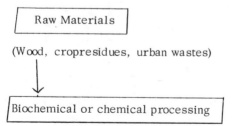

FOOD/FEED	FUELS	CHEMICAL	FIBER	OTHER
Nutrients	Biogas	Polymers	Paper	Liginin
Composts	Pyrolysisgas	Ethanol	Board	Products
Fodder	Ethanol	Fermentation		
Protein	(via sugar)	Chemicals		

Raw Materials

There are in Kenya

Cattle	16.7 million
Sheep & goats	15 million
Poultry	6 million
Pigs	69,400
Camels	607,000
Donkeys	135,000
Rabbits	33,700

(1978 Annual Report for Animal production Branch Ministry of Agriculture).

Daily Manure Production		N. P.	
Dairy Cattle	34.95 kg	0.38	0.10
Swine	25.77 kg	0.83	0.47
Goats	18.18 kg	1.00	0.30
Poultry(100)	28.40 kg	1.20	1.20
Horses	25.46 kg	0.86	0.130

Total Manure Production on any day by

Cattle	34.95 x 16.78	=	586,461,000 kg
Swine	25.77 x 69,400	=	1,788,438 kg
Sheep & Goats	18.18 x 15.0 m	=	272,700,000 kg
Poultry	28.40 x 60,000	=	1,704,000 kg
			862,653,438 kg

The point here is that there is enough raw materials from the alone despite the fact that there are unquantifiable tonnage of coffee waste, pineaple waste and general farm vegetable residues.

Design

The Hutchinson design is basically the Indian design without the middle wall. Another design has been developed based on the Indian model and derives from the popularly constructed stone or steel iron corrugated sheets. It is demonstrated in diagram below:

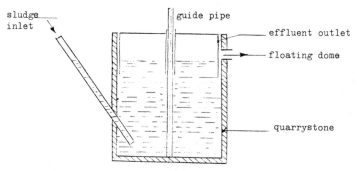

There are 3 sizes being recommended the 6' x 6', the 9' x 9' and 12' x 12' depending on the requirements specified. A do- it-yourself manual has been prepared.

The Slurry

The Slurry fertilizer is being tested for use by farmers. Table 3 shows nitrogen enhancement through digestion.

Table 3: Comparison of losing N between digester manure and farmyard manure

Treatment	Total N		Ammoniacol N	
	Unit wt	%	Unit wt	%
Before treatment	1	100	1	100
Digester manure	0. 99	98. 9	2. 61	260. 7
Open air pool	1	68. 0	1	82. 5
Compost	0. 89	60. 2	0. 22	17. 8

Factorial design experiments are in progress now to demonstrate increased yield of maize while using slurry.

Conclusion:

If all the 2. 5 households were to have 9' x 9' digester each, approximate volume 20 cubic metres, $1\frac{1}{2}$ days would be required to charge
from the animals available.   That would have a lasting effect on the
environment because of the organic-slurry fertilizer.

COMMUNITY BIOGAS PLANTS - IMPLEMENTATION IN RURAL INDIA

R. ROY

Alternative Technology Group, Faculty of Technology,
The Open University, Milton Keynes, U.K.

Summary

Standard designs of biogas plant are only accessible to a
minority in rural India because of their initial cost and input requirements. Two approaches to this problem adopted in India are (a) to
reduce the cost of the plant by developing fixed gas-holder 'Chinese-type'
designs (b) to instal community plants. While Indian fixed gas-holder
designs are still too costly for widespread adoption, a community biogas
scheme installed by the Uttar Pradesh State Government has since late
1978 been supplying inhabitants of a small village in Western U.P. with
cooking fuel, electric lighting and mechanical power for flour milling,
etc. Although the villagers have been involved in the planning and
operation of the system, social, cultural and financial problems have
arisen. In particular these relate to the natural reluctance of the
rural poor to pay cash for fuels formerly obtained free. The results of
this pilot project can help in planning future community energy schemes
both technically and socially. For example, experience here indicates
the need to incorporate income-generating services into future community
energy schemes in order to subsidise household fuel supplies.

1. **INTRODUCTION**

In India some 80,000 *family* biogas plants have been constructed since the 1950s. These however have mainly benefitted a wealthy minority owning the minimum of 4 or 5 cattle required to supply the smallest plant (1). Because a family gas plant is an impossible goal for the rural majority in India, the concept of a *community* biogas plant has evolved. This paper discusses a pilot community biogas scheme which the author visited in late 1979 while working with the Appropriate Technology Development Association, Lucknow (2).

2. **A PILOT COMMUNITY BIOGAS SCHEME IN UTTAR PRADESH**

In 1978 the Planning Research and Action Division (PRAD) of the Government of Uttar Pradesh's Planning Department began work on the installation of a community biogas plant in a small north Indian village called Fatehsingh-ka-Purwa. The plant included two floating gas-holder type digesters of $35m^3$/day and $45m^3$/day capacity, a distribution pipeline connecting the 27 village households to the gas supply, and a machine room with a biogas/diesel generator of 3.5 kVA capacity to supply household electric lighting and a 7 h.p. biogas/diesel engine available for water pumping, chaff-cutting, or flour milling. The capital cost of the scheme of Rupees 160,000 (about £9500) was met by a UNICEF grant.

The system was installed in consultation with the villagers and is managed by government officials who are advised by an 'executive committee' of six villagers, itself nominated by a 'village committee' comprising one representative from each household.

The plant is operated and maintained by two members of the village employed and trained by PRAD. When sufficient is available, gas for cooking is supplied between 7-9 a.m and 7-9 p.m and electricity for lighting from 7-10 p.m. Fifteen minutes before the gas is turned on at the main digester control valves, a bell is rung.

The villagers are traditionally cattle and goat keepers and the ratio of people to cattle (at 1.4:1) is relatively favourable. Dung for the plant is supplied by the owners of animals, who are allowed each month to extract digested slurry for use as a fertilizer in proportion to the amount of dung they have contributed.

3. ASSESSMENT

The installation is well-engineered and, from a technical viewpoint, the plant has been operating quite successfully. The real test of success with a community plant, however, is how well it is functioning within its particular social context. It is apparent from various sources (2,3,4) that many human and social problems have arisen which have so far prevented the scheme from meeting all its original objectives.

(a) *Finance*. Although the capital cost of the plant had been met by a grant, it was the intention of PRAD that its running costs, estimated at Rs 9000 (£500) per year, would be met by the villagers. While most villagers were willing to pay for services such as flour milling and threshing, they have been very resistant to the idea of contributing even nominal sums towards the cost of cooking gas and lighting, claiming they were promised that these services would be provided free. Eventually, under pressure from PRAD, the village committee resolved in October 1979 that each household would pay Rs 5/month (£0.30) towards the cost of cooking gas and Rs 5/month for lighting. Nine months later it was reported that about two-thirds of the families were paying, but only for cooking gas.

(b) *Factionalism*. The villagers are divided into two main factions - a majority group in favour of the plant, and a minority against it. These factions are apparently the result of pre-existing political and family divisions in the village. The 'anti-plant' group, representing about one quarter of the village, views the scheme as symbolising the interests of the other faction and expresses this by withholding their cooperation and complaining that they are being forced to pay for fuels which formerly they could obtain free.

(c) *Dung contributions*. The scheme is highly dependent on the villagers supplying sufficient dung to operate the plant. However in the first eight months of operation, from November 1978, only about 200 kg/day of dung was supplied by all households. This is just 20% of the 1000 kg/day that should be available from 125 cattle (assuming 7.7kg/day/head cattle can be collected). Over the next eight months total dung contributions rose to 350-700 kg/day with an average of 560 kg/day. This implies that, while the 'anti-plant' households are refusing to cooperate, the 'pro-plant' families must be contributing most of the dung from their animals while retaining small amounts for traditional purposes, such as the preparation of milk over a slow-burning dung fire.

(d) *Gas production and use.* While the plant was designed to generate $80m^3$ gas per day in summer from 1500 kg/day dung (5), in practice gas output was much lower, partly due to low dung contributions, but also due to overoptimistic assumptions about gas yields and dung availability. Indeed during September 1979, when dung contributions were at their peak, the mean daily gas output was only $17.6m^3$. It is not surprising therefore that there has not been enough gas to supply all the demands for which the plant had been designed. A survey of the villagers has indicated that only 5 households out of 26 cook exclusively with biogas while the remainder still use firewood and agricultural wastes. In particular women with large families and women of the poorer households, who may have to work in the fields during the morning, find the gas supply hours inconvenient (3). Likewise, only 11 householders find their lighting needs fully met from the electricity supply. Of the remainder, 8 households use kerosene lamps as well as electricity, while 7 householders now rely entirely on kerosene lamps for light because they failed to replace their light bulbs after they had fused.

4. CONCLUSIONS

There seems little doubt that a well-installed community biogas plant operated by trained people can be made to function in an Indian village. Whether such plants can succeed economically and socially is still very uncertain and there are still many lessons to be learned.

(a) The rural poor, most of whom are living from hand to mouth, are understandably extremely loath to pay cash for fuels which traditionally they obtained free, simply for the sake of greater convenience. Such considerations are only likely to become important when a family has enough income to meet its basic needs (and when women have a greater say in decision-making). Payment for domestic fuel is most likely to be acceptable where these fuels are already commercialised, but in a community biogas scheme it seems desirable that cooking fuel at least is supplied free. Grants seem essential to meet the capital cost of community plants, although running costs might be met from income-generating services such as irrigation pumping and flour milling or from small-scale industries that require mechanical power.

(b) It is necessary to accept that a variety of fuels will be used for cooking, partly for reasons of tradition and taste, but also because it is unlikely that enough gas would be generated to meet all demands, especially in villages with a less favourable person:cattle ratio than at Fatehsingh-ka-Purwa.

(c) Electric lighting is probably not a strongly felt need, at least among the poorer and less well-educated sections of the rural population. Lighting might better be viewed as a service that could be introduced at a later stage in the development of a community biogas scheme, if the villagers wanted it and gas production was sufficient.

(d) Factionalism in Indian society is a major obstacle to the cooperative spirit required for a communal scheme to succeed. One way to combat this tendency is for communities that are already cooperating in some field to decide for themselves that they wish to develop a communal energy system rather than having a system 'offered' from outside.

(e) 'Chinese-type' fixed gas-holder plants, such as those being developed by PRAD and others (6), seem worth considering for future community biogas schemes, since they cost less and should provide a more constant gas output over the year, particularly if the waste heat from biogas engines is used to warm the water used to mix the slurry. To further boost gas yields it should be possible to train plant operatives to experiment with adding agricultural wastes to the materials traditionally used to generate biogas in India.

5. REFERENCES

1. Moulik, T.K. and Srivastava, U.K. (1975) *Biogas plants at the village level: problems and prospects in Gujarat,* Centre for Management in Agriculture Monograph No. 59, Indian Institute of Management, Ahmedabad.

2. For a fuller assessment see Roy, R. (1980) *Family and community biogas plants in rural India and China,* ATG 9, Alternative Technology Group, The Open University, Milton Keynes.

3. Bahadur, S. and Agarwal, S.C. (1980) *Community biogas plant at Fatehsingh-ka-Purwa - An evaluation,* Planning Research & Action Division, State Planning Institute, Lucknow, Uttar Pradesh.

4. Vidyarthi, V. (1979) Appropriate technology for meeting rural energy

needs: an appraisal of biogas plants and their scope, *Decision,* Indian
Institute of Management, Calcutta, Oct., pp 443-449.

5.   Ghate, P.B. (1978) *Biogas: a pilot plant to investigate a decentralized*
energy system, Planning Research and Action Division, State Planning
Institute, Lucknow, Uttar Pradesh.

6.   Garg, H.P. et al. (1980) Designing a suitable biogas plant for India,
Appropriate Technology, Vol. 7 No. 1 pp 29-31.

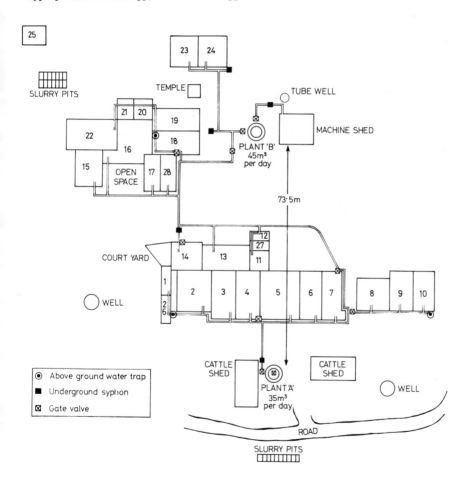

Figure: Layout of the community biogas scheme at Fatehsingh-ka-Purwa

A 4 TO 6-MW WOOD GASIFICATION UNIT: SCALING-UP THE
GASIFICATION TECHNOLOGY INTO AN UNKNOWN SIZE CATE-
GORY (CASE STUDY GUYANA)

G.D. LINDAUER and T. KRISPIN

AGRAR- UND HYDROTECHNIK GMBH, ESSEN, GERMANY

Summary

A power plant of 4-6MW based on wood gasification technology is presently under construction in the framework of
a sawmilling project in Guyana. The plant is the first of its
size and will start operation in 1981 at a capacity level of
about 4MW. Technical performance will have to be monitored
closely.

Where steam is not needed gasification is the more economic
solution. As long as the raw material (in this case wood
wastes) is available "free of cost", this technology is
even a financially viable proposition.

1. INTRODUCTION

The biggest power plant, known to be run on wood gasification technology is currently being constructed as part of a forestry and sawmilling project in the Republic of Guyana. The project comprises the following major components: (a) logging and transport of about 94,000 cbm of roundwood per year, (b) a sawmill including repair shops and transport fleet, with an annual output of 42,000 cbm of sawn lumber and poles, (c) the power plant, (d) a township of, in the first phase, about 1,500 inhabitants (sawmill employees and their families), and (e) port facilities and a 25 km access road.

The nearest settlements are 25-50 miles away from the mill and the power plant with its mill and township is a typical example of an isolated insular grid. The considerable volume of wood waste which will result from the sawmill operations suggested the use of these wastes as fuel for the generation of electric power and of other energy required for the processing of timber products and the power supply of the township. The new technology makes the sawmill, its transport fleet equipped with mobile wood gas units, and the township fully self-sufficient in energy. It saves foreign currency, reduces environmental problems and lays the base for a future development nucleus with ascertained and independent, locally generated power supply.

2. DESIGN CAPACITY OF THE POWER PLANT

Average electric power requirements of 3,2oo kW during the two 8-hour shifts, 1,4oo kW during the non-working days, and some 6oo kW during the night hours have been estimated for the sawmill complex and the township. Considering some additional 3o per cent for peak load periods and provisions for maintenance and stand-by sets, the power plant is planned to have an installed capacity of 4 ooo to 6 ooo kW from up to 8 generators of 8oo kW each.

3. PLANT LAYOUT AND TECHNICAL SPECIFICATIONS

All wastes originating within the factory precincts are transported to the site of the power plant by a conveyor system. They are chipped down and their moisture content is reduced from 50 per cent to a maximum of 25 per cent in driers of 11t/h capacity supplied with heat from the exhaust of the combustion engines. Additional heating can, if necessary, be supplied from wood gas burners. The dry chips are then stored (capacity under roof: 250 cbm). They enter the gasifiers via hoppers; the solid fuel is placed in the gasification tanks according to demand and distilled into charcoal and finally gasified to lean gas.

The equipment is capable of producing wood gas free of dust, tar components, phenols, or other heavy hydrocarbons. After being processed, the gas is cooled to less than 20°C above the ambient temperature, and purified from mineral dust, so as to be adequate as fuel for combustion engines.

The equipment further allows the use of the exhaust gas for chip drying purposes, and the transfer of heat from the engine cooling system to the site of the lumber drying kiln.

Since definite orders have not been placed yet with the suppliers, only some tentative data can be given for two alternatives in the lower capacity range around 4 MW with which the power station is expected to start operation in 1981:

	Alternative	
	I	II
Type: Air-gasifying type, for wood chips and sawdust (max.moisture content 25%)		
Quantity of gasifying units:	5	4
Performance:		
a) Weight of wood residues to be gasified yearly:	43 800 t	35 000 t

Alternative

	I	II

b) The max. hourly performance depends on the requirements of the gas motors and generator group
   with an installed electric
   capacity of: 4 54o KW   3 86o KW

c) Gas supply to the drum-
   drier, when exhaust gases from the gas-motor
   not available, i.e. 274
   days x 8 h and 91 days
   x 24 h per year

Quality of wood-gas

a) lower heating value: $\quad$ 4 6oo KJ/Nm^3 (4 62o to
   5 46o KJ/Nm^3)
   (ca. 1 1oo $Kcal/Nm^3$ to
   1 3oo $Kcal/Nm^3$ = 4 366 to
   5 16o BTU/Nm^3
   = 124 to 146 BTU/cu .ft.)

b) Hydrogen content less than   24% (15-20%)

c) CO   2o to 23%

d) CH_4 1,5 to 2%

e) CO_2 10 to 14%

f) N_2 48 to 52%

g) Temperature of gas behind the
   cooler less than: $2o^\circ C$ above the air-tem-
   perature

h) Absolutely free from tar and
   other hydrocarbons, which
   have to be cracked during
   the gasifying process.

i) Absolutely free from ash,
   dust and grid particles.

Control of gasifiers:

Flexibility of control of gas-volume: 1:1o
i.e. for operations between 386 KW
and 3 86o or between 45o KW and
4 54o KW, depending on alternative, according to the weekly,
daily and nightly power consumption patterns of the consumers.

Control Station for Wood Gasification Plant

Switchboard-type. Control devices for: fuel supply, gas
pressure, automatic ash removal, condensed water removal,

infeed conveyor, etc.. All control units are equipped with optical and accustical signals.

Alternative

	I	II

Wood-Gas Motors and Electric Generators,
Gas-Otto-Motors:

a) Type: Gas-Otto-Motor for clean gas combustion (4 6oo KJ/Ncbm), with turbo-loader and loading-air cooler and air/water-radiator cooling system for the engine.

	I	II
b) Capacity of all motors(elt.)	3 76o KW	2 82o KW
c) Quantity of motors:	4	3
d) Revolutions per min.	1 2oo KW	1 2oo KW
e) Electr.Power 4(or3) x 94o KW (elt.)	3 76o KW	2 82o KW

Gas-Diesel-Motors:

a) Type: Gas-Diesel-Motor for dual-fuel combustion, max. 1o % diesel oil and min. 9o % clean gas, with turbo-loader and loading-air cooler and air/water-radiator cooling system for the engine.

	I	II
b) Capacity of all motors, dual fuel(elt.):	78o KW	1 o4o KW
c) Quantity of motors:	1	1
d) Revolutions per min.	1 2oo	1 2oo
e) Electr. Power Gas-Diesel:	78o KW	1 o4o KW
f) Without gas, 1oo % Diesel	1 2oo KW	1 5oo KW

Electric generators

	I	II
a) Capacity of all generators:	4 96o KW	4 32o KW
b) Voltage (Volts)	46o	46o
c) Frequence(cycles)	6o	6o
d) Revolutions per min.	1 2oo	1 2oo
e) Quantity of generators:	5	4
f) Cooling medium:	air	air
g) Tropical protection:	VDE o53o,	IP 23

Control Station for Gas-Motors and Generators

Switchboard-type. Control devices for motors are for:
revolutions, lubrication oil pressure and temperature, water temperature and filling level of radiator, gas
pressure, fan control. Control devices for electric generators comprise: frequency, voltage, ampere meters, un-
derload, malfunction.

4. COMPARATIVE ANALYSIS

Motors or engines running with lean gas produced from
wood waste will convert up to 30 % of the heat supply from the
gas into mechanical or electrical energy. About 15 % are lost
through general radiation, the remaining 55 % escape with the
cooling air/cooling water and with the exhaust gases.

In stationary plants, particularly in plants designed for a
higher energy gain, such as the power station discussed, the
waste heat can be used effectively for both drying and heating
processes. The exhaust gases are clean and do not contain
any noxious materials polluting the environment. They may
even be used directly for drying and heating by simply adding
air to them. Thus oven-fried wood chips can be produced
"gratis" utilizing the waste heat of the cooling equipment and
of the exhaust of the power station.

The capital cost of such a plant are about 15 % to 20 % lower
than those of a steam power generation plant.

One kilogramme of wood with a 20 % to 25 % moisture content
yields about 2.3 cbm of generator gas, meaning that, e.g. one
litre of petrol can be replaced by 3 to 3.5 kg of wood. With
oven-dried wood chips the latter ratio is reduced to 2.5 to
3 kg of wood for the equivalent of one litre of diesel oil, or
0.2199 British Imperial Gallons.

5. FURTHER POTENTIAL APPLICATIONS AND LIMITS

While the technology of smaller units is already fairly
well known, bigger plants, as the one described above, will
still need observation and improvements.

Problem areas where improvements may be expected soon are:

- fuel supply for mobile units. Bigger units may only prove commercially viable in the form of stationary units as the one in Guyana. Smaller units will certainly be viable as mobile entities, too, e.g. cars, tractors, trucks etc.

- continuity of gas production. At present loading of wood and particularly the unloading of char are still carried out more or less batchwise.

- further improvements in the energy conversion ratio should be possible although the present ratio is already better than that for simply burning wood.

The financial price of the wood fuel is, as long as it is waste, very low indeed; its economic price is practically zero (assuming that the avoidance of environmental harm equals economic transportation costs of the wastes to the burner). This shows already the present financial crux of the matter: at the present level of petrol prices this technology is only superior as long as the fuel wood has no or a very low financial cost price; this may, however, change quickly with rising petrol prices in future. In economic terms, however, this technology is already now competitive in almost any soft-currency country, even under wood or coppice "farming" conditions, i.e. deliberate wood production exclusively for fuel purposes, if the primary material originates in the country and petrol has to be imported. The high shadow price for foreign currency necessary for petrol imports on the one hand, and the high shadow value of additional jobs created in the country's forest economy on the other, make this technology a worthwhile alternative today under economic and social benefit aspects, even if the solid fuel (wood) has to be specially purchased.

This is even more the case where electricity and heat produced can be utilized. The second phase of the Guyana project therefore envisages the attraction of wood processing industries, especially for prefabricated timber houses, creating at a time further value added and jobs in the area and utilizing some of the excess wood wastes and electricity still available.

A DESIGN STRATEGY FOR THE IMPLEMENTATION OF
STOVE PROGRAMMES IN DEVELOPING COUNTRIES

S. JOSEPH and Y.J. SHANAHAN

Intermediate Technology Development Group Ltd.
9 King Street,
London WC2E 8HN.

Summary

Fuelwood is the primary energy source for the great majority of households in developing countries, but de-afforestation is having grave ecological and social consequences. As one approach to reducing fuelwood consumption, I.T.D.G., through its two year collaborative programme with indigenous organisations, has developed a systems design strategy for the widespread adoption of improved cooking stoves. The reasons why previous stove programmes have failed are analysed to show that the design strategy should consist of:

1. Survey of traditional cooking practices to ascertain socio-cultural criteria that stoves must meet.

2. Field testing of existing stoves.

3. Assessment of alternative designs.

4. Laboratory testing of alternative designs and modification or design work on existing stoves.

5. Limited and extended field testing.

6. National or Regional Extension programmes and formulating national stove policies.

This paper presents the results of the first four steps of the strategy, viz. the survey, design, and laboratory and field testing. It is shown that the design and testing steps have considerably increased the knowledge base for engineers to design efficient stoves for varied socio-cultural settings. Initial experience of dissemination is discussed and further research areas delineated.

1. <u>INTRODUCTION</u>

Fuelwood is the primary source of energy for over 90% of households in developing countries. In many areas, population pressures have led to demand consistently exceeding supply, resulting in higher costs whether measured in money, in labour costs or both. This increased demand is having grave ecological consequences; namely reduced soil fertility (even desertification), and severe flooding.

It is likely that wood and other biomass fuels will continue to be the main fuel for most people in developing countries for some time. The problems of excessive demand needs to be attacked on two fronts: on the one hand, available stocks should be used more efficiently, and on the other hand, reafforestation and agroforestry programmes should be initiated. Widespread use of more efficient wood stoves could markedly reduce the demand on biomass fuels. At least ten stove programmes in Third World Countries have been initiated between 1955-75. None of these programmes have led to the widespread use of better stoves.

2. <u>REASONS FOR FAILURE OF STOVE PROGRAMMES</u>

Political, cultural, historic, economic and technical factors all contributed to the failure of stove programmes. The most important factors are summarised as follows:

1. <u>Political factors</u>: lack of commitment by central and local governments, eg. in Indonesia the kerosene was subsidised so that it was within the means of most people; opposition by vested interests who control the production or distribution of fuelwood.

2. <u>Socio-cultural factors</u>: incompatibility of stoves with requirements governing the preparation and distribution of food, eg. if most of the cooking is done outside, then a fixed indoor stove will be unsuitable; failure to involve the existing social structure in the introduction of the innovations, eg. women were generally excluded, until recently, from the design and training stages of stove programmes.

3. Historical factors: lack of knowledge of the historical forces governing the successful introduction of innovations into a society, eg. in most programmes, stove introduction has been based on principles of cooperative developments with stoves being built by the user after receiving training by extension workers. However, in many societies most innovations have difussed through craftspeople/entrepreneurs, informal contacts and trade stores.

4. Economic factors: the cost of stoves which involve metal, asbestos or clay for chimney pipes or dampers have often been prohibitive to the poor; the import of additives for stove construction which are not locally available requires provision of a suitable import infrastructure.

5. Technical factors: stove programmes have failed due to poor design, lack of laboratory and field testing and inadequate extension programmes. The main design failure has been that under field conditions, the stoves actually used more wood than traditional fireplaces. This is due to the improper use or loss of dampers; leaving cooking ports uncovered when not being used for cooking; incorrectly placed chimneys; insufficient maintenance of stove body. A secondary design failure has been the unacceptability of smokeless stoves where the user required a certain amount of smoke for the reduction of insect pests or keeping a thatched roof dry. Inadequate extension programmes for stove builders and users has led to poorly built stoves, short lifetime and incorrect use. Most programmes have focussed on one design, often taken from another country. Because of this a minimum of original design work and modifications were made to suit the local conditions. Alternative stove designs have rarely been presented for user evaluation. This resulted from lack of detailed information on the principles of stove design and the relative performance of alternative stoves. Recent work carried out in laboratories in both First and Third World countries, has shown that simple modifications to traditional designs can have a greater impact than modified First World designs.

3. DESIGN STRATEGY

A design strategy has been developed from the analysis of

the problems of stove programmes. This strategy should facilitate the widespread introduction of improved traditional cooking stoves. This strategy consists of 6 stages:

Stage 1: survey of cooking practices and fuel usage;
information on local cooking practices must be obtained if new
stoves are to be designed that will perform all the functions
of the old stoves as well as using less firewood. The survey
should broadly establish: a) the local cooking practice (type
and quantity of food used, time to cook and method of preparation, where the cooking is done); b) the cultural rules
associated with cooking, (e.g. meal times and any taboos and
sex roles associated with cooking, significance of fireplace);
b) the type and size of pots, stoves and fuel used; d) who
constructs the stove, how it is carried out and time devoted
to repair and maintenance; e) the method of obtaining firewood, (i.e. is the wood gathered in the field or is it obtained
from wood dealers? What is the price of the wood and the time
required to collect it?)

Stage 2: field testing of stoves already in use in different localities; field tests of the stove currently in use
should be conducted to determine the type and amount of wood
used in cooking and time taken to cook food or boil water.

Stage 3: assessment of alternative designs to see if they
meet requirements found in the initial survey; the data obtained from Stages 1 and 2 should be used by the designer to decide
whether an alternative stove design will have a better performance than the indigenous stoves but will be culturally acceptable. Initial assessment of alternative designs is only possible if information is available on: a) the dimensions of the
stove; b) the method of construction and type amd amount of
material needed; c) the labour required for construction;
d) estimation of lifetime; e) detailed instructions on how to
use the stove efficiently; f) sort of fuels that the stove
can burn and methods of cooking for which it is suitable; g)
the amount of fuel used in either cooking standardised representative meals or some standardised water boiling experiments;
h) time required to bring different quantities of water to
the boil.

Stage 4: laboratory testing and design work: If necessary detailed information on alternative designs is not available then laboratory testing will have to be undertaken. If the results of these tests indicate that the stoves are not suitable, modification to the indigenous stoves must be undertaken or new stoves designed.

Stage 5: limited and extended field testing: having ascertained the possible suitability of a stove design from the results of the laboratory tests, performance data under field conditions must now be obtained. We have found that the best way to do this is to provide a limited number of different stoves to at least 50 families who have had previous involvement with the extension agency. Performance of these stoves should be measured at the start of the trials, and at periods of three and six months. The results of the experiment should provide information on: a) user reaction to the new stove; b) how crucially affected the stove performance is by the accuracy of construction, efficiency of operation and stove deterioration; c) possible design modifications that could improve the acceptability or performance of the stove. Once designs have proven successful testing can be extended to other villages.

Stage 6: National or Regional Extension Programmes: Once a range of designs have been proven, extension on a large scale can be undertaken. For successful dissemination, widespread publicity, encouragement and assistance from the government is necessary. Stoves may be fabricated in small industries and sold through the market place or built by extension workers and owners. All require finance, organisation and government involvement at a policy level.

5. IMPLEMENTATION OF DESIGN STRATEGY

ITDG has established collaborative programmes with organisations in India and Sri Lanka (further programmes are to be established in Africa and Asia) in order to determine if the design strategy will lead to widespread dissemination of fuel conserving stoves.

ITDG's role is to provide information on test procedures, principles or heat and mass transfer that govern the design of

efficient stoves, a procedure for optimising the design and
testing of stoves and information on the test results of alternative stove designs. The collaborators use this information
within the strategy framework outlined.

During the last year field surveys and test work have been
carried out in Sri Lanka and Indonesia. In both countries,
tests on the different types of local stoves have indicated
that open fires and small ceramic stoves use very little wood
to bring the water to the boil but use much more wood for longer cooking periods. This indicated that there was a need for
small monolithic mud stoves and improved ceramic stoves.

While this survey and test work was being undertaken, ITDG
carried out a series of laboratory tests in England on small
monolithic mud stoves to determine factors that influence their
performance. The results were fed back to the collaborators
for use in their design work. The combined information from
the laboratory and the field was used to design a range of mud
stoves. These stoves were evaluated in field test centres
before field testing began in the villages.

In Sri Lanka, a number of problems have been encountered
with the new stove designs. Further remedial design and test
work will be carried out in the test centres before proceeding
with the field testing.

In contrast, the initial field trials in Indonesia were
so positive that a more extensive programme was undertaken.
Over 500 mud stoves were built in a six month period. However,
the evaluation of the field trials indicated that further
design and test work was necessary in order to optimise the
stove designs even further.

Design and laboratory test work was also carried out on
ceramic stoves by ITDG. These results were used by the collaborators to design, build and test two ceramic stoves in Indonesia and one in Sri Lanka. The stoves were made by local
potters and to date only the Indonesian stoves have been tested in the field. The initial results indicate that the stoves
have a better performance than the existing ceramic stoves in
Indonesia.

The combined results of field and laboratory test work on

mud stoves showed that the main cause of poor stove performance was inaccurate construction. Thus there is a clear need to provide training in the proper construction of mud stoves for the stove builder. In addition, the extension worker needs field aids to check and test the mud stoves. Poor operation and insufficient stove maintenance was also shown to affect the stove's performance, therefore the user must be trained to operate and maintain the stove correctly. Thus recourses must be available for the implementation of successful extension programmes.

6. CONCLUSIONS

It would appear that this stove strategy has allowed ITDG's collaborators to overcome problems found in other stove programmes. Obviously, final assessment is not possible until these and other stove programmes have been underway for 3-5 years. Further research in areas of: a) survey techniques; b) heat mass transfer in stoves, and c) extension techniques for stoves is needed to further ensure the success of national stove programmes.

THE UNITED NATIONS SYSTEM AND BIOGAS ACTIVITIES

by

E.J. DaSILVA, J.F. McDIVITT and V.A. KOUZMINOV

Unesco, Place de Fontenoy
Paris 7°, France.

Abstract

The report outlines recent activities in the United Nations System concerning bioconversion applications, particularly for the benefit of developing countries. Reference is also made to some activities carried out by organizations outside the U.N. system.

The development of low-cost, low-waste technologies that
resort to recycling and complementary processes (in which the
waste of one level is used as raw material in the next) is
recommended in RIO (1). Identified as the cornerstone of many
such industries is the bioconversion of solar energy. Given the
potential of deoxyribonucleic acid (DNA), enzyme engineering,
and the economic competitiveness of the process, the production
of bio-energy-methane (CH_4) methanol (CH_3OH) and ethanol (C_2H_5OH)
- from forest wastes residues from sugar-cane and cassava 'energy
plantations', and urban waste, is identified by RIO in the search
for environmentally prudent technologies. Their application also
helps counteract the energy crisis.

Wastes from non-energy sources, such as food and paper production and crops grown explicitly for their energy value, are
known to be valuable sources of organic fuels. Defined as "ressources out of place", organic biodegradable wastes contain
energy that is recoverable either by physical, chemical or microbiological means. Energy is physically recovered through incineration of sewage sludges, municipal refuse and solid wastes of
animals. Chemical processes involve the use of pyrolysis and
gasification. The most common method - a microbiological non-
waste producing technology - is biogas production, the problems
and prospects of which have recently been reviewed (2, 3) and the
technology of which ranges from simple, small-farmer lagoons
through the medium technology of bag-digesters to the high technology of high-rate biogas digesters (4).

The value of biogas technology is established in its multi-
purpose benefits (Table I). It produces non-polluting energy at
much greater efficiencies than do dung or wood fires; it retains
nitrogen, phosphorus and potassium in its residual sludge; it
kills most pathogenic bacteria and protozoa of intestinal origin,
and is environmentally sound.

Biogas technology involves the anaerobic bacterial methanization of organic agricultural and industrial wastes. The
anaerobic process is highly dependent on ambient conditions : a
suitable decomposition period optimum temperature of 30-35°C

with a pH ranging between 6.8 and 7.4. The technology is very simple under tropical conditions : organic wastes are decomposed in a gas-tight container and the resulting gas is collected in gas tanks. The principles and in-depth consideration of various parameters encountered with biogas production have been dealt with in detail (2, 3) as well as its potential as a fuel for the future (5). Furthermore, the use of biogas technology has been favoured as an effective weapon in counteracting desertification (6).

Several agencies in the U.N. System are involved in activities concerning biomass conversion and biogas production. A brief description is attempted below :

1. THE ECONOMIC AND SOCIAL COUNCIL (ECOSOC)

(a) In 1976, the Council adopted a resolution 2031 (LXI) entitled "Research and Development in non-conventional sources of energy" in which it requested the Secretary-General of the United Nations, with the assistance of the UN Advisory Committee on the Application of Science and Technology to Development (ACAST) to prepare surveys of on-going Research and Development activities in the field of non-conventional sources of energy (see Table II).

(b) In 1977, the Council adopted a resolution on "new and renewable sources of energy" (2119 (LXIII)) calling for the preparation of a report on the feasibility of holding an international conference on new and renewable sources of energy. The General Assembly, at its last session, has approved the convening of this Conference which will be held in Nairobi in 1981.

2. THE DEPARTMENT OF INTERNATIONAL ECONOMIC AND SOCIAL AFFAIRS (DIESA)

In the Department of International Economic and Social Affairs, surveys have been prepared on Research and Development activities in the field of non-conventional sources of energy, with a view to identifying gaps in Research and Development activities in the countries.

3. THE DEPARTMENT OF TECHNICAL COOPERATION FOR DEVELOPMENT (DTCD), AND THE UN CENTRE FOR NATURAL RESOURCES, ENERGY AND TRANSPORT (CNRET)

In the Department of Technical Co-operation for Development

(DTCD), the Centre for Natural Resources, Energy and Transport
(CNRET) is responsible for operational projects of technical co-
operation in the field of natural resources.

The Centre is also responsible for monitoring Rural Energy
Centres in Africa and Asia, to serve as demonstration units with the
object of encouraging developing countries to use alternative sources of energy. The main concept of the Rural Energy Centres (REC)
is the "energy mix", whereby solar energy, wind and bio-gas obtained
from waste, will complement each other, through storage units to
supply energy to a village of about 200 families (1,000 persons) for
cooking, lighting, pumping water and running agro-based village
industries. Sites have been selected in Senegal, at Ndia Gorey, in
Sri Lanka, at Pattiyapola, and in Mexico. Feasibility studies have
been prepared by the Brace Research Institute, McGill University
(Canada) for the Senegalese village, and by Oklahoma State University for Sri Lanka. Demonstration projects on non-conventional sources of energy had been recommended by ACAST and initiated by the
United Nations Environment Programme (UNEP).

4. THE REGIONAL ECONOMIC COMMISSIONS

4.1 Europe

The Economic Commission for Europe (ECE) has two current programmes in the field of non-conventional sources of energy. The
first project, entitled, "Environmental aspects of energy production
and use, with particular reference to new technology" was started in
1975 and is still in progress with the cooperation of the United
Nations Environment Programme (UNEP). The second project, entitled,
"Review of new energy technologies" would identify areas for possible international cooperation research.

4.2 Western Asia

The Economic Commission for Western Asia (ECWA) has a programme
of non-conventional sources of energy which consists of a feasibility study and an evaluation of utilizing these sources for application in the countries of the ECWA region.

4.3 Africa

The Economic Commission for Africa (ECA) has been interested in
the development of energy for a number of years. For example,
technical assistance was provided for the development and utiliza-

tion of solar and biogas energy to the United Republics of Tanzania
and of Cameroon, Liberia, Sudan, Malawi and Cape Verde.

4.4   Asia and the Pacific

The Economic Commission for Asia and the Pacific (ESCAP) has
been particularly active in organizing seminars and workshops on
Bio-gas technology. For instance, ESCAP convened a Workshop on
Bio-gas and other rural energy resources (Suva, June-July 1977) and
conducted a roving seminar on rural energy development (Bangkok,
July-Aug.1977; Manila, Aug.-Sept.1977; Jakarta, Bandung, Oct.1977).
A field mission has been undertaken by ESCAP in 14 developing coun-
tries of the region.

4.5   Latin America

The Economic Commission for Latin America (ECLA) has promoted
studies on the prospects for using non-conventional sources of
energy in Latin America.

5.   THE FINANCIAL INSTITUTIONS

5.1   The United Nations Development Programme (UNDP)

The United Nations Development Programme (UNDP) funds feasibility studies and technical cooperation projects at the request of
Governments.

5.2   The World Bank (IBRD)

The International Bank for Reconstruction and Development
(IBRD, World Bank) does not generally fund technological research.
However, the Bank follows current research and development activities related to non-conventional sources of energy, with a view to
assess their potential use in projects financed by the Bank, particularly for rural development.

6.   THE UN ENVIRONMENT PROGRAMME (UNEP)

The United Nations Environment Programme (UNEP) is preparing a
proposal for a study on the impact on the environment of the various
sources of energy.

A study on the production and use of alternative energy sources
is to be reviewed.

7.   THE UNITED NATIONS INDUSTRIAL DEVELOPMENT ORGANIZATION (UNIDO)

UNIDO provides technical assistance to developing countries,
on their request.

In 1974, UNIDO and UNDP initiated a five-year project on biogas plants, for the production of methane gas from animal manure.
In cooperation with the Economic and Social Commission for Asia and
the Pacific (ESCAP), UNIDO has prepared several workshops on biogas.

8. THE UNITED NATIONS INSTITUTE FOR TRAINING AND RESEARCH (UNITAR)
AND UNITED NATIONS UNIVERSITY (UNU)

8.1 UNITAR

The UN Institute for Training and Research (UNITAR) has a programme of studies concerning the future. UNITAR in cooperation with
the Federal Republic of Germany and the UN Office for Science and
Technology organized a Seminar on Microbial Energy Conversion. (10)

8.2 UNU

The UN University (UNU), founded in 1973, has its headquarters
in Tokyo and a network of research and advanced training operations
in some sixty countries. It has undertaken pilot projects on solar,
bio-gas and wind energy in rural communities in developing countries.

9. THE SPECIALIZED AGENCIES

9.1 The United Nations Educational, Scientific and Cultural Organization (UNESCO)

The UN Educational, Scientific and Cultural Organization co-
sponsored with ISES and COMPLES, the International Conference on
"Sun in the Service of Mankind".

UNESCO is now pursuing a programme of work in the field of non-
conventional sources of energy, which comprises :

(a) Promotion of research in the scientific and technological
fields.

(b) Assistance in the creation or development of regional or
sub-regional centres for research and training in the application of
non-conventional energy sources.

Unesco has sponsored and organized several research and training
meetings relating to biogas production and biomass conversion. A
few of these are :

Bangladesh	:	Biogas Utilization and Technology
Brazil	:	Regional Seminar on Biomass Conversion and Combustion

Guatemala	:	Fuels and Chemicals from Biomass
India	:	Symposium on Bioconversion and Biochemical Engineering
Kuwait	:	Food-producing and Bioconversion Systems for Arid Lands
Nigeria	:	Biofuels and Biofertilizer Symposium
United Kingdom	:	Renewable Energy Systems for Development
Uruguay	:	Biomass Workshop
Yugoslavia	:	Bioproductivity and Photosynthesis

In addition, planned or released publications are :

The Energy Potential

Harnessing Ocean Energy

Energy Perspectives

Renewable Energy Sources for Developing Countries

The Potential for the Application of Microbiology within the Third World Rural Communities.

9.2 The Food and Agriculture Organization (FAO)

The Food and Agriculture Organization (FAO) activities in the field of non-conventional sources of energy have been limited to monitoring developments in the area and providing technical advice concerning applications and use in agriculture and rural areas on request from Member States. FAO is involved in the practical application of non-conventional energy resources at farm and village levels.

9.3 The World Health Organization (WHO)

The World Health Organization (WHO), Division of Environmental Health, has published a report on "Health hazards from new environmental pollutants".

10. THE GENERAL ASSEMBLY AND THE UNITED NATIONS CONFERENCE ON NEW AND RENEWABLE SOURCES OF ENERGY

The General Assembly at its last session in December 1978 approved the convening of a United Nations Conference on new and renewable sources of energy to be held in 1981. The scope of this Conference will be to promote the development and utilization of new and renewable sources of energy such as solar, geothermal and wind power, tidal power, wave power and thermal gradient of the

sea, biomass conversion, fuel-wood, charcoal, peat, energy from
draught animal, oil shale, tar sands and hydro-power.

Developed and developing countries (Table III a) and several
international organizations (Table III b) have shown interest in
biogas systems with respect to various objectives: a renewable
source of energy, biofertilizer, waste recycling, rural development,
public health and hygiene, pollution control, environmental management, appropriate technology and technical co-operation. Over
the past five years several biogas plants have been set up in
China and India, and several more are to be found elsewhere
(Table III a).

In Nigeria, experimental evidence (7) has shown that the weed
Eupatorium odoratum makes a useful substrate for bio-gas production.
Likewise in Singapore, successful experiments (8) with small scale
bioconversion units for methane production from pig and palm oil
wastes and water hyacinth have been carried out.

The National Institute of Science and Technology (Philippines),
the Small Industries Development Organization (Tanzania) the Khadi
Village Industries Commission (India), and the Choqui Cooperative
(Guatemala) are but a few of the many national agencies that are
involved in promoting research and development and demonstration
projects aimed at rural electrification, rural development and rural
energy centers.

Preliminary work on a small scale has begun at the College of
Agriculture of the University of the Philippines. Maya Farms, 40
miles south of Manila, is the largest biogas establishment in Asia.
The complex consists of 48 batch plants based on 15,000 pigs. The
gas is utilized in a canteen, a meat processing plant and a soup
cannery. It runs the diesel pumps and generators which provide
water and power for the farm. Worker dormitories receive electricity from gas-powered generators, and lighting, refrigeration,
cooking and laundry facilities are provided in an on-site demonstration system : the foreman's living quarters. Moreover, analysis
of the sludge material has shown high concentrations of vitamin B_{12},
justifying the experimental incorporation of sludge material in pig
feed.

In Africa, there is increasing interest in the utilization of biogas plants. In Kenya, biogas plants have been used since 1954. Furthermore, coffee has been grown entirely on the fertilizer residues of the methane plant. Active experimentation in harnessing the multiple utility benefits of biogas technology are currently being carried out in the United Republic of Cameroon, Ethiopia, Rwanda, Senegal, United Republic of Tanzania, Upper Volta, Zaire and Zambia. In Botswana, biogas has been deployed to substitute 80 per cent of diesel fuel in diesel engines.

In Cape Verde, the primary use of energy in the rural sector (about 80 per cent of the population) is for the basic necessities of cooking, drying of agricultural produce, and provision of fresh water. An assessment of the utilization of alternative energy sources has focused on the contributions of wind, sun and biogas.

Microbiology can sustain rural development even in its early stages, provided that it contributes to meeting the basic needs through manpower utilization. In other words, its involvement in agricultural production, and in processing agricultural produce, should envisage a multitude of small production units. In its extreme consequence, microbiology should be able to prescribe guidelines for the smallest production unit, in an appropriate socioecological background - which in traditional societies is the rural family household.

References

1. Tinbergen, J. Reshaping the International Order: A Report to the Club of Rome. New York, N.Y., E.P. Dutton, 1976.

2. National Academy of Sciences, Methane Generation from Human Aminal and Agricultural Wastes, Washington, D.C., National Academy of Sciences, 1977.

3. DaSilva, E.J. "Biogas Generation: Developments, Problems and Tasks (An Overview)". Proceedings of the joint WH-NR United Nations University Conference: Bioconversion of Organic Residues for Rural Communities, Guatemala, November 1978.

4. DaSilva, E.J. (1980). Microorganisms as tools for biomass conversion and energy generation, Impact of Science on Society, Vol. 29; 361-374.

805

5. DaSilva, E.J. (1980). Biogas: Fuel for the Future? Ambio, Vol. 9, pp. 2-11.

6. Schechter, J. (1977). Desertification processes and the search for solutions, Interdisciplinary Sci. Revs., Vol. 2: pp. 38-53.

7. Odeyemi, O. (1980). Biogas from Eupatorium odoratum: an alternative cheap energy source for Nigeria, GIAM VI Conference Abstracts, Lagos, Nigeria.

8. Chin, K.K. (1980). Potential Energy Production from Biomass and Waste in Rural Areas, Journal Eng. Educ. in Southeast Asia, Vol. 10, pp. 32-36.

TABLE I. The Biogas Ledger

Present problems	Benefits of Biogas
Depletion of forests for firewood and causation of ecological imbalance and climatic changes.	Positive impact on deforestation; relieves a portion of the labour force from having to collect wood and transport coal; helps conserve local energy resources.
Burning of dung cakes: source of environmental pollution; decreases inorganic nutrients; night soil transportation a hazard to health.	Inexpensive solution to problem of rural fuel shortage; improvements in the living and health standards of rural and village communities; provides employment opportunities in spin-off small-scale industries.
Untreated manure, organic wastes, and residues lost as valuable fertilizer.	Residual sludge is applied as top-dressing; good soil conditioner; inorganic residue useful for land reclamation.
Untreated refuse and organic wastes a direct threat to health.	Effective destruction of intestinal pathogens and parasites; end-products non-polluting, cheap; odours non-offensive.
Initial high cost resulting from installation, maintenance, storage, and distribution costs of end-products.	System pays for itself.
Social constraints and psychological prejudice to use of human waste materials.	Income-generator and apt example of self-reliance and self-sufficiency.

TABLE II.   Research and Development Activities in
Non-conventional Sources of Energy

	1	2	3	4	5	6	7	8	9
Argentine	X	X	X			X	X		
Australia	X	X	X		X	X	X	X	
Austria	X		X		X	X	X	X	
Barbados			X	X			X		
Belgium	X	X		X		X		X	
Brazil	X	X	X	X	X	X	X	X	
Bulgaria	X					X	X		X
Canada	X	X	X	X		X	X	X	
Chad	X	X		X	X		X	X	
Chile	X			X	X			X	
Costa Rica	X	X	X				X		
Cuba	X	X	X				X		X
Denmark	X					X	X	X	
Ecuador	X	X							
Egypt	X	X	X	X	X	X	X	X	
Finland						X			
France	X	X	X	X	X	X	X	X	X
Germany, Federal Republic of	X	X	X	X	X	X	X	X	
Greece	X	X			X	X			
India	X	X	X	X	X	X	X	X	
Iran	X	X	X	X	X	X	X	X	
Iraq	X						X	X	X
Ireland	X	X	X		X	X	X	X	
Israel	X	X	X	X	X	X	X	X	
Italy	X	X		X	X	X		X	
Japan	X	X	X	X	X	X	X	X	X
Jordan	X	X	X						
Kuwait	X				X	X			
Malaysia	X	X	X	X	X	X		X	
Mali	X	X	X	X	X				
Malta	X						X	X	
Mauritius	X	X							
Netherlands	X			X		X	X	X	
New Zealand	X								
Niger			X	X	X				
Nigeria			X						
Norway	X	X							X
Pakistan	X	X	X		X	X		X	
Papua New Guinea			X	X					
Philippines	X	X	X	X			X	X	
Portugal	X	X							
Saudi Arabia	X		X	X	X	X		X	
Senegal	X		X		X		X	X	
Singapore	X	X			X			X	
Spain	X		X		X	X	X		
Sri Lanka	X		X	X			X		
Sweden	X				X	X	X	X	
Turkey	X		X	X		X	X		
United Kingdom	X			X	X	X	X	X	X
United States of America	X	X	X	X	X	X	X	X	X
USSR	X	X	X	X	X	X	X	X	

Legend :   1 = solar heating
          2 = solar cooling of buildings
          3 = crop drying
          4 = water pumping
          5 = solar electricity (thermal)
          6 = solar electricity (photovoltaic)
          7 = wind energy
          8 = biological energy
          9 = energy from the sea

Source:
ECOSOC Documents
E/C. 8/56 and
E/C.8/56/Corr.1,
United Nations,
January 1978.

TABLE III a.

The Status of Biogas in Developing Countries

Countries	Remarks
Afghanistan	Demonstration project in biogas technology
Algeria	R & D projects in experimental stage
Bangladesh	Development of rural community-size biogas plants
Barbados	Bioenergy and fertilizer production
Botswana	Production of biofuel for diesel engines
Brazil	Development of integrated biogas systems for rural areas
Cameroon	Rural development
China, People's Republic	Fertilizer and energy production; family-size plants
Congo	R & D projects underway
Cook Islands	Integrated biogas farming systems
Costa Rica	R & D projects underway
Ecuador	Biogas utilized in cooperative farms
Egypt	Biogas R & D programme
El Salvador	Biogas generation from coffee wastes
Ethiopia	Rural development scheme
Fiji	Integrated farming systems with algal and fish ponds for protein and fertilizer production
Guatemala	Biogas from piggery wastes and straw
Guyana	Bioenergy production
Honduras	Biogas as alternative to wood fuel
India	Biogas in rural areas
Indonesia	Regional training network in biogas plant operation.
Iran	Promotion of public health programmes via biogas plants.
Jamaica	Bioenergy production
Korea, Republic of	Feed and fertilizer production
Lesotho	Rural development
Malaysia	Biogas technology for rural electrification
Mauritius	Cane sugar residue used for biogas
Mexico	Rural energy integrated system
Nepal	Bioenergy and fertilizer production
Nicaragua	Biogas used as fuel
Nigeria	Utilization of weeds for biogas production
Pakistan	Fertilizer production
Papua New Guinea	Fertilizer production
Philippines	Algal oxidation ponds and fertilizer production
Rwanda	Biogas for domestic use
Senegal	Rural electrification and domestic use
Singapore	Experimental and training activities
Sri Lanka	Rural development and energy centres
Thailand	Rural cooking and electrification
Trinidad	Energy and fertilizer from biogas
Tanzania	Fuel and fertilizer production
Upper Volta	Village development scheme
Zambia	Biogas on family farms
Zaire	Laboratory experiments in progress

TABLE IIIb - International Agencies working with bio-energy and area
of involvement *

UN Agencies	Areas and Remarks	Co-operating Organization
United Nations (UN)	Prospects and Trends: UN Conference on New and Renewable Sources of Energy	
	Technical Meetings of Panel on Biomass (New York 1979, Geneva 1980)	
	Survey Report: Research in Conventional and Non-Conventional Sources of Energy	UNESCO
Food and Agricultural Organization (FAO)	Exploration and Production: Promotion of Technology for biogas production from wastes and residues and use of effluents as fertilizers	UNEP, UNIDO, ESCAP, UNDP, SIDA
United Nations Development Programme (UNDP)	Technical co-operation and demonstration of biogas and allied technologies in the People's Democratic Republic of Yemen, Lesotho, Philippines, Tanzania	UNESCO, UNIDO
United Nations Environment Programme (UNEP)	Utilization of Biogas Technology for environmental management; Rural energy centres in Mexico, Senegal and Sri Lanka; Integrated systems	
United Nations Environment Programme (UNEP) (cont'd)	Information dissemination through INFOTERRA	
	Support to Training Centre in China	
United Nations Educational, Scientific and Cultural Organization (UNESCO)	Regional Workshops and demonstration projects in biomass conversion into fuel in Brazil and Uruguay	
	Regional Training Courses within the framework of MIRCEN networks	UNEP, IFIAS, ICRO, ESCAP, Govt. of Japan
	Expert meeting on use of Biogas	BCSIR
	Special Symposium "Biofuels and Biofertilizers", Lagos, Nigeria, 1980	UNEP, CFTC, OAU
	Survey on Education and Training in New and Renewable Sources of Energy	
	Study on International Information System relating to New and Renewable Energy Sources	UNISIST
United Nations Children's Fund (UNICEF)	Provision of basic services to children via village and rural biogas systems	

*) based in part on data available in ECOSOC document E/AC.51/99/Add. 1

TABLE IIIb (continued)

UN Agencies	Areas and Remarks	Co-operating Organizations
United Nations Industrial Development Organization (UNIDO)	Provision of information and assistance in mobilization of existing technology for integrated development in Tanzania, Upper Volta, Afghanistan	UNIDF
United Nations Institute for Training and Research (UNITAR)	Provision of specialized training	
United Nations University (UNU)	Programme on bioconversion of organic residues for rural development through biomass and biogas production (Tanzania, China)	UNESCO, FAO
World Health Organization (WHO)	Specialized technical bulletin in sanitation and composting technologies	
Commonwealth Science Council (CSC)	Rural Development and alternative energy programme	
Commission of the European Communities (EEC)	High-level research programmes in microbial fermentation to Ethanol and Biogas	
International Development Bank (IDB)	Intermediate Technology programme on fuel production from agricultural and animal wastes for Central America	
International Development Research Centre (IDRC)	Support of Research Projects; Social and Economic Evaluation of Biogas Technology	
The Arab Educational, Cultural and Scientific Organization (ALECSO)	Survey report on the present state of activities and potential uses of renewable energy sources in all Arab States	

SESSION VII : IMPLEMENTATION - DEVELOPED COUNTRIES

Chairman: Dr. G.H. KING, Energy Technology Support
Unit, Harwell, United Kingdom

Summary of the discussions

Invited papers

Alternative fuels from biomass and their use in transport

The French bioenergy programme

Planning for transport fuels from biomass - The New Zealand
experience

Poster papers

Possibilities and limits of using biomass as substitutes for
exhaustible resources - A systems-analysis approach of producing
fuels from rape-seed in the Federal Republic of Germany

Biomass and hydrogen: an answer to the European liquids fuels
crisis in the 21st century

Encouraging biomass use in California

Optimization of an integrated renewable energy system in a dairy
farm

Heat supply system for the Community of Sent

Realistic assessment of biomass energy contributions

Potential for energy cropping in Swedish agriculture

Food and fibre or energy-implications of biomass energy for
Canada

The use of biogas for thermal, electrical & mechanical power
generation

Generation of current from biogas

Studies of the potential for ethanol production from selected
biomass crops in temperate climates: an engineering approach

Use of 95%-ethanol in mixtures with gasoline

The utilization of sunflower seed oil as a renewable fuel for Diesel engines

The economics of improving octane values of gasoline with alcohol additives

Design and advance of the bioenergy in french sugar industry

813

SUMMARY OF THE DISCUSSIONS

Session VII   IMPLEMENTATION - DEVELOPED COUNTRIES

Rapporteur :   B. BERGER, U.S.A.

Speakers :   W. BERNHARDT, Germany
                  H. DURAND, France
                  G.S. HARRIS, New Zealand
Chairman :   G. KING, U.K.

Poster Session : 16 papers presented

The papers presented in the poster session all represent important considerations in increasing the use of biomass for energy. Not only economic
and technical aspects were presented, but also arguments for a secondary
concern viz. energy efficient use. As DOHNE pointed out, cogeneration of
electricity and steam for heating is desirable though often not economical. There is also the question of the overall energy balance of the systems, mentioned by SCHMIDT, a question frequently raised in discussions
on grain-based ethanol. BERGMAN pointed out that solid fuels from biomass
are the most economical : it may be advantageous to substitute biomass
directly in order to displace oil use. Production of high value "petrochemicals" from biomass could save oil and avoid the losses in conversion
associated with the petroleum feedstock. Such uses of biomass appear both
thermodynamically and economically attractive.

The session's papers highlighted the continued need for technological
advances. BERGMAN mentioned a key parameter - the increase in production
per unit area of land. With increased yields from improvements in plant
genetics and cultural practices, more biomass is available for energy
without disrupting traditional farm and forest practices. With further
efforts on oil seed crops, diesel oil may be replaced by, or extended
with, the vegetable oils - a point made by BRUWER.

The future contribution of energy from biomass depends on R&D. In the
near term, as noted by WHITE and PLASKETT, the most available and economical resources are wastes from forests, farms, industries and municipa-
lities. BUCHLI and STUDACH cite the attractions of using waste wood for

community heating needs (1-5Mw) in Switzerland. PELLIZZI et al have illustrated the superior economics of integrated (direct combustion, digestion and solar heating) systems over traditional ones. In all applications possible environmental consequences must be considered.

In implementing commercial biomass ventures, once chemical and economical hurdles have been overcome, institutional factors become the major obstacles. Businessmen and financial institutions are traditionally cautious in investing in new business ventures. To this end, DURAND points out that the French government is providing a subsidy for individuals installing equipment which increase the use of wood. LANG described a revolving loan fund in California, while the US government is now making loan guarantees available for facilities using biomass.

It was clear from the breadth of policy and technical papers presented at this session that nations have a wide diversity of experience in using biomass for energy. This is an advantage if we are to share information on our technical, economic and institutional solutions. As DURAND said "France has biomass ; our real problem is to know how to use it". The international truth of this statement was borne out in this session.

ALTERNATIVE FUELS FROM BIOMASS

AND THEIR USE IN TRANSPORT

DR. W. BERNHARDT

Volkswagenwerk AG Wolfsburg, W. Germany

Energy Research and New Technology Dept.

Summary

The present situation on the crude oil market obliges us to increase our efforts in the search for and development of alternative renewable energy sources.

As far as the automotive industry is concerned the alcohols methanol and ethanol are particularly suited to making a contribution to saving crude oil. In countries rich in coal, gas or wood and provided with highly-developed technology, production of methanol recommends itself. Favourable climatic conditions and an appropriate agricultural structure indicate the production of ethanol from biomass. Even for Europe this possibility should not be ruled out, since here both a high degree of agricultural productivity and a highly developed process technology exist. Taking into account all available raw materials contained in waste and refuse materials about 5 % of current petrol requirements could be substituted by ethanol. The Volkswagenwerk AG research division is already creating the conditions for the use of alternative fuels in our existing vehicles by the development of suitable engines.

The growing crude oil price and the expected future shortage of crude oil has prompted Volkswagenwerk AG, Germany, already in 1973/74 to investigate extensively in alternative automotive fuels for transportation. We focused our activities on alcohols, since these liquid fuels can be produced from natural gas, coal, municipal waste, farm products, wood and other sources. Volkswagen's work in the early 70s has been mainly with the application of methanol/gasoline mixtures and neat methanol to current engines and vehicles. The results of these activities have been so encouraging that research projects on methanol were started in 1975 also in Sweden, Canada, the U.S., New Zealand, South Africa, the Soviet Union and Japan. Though methanol and methanol/gasoline blends may cause material, cold start and driveability problems, the overall results were very promising, especially in the field of emissons, energy consumption and performance.

In 1975/76 suddenly the interest in ethanol from plant material (biomass) increased. Volkswagen was asked to support actively the Brazilian National Alcohol Program which will lead to an alcohol production rate at the end of 1980 of more than 3 million m^3. Volkswagen do Brasil is now playing a dominant role in the field of ethanol fueled cars in Brazil.

Based on the progress and success in Brazil, a number of other countries have quoted for ethanol from biomass as an automotive fuel: The U.S., (at least partially), the Philippines, Paraguay, India and other Latin American and African countries, see Fig. 1, since the biomass potential is very high in these respective regions. The world's biomass production per year by photosynthesis is estimated to be 114 x 10^7 t of Coal Units (CU). The energy content of the biomass used for foodstuffs, heating, etc. is approx. 2 %.

It is Volkswagen's estimate that countries provided with extensive coal or gas resoures and an advanced processing technology will especially develop the methanol fuel technology, whereas countries having excellent conditions for plant growth will adopt or have already adopted ethanol as an alternative fuel.

Alcohol fuels - a short review

Ethanol fuels for automobiles are not new. In 1902 e. g. 3 - 4 HP ethanol engines have been quite common. However, some research work had to be done during the last decade to develop modern alcohol vehicles and proper alcohol fuels.

Fig. 2 shows the major modifications for a straight alcohol fueled car in comparison to a gasoline car. The next figure (Fig. 3) shows the first VW straight ethanol fueled concept car with an air-cooled alcohol engine (1975). Fig. 4 shows a water-cooled alcohol engine, a concept which was developed for regions such as U.S.A. and Europe in 1976. Fig. 5 shows the first test car with that engine.

Based on these research activities VW Brazil developed its
ethanol engines from two of its current gasoline engines:
The water-cooled Passat 1.5-liter straight four-cylinder
engine and the air-cooled "Beetle" 1.3-liter horizontally-
opposed four-cylinder engine.

The next figure (Fig. 6) shows several alcohol cars which
are used in the German alcohol fuel technology program. The
German government sponsored 4 years program titled "Alternative Energies for Road Transport" deals mainly with methanol. However, ethanol will also be investigated.

The next figure (Fig. 7) gives some information on VW vehicles that will be used by "ordinary" customers in the German research and demonstration program in three distinguished
phases.

These three phases are illustrated in Fig. 8:

First phase: Alcohol/gasoline blends
Second phase: Neat alcohol, mainly methanol;
               However, some vehicles will be operated by
               ethanol for direct comparison purposes
Third phase: Application of alcohols to Diesel engines

It is the first time that such a comprehensive program is
conducted by all major oil and all automobile companies in
Germany.

Alcohol fuel potential

Fig. 9 illustrates the most important sources for ethanol
production indicating also the efficiency of conversion.
At the lower part of this figure several application possibilities for ethanol in spark ignition and Diesel engines
are shown.

Table I: Conversion Efficiencies for Alcohol Production
           from Biomass

Biomass	Fuel	Efficiency %
Potatoes	Ethanol	82
Wheat	Ethanol	24
Maize	Ethanol	32
Cassava	Ethanol	18
Sugar Beet	Ethanol	35
Sugar Cane	Ethanol	31
Wood	Ethanol	32
Wood	Methanol	38
Municipal Waste	Methanol	41
Algae	Methanol	28

When alcohol fuels from biomass are discussed, often the
question is raised whether these biofuels have a high potential to be used for automotive purposes in the future
or not. Some researchers are pointing out that the conversion
efficiencies for alcohol production are too low so that other
sources than biomass have to be considered for fuel production.
However, in the following section it will be discussed that
other parameters than the conversion efficiency have to be
considered for the valuation of fuels from biomass.

Table 1 summarizes the conversion efficiencies for some biomass alternates. The starch containing potatoe has the highest efficiency of 82 %, the starch containing cassava root
has the lowest, just 18 %, sugar beet has an efficiency of
35 %. Table 2 gives information on the ethanol yield(litres/ha. year)
for selected countries. The ethanol production rate for the
high-efficiency potatoe is just 3 300 litres/ha. year, for
cassava just 2 800 litres/ha. year, however for sugar beet
the highest yield is obtained, 4 400 litres/ha. year, though
the efficiency is average.

Table 2: Ethanol production from biomass via fermentation

Biomass	Biomass Yield t/ha. year	Ethanol Yield litres/ha. year
Potatoes	27)	3 300
Wheat	4,5) Germany	1 600
Maize	5,6)	2 200
Cassava	15 Brazil	2 800
Sugar Beet	45 Germany	4 400
Sugar Cane	55 Brazil	3 900
Wood*	12 Brazil	3 350

* Wood is converted to sugar via hydrolysis.

Table 3: Methanol production from biomass

Biomass	Biomass Yield t/ha. year	Process	Methanol Yield litre/ha. year
Wood	12 (Brazil)	Pyrolysis - MeOH Synthe-sis	5 570
Algae	80 (Italy)	Digestion - Cracking - MeOH Synthesis	17 470

Table 3 illustrates the situation for methanol production from biomass. The production route wood - pyrolysis - methanol synthesis leads to 5 570 litres/ha. year (equivalent to 4 030 litres EtOH/ha. year). Salt water algae have a biomass yield of 80 t/ha. year; this leads to 17 470 litres MeOH/ha. year equivalent to 12 640 litres EtOH/ha. year when the process route digestion - cracking to synthesis gas - methanol synthesis is used. Hence, proper aquatic plants seem to have a very high potential as renewable energy source ("Aquaculture").

It is a well-known technology to produce ethanol from sugar beets, sugar cane and cassava by the process of fermentation and distillation. However, the usage of wood as raw material for alcohol production is up-to-date limited because the economies are insufficient and alcohol energy yield is often low.

Table 4 illustrates that the conventional hydrolysis - fermentation route leads to an alcohol energy yield of 8.3 MJ/m^2 year; this is less than the ethanol from sugar beet route with 9.4 MJ/m^2 year. The conventional wood pyrolysis process has an energy yield of 8.6 MJ/m^2 year, a recently developed Lurgi pyrolysis process yields 11.7 MJ/m^2 year. This is up to date the best yield of wood to alcohol processing. However, some researchers believe that the energy yield and economics can be improved for the alcohol from wood production.

Fig. 10 illustrates that there are at present new processing routes under investigation to make not only use of the cellulosic portion of wood. One interesting approach is to convert hemicellulose to C_5 sugars and to ferment these sugars to ethanol. Thus, the alcohol energy yield can be improved substantially.

With respect to the automotive application of alcohol fuels the area demand for the operation of one alcohol car during the period of one year is of specific interest. Table 5 illustrates that the annual area demand moves between 0.14 ha/car (for algae) and 0.66 ha/car (for cassava). Here again the high potential of algae culture and silviculture is obvious.

Table 4   Wood as Feed Stuff for Alcohol Production

Assumption: Wood yield 12 t/ha year (Brazil)

Process	Product	Yield 1/ha year	Alcohol Energy MJ/m^2 year
(Conventional) Hydro-lysis - Fermentation - Pyrolysis	EtOH	3 350	8.3
MeOH Synthesis (con-ventional, from Table 3)	MeOH	5 570	8.6
Pyrolysis - MeOH Synthesis (LURGI)	MeOH	7 570	11.7
Delignifaction & combined Sacchari-fication/Fermentation (Penn/G.E.)	EtOH	4 550	9.7
For comparison: Ethanol from Sugar Beets (55 t/ha year), Fermentation & Distil-lation (from Table 2)	EtOH	4 400	9.4

Table 5   Area demand for the operation of one alcohol car
          during one year

Assumption: milage 15 000 kilometres/year, fuel consumption
              10 litres gasoline equivalent/100 kilometres
              energy consumption 20 % less than a gasoline
              car

Biomass	Fuel	Area ha/car
Sugar Beet	Ethanol	0.42
Sugar Cane	Ethanol	0.47
Cassava	Ethanol	0.66
Potatoes	Ethanol	0.56
Wood	Ethanol	0.55
Wood	Methanol	0.34 - 0.46
Algae	Methanol	0.14
Wood (hybrid poplar 30 t/ha year)	Ethanol	0.22

Alcohol from biomass research at Volkswagen

Research work at Volkswagen in the field of alternative fuels from biomass is not only done in application but also in the area of basic research with regard to alcohol and bio-gas production, the improvement of already existing production routes and finding new sources for biofuels. Fig. 11 illustrates 5 different routes from biomass to biofuel which have been selected for further investigation.

The first processing route deals with the conversion of bio-mass into fermentable sugar by using acids (called hydroly-sis). The second route is using pyrolysis processes which lead to methanol via synthesis gas. The third route is not very wellknown. We call this route biomass hydrogenation. This route is a potentially attractive technique due to pre-liminary results from lab tests. During the first steps of this route oily and tarry substances are obtained that can be refined to hydrocarbon fuels.

The fourth route includes the thermochemical and biological gasification. The goal is to improve digestion efficiencies in terms of methane yield. Methane is then cracked to syn-thesis gas which can be converted into methanol.

The last route is one of the most attractive routes because this would lead directly to bio-hydrocarbons via extraction. A couple of useful chemical products including bio-fuels may be obtained. Furthermore, the project includes the se-lection of distinguished plants that grow also under arid or semi-arid land conditions. Up to date these studies in-clude euphorbia, guayule and jojoba.

Besides hydrocarbons from plant as a long term solution, Volkswagen is doing some research work in the field of the application of vegetable oils in diesel engines. Vegetable oils such as peanut oil, sunflower oil, dendê oil, soja bean oil, marmeleiro etc. have cetane numbers around 40 and low ash content. However, since these oils cause gum formation on the surfaces of the combustion chamber, vegetable oils have probably to be treated chemically in order to avoid this effect.

Unfortunately, it is impossible to discuss all of the major aspects with regard to the production and application of fuels from biomass, such as specification and optimization of alcohol fuels, quality control during the production and distribution phases, potential corrosion and ware problems with regard to alcohol engines and vehicles, taxation of alcohol fuels, performance and emission characteristics, environmental and toxicological aspects. However, I would like to point out that all these problems can be solved by joint efforts of the oil industry, automobile industry, government authorities, research institutions and users. Alcohols are excellent automotive fuels, cleanly burning, lead to higher performance (up to 20 % more power) and higher efficiency (up to 30%). And the alcohol fuel technology is already well applied in Brazil in the transport area; other countries will certainly follow.

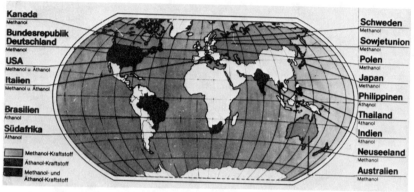

Fig. 1   Countries with interest in alcohol fuels

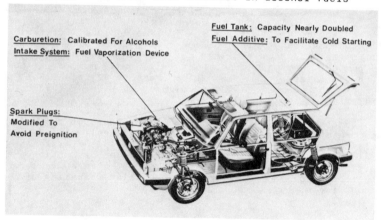

Fig. 2   Modifications for a straight alcohol fueled car

Fig. 3 - First VW straight ethanol fueled concept car developed for VW do Brasil (1975)

Fig. 4 - VW water-cooled alcohol engine (1976)

Fig. 5 - VW alcohol concept car (1976)

Blends:
Higher Alcohol Content
Variation in Base Fuel
Switch System to Gasoline

Neat:
Cold Start Additives
Cold Start Devices
Methanol/Ethanol Fuels
Combustion
Lean burn
Aldehydes

Diesel:
Blends
Cold Start
Combustion

Fig. 8 – The three phases of the German alcohol fuel technology program

Fig. 6 – Alcohol cars used in the present German alcohol fuel technology program (1980)

	Total Number of Vehicles	VW
M 15	1 050	600
Methanol Fuel small cars	130	100
trucks	42	–
Alcohol Diesel Fuel	100	100

Fig. 7 – Volkswagen's participation in the German alcohol fuel technology program

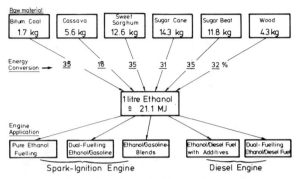

<u>Fig. 9</u> – Sources for ethanol production and routes for ethanol application

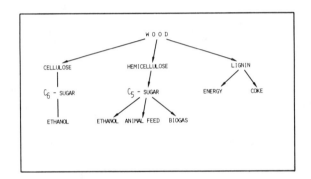

<u>Fig. 10</u> – Wood hydrolysis products

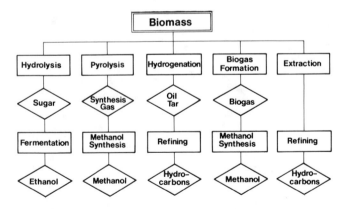

<u>Fig. 11</u> – Routes from biomass to biofuels

THE FRENCH BIOENERGY PROGRAMME

Henry Durand

Président, Commissariat à l'Energie Solaire
(French Solar Energy Authority)

1. RENEWABLE ENERGIES IN THE FRENCH CONTEXT

The French energy situation is presently dominated by the importance of imported oil (domestic production accounts for less than 1% of our oil imports). Thanks to a considerable effort in energy savings, total energy consumption has increased by only 1.5% per year during the 1973 - 1979 period. Furthermore, oil consumption has decreased by 8% during the same period. The present policy of the French Government is to diversify strongly our energy resources, and the French nuclear program is becoming the largest in the world. But gas, coal and renewable energies will play a great role in our future requirements. Table I gives the breakdown on energy consumption in France (in millions of tons of equivalent-oil : "M toe") for past periods and the forecast for the future.

It is interesting to analyse the category "hydro-electricity and renewable energies" more closely. Table II gives the corresponding break-down for hydroelectric, solar, wind and biomass resources up to the year 2000.

2. OUTLINE OF THE FRENCH "BIOMASS AND ENERGY" PROGRAMME

In the case of a country like France, which has agricultural and forest waste resources that to a large extent are currently not being utilized fully, several factors must be brought together in order to pre-pare a programme for biomass. Energy enhancement:

- a close estimate of available resources, broken down by regions and types of products
- accurate forecasts of the use of energy to be produced
- major technological efforts to be made between the biomass production and utilization stages in order to achieve a comfortable balance between raw material supply and energy demand

All three aspects--resources, technology and market development--must
therefore be handled in parallel.

3.   RESOURCES

Unlike other European countries, France should not encounter any
serious problems about the very existence of raw resources, at least for
the next 10 or 15 years. French forests cover 14.5 million hectares, yet
only 60% of their renewable content is thought to be correctly managed.

Similarly, from the 32.5 million hectares of existing agricultural
lands, about 20 million tons of wastes (one-third of which is straw) could
be used for energy purposes.

Animal waste resources (dung and manure) are very abundant in
France and their theoretical potential could amount to 4 million tons of
equivalent oil, a quarter of which could readily be recovered.

The real problem consists of how to collect such unused wastes or
droppings. Major efforts must be made if this biomass is to be recovered
economically, transported at a low cost, and then used conveniently.

Many non-technical obstacles have also emerged. Thus, French
forests, and to a lesser extent French agriculture, are divided up amongst
many landowners. For instance, the 14.5 million hectares of forest belong
to 1.5 million different owners. Another difficulty shows up in terms of
the competition that already exists between the use of wood for industrial
purposes (lumber, paper-making) and for firewood, the latter having caused
all wood prices to have climbed upwards over the past two years.

A different kind of obstacle is involved in implementing a policy
on agricultural alcohol. Sugar prices, and hence sugar-beet prices, are
set by a rather complex Common Market system covering both government
intervention and production quotas. In general, European agriculture is
tightly regulated, so that up to now there has been no room for energy-
oriented crops.

These obstacles lead us to be very cautious and careful to avoid any
excessive speed-up in the biomass energy use programme that might overly

disturb existing agricultural and industrial patterns.

In the first phase of operations, running from 1980 to 1995, the
mobilization of existing resources should indeed make it possible to avert
any major upheaval in current agricultural and forestry structures, with
the exception of an improvement in the waste recovery rate. During this
period, experiments should be conducted with areas planted specifically
for energy production (high rotation forests, energy-oriented crops, even
ocean and lake biomass aquaculture).

In the second phase, which will probably be under way by the year
2000, such specifically planted areas will gradually be fitted into our
system of production. They will thereby bring about a noticeable change
in the distribution and final purposes of species. The transition
between these two phases, however, will probably occur very smoothly, due
to the fact that France has several million hectares of waste land available which could, with proper species selection, be used to provide at
least partial support for such a program.

Specialists vary in their quantitative estimates of what France
might be able to produce in the way of energy biomass during the first
of these phases. Experts do not always agree on what should be meant by
a "programme that disturbs existing structures as little as possible".
Their estimates can vary by a factor of two, depending on how much they
expect authorities to interfere with resource utilization patterns.

As an example, Table III gives two potential estimates of the
residues available in France. The first estimate is based on the work of
a Committee of Experts who helped prepare the French programme[1]. It
covers the current availability of resources that could be readily used
without modifying any of the processes currently employed in agriculture
and forestry.

The second is an evaluation of the total recoverable biomass, taking
into account the physical limits involved in recovery but excluding consideration of specific crops. This Table also shows the official target
for 1990 and the forecasts for the year 2000.

4.   THE SCIENTIFIC AND TECHNOLOGICAL PROGRAMME

Using raw biomass economically to produce energy entails a whole
series of developments, especially technological ones, but also some which
rely on basic research in fields like photochemistry, microbiology, genetics
and agronomy. Among the most urgent technological development figures, as
was mentioned above, the invention of equipment for the economic harvesting
and packaging of biomass so that it can be transported and processed. This
sub-programme includes, in particular, forestry machinery and equipment
for cutting wood into alibrated product sizes, granulating straw, caking
various residues etc.

Naturally, the main emphasis and the greatest stake lie in the
development of liquid fuels. If such processes provided an economically
valid, high-yield product, they would help to simplify the problem of
searching for appropriate energy outlets for biomass inasmuch as a liquid
fuel/gasoline mixture would probably consume any amount available. Nonetheless, the time required to set up a distribution network for a new
fuel should not be underestimated.

After a thorough examination of the problem of liquid fuels, our
programme decided to give priority to methanol production by gasification
with oxygen, despite the toxicity of methanol and its rather poor energy/
volume ratio. Economic studies have indicated that when this new
technology is mastered, the cost of methanol will run close to that for
premium-grade gasoline at the refinery.

This priority, however, has not caused us to overlook methyl and
ethyl fermentation processes. Anaerobic digéstion can provide an
effective contribution towards satisfying the energy needs of rural
communities, provided a local use is found for the methane produced.
Batch production processes are being perfected, whereas continuous
processes still require further study and major trials.

Ethyl fermentation, especially of sugar beets, does not seem attractive
in France under present conditions. First of all, its energy balance is
poor if not nil. In addition, for an equivalent amount of energy, the
cost of alcohol product is about four times greater than that of gasoline.
Conditions in Europe are very different from those prevailing in

countries like Brazil. Nevertheless, active research is continuing in this field since the future is not necessarily as gloomy as it may appear today, especially when one considers the possibility of substituting synthetic alcohol for plant alcohol. The energy balance in this case becomes more favorable, and its economic competitiveness sufficient for the chemical use of alcohol. Furthermore, research in progress throughout the world may perhaps make it possible to make economic use of other raw materials, such as rough wood, by enzymatic catalysis of cellulose. Finally, various species may be selected for their alcohol production value, such as the Jerusalem artichoke or sweet sorghum. Processes for distillation or more generally extraction of alcohol may likewise improve the present situation.

5. MARKET DEVELOPMENT AND MARKETING

The French biomass development programme is emphasizing application research and market research. In the short run, the chief objectives is to increase the use of firewood, for which between now and 1990 the consumption figures should double from three to six million tons of equivalent oil. To accomplish this, a campaign has been undertaken to encourage stove and furnace manufacturers to develop higher-performance products for industrial and home use.

Moreover, the French Energy Conservation Agency is subsidizing equipment that replaces conventional fuels with wood, providing 400 Francs for each ton of equivalent oil thus saved. Although wood consumption is hard to evaluate, especially for households, because it often involves short supply pathways that do not entail any commercial network, various indications suggest that the current growth rate for wood consumption is running about 7 to 10%. Such a rate would indeed allow for the figures to double before 1990 as expected.

Although other biomass uses are less common than the plain burning of firewood, many methyl fermentation installations already exist at the experimental level and plans call for a national bidding competition to induce certain industrial firms to start manufacturing digestors.

Other industrial possibilities have also involved research work and practical achievements. Amongst them, mention can be made of:

- the development and sale of lean gas generators, a French
   speciality. Several manufacturers are selling gas generators
   designed to produce motive power and experimental equipment for
   tractor or truck propulsion.
- the use of alcohol as a fuel when mixed with petrol which is
   going to be experimented with on a fleet of operating vehicles,
   for the purpose of studying the feasibility of using fuels
   containing 10 to 15% of methanol
- applied programmes which are also under way in various French
   departments and overseas territories. Special mention can be
   made of a programme starting to use bagasse to produce electricity on the Island of Reunion, where the power capability
   using leftovers of this product may exceed 100 MW.

6. THE PROGRAM'S CURRENT STATUS

Research on biomass energy applications began in France in 1975
in a few applied research laboratories, such as the National Centre for
Agronomical Research (INRA), or the National Centre for Testing and
Experimentation of Farm Machinery (CNEEMA). In 1976, under the auspices
of the State Secretariat for Research, a first incentive program was
undertaken. Although granted only modest funding (2 to 3 million Francs
per year), this program increased the number of research teams receiving
support by 1979, and established the basis for coherent research work in
this field. During the same period, a few contracts signed by French
research groups with the Commission of the European Communities likewise
contributed to enhancing the scientific and technical potential available
in France. Annual public expenditures in this field by then had reached
about 10 million Francs, including the amount spent within the major
research organizations. On the other hand, very few funds were available
for experimental demonstrations or industrial development projects.

In 1979, the French Government called upon the French Solar Energy
Authority (COMES) to draw up a much more ambitious program for the year
1980. A call for tenders covering the entire field was issued in June
1979 for the purpose of identifying research teams, surveying demonstration
projects, and encouraging industrial developments. More than 200 proposals
were submitted and then examined by some 50 independent experts divided
up into 8 specialities:

(a)   New Concepts and Basic Research

This group looked at projects dealing with photosynthesis,
direct or indirect water photolysis, and reactions designed to produce
hydrocarbons.

(b)   Forest Resources

The proposals studied involve harvesting machinery, species
selection, rapid-rotation forests, economic studies and regional
inventories.

(c)   Agricultural Resources

On topics close to those covered by the preceding heading, projects
were started for the use of currently available agricultural wastes
(straw, corn), as well as for research on new species for energy
purposes.

(d)   Aquatic Biomass Resources

This category includes pilot crops of algae and water hyacinthe,
and research on optimizing their growth, harvesting equipment, etc.

(e)   Thermo-chemical processes Technology

The proposals studied involve gasification with air or oxygen,
pyrolysis, charcoal, etc. This heading includes, in particular, the
"methanol" sub-program in which three different processes for gasification
with oxygen are being studied by French firms.

(f)   Fermentation Technologies

This group includes rather advanced research projects especially
those dealing with the enzymatic catalysis of cellulose, but also
technical work on ethyl and methyl fermentation processes, especially
in continuous operation.

(g)   Market Development for Fluid Fuels

This heading comprises a series of applications dealing with the
use of gas generator, gas or liquid fuel, especially in electrical
power units or vehicles.

(h)   Market Development for Firewood

This sub-programme is being managed jointly with the French
Energy Conservation Agency. It contains projects aiming to bring about
a more effective use of firewood and to encourage its use by the public.

For all of these programmes the French Solar Energy Authority has about
32 million Francs in 1980, to which are added other funds provided by the
Ministry of Agriculture, the Energy Conservation Agency, and the Agency
for Research Enhancement (ANVAR) which is in charge of financing
innovative work. The total amount is nearly 50 million Francs, not
including the in-house funds spent to operate these research organizations.
This gives a grand total of about 60 million Francs as the overall budget
in France for "Biomass and Energy".

Table IV provides a summary of the contracts which exist as of
September 1980 between the French Solar Energy Authority and outside
organisations, divided up in accordance with the headings given above,
and showing the cost for the various projects and the funding share provided by the French Solar Energy Authority.

7.   CONCLUSION

The very sharp increase in public financing, which shot up from
12 to 60 million Francs between 1979 and 1980, is adequate basis for the
assertion that the French "biomass and energy" programme really got
started just this year.

The future of biomass energy in a country like France will
naturally depend upon the cost of other sources of energy which it will
have to compete with, amongst which coal is the most important. The
basic assumption for our work is generally an average estimated cost
of 200 Francs per ton of dry biomass. It should not be forgotten that
unlike other solar energy sources (including wind energy), the primary
energy is not free of charge in this case. Biomass utilization thus
more closely fits the cost pattern of conventional energies than it does
other kinds of solar energy.

Any program involving biomass hence entails an economic challenge
which can be met only under two conditions:

-       either by having a short commercial circuit for raw biomass,
        especially for applications using it as a solid fuel

-       or by offering a rich energy product, produced on a regional basis
        but readily distributed and broadly used, as might be the case
        with a liquid fuel.

These two possibilities clearly underscore the priorities chosen for our activity, since both are in the last analysis tied to the market's capacity to consume products at a profitable level.

For a country like France, which is relatively rich in plant resources, the real bottleneck is located in the search for market outlets. Our program, ranging from basic research to highly practical applications, is very clearly dependent upon a correct evaluation of the methods and means designed to get the marketplace to absorb this national energy resources.

France has biomass; our real problem is to know how to use it.

(1)   This document, entitled "Studies and Recommendations for using
       Green Energy", May 1980, may be ordered from the Commissariat a
       l'Energie Solaire, 208 rue Raymond Losserand, 75014 Paris, France.

TABLE I

ENERGY DEMAND IN FRANCE (1973 - 1990)

	1 9 7 3		1 9 7 9		1 9 9 0	
	M Toe	%	M Toe	%	M Toe	%
COAL	30,5	17	34,5	18	33/28	14/11,5
OIL	115,3	66	108,5	56	68/80	28/33
GAS	15	8,5	23	12	42/37	17/15,5
NUCLEAR	3	1,5	8,5	4,5	73	30
HYDRO AND RENEWABLE	12	7	19	9,5	26/24	11/10
TOTAL	175,8	100	193,5	100	242	100

TABLE II

RENEWABLE ENERGIES RESOURCES (1979 - 2000)

	1 9 7 9		1 9 9 0 FORECAST		2 0 0 0 TARGETS	
	M Toe	%	M Toe	%	M Toe	%
HYDROELEC- TRICITY (x)	16	8	14	6,2	17/18	6
SOLAR HEATING	SMALL	0	1,3	0,5	3/5	1/2
SOLAR ELEC- TRICITY AND WIND	0	0	SMALL	0	0,5	0,3
GEOTHERMAL ENERGY	SMALL	0	0,7	0,3	1,5	0,7
BIOMASS	3	1,5	8/10	3/4	11/14	4,5
TOTAL	19	9,5	24/26	10/11	33/39	12/14

(x) THE 1979 FIGURE FOR HYDROELECTRICITY IS EXCEPTIONALLY HIGH, DUE
TO GOOD CLIMATIC CONDITIONS

TABLE III

FORESTRY AND AGRICULTURAL RESIDUES

AVAILABLE FOR ENERGY IN FRANCE (in Mtoe/year)

	Practical availability (COMES estimate)	Reasonable physical limit (Quoted by Chartier)
Non extracted wood production	0,7 to 1,5	4
Cutting old coppices, per year	1,5	4
Wood wastes	0,8 to 1	4
Present consumption (Firewood	2	(Probably included)
(Industrial residues	1	(in above figures)
TOTAL WOOD	6 to 9	12
Straw	1,5	2,5
Other agricultural residues	3,5 to 4,5	4
Animal wastes	1	3,2
TOTAL AGRICULTURE	6 to 7	9,7
TOTAL BIOMASSE FOR ENERGY	12 to 16	21,7
OFFICIAL TARGET FOR 1990	8 to 10	
COMES FORECAST FOR 2000 (excluding specific plantations)	11 to 14	

TABLE IV

PRESENT STATUS OF THE FRENCH PROGRAMME AT COMES (as of SEPT., 1980)

	Number of contracts	Cost of the Programme (thousands of Fr.Francs)	COMES incentives (thousands of Fr.Francs)
I. NEW CONCEPTS ; FUNDAMENTAL RESEARCH Photosynthesis, direct hydrolysis	8	3 026	1 625
II AGRICULTURE RESOURCES			
Agriculture residues inventory	4	730	530
Specie selection, specific crops	6	1 363	1 165
Harvesting and conditioning	6	890	600
Demonstration	5	8 090	1 275
	21	11 073	3 570
III. FOREST RESOURCES			
Resources Inventory	3	345	345
energetic plantation development	8	1 785	1 600
Harvesting and conditioning (x)	10	5 994	1 667
	21	8 124	3 612
IV. MARINE BIOMASS			
basich research	7	2 557	1 480
growing and harvesting methods	5	918	611
	12	3 475	2 091
V. THERMOCHEMICAL PROCESSES			
basich research	4	1 536	995
new technologies research	3	1 650	1 027
(except liquid fuel)			
development and demonstration	7	3 550	1 515
	14	6 736	3 537
VI. FERMENTATIONS			
- ligno-cellulose hydrolysis			
a - enzymatic	4	937	937
b - fungic	2	244	196
- methanisation			
. research	10	5 610	2 000
. demonstration	15	15 680	4 913
. ethyl fermentation	2	362	362
	33	22 833	8 408
VII. FUEL MARKET DEVELOPMENT			
Solid fuel	2	2 900	400
Liquid fuel	3	9 140	3 548
	5	12 040	3 948
VIII. FIRE WOOD MARKET DEVELOPMENT	4	315	225
TOTAL	118	66 707	27 011 (x)

(x) Plus about 7 Million francs to be awarded from September to December 1980.

PLANNING FOR TRANSPORT FUELS FROM BIOMASS

THE NEW ZEALAND EXPERIENCE

Dr G.S. Harris

Executive Officer
New Zealand Energy Research and Development Committee
Auckland, New Zealand

Summary

This paper discusses a systems study of the potential
of energy farming for transport fuels in New Zealand.
The results of the systems study indicate directions
for on-going research, development and demonstration.
Current and proposed research projects are listed.

1. INTRODUCTION

New Zealand is situated in a temperate region, is blessed
with ample rain and has a considerable amount of expertise and
experience in agriculture and forestry. It was thus only
natural that energy farming should be considered by research
workers as a potential technology for New Zealand.

The concept of energy farming was first mooted as a
serious option for New Zealand in 1974 by research workers
with an interest in the processing of biomass to liquid fuel
and laboratory research commenced at that time. The overall
concept of energy farming was studied during the formulation
of Energy Scenarios for New Zealand (1), this work being
undertaken mostly in 1975 and 1976. Several workshops and
seminars on energy farming were held in the period 1975 to
1979 (2).

The Energy Scenario Research clearly identified transport
and the supply of liquid fuels for transport as being the most
important energy supply problem facing New Zealand. It also
pointed out the potential for energy farming but drew attention
to the many unknowns related to this technology.

In 1977 the New Zealand Energy Research and Development
Committee, which was already funding a considerable proportion
of the research on energy farming (mainly fuel production
research), decided to set up a group to study the technology
as a whole - that is a systems study of energy farming. The
main objective of the study was to answer the question "what
is the potential of energy farming in New Zealand?" This
research commenced with a Workshop in September 1977 attended
by research workers from government, industry and universities.

The systems study has been completed and the final report
published in August 1979 was subjected to an intensive review
by a wide range of organisations and individuals. The review
was published as a separate volume in July 1980 (3).

This paper briefly describes the research which has been
undertaken on processing crops to fuel, the main results of

the systems study with some of the criticisms which have been made, and the present state of the recommendations of the study.

It should also be noted that there is a wide range of New Zealand research on use of alternative fuels in vehicles which is not reported in this paper. In particular, research on 15% methanol blends has advanced almost to the point where the government can make a decision on blending methanol from natural gas with petrol.

2.   OBJECTIVES

The study was concerned with use of energy farming to produce substantial quantities of transport fuel. (The small-scale production of fuel, e.g. on farm methane, was not studied although such technology could well have a part to play in the future). The study was undertaken using data which are fairly readily available, in order to determine the potential of energy farming for New Zealand. The study was from the national point of view. If the concept of energy farming is eventually accepted for implementation, feasibility studies will be required for particular processing plants based on areas of land dedicated to supply biomass for that plant.

The study had the following principal components:

a.   Evaluation of land suitable and available for various crops;

b.   Technical, economic and environmental evaluation of agricultural crops as a source of biomass (sugar beet, fodder beet, maize, lucerne, oats, pasture);

c.   Technical, economic and environmental evaluation of forestry based on radiata pine;

d.   Evaluation of other crops on which there is less information available than for the crops mentioned above.

e.   Technical and economic study of eight processing routes to produce liquid or gas;

f.   Environmental and social implications of energy farming.

As mentioned earlier, study of the use of alcohols in

vehicles is now mainly covered by extensive contracts funded
by the Liquid Fuels Trust Board.

3.   LAND USE IN NEW ZEALAND

New Zealand landscape, landform, soils and climate are
very varied. Yet this diversity is not matched by a diversity
of land use. The bulk of New Zealand's agricultural industry
revolves on the grazing of only two types of livestock - sheep
and cows. Most of the existing forest industry is based on
one species - radiata pine. Or to put the argument in another
form, it is not surprising that given such diversity of landform and landscape, there is potential for an equally varied
range of land uses.

In New Zealand earth and environmental sciences have
advanced to a point where it is now very logical to base land-
use decisions on environmental data. The systems study of
energy farming is probably the first New Zealand attempt to
assemble a full range of data on which a new nation-wide
land-based industry can be made.

The important concept underlying the procedure used in
the study is that land defined as "suitable" for a particular
crop is capable of sustained high yields of that crop. This
concept was indirectly criticised by many agronomists who
failed to perceive that sustained high productivity can be
achieved by matching crop requirements with appropriate
properties of the land and climate.

The concept of terrestrial energy farming involves using
land to produce an agricultural or forest crop. The land so
used needs to be both "suitable" for the crop and "available".
While criteria for "suitability" can be written in terms of
soil type, climate and topography, the criteria for availability depends on definition of an energy farming processing
plant in a particular locality. For instance, only if farmers
have an idea of the price to be paid for a crop will they say
that some or all of their farm is "available". Hence, the

study concentrated on land suitability. Three agricultural crops, maize, beet (sugar and fodder), and lucerne and one tree species, radiata pine, were studied. For the first three crops, most of the suitable land is presently high-value agricultural land. This same land is suitable for radiata pine, but in addition, there is much marginal farmland, scrub and cut-over native forest - land which could conceivably be used for a new, more profitable activity such as energy farming.

In general terms it can be stated that the area of land which is potentially available for energy farming using any of the crops - beet, maize, lucerne or radiata pine - is well in excess of the land area which would be required to supply 100% of the liquid fuel requirements in 2000 (1978 projections).

4. AGRICULTURAL CROPS

The study of agricultural crops proceeded with the philosophy that any crop which is used for energy farming should be integrated into the existing farming situation. If energy farming were to proceed with agricultural crops, then the existing farmers with their knowledge of particular crops and farming techniques would be required to change their method of management in only a comparatively small way in order to produce the energy crops. In many cases it is quite possible that stocking rates could be held at existing levels. In this way energy farming would be more likely to succeed.

The second point of philosophy in the study has been to minimise the energy inputs used in growing the energy crop. Generally, conventional technology and practices have been used, and no attempt has been made to minimise energy inputs through such new practices as low tillage cropping.

A range of crops was studied and energy ratios and cost of production evaluated. As a result of the study and intensive review after publication of the report, it is fair to say that beet (either sugar or fodder) is seen as being the most promising crop. Maize is also seen to have some potential although it is likely to be a more expensive option.

For all crops studied in detail the energy ratio is greater than 10, i.e. at the factory gate there is a satisfactory return on the energy investment in growing, harvesting and transporting the crop.

The cost of crop production was the subject of much debate following the publication of the report. Costs are rapidly changing in New Zealand and any figures can only be valid for a particular date. Table I shows costs of production for certain crops at December 1979.

It is worth mentioning the intense debate that took place as a result of an optimistic yield figure for fodder beet that was used by the research group. Fodder beet has not normally been grown as a well-managed crop in New Zealand and much of the data subsequently produced relates to trials which may not have been on "suitable" land. In any event it is now clear that beet is still one of the most attractive crops but sugar beet is likely to be a better crop than fodder beet. Annual sugar yields of the order of 8 tonnes per hectare rising to 12.5 tonnes per hectare after some time were considered by reviewers to be appropriate for commercial operation.

Any arguments about growing beet or other crops for energy can only be resolved by large-scale trials and commercial growing on a substantial scale. Of particular importance is the yield of fermentable sugars and the storage qualities of the beet in the ground and in stockpiles under various climatic conditions. Results of various trials over the 1979-80 season will be available shortly.

5.   FORESTRY

Much of the native forest of New Zealand has been cleared and what remains is a valuable timber resource. Exotic forest planting commenced in New Zealand on a large scale in the 1920's and mature forests now support a sizeable timber, pulp and paper industry.

Table 1 – National economic analysis

Plant Size†	Capital Cost#	Feed Price	Design Operating Period	Annual Production	Product Cost§			Performance Value‡ (petrol equivalent)	
ODt/d	M$	$/ODt	days/yr	t/yr	$/t	c/l	$/GJ	c/l(blends)	c/l(pure)
Beet ethanol									
200	8.10	80	200	13 000	394	31.1	14.6	31	43
,,	8.00*	,,	150	13 800*	370	29.2	14.4	–	42
,,	8.10	,,	,,	9 700	423	33.4	15.7	34	46
,,	8.00*	,,	,,	10 300*	398	31.4	15.5	–	45
1000	29.37	100	200	64 800	402	31.7	14.9	33	44
,,	,,	,,	150	48 600	432	34.1	16.0	35	47
Maize ethanol									
200	8.32	120	333	22 500	491	38.8	18.2	40	53
1000	29.10	,,	,,	112 400	442	34.9	16.4	36	48
Wood ethanol									
200	13.90	30	333	13 300	444	35.1	16.5	36	48
1000	53.17	50	,,	66 600	441	34.8	16.3	36	48
Wood methanol									
200	24.20	30	333	31 500	232	18.5	11.6	21	33
500	46.64	45	,,	78 800	229	18.1	11.4	21	32
2500	145.30	60	,,	393 800	217	17.3	10.8	20	32
Crop methane (with car convers. and storage costs)									
20	3.04	55	333	1 090	876	–	18.5	–	47
200	28.60	,,	,,	10 900	798	–	16.9	–	42
Crop methane (without car convers. or storage costs)									
20	0.92	55	333	1 090	588	–	12.5	–	31
200	7.38	,,	,,	10 900	510	–	10.8	–	27

* All liquid fuels are anhydrous except for the two rows marked in the 200 ODt/day beet ethanol case. These figures apply to plants producing 94 wt% ethanol. Similar reductions in cost can't be obtained by only producing 94 wt% ethanol in the other plants and should be exploited if possible since anhydrous is only really needed for blending.

\# All capital costs are to December 1979

§ The method of economic analysis is described in Report 46 (Ref.3). Discounted cash flow is over 15 years with a discount rate of 10%. National economic analysis (i.e. no taxes). In all cases 2 years has been allowed for construction except for the large methanol plant (2500 ODt/day) where 3 years was allowed and the small methane plants (20 ODt/day) where 1 year was used.

‡ Ex-refinery petrol cost at December 1979 was 20 c/l. Imported petrol cost at December 1979 was 22 c/l.

† Ex-refinery petrol cost at December 1979 was 20 c/l.

† None of the sizes is necessarily the optimum.

The main forests are radiata pine, which is a resource which could be used for energy farming. Current planting rates exceed 40,000 hectares per year and the total exotic forest area exceeds 700,000 hectares. While the existing mature forest is almost entirely committed to providing feedstock for existing processing plants and for export and domestic demand, the fate of the forest now being planted has by no means been determined. Thus, in the early 1990's a considerable quantity of wood could become available for production of transport fuels. A decision at the present time to implement energy farming based on forest crops would mean that such crops specially planted probably could not be available until well into the 1990's because of the growing time required for any species.

Several forest regimes were studied and for all the energy ratios were high, generally higher than for agricultural crops. It is clear that transport is an important energy cost so that it is desirable for the processing plant to be sited close to the forest.

Studies have been made of the use of forest residues and wood wastes, but it is not possible to obtain sufficient quantities of these to manufacture a large proportion of New Zealand's transport fuel requirements. It is acknowledged, however, that residues and wastes may well be the feedstock of the first processing plants because they are much cheaper than whole-tree feedstock.

The cost of production of wood is shown in Table I for an energy forest required to support a large-scale processing plant and for waste wood supporting a smaller plant.

6. OTHER SPECIES

Many of the reviewers and indeed the research group itself mentioned other species such as sorghum, sugar cane, eucalypts, red alder, etc. The research to date has concentrated on species for which there is reliable data on growth and costs of production under large-scale management systems. Any alternative crop would need to be significantly better in terms of yield and cost of production than "conventional" crops in order to succeed, at least in the initial period.

There was strong support from reviewers for trials of other species and for testing improved varieties.

Eventually diverse crops could be grown for energy and the crops could vary from region to region.

7.   PROCESSING TO TRANSPORT FUEL

The following process routes have been considered:

      Beet to ethanol
      Wood to ethanol
      Maize to ethanol
      Forage to methane
      Forage to ethanol and methane
      Wood to methane
      Wood to hydrogen
      Wood to methanol

In Table I the costs of fuel produced for some of the various routes studied is shown. It should be noted that the cost of feedstock used for the various cases has been chosen as being representative of conditions for the particular sized plant, e.g. higher feedstock cost for greater average transport distance.

The cost of fuel produced is in some cases close to the ex-refinery cost of petrol but it should be noted particularly that the figures all relate to a national accounting viewpoint. It should also be noted that costs are indicative only but that the relativity between processes and fuels is probably correct since all costings were done on the same basis. Subsequent studies from a conservative commercial viewpoint show cost of ethanol from beet in the range of 50-60 c/l. The question of government subsidy or incentive for production of biomass-based fuels has not yet been fully addressed in New Zealand.

The processing of forest products in New Zealand is
conventionally through a highly integrated forest-industrial
system. For conventional agricultural products the degree of
integration is considerably less. For energy farming it would
be essential to develop a fully integrated agro-industrial
system for efficient production of transport fuel. The cost
of transporting feedstock to the processing plant is a
significant proportion of the cost of production. There are
economies of scale in the various process routes, e.g. a large
gasification plant processing 2500 oven dry tonnes (ODt) of
wood per day is more economic than a 1000 ODt per day plant.
However, a larger processing plant intake means a larger area
of land which is required to supply the plant and hence a
larger and more costly feedstock collection system. When
feasibility studies for plants in particular regions are
undertaken, there will be a need to strike a careful balance
between land suitability and availability, transport distance
including relation of the transport network to land availability, and processing plant size and location.

8.   ENVIRONMENTAL AND SOCIAL

In all of the above studies, environmental factors have
been considered. For instance, allowance is made in design
of the processing plants to recycle the waste materials to
farms where this is appropriate. Cropping systems have been
designed to interfere as little as possible with the existing
farm environment and for most, the use of fertiliser has been
minimised.

A small survey of the 10 year-old maize industry in the
Waikato was undertaken. This shows that a diversification
into an energy farming crop will have its main effect at the
farm level, with other consequences likely in allied sections
of the agricultural industry and in the local community. There
is, however, some uncertainty as to the degree of the effects
on farmer/farm operations and beyond until further details on

the energy farming operations are known. Farmers and people associated with the farming industry are very receptive to the idea of energy farming but it is apparent that some additional incentive may be required in the initial phases. This will be sensitive to the scale of operation in the locality.

As part of the review, questionnaires were circulated to a group of 80 people attending a seminar to discuss the report. The farmers in this group saw energy farming as a way of diversifying their farming operations and as a means to come self-sufficient in fuel on their own farms. From a national viewpoint they saw energy farming as a way of saving overseas funds using a renewable source of energy. But they all wish as a group to retain significant control of the overall energy farming system, from farm through to distribution of fuel. Academics and professionals saw diversification of cropping and regional development as important issues as well as the need for reduction of imported transport fuels and a move to renewable sources. All groups were concerned about the difficulties of transferring the new technology to farmers.

9.   ON-GOING PROGRAMME

Over the next 15 years New Zealand is likely to invest substantial amounts of capital and other resources in the manufacture of liquid fuels from Maui natural gas. The transport fuels so produced are likely to be cheaper than comparable fuels using biomass feedstock. However, there are many advantages to be gained by undertaking a complementary energy farming programme at a modest level to test the biomass production and processing technology, and to set the scene for rapid implementation of this option when required.

The report recommends a programme with the following two principal objectives:

a.   to have a single 200 oven dry tonnes per day of feedstock fermentation ethanol plant on line by 1982;

b.   to have a single 500 oven dry tonnes per day of wood
gasification methanol plant on line by 1988.

Only two commercial plants are envisaged in this on-going
programme but if these objectives were achieved, New Zealand
would be well positioned to proceed rapidly with a major
energy farming programme, should the economic and social
climate of the late 1980's require it. To facilitate future
rapid implementation, a necessary adjunct would be a
substantial research, development and demonstration programme
to obtain the required knowledge and experience. It would
also be a pre-requisite that this knowledge and experience
be locally and readily available throughout New Zealand.

The concept of the demonstration ethanol plant has
provoked considerable interest both amongst farmers and
potential processors. A number of studies have been undertaken
by various groups and proposals are to be submitted to the
government during August 1980. The reaction of government
to these proposals has yet to be determined but presumably
depends in part on the level of subsidy required to make a
commercial plant viable.

Action on the demonstration methanol plant is not so
evident since much research and development both in New
Zealand and overseas is probably required. Several
government agencies are commencing research programmes aimed
at developing an understanding of gasification.

Time is needed to research and to fully prove variations
in the technology and to solve some of the problems (e.g.
waste disposal) connected with energy farming. For instance,
the development of new crop species especially suitable for
energy farming may take many years. New Zealand's crops,
climate, farming and forestry differ from those in other parts
of the world and the solution for this country will be unique.
Only if money is invested now in the on-going programme can
the longer-term elements of the programme be brought to a
timely conclusion.

Demonstration plants will only answer some questions about energy farming. Others can only be solved through research and development and therefore it is essential that a well constructed research and development programme be funded by government. As discussed in the next section this programme is now moving ahead, at least in certain areas.

10. RESEARCH AND DEVELOPMENT

The title of Government-funded research projects is given together with the institution(s) undertaking the research.

(a) Land Use
- regional land suitability and availability: (contract not yet let)

(b) Agriculture
- evaluation of sugar and fodder beet species: Ministry of Agriculture and Fisheries
- plant productivity comparisons: Ministry of Agriculture and Fisheries (MAF)
- farm-scale biogas production: MAF
- beet planting trials: Lincoln College
- survey of contract farming: Lincoln College
- controlled environment studies of beet: Plant Physiology Division, DSIR
- virus/vector relationships for beet: Entomology Division, DSIR

(c) Forestry
- sources of wood and multiple use of wood: Forest Research Institute (FRI)
- regional studies of log and/or residue supply: FRI
- predictive models for biomass production for radiata and other species: FRI
- logging and handling: FRI
- nutrient requirements of forests: FRI
- screening and trials of other species: FRI

(d) Processing
  - acid hydrolysis of radiata pine: University of   *
    Auckland
  - ethanol from wood: Forest Research Institute   *
  - fermentation of sugar to ethanol: Lincoln College   *
  - coupled fermentation of crops to ethanol and   *
    methane: Cawthron Institute
  - novel fermentation techniques: Massey University   *
  - leaf protein concentrate and ethanol: Ministry of   *
    Agriculture and Fisheries
  - farm scale demonstration of ethanol production: *
    Contract not yet let
  - processing of beet; sugar extraction and waste   *
    disposal: Contract not yet let
  - comparative economics of coal and biomass as raw
    materials for transport fuel production:
    Contract not yet let
  - wood gasification and liquid fuel synthesis: *
    Physics and Engineering Lab, DSIR
  - direct liquifaction of biomass: Chemistry
    Division, DSIR
  - size reduction of wood with steam: Chemistry
    Division, DSIR

(Projects marked * involve significant experimental or pilot
plant work).

From the above list it can be seen that there is
considerable activity on research and development related to
biomass-based transport fuels. Co-ordination is achieved
through a Biomass Co-ordination Group with representatives
of all the government agencies funding research in this area.
At this stage, funding is fairly evenly spread between
Department of Scientific and Industrial Research, Ministry of
Agriculture and Fisheries, New Zealand Forest Service, Liquid
Fuels Trust Board and New Zealand Energy Research and
Development Committee. The impetus and direction for
research projects is coming largely from the systems study

results. Research workers are keen to take up some of the challenges which arise from the systems study but work on biomass has to be seen in context with other demands for research resources (e.g. agricultural research).

Research in the private sector is limited at this stage and is not well documented. In New Zealand the private sector very much relies on research results from government agencies and from overseas.

11. CONCLUSIONS

The answer to the question "What is the potential for energy farming in New Zealand?" is clear. There is adequate land area to supply New Zealand's transport fuel requirements in the year 2000 by processing agricultural or forest crops specially grown for that purpose. The net energy is high and the cost not greatly above the ex-refinery cost of gasoline at present. The price of oil will continue to rise.

The on-going programme mentioned previously is proceeding, at least to a certain extent, as the basis for further work in energy farming. This goal-oriented programme is giving research workers, farmers and industry an incentive to explore whether energy farming really can contribute substantially to New Zealand's energy supplies in the 1990's. Involvement of the private sector - farmers, forestry companies and other industry - should be an important part of a successful on-going programme. While a commitment for at least five years by government to a research, development and demonstration programme of the nature proposed is essential if it is to be successful, the results obtained from it by the mid-1980's and the changes in world and New Zealand energy supplies and economy will govern the decision whether or not to proceed to a major energy farming programme at that time.

12. ACKNOWLEDGEMENTS

The considerable assistance given by members of the

Energy Farming Research Group and their respective
institutions is acknowledged: Professor J.B Dent and Dr
W.A.N. Brown, Lincoln College; Dr W.B. Earle, Dept of
Chemical Engineering, University of Canterbury; Mr T. Fraser,
Forest Research Institute, Rotorua; Dr M. Leamy, Soil Bureau,
DSIR; Mr J. Gilbert, Commission for the Environment; Mr T.
Fookes, Department of Geography, Waikato University; Mr J.
Lee, Link Consultants.

13.   REFERENCES

1.   "Energy Scenarios for N.Z." by G.S. Harris, M.J.
     Ellis, G.C. Scott, J.R. Wood and P.H. Phillips,
     NZERDC Report No. 19, March 1977.

2.   "The Potential for Energy Farming in N.Z." -
     Proceedings of a Symposium of Physics and Engineering
     Laboratory, DSIR, Lower Hutt, 26 November 1975.

3.   "The Potential of Energy Farming for Transport Fuels
     in N.Z." by G.S. Harris, M.L. Leamy, T. Fraser,
     J.B. Dent, W.A.N. Brown, W.B. Earl, T.W. Fookes, J.
     Gilbert, J. Lee, NZERDC Report No. 46, August 1979
     (Main report & Appendices) and August 1980 (Review).

Note: Copies of references 1 and 3 can be obtained from
New Zealand Energy Research and Development Committee,
University of Auckland, Private Bag, Auckland, New Zealand.
Reference 2 can be obtained from DSIR, Private Bag, Wellington,
New Zealand.

POSSIBILITIES AND LIMITS OF USING BIOMASS AS SUBSTITUTES FOR EXHAUSTIBLE RESOURCES

A SYSTEMS-ANALYSIS APPROACH OF PRODUCING FUELS FROM RAPE-SEED IN THE FEDERAL REPUBLIC OF GERMANY

T. BÜHNER and H. KÖGL

Institute of Farm Economics

Federal Research Center of Agriculture

Braunschweig, Germany (F.R.)

Summary

A systems-analysis approach of production, conversion and demand for rape-seed is done showing the present and future
possible economic contribution of rape-seed oil as a substitute of diesel oil in the Federal Republic of Germany. For
that purpose supply functions of rape-seed (rape-seed oil)
and demand functions of coarse colza meal are estimated.
Afterwards the demand for rape-seed oil as a substitute of
diesel oil and the corresponding required subsidies for the
production of rape-seed are derived. Beyond this the reaction
on food supply and the corresponding government expenditures
are indicated.
It is shown that at present prices of diesel oil a much higher
subsidy for rape-seed will be needed than in the case, when
rape-seed oil is used for other purposes than fuel. Only at
high diesel oil prices the situation changes. But even in
this case an increasing production will require additional
budget.

1. THE OBJECTIVE

The use of rape-seed oil as a substitute of diesel oil
is one alternative of winning energy from renewable resources.
The purpose of this investigation is to find out on what economic conditions rape-seed oil from domestic production can
improve the energy situation in the Federal Republic of Germany.

2. THE CHOOSEN APPROACH

The analysis of an economical use of renewable resources
- here rape-seed oil as a substitute of diesel oil - requires
to regard the whole process of production, conversion and demand as a simultaneous system for existing and future possibly situations. Linear programming models are applied to estimate the main functions of the system. Regression analysis is
only used to specify the farm models for the supply analysis.

3. THE SYSTEM 'PRODUCTION, CONVERSION AND DEMAND FOR RAPE-SEED AND ITS BY-PRODUCTS'

The equations of the system are represented in figure 1.
At first the supply of rape-seed (equation 1) and the demand
of coarse colza meal (equation 2) which is used as a valuable
feeding stuff in animal production are estimated. Further on
we derive the demand for rape-seed oil as a substitute of
diesel oil (equation 12) and the required subsidy for the production of rape-seed (equation 9).

For the instance of simplification cost of transportation are not included. The consumption of rape-seed oil for
other purposes than fuel is regarded as an exogenous variable.
The price instability of protein feedingstuffs in the world
market (2) and the enormous increase in prices of fossil energy are considered by different price levels.

4. THE POSSIBLE SUPPLY OF RAPE-SEED

For estimating supply functions farm models are build
using the results of regression analysis which show a statistically significant influence of price relationship between

Figure 1 – Structure of the system "Production, conversion and demand for rape-seed and its by-products"

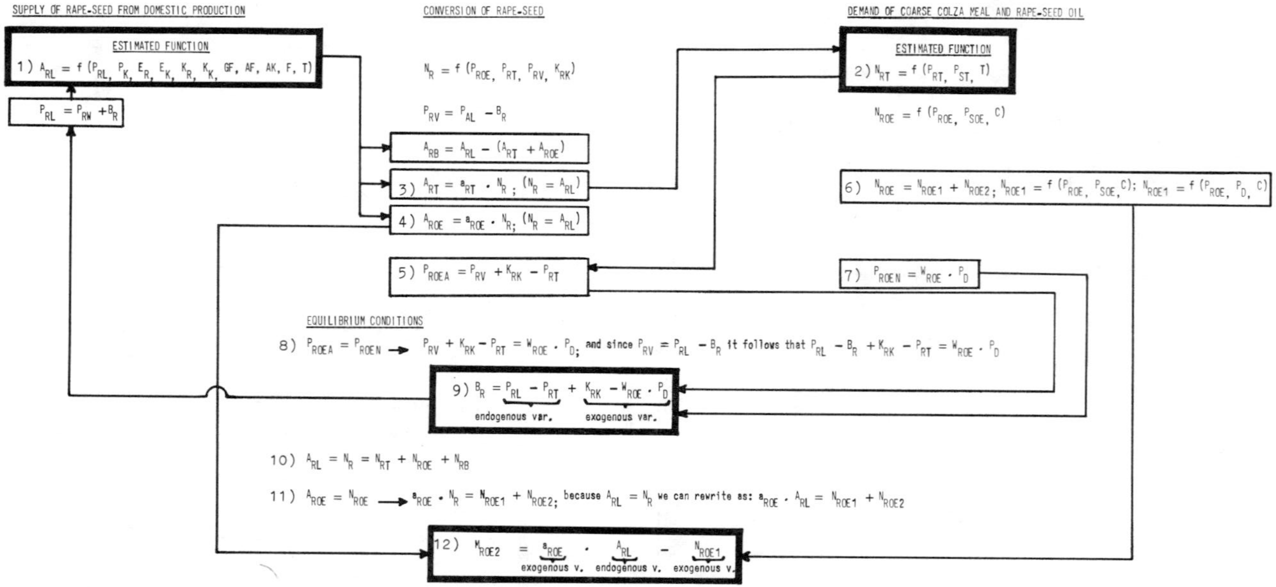

SUPPLY OF RAPE-SEED FROM DOMESTIC PRODUCTION

ESTIMATED FUNCTION

1) $A_{RL} = f(P_{RL}, P_K, E_R, E_K, K_R, K_K, GF, AF, AK, F, T)$

$P_{RL} = P_{RW} + B_R$

CONVERSION OF RAPE-SEED

$N_R = f(P_{ROE}, P_{RT}, P_{RV}, P_{RK})$

$P_{RV} = P_{AL} - B_R$

$A_{RB} = A_{RL} - (A_{RT} + A_{ROE})$

3) $P_{RT} = P_{RT} \cdot N_R; \quad (N_R = A_{RT})$

4) $A_{ROE} = B_{ROE} \cdot N_R; \quad (N_R = A_{RL})$

5) $P_{ROEA} = P_{RV} + K_{RK} - P_{RT}$

DEMAND OF COARSE COLZA MEAL AND RAPE-SEED OIL

ESTIMATED FUNCTION

2) $N_{RT} = f(P_{RT}, P_{ST}, T)$

$N_{ROE} = f(P_{ROE}, P_{SOE}, C)$

6) $N_{ROE} = N_{ROE1} + N_{ROE2}; \quad N_{ROE1} = f(P_{ROE}, P_{SOE}, C); \quad N_{ROE1} = f(P_{ROE}, P_D, C)$

7) $P_{ROEN} = W_{OE} \cdot P_D$

EQUILIBRIUM CONDITIONS

8) $P_{ROEA} = P_{ROEN} \rightarrow P_{RV} + K_{RK} - P_{RT} = W_{ROE} \cdot P_{ROE} \cdot P_D;$ and since $P_{RV} = P_{AL} - B_R$ it follows that $P_{RL} - B_R + K_{RK} - P_{RT} = W_{ROE} \cdot P_D$

9) $B_R = P_{RL} - P_{RT} + K_{RK} - W_{ROE} \cdot P_D$
   endogenous var., exogenous var.

10) $A_{RL} = N_R = N_R + N_{RB}$

11) $A_{ROE} = N_{ROE} \rightarrow B_{ROE} \cdot N_R = N_{ROE1} + N_{ROE2};$ because $A_{RL} = N_R$, we can rewrite as: $B_{ROE} \cdot A_{RL} = N_{ROE1} + N_{ROE2}$

12) $N_{ROE2} = \dfrac{B_{ROE}}{A_{RL}} \cdot A_{RL} - N_{ROE1}$
    exogenous v., endogenous v., exogenous v.

A_{RL} : supply of rape-seed from domestic production
P_{RL} : producer price of rape-seed
E_K : producer price of competing crops
E_R : yield of rape-seed
E_K : yield of competing crops
K_R : cost of rape-seed production
K_K : cost of competing crops
GF : permanent grassland
AF : arable land
AK : labour
F : crop rotation
T : quantity and structure of animal production
P_{RW} : world price of rape-seed
B_R : required subsidy for rape-seed

N_R : demand for rape-seed from domestic production
P_R : price of rape-seed oil
P_{ROE} : price of coarse colza meal
P_{RT} : cost price of coarse colza meal
P_{RK} : cost price of oil mills
P_{RV} : cost of conversion
A_{ROE} : supply of rape-seed oil from domestic production
A_{RT} : supply of coarse colza meal from domestic production
A_{RB} : supply of other by-products
B_{RT} : yield of coarse colza meal
A_{RT} : yield of rape-seed oil
P_{ROEA} : minimum price of rape-seed oil
N_{RB} : demand for other by-products
M_{ROE} : quantity of rape-seed oil from domestic production available as fuel

N_{RT} : demand for coarse colza meal
P_{ST} : price of substitutes of coarse colza meal
P_{ROE} : total demand for rape-seed oil from domestic production
P_{SOE} : price of substitutes of rape-seed oil
C : level of consumption
N_{ROE1} : demand for rape-seed oil from domestic production for other purposes than fuel
N_{ROE2} : demand for rape-seed oil from domestic production as fuel
P_D : price of fuel
P_{ROEN} : ceiling price of rape-seed oil
W_{ROE} : efficiency of rape-seed oil compared to traditional fuels

rape-seed and wheat and the importance of farm structure for the supply of rape-seed. The formulation of the models regards the specific resource equipment of sample farms with more than 10 hectars of agricultural land in the Federal Republic of Germany. Starting from the present yield and price relation the potential supply response is estimated regarding prices of rape-seed from 950 DM to 1 500 DM per ton (this means a change of price ratio between rape-seed and wheat from 2:1 to 3:1). The results are projected corresponding to the distribution of farm types and sizes in single states (Länder) of the Federal Republic of Germany.

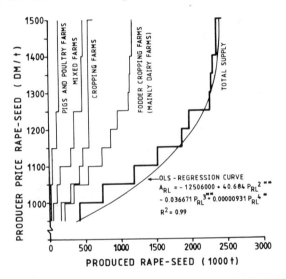

FIGURE 2: AGGREGATE STEPPED AND SMOOTHED SUPPLY FUNCTIONS FOR RAPE-SEED IN THE FEDERAL REPUBLIC OF GERMANY, SHOWING THE CONTRIBUTION OF EACH TYPE OF FARM FOR VARIOUS PRICES OF RAPE-SEED

Figure 2 shows the projected supply functions. Each type of farm has a specific supply elasticity. But regardless of this fact, the total supply of rape-seed (2,4 million tons) is mainly limited for reasons of crop rotation by the production of sugar beet.

5. THE POSSIBLE DEMAND FOR BY-PRODUCTS OF RAPE-SEED OIL

Oil and fat for human consumption requires a full refinement to remove undesired attributes(above all dyes, smell and taste). For using oil as fuel perhaps a partly refinement will do (1). But since information about this is still insufficient we assume in our investigation a full refinement of rape-seed oil.

The cost of separating coarse colza meal and oil amounts to 120 DM per ton of oil. For full refinement additionally about 300 DM per ton of oil is needed (partly refinement only requires 150 DM per ton of oil). The yield of oil is 38 per cent and the yield of coarse colza meal is 60 per cent of rape-seed. The rest of 2 per cent are other components which are partly processed.

There are no decisive differences between rape-seed oil and diesel oil per unit of volume as far as their power is regarded. But not yet solved technical problems arise at low temperatures. Further on residues of rape-seed oil do not allow a continuous working of engines.

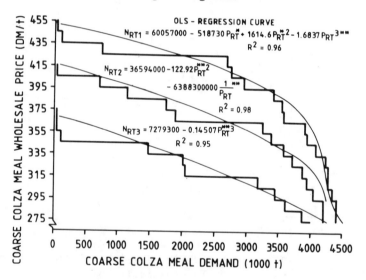

OLS - REGRESSION CURVE

$$N_{RT1} = 60057000 - 518730\,P_{RT}^* + 1614.6\,P_{RT}^{**2} - 1.6837\,P_{RT}^{3**}$$
$$R^2 = 0.96$$

$$N_{RT2} = 36594000 - 122.92\,P_{RT}^{**2}$$
$$- 6388300000\,\frac{1}{P_{RT}}^{**}$$
$$R^2 = 0.98$$

$$N_{RT3} = 7279300 - 0.14507\,P_{RT}^{**3}$$
$$R^2 = 0.95$$

COARSE COLZA MEAL WHOLESALE PRICE (DM/t)

455 — 435 — 415 — 395 — 375 — 355 — 335 — 315 — 295 — 275

0 500 1000 1500 2000 2500 3000 3500 4000 4500

COARSE COLZA MEAL DEMAND (1000 t)

FIGURE 3: AGGREGATE STEPPED AND SMOOTHED DEMAND FUNCTIONS FOR COARSE COLZA MEAL IN THE FEDERAL REPUBLIC OF GERMANY AT DIFFERENT PRICES (P_{ST}) OF IMPORTED PROTEIN FEEDING STUFFS

N_{RT1} INCLUDES PRICES $\bar{P}_{ST}$)
N_{RT2} INCLUDES PRICES $\bar{P}_{ST}-s/2$ } 1969/70-1978/79
N_{RT3} INCLUDES PRICES $\bar{P}_{ST}-s$)

PRICES OF OTHER FEEDING STUFFS AND LIVESTOCK SITUATION IN 1977/78

Because coarse colza meal is a by-product of oil production its economic value as a feeding stuff influences the competitiveness of conversion. For estimation of the demand function for coarse colza meal a sectoral linear programming model of least-cost feed mixes is used. This model takes into account the various feeding values and the possible demanded quantities by several branches of livestock production (extent and structure in 1977/78).

The estimated demand functions for coarse colza meal (figure 3) indicate a strong decline from 2.5 to 3 million tons. The maximum possible supply (1.4 million tons at a price of 1500 DM/t rape-seed) faces a very elastic demand. At wholesale prices of coarse colza meal from 385 DM per ton to 455 DM per ton the potential demand amounts to more than 1.7 million tons depending on the prices of imported protein feeding stuffs.

6. THE EVALUATION OF CONVERSION

6.1 COMPETITIVENESS OF RAPE-SEED OIL AS A SUBSTITUTE OF DIESEL OIL

Table 1 shows the required subsidy for rape-seed oil as fuel (demand function N_{RT2} of coarse colza meal used) at different production levels of rape-seed and prices of diesel oil. At the present price of diesel oil (about 800 DM per ton without tax) a much higher subsidy for rape-seed (611 DM per ton) will be needed than in the case, when rape-seed oil is used for other purposes than fuel (423 DM per ton in 1979). Only if the price of diesel oil is higher than 1200 DM per ton the required subsidy of rape-seed oil will be nearly at the same level as today. It lessens, if prices of imported protein feeding stuffs increase.

Assuming the same price advance of petroleum in future as was observed in the past (from 1970 to 1979 about + 250 per cent), the use of rape-seed oil can become important as a substitute of diesel oil. But there are two restrictions:

TABLE 1 : INFLUENCE OF FUEL PRICES ON THE SUBSIDY FOR RAPE-SEED AT VARIOUS LEVELS OF
RAPE-SEED OIL PRODUCTION

Producer Price of Rape-Seed	Usable Production of Rape-Seed Oil	Price of Fuel (without Tax)	Required Subsidy for Rape-Seed	
DM / t	1000 t	DM / t	DM / t	DM/hectare
1 000	310	800 1 200 1 600	611 459 307	1 460 1 097 734
1 100	586	800 1 200 1 600	715 563 411	1 637 1 289 941
1 200	766	800 1 200 1 600	818 666 514	1 832 1 492 1 151
1 300	864	800 1 200 1 600	920 768 616	2 024 1 690 1 355
1 400	902	800 1 200 1 600	1 020 868 716	2 234 1 901 1 568
1 500	911	800 1 200 1 600	1 121 969 817	2 455 2 122 1 789

- the mentioned technical problems have to be solved and
- the maximum quantities of rape-seed oil shown in table 1 can only get up to about 1 per cent of the present demand for diesel oil.

6.2 THE REACTION ON SUPPLY OF FOOD AND GOVERNMENT EXPENDITURES

In the farm models an increase of rape-seed production causes a decline only of cereals (table 2), whereas the production volume of other crops and livestock does not change.

TABLE 2 : RELATION BETWEEN THE EXTENT OF RAPE-SEED PRODUCTION AND THE VOLUME OF SUBSTITUTED CEREALS

Producer Price of Rape-Seed	Acreage of Rape-Seed	Average Yield of Rape-Seed	Production of Rape-Seed	Average Yield of Cereals substituted by Rape-Seed	Cereals substituted by Rape-Seed
DM / t	1000 hecta.	t/hectare	1000 t	t/hectare	1000 t
950	141	2,50	353	4,20	592
1 000	342	2,39	817	"	1 436
1 050	523	2,32	1 213	"	2 197
1 100	674	2,29	1 254	"	2 831
1 150	804	2,25	1 809	"	3 377
1 200	900	2,24	2 016	"	3 780
1 250	985	2,20	2 168	"	4 137
1 300	1 033	2,20	2 273	"	4 339
1 350	1 062	2,20	2 336	"	4 460
1 400	1 084	2,19	2 374	"	4 553
1 450	1 091	2,19	2 389	"	4 582
1 500	1 095	2,19	2 398	"	4 599

Recent investigations (3) show that the EC will become a net exporter of cereals in the next years, if the level of grain prices is not reduced and the relations of prices are not adjusted to their feeding value. If we assume that the worldmarket price of cereals in the long run is lower than the price in EC, aids have to be paid for exports. In this situation an increase in rape-seed production not only would require additional subsidies but also would lower the total amount of aids for grain exports. Therefore we can do the following calculation:

cost of intervention $^{+)}$	370 DM/hectar
aids for exports $^{++)}$	372 DM/hectar
total cost	742 DM/hectar

+) yield of grain: 4,2 t/hectar ; average price in 1970/71 to 79/80; ++) arithmetical mean for feed wheat and barley

The saving of money in the case of a lower grain production amounts to 742 DM per hectar only if surplus is intervened and exported. It depends on the intervention prices, worldmarket prices and the yield of grain and rape-seed.

TABLE 3 : ADDITIONAL BUDGET REQUIRED FOR EXTENDING THE RAPE-SEED PRODUCTION AT VARIOUS LEVELS OF FUEL PRICES

Production of Rape-Seed		Price of Fuel (without Tax)	Required Subsidy for Rape-Seed	Maximum Expenditures for Market Policy in Cereals	Additional Budget Required
1000 hectare	1000 t	DM / t	DM/hectare	DM/hectare	Mio DM
342	817	800	1 460	742	245,6
		1 200	1 097	"	121,4
		1 600	734	"	- 2,7
674	1 254	800	1 637	"	603,2
		1 200	1 289	"	368,7
		1 600	941	"	134,1

By comparing the maximum saving money for exports and the required subsidies for using rape-seed oil as a substitute of diesel oil, we see (table 3) that the total budget decreases only if prices of diesel oil are higher than 1600 DM per ton and production of rape-seed is lower than 800 000 tons. In any case increasing production requires additional budget.

This is - regardless of the possibility that other resources might be more competitive - the price for a little bit

more independence of fossil energy and imported protein feeding stuffs.

References:

1. BATEL, W., GRAEF, M., MEYER, G.-J., MÖLLER, R. and
   SCHOEDDER (1980).: 'Pflanzenöle für die Kraftstoff- und
   Energieversorgung'. Grundlagen der Landtechnik, p.40-51.

2. BÜHNER, Theo (1980). 'Möglichkeiten und Grenzen der Er-
   schließung und wirtschaftlichen Nutzung inländischer
   Proteinressourcen für die tierische Veredlung in der
   Bundesrepublik Deutschland'. Agrarwirtschaft, SH 82,
   Hannover.

3. UHLMANN, F. (1980). 'Langfristige Vorschätzung der EG-Ge-
   treideversorgung'. Agrarwirtschaft, H.8, p.235-244.

BIOMASS AND HYDROGEN: AN ANSWER TO THE

EUROPEAN LIQUIDS FUELS CRISIS IN THE 21ST CENTURY

Michael Messenger

International Institute for Applied Systems Analysis

ABSTRACT

An all out effort to collect biomass could yield 4732 Twhrs (2.7 billion boe) of primary energy from waste streams and biomass plantations at an average delivered cost of $3.67/GJ($22/bbl) in Western Europe. Future land use conflicts and uncertainties in projected collection factors suggest a more practical "limit" of 3105 Twhrs could be harvested using only 4.5% of the land area at a cost of $20/boe.

This practical limit of biomass collection could produce sufficient methanol to meet the 1975 European motor fuel demand of 1761 Twhrs. The production cost of this methanol is estimated to range from $40 to $70/boe using current technology. Improvements in process efficiency and reductions in capital costs over time could reduce this cost in half and meet a fuel demand of up to 1.5 times today's before serious land use conflicts are expected.

A greater harvesting effort to meet higher fuel demands would press the biomass stock to its physical and ecological limits. If 5% is considered to be a reasonable limit of land area devoted to bioenergy farms, the use of hydrogen blending in methanol synthesis would increase production to levels to roughly three times today's demand. This process would reduce the biomass requirement by 60% but increase the expected cost of methanol to $60 to $100 per boe. Dramatic breakthroughs in technology could reduce this cost to $30/boe in the 21st century.

This study considers the eventual shift by the nations of Europe from fossil fuels to biomass to meet the liquids fuel demand in the 21st century. The primary objective is to estimate the potential for collecting biomass and transforming it into methanol to meet a range of projected demands for automobiles and ground transportation. To accomplish this task both the cost and quantities of biomass that could be harvested in the medium run has been estimated for three major categories; waste streams, agricultural energy farms and silvicultural forest plantations.

Within each category the harvestable energy content of biomass is estimated as a function of national conditions, expected yield and delivery cost to the plant door on a national and regional level. These estimates include not only collection and transport costs but also the opportunity cost of not selling the given product in another market (e.g. wood chips to the pulp industry). Due to uncertainties in the future demand levels and the cost of energy, three separate cost categories (cheap, moderate and expensive) have been constructed on a regional basis for each type of biomass feedstock.

All of the biomass is transformed to methanol or an intermediate methane gas even though it is recognized the production of methanol may be significant during the transition. Methanol is the fuel of choice in the long run due to higher process efficiencies and greater flexibility in the feedstocks.

Two alternative processing routes to produce methanol are examined. The first uses biogasification and a methanol synthesis reaction using only biomass feedstock, while the second uses available hydrogen to increase the yield of biomass to methanol by roughly a factor of two. Using the expected evolution of the capital costs of the processes a comparison of the cost of producing methanol with and without hydrogen is constructed for a range of fuel demand from 1.5 to 3 times the 1975 motor fuel demand for Western Europe.

The results of a detailed investigation of the yields,

feasible collection factors and cost of biomass streams are
presented here as weighted averages of nations within one of
the three regional groupings in this study*. Much greater
detail at the national level and a full discussion of the
methodology is available in the complete version of this
paper (Messenger, 1980).

Table 1. COLLECTION FACTORS AND YIELDS IN 2050 (Central region)

Waste Stream	Coll.Factor	x Yield	Energy	Cost
units	%	waste/product	TWhrs	$/GJ
Straw	40	.9 straw/grain	84	3.10
Mill Wastes	40	1.5 waste/boardft	102	2.18
Logging Wastes	30	.27 w./harvest bft	76	2.82
Urban Wastes	65	330 kg/cap/yr	144	3.14
Livestock Man.	50	6 GJ/cattle/yr (gas)	22	6.5
Human Waste	50	60 kg/cap/yr (gas)	12	5.0
TOTAL	−	−	440	3.04 (wtd)

The estimated energy from waste stream for all of Europe
at a weighted cost of roughly $3.0/GJ is shown below;

Table 2. ENERGY FROM WASTE STREAMS IN EUROPE (TWhrs)

	Straw	Wood	Urban Waste	Biogas from Manure	
North	33	370	15	16	
Central	84	178	144	37	
South	224	247	130	93	Grand Total
TOTAL	341	847	289	136	1655

The estimated yields and delivered cost for catch crops,
agricultrual energy plantations and wood energy farms are shown
below for the Central regions. The yields are averages of more
country specific data taken from Sweden, France and New Zealand
for pilot plantations.

---

*3 Regions = North (Denmark, Finland, Norway, Sweden), Central
(FRG, U.K., Holland, Belgium, Ireland, Switzerland, Austria)
and South (Italy, Spain, Portugal, Greece, Yugoslavia and
Turkey).

Table 2. Bioenergy Farms in the Central Region (medium cost)

Type	Yield tons/ha	Land Use 10^6h	Energy TWhrs	D. Cost $/GJ
Catch Crops	5.2	6.1[a]	210	2.05
Agricultural (fodder beet)	16.0	5.4	432	3.2
Wood (European alder)	18.0	1.54	138	3.3
TOTAL	-	6.94[a]	780	2.91 (wtd)

[a]The joint use of land for catch crops and cereals productions is not double counted as land for energy farms.

Biomass Resource Summary

The maximum amount of collectable biomass feedstocks is shown in Table 3. These estimates imply an allout effort to collect biomass for liquid fuels. Given the likelihood that some of these projects will fall short of the yields projected and the possibility that some of the governments will opt for different solutions to the liquids demand (investments in conservation, mass transit, coal liquefaction, etc.) an alternative table has been juxtaposed to the left to estimate the "practical" or realizable biomass potential. The reduction from maximum to practical stems primarily from the potential land use conflicts the maximum strategy might generate in a world pressed for food and to a lesser extent from uncertainties in collection factors and yields.

Table 3. Biomass Sources at Realistic and Maximum Potential

Source	Realistic Potential Land 10^3km^2	% Current Streams	Energy TWhr	Maximum Potential Land 10^3km^2	% Current Streams	Energy TWhr
Wastes	-	25%	1173	-	40%	1655
Catch Crops	-	9%	307	-	14%	442
Pastures	121	16%	635	210	22%	952
Marginal Farms	-	-	-	46	4%	318
Energy Forest	104	5.8%	990	148	8%	1395
TOTAL (TWyr/yr)	225	4.8[1]	3105 (0.39)	404	9.7%[1]	4732 (0.54)

[1]Percentage of total land area.

Biomass resource availability as a function of price is shown below. Three discrete cost categories are defined for convenience.

Table 4. BIOMASS RESOURCE SUMMARY

	Total Energy TWhrs	Average C	Marginal C	Subtotal
Cheap	1675	1.8	3.2	–
Moderate	3057	3.67	5.2	4732
Expensive	2850	6.25	10.5	7582

Estimates of the capital costs of biomass to methanol plants designed in the United States and New Yealand were used for the year 2000 (1,7,11). Capital costs of plants built before then are assumed to be 25% higher due to cost overruns. The evolution of these costs over time are projected to decline and approach twice the cost of refining crude oil for gasoline ($100/kW) if hydrogen is available. Reduction in biomass alone costs are limited to 1.5 times the hydrogen process cost or $300/kW.

To simplify the analysis we have assumed only two demand levels for Europe in the 21st century. The high demand is 3.5 times the 1975 demand for gasoline in ground transport while the low demand is 1.75 X.

Meeting the Low Demand (3010 TWhrs)

Using the practical biomass collection limit of 3105 TWhrs, 74% of the low demand could be met using biomass alone to methanol synthesis plants. The cost is projected to range from 40 to 70 dollars per boe of methanol (in the near term).

To completely meet the medium demand would require full use of the maximum biomass potential and dedication of 10% of Europe's land to energy farms. This could trigger serious land use conflict between the traditional forestry and agricultural sectors and the new energy feedback business.

An alternative to relying on increased land use for biomass would be to use hydrogen blending. This would reduce the feedstock requirement by 55% to meet the same demand but increase the cost of methanol to roughly $60-100 by 2030.

Meeting the High Demand (6020 TWhrs)

Within the assumed constraints of land availability
(15% of the commercial forest, 5% of agricultural land) a
biomass alone to methanol system could not meet the high
liquid fuel demand (roughly three times today's demand level).
The use of hydrogen blending could produce 100% of the high
demand at a minimum cost of $50/boe. This optimistic estimate
assumes a production cost for hydrogen of 30 mills/kWhr from
electrolysis, whereas the actual cost of "solar hydrogen" may
range from 40 to 80 mills from solar towers in the southern
region (Caputo).

Dramatic breakthroughs in hydrogen production technology
(e.g. thermochemical splitting at high temperature) and
hydrogen blending technology could conceivably bring the cost
of methanol down to $40/boe using very low cost biomass feed-
stocks ($6/boe or $15/ton). However, this must be considered
an absolute lower limit since even the existence of "free
carbon resources" and low cost hydrogen with capital costs
approaching those of the oil refining industry would yield
a methanol production cost of $25/boe.

Cost/Benefits of Using Methanol Synthesis Plants

The primary benefit of developing a biomass-methanol
system is economics since it could be a competitor to oil by
2000 and thus set a price ceiling on future fossil fuel price
rises. Additional environmental and health benefits accrue in
comparison to methanol from coal liquefaction. The potential
costs of the system are harder to define because they involve conflicts over land use, foregone opportunities, and the
longer term environmental effects of heavy fertilizer
application and the subsequent freezing of forestry succession
cycles in the growth phase. These costs are not sufficiently
lare to outweigh the benefits of proceeding to develop
a pilot plant to prove the biogasification-methanol synthesis
concept. The future commitment to full commercialization
should rest on a better understanding of the total social
costs of the system relative to coal liquefaction or the continued dependence on imported oil.

Towards the Transition to a Biomass-Methanol System

The results of this study suggest considerably more
support should be devoted to planning and building the first
generation of biomass plants in the 1980's. With an all-out
effort to mobilize biomass resources and commercialize
gasification designs up to 15% of the 1975 gasoline demand
in Europe could be displaced by methanol in the year 2000.
It is my hope that the potential for fuels from biomass outlined here may serve as a catalyst to stimulate more
government support.

References
1. Caputo, R. Solar Energy for the Next Five Billion Years
        (IIASA, in press).
2. Harris, G. A. The Potential of Energy Farming for Trans-
        port Fuels in New Zealand,(New Zealand Energy Research
        and Development Comm. Report No. 46, 1979).
3. Inman, R. E. Silvicultural Biomass Farms Vol. 1 Summary,
        (Mitre Corp, Mitrex Division, U.S.A., May 1977).
4. Messenger, M. Biomass and Hydrogen: An Answer to the
        European Liquid Fuels Crisis in the 21st Century?
        (IIASA Working Paper, in press, 1980).
5. Salo, D. J. Pilot Silvicultural Biomass Farm Layout
        and Design (Mitre Corp., June 1979).

ENCOURAGING BIOMASS USE IN CALIFORNIA

RICHARD C. LANG

Energy Fuels Specialist

California Energy Commission

1111 Howe Avenue

Sacramento, California   95825

Summary

The State of California has taken an active role in encouraging the implementation of alternative energy systems (i.e., solar, wind, and geothermal energy, and biomass). The California Energy Commission has programs which total over 20 million dollars to encourage the use of biomass systems for energy production. This paper describes these programs and discusses the rationale behind them. The Energy Commission projects that biomass can contribute 295×10^{12} Btu/year of California's energy needs.

The three major areas that this paper addresses are:

o The wide range of biomass residues available in California.

o Examples of existing biomass conversion projects.

o California's programs to encourage the implementation of biomass systems for energy production.

California's biomass resources include forestry residues (e.g., sawmill and woodworking waste, slash and thinnings), and agricultural residues (e.g., food processing wastes, orchard prunings, field residues, and manure). Potential resources include energy farms and kelp.

Biomass residues can be converted to useable energy through many types of conversion technologies including direct combustion, thermal gasification, pyrolysis, and methane/alcohol fermentation. This paper presents examples of projects which use these technologies in California.

1. CALIFORNIA'S BIOMASS RESOURCE

Two of California's largest industries, forest products and agriculture, generate over 90 million tons of biomass residues annually. These residues include forest slash, sawmill wastes, food processing wastes, field residues, and manure. Table I summarizes California's biomass resource, the amount of energy currently produced from biomass, and an estimate of the current available energy potential. The total available potential of 238×10^{12} Btu is now equal to approximately 5 percent of California's 1978 energy consumption and is more than four times the state's current energy production.

Table I. 1978 California Biomass Resource* $(10^{12}$ Btu)

	Gross Amount	Available Amount	Produced
Agricultural	205	75	0
Forestry	636	103	50
Municipal Solid Waste	305	60	5
TOTAL	1,146	238	55

*1978 energy consumption in California was approximately $5,086 \times 10^{12}$ Btu.

In 1979, three major developments improved the attractiveness of using biomass for energy production in California. First, the California Public Utilities Commission ruled that utilities pay for cogenerated electricity based on their avoided cost of electrical production (i.e., the cost that the utility would have to pay if the cogenerated electricity had not been purchased. This is currently the cost of generating electricity from an existing oil-fired plant). It is anticipated that this ruling will raise the price of purchased electricity from 2 cents/kWh to 6 cents/kWh. Second, the cost of natural gas from Canada was raised from 23 cents/therm to 45 cents/therm. Third, the cost of gasoline approximately doubled.

The increasing cost of conventional fuels and the more favorable purchasing arrangement with the utilities have sharply increased the number of projects which are economically attractive. Table II is an estimate of the increased use of biomass through the year 2000. The total represents about 5 percent of the projected energy consumption for that year.

Table II. Biomass Projections (1) (10^{12} Btu/Year)

	1978	1984	1991	2000*
Agricultural	0	20	50	95
Forestry	50	90	120	150
Municipal Solid Waste	5	20	30	50
TOTAL	55	130	200	295

---

*Projected energy consumption in the year 2000 is 5,818 to 6,646 x 10^{12} Btu

The energy potential from biomass may seem small in terms of the state's total energy needs; however, this contribution, if included in an overall program of conservation and energy production from other alternative sources such as geothermal steam, wind, and solar energy, could make a sizeable contribution to reducing the need for oil.

2. THE CALIFORNIA ENERGY COMMISSION'S BIOMASS PROGRAM

The purpose of the California Energy Commission's biomass program is to promote the use of biomass for energy production. Although recent events make biomass projects more attractive to industry, many companies still consider these projects to be risky because they differ from past operating practices. The main thrust of the commission's 20 million dollar biomass program is therefore to demonstrate various conversion technologies on a variety of biomass feedstocks. To achieve this, the commission co-funds conversion projects during the construction and shake-down phases of the project. This technique of sharing the financial risk with industry has proven to be an excellent way to establish the reliability and economics of biomass technologies.

In the past four years, the commission has participated with industries and others in many demonstration projects. The following are examples of completed projects.

Incinerator-Air Heater: With J.G. Boswell, Inc.
Description: Incinerator-Air Heater System for Drying Cotton
Fuel: Cotton Gin Trash
Location: Corcoran, California
Size: 24 million Btu/Hour
Cost: 400,000 dollars (commission, 76 K; Boswell, Inc., Norman Pitt, Inc., and Cotton, Inc., 324 K)

Suspension Burner: with Tri-Valley Growers
Description: Retrofit of Oil-fired Boiler
Fuel: Fruit Pits
Location: Modesto, California
Size: 60 million Btu/Hour
Cost: 530,000 dollars (commission, 115 K; Tri-Valley, 415 K)

Methane Fermentation: with the University of California at Davis
Description: Trailer-mounted Methane Fermentor
Fuel: Manure, Cannery Waste
Location: Four Sites in California
Size: $.5 \times 10^6$ Btu/Day
Cost: 50,000 dollars

Alcohol Blend Test Fleet: with Santa Clara University
Description: One-year Fleet Test of Alcohol Blend Fuels in Stratefied-Charge Engine
Fuel: Methanol and Ethanol/Gasoline Blends (5 percent, 10 percent, 15 percent)
Location: Sacramento, California
Size: Four-car Fleet
Cost: 84,000 dollars

Low-Btu Gasifier: with the University of California at Davis
Description: Trailer-mounted Air-Gasifier
Fuel: Wood, Nuts, Fruit Pits
Location: Three Sites in California
Size: 8×10^6 Btu/Day
Cost: 285,000 dollars

Based on its successful demonstrations, the California
Energy Commission worked with other state agencies and the state
Legislature to establish an aggressive biomass commercialization
and development program. Senate Bill 771 (State Agricultural
and Forestry Residue Utilization Act of 1979) addressed biomass
conversion. A special section of Senate Bill 620 (Alternative
Transportation Fuels Program) addressed alcohol fuels.

The State Agricultural and Forestry Residue Utilization
Act allocates 10 million dollars to develop 20 or more biomass
conversion projects to be co-funded with members of the agricultural and forestry industries, public utilities, and investor
owned utilities. The program demonstrates three technologies:
direct combustion, gasification, and methane fermentation.

These technologies are being demonstrated at different
sites using a variety of residues. Projects are selected under
specific criteria that consider the feasibility of the particular process, availability of markets, economics, local employment, environmental quality and conformance with local land use
plans. The State of California will fund 10 to 50 percent of
project cost.

State funds will be used to reimburse the applicant for
the purchase of equipment.If the equipment meets the established
performance criteria, the funds will be repaid by the applicant.
If the equipment does not meet the performance criteria, the
state will assume ownership of the designated equipment and
salvage or reuse it. All recovered funds will be returned to
the program's revolving account and will be used to fund additional projects over the five-year program.

Senate Bill 620 provides funds to investigate the practi-
cality and cost-effectiveness of alternative motor vehicle fuel.
The Energy Commission will conduct fleet tests of pure ethanol
and methanol fuels and co-fund demonstrations of commercial
ethanol production from agricultural residues and livestock
feeds.

The alcohol fleet test program involves the development
and demonstration of 75 to 100 specially modified automobiles
which represent new vehicle technology. Both methanol and

ethanol fuels will be used in the tests. The automobiles must
be modified to achieve required emissions, performance, and
fuel economy goals and must be suitable for future mass production. After the modified technology passes all the necessary
tests, vehicles will go into daily use with continual monitoring and with periodic emissions and durability checks. The
successful conclusion of this program will lead to a broader
application of the vehicles in the state and in other captive
fleets of vehicles.

The commercial ethanol production demonstration program
was designed as an extension of the Senate Bill 771 program
and develops ethanol fermentation as an additional technology. Funding for this activity is approximately 7 million
dollars.

Table III summarizes the California Energy Commission's
biomass program budget.

Table III. Summary of the Energy Commission's Biomass Budget
(Million 1980 Dollars)

	Funding
Agricultural and Forestry Demonstrations (SB 771)	10.5
Alcohol Fuels (SB 620)	
. Alcohol Production	7.0
. Fleet Tests	2.1
Other	.5
TOTAL	20.1

California's program of sharing the risk of projects with
industry has proven to be very successful and should be considered by other state and national governments as a way to encourage the use of alternative energy resources.

Reference

(1) California Energy Commission, 1979 Biennial Report to the
Legislature. This report may be obtained by writing to the
California Energy Commission, Publications Unit, 1111
Howe Avenue, Sacramento, California 95825.

OPTIMIZATION OF AN INTEGRATED RENEWABLE ENERGY SYSTEM IN A DAIRY FARM

L.BODRIA, G.CASTELLI, G.PELLIZZI and F.SANGIORGI

Istituto di Ingegneria Agraria dell'Università degli Studi di Milano
Via Celoria, 2 - 20133 - MILANO

Summary

This study aims at the optimization of an alternative energy system to meet the energy requirements of a farm. It is based on a typical Po Valley dairy farm, having 60 ha surface and 120 equivalent head of cattle. The energy demand profile and the available renewable energy sources (solar, anaerobic digestion and combustion) are established; then the possible utilization of the several energies is analysed on the basis of the electricity and heat needs of the farm. The different technical solutions are determined and an optimization model is developed. Afterwards, the dimensions of the individual components are established so as to define a solution which will minimise the cost of the produced energy and meet the farm's requirements. This model makes it possible to assess the economic incidence of the various possible solutions and the unit costs of renewable energy, in the framework of a closed "input/output" system, as represented by the farm.

1. FOREWORD

The study of the possible utilisation of renewable energies involves
two quite separate approaches, pertaining to different, specific activities. Indeed, if the first step in such a study entails assuring the avail
ability of the new energy sources, the next step will be the optimisation
of their use.

This principle is generally valid and becomes especially significant
in the agricultural field. Here, the exigences to be met are those of a
number of time-discontinuous, rather rigid uses – we only need mention the
period of time in which certain operations have to be performed, and the
consequent promptness required –; moreover, these uses are widely differ-
entiated because of the complexity of the "farm" system.

Therefore, the Institute of Agricultural Engineering of Milano, in
the framework of the research started by the It.Nat.Res.Council Coordinat
ed research programme on Agricultural Mechanization on the energy problem
in agriculture, besides investigating technologies and processes also
tackled the study of integrated energy-user systems at farm level, aimed
at minimizing the cost of the farm's energy requirements.

We feel in fact that, while the technical aspects come foremost in
basic research and for research applied to the pilot plants, the economic
aspect cannot be neglected at the application stage.

On the other hand, the search for simple, low cost solutions for farm
ing applications of renewable energy – and chiefly of low temperature solar
energy – has yielded quite satisfactory results.

A first simulation mathematical model was therefore developed, to com
pare the energy requirements of a farm and the various available energies
considered. By this model, the economic validity limits of the various solutions were pointed out, and the "indifference level" for a competitive
cost of the facilities making use of the renewable energies was identified.

Since the survey should be limited to those integrated systems that
are of actual, short-term interest to the agricultural world, powering mo
bile equipment was not considered, as the corresponding energy sources
need further investigation and testing.

2. THE FARM CONSIDERED

The farm considered is located in the Lombardy plain; on a surface of 65 ha 100 head of cattle are raised and the crop pattern is such as to ensure the required livestock fodder. For hay, artificial drying by preheated air was assumed. This known technique entails harvesting the half-dried produce at 45–50% moisture content, and finish-drying in suitable barns by means of preheated air and with 7 to 10°C temperature differential.

The advantages of this method are known to be much lower harvesting losses and better forage quality. The actual quantity of heat required on a farm will vary widely, depending on the cattle raising technique; however, certain indicative mean values can be assumed as a basics for a first assessment.

Hot water requirements for the stable were evaluated according to the following parameters : for udder washing : 5 litres of 40°C water per head and per day; for washing the plant, inside and outside : 250 litres of 70°C water per day; for calf weaning : 7 litres of 40°C water per head and per day. For forage drying, a yearly requirement of 300 t of hay could be assumed.

Residential need, for the two families living on the premises, were taken as 1,000 m^3 heated volume and 300 litres of hot water per day.

On the basis of the above, and taking into account the climatic conditions of northern Italy and the forage production over the various months, the monthly desaggregated hear requirements were established.

Electricity requirements were determined in a similar way.

The result is 980,000 MJ/year of head and 65,000 kWh/year of electricity.

3. THE RENEWABLE ENERGIES CONSIDERED

Starting from the above remarks on the energies to be investigated, it was assumed to use air or water-type, simplified solar collectors as well as cattle manure-biogas production facilities, and facilities for direct burning agricultural byproducts.

As concerns availability, we assumed : the mean insolation values of northern Italy, 0.45 efficiency for the air collectors and 0.55 for the water collectors; variable biogas output, averaging 600 MJ per month and equivalent head of cattle; 3 t/ha of straw, having 16 MJ/kg heat value and at 0.6 combustion efficiency. Unit investment costs were takes as L. 150 per equivalent head of cattle for the biogas plant; L. $25/m^2$ and L.$50/m^2$ for the air type and water type solar collectors; straw was evaluated at L. 40 per t.

The following prices of the conventional energies were assumed : fuel oil, 1.75 p/kg; electricity, 0.23 p/kWh.

The plant write-off period rate was supposed to be ten years 7% reduc ed borrowing rate was taken for the investment capital, and maintenance and upkeep assessed at 6% of plant value per year.

4. ANALYSIS OF THE RESULTS

A first calculation was made considering the separate use of air type and water type solar collectors and of the biogas obtained from anaerobically digested cattle manure.

If only water solar collectors are installed, the heat load requirement can be theoretically covered by 100%, provided the plant is overdesigned and its utilisation factor allowed to drop below 50%.

The optimum solution (fig. 1-A) corrisponds to 500 m^2 approx. of col lector surface.

The performance of the air solar collectors system was investigated in the same way (fig. 1-B). The load is never covered by more than 55% since, because of their intrinsic characteristics, the collectors considered can only be used for forage drying, and only for 75% of the process heat needed. Here again the optimum solution is not the one ensuring maximum load coverage, but lies in the neighbourhood of 500 m^2 of collector surface.

Lastly, the graphs for a biogas generating plant, with subsequent con version into heat and electricity by a total energy system are given in fig.1-C.

As the size of the biogas plant is limited by the number of cattle bred on the farm (120 eq.head), it appears immediately that the generated energy will meet max. 45% of the global heat demand, and approx.90% of the electricity demand. This means that – any economical consideration apart – biogas by itself cannot fully meet the energy demand of the farm, and least of all its heat demand.

An analysis of the economic balance shows the optimum solution to be around 80 head; in this case the output covers 35% of the heat load and 70% of the electric load.

It appears from the examination of the three solutions above that one renewable energy, if taken separately, could not possibly represent an eco nomically sound way of meeting the global energy demand of the farm. There fore, integrated energy systems, making use of several sources of renewa- ble energy, were considered. An evaluation was made of two tentative inte- grated systems. The first is based on biogas, water and air type solar collectors, and fuel oil. With the cost data mentioned before, the opti- mum solution would consist of : 70 head biogas plant; 120 m^2 of water type solar collectors; 350 m^2 of air type solar collectors.

The heat load supplied by the renewable energies equals 83%, and the electric supply equals 64%. The balance heat (17%) is supplied by fuel oil, and 36% electricity is purchased externally (fig. 2).

The degree of utilisation of the renewable energies is 91% for heat and 99% for electricity for the biogas plant, 75% for the water solar plant and drops to 54% for the air solar plant since the last operates in the summer months only and only for forage drying. With this first assumption, the cost of the energy supplied to the farm is cut by 23% as against the conventional energy sources.

It it immediately apparent from fig. 2 that, when using "rigid" re- newable energies such as the sun and biogas, overdesigning to meet the de- mand peaks is an economical nonsense, as the utilisation factor would plum met as a consequence. It is more convenient to meet the peaks by flexible energy, available at any time and in any quantity (from this point of view, fuel oil is extremely flexible). A confirmation of the above was ob- tained by running the programme under the same assumptions, but with the further condition that the farm be self-sufficient : the cost of the ener- gy produced increases by 20% above the cost of conventional energy.

Consequently a further system was investigated; this system includes the combustion of agricultural byproducts, and specifically straw, as this exhibits a flexibility akin to fuel oil's.

Thie further system them considers the use of biogas, water and air type solar collectors and straw combustion.

For this second assumption, the optimum solution is : 70 head biogas; 30 m^2 water type solar collectors; 320 m^2 air type solar collectors; 32 t straw combustion.

The heat demand is fully met by the renewable energies; while the electric load is covered by biogas for 64% (fig. 3).

The utilisation factors of the various renewable energies are : 91% for heat and 99% for electricity form biogas; 83% for the water type solar plant 54% for the air type solar plant and 100% for straw combustion. The energy supply costs of the farm drop by an overall 28% approx.

Note that in this last solution fuel oil combustion was not envisaged; this not for a matter of principle, but because the other energy sources, and specifically straw combustion, are economically more convenient.

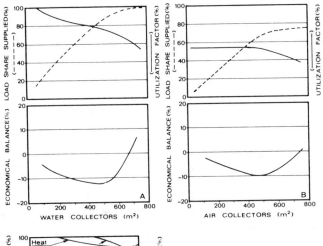

Fig. 1 – Load coverage, utilisation factor profiles and incidence on the energy supply cost assuming : water type solar collectors (A); air type solar collectors; biogas generating plant (C).

882

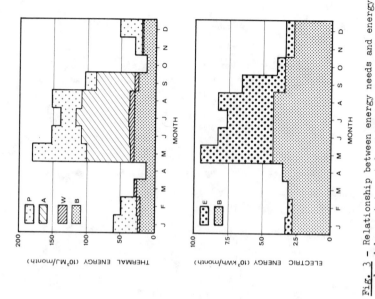

THERMAL ENERGY (10^3MJ/month)

ELECTRIC ENERGY (10^3kWh/month)

MONTH

MONTH

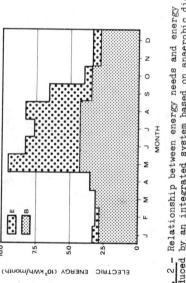

Fig. 2 - Relationship between energy needs and energy produced by an integrated system based on anaerobic digestion and solar energy (water+air). The total energy cost is 23 % less than for conventional energies. (G: fuel; A: air; W: water; B: biogas; E: electricity from network)

Fig. 3 - Relationship between energy needs and energy produced by an integrated system based on: anaerobic digestion; solar energy (water+air); straw combustion. The total energy cost is 28 % less than for conventional energies. (P: straw; A: air; W: water; B: biogas; E: electricity from network

HEAT SUPPLY SYSTEM FOR THE COMMUNITY OF SENT

J. BUCHLI and J. STUDACH

 **IGEK** Engineering consultants

Summary

The project aims at a central heat supply by wood gas-operated thermal power plants, combustion of wood, and biogas plants. The combustion wood
yearly available, which will be procured according to a special plan, will
be used optimally in combinations including wood gas generators, gas-powered Otto-engines, and heat pumps. Ambient air and solar radiation serve as
heat sources in the heat pump circuit. These sources will be exploited by a
collector/air heat exchanger element, designed to work both as solar collector and as direct evaporator element. These elements form the roof of
the plant.

A 4-tube long-distance grid provides an independent supply of hot water for
domestic and heating purposes. This allows to keep the temperature in the
grid adapted to the outdoor temperature, which means extremely low throughout the heating period.

This integral project includes a system-oriented tariffing, building insulation in order to reduce the heat demand, as well as the proposals for the
choice of an institution for the maintenance and operation of the plant.
Based on detailed calculations of the availability of combustion wood and
of the demand, conclusions can be drawn in terms of the energy autarchy of
rural areas and of the transferability of the chosen system. The existing
wood-burning stoves are sensibly integrated in the concept. Detailed results are condensed in a report of IGEK.

1. Introduction

There are a surprisingly high number of rural communities whose proportion of wooded area per inhabitant and visitor is large enough to provide sufficient fuel wood to reach a high degree of, possibly even an absolute self-sufficiency in terms of heating capacity.

Consequently, the idea of a "fully effect-optimized application of fuel wood" by an integral heat supply system, accurate in exergy, ecologically beneficial, and economical on a long-term basis, is not only applicable to particular cases. On the contrary, it corresponds to desired conceptions of an optimal exploitation of this abundant source of energy, which is not limited to specific areas.

2. Procurement of a maximum quantity of fuel wood

Within the bounds of the sustained yield method granted by forest-law and without affecting the natural balance of nutritive substances, as much of the biomass as economically possible shall be extracted from the forest. New or rediscovered kinds of fuel wood accounting for 16 - 17 % of the annual yield can be found in the form of branches and bark.

The intensified cutting of log timber would not only provide more jobs and transport advantages, but also increase the yield of structural timber and fuel wood by 20 - 25 %.

Potential fuel wood shall be consistently separated from the waste arising from domestic usage and from demolished buildings.

In this way, a certain area would have some 40 - 50 % of the cut at disposal as fuel wood on a long-term basis, without reducing the present yield of structural and straight timber. These kinds of lumbering are known but not established.

3. Generating heat with optimal efficiency

Wood-gas: Power plant

To maximize the energy outpout, more sophisticated technology is required and, therefore, central plants with an appropriate distribution network are the rule. This can also be said for the use of fuel wood, which shall be burned in commercial plants involving a combined heat and power station. In a plant, where all the machinery necessary to generate thermal energy is installed in the same building, the wood will be stored to reach its optimal content of moisture and will then be used with the best degree of exergy. This is done by wood gasification and subsequent gas combustion in the gas-powered Otto-engine, making use of all waste heat resulting from the cooling of wood-gas, cylinder head, and exhaust fumes. The required efficiency of such a plant has been established by tests; long-term experiences are being gathered.

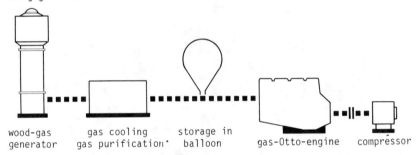

wood-gas   gas cooling   storage in
generator   gas purification· balloon   gas-Otto-engine   compressor

By coupling the plant with a heat pump, it is possible to obtain an optimal thermodynamical exploitation of the generated power. At the same time, this leads to the desirable side-effects:
- availability of a suitable source of heat,
- relatively low temperatures of heating water.

Apart from the planned heat pump units, wood gas-powered boilers and wood-burning boilers will be installed to meet peak demands and to build up an energy reserve.

Source of heat: Ambient air and solar radiation

Since other sources of heat are lacking, which is rather common in higher locations, a new type of evaporator element has been designed to exploit both, ambient air and solar radiation as heat sources. Under a glass cover, conventional, laminated evaporator elements coated with a black varnish are arranged in such a way that they work as solar collectors and, at the same time, - that is to say when solar radiation is insufficient or when demand is high - they can also function as air evaporators, the air being circulated by two fans.

With this construction it is possible to let the evaporator temperature rise during summer, when the radiation intensity is high, and release the evaporation heat in the condenser, based on the principle of the "heat pipe", without using a compressor. According to computer calculations with hourly solar radiation data for a reference community obtained from the MZA (Swiss Meteorological Institute), this system appears to be far more efficient than ordinary collectors, while thermal energy is generated at low temperatures.

These elements will be structurally integrated in the roofing of the plant and wood storage building.

4. Adaptation of the network and the temperature of the heat-transfer medium

The use of heat pumps requires lower operating temperatures than those common today (max. 90°C/194°F), as well as flexible adaptation of the temperature in the network throughout the whole heating period and a widening of the temperature range between the ongoing and returning flow of water. Another consequence is the separation of hot-water- and heating-water-circuit in a 4-tube network.

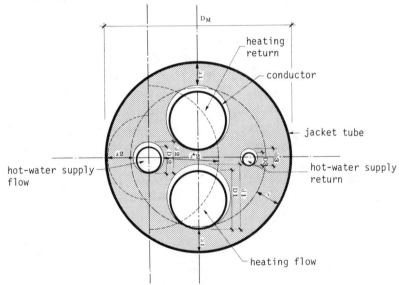

The slightly higher costs are compensated by a single, space-saving insulation cover, average heating-water temperatures (forward and backward flow) lower than 50°C/122°F, and thus smaller losses within the network, as well as by more simple kind of distributing stations needing less space (no hot-water treatment , control system nor pump). A differential pressure-controller provides a constant flow of heating-water in the building.

5. System-oriented tariffing

By taking the advice and the support of the heating station, and by applying appropriate insulating methods, the consumption of hot-water and heating energy shall be reduced. By means of hot-water meters (no calorimetry),
the station records the flow of heating-water and hot-water. This simplified measuring system favours customers with total or partial low-temperature heating systems and thus leads to the best possible cooling of the
heat-transfer medium, which results in improving the capacity of the distributing system and creating more favourable conditions for the heat pumps.
Since the metering is limited to the rate of flow, it is necessary to take
seasonal differences of demand into account by establishing suitable
accounting periods with tariffs varying according to the flow/return temperature range of the respective period.

6. An economical power supply

According to competent estimates, the procurement of the anticipated quantity of fuel wood, combined with the described cutting method for commercial timber will cost between 45.-- and 55.-- Swiss francs per cubic meter
and will hence lie below the price parity for crude oil. Because of the
optimized efficiency intended, the investments for the described system for
generating and distributing thermal energy, amounting to 1,5 million Swiss
francs per MW output, are relatively high. The operating cost per unit,
however, will be lower.

To many communities it may appear well worth the effort to try to regain
their self-sufficiency in energy by installing heat generating plants with
an output capacity ranging from 1 to 5 MW. For them, the operation of a
communal thermal station and the maintenance of the heat-distributing grid
will be a feasible task, just as their water supply and the sewage disposal.

Energy balance

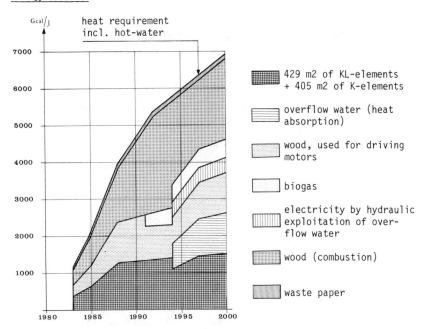

429 m2 of KL-elements + 405 m2 of K-elements	
overflow water (heat absorption)	
wood, used for driving motors	
biogas	
electricity by hydraulic exploitation of over-flow water	
wood (combustion)	
waste paper	

Yearly operating cost

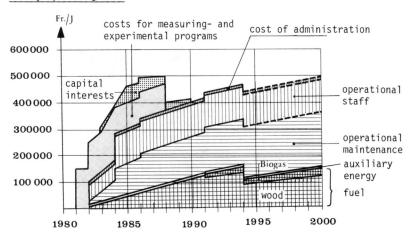

REALISTIC ASSESSMENT OF BIOMASS ENERGY CONTRIBUTIONS

L. P. WHITE and L. PLASKETT

General Technology Systems Ltd. & Biotechnical Processes Ltd.

Summary

In estimating energy contributions from biomass calculations often assume yields based on measured growth rates that are much higher than those demonstrated as harvestable and views on engineering costs, land acquisition and competition for space tend to be optimistic.

In making calculations for the European Community estimates of quantities of wastes, crop production figures and areas of available land set firm upper limits strongly modified by practicability.

Studies for the Community indicate quite modest contributions unless dramatic scenarios are envisaged, involving major changes in land-use or reassignment of biomass resources. In the case of marine resources there is only small potential. For this and other more drastic scenarios engineering problems impose an extended time-scale on any large-scale up-grading of estimates.

In biomass plantation/energy farm schemes the engineering and management costs outweigh the others and they involve new approaches and new equipment which are notoriously prone to under-estimation. A further difficulty is that the biomass resources may be successfully competed for by alternative uses, for materials, foodstuffs and chemical feedstock.

On the other hand, there may be considerable add-on benefits.

1. <u>INTRODUCTION</u>

Assessments of how much energy can be contributed from biomass are based on the following:-

i. In the case of waste biomass, the amount produced is governed by the initial consumption of the original material. In the case of animal wastes it is the number of animals.

ii. In the case of biomass produced directly for fuel use quantities depend directly upon the land areas available and the potential production levels.

Each of these can be readily determined for current situations and assumed for future scenarios, establishing the broad limits for potential contributions.

The amount of land that can be made available for biomass production has clearly recognisable limits while factors like maximum solar conversion efficiency and energy conversion to heat or intermediate fuels are to all interests and purposes immutable, and set firm limits to what can be extracted from a particular resource base.

Within these physical constraints, possibilities are further limited by the economics of the mobilisation and processing of the resource and these themselves are influenced by pressures for alternative uses for the raw materials, and for land.

2. <u>PRODUCTION FIGURES</u>

Claims for biomass yields are ranged in an "envelope" extending from observed natural or cultivated production figures; generally from an order of no more than 10 t/ha/y (representing a solar conversion efficiency of 0.5 per cent) in temperate zones to over 100 t/ha/y (5 per cent plus) for some aquatics. Though the maximum solar conversion efficiency of 5 to 11 per cent, (depending on the authority), seems to indicate great scope for very high yields, all the factors of actual field conditions conspire against this and with "optimum" field condition in temperate zones the annual efficiency rarely appears to be better than 1 per cent.

In most cases the highest quoted figures seem to be obtained by extrapolation of maximum observed yields and are unlikely to be achieved over large areas over the whole year. Further confusion arises from the use of terms; dt may mean "bone dry" with 0 per cent water content of the material, or "air dry," which varies with the material but which is usually between 15 and 25 per cent.

In the case of of aquatics however, "air dry" material can include up to 50 per cent water. Wood harvested "green" may contain 50 per cent water, and drained-off seaweed 90 per cent. These variable water contents not only affect true yield calculations, but also the energy content of the raw materials most "bone-dry" organic materials have more or less the same energy content of about 18 GJ/t.

Though the proportion of this that can be converted to useable energy varies with the method and different materials have different optimum methods, there are clear conversion efficiency limits of generally no more than 60 to 70 per cent.

3. CALCULATIONS OF POTENTIAL

Figures on the poster display are calculations of the order of size of the potential biomass energy contributions from various sources from the nine member European Economic Community (Euro-9). Calculations use basic statistics and a number of generalised assumptions. The current readily realisable potential and some limited scenario options are given. Some of these have overlapping demands which affect the totals.

The poster also includes comments on what are seen as the principal advantages and obstacles to the realisation of current potentials and what are regarded even as modest scenarios. These include add-on benefits in the form of cost saving in waste disposal.

4. SOURCES

Animal Wastes (including bedding)

Wastes from:-

	Total	Collectable Waste	Gross Energy Content
77.1 M Cattle	88.8Mt/y	25% = 22.2Mt/y	8.6 Mtoe
72.1 M Pigs	8.1	90% = 7.29	2.6
316.5 M Chickens	6.1	90% = 5.49	1.8
Plus bedding straw	29.3	90% = 26.37	8.1
	132.3Mt/y	61.35Mt/y	21.1 Mtoe

If all cattle were housed indoors for the winter and zero grazing systems were introduced for 10 per cent of the animals, there would be a further 9.2 Mtoe in collectable arisings.

Crop Wastes

	Total Waste	Available	Energy Content
Cereal straw	81.5Mt/y	25% = 20.3Mt/y	7.0 Mtoe
Maize and rice straw	15.6	60% = 9.36	3.4
'Green' vegetable matter	18.8	60% = 11.28	4.5
Woody arisings	6.8	50% = 3.4	1.4
	122.7 Mt/y	44.34Mt/y	16.3 Mtoe

Assuming that agriculture cannot be greatly expanded or yields greatly increased or useful wastes diverted, these figures are essentially inflexible.

Field Energy Crops

The area theoretically available for catch crop cultivation not involving new techniques is 28.8 Mha, with a yield of 107 Mt/y with gross energy content of 42.6 Mtoe. In practice only 50 per cent of this is likely to be available giving a 21.3 Mtoe/y contribution.

The area available for plantation crops on land not currently used by agriculture is estimated at 4.3 Mha. This would give a yield of 64.2 Mt/y with a gross energy content of 25.5 Mtoe. In addition the utilisation of some 8 Mha contributing about 25 per cent of permanent pasture land in the Community would produce an additional 129.4 Mt/y with an energy content of 51.5 Mtoe.

A further more drastic scenario envisage the adoption of whole crop harvesting of 50 per cent of the cereal crop combined with aerial sowing of catch crops. This could give rise to an additional 53.6 Mt/y material with a gross energy content of 10.7 Mtoe.

Wood Waste

Figures assume a continued Euro-9 wood consumption of 195 m^3/y with 38 per cent wastage in processing plus a further 25 per cent waste recovery from a domestic European timber production of 78 m^3/y. The 31.5 Million t or so of wood waste that this represents, with an energy content of 16 GJ/t, give a potential of 11.5 Million tonnes oil equivalent.

The only way that the energy contribution from this source could be substantially increased is by an increased domestic European forest production on 4 Million hectares of marginal land as suggested by the Manshold Plan. As even high yielding commercial timber species would not be expected to give more than

3 t/ha/y the total new production of 12 Million t/y would give additional harvested waste of 3 Mt/y giving a further 1 Mtoe of energy raw material.

Short Rotation Forestry

SRF is a similar scenario concept to field energy crop operations. If the 4 Million hectares of marginal land are used for this instead of for conventional forestry much higher yields could be obtained. Assuming a 14 dt/ha/y figure used in some less ambitious projections for the United States, at 16 GJ/t, 56 Mt of wood would represent about 20 Mtoe.

Seaweed

The scenario envisages harvesting 1 Mt of material with an energy content of 10 GJ/t by a variety of methods. This very large increase in seaweed production would still only represents 0.227 Mtoe. Effective methods of harvesting and farming seaweed for energy supply have not yet been developed.

Algal Production Schemes

There is some evidence for high potential yields of 50 dt/ha/y of material at 20 GJ/t, to produce enough to contribute even 1 Mtoe would require 44,000 ha of ponds or their equivalent.

Apart from the questions of land availability the capital costs and economics of such highly engineered systems are very much an unknown quantity at present.

These calculations to a large extent ignore the fact that a part of, or even all of a particular resource very well be diverted to more profitable uses. Any of the figures can be altered by varying the assumptions but within the constraints of the physical laws. Despite this only major scenario changes will significantly effect the overall size of the estimates.

5. DISCUSSION

The figures represent the limits only of what could be available. Realisation of the various biomass energy potentials must be below these figures and will be governed by the effectiveness of the techniques employed. These are at various stages of development but it is apparent that the chief obstacles are in economically effective operation rather than in the need for fundamental technological break throughs, the scope for which is limited. The most effective routes at present for achieving energy gains from biomass, therefore, appear to be

where there is a complimentary or secondary benefit e. g. :
- where waste can be readily and cheaply assimilated into an existing industrial set-up as in the timber industries.

As well as capital cost subsidies, integration in this way has the advantage of flexibility of opportunity in the use of energy. For instance in an industrial complex digesters used primarily for BOD control can utilise low grade waste process heat as well as contributing to the premium fuel supplies of the plant.

"Energy dedicated" schemes like energy plantations do not have this advantage. They also usually propose using large areas of land where, in the case of Europe, any major land-use change creates major socio-political problems.

In the case of short rotation forestry however, it is possible that a combined production of paper pulp and energy could be a more attractive solution. In the case of field crops a combination of fuel and fodder production could be considered.

The figures displayed represent some sort of 'reasonable' limit to potential contributions of energy from biomass. Large contributions in Europe would necessitate drastic scenarios like millions more hectares of changed land use, fundamental changes in diet, or marine engineering enterprises of the magnitude of the nuclear power programmes.

The scenarios quoted in this paper are taken from a study by the authors for the European Commission Solar Energy Programme. The results of this work are being published by Applied Science Publishers under the title 'Energy from Biomass in Europe', Ed. W. Palz and P. Chartier.

Conversion factors used

$$\text{Wood: 1 tonne} = 3m^3$$
$$= 16 \text{ GJ/t}$$
$$1 \text{ Mtoe} = 44 \times 10^6 \text{ GJ}$$

POTENTIAL FOR ENERGY CROPPING IN SWEDISH AGRICULTURE

K.G. BERGMAN

Department of Economics and Statistics,

Swedish University of Agricultural Sciences

Summary

The swedish energy import has during the last years grown to a very serious economic problem for the country. At the same time swedish agriculture is producing an increasing surplus of agriculture products, sold on the international market at low prices. Some of this surplus can be converted to fuel and thus replace imported petroleum based oil.

Energy can be produced in agriculture in many forms. Liquid fuels can be produced in the form of ethanol and vegetable oil using crops rich in sugar, starch or oil. Solid fuels can be produced using straw and cellulose crops.

The potential for energy cropping has been estimated to 500 000 ha in 1990 and 700 000 ha (25% of total hectarage) in year 2000. Corresponding amounts of energy in harvested crops is 30-55 TWh per year (= 10% of national energy consumption in 1980).

Energy efficiency measures show good result in all energy production technics provided biproducts are considered.

Solid fuels from agriculture used for heating purposes seem to compete well economically with petroleumoil, especially for straw. Liquid fuels, however, need a substantial support to be able to compete with petrol and diesel fuel.

INTRODUCTION

Swedish energy policy aims at replacement of imported non-renewable energy sources with domestic renewable energy sources. Available domestic renewable energy sources are residuals from forestry and agriculture, energy forestry, energy cropping, methane from waste products, wind energy and direct sun energy. Among the alternatives available energy from biomass is of great importance.

Swedish agriculture has the capacity to produce more than is demanded on the swedish market in terms of food raw material. Surplus production therefore is sold on the international market at prices lower than the domestic production cost. It is therefore of interest, both from energy policy point of view and from agriculture policy point of view, to investigate wheather the agriculture surplus production can be converted to fuels, which can replace imported fuels, especially petroleum fuels.

ENERGY MARKETS, ENERGY CARRIERS AND ENERGY CROPS

The market for energy carriers consists of submarkets for heat, engine fuels and electricity. All these submarkets can be supplied with energi produced from agriculture raw material. In addition to this, petroleum oil can be replaced by biomass in the production of chemicals.

The various kinds of energy carriers of interest for energy production in agriculture are in the short run ethanol, solid fuels, methane and vegetable oil (figure 1). In a longer perspective other gaseous energy carriers and methanol can be of interest. Figure 1 gives a broad overview over the technics used in the production of the various energy carriers. The figure also tells us the role of primary agriculture in the energy production: to produce sugar crops, starch crops, cellulose crops and/or oil crops.

Sugar crops of interest in swedish agriculture are mainly sugar beet, fodder beet and turnip. Starch crops are wheat, barley, oat, rye and potatoes. Cellulose crops are first of all grasses of various kinds. Oil crops are rape and white mustard. The same varieties and production technics can be used in energy cropping as are used in food production. However, in a longer perspective, it is expected that varieties and technics adjusted to energy cropping demands will increase energy yield and efficiency.

898

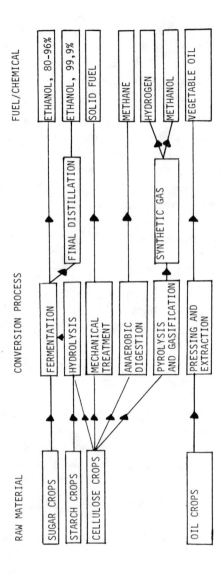

Figure 1: CONVERSION OF RAW MATERIAL TO ENERGY CARRIERS

RAW MATERIAL   CONVERSION PROCESS   FUEL/CHEMICAL

POTENTIAL FOR ENERGY CROPPING

There will be room for energy cropping in swedish agriculture only if this use of agriculture resources will contribute to national goals more than does export of cereals. This issue has so far not been satisfactory analysed. However, this question can not be fully answered until technical and economical aspects of energy cropping are better known than today.

The average year during the three year period 1976-1978 cereals were exported equivalent to 277 000 ha and vegetable oil equivalent to 54 000 ha, together 331 000 ha. This hectarage therefore can be used for other pruposes, for instance energy cropping, without impacts on national self sufficiency in food production. The area which under these assumptions can be made available for energy cropping is growing over time because of increasing yields per ha.

During the seventies the yields for breadcereals increased by 2,3% per year and the yields for fodder cereals by 1,7% per year. Demand increased by 1,5% in the first case and 0,5% in the latter case. Net increases in surpluses therefore correspond to 2 500 ha per year for breadcereals and 15 600 ha per year for fodder-cereals, together 18 100 ha per year. This means that 18 100 new hectares could have been made available for energy cropping every year.

The future need of agriculture land for self sufficiency in food is hard to predict. However, if development under the seventies is continued the following two decades, hectarage made available for energy cropping will be as follows:

Year	Hectares	% of total ha
1977	331 000 ha	11
1980	385 000 ha	13
1990	566 000 ha	19
2000	747 000 ha	25

Total agriculture land in Sweden amounts to 2 950 000 ha. The "surplus" land, according to the calculations above, will raise from 11 to 25% of total agriculture land until year 2000.

Real increases in yields may be lower than is estimated above. This does not reduce interest for energy cropping. Already the area available today can play an important role in fulfilling the energy policy goals.

The amount of energy produced in agriculture can be estimated if area, yield in dry matter per ha and energy content per unit dry matter is known. The yield in dry matter is estimated to 10, 11 and 12 tons per ha in years 1980, 1990 and 2000 respectively when conventional food production technics are used. Using adjusted energy cropping technics, the yields are estimated to 11, 13 and 15 tons dry matter for the same years.

Energy content for most biomasses is about 5 KWh per kg dry matter, provided that the stuff is dry.

The energy yield in 1990 can now be calculated to 566 000 ha x 11 tons dry matter x 5 KWh per ton dry matter, which is 31 TWh per year (37 TWh when adjusted energy cropping technic is used). For year 2000 the same figures will be 45 TWh and 56 TWh respectively. Total energy consumption in Sweden 1980 is 500 TWh.

SOLID FUELS

Solid fuels from agriculture can be either straw from cereal crops or oil crops or crops grown specifically for this purpose, f.i. grass.

Straw can be used either at the farm itself or on a large scale in the district heating system. It is estimated that under swedish conditions 2 tons dry matter of straw per ha can be used in this way without reducing the long term fertility of land.

It is also estimated that straw from one ha contains 15 MWh as an average. Energy inputs in harvesting and handling process require about 0,5 MWh per ha. The annual net energy gain per ha therefore amounts to 14,5 MWh.

The cost for harvesting and handling of straw for energy purposes have been estimated to 0,07 -0,09 Scr per KWh (1 Scr= 0,1 £) if the quantity handled is 300-600 tons of straw per year.

Trials with use of grass and other cellulose crops for direct burning have so far not been carried out. Some very preliminary calculations show that 35 MWh could be harvested per ha per year. Energy input requirements amount to 5 MWh per ha and year for growing, harvesting and handling, provided that no energy is used for drying of the stuff, that is if sun can be used in the field to bring down the water content to an acceptable level.

Costs for growing, harvesting and handling have been estimated to
0,13 Scr per KWh.

Growing of solid fuel crops in agriculture therefore can be an
interesting technic if problems with high water and ash contents can be
overcome.

ETHANOL

Ethanol can under certain conditions be used as an engine fuel in
both petrol and diesel engines. Water free or water mixed ethanol can be
used depending on engine construction and fuel mixture.

Literature shows different figures for energy consumptions and production cost for ethanol. Both technic, energy comsumption and costs therefore still are not known by enough certainty.

Energy content in ethanol amounts to about 6,5 KWh per liter. For
growing of raw materials and for the conversion process is required roughly
the same amount of energy. However, even if ethanol production
is not a net producer of energy it may very well be a technic for raising
the energy quality from f.i. solid to liquid fuels.

Biproducts are important both for energy efficiency and for costs.
If biproducts are included in energy efficiency calculations, ethanol may
very well be a net producer of energy.

Costs for growing of raw material and for conversion of ethanol have
been estimated to 3,20-3,80 Scr per liter. Returns from biproducts, used
as cattle feed, have been estimated to 1,20-2,10 Scr per liter of ethanol.
Net cost therefore amounts to between 1,10 Scr (very favourable conditions)
and 2,60 Scr per liter ethanol.

VEGETABLE OIL

Vegetable oil can replace diesel oil in diesel engines. Only minor
adjustments of the engines seem necessary. Trials have shown results near
to those of diesel oil. Together with ethanol, vegetable oil is the only
renewable engine fuel available during the eighties.

One liter of vegetable oil contains roughly 9 KWh. For growing and
extraction of oil is required 7 KWh. Net energy production therefore
amounts to around 2 KWh per liter vegetable oil. In addition to that straw

is received as a biproduct containing 12 KWh per liter produced vegetable oil.

Costs for growing of oil seeds and for oil extraction have been estimated to 3,50 Scr per liter vegetable oil. Returns from biproducts, used as feed, are estimated to 1,30 Scr per liter vegetable oil. Net cost therefore is estimated to 2,20 Scr per liter vegetable oil.

ECONOMIC FEASIBILITY

Although the new renewable domestic energy sources fulfil the national energy policy goals, they have to compete economically with conventional non-renewable and in most cases imported energy sources.

Economic feasibility seems most favourable for solid fuels from biomass. Costs for agriculture solid fuels have been estimated to 0,07-0,13Scr per KWh. Present cost for petroleum oil, used for heating purposes, is about 0,13 Scr. Costs for coal and peat on the other hand have been estimated to 0,03-0,05 Scr per KWh. These two energy sources, however, have negative effects on the environment, especially coal. Neither are they renewable. Only peat is domestic.

Costs for ethanol and vegetable oil have been estimated to well above production costs for petrol and diesel oil. Cost for ethanol has been estimated to about 2,00 Scr per liter. To replace one liter of petrol 1,5 liter of ethanol is needed. Production cost for petrol is about 1,20 Scr per liter.

Vegetable oil cost has been estimated to 2,20 Scr per liter. To replace one liter of diesel oil 1,1 liter of vegetable oil is needed. Production cost for diesel oil is about 1,00 Scr per liter.

To enable ethanol and vegetable oil to compete on the market, government support is needed. Such support can be motivated by the advantages of having domestic, renewable engine fuels available in case of disturbances or interruption in international trade.

FOOD AND FIBRE OR ENERGY-IMPLICATIONS OF BIOMASS ENERGY FOR CANADA

STELIOS M. PNEUMATICOS
Conservation and Renewable Energy Branch,
Energy, Mines and Resources Canada

Summary

Biomass is currently the source for over 3% of Canada's total primary energy. This is achieved primarily through combustion of mill residues in the forest industry system for process heat. Government policies and programs have aimed to accelerate this energy option by providing financial incentives for retrofitting. At the same time, a moderate scale R&D program has been launched to evaluate biomass resources and to develop conversion technologies and more desirable fuels. However, because of an increasing deficit of oil there is growing interest for substitution of liquid fuels with more abundant energy sources. The possibility of alcohol fuels production from biomass feedstocks is a subject of intense public interest; particularly since biomass resources are available in all regions of the country. Carbohydrate containing crops such as grains could be easily converted to ethanol to substitute a significant portion of gasoline. The implications of such a strategy however in terms of food production and exports may be very serious. The forest resource on the other hand seems to hold the potential for more energy supplies. Mill and logging residues, currently unmerchantable wood and ultimately biomass energy plantations can provide more than 10% of Canadian primary energy without adverse effects on forest products production and trade.

Background

Canada is a large and resource rich country with a relatively small population. Compared to the rest of the world, Canada has some of the highest per capita energy resources with extensive reserves of fossil fuels, uranium, and renewable resources such as hydroelectric power, sunlight and biomass. Few of these resources, however, are evenly distributed throughout the country. Most of the known reserves of fossil fuels are located in the western part of the country, mainly in the province of Alberta. Over two-thirds of the Canadian people live over 3000 km away, in the eastern part of the country.

Energy consumption per capita in Canada is among the highest in the world because of a severe climate, dispersed population with a high standard of living, and a high degree of primary manufacturing. Canadians use approximately 9 EJ of primary energy derived from various sources as indicated in Figure 1. Biomass derived energy accounts for 3.1% of primary energy; almost the same contribution to that of nuclear. The important point, however is that over 40% of Canada's primary energy requirements come from oil.

Canada is not immune to the oil crisis brought about since 1973 for several reasons. The most important one, is that conventional oil reserves are rapidly diminishing and the country can no longer rely on them to supply the same high proportion of energy as now. Second, even though the country has extensive reserves of heavy oils, oil sands and, perhaps, frontier oil, the production of oil from those sources and it's delivery to markets is extremely capital and time intensive. A plant to extract oil from Alberta oil sands has a capital cost of over $350,000 per tonne of daily capacity. Production costs for such oil are accordingly very high. In addition to expensive production facilities there is the need for extensive pipelines to deliver the resource to consumers. Thirdly, and just as important, even if Canada were to become self sufficient in oil, it could not ignore the global shortage because of it's major trading interests. If Canada's trading partners were to suffer economically because of energy shortages, Canadians would suffer too since they would not be able to export their commodities.

The outlook for Canadian oil supplies and requirements is not clear. The country is presently a net importer of oil; further, the gap between domestic supplies and demand is expected to widen in the next few years. Energy conservation, oil substitution and heavy oil and oil sands developments, are all high priority policies of the Canadian government.

Present Contribution of Biomass Energy

The potential role of renewable resources, and particularly biomass is well recognized in Canada. The total contribution of renewables to Canada's primary energy is 26.2%; hydro-electricity accounts for 23.1%, and biomass for 3.1%. Energy from biomass in Canada is synonymous with wood-derived energy. Further, it is primarily energy derived through combustion of mill residues by the forest industry for heat, steam or electricity (through cogeneration). Another significant biomass energy user is the residential sector in rural areas. Many Canadians have been purchasing wood stoves, or furnaces and they use them either as a supplementary source or in a few cases as the only source for space heating.

In order to stimulate biomass energy in Canada, the government has adopted a series of programs which aim to directly substitute fossil fuels with wood and other residues. They include direct financial incentives for retrofitting industrial energy systems to use wood residues, and R&D programs to evaluate the resource and improve the various biomass energy technologies. These programs are well received by both the industry and the public in general, because they lessen the country's dependency on oil imports.

Future Role

Although biomass energy in Canada is advancing very rapidly, the continuous public interest in oil and oil product substitution raises the issue of potential liquid fuels from a renewable resource. Alcohols in particular have been viewed as an alternative to gasoline and diesel in the transportation system. This interest is amplified by the fact that

other countries with extensive biomass resources have adopted alcohol
fuels (mainly ethanol) in order to conserve oil products. A brief
examination of the Canadian biomass resources, however, points out the
particular problems and opportunities.

Production of ethanol from grains and other carbohydrate containing
agricultural crops is technologically straightforward. Plants could be
built relatively fast and the product could be used in blends with
gasoline. Experience in the United States does not indicate any serious
problems even in cold climates. Canada's agricultrual crops which would
be available in large quantities for ethanol production are principally
wheat and barley. Canada produces annually about 20 million tonnes of
wheat and 11 million tonnes of barley (4). If all the exported Canadian
wheat were converted to alcohol fuel, the gasoline replaced would be less
than 16% of current national consumption. This percentage would increase
to 33%, if the total production of both wheat and barley were allocated
to alcohol fuel production (Figure 2). Projected production economics
and energy balances do not seem very attractive. The moral and social
implications of such diversions of food or feed resources from the
domestic or export markets are too important to be ignored. It is
considered therefore, highly unlikely that Canada would embark in such
large scale distortion of grains trade. It may be probable however, that
small quantities of ethanol fuel may be produced from food processing
residues, cull crops and generally carbohydrate-containing wastes;
particularly in situations where there exist environmental credits, low
cost non-petroleum energy, and markets for process byproducts.

The use of lignocellulosic biomass (primarily wood) for alcohol fuel
production is not as easy a technological process as in the case of a
sugar or starch containing feedstock. Both the gasification-methanol
syshesis process and the cellulose hydrolysis-fermentation to ethanol
process are at the development level. Canada is participating actively
in the worldwide search for biomass to liquids conversion processes.
Canadian research organizations are also investigating the ways and means
of utilizing alcohol. The methanol option is attractive in Canada
because of the availability of a variety of resources for its

manufacture. In addition to biomass, Canada could use natural gas, coal and peat to complement methanol production capacity or to optimize the production process. Hybrid methanol production systems using biomass and natural gas or biomass and electricity have been discussed. The option of ethanol production from cellulosic biomass is also under investigation.

The supply potential of forest biomass for energy does not present the limits associated with agricultural crops. If residues from all Canadian forest operations were converted to alcohol, and if the product was suitable for direct substitution of gasoline, the amount of gasoline which could be replaced would be over 90% of current consumption (Figure 3). This estimate of forest biomass is additional to the current harvest or projected future requirements for industrial wood by the forest industries. Present harvest of industrial wood is 51 million oven dried tonnes ($155 \times 10^6 m^3$) while the annual allowable cut is estimated at 97 million tonnes (3). In addition, Canada could draw from its remote forest lands and ultimately from forest energy plantations.

Recognizing that the lignocellulosic resource is the most important biomass feedstock for future energy supply, the Canadian government is assessing biomass availability by type, quantity and projected harvesting costs. Preliminary data is becoming available and it is summarized in Figure 4 (1-3). It is generally agreed that residues from forest, farm and human activities represent the best marginal candidates for increasing the energy contribution of biomass. They are available for energy production at costs of $1-$2/GJ ($20-$40/oven dried tonne). Mill residues, in particular are available on site at less than $1/GJ and they are used for combustion by the forest industry as mentioned earlier. Logging residues normally left on the forest floor, during harvesting for industrial wood, are estimated to be the next major cost effective source of biomass derived energy. Roundwood harvested specifically for energy could come from currently unmerchantable forests or from energy plantations established under an intensive forest management system.

Conclusions

At costs which are significantly lower than equivalent world priced oil, Canada can draw from its biomass resources at the rate of 100 million tonnes per year or about 2 EJ of primary energy. The most cost effective source of such biomass supply is the residue generated primarily from forestry operations and to a much lesser extent from agricultural and residential activities. Utilization of such residues for energy often provides environmental benefits. Although the most efficient method of utilizing biomass for energy is usually through combustion, the potential production of liquid fuels ranks high in the public interest. In order to keep its options open, Canada is researching the various conversion technologies for turning biomass into a liquid fuel. The forest resource could provide the feedstocks for the production of significant quantities of liquid fuels without adversely affecting current production and trade patterns for forest product commodities.

References

1. InterGroup Consulting Economists Ltd., 1978 Liquid Fuels from Renewable Resources: Feasibility Study, Environment Canada, Ottawa

2. Love, P., 1980. Biomass Energy in Canda, Its Potential Contribution to Future Energy Supply. Prepared for Energy, Mines and Resources Canada.

3. Overend, R.P., F.L.C. Reed and C.R. Silversides, 1980. Energy and Forestry in Canada. National Research Council (N.R.C.C. 18449), Ottawa.

4. Statistics Canada, 1979, Grain Trade of Canada: 1977-78 Statistics Canada (22-201), Ottawa.

SOURCES OF PRIMARY ENERGY IN CANADA

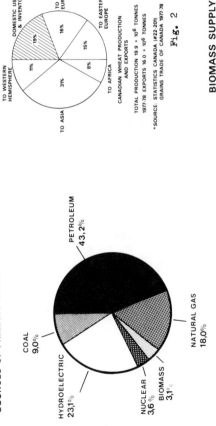

Fig. 1

PETROLEUM 43,2%

COAL 9,0%

HYDROELECTRIC 23,1%

NUCLEAR 3,6%

BIOMASS 3,1%

NATURAL GAS 18,0%

CANADIAN WHEAT PRODUCTION* AND ITS POTENTIAL FOR LIQUID FUEL SUBSTITUTION

IF ALL EXPORTED CANADIAN WHEAT WERE CONVERTED TO ALCOHOL THE GASOLINE REPLACED WOULD BE LESS THAN 16% OF CURRENT NATIONAL CONSUMPTION

POTENTIAL ALCOHOL YIELD FROM EXPORTED WHEAT IS 6×10^6 M³

Fig. 2

TO WESTERN HEMISPHERE

DOMESTIC USE & INVENTORY 19%

TO WESTERN EUROPE 16%

TO EASTERN EUROPE 15%

TO AFRICA 8%

TO ASIA 31%

11%

CANADIAN WHEAT PRODUCTION AND EXPORTS

TOTAL PRODUCTION 19.9 × 10⁶ TONNES
1977-78 EXPORTS 16.0 × 10⁶ TONNES

*SOURCE: STATISTICS CANADA (#22 201)
GRAINS TRADE OF CANADA 1977-78

**IF ALL CANADIAN WHEAT AND BARLEY WERE CONVERTED TO ALCOHOL THIS QUANTITY WOULD INCREASE TO 33%

BIOMASS SUPPLY CURVE FOR CANADA

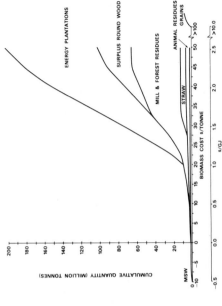

CUMULATIVE QUANTITY (MILLION TONNES)

ENERGY PLANTATIONS

SURPLUS ROUND WOOD

MILL & FOREST RESIDUES

STRAW

ANIMAL RESIDUES

GRAINS

MSW

BIOMASS COST $/TONNE

$/GJ

SOURCE: OVEREND

Fig. 4

FOREST BIOMASS POTENTIALLY AVAILABLE FOR ENERGY*

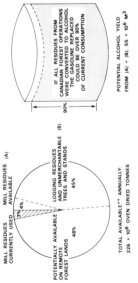

Fig. 3

SOURCE: EMR. UPDATE 1979

MILL RESIDUES CURRENTLY USED 3%

MILL RESIDUES AVAILABLE (A) 4%

POTENTIALLY AVAILABLE ON REMOTE FOREST LANDS 48%

LOGGING RESIDUES AND UNMERCHANTABLE (B) TREES AND STANDS 45%

IF ALL RESIDUES FROM CANADIAN FOREST OPERATIONS WERE CONVERTED TO ALCOHOL THE GASOLINE REPLACED COULD BE OVER 90% OF CURRENT CONSUMPTION

POTENTIAL ALCOHOL YIELD FROM (A) + (B): 55 × 10⁶ M³

TOTAL AVAILABLE** ANNUALLY:
226 × 10⁶ OVEN DRIED TONNES

* SOURCE: OVEREND, REED AND SILVERSIDES

** THIS ESTIMATE DOES NOT INCLUDE THE ALLOWABLE CUT FOR INDUSTRIAL WOOD

THE USE OF BIOGAS FOR THERMAL, ELECTRICAL

& MECHANICAL POWER GENERATION

D.J. PICKEN and M.F. FOX

Leicester Polytechnic

Summary

The balance of energy output and energy demand of a range of digester sizes is considered in detail. It is shown that the best use of the gas is determined at least partly by the size of digester and that considerations of waste heat recovery from an engine can be at least as important as thermal efficiency. A distinction is made between the low grade (heating) energy requirement of a digester and the high-grade (mechanical) requirement.

A low technology device is described which is capable of supplying the mechanical power needs in terms of pumping slurry, stirring and pumping in digester heating water, as well as supplying the low grade heating through a self-contained heat recovery system.

1. INTRODUCTION

Anaerobic digester systems may be employed in any of three modes – as waste material treatment, as energy generators, or as a combination of both.   In all cases a methane-rich gas is produced, and the economics of the system may be influenced by the manner in which the gas is used.   The research at Leicester Polytechnic has concentrated on the use of biogas in engines and other prime movers, but it has been considered appropriate to undertake the research in the knowledge that to be relevant it had to be undertaken with an understanding of the requirements of the digester and the needs of the farm or process unit in which the digester is situated.

2. BALANCE OF GAS PRODUCTION AND ENERGY NEEDS

Digester gas production is often quoted as net production, i.e. gas yield after some of the gas has been used to heat the incoming slurry to the digester and overcome heat losses to the atmosphere.   In practice it is known that this heat can often be supplied from waste heat from an engine driven by the gas.   This balance will of course be a function of the type of digester, the feed biomass in use, and the retention time.   As an example we have chosen to use the figures supplied by BABA* relating to pig slurry with a 10 day retention time.   Figure I is an extension of these results assuming a linear relationship between total gas yield and digester size.   The upper line represents the gross gas yield, which, to be of any value, must all be burnt in a heater, an engine or some other thermodynamic device.

Since it is common to consider electrical power as the most desirable product from gas, line 2 has been drawn so that the difference between line 1 and line 2 is the electrical power available (assuming 25% efficiency). The distance between line 2 and the base line then represents the waste heat from the engine-generator unit, of which up to about 70% is recoverable, depending on the sophistication of the waste heat recovery system.

The heat requirement of the digester has again been taken from the BABA figures, the distance of line 2 from the base representing 'worst case' feed heating, and from 3 to the base 'worst case' feed heating and digester cooling (with lagging assumed for smaller digesters).

This graph therefore shows clearly that the digester heating can be achieved by waste heat recovery from any system which is at least 50% efficient.

The mechanical power requirement of the digester is more complex. Power is normally required for slurry feed pumping and for some form of stirring. Whereas pumping requirement is a fairly easy calculation, there appear to be many different methods in use and ideas on the total need for stirring.

The solution used for Fig.1. is that stirring would be on a slurry circulation basis with a two hour slurry movement cycle suggested by Mills. More commonly gas circulation would be used but this would also come within the power range suggested.

The stirring and feed power must come from the high grade electrical or mechanical power and is thus shown as the shaded area between lines 1 and 5, decreasing the net high grade energy output.

3. CHOICE OF POWER UNIT

Fig.1. has deliberately been drawn to represent the smaller sized digester plants - larger plants become progressively easier to choose the correct engine/generator system. Three bands of operation are shown:-

(i) Digesters of above about $90m^3$ (pig slurry).

These can operate engines of 10 kW output and above (continuous rating). Water cooled engines are available and have, or can be modified to have, whatever degree of heat recovery is required. It is suggested that in these cases, unless there is a need for heat energy in the immediate vicinity of the plant, an internal combustion engine providing electrical power will be used to burn all the gas produced by the plant.

There may then, for very large plants, be some need to decide whether the engine should be spark or compression ignition.

(ii) Digesters in the range 45 to $90m^3$.

The power range of engines which can be operated continuously is from 5 kW to 10kW. The majority of engines in this range

will be air-cooled with a consequent increase in the
difficulty of getting sufficient waste heat recovery.

The choice may then be between the comparatively high unit
cost of a continuously operating engine with waste heat
recovery and using an air cooled engine intermittently with
a separate water heater to provide digester heating.
Perhaps the most desirable choice is to use the gas in this
range for heating only, if a suitable use for the heat energy
is nearby.   In this latter case it should be noted that the
penalty to be set against the supply of 5 to 10 kW of heat
power is the need to buy in between $\frac{1}{2}$ kW and 1 kW of high
grade (electrical) power.   This penalty may be increased
because of the need to use pumps of a higher power to avoid
operating on very small sizes of pumps, motors and pipework.

(iii)   <u>Digesters below 45m^3 capacity</u>

Although capable of powering engines, it is doubtful whether
it is worth the investment of installation, particularly as
waste heat recovery would be difficult and the engine would
therefore have to share the available gas with a digester
heating unit.   Once again any heat energy output would have
to be set against the electrical energy needed to operate
pumps, etc.

The Leicester Polytechnic gas operated slurry pump, Fig. 2   has
been developed to provide the mechanical power required by a digester in
a simple and reliable form, using only the heating value of the gas
which is required for digester heating anyway.   It has an extremely low
efficiency of converting heat energy to mechanical work, but the
advantage of a very good, self-contained heat recovery system.   Since the
mechanical work requirement of the digester is only of the order of 2% of
the gross energy output (Fig.1.), the total energy utilisation is much
more important than mechanical efficiency.

The action of the pump is described in ref (2) and its operation is
illustrated in Figure 2.

A prototype of this pump has been operated without failure in conjunction with an experimental digester for about 400 hours over 4 months at Brooksby College of Agriculture.

References

1. News letter of British Anaerobic & Biomass Association
    (No.1. March 1980).

2. Analysis of a low technology steam operated water pump.
    Polytechnics Symposium on Thermodynamics & Heat Transfer. Nov. 1979.
    D.F. Brewin and D.J. Picken.

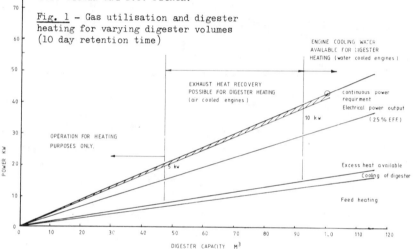

Fig. 1 - Gas utilisation and digester heating for varying digester volumes (10 day retention time)

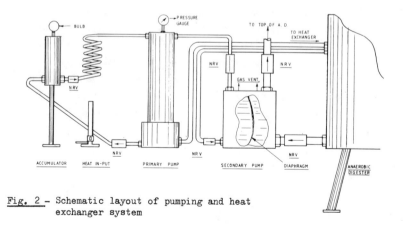

Fig. 2 - Schematic layout of pumping and heat exchanger system

GENERATION OF CURRENT FROM BIOGAS

E. DOHNE

Kuratorium für Technik und Bauwesen in der Landwirtschaft (KTBL), Darmstadt, Federal Republic of Germany

Summary

When planning biogas plants intended for farms, it appears politico-economically wise to design them in accordance with the overall livestock. One of the key problems will consist then in actually making full use of the producible gas. Subject of fierce discussion is at present that full use for the generation of current by means of cogeneration sets with the possibility that one part of the waste heat might serve as process energy for the digester as well as for comfort heating, hot-water heating and drying of agricultural products. A rough calculation shows that in this case the number of LU required for heating a dwelling house amounts to about 80 to 100. In addition, it shows that the production cost of one kilowatthour of current (kWh_{el}) on an average farm is higher than supplied by the electric power companies. And surplus energy can be sold only for a fraction of the production cost. Hence, the generation of current from biogas under conditions as are prevailing in the Federal Republic of Germany is normally no solution unless in particular cases. Detailed preliminary studies are imperative.

1. Introduction

When planning agricultural biogas plants it becomes apparent that only
parts of the biogas susceptible to be produced in the course of one year
on an average farm can be used on the farm itself. From about 30 LU on,
there will be produced a more or less large amount of surplus gas, the
so-called problem gas, which must be burnt off. A "biogas balance" diagram shows the situation of a specific farm. The generation of current
by means of a cogeneration set often appears as an ideal utilization, the
amounts of waste heat produced in the course of generation being used
in all those ranges of application where normally biogas would have been
used directly (process energy, comfort heating, hot-water heating, drying). From the angle of social economics, this scheme makes savings
in primary energy possible. The point is whether this is also useful from
the angle of private economics, as only parts of the current and heat produced can be consumed on the farm itself. Surplus current may be fed
into the mains. This issue can be considered from the legal, technical,
energetic, social-economic and cost-effective point of view.

2. Problems involved by the generation of current from biogas

The legal aspects of the generation of current from biogas vary from
country to country. The electric power supply companies in the Federal
Republic of Germany are bound to accept the excess current produced
from biogas, pay it however only at a price between 0.03 and 0.06 DM/
kWh_{el} and make the avoidance of any detrimental effects on the public
supply a condition. The pertinent safety disconnecting devices cause
additional costs.

Cogeneration sets from 15 kVA on are on the market. Sets between 15
and 60 kVA are being prevailingly in consideration for farms (Tab. 1).
In that range it is almost always possible to make use of rather economy-
priced asynchronous generators. The technical problems rather lie in the
today's short life of the driving engines - modified vehicle engines. By
full-service contracts, involving the corresponding charges (0.02 to

0. 06 DM/kWh$_{el}$), this problem can be obviated in part without however preventing down-times.

Economically speaking, the generation of current from biogas for farms can only be good policy if the foreign energy saved by biogas should cause higher costs than the own production cost for this purpose with the use of biogas. In short, higher charges for foreign current than for own current. Future increases in energy prices may obviously be taken into consideration. Detailed studies would necessarily determine a mixed price between the kWh$_{el}$ produced on site and the kWh$_{therm}$.

In the following it will be tried to determine the cost price of the kWh$_{el}$ on the basis of the "machinery cost fraction" of the cogeneration set and the "energy cost fraction" of biogas (biogas price) according to the investment charges for the biogas plant and the given operating conditions. (Another approach to the calculation would place the annual costs of the biogas plant and the cogeneration set together to the debit of the usable effective energy.)

The evaluation of the usable "thermal" energy is analogous to that of the utilization of a regular gas heating boiler. The value is compared with the today's price of current paid to the power supply company and the company's acceptance price for surplus current.

It must be stated that the generally suitable "small power units" compared to high-power sets are relatively expensive in purchase cost (in part over 100 per cent more expensive). The waste heat of the set can be made use of for the process energy required for the biogas plant and, moreover, only up to a certain output (or number of LU) and, to a large extent, only in winter. To fully supply a dwelling house heating system from waste heat the number of LU required amounts to about 80 to 100. The full-service contracts as usually signed charge about 0.02 to 0.07 DM kWh$_{el}$. A certain insecurity in this connection is to be explained by the still lacking experience. In the fraction of current used for one's own requirements it is only the prevailing kWh-price that can be saved, as nobody will be inclined to work without a standby supply by the power supply company. On the federal average this amounts to about 0.12 DM/kWh. The price charged to the respective power supply company for surpluses of current does not exceed 0.03 to 0.06 DM/kWh.

Calculations of profitability have to take into account the local conditions. The prime cost of the current is composed of the proportion of machinery cost for the cogeneration set (small power unit) and the energy cost fraction, i.e. the biogas production cost. The latter is determined by the question as to what extent the exploitation of gas with the cogeneration set being used is allocated to the use of current and heat. The price of the self-produced current rises in the same measure as heat cannot be utilized.

The data of the different manufacturers of biogas power units concerning the electric and thermal output as well as the power loss of their units vary largely.

Biogas energy content	100 %	Simplified calculation of average value for biogas
electric output	20 to 37 %	25 %
thermal output	46 to 66 %	55 %
losses	8 to 25 %	20 %

Thus the extent of the variations presented by the yield of current from 1 cu. m of biogas (which also varies in its energy content, an average of 21 MJ/cu. m) is relatively large, of the order of 1.1 to 1.9 kWh_{el}/cu. m and 2.9 to 3.8 kWh_{therm}/cu. m. That is to say, with a virtual gross range of variation of 1 to 2.0 cu. m of biogas/LU · d, the yield of current may vary between 1.1 and 3.8 kWh_{el}/LU · d. As mean value is calculated:

$$2.3 \ kWh_{el}/LU \cdot d \qquad \text{and} \qquad 5 \ kWh_{therm}/LU \cdot d$$

On the basis of a usual process energy demand of 15 to 30 % of the gas production - produced via a gas heating boiler -, and taking into account that combustion efficiency may be left unconsidered in this case, the amount of biogas required for the digester heating is of the order of 0.7 to 1.4 kWh_{therm}/cu. m so that 1.6 to 2.9 kWh_{therm} of biogas are still available for comfort heating and others (1.6 to 5.8 kWh_{therm}/LU · d). For an estimate, after 25 % of net process energy (=about 30 % of gross process energy) being deducted, there are still 2.7 kWh_{therm}/LU · d out of the 5 kWh_{therm} "freely" available for whatever use, for example, heating in winter.

A medium one-family house needs about 80 to 100 LU to meet the comfort heating requirements to the full. The use of the free heat in summer is problematic. In small plants the amount of the available heat that can be made use of throughout the year is estimated to be as little as 50 to 70 %. Practical tests have been lacking so far.

The purchase price for the units, which varies between 1000 and 3000. - DM/kVA, must be added to the costs for the foundations, the electrical and thermal installation including possible special directions, as well as extra sound insulation (in part up to 1000. - DM/kVA). The evaluation of the "freely" disposable "thermal energy" can be made only by analogy with the production in a regular gas heating boiler. On the basis of a 20-year life, an operating time of 2000 h/year, a combustion efficiency of 90 %, the machinery cost fraction to be reckoned with is about 0.005 DM/kWh_{therm}.

The machinery cost for some cogeneration sets can be taken from Table 2. It shows that the machinery cost fraction for the kWh_{el} alone is of the order of 0.09 to 0.19 DM. On the other hand, full utilization of the waste heat reduces the machinery cost fraction only by about 5 %.

To the machinery cost fraction must be added the energy cost fraction. According to Fig. 1, this cost fraction can be limited. It is primarily dependent on the capital invested into the biogas plant, the gas yield and the proportion of process energy and the use of the "free" heat. Calculating investment charges in the case of a thermophile plant on a realistic

basis of presently 1,400.- DM/LU, a gross yield of biogas from cattle
of about 1.5 cu.m/LU (= 0.3 cu.m/kg of dry matter), a gross process
energy demand of 30 %, the energy cost fractions,according to Fig. 1,

amount to about:	with the utilization rate of the "freely" disposable heat being:
0.09 DM/kWh$_{el}$	100 %
0.13 DM/kWh$_{el}$	50 %
0.20 DM/kWh$_{el}$	0 %

That is to say, the possibility of generating on site the kWh$_{el}$ as a whole
(machinery cost and energy cost) for less than 0.20 DM is limited to
particular cases, while the kWh$_{el}$ supplied by the power supply company
is being charged on an average at about 0.12 DM. Consequently the generation of current from biogas is profitable only in particular cases.

Table 1: Livestock (LU) allocated to size of generator

LU	daily yield of current kWh/d	average daily/annual operating hours with units of		
		15 kW (19 kVA)	30 kW(37 kVA)	45 kW (56 kVA)
20	45	3 / 1100	-	-
40	90	6 / 2200	3 / 1100	-
80	180	12 / 4400	6 / 2200	4 / 1500
120	270	18 / 6600	9 / 3300	6 / 2200
200	450	(24 / 8760)	15 / 5500	10 / 3700
300	675	-	22 / 8000	15 / 5500

Tab. 2 - Machinery cost for biogas-generated current

driving engine	generator		high life synchron	
	mean life			
	asynchronous	synchronous		
kVA		18	35	
kWel	15	15	30	
kWtherm	30	30	55	
h/a	3000	2000	2000	3000
kWh el/a	45000	30000	60000	90000
purchase price complete	25000	30000	43000-	50000
full service DM/kWhel	0,05	0,05	0,03	0,03
annual costs				
depreciation %	13	10	5	7,7
DM	3250;	3000,	2150,-	3850,-
interests (4%) DM	1000	1200,	1720,-	2000,-
full service DM	2250,-	1500,-	1800,-	2700,-
total DM	6500,-	5700,-	5670,-	8550,-
machinery costs DM kWhel				
use of free waste heat 0%	0,144	0.19	0.095	0,095
100%	0,138	0,184	0.089	0.089

Fig. 1 - Energy cost fraction of biogas-generated current

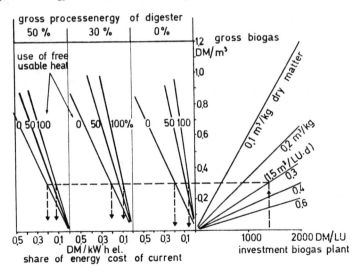

STUDIES OF THE POTENTIAL FOR ETHANOL PRODUCTION
FROM SELECTED BIOMASS CROPS IN TEMPERATE CLIMATES:
AN ENGINEERING APPROACH

C.E. Dodson,
Helix Multi-Professional Services
and
S.R. Martin
Stone & Webster Engineering Limited

SUMMARY

A scheme is described for the production of 100,000 m^3/year of fuel ethanol from wheat, with straw serving as the source of process energy.

Based on this scheme, production cost from wheat at £100/ton and straw at £15/ton is estimated at 29 pence/litre which on an equal energy basis is over $2\frac{1}{2}$ times the present wholesale price of premium gasoline in Europe.

Cost of feedstock is the dominant element in ethanol cost. It is likely that selected non-cereal crops can provide fermentable carbohydrate at significantly lower cost than cereals, however, their high water content limits the possibilities of their transport and storage, hence also their potential for lowering the cost of fuel ethanol.

INTRODUCTION

In recent years the large-scale production of ethanol as a vehicle fuel has come to be associated mainly with low-cost sugar cane (as in Brazil) or with a combination of a local grain surplus and low-cost coal (as in the USA).

Three general factors will however operate to make the fuel ethanol option more widely attractive as time goes on:

    a)   increase in cost of fossil-derived vehicle fuels
    b)   improvements to agricultural technology
    c)   improvements in processing technology as compared with the current technology based largely on traditional beverage alcohol practices.

It is of particular interest to make estimates of likely production costs in the cool climates which predominate in the developed countries, since they have the greatest consumption of vehicle fuels.

It should also be remembered that cool climates are not at any serious disadvantage against the tropics in respect of potential for biomass production since a) annual incident solar energy falls off quite slowly with latitude and b) water supply is more generally adequate at higher latitudes.

Nor should it be too readily accepted that agricultural potential has reached a limit just because all the land is already being farmed. Large areas given over to pasture have very low productivity; and it is questionable whether production of meat and dairy products should receive large subsidies in preference to "energy farming."

Wheat was selected as feedstock for initial study for the same reasons that have made it the staple food in many parts of the world - adaptability to a wide range of soil and climate, ease of storage, and ready conversion of starch to sugar by enzymes.

To minimise use of of fossil fuel, crop residues and unfermented residue from the process are considered as sources of process power and heat. Anaerobic fermentation was selected for the scheme studied as a means of conversion of these materials to fuel gas because it offers interesting prospects of providing protein feed and organic fertiliser as byproducts.

PROCESS DESCRIPTION AND BASIS (WHEAT)

Assumed grain composition and the derived hourly quantities are shown in Table 1 and the estimated energy flows in Table 2.

Storage on the plant is assumed for grain (12 months' supply) and for straw, protein, compost, and product ethanol (30 days' throughput each).

Grain Milling and Cooling

Grain is metered by weigh feeders to hammer mills. The mills are protected by guard screens against entry of stones or tramp metal.

The milled grain is mashed at a controlled temperature with return stillage, fresh process water, and previously prepared liquefying enzymes. The mash is heated by steam injection and passed through continuous tubular cookers to solubilise the starch. The cooked mash is cooled to brewing temperature (35°C) by flashing at reduced pressure. Enzymes for conversion of starch to sugar are added, and the solids content is adjusted by water addition.

Fermentation, Enzyme and Yeast Preparation, and CO_2 Production

The prepared mash and previously prepared yeast are charged to one of a number of batch fermenters in which mixing and temperature control is maintained by an external pump and cooler. The carbon dioxide evolved during fermentation is scrubbed with a small flow of process water to recover ethanol. When conversion of starch to sugar to ethanol is approaching completion, the contents of the fermenter are pumped to a holding vessel. The enzyme-producing organism and the yeast are each cultured in batch equipment, using broths prepared from portions of the cooked mash, added nutrients, and inoculum recovered from previous batches. Provided that an assured outlet can be found, e.g. in the carbonation of beverages, it will prove highly profitable to install the additional compression, purification, storage and despatch facilities required to produce liquid carbon dioxide.

Distillation and Denaturing

Rectified spirit (approx. 95 wt% ethanol) is separated overhead from the beer still.

The rectified spirit flows to the drying column to the top of which a liquid hydrocarbon stream is returned as reflux. Ethanol product (approx. 99.5 wt%) is withdrawn from the base of the drying column. The mixed vapours from the drying column when condensed separate into a hydrocarbon phase, which is the reflux for this column, and an aqueous phase. Hydrocarbon and ethanol are recovered from the aqueous phase in the recovery column, the vapours from which are condensed and recycled via the liquid separator.

The ethanol product is denatured by addition of gasoline in a controlled proportion.

Digestion and Treatment of Digester Products

After separation of a proportion of liquor which is recycled to cooking, the net residue from the beer still is mixed with a controlled proportion of chopped straw and fed to the digester. Temperature of the liquor is adjusted to maintain the digester contents at about 35°C. In a continuous operation with a mean residence time of more than 20 days, carbonaceous matter is to a great extent converted to methane, CO_2, and microbial biomass.

Treatment of Digester Products

Unconverted fibre is recovered by a screen, mechanically
dewatered, composted over a period of about 3 weeks and transported to
the farms for spreading as a fertiliser or use as a horticultural growing
medium.

Next, a dry product of about 37% protein content is obtained by
flocculation, screening, dewatering, and drying. It is anticipated that
this material will be useful as an ingredient of prepared animal feeds,
approximating in value to yeast on the basis of equivalent total nitrogen
content. The residual water is given conventional aerobic and clarification treatments to meet all requirements for discharge and to
recover fertiliser in slurry or solid form.

The evolved biogas is collected by floating caps which act as
gasholders and is burned in the boilers. When necessary for safety,
excess gas is flared.

Process Heat and Power

Steam is raised at high pressure in boilers fuelled by biogas.
The high pressure steam is expanded in turbines to provide medium and low
pressure steam and power for process requirements, and a surplus of
electrical power for export.

The cooking, distillation, and drying steps include process
features which contribute to a high overall fuel efficiency.

TABLE 1 - FEED AND PRODUCT MASS FLOWS - ETHANOL FROM WHEAT
(kg/hr)

	Grain	Straw	Ethanol kg/hr	CO_2	Compost	Protein
Component						
Water	4,690	4,020	50	9	2,080	225
Starch + Sugar	21,976					
Protein	4,020	959				1,665
Fat	603	369				
Fibre	770	9,102			820	200
Other volatile	838	10,308			100	2,410
Ash	603	2,042			200	
Ethanol			9,950			
CO_2				10,226		
TOTAL	33,500	26,800	10,000	10,235	3,200	4,725

TABLE 2 - ENERGY FLOWS- ETHANOL FROM WHEAT

	MW	Ratio		MW	Ratio
Grain	136.05	165.4	Ethanol	82.25	100.0
Straw	100.42	122.1	Electricity Export	14.82	18.0
			Fibre	4.34	5.3
			Protein	17.81	21.6
			Flue Gas	13.34	16.2
			Losses	14.30	17.4
			To Cooling Water	89.61	109.0
	236.47	287.5		236.47	287.5

COST ESTIMATE, ETHANOL FROM WHEAT

It is envisaged that a number of farms within a radius of about 30 km would be linked to the plant by a profit sharing or long-term contract arrangement for supply of crops and the recycling of organic fertiliser. Within such an organisation it should be possible to deliver feed to the plant at slightly below general market prices and on this basis prices of £100/ton for grain and £15/ton for straw are assumed.

Protein product is assumed to realise £200/ton at the plant gate. Organic fertiliser is conservatively assumed to have zero value after deduction of cost of its transport to the farms.

Credit at a rate of 2.5 p/kWh is assumed for exported electrical energy.

Fixed and operating capital requirements are estimated at £30 million and £20 million respectively.

It is estimated that 100 people (14 operators and 5 others on shift, plus 24 on day) would be directly employed at the plant, at an average salary of £6,000 p.a.

All price and cost figures are in terms of constant 1980 values, i.e. ignoring inflation.

Item	£000 Per Year	Pence Per Litre Ethanol
Consumption:		
Grain 268,000 t @ £100/t	26,800	
Straw 214,400 t @ £15/t	3,216	
Water 1.34 million m^3 @ £0.1/m^3	134	
Chemicals	300	
Subtotal	30,450	
Byproducts		
Protein 37,800 t @ £200/t	7,560	
Electricity 118.6 million kWh @ £0.025	2,964	
	10,524	
NET VARIABLE	19,926	19.93
Salaries 100 x £6,000/yr	600	
Overheads 50% x salaries	300	
Maintenance 3%/yr x £30 million	900	
Rates and Insurance	300	
Depreciation 10%/yr X £30 million	3,000	
Interest 5%/yr x £20 million	1,000	
FIXED	6,100	6.10
10%/yr return on £30 million	3,000	3.00
PRODUCTION COST	29,026	29.03

OTHER CROPS

General

There is probably no single 'best' crop for ethanol production. Most likely, crops will be selected and developed to suit particular environments. Plants may also be designed to suit more than one crop, to take advantage of different cropping seasons or, less predictably, occasional surpluses.

Roots and Tubers

Sugar beet and fodder beet are highly productive in some areas and the methods of handling and processing to produce sugar solution are well developed.

Crop storage is the main problem with beet. Leaving the crop in the ground is possibly the most effective storage method where the climate allows digging to continue through the winter. It has been suggested that the processing season could be extended to half the year or more by this means, however it should be noted that in sugar making practice the processing season is very much shorter than this.

The potato shares with beet the disadvantages of high costs of cultivation, harvesting, transport and cleaning. Ethanol yield per acre tends to be less than for beet, and processing cost is higher since cooking and hydrolysis steps to solubilise and convert the starch are required. Ethanol production from potatoes may be interesting mainly as an outlet for spoiled tubers and occasional surpluses.

The Jerusalem artichoke is interesting because it crops well in a wide variety of situations, requires less cultivation than the potato, and enjoys relative freedom from pests and diseases. Processing of the tubers is said to be simple; their carbohydrate (inulin) being hydrolysed under mild conditions with the aid of enzymes already present. The tubers do not store well, however, once they have been lifted from the ground.

Green Crops

A wide variety of crop plants from grasses to Jerusalem artichoke "tops" have a high yield of easily fermentable carbohydrates if cut at the optimum, unripe, stage. Cultivation and harvesting costs are potentially very low for such a crop system. A major problem with these crops however is that they deteriorate rapidly once they are cut. It seems unlikely that their harvesting could be phased to provide a continuous supply to a processing plant, therefore means of preservation are desirable.

The use of wet (i.e. non-cereal) crops for ethanol would be greatly facilitated if an economic method of preservation could be developed. Known methods are:-

a)   Silage. In making silage, fermentable carbohydrates are largely converted to lactic acid. The method therefore does not appear to be applicable to ethanol production.

b)   Sun drying (hay). A major proportion of the fermentable carbohydrate content is lost during natural drying, so this does not appear to be practicable as a route to ethanol production.

c)   Rapid drying. A drying and compacting machine located at or near the farm could in principle produce a material which would be easy to store, transport and eventually process to produce ethanol. Unfortunately, the mass of water to be removed is 20 or more times the potential ethanol product. Even with very sophisticated (and therefore expensive) machinery, it is almost inconceivable that the consumption of fuel by the dryer/compacter could be reduced enough to produce a useful energy balance from the operation.

Cellulosic Materials
 The use of cellulose materials as feedstocks for ethanol production awaits an economic process for cellulose hydrolysis to produce a sugar solution of, say, 5 to 10 wt% concentration. This potentially very important line of development is outside the scope of the present paper.

DISCUSSION
 The estimate, which is based on technology which could be used for plants designed now or in the near future, leads to a possible wholesale price of 29 pence/litre of fuel ethanol, equivalent on the basis of energy content to 42 pence/litre of premium gasoline.
 For comparison, in September 1980 retail prices of premium gasoline, converted at current exchange rates, were:

	pence/litre	
	ex-tax	with tax
U.S.A.	12.6	13.9
U.K.	16	30
France	15	34.5
W. Germany	16	29
Belgium	15	34
Austria	14.5	30
Sweden	15	29

 Tax structures could certainly be devised which would provide an adequate incentive to producers of fuel ethanol with little or no loss of revenue to the government, but at the cost of substantially higher price at the pump.
 The cost to a country's economy has to be considered in terms of the extent to which ethanol production would divert resources of land, labour, and capital away from production of food, etc.
 Benefits of ethanol production would include lessening of dependence of transport upon fossil-derived fuels. As a bonus, given proper attention to techniques, improvements may be expected to the fertility of many types of soil.
 Use of crops and process residues as a source of energy is a crucial aspect of the type of scheme described. Anaerobic digestion of straw etc. is interesting because a) moisture content of the stored biomass is not critical and b) nitrogen can be recovered as feed protein and fertiliser. Much development work remains to be done before a valid comparison can be made with other possibilities, including direct combustion.

CONCLUSIONS AND RECOMMENDATIONS
1.　　Fuel ethanol can be produced from cereal grains priced at £100/ton as feedstock and straw at £15/ton as a source of energy, at a cost of about 30p/litre.

2.　　Other food crops are restricted by their high water content and the associated storage problems to a limited season of availability, during which they might usefully substitute for cereals.

3.　　Anaerobic fermentation of crops and process residues merits continuing investigation and comparison with other means of conversion and with direct combustion.

USE OF 95 %-ETHANOL IN MIXTURES WITH GASOLINE

A. SCHMIDT

Institute of Fuel Technology,

Technical University Vienna, Austria

Summary

Ethanol specifications for use as a gasoline extender have usually required 99.5 % ethanol min. or 0.5 % water max. As ethanol forms an azeotropic mixture with water a considerable quantity of energy is required to dehydrate technical grade ethanol (95 %) to this low water content. The reason for the low water specification is possible phase separation of gasoline-ethanol-water mixtures at low temperatures.

The present investigation shows that depending on climatic conditions 95 % ethanol can be used in mixtures with gasoline alone, or mixtures of 95 % and 99.5 % ethanol must be used, or small amounts of co-solvent (propanols or butanols) are required with 95 % ethanol. The effect of the addition of lubricating oils to the fuel mixtures, as required for two-stroke engines, has also been determined.

Ethanol as a gasoline extender is already used in a few countries, others are discussing its use. A major obstacle is in most cases its higher cost as compared with gasoline. Ethanol specification for use as gasoline extender usually requires 99.5 % C_2H_5-OH min., or water 0.5 % maximum. As ethanol forms an azeotropic mixture with water, simple fractionation will yield a product with 95 % max., higher concentrations require special processing and equipment (dehydration).

The reason for setting the ethanol specification at 99.5 % is the phase behaviour of gasoline-ethanol-water mixtures. At low temperatures mixtures can separate into a high-density water-ethanol and a low-density gasoline-ethanol phase. Obviously, phase separation in the fuel system of a motor-car would seriously impair the normal functioning of the combustion engine.

Efforts to reduce cost and energy requirements of ethanol production have led to the question whether 95 % ethanol as produced by normal fractional distillation can be used as a gasoline extender.

Visual observation shows that gasoline-ethanol-water mixtures have on cooling a definite cloud-point temperature; the two-phase emulsion formed is fairly stable at that temperature. On further cooling phase separation starts, but this temperature is not so well reproducible as the cloud-point temperature: The interval is usually four to seven degrees. In the present investigation cloud-point temperatures have been determined.

The investigation consisted of two parts: In the first part cloud-point temperatures of the systems toluene-ethanol water, n-heptane-ethanol water, regular gasoline-ethanol-water and premium gasoline-ethanol-water were determined; in the second part cloud-point temperatures of regular and premium gasoline-ethanol-water mixtures in the presence of co-solvents were investigated.

Four climatic zones have been defined:

Tropical zone	minimum temperature	$+ 10^{o}C$
Subtropical zone	- " -	$0^{o}C$
Moderate zone	- " -	$- 10^{o}C$
Continental zone	- " -	$- 20^{o}C$

Permissible cloud-point temperatures for gasoline-ethanol mixtures have been set equal to the temperature minima of the climatic zones mentioned. This means that a safety margin of five to seven deg. C is available until phase separation would actually occur in a motor-car.

Gasolines used in the test are

regular grade	paraffines	72.3 %
(ROZ 88)	olefines	9.6 %
	aromats	18.1 %
premium grade	paraffines	49.5 %
(ROZ 98)	olefines	14.8 %
	aromats	35.7 %

Results of the investigation show that under certain conditions 95 % ethanol can be used in the gasoline-ethanol mixtures, tab. 1 and 2.

If this is not possible, only part of the ethanol used must be dehydrated to 99.5 %, the rest can be 95 % quality; the relative quantity of 95 %-quality necessary is also shown in tab. 1 and 2 for various conditions.

If no 99.5 % ethanol is available 95 % quality can be used if propanols or butanols are added as co-solvents. The quantities necessary are shown in tab. 3 and 4.

Tab. 1   Use of 95 % ethanol and 95%/99.5%
          ethanol in ethanol-gasoline mixtures

regular   grade gasoline; figures show relative
amounts of 95%-ethanol

	Ethanol in the mixture		
Climate	5 %	10 %	15 %
Tropical	77.5	100	100
Subtropical	66.7	84.7	100
Moderate	60	73.6	84.7
Continental	50	62.1	70.9

Tab. 2   Use of 95 % ethanol and 95%/99.5%
          ethanol in ethanol-gasoline mixtures

Premium grade gasoline; figures show relative
amounts of 95%-ethanol

	Ethanol in the mixture		
Climate	5 %	10 %	15 %
Tropical	92.6	100	100
Subtropical	83.3	100	100
Moderate	75	100	100
Continental	66.7	88.9	100

Tab. 3   Use of co-solvents in ethanol-
          gasoline mixtures

regular   grade gasoline; figures show quantity
of co-solvents necessary when using 95%-ethanol

| | Ethanol in the mixture | | |
Climate	5 %	10 %	15 %
Tropical	1	O	O
Subtropical	1.4	1	O
Moderate	1.8	1.8	1.2
Continental	2.2	2.6	2.5

Tab. 4   Use of co-solvents in ethanol-
          gasoline mixtures

Premium grade gasoline; figures show quantity
of co-solvent necessary when using 95%-ethanol

| | Ethanol in the mixture | | |
Climate	5 %	10 %	15 %
Tropical	0.3	O	O
Subtropical	0.7	O	O
Moderate	1.0	O	O
Continental	1.4	0.9	O

The addition of 5 % (vol.) of lubricating oil to the fuel mixtures - as necessary for two-stroke-engines - changes the cloud-point temperatures as shown in tab. 5.

Tab. 5   Changes of cloud-point temperatures
          caused by adding 5 % (vol.) lubricating
          oil, deg. C.

Gasoline	Ethanol in the mixture		
	5 %	10 %	15 %
regular grade	+ 2	+ 4	+ 7
Premium grade	- 4	+ 3.5	+ 8.5

THE UTILIZATION OF SUNFLOWER SEED OIL
AS A RENEWABLE FUEL FOR DIESEL ENGINES

J.J. BRUWER
B. v. D. BOSHOFF
F.J.C. HUGO
J. FULS
C. HAWKINS
A.N. v.d. WALT
A. ENGELBRECHT

Division of Agricultural Engineering,
Department of Agriculture & Fisheries, Pretoria, RSA;

and

L.M. du PLESSIS

Council for Scientific and Industrial Research, Pretoria, RSA.

Summary

Prior to and during World War II, some detailed but short-term research work was carried out on the use of sunflower seed oil in diesel engines; but little in the way of thorough investigations in this field has since been published. Escalating prices of petroleum-based fuels re-awakened interest in possible alternative renewable fuels in South Africa, especially diesel fuel replacements for use in agricultural applications.

Because virtually no literature was available on the utilization of vegetable oils in *modern* diesel engines, a project was initiated to study aspects concerning the use of such oils in agricultural tractors. Research, using several makes of diesel engine, has shown that sunflower seed oil, and particularly an ethyl ester mixture, has the potential to extend diesel fuel provided solutions are found for practical problems encountered.

1. PROPERTIES OF SUNFLOWER SEED OIL

Some properties of sunflower seed oil and diesel are compared below.
The greatest single difference is in the kinematic viscosity.

Fuel	Cetane Value	Calorific Value (MJ/kg)		Kinematic Viscosity (20°C)
		Gross	Nett	
Diesel	48	45.5	42.7	6,03
Sunflower oil	37	39.2	36.7	73,18

2. ENGINE PERFORMANCE AT MAXIMUM POWER

Tests were made on tractor engines running on 100% and blended sunflower seed oil, and results compared with performances on diesel fuel.
For example, the conventional O.E.C.D. Nebraska engine performance tests
were conducted on nine tractor models, each with direct injection. On
100% sunflower seed oil they all operated normally, and delivered almost
full power without modifications. However, extended studies disclosed
practical problems which make fuel/engine modifications necessary before
the technology can be applied as general practice.

A number of blends were also tested, mainly to determine the effects
of lower viscosity (see also results obtained under "Injection Equipment
Performance"). Tables I and II show *some* results obtained at maximum p.t.o.
power for each tractor, operated on diesel fuel, 100% sunflower seed oil
(degummed), and on blends of 90% sunflower seed oil with 10% gasoline, and
80% diesel fuel with 20% gasoline.

3. ENGINE SERVICE LIFE

At the outset of the experiments, one of the tractors (a Ford 5000)
completed a 100-hour maximum power dynamometer test on 100% sunflower seed
oil without any noticeable adverse effects. The same tractor running on a
moderate extender of 20% sunflower seed oil to 80% diesel fuel subsequently
completed 1004 engine meter hours of trouble-free operation on a farm. At
the end of this period an 8% power loss was measured at the power take-off.
The injector nozzles were then replaced and the injector pump recalibrated
to specification and this reduced the power loss to only 4%.

Following that field test, the tractor was coupled to a power take-off
dynamometer and operated at an arbitrary continuous load of 70% maximum
power using the same fuel. It was run for 24 hours a day on this 20:80

blend for a further 278 engine meter hours before the exhaust smoke increased noticeably. The reason was that the injector nozzles had started to carbon up around the orifices.

Incomplete combustion, which leads to lubricating oil contamination, has more serious consequences in engines operating on 100% or blended sunflower seed oil than in diesel-fuelled engines. The anti-oxidant additives in mineral lubricating oils rapidly become depleted, resulting in oil polymerization. In extreme cases (e.g. 20% crankcase dilution), this could result in a catastrophic solidification of the oil throughout the engine.

Experience with some of the tractors in the experiments has shown that one of the reasons for incomplete combustion is that the injector nozzles coke up excessively under prolonged part-load conditions. At part-load, and especially during idling conditions using 100% and blended sunflower seed oil, incomplete combustion is more pronounced than with diesel fuel, even manifesting unburned liquid fuel in the exhaust system.

At the end of the total 1382 hour period, deposits in the combustion chamber, cylinders and piston ring grooves were indistinguishable from those when operating normally on diesel fuel. However, another make of test tractor operated continuously at 70% of maximum power with 80% sunflower seed oil and 20% gasoline for about 300 hours *did* exhibit symptoms of piston ring sticking, deposit-forming on the injector tips and crankcase dilution. This was probably due to incomplete combustion associated with inadequate injector atomization.

4. INJECTION EQUIPMENT PERFORMANCE

A specially constructed test apparatus was used to study the atomization characteristics of different fuel systems. The apparatus is capable of separating small and large droplets in the spray cloud during injection in the test apparatus. The percentage of the volume of the fuel injected which is carried in large droplets could then be compared for different fuels. The quality of atomization relative to that of diesel fuel could thus be assessed. It was clear from these bench tests that the atomization spectrum of 100% sunflower seed oil was much inferior to that of diesel fuel, and that improved atomization resulted from any additive to sunflower seed oil that reduced viscosity. Preheating to reduce the viscosity of the sunflower seed oil also improved atomization.

5. SUNFLOWER SEED OIL ESTERS

Another way to reduce the viscosity of sunflower seed oil is to change its chemical composition by a very simple and well-established process, which involves chemically changing the pure sunflower seed oil to, say, an ethyl or a methyl ester mixture. This has the effect of bringing the distillation curve of the fuel nearer to that of diesel fuel (Fig. 1). Further, it brings about a significant decrease in the viscosity to about the same order as diesel fuel (Fig. 2).

Preliminary dynamometer tests on tractor engines revealed that, after 100 hours of operation at 80% of maximum power, the ester-based fuel:

- Caused less coking than diesel fuel (Fig. 3);
- Produced much less exhaust smoke;
- Increased engine thermal efficiency to a level higher than that of diesel fuel (Table III); and
- Showed no difference in combustion properties compared with diesel (from the cylinder pressure diagrams, Fig. 4).

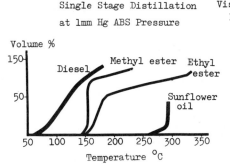

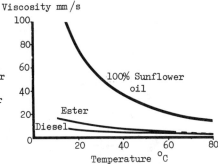

Fig. 1 Distillation curves showing percentage volumes distilled at various temperatures.

Fig. 2 The relationship between viscosity and temperature: sunflower seed oil, ester and diesel fuel.

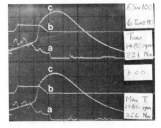

Fig. 3 Injector tips after 100 hours operation in an engine on sunflower seed oil based ester(left) and on diesel fuel(right)

Fig. 4 Oscilloscope traces for an engine operating on ethyl ester mixture(above) and diesel fuel(below)
a) injection line pressure   b) needle lift   c) cylinder pressure

TABLE I

PERFORMANCE OF VARIOUS TRACTOR MAKES WITH DIESEL, SUNFLOWER SEED OIL AND PETROL BLENDS
(Maximum Power, Brake Thermal Efficiency and Smoke Values)

Tractor	Engine	Maximum Power (kW)				Difference in power compared to diesel			Brake Thermal Efficiency				Difference in Brake Thermal Efficiency compared to diesel			Hartridge Smoke Value			
		Diesel	SN100	SN90/10P	D80/20P	SN100	SN90/10P	D80/20P	Diesel	SN100	SN90/10P	D80/20P	SN100	SN90/10P	D80/20P	Diesel	SN100	SN90/10P	D80/20P
MF 240	Perkins AD3-152	27.8	28.2	27.9	26.5	+1.4%	+0.4%	−4.7%	30.7%	30.6%	32.6%	31.0%	−0.3%	+6.2%	+1.0%	39	43	37	27
MF 285	Perkins A4-248	41.6	40.0	39.8	40.8	−3.8%	−4.3%	−1.9%	29.8%	29.1%	29.3%	29.3%	−2.3%	−1.7%	−1.7%	56	51	46	58
Landini 8500 2WD	Perkins A4-248	51.3	49.3	49.8	50.0	−3.9%	−2.9%	−2.5%	30.0%	29.3%	29.6%	31.3%	−2.3%	−1.3%	+4.3%	63	71	71	53
Landini 8500 4WD	Perkins A4-248	47.8	46.2	46.6	48.6	−3.3%	−2.5%	+1.7%	31.4%	30.5%	31.2%	32.1%	−2.9%	−0.6%	+2.2%	57	54	61	59
Fiat 780 4WD	Fiat 8045	43.4	41.7	41.9	45.2	−3.9%	−3.5%	+4.1%	28.6%	28.2%	27.7%	30.6%	−1.4%	−3.1%	+7.0%	62	79	78	59
Fiat 880	Fiat OMCO 3/75	61.0	59.8	60.4	59.7	−2.0%	−1.0%	−2.1%	30.8%	30.7%	31.3%	30.7%	−0.3%	+1.6%	−0.3%	60	42	57	50
John Deere 2030	John Deere	41.5	40.2	39.7	39.7	−3.1%	−4.3%	−4.3%	28.0%	26.9%	26.6%	29.1%	−3.9%	−5.0%	+3.9%	67	70	59	54
IH 844-S	I H Neuss D268	54.6	51.1	51.1	53.4	−6.4%	−6.4%	−2.2%	30.1%	30.5%	30.4%	31.8%	+1.3%	+1.0%	+5.6%	64	57	54	43
IH 1066	IH/DT414 (Turbo-charged)	81.3	79.2	78.1	76.2	−2.6%	−3.9%	−6.3%	28.4%	31.2%	31.2%	31.5%	+9.9%	+9.9%	+10.9%	34	20	23	34

NOTE:

D = Diesel fuel
SN = Sunflower seed oil
P = Petrol (Gasoline)

Altitude = 1 400 m

TABLE II

PERFORMANCE OF VARIOUS TRACTOR MAKES WITH DIESEL, SUNFLOWER SEED OIL AND PETROL BLENDS
(Fuel consumption and Torque)

Tractor	Engine	Fuel Consumption (l/h)				Difference in fuel Consumption compared to diesel			Specific fuel consumption (ml/kW-h)				Max. Torque (N-m)				Difference in Max. Torque compared to diesel		
		Diesel	SN100	SN90/10P	D80/20P	SN100	SN90/10P	D80/20P	Diesel	SN100	SN90/10P	D80/20P	Diesel	SN100	SN90/10P	D80/20P	SN100	SN90/10P	D80/20P
MF 240	Perkins AD3-152	9.02	9.78	9.12	8.68	+8.4%	+1.1%	−3.8%	324	347	327	328	139	137	136	134	−1.4%	−2.2%	−3.6%
MF 285	Perkins A4-248	13.90	14.61	14.48	14.14	+5.1%	+4.2%	+1.7%	334	365	364	347	214	198	202	217	−7.5%	−5.6%	+1.4%
Landini 8500 2WD	Perkins A4-248	17.04	17.89	17.93	16.25	+5.0%	+5.2%	−4.6%	332	363	360	325	240	229	229	238	−4.6%	−4.6%	−0.8%
Landini 8500 4WD	Perkins A4-248	15.15	16.10	15.92	15.41	+6.3%	+5.1%	+1.7%	317	348	342	317	239	229	231	242	−4.2%	−3.3%	+1.3%
Fiat 780 4WD	Fiat 8045	15.09	15.66	16.11	15.00	+3.8%	+6.8%	−0.6%	348	376	384	332	183	161	160	192	−12.0%	−12.6%	+4.9%
Fiat 880	Fiat OMCO 3/75	19.74	20.66	20.60	19.78	+4.7%	+4.4%	+0.2%	324	345	341	331	271	249	249	275	−8.1%	−8.1%	+1.5%
John Deere 2030	John Deere	14.73	15.88	15.93	13.88	+7.8%	+8.1%	−5.8%	355	395	401	350	183	171	169	176	−6.6%	−7.7%	−3.8%
IH 844-S	I H Neuss D268	18.03	17.77	17.93	17.04	−1.4%	−0.6%	−5.5%	330	348	351	319	259	258	257	249	−0.4%	−0.8%	−3.9%
IH 1066	IH/DT 414 (Turbo charged)	28.48	26.92	26.70	24.55	−5.5%	−6.3%	−13.8%	350	340	342	322	418	386	386	400	−7.7%	−7.7%	−4.3%

TABLE III

PERFORMANCE OF TWO DIESEL ENGINES OPERATING ON SUNFLOWER SEED OIL BASED ESTER AND DIESEL FUEL

Engine	Max. Power (kW)		Brake Thermal Efficiency (%)				Hartridge Smoke Value			
			Diesel		Ester		Diesel		Ester	
	Diesel	Ester	Max. Power	35 kW	Max. Power	35 kW	Max. Power	35 kW	Max. Power	35 kW
Perkins 4.236	45.1	41.6	30.7	33.3	32.1	35.8	45	8	11	9
Ford 6 600	41.7	37.2	29.6	31.7	32.8	33.4	66	42	18	21

Diesel injector needles tended to stick after the engine had cooled off following shut-down, and it is suspected that this problem is related to the catalyst used in the trans-esterification process.

Although extended tests still remain to be completed, the concept is bright with promise.

6. SUMMARY

1. Unmodified direct-injection diesel engines fuelled with 100% sunflower seed oil (degummed) deliver approximately the same maximum power with slightly higher specific fuel consumption.

2. When using 100% sunflower seed oil as a fuel for diesel engines over an extended period of time, coking of injector tips is a major problem.

3. Sunflower seed oil blends with 10% gasoline, with 20% diesel fuel, and with 40% kerosene, were evaluated in diesel engines in attempts to solve the injector coking problem. Although in each case engine performance was quite satisfactory, the only blend which did *not* create injector problems was the 20% sunflower seed oil : 80% diesel fuel blend used in the Ford 5000 tractor during field trials.

4. It would thus appear that attempts to modify the physical charac-teristics of sunflower seed oil by blending with other fuels would not successfully resolve injector coking problems; although modi-fications to certain injector systems could possibly obviate them.

5. A chemical modification of sunflower seed oil, yielding an ester compound, was assessed for possible application as a fuel. After 100 hours of its use in a diesel engine, no coking on the injector tips could be noticed. However, a problem of injector needle sticking after engine shut-down remains to be solved.

6. Considering the promising results achieved, it is evident that sunflower seed oil, particularly in the ester phase, is potentially a highly suitable renewable fuel for diesel engines.

THE ECONOMICS OF IMPROVING OCTANE VALUES OF GASOLINE
WITH ALCOHOL ADDITIVES

P. JAWETZ

Independent Consultant on Energy Policy
425 E. 72 St. New York,N.Y. 10021,U.S.A.

Summary

Mixing one part ethanol or methanol with nine parts low
octane gasoline, the so called gasohol or its methanol counterpart, results in an octane improvement of the mixture by
over three octane numbers. If one were to achieve a similar
octane improvement of the base gasoline via energy intensive
processes in reforming or isomerizing of hydrocarbons, one
had to use in the production of a gallon of the higher octane unleaded gasoline 6-10 percent more crude than in the
production of the base low octane gasoline. A six percent
saving on the full gallon of the mixture translates then to
a 60 percent saving for the one tenth gallon of ethanol
which can achieve an equal octane value. The economical effectiveness of using fermentation ethanol is thus influenced
by three factors: the first factor was dealing with the
crude displaced at the refinery, the second factor deals
with the inadequacy of using Btu measurements and the third
factor is determined when comparing the miles per gallon
yields of different fuels. These calculations are independent of any energy balance in the production of the alcohol.

1. PREFACE

In order to avoid misconceptions the author wishes to
state first that he deals with ethanol and methanol not as a
replacement for gasoline, and not merely as an extender to
gasoline, but rather as addition to low octane gasoline for
the production of unleaded gasoline with an octane value acceptable for use in present day motor vehicles.

The economical thesis of this paper is that gasohol, as
defined in the United States arbitrarily - a mixture of 10
percent fermentation ethanol and 90 percent low octane gasoline - made economical sense for the United States already in
1978.

2. INTRODUCTION

The inclusion of three percents ethanol or methanol to
gasoline stock (unleaded and of the 80-90 average octane value) will increase the average octane value of the mixture
(that is the average of the road octane value and the motor
octane value) by one point. This increase in the range mentioned is nearly propotionally to the amount of alcohol present in the mixture. 10 percent alcohol mixtures (as used in
the U.S.) will increase the octane value by about three
points and 20 percent alcohol mixtures (as used in Brazil)
will increase the octane value by somewhat less than 6 points
or about 5 points.

As shown in Brazil the inclusion of 20 percent nearly
anhydrous ethanol in gasoline has no negative effect on the
use of the motor vehicle and in Brazil such a mixture is in
effect called "gasoline" and quite naturally the ethanol is
regarded as (1) an additive to the gasoline and (2) an "extender" to the supply of the original petroleum crude product.

In the U.S., where the oil industry and even the automobile manufacturers went to work to find arguments and ways
to deny the use of ethanol one had to invent the mistifying
new product "gasohol" rather than to allow for the use of

ethanol as one more additive in the long list of gasoline additives.

The use of the "gasohol" definition in present U.S. legislation does not allow for the addition of ethanol to unleaded gasoline according to octane value requirements and these regulations thus present a barrier to a more rational use of the alcohol-gasoline fuel mixtures.

The addition of methanol to gasoline was practiced by the oil industry in the past and no negative effects were known. Now after this subject became public knowledge and different tests were performed, one can say that methanol can be added to gasoline at percentages lower than ethanol in order to avoid significant corosion effects though mixtures with butanol and higher alcohols would improve the usability of both methanol and ethanol.

One interesting comment that was brought to our attention lately is that South Africa does in effect use a gasohol-type gasoline as its synthetic gasoline, as produced at the SASOL plants, includes alcohols that are a byproduct of the synthetic process .

3. THE PETROLEUM CRUDE DISPLACEMENT VALUE OF ETHANOL

When studying the usefullness of alcohols one has to keep in mind that it is not the energy content of the alcohol that is of importance but rather we should focus on the petroleum crude displacement value of the ethanol. As such it is the use of the alcohol that is important as well as the energy balance in the production of the alcohol.

(a) When adding 10% ethanol to low octane gasoline we displace not only 10% gasoline but an additional 6-10% crude that would have been otherwise needed for the energy intensive further reforming. Assuming the lower figure, 6% additional crude is used up in the production of the full gallon of gasoline, then we could credit the one tenth of gallon of ethanol that obtains the same end result with an additional 60 percent effective displacement of petroleum. The ethanol displays therefore an efficiency factor 1.6 for its energy content.

(b) a second factor results when viewing the energy content of ethanol as measured in Btus. - a unit based on calorimetry rather than on the transformation of chemical energy to mechanical energy. If ethanol has only 2/3 the value in Btus as compared to gasoline - but in a mixture with gasoline its effectiveness as a fuel is equal to that of gasoline - then one can say that its energy when measured in Btus has a 1.5 higher usefulness.

(c) a third factor results when comparing in an actual fleet test the fuel efficiency as measured in miles per gallon gasoline versus gasohol (or any other gasoline-alcohol mixture). Averaging over fleet tests results performed in the States of Nebraska and Illinois in the U.S.A. an increase in miles /gallon of 5.7% for gasohol, at a 89-90 average octane value, as compared to gasoline, at 87 average octane, was obtained. This allows for a third efficiency factor equal in this case to 1.57.

The three factors developed are independent of each other and therefore the total effect results in <u>an efficiency factor of 3.77 for ethanol when used in gasoline at the gasohol composition</u>. Each Btu of ethanol <u>used</u> displaces 3.77 Btus of petroleum crude or its products and this value has now to be multiplied by the appropriate factor for the <u>production</u> of ethanol, as measured in conventional energy balance calculations, in order to obtain the overall energy balance that relates all the inputs in the manufacture of the alcohol (fertilizers, agricultural machinery, the plant equipment, distillery energy use, transportation etc.) to the effective displacement of the crude at the refinery as well as in the motor vehicle.

4. <u>ALCOHOL AS AN OCTANE BOOSTER VERSUS ALCOHOL AS A NEAT FUEL</u>.

Our present paper compares the economics of using alcohol as an octane boosting additive with the use of alcohol in a neat alcohol engine.

Considering the factor developed above, when using ethanol in the "gasohol" context, recognizing the lower Btu content of

ethanol, one obtains that each gallon of ethanol displaces 2.5
gallons of a petroleum product and one can therefore state
that ethanol when used in this context could cost the national
economy 2.5 times as much as gasoline without this having any
negative side effects upon oil importing countries, but this
contention has no validity if the ethanol is not used as an
octane boosting additive but only for its energy content. Furthermore, recognizing that there is some debate about the actual performance of gasohol in a car as measured in miles per
gallon we will also point out that the value developed here
for the savings of crude at the refinery were quite conservative and also that no externalities important to the national economy were reviewed in this context either (e.g. employment factor, decrease of the outpouring of currency etc.).

Considering the savings of crude at the refinery alone,
it should be pointed out to the refiner when blending an alcohol-gasoline mixture that his costs do not increase even if
the cost of alcohol is higher than his costs of gasoline by
over 50%. In October 1980 values one can thus say that the refiner should be asked to absorb a 50 cents cost differential
as this represents his savings when using alcohol octane
boosters for the production of high octane unleaded gasoline
while phasing out the metal containing octane boosters.

The consumer on the other hand can be asked to pay for
the increased effectiveness of the fuel in use in his motor
vehicle thus bringing up the cost of motor vehicle fuel to the
point that,when used as an octane boosting additive, the production of ethanol may not need indeed any further governmental subsidies beyond incentives for the construction of new
distilleries and a legislative move requiring the refiner to
use this particular octane boosting additive. The gains to the
environment by improving air quality from phasing out the lead
octane boosters are an added bonus to such legislative moves.

5. POLICY CONCLUSIONS.

The initial use when phasing in alcohol fuels as part of
a national energy policy should be via octane boosting additives. When a country is in a position to produce more

alcohol than needed for the use as an additive two alternate paths may become available:

(1) the development of an international trade in alcohol to countries in need of this octane booster and (2) the development of neat alcohol engines. The second possibility, disregarding the conclusions of this paper, may be recommended nevertheless to policymakers who are intent to decrease the importation of crude because of their overdependence on this commodity in an insecure market and because of the high cost of petroleum crude.

DESIGN AND ADVANCE OF THE BIOENERGY IN FRENCH SUGAR INDUSTRY

J.P. LESCURE

Institut de Recherches de l'Industrie Sucrière - VILLENEUVE d'ASCQ, France

Presented by R. MOLETTA

Institut National de la Recherche Agronomique
Villeneuve d'Ascq - France

Summary

The first IRIS programme, begun in 1975, had for object to treat the wastewater pollution by the anaerobic way. It included three successive steps : analytical study of water composition, biological study of degradation of the identified components and chemical engineering study for industrial application of the project. This first programme reached the fixed object from 1978 : need of a minimum energy and production of an easily usable gas at 80 % methane. The experience is then used in a second research programme for treatment of particularly polluted waters, the most important ones as far as the worked volumes are concerned, are distillery stillage. This research succeeded in part and is continued by study of the inhibitions mainly due to inorganic components or to nitrogen. For the wastewater treatment IRIS designed and developed an original system permitting to reduce considerably the cost of the treatment plants. From 1979 the energy crisis leads to intensify in priority research related to biogas production. A new research programme is then launched to produce biogas from pulp and beet trash. In laboratory a gasification yield near 80 % is achieved with loadings of 6 kg dry substance per m^3 and per day. This programme is continued in 1980 by a pilot-experiment in a sugar factory with assistance of the "Commissariat à l'Energie Solaire".

L'Institut de Recherches de l'Industrie Sucrière, IRIS, s'intéresse
à la bioénergie appliquée depuis 1975. Son objectif initial, ou premier
programme de trois ans, est très modeste : rechercher une voie économique et
nouvelle pour traiter rapidement la pollution sans gaspiller l'énergie électrique en pure perte.

I - PREMIER PROGRAMME (1975-1978) : TRAITEMENT des EFFLUENTS AQUEUX

On s'efforce de suivre la méthode suivante :

1°) Connaître la nature exacte des polluants à dégrader dans les
effluents de sucrerie,

2°) Connaître les mécanismes biochimiques des dégradations du sucre
jusqu'au méthane.

3°) Mettre au point une installation de traitement adaptée à la sucrerie sans faire appel aux technologies onéreuses du traitement anaérobie
à faible charge, proposées par certains constructeurs.

Les principaux résultats de l'étude sont connus et publiés depuis
l'exposé qui en a été fait à la 16ème Assemblée Générale de la Commission
Internationale Technique de Sucrerie en Juin 1979. On en rappellera ici les
principaux enseignements.

1 - ANALYSE de l'EAU

La pollution de sucrerie qui provient essentiellement de pertes de
jus sucrés en différents points de la fabrication et particulièrement au
lavage des betteraves, se transforme très rapidement en acides gras volatils
ou A G V, ce qu'avait reconnu A. CARRUTHERS (1960). J.P. LESCURE et P.
BOURLET précisent que le saccharose est complètement dégradé en 1 à 2 jours
dans les bassins de décantation et que par leur caractère plus ou moins anaérobie, les fermentations produisent une plus ou moins grande proportion
de butyrate et de n.valérate dans les eaux en plus d'acétate et propionate
que l'on rencontre toujours. Le lactate se trouve aussi en quantité plus ou
moins importante. Seuls 10 % des composants carbonés échappent à cette analyse, où l'on n'a pas recherché les composés organiques azotés.

2 - BIOCHIMIE de la DEGRADATION ANAEROBIE

Pour étudier la biochimie de la dégradation anaérobie des composés identifiés, on utilise un petit appareillage très simple qui permet de suivre en batch la dégradation d'un composé par une flore extraite d'un pilote de traitement d'eau résiduaire de sucrerie. La dégradation anaérobie du saccharose dans ces conditions est très complexe et il se forme un polyglucose à longue chaîne par transglycosidation du saccharose, de plus les anions formés rendent le pH très instable. La dégradation du lactate est beaucoup plus simple avec production de propionate. Les A G V supérieurs sont dégradés en acétate par des coupures oxydatives tout à fait en accord avec le mécanisme de la β.oxydation décrit par J.S. JERRIS et P.L. Mc CARTY (1965). La majeure partie du méthane produit vient finalement de la dégradation d'acétate.

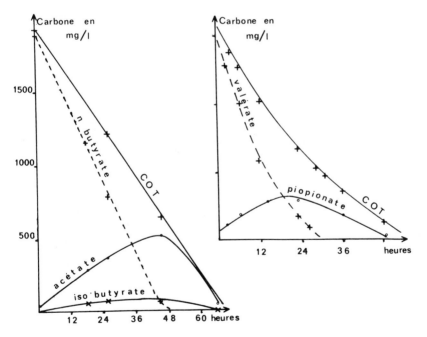

DIGESTION ANAEROBIE DE BUTYRATE ET VALERATE

3 - <u>GENIE CHIMIQUE</u>

Si la biochimie est relativement simple, la mise au point du procédé industriel demande une étude particulière de génie chimique. L'IRIS a collaboré avec une société sucrière à la mise au point d'un nouvel appareillage en réalisant un pilote puis un premier prototype industriel. Il s'agit d'un bassin, de forme pyramidale ou cônique, tronquée et inversée. Les bords sont des digues à 45° dont le revêtement peut être très léger, film PVC, ou plus sophistiqué, au gré de l'utilisateur ou selon les exigences des sols. L'équipement est très simple : un trépied central surmonté d'une petite cloche métallique de faible volume porte un agitateur vertical; un film entièrement immergé en marche normale sert d'entonnoir pour le recueil vers la cloche métallique des gaz produits. Ce film hermétiquement fixé à la cloche centrale est tendu vers le milieu des digues; il sépare le bassin en deux secteurs et la partie située sous le film constitue le fermenteur proprement dit. Les eaux à traiter sont introduites dans le fermenteur sensiblement sous l'hélice d'homogénéisation et s'évacuent progressivement par la périphérie du film immergé. Elles sont alors recueillies dans une nochère ceinturant le dispositif et évacuées hors de l'appareil.

Le dispositif décrit, couramment appelé procédé IRIS, a fait l'objet d'un dépôt de brevet et connaît actuellement 7 applications en Europe.

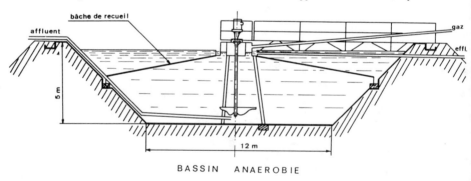

BASSIN ANAEROBIE

En 1978, l'objectif fixé est atteint. Le premier prototype fonctionne avec une puissance installée de 7,5 kW pour l'agitation et traite 3 000 kg/j. de D C O avec un rendement de 80 à 90 %. On produit environ 900 m^3/j. de gaz à 80 % de méthane.

II - DEUXIEME PROGRAMME

Partant de l'expérience déjà acquise, les sucreries confient à l'IRIS un second programme de 3 ans pour le traitement d'eaux résiduaires plus fortement polluées :

- Eluats de divers procédés d'échangeurs d'ions en sucrerie,
- éluats de résines décolorantes,
- eaux de pressage d'herbes et de radicelles,
- vinasses de distilleries agricoles : betteraves et mélasses.

Industriellement, tant par leur charge polluante que par les volumes mis en oeuvre, les vinasses des distilleries et sucreries-distilleries sont les plus importantes. On constate que la biogazéification des vinasses de jus de betteraves se fait avec un rendement d'environ 75 %, pour une épuration des matières organiques de l'eau d'environ 90 %. Le gaz produit titre entre 65 et 70 % de méthane mais contient aussi des teneurs élevées en sulfure d'hydrogène.

Ces essais ont été corroborés en 1978 par un essai industriel en vraie grandeur. Cette année, trois installations anaérobies de respectivement 450 m^3, 1000 m^3 et 2 000 m^3 utiles seront en service en FRANCE.

Les premières estimations permettent d'espérer produire environ 1/3 du combustible nécessaire en distillerie classique ou une plus grande proportion si l'on effectue une recompression de vapeur. Les résultats sont plus aléatoires en ce qui concerne la méthanisation des vinasses plus concentrées des distilleries de mélasses. Le principal inhibiteur est alors l'ammoniaque libérée par désamination.

L'application du traitement anaérobie aux éluats de déminéralisation des jus sucrés ou des résines décolorantes très chargées en sels minéraux pose des problèmes d'inhibition actuellement à l'étude.

III - METHANISATION des DECHETS VEGETAUX

Les sucreries recherchent activement des voies nouvelles pour utiliser les déchets de lavoir et même éventuellement les pulpes comme source d'énergie de remplacement. Ajoutée à ce qui est déjà réalisé pour le traitement des eaux, l'opération permettrait d'obtenir la quasi-autonomie énergétique des sucreries.

Deux voies sont envisagées :

1 - surpressage et combustion directe,

2 - digestion anaérobie et combustion des gaz.

La première nécessite la création de presse à très haute teneur pour obtenir 50 % de matières sèches (contre 25 % M.S. obtenues actuellement). En supposant ce pressage réalisé, les chaudières à fuel devraient être remplacées, car elles n'admettent pas ce type de combustible solide. La deuxième solution, moins avantageuse sur le plan énergétique, présente l'avantage de fournir un combustible plus facile à utiliser.

Les sucreries poursuivent leurs recherches dans les deux voies, mais nous n'évoquerons ici que la méthanisation. La méthodologie suivie par l'IRIS est la suivante :

1ère étape : On recherche au laboratoire la voie fermentaire techniquement la plus intéressante. Très rapidement, la voie thermophile est abandonnée car elle ne permet pas d'obtenir une vitesse d'hydrolyse assez rapide ni des rendements suffisamment élevés. De plus, la possibilité d'ensemencement d'une installation industrielle paraît douteuse.

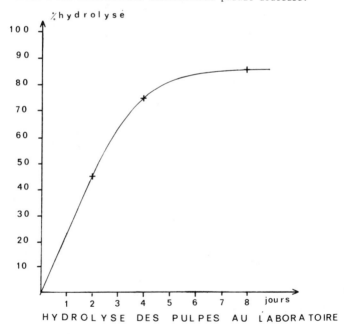

HYDROLYSE DES PULPES AU LABORATOIRE

La voie mésophile est la plus intéressante. On mesure en particulier la vitesse d'hydrolyse des pulpes et le rendement qui est supérieur à 80 % en 8 jours.

On recherche ensuite le schéma qui permet d'obtenir la vitesse de digestion la plus rapide et le choix se porte sur une fermentation avec un ensemble de sous-populations microbiennes permettant la digestion en une étape avec une charge de 6 kg de M.S. par m^3 et par jour.

2ème étape : Cette étape actuellement en cours de réalisation est une démonstration à l'échelle pilote destinée à permettre de déterminer le rendement énergétique réel du processus industriel.

Ce pilote est un ensemble de 2 cuves de fermentation de 30 m^3 chacune.

Performances obtenues

	Réduction de la DCO	Charge volumique kg/m^3.d	Energie récupérable en kg équivalent pétrole par t. de betteraves
Eaux décantées (prototype à VAUCIENNES)	90 %	5 (DCO)	0,4
Vinasses (jus de betteraves) (prototype à GOUSSAINVILLE)	80 %	4 (DCO)	3
Pulpes et déchets de sucrerie (essais de laboratoire).	hydrolyse 80 %	6 (M.S.)	11,5

Références :

A. CARRUTHERS et al. (1960) - the composition and treatment of sugar beet factory waste waters - 13 th Annual Technical Conference of the British Sugar Corporation Limited.

J.S. JERIS, P.L. Mc CARTY (1965) - the biochemistry of methane fermentation using C^{14} tracers - J. Water Pollution control fed. 37 p.178

J.P. LESCURE et P. BOURLET (1979) - Epuration des eaux de sucrerie par fermentation méthanique mésophile - C.R. 16ème Assemblée Générale de la C.I.T.S. - Amsterdam 1979 - p.29 - Secrétariat Général, Aandorenstraat 1 - 3300 TIENEN Belgique, Editeur.

ABOU-ELMASR, Dr. T.   Physics Department, Al-Azhar University, Cairo, Egypt.

ADAIR, Dr. D.   Tropical Products Institute, Culham, Nr. Abingdon, Oxon, UK.

ADAMS. Dr. M. A.   Tropical Products Institute, 56 Grays Inn Road, London, UK.

ADER, G.   George Ader & Associates Ltd., Croydon, UK.

ALFANI, Dr. F.   University of Naples, Ple Tecchio, 80125, Naples Italy

ALLIRAND, Mr.   INRA, CNRA, Bioclimatologic, Route de St. Cyr, 78000 Versailles, France.

ALMOND, F. R.   Intermediate Technology Dev. Group, Myson House, Railway Terrace, Rugby, UK.

ANDERSEN, Dr. J. R.   Riso National Laboratory, Riso, DK- 4000 Roskilde, Denmark.

ANDERSON, Mr.   RHM Research Ltd., Lincoln Rd., High Wycombe, UK.

ANDERSSON, K.   University of Goteborg, Botanical Institute, Carl Skottsbergs Gata 22, S-413 19 Goteborg, Sweden.

ANDERTON, Miss M.F.   International R&D, Co. Ltd., Fossway, Newcastle-upon-Tyne, UK.

ANTUNES, S. D.   UCL-FSA-Unite Auto, Place du Levant 3 - B.1348, Louvain-la-Neuve, Belgium.

ARLIE, Dr. J. P.   Institut Francais de Petrole, IFP, 1 Avenue de Bois Preole 92500, Rueil Malmaison, France.

ARNOLD, J. H.   MIT/META, Meta Systems Inc. 10 Holworthy, Cambridge, Mass., USA.

ARNTZEN, Dr. C. J.   Michigan State University, MSU/DoE Plant Research Lab., E. Lansing, MI 48824, USA.

ASINARI DI SAN MARZANO, Mrs. C.   University of Louvain, Unite de Genie Biologie, Place Croix du Sud 119, B1348 Louvain-la-Neuve, Belgium.

AUBART, C.   PEC Engineering, 62 Rue Jeanne D'Arc, 75646 Paris Cedex 13, France.

AUCLAIR, D.   INRA, Ardon, 45160 Olivet, France.

AVELLO, Dr. R.   CNEN, CSN, Cosaccia v. Auguillatere, Rome, Italy.

BAHUREL, Mr.   Michelin Bibliotheque Technique, 63040, Clermont-Ferrand, Cedex, France.

BAILEY, P. H.   SIAE, Bush Estate, Penicuik, Midlothian, Scotland, UK.

BARFOED, S.   DDS, 1 Langebrogade, Copenhagen 1411, Denmark.

BARILLI, F.   European Investment Bank, 100, Bd. Konrad Adenauer 1020 Luxembourg.

BARNARD, G.   28 Fairholme Road, London, W.14

BARNES, P. M. L.   Shell International Petroleum Co., SIPC (PLE/21) Shell Centre, London, SE1, UK

BARNETT, J. A.   SPRU, Mantell Building, University of Sussex, Brighton, UK.

BARREVELD, W. H.   AGS Division, FAO Hq, Rome, Italy.

BARRY, Ms. M.   Department of Microbiology, University College, Galway, N. Ireland.

BAUBION, P. A.   CEM, 85 Avenue Aristide Brianol, 93240 Stains, France.

BEBIN, J.   Ste Degremont, B.P.46 92151 Suresnes, Cedex, France.

BEENACKERS, Dr.   Twente University, P.O. Box 217, Enschede, Netherlands.

BELAICH, Dr. J. P.   CNRS, Lab. de Chimie Bacterienne, BP 71, Marseille Cedex 9, France.

BELL, T. A.   Caltex (UK) Ltd., 30 Old Burlington St., London, UK.

BELL, Dr. W. J.   Atkins Group Ltd., Woodcote Grove, Ashley Rd., Epsom, Surrey, UK.

BENTE, Dr. P. F.   Bio-Energy Council, 1625 Eye St., N. W.,Suite 825A, Washington DC 20006, USA.

BERARD, Mrs. H.   Elf-Aquitaine, 7 Rue Nelalou - Paris 15 eme, France

BERGMAN, K. G.   Swedish University of Agric. Sciences, S-75007 Uppsala, Sweden.

BERNHARDT, Dr. W.   Volkswagen AG, Wolfsburg, Germany.

BERTIN, G.   CEVA SA, 14 Rue Garnier, 92200 Neuilly S/Seine, France.

BEST, Dr. G.   United Nations-CEPAL, Casilla, 179-D, Santiago, Chile.

BESWICK, R. H. D.   GIRA-SA Switzerland 142-144 Oxford Street, London, W.1.

BIANCHI, A.   CNRS University de Provence, 3 Place Victor Hugo, 13331 Marseille Cedex 3, France.

BIGG, S. D.   Shell International Petroleum Co. Ltd., Shell Centre, Waterloo, London, UK.

BINDER, Dr. H.   Inst. F. Radiochem, University of Innsbruck, Innrain 52 a, A-6020 Innsbruck, Austria.

BLACKADDER, W. H.   Studsvik Energiteknik AB, S.61182 Nykoping, Sweden.

BLANC, J. H.   SNEA(P), 64018 PAU, Cedex France.

BLANC-MUESSER, Mrs.   CNRS, 53 X 38041, Grenoble, France.

BODO, L. B.   Lucas CAV Ltd., P.O. Box 36, Warple Way, London W3, UK.

BODRIA, Dr. L.   University of Milano, via G. Celoria, 2-20133, Milano, Italy.

BOL, Dr. W.   Eindhoven University of Technology, Papenvoort 88, Geldrop, Netherlands.

BOLHAR-NORDENKAMPF, Dr. H. R.   University of Vienna, Franz-Josefs-Kai 21/35, A1010 Vienna, Austria.

BOND, Dr. R. P. M.   Shell Research Ltd., Sittingbourne, Kent, UK.

BOND, D.   Desmond Bond, Inc., 2416 Luckett Ave., Vienna, Virginia 22180, USA.

BONDI, Sir Herman,   NERC, Polaris House, North Star Avenue, Swindon, Wilts, UK.

BONDUELLE, Dr. B.   CNRS-GESER, E.C.P. grande voie des vignes - 92290 Chatenay-Malabry, France.

BONINO, G.   SES 20 Via Cuneo, Turin, Italy

BONNER, Prof. T. G.   26 Hadley Gardens, Chiswick, London W44 NX, UK.

BOSSUYT, J.   NV Bossuyt, 81 Nijverheidstraat-8749 Waregem-Beveren, Belgium

BOTTAI, Dr.   Termomeccanica, via Del-molo No. 1, Laspezia, Italy

BOURGEOIS, B.	CNRS, SPRU, Mantell Building, University of Sussex, Brighton, UK.
BRADLEY-SMITH, C.	Commonwork Construction, Edenbridge, UK.
BRANDON, O.	Energy Technology Support Unit, Building 156, Harwell, UK.
BRIDGWATER, Dr. A. V.	Chemical Engineering, Aston University, Birmingham 4, UK
BROWN, Dr. A.	Energy Technology Support Unit, Building 156, Harwell, UK.
BROWN, O. M. R.	United Molasses Co., Sugar Quay, Lower Thames St., London, UK.
BRUGERE, D.	INRA, INRA-SAD, Route de St-Cyr, 78000 Versailles, France.
BRYAN-JONES, Dr. D. G.	Distillers Co. Ltd., Glenochil Research Station, Menstrie, Clacks, Scotland, UK.
BRYANT, M. W.	Foster Wheeler Ltd., Station Rd., Reading, UK.
BUHNER, Dr. T.	Group VII, Institute for Betriebswirtschaft, Bundesallee 55/D3300 Braunschweig, Federal Republic of W. Germany.
BUNCE, Dr. R. G. H.	NERC, Merlewood Research Station, Grange-over-Sands, Cumbria, UK.
BURGESS, D.	Petawawa National Forestry Institute, Canadian Forestry Service, Chalk River, Ontario, Canada, KOJ 1JO.
BURTSCHER, Mr.	University of Innsbruck, Austria
CABANETTES, A.	INRA, Station Foret-Environnement, Ardon 45160 Olivet, France.
CADE, A.	TOTAL, 84 Rue de Villiers, 92 Levallois Penet, France.
CALLAGHAN, Dr. T . V.	NERC, Merlewood Research Station, Grange-over-Sands, Cumbria, UK.
CAMPAGNA, Dr. R.	Instituto Guido Donegani, Via Fauser 4, 28100 Novara, Italy.
CANNELL, Dr. M.G.R.	Institute of Terrestrial Ecology, Bush Estate, Penicuik, Midlothian, Scotland, UK.
CANTARELLA, M.	Via Cerbone 51, NA, Italy

CARAMELLE, Mrs. D. STMI, 9 Rue F. Leger, 'Val Courcelles', 91190 Gif-S/Yvette, France.

CARIOCA, Dr. J. O. University Federal of Ceara, Energy Nucleus 60 000 Fortaleza/Ceara - Brazil. Caixa Postal 935

CARNEIRO DE MENDONCA, P. L. Brazilian Embassy, 32 Green St., London W1, UK.

CARRARA, Dr. P. IREN, Universite d Abigjan - 08BP V34, Abidjan, Ivory Coast

CARRASSE, J. ZAZE, 15 Rue Louis Barthou, 92160 Antony, France.

CARRUTHERS, S. P. Department of Agric. & Horticulture, University of Reading, UK.

CASADEVALL, Mrs. E. CNRS, ENSCP, 11 Rue P. et M. Curie, 75231 Paris Cedex 05, France.

CATHELINAUD, Y. OECD, 2 Rue Andre-Pascal, 75016, Paris, France.

CATHONNET, Dr. M. CNRS, CRCCHT, 45045 Orleans Cedex, France.

CATTEL, G. S. APV Co. Ltd., PO Box 4, Crawley, Sussex, UK.

CAUBET, M. Universite Paul Sabatier, 31062 Toulouse, Cedex, France.

CHADWICK, Dr. A. T. ETSU, Building 156.3, AERE Harwell, Oxon, OX11 ORA UK.

CHASSANY, Mr. ENSAM-CRAM, Place Viala, 34060 Montpellier, France.

CHASSIN, P. INRA, Science du Sol, Rte de St. Cyr 78000, Versailles, France.

CHAUMONT, D. SFBP, Heliosynthese, BP No. 1, 13117 Lavera, France.

CHIESA, G. Castagnetti SpA, Via Fabbrichette 65, 10095 Grugliasca, Torino, Italy.

CHILDS, Dr. A. F. Albright and Wilson, Ltd., P.O. Box 3, Hagley Rd. West, Oldbury, Warley, W. Midlands, UK.

CHRISTOPHERSEN, Dr. K. A. US Forest Services/SEA, 348 St. James St., Falls, Ch VA 22076, USA.

CIBO, Dr. C AGIP-FORE, Via Del Giorgiove 59, 00147 Rome, Italy.

CLAIRE, M. INRA, CNRF Champenoux, 54280 Seichamps, France.

CLAUDON, Dr.	SFBP, Boite Postale h-519, 59381 Dunkerque Cedex, France.
COCQUEMPOT, Mrs. M-F	University of Compiegne, BP 233, 60200 Compiegne, France.
COLEMAN, Dr. A.	ICI Corporate Laboratories, P.O. Box 11, The Heath, Runcorn, Cheshire, UK.
COLLERAN, Dr. E.	Department of Microbiology, University College, Galway, Ireland.
COOMBS, Dr. J.	Tate and Lyle Ltd., P.O. Box 68, Reading, UK.
CORTE, Mr.	Universite Paul Sabatier, 31062 Toulouse Cedex, France.
COURTINAT, M.	Creusot Loire Enterprises, Tour Gen Cedex 13-92082 Paris La Defense, France.
CRAMER, Dr. H. H.	Bayer AG PF-AT, Geb. W 7, D509 Leverkusen, Germany.
CREUZET, Dr. N.	CNRS, LCB - 31 Chemie Joseph Aiguier, 13274 Marseille Cedex 2, France.
CRIDEN, Dr. J.	R&D Institute, 27 Hayarden St. Ramat Gan, Israel.
CROWTHER, R. E.	Forestry Commission, Alice Holt Lodge, Wrecclesham, Farnham, Surrey, UK
CUPP, Dr. C. R.	Inco R & D Center Inc. Sterling Forest, Suffern, New York, 10901, USA.
CUSHMAN, Ms. J. H.	Oak Ridge National Lab., P.O. Box X, Oak Ridge, Tennessee 37830, USA.
DALGLIESH, Dr. C. E.	24 Ringley Park Avenue, Reigate, Surrey, UK.
DASILVA, E. J.	Scientific Research and Higher Education Division, UNESCO, Paris, France.
DAWSON, L. G.	19 Piper Drive, Long Whatton, Loughborough, UK.
DEFAYE, Dr. J.	CNRS, CERMAV 53X, 38041 Grenoble Cedex, France.
DEFAYE, S.	Chambre Regionale d'Agriculture de Normandie, 4 Promenade Mme de Sevingne, 14039 Caen Cedex, France.
DEGLISE, X.	University of Nancy, France.
DEL CAMPO, Dr.	Instituto de Biologia Celular, Velazquez No. 144, Madrid 6, Spain.

DELMON, Prof. B. University of Louvain, Place Croix du Sud 1, 1348 Louvain-la-Neuve, Belgium.

DENNINGTON, Dr. V. N. University of York, 20 Briston Grove, London, UK.

DESCHAMPS, Dr. F. IRCHA, BP No.1, 91710 Vert-le-Petit, France.

DE SILGUY, Mrs. IGER, 21 Rue Chaptal, Paris 75009, France.

DESMET, E. ABR Engineering, 7 Square Frere Orban, 1040 Brussels, Belgium

DE TURCKHEIM, N. Bertin and Cie., B.P. 3, 78370 Plaisir, France.

DE WAART, Dr. J. TNO, Utrechtseweg 48, 3704 HE Zeist, Netherlands.

DE WAEL, I. R. Agri-reizen, Oude Markt 24, Postbus 102, B-3000 Leuven, Belgium.

DI GIOACCHINO, S. Assoreni, Via E. Ramarini, 32, 00015 Monterotondo, (Rome), Italy.

DIGIORGIO, Dr. G. CNEN, Rome, Italy.

DOAT, Mrs. J. CTFT, 45 bis Avenue de la Belle Gabrielle, 94130 Nogent sur marne, France.

DOHNE, E. Kuratorium fur Technik and Bauwesen in der Landwirtschaft, Privat: Elisabethenstr. 27, D-6106 Erzhausen, Germany.

DRIGUEZ, H. CNRS, CERMAV, 38041 Grenoble Cedex 53, France.

DUBOIS, P. Laboratoires de Marcoussis, Route de Nozay – 91460 Marcoussis, France

DUBOURGUIER, Dr. H. C. INRA, 63110 Beaumont, France.

DUCROS, J. M. TECHNIP, Cedex 23 – 92090, Paris la Defense, France

DUJARDIN, Dr. E. C. Liege University, B22, 4000 Sart Tilman, Belgium.

DURAND, Dr. H. Solar Energy Commissariat, 208 Rue-Raymond Losserand, 75014, France.

DUVERNET, F. Institut Auguste Counte Paris, 5 Rue Descantes 75005, Paris, France.

ECHARD, J. PEC Engineering, 62 Rue Jeanne D'Arc, 75646 Paris Cedex 13, France.

EDELINE, Prof. F. CEBEDEAU, 2 Rue A. Stevart, 4000 Liege, Belgium.

EGNEUS, Dr. H.

University of Goteborg, Botanical Institute, Carl Skottsbergs Gata 22, S-413 19 Goteborg, Sweden.

EHLE, Dr.

ZZG Strube-Dieckmann, D-3065 Nienstadt, West Germany.

ELSEN, Mr.

Usine de Wecker, Wecker, Luxembourg.

ENGLISH, M.

W. S. Atkins Group Ltd., Woodcote Grove, Ashley Rd., Epsom, Surrey, UK.

ERCKEL, Dr.

Hoechst Aktiengesellschaft, Postfach 80 03 20, 6230 Frankfurt am Main 80, Germany.

ESNOUF, Miss

ENGREF, 19 Ave. du Maine, 75015 Paris, France.

ETIEVANT, C.

CNRS-GESER, Grande Voie des Vignes, 92290 Chatenay-Malabry, France.

FABRE, M.

SERST, BP 3218 Dakar, Senegal.

FALLOWFIELD, H. J.

Department of Agric. & Food Chemistry, Queens University, Belfast, N. Ireland. UK.

FARDEAU, Dr. M-L

CNRS, 31 Chemin Joseph Aiguier, 13274 Marseille Cedex 2, France.

FARGET, Mrs. M. A.

INRA, CEE, DG VI, 200 Rue de la Loi, 1049 Brussels, Belgium.

FARLEY, I.

Baker Perkins Holdings, Westfield Rd., Peterborough, UK.

FEDTKE, Dr. C.

Bayer AG, Biolog. Forschung, 5090 Leverkusen 1, West Germany.

FERENCZI, Mr.

Groupe Energie et Biomasse, 40 Rue du Chateau, 92 Boulogne, France.

FERNANDEZ, Dr. J.

JEN. Isotopos-Biosintesis, Apartado 3055, Madrid, Spain.

FERRERO, Dr. G. L.

CEC, 200 Rue de la Loi, 1049 Brussels, Belgium.

FINCK, J-D

Elf Aquitaine, Paris, France.

FINNEY, G. W.

Building 351.28, UKAEA Harwell, UK

FIORITO, G.

CESEN, Via Serra 6 - 16122, Genova, Italy.

FLETCHER. R.

Foster Wheeler Power Products Ltd., PO Box 160, Greater London House, Hampstead Road, London, UK.

FLOYD, J. S.

The Polytechnic of Wales, Llantwit Rd., Treforest, Pontypridd, Mid. Glamorgan, Wales, UK.

FLURI, P   Inventa AG, CH 7013 Domat/Ems, Switzerland.

FOIDL, K.   Institute for Environmental Research, 8010 Graz,
                            Austria.

FORRAT, P.   Laboratoire de la Roquette, 34190 St Bauzille de
                            Putois, France.

FOUCHE, Dr. S.   Societe Europeenne de Propulsion, BP 802, 27267
                            Vernon, France.

FRAISSIGNES, B.   British Petroleum, 10 Quai Paul Doumer, 92401
                            Courbevoie, France.

FREEMAN, Dr. R. F.   Bass Ltd., Burton-on-Trent, UK.

FREROTTE, J.   Electrobel, 1, Place du Trone, B 1000 Brussels,
                            Belgium.

FRICKER, R.   Allied Breweries Eng. Services Ltd., Station St.,
                            Burton-on-Trent, UK.

GADELLE, Mrs. A.   CERMAV-CNRS, Domaine University, St. Martin
                            d'Herces, 38041 Grenoble Cedex, France.

GALLARATI, Dr. Eng. E.   SIDI, Viale Mentanagz - 43100 Parma, Italy.

GANDAR, M.   Institute of Natural Resources, PO Box 375,
                            Pietermaritzburg, South Africa

GASSER, Dr. J. K. R.   Agricultural Research Council, 160 Great Portland
                            St., London, UK.

GAY, Dr. R.   Faculte des Sciences, University Nancy C0140,
                            54037 Nancy Cedex, France.

GEHRMANN, J.   KFA Julich Gmbh, Postfach A913, D-5170 Julich

GENNER, C.   Dept. of Microbiology, University College, Newport
                            Rd., Cardiff, UK.

GERSTER, Mrs. D.   CEA, DP2/EC BP 16, Pierrelatte 26700, France.

GEYNET, Dr. P.   Innovation 128, 38 rue des Mathurins - 75008,
                            Paris, France

GIANFREDA, Prof. L.   University of Naples, Ist. di Principi, di Ing.
                            Chim. P. le Tecchio, Napoli, Italy.

GIBBENS, J. G.   Foster Wheeler Power Products Ltd., PO Box 160,
                            Greater London House, Hampstead Road, London, UK.

GIBBONS, T.   Trinity College, University of Dublin, Eire.

GILBY, Mr.   Electrowatt Eng. Services, Ltd., 20 Harcourt House, Cavendish Square, London, UK.

GILL, L. L.   Shell UK Admin. Service, Shell Centre, London SE1 UK.

GILLOT, J. L.   CEC, 86, Rue de la Loi, 1040 Brussels, Belgium.

GIOT-WIRGOT, P.   Labo-Forestier, Place Croix -du-Sud, 2-1348 Louvain-la-Neuve, Belgium.

GLORIA, Dr. C   SES SpA, 20 Via Cuneo, Torino, Italy.

GOEBEL, O. H.   Cora Engineering, CH-7001 CHOR, Sagenste 97, Switzerland.

GOMEZ-MORENO, Dr. C.   DPTO Bioquimica, Facultad de Biologia, Av. Reina Mercedes, Spain.

GOSLING, Dr. D. L.   University of Hull, 266 Northgate, Cottingham, N. Humberside, UK.

GOUDEAU, Dr. J-C   GRCPC, Le Defend, Mignaloux Beavoir, 86800 St. Julien L'Ars, France.

GOUPILLON, Mr.   CNEEMA, Parc de Tourvoie, 92160 Antony, France.

GOYVAERTS, A.   KIH Limburg, Torenplein 6 bus 9, B-3500 Haselt, Belgium.

GRANGE, Dr. P.   University of Louvain, Place Croix du Sud 1, 1348 Louvain-la-Neuve, Belgium.

GRASSI, Dr. G.   CEC, Rue de la Loi 200, Brussels, Belgium.

GRECO, Prof. G.   University of Naples, Ist, di Principi di Ing. Chim. p le Tecchio, Napoli, Italy.

GRIEDER, Dr. E. P.   BFF, Postfach 1987, 3001 Bern, Switzerland.

GROENEVELD, Dr. M. J.   Energy Equipment Engineering, P.O.B. 316, 7570 AH Oldenzaal, Netherlands.

GROSSIN, R.   Bertin and Cie, B.P. 3, 78370 Plaisir, France.

GROUT, H. J.   W. S. Atkins Group Ltd., Woodcote Grove, Ashley Rd., Epsom, UK.

GUDIN, Dr. C.   SFBP, Heliosynthese, BP no.1, 13117 Lavera, France.

GUYMER, D. A.   Unilever Research Laboratory, Sharnbrook, Bedford, UK.

HANSEN, Dr. A. B. Risø National Laboratory, Risø, DK-4000 Roskilde, Denmark.

HANSSON, L. R. NE, Box 1103, S-163 12 Spanga, Sweden.

HARPER, S. H. T. Agric. Research Council, Letcombe Laboratory, Letcombe Regis, Oxon, UK.

HARRIS, Dr. G. S. University of Auckland, Private Bag, Auckland, New Zealand.

HARTL, H. Tullner Zuckerfabrik A.G., Reitherstrasse 21-23, Austria.

HAVE, Dr. H. Royal Veterinary and Agricultural University, Agrovej 10, DK-2630 Taastrup, Denmark.

HAWKES, Dr. D. L. Polytechnic of Wales, Dept. of Mech. Eng. Mid Glamorgan, Wales. UK

HAWKES, Dr. F. R. The Polytechnic of Wales, Department of Science, Pontypridd, Mid Glamorgan, Wales, UK.

HAWLEY, Dr. M. C. Michigan State University, East Lansing, Michigan, USA.

HAYES, Dr. P. Queens University, Newforge Lane, Belfast, N. Ireland, UK.

HEATH, W. M. Stone and Webster Engineering Ltd. London. UK.

HEDEN, Dr. C-G Karolinska Institutet, 10401 Stockholm, Sweden.

HEDGER, Mrs. M. Penddolfawr, Pontrhydfendigaid, Ystrad Meurig, Dyfed, Wales, UK.

HEIKAL, Prof. Dr. Helwan University, Cairo-Heliopolis, Almasastr 25, Egypt.

HEPHERD, R. Q. National Institute for Agricultural Engineering, Wrest Park, Silsoe, Bedford, UK.

HERINCKX, DR. C. Commission of the European Communities, Brussels, Belgium

HERINCKX, Dr. J. Faculte Agronomique De L'Etat, 5800 Gembloux, Belgium.

HICK, Dr. P. C. University of Cordoba, Calle 3, No.528 P. V. Sarfield, 5000 Cordoba, Argentina.

HILL, E. C. University College, Newport Rd., Cardiff, Wales, UK.

HITZHUSEN, Prof. F. J.   Ohio State University, 2120 Fyffe Rd., Columbus, Ohio 43210, USA.

HOFFMANN, G.   TUV-Rheinland, 5 Koln 1, Postfach 101750, Institute for Energietechnik, Germany.

HOFFMANN, Dr. L.   Dornier System GmbH, Postfach 1360, D-7990 Friedrichshafen 1, Germany.

HOLT, T. J.   University of Liverpool, Port Erin, Isle of Man, UK.

HOLVE, W. A.   Badger Ltd., Turriff Bldg., Great West Road, Brentford, UK.

HOLZAPFEL, K-O.   Fachinformations-Zentrum, D 7514 Leopoldshafen, Germany.

HOS, I.   Twente University, P.O. Box 217, Enschede, Netherlands.

HOWELLS, DR. E. R.   ICI Ltd., ICI House, Millbank, London, SW1. UK

HUFFMAN, D.   Forintek Canada Corpn., 800 Montreal Rd., Ottawa, Ontario, Canada.

HURARD, H.   Solvay & Cie SA., Rue Prince Albert, 33, B-1050 Brussels, Belgium.

IMARISIO, Dr. G.   CEC, DG XII/C1, Rue de la Loi 200, B-1049 Brussels, Belgium.

INGLE, Dr. J.   Agricultural Research Council, 160 Great Portland St., London, UK.

JACQUEMIN-MAHY, Mrs.   University of Namur, Rempart de la Vierge, 8, B5000 Namur, Belgium.

JAMES, P. A.   Shell International, MKBE/1, Shell Centre, York Rd., London, UK.

JAMES, R. N.   New Zealand Forestry Service, C/O Dept. of Forestry, Oxford University, UK.

JANNUZZI, G.   Cavendish Laboratory, Madingley Rd., Cambridge, UK.

JAWETZ, P.   Consultant on Energy Policy, 425 East 72 St., New York, NY 10021, USA.

JENSEN, Dr. A.   University of Trondheim, N-7034, Trondheim-Nth., Norway.

JOB, Dr. B.   ICI (Europa) Ltd. Everslaan, 45, 3078 Everberg, Belgium

JONES, B. A. J.   CDC, 33 Hill Street, London, W.1. UK.

JONES, Dr. J. M.   University of Liverpool, Dept. of Marine Biology,
                             Port Erin, Isle of Man, UK.

JONES, M. R.   Dept. of Agriculture and Horticulture, University
                             of Reading, UK.

JOSEPH, S.   Intermediate Technology Development Group Ltd., 9
                             King Street, London, WC2, UK.

JUNGE, Dr. D. C.   Oregon State University, Corvallis, Oregon 97331,
                             USA.

JUNGSCHAFFER, Dr.   Universitat - Graz, Inst. for Biochemie,
                             Halbartg.5, A-8010 Graz, Austria.

KANDLER, Prof. O.   Botanisches Institut der Universitat, 8 Munchen 19,
                             Menzingerstrasse 67, Germany.

KARIUKI, Dr. P. N.   Ministry of Energy, Box 30582, Kenya.

KEENE, R. M.   BOC Ltd., Deer Park Rd., London SW19, UK.

KEREVER, A.   Commissariat l'Energie Atomique, DPR-STEP, SPT, BP
                             No. 6, 92260 Fontenay aux Roses, France.

KEUNE, Dr.   BMFT, D-5300 Bonn, Germany

KHALIL, Dr. M. A. K.   Canadian Forestry Service, P.O. Box 6028, St.
                             John's, Newfoundland, Canada.

KNOEPFFLER-PEGUY, Dr. M. Universite Paris VI, Laboratoire Arago, 66650
                             Banyuls/Mer, France.

KOLSTER, H. W.   Stichting Industrie - Hout, Postbus 253, 6700 AG
                             Wageningen, Netherlands.

KREUZBERG, K.   University of Bonn, Inst. of Botany, Kirschallee 1,
                             D53 Bonn 1, Germany.

KRISPIN, T.   Agrar-und-Hydrotechnik, P.O.B. 10 01 32, 4300
                             Essen 1, Germany.

LAAN, C. J.   Humphreys & Glasgow, 22 Carlisle Place, London,
                             SW1, UK.

LAMPORT, Dr. D. T. A.   US Department of Energy and Michigan State
                             University, MSU/DOE Plant Res. Lab., East Lansing,
                             Michigan 48824, USA.

LANG, R. C.   California Energy Commission, 1811 Castro Way,
                             Sacramento, California 95825, USA.

LARGEAU, Dr.	CNRS, 54 Allee de Mocsouris, 91190 Gif sur Yvette, France.
LARKIN, S. B. C.	National College of Agricultural Engineering, Silsoe, Bedford, UK.
LARROQUE, F.	SEMA metra International, 16-18 rue Barbes, 92126 Hontroupe-Cedex, France.
LASSON, Dr. A.	Institute of Technology, Trondheim, Norway.
LAVAGNO, E.	Politecnico, Corso Duca, Degli Abruzzi 24, Torino, Italy.
LAWSON, G. J.	ITE, NERC, Merlewood, Grange-over-Sands, Cumbria, UK.
LAWSON, V.	Waste Management Consultants, 2 Eaton Crescent, Bristol 8, UK.
LE TIEC, J.	CEM, 85 Avenue Aristide Buand, Stains 93240, France.
LEAKEY, Dr.	Wimpey Laboratories, 15 Cambridge Rd., Girton, Cambridge, UK.
LEMASLE, Mr.	Creusot-Loire, B.P. 31, F-71208 Le Creusot, France.
LEPIDI, Dr. A.	University of Pisa, Istituto Microbiologia Agraria, Italy.
LERNER, P.	CGE Lab. de Marcoussis, Route de Nozay, 91460 Marcoussis, France.
LEVERT, Prof. J. M.	Institut de Chime et de Metallurgie Faculte Polytechnique de Mons, Rue de l'Epurgne, Mons. 7000, Belgium.
LEVI, Dr. J. D.	The British Petroleum Co. Ltd., Chertsey Rd., Sunbury-on-Thames, UK.
LEVY-LAMBERT, Mr.	Groupe Energie et Biomasse, 5 Rue Ste, Croix de la Bretonneric, 75004, Paris, France.
LEWIS, Dr. C.	University of Strathclyde, 100 Montrose St., Glasgow, Scotland, UK.
LIEN, Dr. S.	Solar Energy Research Institute, 1617 Cole Blvd., Golden, Colorado 80401, USA.
LIINANKI, L.	Kemisk Teknologi, Tekniska Hogskolan, K91, Kameralsekhonen, Stockholm, Sweden.

LINDNER, Dr. C. Lurgi, Kohle u. Mineraloltechnik GmbH, Bockenheimer Landstr. 42, Postfach 11 91 81, Germany.

LINNEBORN, Mr. Fritz Werner Industrie-Ausrustungen GmbH, D-6222 Geisenheim/Rhg, Germany.

LONG, Dr. G. Energy Technology Support Unit, Building 156, AERE, Harwell, UK.

LONGIN, Dr. R. Institut Pasteur, 28 Rue du Dr Roux, 75015, Paris.

LOPEZARIAS, Dr. Av. General Peron, Y-9-E, Madrid, Spain.

LORENC, G. Sirycon Ltd., Regal House, Twickenham, Middx., UK.

LUCA, Dr. S. Institut fur Garungsgewerbe und Biotechnologies, Seestrasse 13, 1000 Berlin 65, Germany.

LYONS, G. Irish Agricultural Institute, Oak Park Research Centre, Carlow, Eire.

MACBRAYNE, C. Forestry Department, Aberdeen University, Scotland, UK

MADDEN, Dr. R. H. CNRS Marseille, c/o 14 Broomhall Gdns., Edinburgh, Scotland, UK.

MAGAUDDA, Dr. CNEN-FARE, B.P. 2400, Rome, Italy.

MAHAIM, Miss I. Universite de Mons, 39 Route du Tonnelet, 4880 Spa, Belgium.

MALLALIEU, B. D. Ewbank & Ptnrs Ltd., Prudential House, North Street, Brighton, UK.

MARIEN, Mr. AFOCEL, 5 Rue des Palombes, 34000 Montpellier, France.

MARSTRAND, Mrs. P. K. SPRU, University of Sussex, Falmer, Brighton, UK.

MARTIN, S. R. Stone & Webster Engineering Ltd., 236 Grays Inn Rd., London, UK.

MARTY, D. CNRS, University de Provence, 3 Place Victor Hugo, 13331 Marseille Cedex 3, France.

MARWOOD, P. Nestle Products Technical Assistance Co. Ltd., CH-1814 La Tour-de-Peilz, Switzerland.

MASONI, Dr. A. University di Pisa, Istituto of Agromomia-Pisa Italy

MASSANTINI. Prof. F.	Universita di Pisa, Istituto di Agronomia, Pisa, Italy.
MASSON, C.	GRCPC, Domaine du Deffend, Mignaloux Beauvoir, 86800 St Julien L'Ars, France.
MATARASSO, Mr.	CNRS-PIRDES, 282 Bvd. Saint Germain, 75007, Paris, France.
MATTHEWS, Prof. J. D.	Department of Forestry, University of Aberdeen, Scotland. UK.
MAUER, Dr. G.	University of Dar-es-Salaam, Tanzania.
MAURIN, Dr. J.	Centre de Recherches Total, BP 27 76700 Harfleur, France.
McELROY, G. H.	Department of Agriculture, Loughgall, N. Ireland. UK
McFARLANE, Dr. N.	Shell UK Admin. Service, Shell Centre, London, SE1, UK.
McLAIN, Dr. H. D.	Department of Agriculture, Loughry College, Cookstown, Co. Tyrone, N. Ireland, UK.
MELLOTTEE, Dr. H.	CNRS, 1c Av. de la Recherche Scientifique, 45045 Orleans, France.
MENDELSOHN, H. R.	Inventa AG, 7013 Domat-Ems, Switzerland.
MERCENIER, J.	CEME, CP 139, ULB, 50 Av. F. Roosevelt, Brussels, Belgium.
MERTEN, Dr. D.	A-7000 Eisenstadt, Leserg 8, Austria
MESSENGER, M.	IIASA, 2361 Laxenburg, Austria.
MICHEL, A.	Belgonucleaire, 25 Rue du Champ de Mars, 1050 Brussels, Belgium.
MICUTA, W.	Bellerive Foundation, 5 Rue Vidollet, 1202 Geneve, Switzerland.
MIGLIAVACCA, Dr. M.	CISE, C. P. 12081, 20100 Milano, Italy.
MILLER, E.	Wimpey Laboratories Ltd., Beaconsfield Rd., Hayes, Middlesex, UK.
MILLS, J. M	J. M. Mills (Industrial Alcohols) Ltd., 59-61 Sandhills Lane, Liverpool, UK.
MILLS, P. J.	Distillers Co. Ltd., Glenochil Res. Stn., Menstrie, Clacks, Scotland, UK.

MITCHELL, Dr. C. P. Aberdeen University, St. Machar Drive, Aberdeen, Scotland, UK.

MOLETTA, Dr. R. 369 Rue J. Guesdes, 59650 Villeneuve D'Asco, France.

MOLLIERE, Ms. C. Techno-Foret, 17 Allee Piencourt, 48000, France.

MOLYNEUX, P. H. Lubrizol Ltd., 57-63 Old Church St., London SW3, UK.

MOORE, J. Parliamentary Under Secretary of State for Energy, UK, Thames House South, London SW1

MORLEY, Prof. J. G. WHT, University of Nottingham, UK.

MORRIS, Dr. J. South African Scientific Liaison Office, 278 High Holborn, London WC1, UK.

MORRIS, Dr. R. M. Open University, Milton Keynes, UK.

MOULLIN, Dr. D. T. ENSA-INRA, Chaire de Genetique & Microbiologie, 34060 Montpellier, France.

MULHEIRN, Dr. L. J. Shell Research Ltd., Broad Oak Rd., Sittingbourne, Kent, UK.

MUNIR, Dr. M. Sudd, Zucker-AG, Zentrallabor, D-6718 Grunstadt 1, Postfach 1127, Germany.

NADAL-AMAT, Dr. A. INIA (Mo. Agric.), Crida 4, Cabrils, Barcelona, Spain.

NEENAN, Dr. M. Irish Agricultural Institute, Oak Park, Carlow, Eire.

NICKEL, Dr. D. European Parliament, P.O. Box 1601, Luxembourg.

NIEDER, Dr. H. Fachverband Stickstoffindustrie, Sternstr. 9-11, 4000 Dusselforf, Germany.

NILSSON, Prof. P. O. University of Agricultural Sciences, S-770 73 Garpenberg, Sweden.

NORTON, J. F. CEC/JRC Petten, Postbus 2, 1755 ZG Petten, Netherlands.

NYNS, Prof. E. J. University of Louvain, Unite GEBI, Pl. Croix du Sud 1-9, 1348 Louvain-la-Neuve, Belgium.

OBEN, G. Economische Hogeschool Limburg, B-3610 Diepenbeek, Belgium.

OLIVIER, D. Earth Resources Research, 40 James St., London W1, UK.

OLSEN, G.	G. V. Olsen Associates, 170 Broadway, New York, NY 10038, USA.
OSWALD, Prof. W. J.	University of California, Berkeley, CA 94720, USA.
OTZEN, Dr.	Verbindangst Landwirt/Industric, Lindenallee 56, 43 Essen, Germany.
OUDINOT, Mrs.	c/o Credit Agricole, France.
OVEREND, Dr. R. P.	National Research Council, Bldg. M-50, Montreal Road, Ottawa, Ontario, Canada.
OWSIANOWSKI, Mr.	GTZ Abt. 21, 6236 Eschborn, Germany.
PAAVILAINEN, Dr.	Finnish Forest Research Institute, Unionine, 40A, 00170 Helsinki, Finland.
PANKHURST, Dr. E. S.	British Gas Corporation, London Research Station, Michael Rd., London, SW6, UK.
PANSOLLI. Dr. P.	ASSORENI, Via E. Ramarini No. 32, 00015 Monterotondo, Rome, Italy.
PARE, Mrs. C.	INPL, 1 Rue de Grandville, 54042 Nancy Cedex, France.
PARTOS, G.	CNRS-LCB, 31 Chemin Josef Aiguier, 13274 Marseille, Cedex 2, France.
PEARCE, M. L.	Forestry Commission R&D Div., Westonbirt Arboretum, Tetbury, Glos., UK.
PEDERSEN, T.	Ministry of Petroleum and Energy, P.O. Box 8178 Dep, Oslo 1, Norway.
PEGURET, Mrs. A.	CERNA, ECOLE des MINES, 60 Bd. St. Michel, 75006 Paris, France.
PELLIZZI, Prof. G.	University of Milano, Via G. Celoria, 2 - 20133 Milano, Italy.
PEREZ-ARANDA, Dr. L.	V. Libre de Bruxelles, 28 Av. Paul Heger, 1050 Brussels, Belgium.
PERNKOPF, J.	Federal Research Institute, Rottenhauser str. 1, A-3250 Wieselburg, Austria.
PETAZZONI, G.	AGIP SpA, S. Donato Milanese, Milano, Italy
PETERS, A.	International Forest Science Consultancy, 21 Biggar Rd., Silverburn, Penicuik, Midlothian, UK.
PEZE, H. F.	Gist-Brocades N. V., P. O. Box 1, 2600 MA Delft, Netherlands.

PICKEN, Prof. D. J.   Leicester Polytechnic, P. O. Box 143, Leicester, UK.

PIERROT, Dr. F.   Rhone Poulenc S.A., 26 Rue Roussy, 69004 Lyon, France.

PIRET, Dr. E.   Energy Resources Company, 185 Alewise Brook Cambridge, Mass., USA.

PLODER, Dr. W.   A-2320 Schwechat, Postfach 7, Austria.

PLUCHET, J.   PEC Engineering, 62 Rue Jeanne D'Arc, 75646 Paris Cedex 13, France.

PNEUMATICOS, Dr. S. M.   Energy, Mines & Resources, Canada, 580 Booth, Ottawa, Ontario, Canada.

POHJONEN, Dr. V.   Finnish Forest Research Institute, SF-69100, Kannus, Finland.

POLLARD, B.   Sirycon Ltd., Regal House, Twickenham, Middx., UK.

PONCELET, Dr. G.   University of Louvain, Place Croix du Sud 1, 1348 Louvain-la-Neuve, Belgium.

POOLE, Dr. J.   ICI Ltd., Jealotts Hill, Bracknell, Berks., UK.

PORTER, Sir George   The Royal Institution, London, UK.

POURBAIX, Mr.   Electricite de France, 6 Quai Watier, 78400 Chatou, France.

PRATT, D. C.   University of Minnesota, Minneapolis, Minnesota 55455, USA.

PRESTON, Dr. G.   Department of Energy, Thames House South, Millbank, London SW1, UK.

PRESTON, Dr. T. R.   13 Cordemex, Yucatan Mexico,

PUCCIONI, Mrs. M. C.   Forestry Department, Aberdeen University, Scotland, UK.

PULEJO, M.   CEC, Rue de la Loi 200 B-1049, Brussels, Belgium.

PULS, Dr. J.   Institute for Wood Chemistry, Leuschnerstr 91, 205 Hamburg 80, Germany.

PURCHASE, Dr. B. S.   SMRI University of Natal, Durban 4001, South Africa.

PYM, D.   W. S. Atkins Group Ltd., Woodcote Grove, Ashley Rd., Epsom, Surrey, UK.

RABSON, Dr. R. U.S. Department of Energy, ER-19, G-256, GTN Washington, D. C. 20545, USA.

RANALLI, Dr. L. AMN SpA, c/o 14/18 High Holborn, London, WCI, UK

RAVETTO, P. Politecnico, Corso Duca, Degli Abruzzi 24, Torino, Italy.

RECHOU, J. Leroy-Somer, 16015 Angouleme Cedex, France.

REDDY, Dr. A. K. N. Indian Institute of Science, Bangalore, India.

REED, Dr. T.B. Solar Energy Research Institute, 1617 Cole Blvd., Golden, Colorado 80401, USA.

RENARD, J-C. Laboratoires de Marcoussis, Route de Nozay, 91460 Marcoussis, France.

REXEN, Dr. F. Bioteknisk Institut, Denmark.

REYNIEX, Mrs. CNEEMA, Parc de Tourvoie, 92160 Antony, France.

RICHARD, Dr. C. Universite de Nancy 1, C.O. 140 54037 Nancy Cedex, France.

RICHARD, Dr. J. R. CNRS, CRCC 4T, 45045 Orleans Cedex, France.

RICHARDS, Dr. K. M. Energy Technology Support Unit, Harwell, UK.

RICHTER, DR. CIBA-GEIGY, BASLE, R-1093, 4-41, Switzerland.

RIEDINGER, Dr. KFA Juelich, PLR, Postfach 1913, D-517 Juelich, Germany.

RIJKENS, B. A. Instituit voor Bewaringen en Verwerking van Land bouwprodukten, Netherlands.

RIVA, Dr. G. University of Milano, Via G. Celoria, 2-20133 Milano, Italy.

ROBERTSON, E. E. The Biomass Energy Institute, P.O. Box 129, P.S. "C", Winnipeg, Canada.

ROBERTSON, R. Foster Wheeler Power Products Ltd., PO Box 160, Greater London House, Hampstead Road, London, UK.

ROBINSON, Dr. K. National Board for Science and Technology, Shelbourne Road, Dublin 4, Ireland.

RODE, INRA, CNRA, Route de St. Cyr, 78000 Versailles, France.

ROKITA, K. Entsorgungsbetriebe Simmering, 11 Haidequerstr, 1110 Vienna, Austria.

ROUET, L.	L'Air Liquide, 57 on Carnot, 94500 Champigny s/Marne, France.
ROUGE, B.	Comes, 208 Rue Raymond Losserand, Paris, France
ROY, Dr. R.	Open University, ATG, Walton Hall, Milton Keynes, UK.
RUSSELL, B. J.	Foster Wheeler Ltd., Station Rd., Reading, Berks., UK.
RUTHNER, Prof. O.	Ruthner Pflanzentechnik AG, Sieveringerstrasse 150, 1190 Vienna, Austria.
RUYSSEN, O.	CEC, 200 Rue de la Loi, 1049 Brussels, Belgium.
SAKELLARIOS, J.	National Energy Council, 42 Akademias St., Athens, Greece.
SAMOOTSAKORN, Miss. P.	Engineering Department, University of Reading, Berks. UK.
SANNA, Dr. P.	ASSORENI, Via Fabiani, 1, S. Donato, Milan, Italy.
SCARAMUZZI, Prof. G.	ENCC – SAF, Casella Postale 9079, Rome, Italy.
SCHLICKEN, Dr. P.	Centre de Recherches de Pont-a-Mousson, BP-28 54700 Pont-a-Mousson, France
SCHOENSTEIN, R.	WIST, Berggasse 16, A-1090 Vienna, Austria.
SCHOERNER, Dr. G.	WIST, Berggasse 16, A-1090 Vienna, Austria.
SCHWEIGER, Dr. G.	Universitat Duisburg, D41-Duisburg, Germany.
SCOTT, R.	ITE Merlewood, Grange-over-Sands, Cumbria, UK.
SCOTT-KEMMIS, D.	Science Policy Research Unit, Sussex University, Falmer, Sussex, UK.
SELLDEN, Dr. G.	University of Goteborg, Botanical Institute, Carl Skottsbergs Gata 22, S-413 19 Goteborg, Sweden.
SEYMOUR, Mr.	CEC, Translation Division, Luxembourg.
SHAMA, G.	Imperial College, London, SW7, UK.
SIMON, Mr.	Bayer AG, 5090 Leverkusen, Germany.
SINGLETON, F. H.	Babcock Woodall Duckham Ltd., The Boulevard, Crawley, Sussex, UK.
SINNER, Dr.	Fritz Werner Industrie-Ausrustungen GmbH, D-6222 Geisenheim/Rhg, Germany.

SINTUNAWA, C.	Strathclyde University, Chesters Road, Bearsden, Glasgow, UK.
SMITH, Dr. P.	University of Otago, New Zealand
SOMMERMANN, Mrs. J.	˙Ministry of Agriculture, Bonn 1, Rachusstr 1, (BML), Germany.
SOYER, N.	ENSCR, Av. General Leclerc, 35000 Rennes, France.
SPANJAARD, Mrs.	Alsthom-Atlantique, 38 Avenue Kleber, 75016, Paris, France.
SPEYER, P	ARES, 37 Rue Croix Baragnon, 31 000 Toulouse, France.
STAFFORD, Dr. D. A.	University College, Newport Rd., Cardiff, Wales, UK.
STASSEN, H. E. M.	T.H.T. P.O. Box 217, 7500 AE Enschede, Netherlands.
STEINBECK, Dr. K.	University of Georgia, School of Forest Resources, Athens, GA, USA.
STELWAGEN, P.	Oy ALKO Ab, Rajamaki Factories, SF-05200 Rajamaki, Finland.
STEPHENSON, R.	Brighton Polytechnic, Grand Parade, Brighton, UK.
STOCKDALE, J. W.	Witch Chips Ltd., Bar Lane, Boroughbridge, York, UK.
STOTT, K. G.	A.R. Council, Research Station, Long Ashton, Bristol.
STOUT, Prof. B. A.	AGR Eng. Dept., Michigan State University, East Lansing, Michigan, USA.
STREHLER, Dr. A.	Landtechnik, Weihenstephan D805 Freising, Germany
STRIXIOLI, DR. P.	CESEM, Ansaldo Spa, c/o Furnival House, 14/18 High Holborn, London, WCI, UK.
STUCKEY, Dr. D.	SINTEF-NTH, N-7034, Trondheim-Nth, Norway.
SURYANARAYANA, Dr. G.	Department of Chemical Engineering, Imperial College of Science and Technology, London SW7, UK.
SUTHERLAND, D. J. J.	Devon County Council, Landscape Institute, Craxon, Diptford, Nr. Totnes, Devon.
SWAAIJ, Prof. W. P. M.	Twente University, P.O. Box 217, Enschede, Netherlands.

TATE, H. S.	Tate & Lyle Ltd., PLMRL, Whiteknights, P.O. Box 68, Reading, Berks., UK.
TAYLOR, J. D.	Stake Technology Ltd., 20-A Enterprise Ave., Nepean, Ontario, K29 0A6, Canada.
TEGTMEIER, U.	Institut fur Garungsgewerbe und Biotechnologie, Seestrasse 13, 1000 Berlin 65, Germany.
THOMA, Dr. H.	Technical University of Munchen-Weihenstephan, Lehrstuhl f. Angew. landw. Betriebslehre D 8050 Freising-Weihenstephan, Germany.
THOMSON, Dr. A. R.	UKAEA Harwell, Oxfordshire, UK.
THORESEN, Dr. P.	Defence Research Establishment, P.O. Box 25, N-2007 Kjeller, Norway.
THURSFIELD, Dr. G.	Tube Investments Ltd., Hinxton Hall, Hinxton, Essex, UK.
TISSIERES, J-N.	Electrowatt Eng. Services Ltd., P.O. Box, CH-8022 Zurich, Switzerland.
TITL, A. J.	MAN, Neye-Technologie, Docheur Strete 667, D-8000 Munich 50, Germany.
TJERNSHAUGEN, Dr.	Agricultural University, Box 15, N-1432 AA5-NLH, Norway.
TOMINAGA, K.	Idemitsu Kosan Co. Ltd., 25 St. James St., London SW1, UK.
TORRENTI, R.	Ecole des Mines, Sophia Antipolis, 06560 Valbonne, France.
ULLMANN, O.	Arge Systemtechnik, c/o MBB, Postfach 80 12 20, D-8000 Munchen 80, Germany.
VAN DEN AARSEN, F.G.	Twente University, Department of Chemistry, P.O. Box 217, Enschede, Netherlands.
VAN DER KELEN	FAC, Genbloux, Belgium
VAN DER MEIDEN, H. A.	Stichting Industrie - Hout, Postbus 253, 6700 AG Wageningen, Netherlands.
VAN WERSCH, Mr.	FN Herstal, 33 Rue Voie de Liege, 4400 Herstal, Belgium.
VANACKER, L.	Seminare voor Ruimtelijke Planning, Onafhankelijkheidslaan, 17-18 B 9000 Gent, Belgium.

VANCE, Dr. I.
British Petroleum Co. Ltd., Chertsey Rd., Sunbury-on-Thames, Middx., UK.

VANLANDUYT, Dr. E.
ACEC, 393 LR/CMA, BP 4, 6000 Charleroi, Belgium.

VECLI, Prof. A.
Istituto di Fisica, Via M. d'Azeglio 85, 43100 Parma, Italy.

VEEGER. Dr. C.
Department of Biochemistry, Agricultural University, 6703 BC, Wageningen, Netherlands.

VERHOEVEN, Dr. W.
Boetele 29, Raalte, Holland.

VERKOREN, J.
Ministry of Agriculture and Fisheries, Bezuidenhoutseweg 73, The Hague, Holland.

VERMOOTE, D.
SES, Industrie Park 15 - Tienen, Belgium

VERSTRYNGE, L.
University of Ghent, Coupure Links, 533, 9000 Ghent, Belguim.

VIGNAIS, Dr. P.
CNRS, Biochimie CEN-G, 85X, 38041 Grenoble Cedex, France.

VILLET, Dr. R.
Solar Energy Research Institute, 1617 Cole Blvd., Golden, Colorado 80401, USA.

VINCENT, Dr. D.
Department of Energy, Thames House South, Millbank, London SW1, UK.

VLITOS, Prof. A. J.
Tate & Lyle Ltd., PLMRL, Whiteknights, P.O. Box 68, Reading, Berks., UK.

VOESTE, T.
Lurgi Technik, 6000 Frankfurt, Germany.

VOETBERG, J. W.
Institut voor Bewaringen en Verwerking van Landbouw produkten, Netherlands.

VON HORNSTEIN, H. H. F.
Grueningen, 7940 Riedlingen, Germany.

VON SONNTAG, Prof. C.
Institut fur Strahlenchemie im MPI fur Kohlenforschung, Stiftstr. 34-36, D-4330 Mulheim, Germany.

VORAGEN, Dr. F.
Department of Food and Science, Agricultural University, De Dreyen 12, 6703 B C Wageningen, Netherlands.

WAHL, K. L.
Dornier System GmbH, Abt. Ntu, 7990 Friedrichshafen, Pf. 1360, Germany.

WARD, Dr. R. F.
United Nations, D.C. 810, New York 10017, USA.

WATSON, Dr. H. C.	National Energy Research Development and Demonstration Program for Australia, , c/o 25 Rowlands Ave., Hatch End, Middx., UK.
WEICKMANS, L.	International Institute for Sugar Beet Research, 47 Rue Noutoyer, B-1490, Brussels.
WEISMANN, A.	Bundesministerium fur Ernahrung, D 53 Bonn 1, Rochinsstr 1, Germany.
WEISS, E.	Inventa AG, 7013 Domat/Ems, Switzerland.
WELLARD, Dr. N. K.	Shell Research Ltd., Biosciences Lab., Sittingbourne, Kent, UK.
WEYMOUTH, Dr. F. J.	Courtaulds Ltd., Foleshill Rd., Coventry, UK.
WHITE, Dr. D. J.	MAFF, Rm 120, Great Westminster House, Horseferry Rd., London SW1, UK
WHITE, L. P.	General Technology Systems Ltd., Forge House, 20 Market Place, Brentford, Middx., UK.
WHITTAKER, Miss H.	ITE, Merlewood, Grange-over-Sands, Cumbria, UK
WHITTLE. E. J. R.	Biomechanics Ltd. Caxton House, Wellesley Road, Ashford, Kent, UK.
WILHELMSEN, G.	NLVF, P.O. Box 61, N-1432 A5-NLH, Norway.
WILKIE, Ms. A.	University College, Galway, Ireland.
WILLIAMS, L.	Director General for Energy Commission of the European Communities.
WITHERS, Mrs. A.	Dept. Agric. & Horticulture, University of Reading, Earley Gate, Reading, Berks., UK.
WOLF, E.	Am Stockener Bach 7B, 3000 Hannover 21, Germany.
WOOD, Dr. T. M.	The Rowett Research Institute, Bucksburn, Aberdeen Scotland, UK.
WRAY, S. W.	Hamworthy Engineering Ltd., P & C Division, Fleets Corner, Poole, Hants. UK.
ZEEMAN, G.	Water Pollution Control, Agricultural University, Brotechnion, De Dreyen, Weigeningen, Netherlands.
ZERBIN, W.	IMBERT, Bonnerstr 49, 5354 Weilerswist, Germany.
ZUERRER, Dr. H.	Institute of Plant Biology, Zollikerstrasse 101, CH-8008 Zurich, Switzerland.

Index of authors

Abdallah, M.   610
Abernethy, W.   198
Ader, G.   598
Ait, N.   324
Alfani, F.   144
Andreoni, P.   337
Andrews, N.J.   75
Antunes, S.D.   372
Asinari di San Marzano, C.M.,   392
Auclair, D.   216
Avella, R.   337

Balachandran, B.N.   719
Balakrishna, M.   719
Barry, M.   416
Beenackers, A.A.C.M.   616
Belaich, J.P.   380
Berger, B.   813
Bergman, K.G.   896
Berkaloff, C.   653
Bernhardt, W.   815
Berra, E.   659
Bianchi, A.   386
Binder, A.   554
Blanc-Muesser, M.   312
Bobleter, O.   554
Bodria, L.   876
Bondi, H.   19
Booth, A.   416
Boshoff v. D., B.   934
Bourreau,   588
Bouvarel, P.   172
Bridgwater A.V.   598, 607
Bruwer, J.J.   934
Buchli, J.   883
Buehner, T.   854
Bunce, R.G.H.   103
Burtscher, E.   554

Callaghan, T.V.   83, 90
Cantarella, M.   144
Carruthers, S.P.   97
Casadevall, E.   653
Castelli, G.   876
Cathonnet, M.,   529
Cattaneo, J.   324
Caubet, S.   542

Chadwick, M.J.   251
Channeswarappa, A.   719
Chartier, P.   1, 2, 22
Chassany de Casabianca, M.L.   672, 678
Chassin,   138
Chaumont, D.   659
Chrysostome, G.   594
Codomier, L.   677
Colleran, E.   416
Concin, R.   554
Coombs, J.   279
Corte, P.   542
Creuzet, N.   324
Crookes, R.J.   516

Da Silva, E.   717, 796
Defaye, J.   312, 319
Deglise, X.   548, 569
De la Rosa, F.F.   647
Dennington, V.N.   251
Dietrichs, H.H.   348
Dif, D.   653
Di Giorgio, G.   337
Divry, A.   583
Doat, J.   535
Dodson, C.E.   922
Dohne, E.   915
Driguez, H.   312
Dubois, P.   582
Duff, J.W.   228
Dujardin, E.   703
Dunican, L.K.   416
Durand, H.   826

Engelbrecht, A.   934
English, M.   360
Elliott, R.   763

Fahim, C.   542
Fallowfield, H.J.   691
Fardeau, M.L.   380
Fontes, A.G.   647
Forget, P.   324
Fox, M.F.   910
Flloyd, J.R.S.   398
Fuls, J.   934

Gadelle, A. 319
Galzy, P. 344
Garcin, J. 386
Garrett, M.K. 691
Gely, A. 677
Ghose, T.K. 261
Gianfreda, L. 306
Goma, G. 298
Gomez-Moreno, C. 647
Goudeau, J.C. 588
Greco, G. jr. 306
Grout, H.J. 360
Gudin, C. 659
Guerif, 138

Hall, D.O. 1, 2, 34
Hardt, H. 292
Harris, G.S. 838
Hathaway, R.L. 244
Hatt, B.W. 598, 607
Have, H. 563
Hawkes, D.L. 398, 429
Hawkes, F.R. 398
Hawkins, C. 934
Hayes, D.P. 198
Hitzhusen, F.J. 610
Hos, J.J. 616
Hounam, I. 752
Hughes, D.E., 406
Hugo, F.J.C. 934

Installe, M. 372
Irlam, G.A. 607

Jawetz, P. 941
Jones, J.M. 681
Joseph, S. 789
Juste, 138

Kandler, O. 472
Kariuki, P.N. 769
Kleinhanss, W. 466
Koegl, H. 854
Kolster, H.W. 193
Kothandaramaiah, P. 719
Kouzminov, V.A. 796
Kreuzberg, K. 709
Krispin, T. 782

Lallemand, M. 588
Lamport, D.T.A. 292

Lang, R.C. 870
Largeau, C. 653
Larkin, S.B.C. 118, 124
Lavagno, E. 423
Lawson, G.J. 83, 90
Lescure, J.P. 947
Lettinga, G. 264
Lewis, C.W. 752
Lien, S. 697
Lindauer, G.D. 440, 782
Lopolo, M. 337
Lyons, G. 232

MacBrayne, C.G. 181
Madden, R.H. 366
Mancini, A. 337
Margaris, N.S. 47
Marien, J.N. 257
Martin, S.R. 922
Marty, D. 386
Masson, C. 588
Matthews, J.D. 181
Mauer, G. 758
McDivitt, J.F. 796
McElroy, G. 198
McLain, H.D. 228
Mellottee, H. 523
Messenger, M. 863
Meynell, P.J. 448, 459
Minier, M. 298
Mitchell, C.P. 103, 181, 239
Mnzava, E.M. 735
Mohrlok, S. 292
Molle, J.F. 574
Monnier, 138
Monties, B. 523
Moore, J. 11
Morandini, R. 159
Morley, J.G. 681
Morliere, P. 569
Morris, R.M. 118, 124
Moulin, G. 344
Muller, 138

Nagaraju, S.M. 719
Naveau, H.P. 392
Neenan, M. 232
Newell, P.J. 416
Noble, D.H. 124
Nyns, E.J. 392

Ofoli, R.Y. 354
Ohleyer, E. 312

Oswald, W.J.   633
Overend, R.   481

Pare, C.   187
Partos, J.   380
Paul, F.   665
Pearce, L.H.   103
Pearce, M.L.   210
Pedersen, C.   319
Pellizzi, G.   876
Petroff, G.   535
Picken, D.J.   910
Piron-Fraipont, C.   703
Plaskett, L.   890
Plaskett, L.G.   110
Plasse, F.   380
Pneumaticos, S.M.   903
Porter, G.   627
Posselius, J.H.   132
Prakash, C.S.   719
Pratt, D.C.   75
Preston, T.R.   763
Proe, M.F.   181
Puls, J.   348

Radley, R.W.   124
Ramaiah, C.   719
Ravetto, P.   423
Ravindranath, N.H.   719
Reddy, A.K.N.   719, 727
Reed, T.B.   496
Renard, J.C.   582
Remy, 138
Rexen, F.P.   50
Richard, C.   548
Richard, J.R.   523, 529
Rijkens, B.A.   435
Rolin, A.   548
Rolin, C.   187
Roy, R.   776
Ruggeri, B.   423

Sanchez, M.   763
Sangiorgi, F.   876
Sauze, F.   672, 678

Scaramuzzi, G.   222
Scardi, V.   144
Schlicklin, Ph.   569
Schmitdt, A.   928
Scott, R.   90
Shanahan, Y.J.   789
Sironval, C.   703
Sixt, H.   440
Slesser, M.   752
Smith, E.L.   607
Smith, G.   292
Somashekar, H.I.   719
Souil, F.   588
Srinath, P.N.   719
Stafford, D.A.   406
Steinbeck, K.   163
Stengel, 138
Stott, K.G.   198
Stout, B.A.   131, 354
Strehler, A.   509
Strub, A.   9
Studach, J.   883

Tauzin, 138
Taylor, J.D.   330
Temper, U.   472
Thring, M.W.   516
Tjernshaugen, O.   150
Traverse, J.P.   542
Trindade, S.C.   59

Van den Aarsen, F.G.   616
Van der Meiden, H.A.   193
Van Kraayenoord, C.W.S.   244
Van Swaaij, W.P.M.   485, 616
Vaseen, V.A.   411
Vignais, P.M.   665

Wagener, K.   625, 685
Walt v.d., A.N.   934
Whittaker, H.A.   90
White, LP.   890
Wilkie, A.   416
Williams, L.   14
Winter, J.   472